AF658398

Advanced Structured Materials

Volume 75

Series editors

Andreas Öchsner, Southport Queensland, Australia
Lucas F.M. da Silva, Porto, Portugal
Holm Altenbach, Magdeburg, Germany

More information about this series at http://www.springer.com/series/8611

Vijay Kumar Thakur · Manju Kumari Thakur
Editors

Eco-friendly Polymer Nanocomposites

Processing and Properties

Editors
Vijay Kumar Thakur
Mechanical and Materials Engineering
Washington State University
Pullman, WA
USA

Manju Kumari Thakur
Division of Chemistry
Himachal Pradesh University
Shimla, Himachal Pradesh
India

ISSN 1869-8433 ISSN 1869-8441 (electronic)
Advanced Structured Materials
ISBN 978-81-322-2469-3 ISBN 978-81-322-2470-9 (eBook)
DOI 10.1007/978-81-322-2470-9

Library of Congress Control Number: 2015940723

Springer New Delhi Heidelberg New York Dordrecht London

Printed on acid-free paper

Springer (India) Pvt. Ltd. is part of Springer Science+Business Media
(www.springer.com)

To my parents and teachers who helped me become what I am today.

Vijay Kumar Thakur

Preface

Nanotechnology has proven to be a potential technology that is rapidly emerging in wider areas such as medicine, electronics and food technology by manipulating nanomaterials for various applications. Nanotechnology is the engineering and art of developing new materials on a nanoscale. A nanomaterial is defined as a discrete entity that has at least one of its dimension as 100 nm or less. It may include any of the following nano forms: nanoparticles, nanotubes, fullerenes, graphene, nanoclays, nanofibres, nanowhiskers and nanosheets to name a few. The applications of nanomaterials are quite broad. Due to their unique properties, the application of nanomaterials in various scientific fields, including environmental sciences, has increased greatly. It is well known that the performance of polymer materials can be greatly enhanced by dispersion of nanometre-size materials. Such materials are called polymer nanocomposites and have the interesting characteristic that the mechanical properties; the barrier properties; the thermal properties and some others such as flammability and water adsorption can be greatly enhanced with addition of a small amount of filler (usually less than 10 wt %).

Indeed, polymer nanocomposite (PNC) is a promising multidisciplinary research activity in the field of material research that might expand the utilization of polymers in various industrial applications. The use of polymers has been extended in different fields as PNC enhances the properties of polymer to obtain a product with essentially a new set of properties. It is expected that the transition from micro to nano materials increases the surface area-to-volume ratio. This in turn results in a prominent increment of the behaviour of the atoms on the surface of the nanomaterials. It affects the properties of the nanomaterials when they react with other nanomaterials. Due to the higher specific surface area of nanomaterials, interaction with other nanomaterials within the mixture becomes more intense. This consequently results in positive properties, such as high temperature capability, resistance against corrosion, noise damping, low in cost/manufacturer, ductile, high specific stiffness and strength high thermal conductivity, and low coefficient of thermal expansion. Another advantage of polymer nanocomposites is that it could be fabricated via rapid and precise manufacturing methods, such as injection molding, compression molding, vacuum bag molding, contact molding, and resin transfer molding. Therefore, polymeric nanocomposites are speculated as appropriate options in overcoming the inherent restrictions of microcomposite and monolithic, while posing preparation challenges related to the control of elemental composition and stoichiometry in the nano-cluster phase.

Within the development of polymer nanocomposites, there are challenges and limitations. Today, mankind is aware of eco-friendliness that has led to growing attention in reducing environmental impact cause by traditional polymer nanocomposites. The rise in environmental awareness has further led governments to make more stringent regulations as well as researchers to explore new eco-friendly polymer nanocomposites. Such nanocomposites obtained by using eco-friendly materials and techniques as well as incorporating nanofillers to biopolymers, are extremely promising products because they provide better properties with conservation of the material biodegradability, environmental friendliness, easy processing, impressive physico-mechanical properties, avoiding eco-toxicity. This assists in evolution of simpler chemical processes or innovative designed product for future generations by the chemical industries that should create least environmental impact. An interest in naturally available renewable materials has been developed due to the global environmental concern. Numerous studies are underway on the preparation and applications of eco-friendly polymer nanocomposites.

Keeping in mind the immense advantages of eco-friendly polymer nanocomposites, this book primarily focuses on the processing and properties of different eco-friendly polymer nanocomposites procured from diverse sustainable resources and techniques. Several critical issues and suggestions for future work are comprehensively discussed in this book with the hope that the book will provide deep insight into the state of art of "Eco-friendly Polymer Nanocomposites". We would like to thank the Publisher for the invaluable help in the organisation of the editing process.

Finally, we would like to thank our parents for their continuous encouragement and support.

Vijay Kumar Thakur
Manju Kumari Thakur

Contents

About the Editors

Dr. Vijay Kumar Thakur has been working as Research Faculty (Staff Scientist) in the School of Mechanical and Materials Engineering at Washington State University, USA, since September 2013. His former appointments include being a Research Scientist in Temasek Laboratories at Nanyang Technological University, Singapore, and a Visiting Research Fellow in the Department of Chemical and Materials Engineering at LHU-Taiwan. His research interests include the synthesis and processing of bio-based polymers, nanomaterials, polymer micro/nanocomposites, nanoelectronic materials, novel high dielectric constant materials, electrochromic materials for energy storage, green synthesis of nanomaterials and surface functionalization of polymers/nanomaterials. He did his post-doctorate in Materials Science at Iowa State University and his Ph.D. in Polymer Science (2009) at the National Institute of Technology. In his academic career, he has published more than 80 SCI journal research articles in the field of polymers/materials science and holds one United States patent. He has also published 15 books and 35 book chapters on the advanced state-of-the-art of polymers/materials science with numerous publishers. He is an editorial board member of several international journals and also is member of scientific bodies around the world. In addition to being on the editorial board of journals, he also serves as the guest editor for the *Journal of Nanomaterials,*

International Journal of Polymer Science, Journal of Chemistry, and American Journal of Applied Chemistry. e-mail: vijayisu@hotmail.com

Dr. Manju Kumari Thakur has been working as an Assistant Professor of Chemistry at the Division of Chemistry, Government Degree College Sarkaghat Himachal Pradesh University—Shimla, India since June 2010. She received her B.Sc. in Chemistry, Botany and Zoology; M.Sc., M.Phil. in Organic Chemistry and Ph.D. in Polymer Chemistry from the Chemistry Department at Himachal Pradesh University—Shimla, India. She has rich experience in the field of organic chemistry, biopolymers, composites/nanocomposites, hydrogels, applications of hydrogels in the removal of toxic heavy metal ions, drug delivery, etc. She has published more than 30 research papers in several international journals, co-authored one book and has also published 25 book chapters in the field of polymeric materials. e-mail: shandilyamn@gmail.com

Eco-friendly Polymer Nanocomposite—Properties and Processing

Pei Dong, Raghavan Prasanth, Fangbo Xu, Xifan Wang, Bo Li and Ravi Shankar

Abstract This chapter mainly reviews the concept, properties and processing, and design method of the eco-friendly polymer nanocomposite (EPN), which is generally biodegradable and renewable. The major attractions of EPN are that they are environmentally friendly, sustainable, and degradable. These polymer composites can be easily composted or disposed without harming the environment. Some efforts have been made on attaining biodegradable reinforcing fillers giving improved performance of composites. Another concern is focused on employing recyclable synthetic fibers with thermoplastic composites to reduce the waste of fillers, and also some research is devoted to reusing or recycling the whole composites for the similar purpose. Simultaneously, people also would like to make composites manufactured with traditional production process become eco-friendly by extra reprocessing. Throughout the stages of development—design, appraisal, manufacture, use, reuse–recycling, and disposal—researchers are supposed to be fully engaged in reducing waste as much as possible, keeping in mind the environment all the time. A series of natural or synthetic materials have been used, such as cellulose, thermoplastic starch, etc. The challenge posed by eco-friendly composites also needs considerable attention in terms of poor bonding between matrix and fillers, loose control of fiber orientation, and difficulty in shaping nanoscale particles.

Keywords Eco-friendly · Polymer · Nanocomposite · Nanoclay · Montomorillionite

P. Dong · R. Prasanth (✉) · F. Xu · X. Wang · B. Li
Department of Materials Science and NanoEngineering, Rice University, Houston, TX 77005, USA
e-mail: prasanth.raghavan@rice.edu

R. Prasanth
Department of Mechanical Engineering and Materials Science, Rice University, 6100 Main MS-321, Houston, TX 77005, USA

R. Shankar
Fujifilm Imaging Colorants, Inc., 233 Cherry Lane, New Castle, DE 19720, USA

V.K. Thakur and M.K. Thakur (eds.), *Eco-friendly Polymer Nanocomposites*, Advanced Structured Materials 75, DOI 10.1007/978-81-322-2470-9_1

Abbreviations

EPN	Eco-friendly polymer nanocomposite
PLA	Polylactic acid
PHB	Polyhydroxylbutyrate
EPC	Eco-friendly polymer composites
PEG	Polyethylene glycol
MC	Methyl cellulose
MMT	Montomorillionite

1 What Is Eco-friendly Polymer Nanocomposite and Why We Need Them?

Generally, biodegradable and renewable polymer composites are called eco-friendly polymer nanocomposites (EPN) or green polymer composites. The major attractions of EPN are that they are environmentally friendly, sustainable, and degradable. These polymer composites can be easily composted or disposed without harming the environment (Baillie 2004; Adeosun et al. 2012).

Biodegradation indicates degradation of a polymer in natural environment. This implies loss of mechanical properties, changing in the chemical structure, and into other eco-friendly compounds (Jamshidian et al. 2010). Degradable polymers from natural sources (such as lignin, cellulose acetate, starch, polylactic acid (PLA), polyhydroxylalkanoates, polyhydroxylbutyrate (PHB)), and some synthetic sources (polyvinyl alcohol, modified polyolefins, etc.) are classified as biopolymers (John and Thomas 2008). It is noticeable that the nanocomposite from nonrenewable synthetic sources is neither wholly degradable nor renewable.

Previously, the major interest was in using synthetic materials such as aliphatic polystyrene, polyesters, glass, nanoclays, carbon fibers, carbon nanotubes, etc., for the production of nanocomposites. The use of these materials is challenging (Leja and Lewandowicz 2010): (A) Shortage of organic compounds due to increasing oil and gas prices, and declining oil and gas resources. (B) Other disadvantages include global warming, environmental concerns, cross-contaminations, and uneconomical costs (Amass 1998; Chandra and Rustgi 1998; Jamshidian et al. 2010). Therefore, research efforts are geared to the quest for materials that can overcome these challenges.

Compared with synthetic sources, renewable polymer composite sources have a lot of advantages, particularly as a solution to the environmental concerns of plastic waste. Renewable polymer composites can maintain sustainable development of ecologically attractive technology that is also economic (Okamoto 2004). Eco-friendly polymer composites (EPC) are widely studied, due to the need for innovations in the development of materials from biodegradable polymers and renewable raw materials. EPC has also attracted tremendous research interest, due to the potential application of agricultural resources, including product wastes (Okamoto 2004). Recycling process with enhanced environmental compatibility can also be improved by using these biocomposites.

Adding nanoparticles can directly maximize the interfacial adhesion inside the composite. Therefore, recently it attracts more and more research works focusing on the development and application of EPN, instead of EPC, for use in construction, packaging, automotive, and medical fields, due to their enhanced mechanical and permeable properties.

2 Design and Processing of Eco-friendly Polymer Nanocomposites

2.1 Challenges

The incompatibility of the hydrophobic (water repelling) polymer matrix and hydrophilic (water absorbing) fibers may give rise to nonuniform dispersion of fibers across the matrix leading to poor mechanical strength, posing the greatest challenge in the fabrication of nanocomposites over the past decades (Ashori 2008; John and Thomas 2008). To arrest this problem, large amounts of additives and techniques are being employed. Various chemical coupling agents or compatibilizers such as maleated polyethylene, carboxylated polyethylene, titanium derived mixture, maleic anhydride polypropylene, corona discharge, etc., calendaring, stretching, thermal treatment, reaction with methanol-melamine, isocyanates, triazine, silane, and mercerization of the matrix have been used (Rong et al. 2001; Kim et al. 2006; Lei et al. 2007; Lee et al. 2008; Qin et al. 2008; Reddy and Yang 2009a, b; Teixeira et al. 2009; Qu et al. 2010), although they all fail to demonstrate significant results. Efforts on the molecular structure and interfacial interaction between the matrix and the fibers may be required to achieve major progress in this research area.

The quest for matrix materials with superb mechanical strength and also being eco-friendly poses another challenge. Biopolymers like starch usually suffer from poor water resistance, lousy tensile strength, and high brittleness when the particle size is large, and therefore amounts of plasticizers (glycerol, etc.) are necessary to guarantee the usefulness of the biopolymers in the environment of considerable heat and pressure (Ke and Sun 2001). No significant effects have been observed in addition to the ductility arising from plasticizers, which, however, may undermine adhesion between the matrix and natural fillers by virtue of the residual sugar when used in starch (Zabihzadeh 2010). Alternatively, PLA exhibits excellent mechanical strength, low toxicity, and barrier properties, while its low temperature of glass transition, above which the weak thermal stability, poor ductility and roughness, and muffled modulus start to prevail, narrows the range of its application (Jamshidian et al. 2010). PLA is considered the most promising biopolymer currently in use, while other biopolymers extracted from cellulose, gelatin, chitosan, and plant-based oils undergo a bunch of disadvantages like the scarcity of sources, as well as the tedious and uneconomic manufacturing process.

To realize the optimal properties, the filler particles need to be shaped quite thin (~ 1 nm) and wide (~ 500 nm), so as to achieve layered structures. High-pressure homogenizers and inline dispersers may perform wonderful work in this tough job. The range of nanoparticles sizes, however, has been found to be unfavorably broad (Kaushik et al. 2010; Tunc and Duman 2011). Moreover, the orientation of particles also comes into play in the tensile properties of nanocomposites, and in macromechanics the dominant influence of the orientation of fibers on the mechanical properties of the composite materials is observed, but there is as yet no control on how nanofibers orient in the matrix. Furthermore, reaggregation, where the particles are lumped together, is likely to take place, and therefore appropriate processing techniques preventing from its occurring should be taken into account.

2.2 Methods of Design and Processing

Currently, there exist a great amount of approaches and schemes aiming at producing nanocomposites with optimal dispersion of fibers across the matrix. Extrusion followed by injection molding is believed to give rise to better exfoliation of fibers in the matrix, although its effect on the properties of product has not been reported (Averous and Boquillon 2004; Averous and Le Digabel 2006). Another frequently used method is melt extrusion, with which the extrusion speed is found to play a significant role in determining the thermal properties, radial expansion, specific energy requirement, and compressibility of the resulting polymer foams (Guan and Hanna 2006; Majdzadeh-Ardakani and Sadeghi-Ardakani 2010). Also, the film stacking method, which compresses a stack of polymer films and fibers, has been reported to inflict a pronounced impact on the tensile strength of the composites (Bodros et al. 2007). Other processing approaches such as melt compounding followed by compression molding (Vignon et al. 1996; Lei et al. 2007), direct melting, solidification one-step in situ intercalative solution polymerization (Huskic and Zigon 2007), injection molding, and solution casting after gelatinization (Fama et al. 2009; Liu et al. 2010), have also been employed. The one-step in situ polymerization method has been reported to exhibit advantages like easy handling and better performance of products. This approach disperses nanofillers in monomer(s), modified by functional groups to enhance interaction between the nanofillers and matrix polymers, or to improve dispersion of nanofillers across the polymer matrix, and then bulk or solution polymerization is performed (Zou et al. 2008; Wei et al. 2010).

New ways of exploring and developing techniques to fabricate and assemble eco-friendly composite materials, through the entire technical flow process from source of raw materials, manufacturing, to processing technology are now undertaken. Some efforts have been made on attaining biodegradable reinforcing fillers giving improved performance of composites. These fillers usually come from natural fibers like cellulose. Another concern is focused on employing recyclable synthetic fibers with thermoplastic composites to reduce the waste of fillers, and also some research is devoted to reusing or recycling the whole composites for the similar purpose. Simultaneously, the current concern is to manufacture composites using traditional production process become eco-friendly by extra reprocessing. Throughout all the stages of development—design, appraisal, manufacture, use, reuse–recycling and disposal—researchers should be fully engaged in reducing waste as much as possible, keeping in mind the environment all the time.

3 Current Eco-friendly Polymer Nanocomposite (EPN)

Here we introduce six types of the most popular EPN, including: (1) EPN from green fillers and (2) EPN from green-base-composite.

3.1 EPN with Green Fillers

Cellulose fibers offer excellent advantages of mechanical strength, low mass density, safe handling, and extensive working conditions, holding great promise as eco-friendly and renewable reinforcement in the future EPN. While supporting the structural rigidity of plants, cellulose can be produced from agricultural waste and thus becomes the most abundant renewable fiber and the most economically feasible source of fibers contemporarily. In the past few years, natural fibers have been serving the purpose of bearing applied load in composite materials.

As a typical example of using natural fibers in the production of composites, the steam exploded hemp fiber was widely used as reinforcement in polypropylene-based composites in the early 1990s (Vignon et al. 1996). In recent years, natural cellulose is extracted from the bark of cotton stalks using alkaline extraction. Reinforced by these natural fibers, of single cell length and width of 9.7 μm, the composites were formed by molding compression with polypropylene matrix. The mechanical strength was also tested on an Instron tensile tester. It was found that the modulus of elasticity of composites to be 502 MPa; while the tensile modulus to be 806 MPa.

Simple alkaline treatment was used to extract natural fibers from milkweed stem and the percentage of cellulose in the fibers was measured as the acid detergent fiber with AOAC method; their performance was also investigated in a variety of composites (Reddy and Yang 2009a, b). X-ray diffractometer was used to identify the shape and position of the peaks representing cellulose, and physical structure of the fibers was determined by percentage crystallinity. With varying pressure, SEM was employed to observe the morphological features of fibers. Using Instron tensile tester, the tensile properties were also determined, such as breaking tenacity, percentage breaking elongation, and Young's modulus. The test results show that milkweed stem contains 75.4 % cellulose fibers of average length of 0.9 mm and diameter of 13.0 μm, and these fibers take a crystallinity of 39 %, breaking tensile strain of 4.7 %, and modulus of 122 g/denier (15.8 GPa). The properties of these milkweed stem fibers bear resemblance to those of prevalent cellulose fibers extracted from cotton and linen, which make possible their application in textile, composite, automotive, and many other industries.

Using three different fractions of wheat bran fibers as fillers: 1.5, 13.5, and 27.1 mg/g, to cast composite films, their effect in biodegradable composites based on cassava starch constituted by glycerol and potassium was studied (Fama et al. 2009). Work carried out using dynamic and quasi-static test to measure the mechanical properties of these fibers showed that with introduction of fibers, the storage modulus and hardening of films are enhanced and can be further improved by increasing the content of fibers; also they can sustain huge deformation before rupture (70 %). Wheat bran contains water-insoluble fibers which are considered responsible for reinforcement. In addition, however, the composites exhibit downgraded moisture content, while the bran content is found increased, and therefore water vapor barrier properties are found to be improved thanks to the increased percentage of wheat bran.

Natural polymers and agro waste cellulose fibers were also reported in research on EPN (Kaushik et al. 2010). The cellulose nanofibrils were separated from wheat straw and plasticized maize starch using steam explosion, acidic treatment, and high shear mechanical treatment. Then they were used to fabricate novel composites based on nanofibril/thermoplastic starch (TPS) by dispersing the nanofibrils into TPS with a high shear mixer in various proportions. The produced nanocomposites exhibited the highest tensile modulus (around 220 MPa) and the highest yield strength (around 6.5 MPa) at a fiber content of 15 %.

Bleached wood pulp was also employed to serve as fibers dispersed into the matrix of commercial grad polyactic to prepare cellulose fibril nanocomposites (Qu et al. 2010), with a chemomechanical method extracting cellulose nanofibrils, which were then dispersed into an organic solvent in a uniform way. Interfacial bonding between the matrix and the fibers is reinforced by adding polyethylene glycol (PEG) as a compatibilizer, and N,N-Dimethylacetamide was used to obtain the produced composite with the solvent casting method. The tensile strength of pure PLA matrix was found to be lowered (30 MPa with 2.5 % tensile strain) after adding cellulose nanofibrils, while adding PEG to the PLA with cellulose nanofibrils led to a significantly improved tensile strength and % elongation, by 56.7 and 60 %, respectively, and by 28.2 and 25 %, respectively, relative to pure PLA. The improvement is ascribed to intermolecular hydrogen bonding, which was found among PLA, PEG, and cellulose nanofibrils by FT-IR analysis.

Bamboo cellulose crystals (BCCs) were also found to be able to reinforce starch composites (Liu et al. 2010), in particular, glycerol plasticized starch. The BCCs were obtained by HNO_3–$KClO_3$ treatment and sulfuric acid hydrolysis, and the structure and morphology were revealed by X-ray diffraction, electron microscopy, and solid-state 13C NMR; Instron tensile tester was employed to investigate the mechanical properties. BCCs were found to possess typical cellulose I structure with diameter ranging from 50 to 100 nm, but morphology depends on the concentration in the suspension. The decrease of BCC content leads to increased breaking elongation, and a high relaxation peak was observed when a considerable amount of cellulose crystals were incorporated. Additionally, the reinforcing efficiency of BCCs is found to be the highest at 8 % loading level.

Corn starch with banana and sugarcane fibers was also reported for production of EPN (Guimaraes et al. 2010). Corn serves as the source of commercial grade starch (about 28 % amylose) and the fibers were extracted from sugarcane bagasse and banana pseudo stem by means of chemomechanical methods, and subsequently, mixing laminates of starch and fibers in a ball mill was carried out. The produced composites exhibited great thermal stability, and in the composite of 70 wt% and 30 % glycerol matrix cracks were found between smooth and rough surfaces by fractographic studies (Fig. 1); also dimples exist in rough regions, indicating the sample was ductile. Adding banana fibers to the starch-glycerol matrix, the Young's modulus of the matrix was increased by 186, 294, and 201 % at fiber contents of 20, 25, and 35 %, respectively. The fiber contents of 20, 25, and 35 wt% give rise to around 129, 141, and 133 % increase of yield strength, while the percent elongation was found to be weakened fivefold, sixfold, and sevenfold for 20, 25, and 35 wt% fiber content, respectively, and the ultimate tensile strength kept invariant relatively. The appreciable modification in yield strength and Young's modulus is attributed to the compatibility between the reinforcement (cellulose chains) and the matrix (starch-glycerol), both chemically and structurally. Researchers also propose that deplastification of starch owing to partitioning glycerol between the matrix and fibers must take responsibility of the improved Young's modulus.

3.2 EPN with Green-Base-Composite

3.2.1 EPN from Cellulose

Cellulose has been identified as a source of biopolymer that can be used as a substitute for petroleum polymers. EPN have been successfully synthesized from cellulose acetate, triethyl citrate plasticizer, and organically modified clay (Misra 2004). The polymer matrix for nanocomposite contains 80 wt% pure cellulose acetate and 20 wt% triethyl citrate plasticizer. Results show that better exfoliated and intercalated structure were obtained from nanocomposites containing 5 and 10 wt% organoclay compared with that of 15 wt% organoclay. Tensile strength of cellulosic plastic reinforced with 10 wt% organoclay improved by 180 % and thermal stability of the cellulosic plastic also increased.

Recently, an active antimicrobial packaging material has been successfully synthesized using methyl cellulose (MC) as the matrix with montomorillionite (MMT) as reinforcement (Tunc and Duman 2011). Carvacrol was then added to the as-prepared MMT/MC composite material to form nanocomposites.

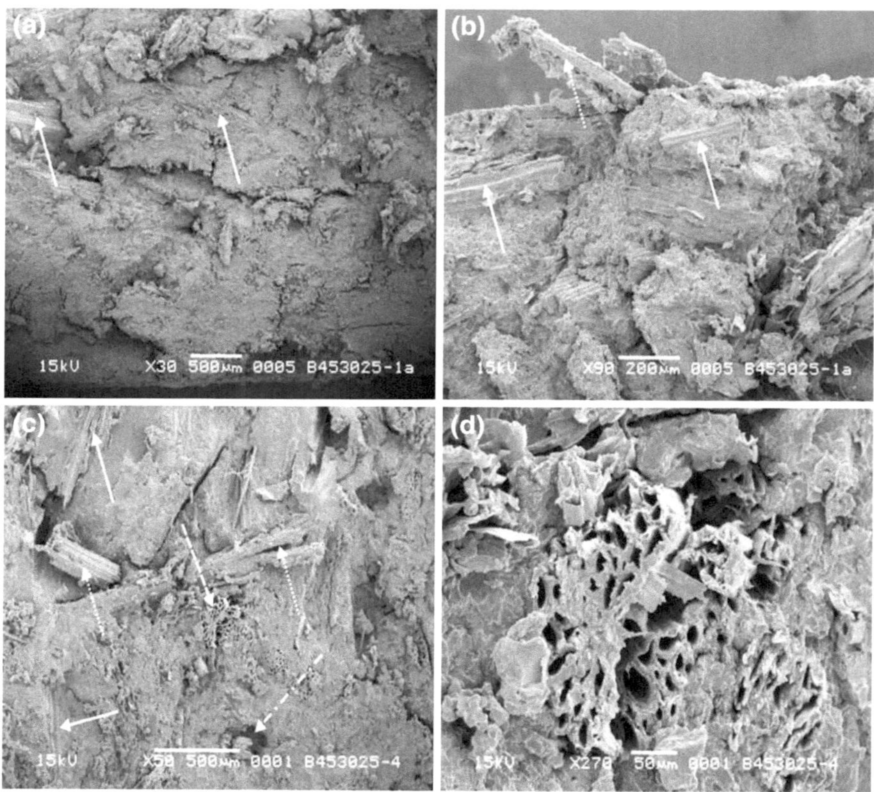

Fig. 1 SEM tensile fractographs of fractographs of 45 % Starch (Amidex-3001)–30 % crude glycerin–25 % banana fiber composite. **a** rough fracture surface with lot of cracks and fibers lying parallel to the crack surface; **b** same a, pulled out fiber and fiber fracture, pull-outs and some voids; **c** fibers lying parallel to the crack surface, some pullouts, ruptured fibers

Thermal analysis result showed the thermal stability of this nanocomposite was found to increase with increase in concentration of MMT. However, this study did not assess the structure of this cellulose-based nanocmposite. Further works in this area by Zimmermann et al. (2011) and Zadegan et. al (2011) studied the cellulose-based nanocomposite for the medical applications. Saska et al. evaluated EPN based on bacterial cellulose-hydroxyapatite prepared by electrospinning for bone regeneration. Fig. 2 shows the SEM images of EPN based on electrospun bacterial cellulose-hydroxyapatite (BC-HA). The authors reported that the BC-HA membranes were effective for bone regeneration in bone defects of rat tibiae. It was found that EPN membranes accelerated new bone formation at the defect sites interestingly the reabsorption of the EPN membranes was slow enough, suggesting that this composite takes longer time to be fully reabsorbed (Saska and Barud 2011).

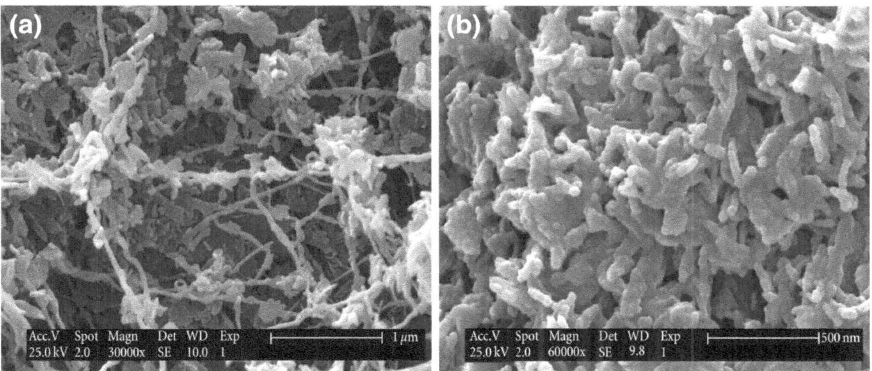

Fig. 2 SEM image of EPN based on bacterial cellulose-hydroxyapatite prepared by electrospinning at 25 kV: **a** 30000x and **b** 60000x (adapted from Saska and Barud 2011)

3.2.2 EPN from Thermoplastic Starch

The early research stage of thermoplastic starch-based composites was focused on the use of plasticized starch as matrix for EPN. The preparation of thermoplastic starch for EPN by melt intercalation in twin screw extruder was first reported by de Carvalho et al. (2001). The composites were prepared with regular corn starch plasticized with glycerin and followed by reinforced with hydrated kaolin. The research result showed a significant incensement in the tensile strength from 5 to 7.5 MPa for the composite from matrix only up to 50 % clay composition. The result also indicates the maximum value of elasticity modulus and tensile strength incorporated in the matrix.

Later, Pandey and Singh (2005) proposed the mechanical and structural properties of resulting composites determined by varying the sequence of addition of plasticizers via solution method. The experiment result deduced that the sequence of addition of components (starch/plasticizer (glycerol)/clay) had a significant effect on the composite formation and the properties of composites were altered without a well-established sequence to determine the appropriate process method. As a result, filler dispersion was found to become highly heterogeneous and highly brittle when starch was plasticized before filling with clay. The authors suggested that improvement in all properties of the composites could be achieved by better dispersion of the clay filler into matrix and changing the sequence of addition of plasticizers is not necessary. Guan and Hanna (2006) produced biocomposites with cellulose fibers by modifying the starch with acetate. X-ray diffraction spectrum showed losses in crystallinity of starch and cellulose after acetate treatment. FTIR spectra indicate the functional groups were maintained after processing. The melting temperatures of the composites change significantly with variation of starch acetate degree. Processing variables such as cellulose content, barrel temperature, and screw speed have a significant effect on thermal and mechanical properties of extruded foams. For example, Tg and Tm of the extruded foams decreased with cellulose content increase.

Further study in starch modification via photo-induced cross linking was done by Kumar and Singh (2008). In this method composite films were prepared from aqueous dispersions of starch with microcrystalline cellulose using glycerol as plasticizer, followed by irradiation under ultraviolet light using sodium benzoate as photo-sensitizer via casting. Consequently, Young's modulus of composites with 5, 10, and 15 wt% of cellulose were improved by 72.41, 42.5, and 32 % respectively.

However, increasing cellulose fiber content and time of photo-irradiation led to decreasing elongation (%) values. Other research on the use of thermoplastic starch without further modification (i.e., changes in experimental conditions) include the work of Lu et al. (2006), Ma et al. (2007), Fama et al. (2009), Kaushik et al.(2010), Liu et al. (2010), Guimaraes et al. (2010), and Kaith et al. (2010). These studies show a significant increase in tensile and thermal properties of thermoplastic starch when the matrix reinforced with nanofibers.

3.2.3 EPN from Polylactic Acid

Ogata et al. (1997) prepared polylactic acid (PLA) based nanocomposites. PLA/organically modified clay mixture was synthesized by dissolving the polymer in hot chloroform adding dimethyl distearyl ammonium modified MMT. As a result, clay existed in the form of tactoids. Tactoids play a key role in the formation of geometrical structures in the blends, which lead to the formation of super-structures in the thickness of the blended film. This structural feature promotes increase in Young's modulus. Bondeson and Oksman (2007) used commercial PLA as matrix with anionic surfactant treated cellulose whiskers as reinforcement. The compounded materials were processed in three steps extrusion. Extruded nano-composites were characterized by compression molded. Compared to PLA, the value of tensile strength and elongation at failure decreased for this composite. This phenomenon was attributed to poor adhesion between the fiber and the matrix (Fig. 3). The reason is directly related to the nature of the fiber and the type of mechanical treatment given.

The thermal, mechanical, and morphological properties of PLA-based composites were investigated and reported by Lee et al. (2008). Compared to the previous study without any improvement, the tensile modulus increased from 62.5 to 169.5 %. More recently, Qu et al. (2010) developed nanocomposites of PLA with a compatibilizer and cellulose fibrils. Bleached wood pulp was used as fiber source and commercial grade PLA as matrix. A hybrid method was used to prepare cellulose nanofibrils dispersed uniformly in an organic solvent. On well dispersion, polyethylene glycol (PEG) was added to the matrix as compatibilizer to improve interfacial bonding/adhesion between the matrix and the fiber. The composites were obtained by solvent casting methods using N,N-Dimethylacetamide. The characterization of PLA reinforced with cellulose nanofibrils resulted in no improvement in both tensile

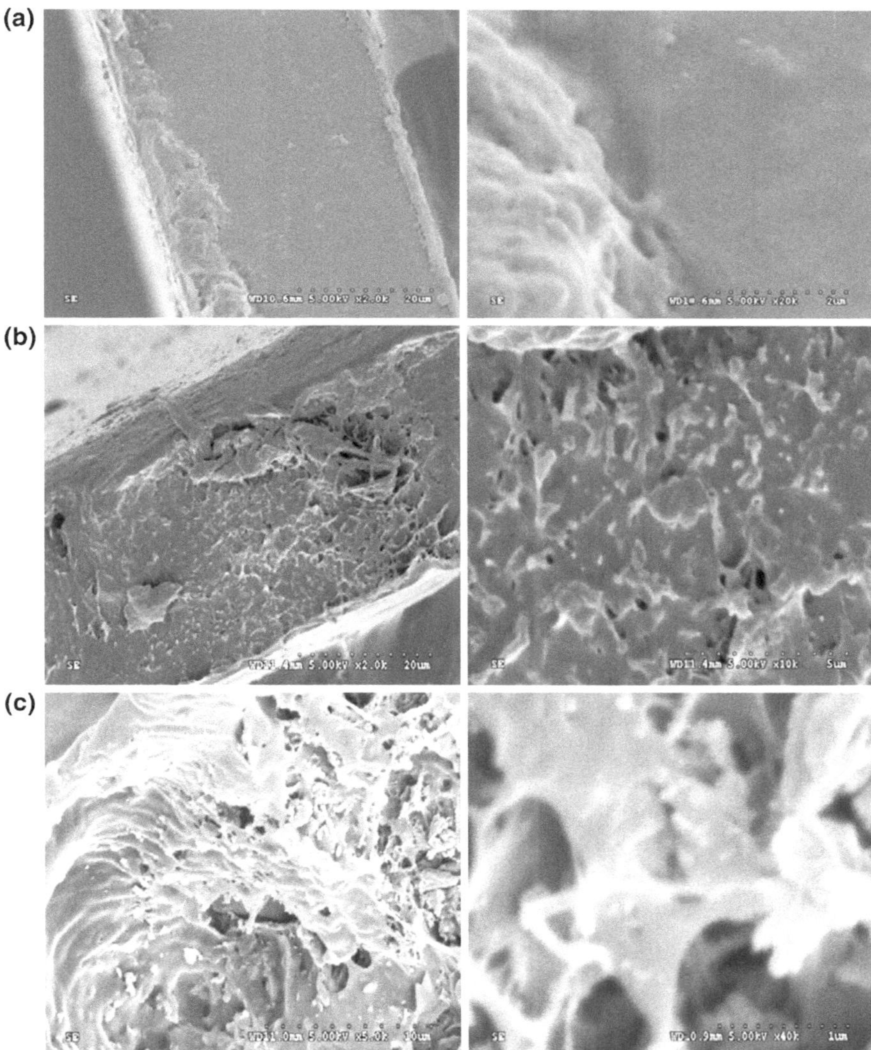

Fig. 3 SEM micrographs of fracture surfaces of pure PLA (**a**); PLA/cellulose nanofibrils (**b**); PLA/cellulose nanofibrils/PEG (**c**). Adapted from Qu et al. (2010)

strength (30 MPa compared with pure PLA) and elongation (2.5 % compared with pure PLA). This is due to poor interfacial bonding between cellulose nanofibrils and the PLA matrix. By adding PEG to the blend of PLA, a significant improvement in tensile strength (28.2 %) and percentage elongation (25 %) can be observed. The mechanism of this enhancement is that PEG covers the surface of cellulose nanofibrils and acts not only as a plasticizer for PLA to improve elongation, but also as a compatibilizer between the hydrophobic PLA and the hydrophilic cellulose nanofibrils. The authors also posited 3 % was the optimum composition of cellulose nanofibrils to obtain better percentage elongation and tensile strength.

3.2.4 EPN from Polymer Mixture

A surge of interest is observed in the biodegradable polymer fabricated by mixing two or more polymers. A great number of polymer composites of this type have been reported, such as starch/PLA, polycarprolactone (PCL)/polyvinyl alcohol, and thermoplastic starch/polyesteramide (DeKesel et al. 1997; Uesaka et al. 2000; Ke and Sun 2001; Martin and Averous 2001; Willett and Shogren 2002). These degradable polymers are regarded as promising candidates for composite applications, due to their widely tested degradability and excellent mechanical properties.

Recently, Sarazin et al. prepared binary and ternary composites formed by polylactide (PLA), PCL and thermoplastic starch (TPS) using a one-step extrusion process and reported their application (Sarazin et al. 2008). At 36 % glycerol, the transition temperatures of TPS/PLA-PCL composites are found to be around −57 °C. The strain at break is also found to increase by tuning up the concentration of TPS in PLA, while the elastic modulus and maximum strength are both weakened at higher percentage of TPS. Nanocomposites formed by mixing natural rubber and starch have been reported. The fractured surface of the starch/natural rubber/clay nanocomposite fabricated with undiluted natural rubber latex, as well as 3 wt% clay, has been presented. Owing to the lack of interfacial adhesion, the thermoplastic starch matrix (gray areas) is apparently separated from dispersed rubber phase (white areas). By contrast, the modified natural rubber leads to finer dispersion and further improves the interfacial adhesion.

3.2.5 EPN from Others

Other biopolymers useful for synthesis of nanocomposites include: (i) gelatin—a water-soluble protein obtained by extracting collagen from animal skin and bones and thermal denaturation. (ii) PHB—a natural product of biosynthesis performed by bacteria in nature. (iii) Chitosan—a natural polymer widely found in exoskeletons of crustaceans and insects, as well as in the cell walls of microorganisms (Maiti et al. 2003; Zheng et al. 2002; Takegawa et al. 2010). Moreover, the mechanical and water vapor barrier properties of chitosan-based nanocomposites with cellulose nanofibers could be enhanced.

4 Conclusions

Thanks to their numerous advantages such as lightweight, strong mechanical strength, excellent barrier properties, chemical resistance, etc., nanocomposites have been broadly used in a variety of applications. The nanocomposites fabricated by biodegradable polymers reinforced with natural fibers are highly appreciated as the most environmentally friendly and, accordingly, diverse research efforts have been devoted to exploring manufacturing "green" nanocomposites with "green" process. Novel polymers with inherent eco-friendly features like renewability and biodegradability are desirable, and a series of natural or synthetic materials have been used, such as thermoplastic starch, gelatin, PLA, chitosan, etc. As for reinforcing fibers, natural fibers are more favorable than synthetic fibers duo to the biodegradability and abundance of their sources: coconut coir, bamboo, flax, etc. The challenge posed by eco-friendly composites also needs considerable attention in terms of poor bonding between matrix and fillers, loose control of fiber orientation, and difficulty in shaping nanoscale particles.

References

Adeosun, S.O., Lawal, G.I., Balogun, S.A. & Akpan, E.I. (2012) Review of green polymer nanocomposites. JMMC 11:385–416(2012)

Amass W, Amass A, Tighe B (1998) A review of biodegradable polymers: uses, current developments in the synthesis and characterization of biodegradable polyesters, blends of biodegradable polymers and recent advances in biodegradation studies. Polym Int 47(2):89–144

Ashori A (2008) Wood-plastic composites as promising green-composites for automotive industries! Bioresour Technol 99(11):4661–4667

Averous L, Boquillon N (2004) Biocomposites based on plasticized starch: thermal and mechanical behaviours. Carbohydr Polym 56(2):111–122

Averous L, Le Digabel F (2006) Properties of biocomposites based on lignocellulosic fillers. Carbohydr Polym 66(4):480–493

Baillie, C. (2004) *Green composites. Polymer composites and the environment.* (Woodhead Publishing Limited & CRC Press LLC)

Bodros E, Pillin I, Montrelay N, Baley C (2007) Could biopolymers reinforced by randomly scattered flax fibre be used in structural applications? Compos Sci Tech 67(3–4):462–470

Bondeson D, Oksman K (2007) Dispersion and characteristics of surfactant modified cellulose whiskers nanocomposites. Compos Interfaces 14(7–9):617–630

Chandra R, Rustgi R (1998) Biodegradable polymers. Prog Polym Sci 23(7):1273–1335

de Carvalho AJF, Curvelo AAS, Agnelli JAM (2001) A first insight on composites of thermoplastic starch and kaolin. Carbohydr Polym 45(2):189–194

DeKesel C, VanderWauven C, David C (1997) Biodegradation of polycaprolactone and its blends with poly(vinylalcohol) by microorganisms from a compost of house-hold refuse. Polym Degrad Stab 55(1):107–113

Fama L, Gerschenson L, Goyanes S (2009) Starch-vegetable fibre composites to protect food products. Carbohydr Polym 75(2):230–235

Guan JJ, Hanna MA (2006) Selected morphological and functional properties of extruded acetylated starch-cellulose foams. Bioresour Technol 97(14):1716–1726

Guimaraes JL, Wypych F, Saul CK, Ramos LP, Satyanarayana KG (2010) Studies of the processing and characterization of corn starch and its composites with banana and sugarcane fibers from Brazil. Carbohydr Polym 80(1):130–138

Huskic M, Zigon M (2007) PMMA/MMT nanocomposites prepared by one-step in situ intercalative solution polymerization. Eur Polymer J 43(12):4891–4897

Jamshidian M, Tehrany EA, Imran M, Jacquot M, Desobry S (2010) Polylactic acid: production, applications, nanocomposites, and release studies. Compr Rev Food Sci Food Saf 9(5):552–571

John MJ, Thomas S (2008) Biofibres and biocomposites. Carbohydr Polym 71(3):343–364

Kaith BS, Jindal R, Jana AK, Maiti M (2010) Development of corn starch based green composites reinforced with Saccharum spontaneum L fiber and graft copolymers—evaluation of thermal, physico-chemical and mechanical properties. Bioresour Technol 101(17):6843–6851

Kaushik A, Singh M, Verma G (2010) Green nanocomposites based on thermoplastic starch and steam exploded cellulose nanofibrils from wheat straw. Carbohydr Polym 82(2):337–345

Ke TY, Sun XZ (2001) Effects of moisture content and heat treatment on the physical properties of starch and poly(lactic acid) blends. J Appl Polym Sci 81(12):3069–3082

Kim JP, Yoon TH, Mun SP, Rhee JM, Lee JS (2006) Wood-polyethylene composites using ethylene-vinyl alcohol copolymer as adhesion promoter. Bioresour Technol 97(3):494–499

Kumar AP, Singh RP (2008) Biocomposites of cellulose reinforced starch: improvement of properties by photo-induced crosslinking. Bioresour Technol 99(18):8803–8809

Lee SY, Kang IA, Doh GH, Yoon HG, Park BD, Wu QL (2008) Thermal and mechanical properties of wood flour/talc-filled polylactic acid composites: effect of filler content and coupling treatment. J Thermoplast Compos Mater 21(3):209–223

Lei Y, Wu QL, Yao F, Xu YJ (2007) Preparation and properties of recycled HDPE/natural fiber composites. Compos Part a-Appl Sci Manuf 38(7):1664–1674

Leja K, Lewandowicz G (2010) Polymer biodegradation and biodegradable polymers—a review. Pol J Environ Stud 19(2):255–266

Liu DG, Zhong TH, Chang PR, Li KF, Wu QL (2010) Starch composites reinforced by bamboo cellulosic crystals. Bioresour Technol 101(7):2529–2536

Lu YS, Weng LH, Cao XD (2006) Morphological, thermal and mechanical properties of ramie crystallites—reinforced plasticized starch biocomposites. Carbohydr Polym 63(2):198–204

Ma XF, Yu JG, Wang N (2007) Fly ash-reinforced thermoplastic starch composites. Carbohydr Polym 67(1):32–39

Maiti P, Batt CA, Giannelis EP (2003) Renewable plastics: synthesis and properties of PHB nanocomposites. Abs Pap Am Chem Soc 225:U665–U665

Majdzadeh-Ardakani K, Sadeghi-Ardakani S (2010) Experimental investigation of mechanical properties of starch/natural rubber/clay nanocomposites. Dig J Nanomater Biostruct 5(2):307–316

Martin O, Averous L (2001) Polylactic acid: plasticization and properties of biodegradable multiphase systems. Polymer 42(14):6209–6219

Misra M, Mohanty AK, Drzal LT (2004) Injection molded 'Green' nanocomposite materials from renewable resources. Global plastics environmental conference

Ogata N, Jimenez G, Kawai H, Ogihara T (1997) Structure and thermal/mechanical properties of poly(l-lactide)-clay blend. J Polym Sci Part B-Polym Phys 35(2):389–396

Okamoto M (2004) Biodegradable polymer/layered silicate nanocomposites: a review. J Ind Eng Chem 10(7):1156–1181

Pandey JK, Singh RP (2005) Green nanocomposites from renewable resources: effect of plasticizer on the structure and material properties of clay-filled starch. Starch-Starke 57(1):8–15

Qin C, Soykeabkaew N, Xiuyuan N, Peijs T (2008) The effect of fibre volume fraction and mercerization on the properties of all-cellulose composites. Carbohydr Polym 71(3):458–467

Qu P, Gao YA, Wu GF, Zhang LP (2010) Nanocomposites of polylactic acid) reinforced with cellulose nanofibrils. Bioresources 5(3):1811–1823

Reddy N, Yang YQ (2009a) Extraction and characterization of natural cellulose fibers from common milkweed stems. Polym Eng Sci 49(11):2212–2217

Reddy N, Yang YQ (2009b) Properties of natural cellulose fibers from hop stems. Carbohydr Polym 77(4):898–902

Rong MZ, Zhang MQ, Liu Y, Yang GC, Zeng HM (2001) The effect of fiber treatment on the mechanical properties of unidirectional sisal-reinforced epoxy composites. Compos Sci Technol 61(10):1437–1447

Sarazin P, Li G, Orts WJ, Favis BD (2008) Binary and ternary blends of polylactide, polycaprolactone and thermoplastic starch. Polymer 49(2):599–609

Saska S, Barud HS, Gaspar AMM, Marchetto R, Ribeiro SJL, Messaddeq Y (2011) Int. J. Biomater. Article ID 175362:8

Takegawa A, Murakami M, Kaneko Y, Kadokawa J (2010) Preparation of chitin/cellulose composite gels and films with ionic liquids. Carbohydr Polym 79(1):85–90

Teixeira ED, Pasquini D, Curvelo AAS, Corradini E, Belgacem MN, Dufresne A (2009) Cassava bagasse cellulose nanofibrils reinforced thermoplastic cassava starch. Carbohydr Polym 78 (3):422–431

Tunc S, Duman O (2011) Preparation of active antimicrobial methyl cellulose/carvacrol/ montmorillonite nanocomposite films and investigation of carvacrol release. Lwt-Food Sci Technol 44(2):465–472

Uesaka T, Nakane K, Maeda S, Ogihara T, Ogata N (2000) Structure and physical properties of poly(butylene succinate)/cellulose acetate blends. Polymer 41(23):8449–8454

Vignon MR, Dupeyre D, GarciaJaldon C (1996) Morphological characterization of steam-exploded hemp fibers and their utilization in polypropylene-based composites. Bioresour Technol 58(2):203–215

Wei LM, Hu NT, Zhang YF (2010) Synthesis of polymer-mesoporous silica nanocomposites. Materials 3(7):4066–4079

Willett JL, Shogren RL (2002) Processing and properties of extruded starch/polymer foams. Polymer 43(22):5935–5947

Zabihzadeh SM (2010) Water uptake and flexural properties of natural filler/hdpe composites. Bioresources 5(1):316–323

Zadegan S, Hosainalipour M, Rezaie HR, Ghassai H, Shokrgozar MA (2011) Synthesis and biocompatibility evaluation of cellulose/hydroxyapatite nanocomposite scaffold in 1-n-allyl-3-methylimidazolium chloride. Mater Sci Eng C-Mater Biol Appl 31(5):954–961

Zheng JP, Li P, Ma YL, De Yao K (2002) Gelatin/montmorillonite hybrid nanocomposite. I. preparation and properties. J Appl Polym Sci 86(5):1189–1194

Zimmermann KA, LeBlanc JM, Sheets KT, Fox RW, Gatenholm P (2011) Biomimetic design of a bacterial cellulose/hydroxyapatite nanocomposite for bone healing applications. Mater Sci Eng C-Mater Biol Appl 31(1):43–49

Zou H, Wu SS, Shen J (2008) Polymer/silica nanocomposites: preparation, characterization, properties, and applications. Chem Rev 108(9):3893–3957

Biodegradable Starch Nanocomposites

N.L. García, L. Famá, N.B. D'Accorso and S. Goyanes

Abstract Biodegradable thermoplastic materials offer great potential to be used in food packaging or biomedical industry. In this chapter we will present a review of the research done on starch and starch nanocomposites. In the case of nanocomposites based on starch, special attention will be given to the influence of starch nanoparticles, cellulose whiskers, zinc oxide nanorods, antioxidants, and antimicrobial inclusion on the physicochemical properties of the materials. The discussion will be focused on structural, mechanical, and barrel properties as well as on degradation, antibacterial and antioxidant activities. Finally, we will discuss our perspectives on how future research should be oriented to contribute in the substitution of synthetic materials with new econanocomposites.

Keywords Starch · Nanocomposites · Starch nanoparticles · Cellulose · Layered silicate · Antioxidant nanofillers · Antimicrobial nanofillers

Abbreviations

σ	Tensile strength
Ag-NP	Silver nanoparticles
B-NC	Bamboo nanocrystals
C-NC	Cellulose nanocrystals
C-NF	Cellulose nanofibers
C-NW	Cellulose nanowhiskers
CH	Chitosan

N.L. García · N.B. D'Accorso (✉)
CIHIDECAR-CONICET; Departamento de Química Orgánica, FCEyN-UBA,
Ciudad Universitaria, 1428, Ciudad Autónoma de Buenos Aires, Argentina
e-mail: norma@qo.fcen.uba.ar

N.L. García · L. Famá · S. Goyanes (✉)
Laboratory of Polymer and Composite Materials, Departamento de Física, FCEyN-UBA,
Ciudad Universitaria, 1428, Ciudad Autónoma de Buenos Aires, Argentina
e-mail: sgoyanes@gmail.com

L. Famá · S. Goyanes
IFIBA-CONICET, Ciudad Autónoma de Buenos Aires, Argentina

V.K. Thakur and M.K. Thakur (eds.), *Eco-friendly Polymer Nanocomposites*,
Advanced Structured Materials 75, DOI 10.1007/978-81-322-2470-9_2

CH-NP	Chitosan nanoparticles
CA-S-NP	Citric acid-modified starch nanoparticles
CO_2	Carbon dioxide
CMC	Carboxymethylcellulose sodium
D	Diameter
DAS	Dialdehyde starch
F-CN	Flax cellulose nanocrystals
HC-NC	Hemp (Cannabis sativa) cellulose nanocrystals
k	Aspect ratio
L	Length
MCC	Microcrystalline cellulose
MEO	Pennyroyal
MFC	Microfibrillated cellulose
MMT	Montmorillonite
PLA	Polylactic acid
PO_2	Oxygen permeability
PVA	Polyvinyl alcohol
REX	Reactive extrusion
RTE	Ready-to-eat
S-NP	Starch nanoparticles
SME	Specific mechanical energy
T_d	Decomposition temperature
T_g	Glass transition temperature
TiO_2	Titanium oxide
TiO_2-NP	Titanium oxide nanoparticles
T_m	Melting temperature
TPS	Thermoplastic starch
WVP	Water vapor permeability
UV	Ultraviolet spectroscopy
wt%	Weight percentage
ZEO	Zataria multiflora Boiss
ZnO	Zinc oxide
ZnO-NP	Zinc oxide nanoparticles
ZnO-NP-CMC	Zinc oxide nanoparticles-carboxymethylcellulose sodium
ZnO-NR	Zinc oxide nanorods

1 Introduction

Nowadays, petroleum-derived polymers are the most widely used materials in the packaging industry. However, there are important problems related with their use, such as no renewability, high costs, and potential pollution they can create (Thakur et al. 2012a, b, c, d, e). The extensive degradation time associated to these materials, which take hundreds of years and involves the production of high CO_2 levels, is the main cause of the environmental pollution and residues accumulation produced (Chaudhry et al. 2008; de Azedero 2009; Arora and Padua 2010; Vieira et al. 2011; González Seligra et al. 2013). A new generation of materials based on biopolymers will reduce the polymers and plastics industry dependency on petroleum, creating more sustainable alternatives (Thakur et al. 2014a, b, c, d, e).

Bio-based polymers are derived from renewable resources such as plant and animal mass from CO_2. They can be divided into two groups: natural or synthetic polymers. Natural bio-based polymers are polymers synthesized by living organisms such as animals, plants, algae, and microorganisms. The most abundant bio-based polymers in nature are polymers from biomass (from gro-resources) as polysaccharides: starches (Famá et al. 2005, 2006; Rojas-Graü et al. 2007; Flores et al. 2007; Ma et al. 2008a; Jiménez et al. 2013; Lamanna et al. 2013; Souza et al. 2013), lignocellulosic products (Santiago-Silva et al. 2009; Sayanjali et al. 2011; Pastor et al. 2013) and others as pectin, chitosan/chitin, and gums (Norajit et al. 2010; Martins et al. 2012; Elsabee and Abdou 2013; Rubilar et al. 2013), protein and lipids (Bourtoom 2009; Jiménez et al. 2010; Murillo-Martínez et al. 2011), and plants (Orliac et al. 2003; Bertan et al. 2005). In particular, polysaccharides have been the focus of research in recent decades as base materials for the development of biodegradable products (Xie et al. 2011a; Thakur et al. 2014a; Thakur and Thakur 2014a). These are a good alternative to perform biodegradable formulations due to their "green" connotation (environmentally friendly material): they are biodegradable, edible, and compostable (Ptaszek et al. 2013; He et al. 2014).

Another topic that was exhaustively investigated during the last few years was the development of biorenewable polymers-based hydrogels. They have attracted great interest for miscellaneous applications including biomedical, toxic ion removal, and water purification (Thakur and Thakur 2014a, b, c).

Constant demands for biodegradability in many products lead to the search for materials that can be produced in significant amounts and are versatile in terms of their properties and potential applications (Singha and Thakur 2009a, b, c, d, e). In this context, recent innovations in polymer materials are widely discussed in the literature, presenting improvements in packaging, surgery, pharmaceutical, biomedical, hygiene, and food industrial uses (Bierhalz et al. 2012; Bouyer et al. 2012; Prajapati et al. 2013; Ghori et al. 2014; Suvakanta et al. 2014; Famá et al. 2014). In particular, cellulose and clays are two of the more common fillers used in the development of biodegradable compounds (Tjong 2006; Lu and Mai 2007; Thakur et al. 2010a; Müller et al. 2011; Sadegh-Hassani and Nafchi 2014; Thakur et al. 2014b; Thakur et Thakur 2014b).

One of the most used polymers for the development of biodegradable materials is the starch, mainly due to its low cost and the possibility of obtaining it from several renewable resources (Tharanathan 2003; Lu et al. 2005; Romero-Bastida et al. 2005; Sorrentino et al. 2007; Talja et al. 2007). The use of different starches in their natural form or as modified by chemical cross-linking has been extensively investigated. Some of the advantages of modified starch-based films are the improvement of mechanical and barrier properties, as well as thermal stability

(Ghanbarzadeh et al. 2011; Zuraida et al. 2012; Olsson et al. 2013; Gutiérrez et al. 2014a, b).

In order to help food preservation, several researches have considered the use of many additives or food components into biodegradable starch films as a way of improving food shelf life (García et al. 1998; Petersson and Stading 2005; Famá et al. 2010, 2011; Gutiérrez et al. 2014a, b). The great research boom in the use of starch-based composites is reflected in the significant number of patents and papers in the literature (Arvanitoyannis et al. 1998; Rindlav-westling et al. 1998; Myllärinen et al. 2002; Famá et al. 2007; Hansen and Placket 2008; Ma et al. 2008a; Famá et al. 2009a, b; Goyanes et al. 2010; Li et al. 2011b; García et al. 2011; Famá et al. 2012).

However, so far there are few applications that have been accomplished. Although starches have many advantages over synthetic materials, their use is still strongly limited because they have worse mechanical properties and higher permeation compared to other nonnatural polymers (Hansen and Plackett 2008; Dhakal and Zhang 2012). The incorporation of micro- and nano-sized fillers into starch materials has been the topic of many studies in order to overcome these disadvantages (Ma et al. 2009; Famá et al. 2009a, b, 2010, 2011, 2012; Cheviron et al. 2014; Lopez et al. 2014).

Many packaging industries are trying to implement composites based on starch as new technologies in bags, plates, cups, bowls, and coatings (Wong et al. 2003; Avérous and Boquillon 2004). In addition, the biomedical industry has shown a great interest in the development of starch nanocomposites for their use as sensors or stimulators of bone cells (Torres-Castro et al. 2011; Xie et al. 2011a, b). However, there are increasing scientific evidences reporting that nanoparticles can cross cellular barriers and may lead to oxidative damage and inflammatory reactions (Chaudhry et al. 2008; Lewinski et al. 2008; Bouwmeester et al. 2009).

It has been well established that the filler properties and size greatly influences mechanical properties of the final composite (Lin et al. 2011a, b). In general, smaller sized fillers improve mechanical, electrical, and thermal properties (Biercuk et al. 2002; Kilbride et al. 2002; Wei et al. 2002; Sandler et al. 2003) of the composites.

Besides improving physicochemical properties of biomaterials, the addition of certain nanofillers with antimicrobials and/or antioxidants characteristics can bring important properties such as good resistance to bacteria, mildew, and insect attack; and can also alter nutrition, flavor, texture, heat tolerance, and shelf life of the product (Lu et al. 2006; Zhao et al. 2011; Li et al. 2012; Liu and Kim 2012; Martinez-Gutierrez et al. 2012; Pérez Espitia et al. 2012).

In this chapter, a review of several researches done on the development and characterization of nanocomposites based on starch will be presented. Special attention will be given to the influences of the incorporation of starch, cellulose, layered silicate, and antioxidant and/or antimicrobial nanofillers on the physicochemical properties of the composites. The discussion will be focused on structural, mechanical, and barrel properties as well as on degradation, antibacterial and antioxidant activities.

2 Starch

2.1 Characteristics and Properties

The scientific and business community finds in starch a good alternative for biodegradable packaging applications, as it is a renewable resource widely available that can be obtained from different agricultural surplus and industrial leftovers from raw materials processing. Besides, among all biodegradable polymers, its cost is relatively low. Starch naturally occurs in a variety of botanical sources such as cereal grains (corn, wheat, and rice), seeds, legumes (lentils), and tubers (potato and cassava). In addition it is completely compostable without toxic residues, odorless, tasteless, colorless, nontoxic, and biologically absorbable (Wong et al. 2003).

Starch is a polymeric carbohydrate composed of anhydroglucose units, which is deposited in plant tissues as insoluble semicrystalline granules that vary in shape, size and structure, depending on their origin. Usually, it is mostly composed by a mixture of two biopolymers (glucans): amylose (straight chain) and amylopectin (branched chain). The structure of the starch granule depends on the way in which amylose and amylopectin are associated and distributed (Zavareze and Guerra Días 2011). The ratio between amylose and amylopectin varies depending on the starch source. In regular starches, amylose constitutes about 15–30 % of total starch (Hoover et al. 2010).

Amylose is essentially a linear structure of α-1,4 linked glucose units (Fig. 1a). The large number of hydroxyl groups in its structure confers hydrophilic properties to the polymer, resulting in a material with high affinity for moisture. Because of its linear nature, mobility and the presence of many hydroxyl groups along the

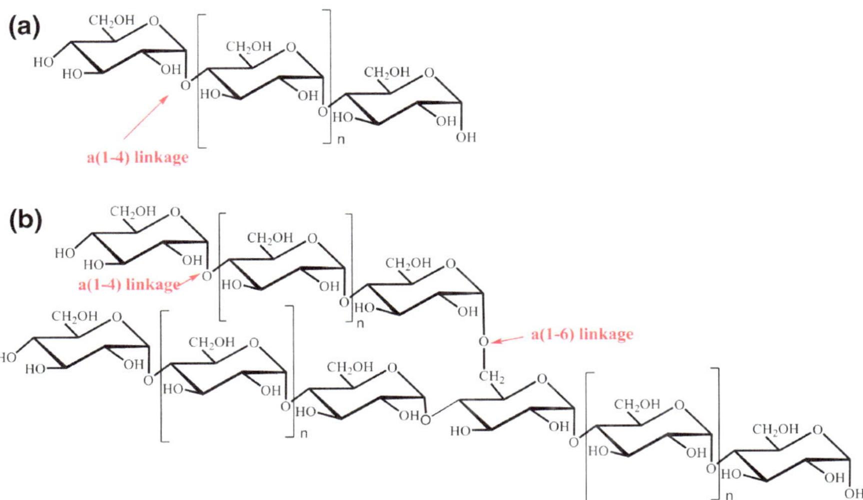

Fig. 1 Chemical structures of amylose (**a**) and amylopectin (**b**)

polymer chains, amylose molecules tend to orient themselves parallel to each other and approach each other near enough to allow formation of hydrogen bonds between adjacent chains. It is widely accepted that the most favorable three-dimensional conformation for amylose is a double helix, which is able to pack into regular arrays (Zobel 1994).

Amylopectin is a highly branched structure of short α-1,4 chains linked by α-1,6 bonds containing 10–60 glucose units and side chains with 15–45 glucose units with an average of 5 % of links α-(1–6) in branching points (Fig. 1b) (Van der Maarel et al. 2002). The molecular weight of amylopectin is about 1000 times the molecular weight of amylose and ranges from 1×10^7 to 5×10^8 g/mol (You et al. 2002). Amylopectin chains are arranged radially within the granule with their nonreducing terminal ends oriented toward the surface. These are composed by alternating crystalline areas (as a double helix) and amorphous areas with regions of branching points (Zavareze and Guerra Días 2011).

Amylose content and branches length and placement in amylopectin are the major determinant factors of starch functional properties, such as water absorption, gelatinization and pasting, retrogradation, and susceptibility to enzymatic attack (Jane 2007; Copeland et al. 2009). According to their origin, starches possess differing amylose and lipid-complexed amylose contents, amylopectin chain-length distributions, relative crystallinity, microstructures, swelling behavior, gelatinization properties, and pasting/rheological characteristics (Bertolini et al. 2003; Kim and Huber 2008; Salman 2009). Due to their strong and flexible structure, transparency (derived from the linear structure of amylose), and their resistant to fats and oils, starch films are useful for numerous applications in food industry. Their functional properties depend on the source but are also affected by other factors like chemical modification, system composition, pH, and ionic strength of the media.

Table 1 shows the amylose and amylopectin content of starches from different sources.

Table 1 Amylose and amylopectin content of starch from different sources

Type of starch	Amylose (%)	Amylopectin (%)	References
Amylomaize	48–77	23–52	Cuq et al. (1977); Takeda et al. (1989)
Banana	17–24	76–83	Cuq et al. (1977); Moongngarm (2013)
Corn	17–25	75–83	Cuq et al. (1977); Sandhu et al. (2007)
Cush-cush Yam	9–15	85–91	Gutierrez et al. (2014a); Pérez et al. (2012)
Chickpeas	30–40	60–70	Cuq et al. (1977); Polesi et al. (2011)
High-amylose corn	55–70	30–45	Sandstedt (1961); Halsall et al. (1948); Cuq et al. (1977)
Potato	17–24	76–83	Rosin et al. (2002); Svegmark et al. (2002); Vasanthan et al. (1999)
Rice	15–35	65–85	Cuq et al. (1977)
Sorghum	25	75	Cuq et al. (1977)
Tapioca (cassava)	19–22	28–81	Cuq et al. (1977); Pérez et al. (2012); Gutierrez et al. (2014a)
Wheat	20–25	75–80	Cuq et al. (1977)
Waxy	<1	>99	Cuq et al. (1977)

Starches also contain minor components such as protein (0.05–0.5 wt%), lipids (0.1–1.7 wt%), inorganic substances (0.1–0.3 wt% of ash), and non-starch polysaccharides (Liu 2005), which can interact with many additive components (Baker et al. 1994; Garcia et al. 1998). This makes them widely used in packaging industry, because in addition to protecting the products against shock and vibration that occur during transport, they can impart antioxidant and/or antimicrobial properties.

In recent years, several attempts have been made to apply starch in the manufacture of films. However, even today, industrial products associated with the production of starch films are scarce, frequently due to problems related to the extrusion of starch, which results much more complex for this polymer than for traditional plastics.

2.2 *Processing*

Similar and adapted techniques to those used for processing conventional synthetic thermoplastics have been used to process starch. Those include solution casting, internal mixing, extrusion, injection molding, and compression molding (Shank and Kong 2012). However, in the case of starch, one important factor to consider in all processes is the presence of water. Water and/or plasticizers are essential for starch processing. The dry starch melting temperature (T_m) is generally bigger than its decomposition temperature (T_d), as it was shown in the works of Liu et al. (2008) and Russel (1987). However, as moisture content increases, water acts as a plasticizer decreasing the T_m gradually, allowing starch treatment. One important problem that could influence the mechanical properties of the final material is the water evaporation during starch processing, which generates instabilities and occluded bubbles. The materials processed exclusively with water are usually brittle. For that reason, not easily evaporated plasticizers, such as polyols, are employed.

More emphasis will be made on the starch processing by casting and extrusion, being them the most used at laboratory and industrial scale.

2.2.1 Casting

The most widely used technique for the laboratorial production of starch-based films is casting (Vicentini et al. 2002; Mali et al. 2005; Famá et al. 2006, 2011, 2012; Müller et al. 2008; García et al. 2009a). The production of films by the classical casting technique consists in pouring on small plates or molds, an aqueous suspension of gelatinized starch, plasticizer and other additives, followed by a drying process where water is evaporated. The starch granule gelatinization is achieved heating the suspension before pouring it (Fig. 2). The average thickness of the resulting films is controlled by the mass of suspension poured on the plate. Several studies report that suspension's drying takes place at room temperature or in

Fig. 2 Pouring suspension of starch gelatinizated

Fig. 3 Film of thermoplastic starch detached from its mold once the drying process is finished

ovens with forced air circulation at moderate temperatures (30–50 °C), requiring drying times of 10, 24, or 48 h (Godbillot et al. 2006; Müller et al. 2009; García et al. 2011 among others). The reported average size of films is diverse, and varies according to the size of the mold used.

Figure 3 shows how these films can be easily detached from its mold once the drying process is finished according to the different techniques of evaporation applied.

The main advantage of this method is its simplicity. Besides, the films obtained by casting are in general adequate testing structures for determination of barrier, mechanical, and other relevant properties.

Different and varied techniques are applied on the preparation of starch suspensions for the casting method. For example, Wang et al. (2014) gelatinized a mild oxidized cornstarch with urea contents between 0 and 50 wt%, relative to the total dry basis, at 90 °C for 90 min. The obtained suspension was poured in glass dishes and dried at room temperature.

López and García (2012) gelatinize ahipa and cassava starch at 90 °C for 20 min; for the filmogenic suspension onto Petri dishes and dry them in ventilated oven at 60 °C for 2 h.

Ghasemlou et al. (2013), transferred a potato starch and glycerol suspension to a water bath at 90 °C for 10 min, and agitated by magnetic stirrer at 500 rpm. After cooling at room temperature, about 70 mL of the sample was spread evenly over a Teflon casting plate (15 cm of diameter) placed on a leveled surface. The drying process continued for 48 h at 30 % RH and 20 °C.

García et al. (2009) and Famá et al. (2006), emphasize that before the casting procedure, the air bubbles contained in the starch suspension must be removed. This can be achieved lowering the pressure in the suspension's container using a vacuum mechanical pump. Once the removal is complete, the suspension is poured in a plastic mold and dried in an air convection oven at 50 °C for 24 h.

In general, optimal film formation, where a film is easily removable from the mold and immediately suitable for further studies, is strongly influenced by the starch source and preparation conditions of the film such as glycerol content, heating time, and heating temperature. Koch et al. (2010) examined the effects of manufacturing time, processing temperature, and plasticizer content on the molecular structure of high-amylose maize starch films. They demonstrated that glycerol played an important role in film formation. In particular, for films prepared at 140 °C for 15 min, when no glycerol was added a non-cohesive, wavy film with curled up areas was obtained (Fig. 4d). In contrast, films prepared under the same conditions but with a glycerol content of 30 % showed cohesivity, but kept exhibiting highly curled up borders (Fig. 4b). Slightly higher glycerol content (34 wt%) was required for smooth, even, and cohesive films (Fig. 4c). On the other hand, the heating time and heating temperature also showed detrimental effects on film formation. In particular, for samples with 34 wt% of glycerol, heating the starch slurry for 30 min instead of 15 min at 140 °C resulted in wavy and non-cohesive films (Fig. 4e), while highly fragmented films (Fig. 4f, g) resulted from heating at 150 °C for 5 and 15 min.

An important disadvantage of the casting technique lies on the fact that it is a batch method that is not able to be industrially replicated which does not allow an easy control on the sample film thickness and uniformity, and rarely allows samples larger than 100–200 cm^2. The sample with the mildest treatment (Fig. 4a), i.e., lower temperature but the same heating time as sample (Fig. 4b), was a cohesive film.

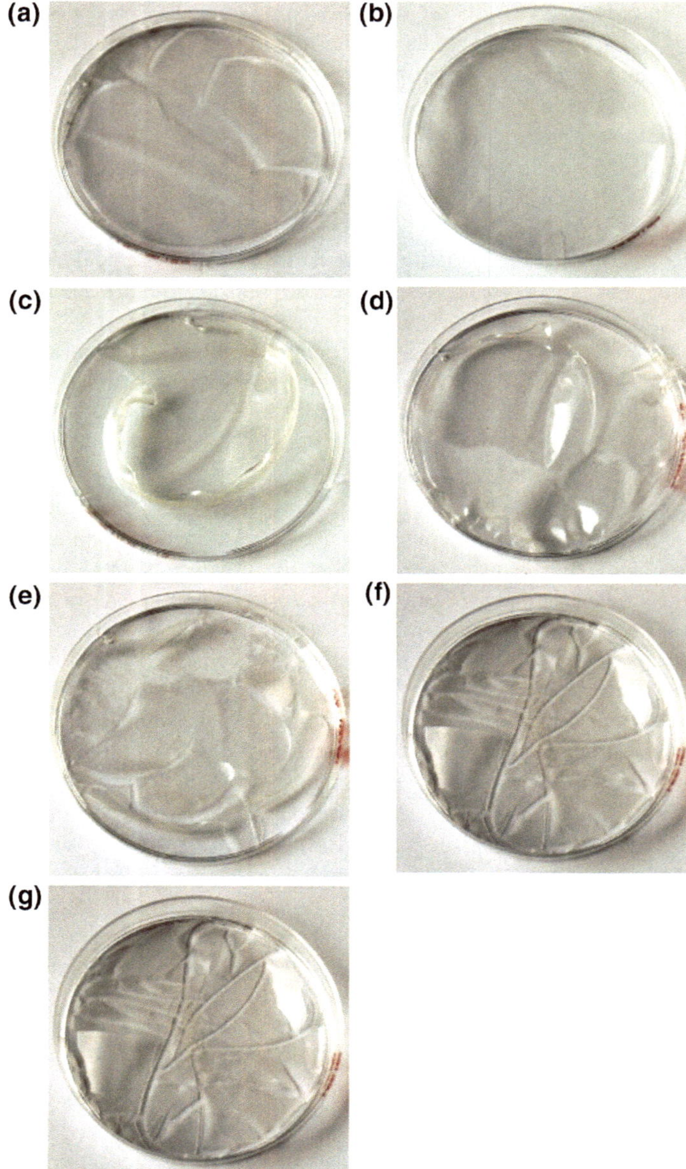

Fig. 4 Digital images of solution-cast high-amylose maize starch films prepared under different experimental conditions. Reproduced with permission from Koch et al. (2010). © 2013, Elsevier Ltd

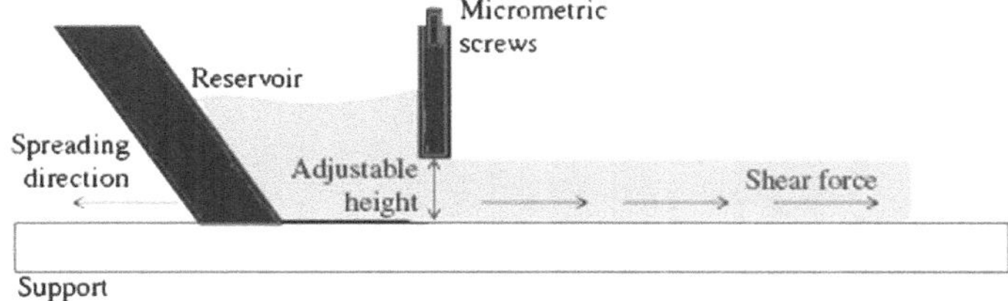

Fig. 5 Sketch of the tape casting process [adapted from Wonisch et al. (2011)]. Two micrometric screws control the gap between the doctor blade and the support. Reproduced with permission from Oliveira de Moraes et al. (2013). © 2013, Elsevier Ltd

An alternative to produce biodegradable films with controlled thickness is a technique named tape casting. This technique (also known as spread casting or knife coating) is well known in paper, plastic, ceramics, and paint manufacturing industries (Richard and Twiname 2000). In tape casting process a suspension is placed in a reservoir with a blade, whose height can be adjusted with micrometric screws (Hotza 1997). The suspension is cast as a thin layer on a support (tape) due to the movement of the carrier tape (continuous process) or the movement of the doctor blade (batch process) (Larotonda 2007). The equipment consists in a constant speed guide driving a knife maintained at an adjustable gap from the film support. This technique allows a very strict control over the thickness and film length. The spreading of the film forming solution (or suspension) can be done on larger supports or on a continuous carrier tape.

The formed film is dried on the support, by heat conduction, circulation of hot air (heat convection) and infrared, resulting in a reduction of its thickness.

A sketch of the tape casting process (adapted from Wonisch et al. 2011) is presented by Oliveira de Moraes et al. (2013) and can be observed in Fig. 5. They obtained, through this technique, cassava–glycerol films reinforced with cellulose fibers. The results showed that tape casting is a suitable technology to scale-up the production of starch-based films.

However, despite these possible methods for forming scalar thermoplastic films, for large-scale manufacture, an extrusion process is usually used.

2.2.2 Extrusion

In gelatinization, when native starch granules are heated in water, their semicrystalline nature architecture is gradually disrupted, resulting in the phase transition from an ordered granular structure into a disordered state in water (Lelievre 1974; Atwell et al. 1988; Ratnayake et al. 2008). This is an irreversible process that includes granular swelling, native crystalline melting (loss of birefringence), and molecular solubilisation in time–temperature sequence (Sullivan and Johnson 1964). Therefore, the gelatinization/melting behavior of starch is quite different when shear treatment is imposed (Xie et al. 2006).

Extrusion is an energy efficient system able to break down the starch granule structure through a combination of high shear, temperature and pressure and can successfully melt starch. Typical single or double screw extruders, in general form consists of a hopper, barrel, feed screw, thermocouples, and dies.

In the bibliography, there are two main types of extrusion: reactive extrusion and extrusion-cooking.

The reactive extrusion (REX) is used primarily in the chemical modification of starch, and to add cross-link agents or to make copolymers. The grafting of monomers from starch as single units, such as ring-opening of epoxides, esterification (with lactones, anhydrides, acids, halides, or vinyl esters), phosphorylation, and silylation; graft polymerization from starch by radical-induced grafting or the ring-opening polymerization of lactones; reactive compatibilization with polyesters and polyolefins by grafting to or from starch; cross-linking of starch with epichlorohydrin or by phosphorylation; and the degradation of starch thermally or catalyzed by acid or enzymes are some of the processes that use REX (Moad 2011). The use of extrusion can be used to produce modified starches in a continuous process with a more consistent product quality. The extruder has the advantage of a being an excellent mixing device and is particularly suitable for processing highly viscous fluids (such as gelatinized starch). Thus, with the use of REX, starch modification can be performed in a homogeneous medium.

There are several studies on the extrusion of modified starches. Chemical starch modifications, such as esterification, etherification, or oxidation before thermoplasticization process, reduce water sensibility and enhance mechanical behavior of the obtained thermoplastic starch (TPS) materials (Gaspar et al. 2005). Likewise, Thunwall et al. (2008) reported that the use of oxidized and hydroxypropylated starches, as well as high plasticizer content could increase the TPS melt tenacity (ability of the melt to deform without rupture), one of the potential limitations in their processing. Different formulations based on native and acetylated cornstarches and glycerol were made by López et al. (2013), who could obtain biodegradable films from thermoplastic native and acetylated cornstarches and glycerol as plasticizer, employing the blowing technique.

Generally, extrusion technology used in food industry is referred as extrusion-cooking. It has been employed for the production of so-called engineered food and special feed. Extrusion-cooking of vegetable raw materials consists in the extrusion of grinded materials at barothermal conditions. With the help of shear energy, exerted by the rotating screw, and additional heating by the barrel, the food material is heated to its melting point or plasticizing point (Moscicki and Van Zuilichem 2011). In this changed rheological status, the food is conveyed under high pressure through a die or a series of dies, and the product expands to its final shape. This results in different physical and chemical properties of the extrudates in comparison to raw materials used.

Nowadays, extrusion-cooking is used for the production of different food staff, ranging from the simplest expanded snacks to the highly processed meat analogues. The most popular extrusion-cooked products are: direct extrusion snacks, RTE (ready-to-eat) cereal flakes and diverse breakfast foods produced from cereal

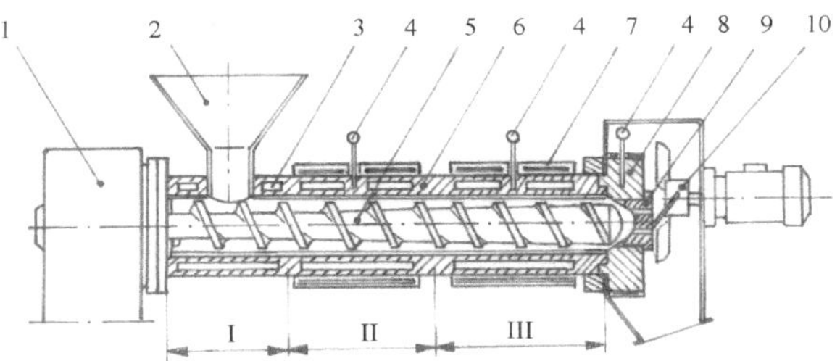

Fig. 6 A cross section of a single-screw extrusion-cooker: *1* engine, *2* feeder, *3* cooling jacket, *4* thermocouple, *5* screw, *6* barrel, *7* heating jacket, *8* head, *9* net, *10* cutter, *I* transport section, *II* compression section, *III* melting and plasticization section. Reproduced with permission from Moscicki et al. (2013). All published manuscripts are licensed under a Creative Commons Attribution 3.0 Unported License

material and differing in shape, color and taste, snack pellets—half products suitable for fried or hot air expanded snacks, precooked pasta, baby food, precooked flours, instant concentrates, functional components, pet food, aquafeed, feed concentrates and calf milk replacers, among others.

A cross section of a single-screw extrusion-cooker is shown in Fig. 6 (Mościcki et al. 2013).

Mitrus y Moscicki (2014) showed, for example, that the application of extrusion-cooking technique to process starch–plasticizer mixtures can be one of the most economical and efficient ways to produce TPS loose-fill foams. Then extrusion-cooking technique can be successfully employed for starch-based foams production.

Application of food extruders gives much better results in processing of starch-based materials than conventional plastic extruders due to the plant origin of the biopolymer. Most of the experimental works which apply extrusion-cooking for the production of starchy loose-fill foams started in the Department of Food Process Eng., Lublin University of Life Sciences in 2012. Their objective is to achieve commercially acceptable biodegradable products based on locally produced potato, corn- and wheat starch, which can replace popular expanded polystyrene loose-fill foam products. Results of the first phase of this study are presented in their work (Mitrus and Moscicki 2014).

More recently, Zhang et al. (2014) presented a new gelatinization technology, the improved extrusion-cooking technology, which is reconstituted from traditional single-screw extruders. Conventional extrusion-cooking is a continuous high-temperature and short-time process, which physically modifies moistened expansible starchy and proteinaceous materials, causing them to swell through the use of the unique combination of high temperature, pressure, and shear forces. Compared to the traditional extrusion-cooking machines, single-screw extruder transformed by Zhang et al. (2014) shows the characteristics of a longer screw (1950 cm), longer residence time (40–68 s), higher die pressure (13.4–19.1 MPa), lower temperature (69.8–120.2 °C), and lower screw speed (20.1–32.6 rpm) than traditional extrusion-cooking extruders. Furthermore, a new forming mold and a rotary cutting knife which are not included in traditional extruders are added (Liu et al. 2011). The retrogradation behavior of rice starch with high amylose content treated by improved extrusion-cooking technology was investigated by morphologic studies, retrogradation enthalpy, percentage of retrogradation, relative crystallinity, and FTIR absorbance ratio.

It is well know that in the plasticization of starch, a phase transition occurs and the phase transition degree determines the process ability and the final product properties. Xie et al. (2014) claim that during the processing of starch, it is also important to know and control the rheological behavior of plasticized starch to prevent flow engineering problems and maintain the final product quality (Xie et al. 2012). Then, in processed plasticized starch-based materials, the phase transition and rheology are the two most important aspects to take into account, and the understanding of the materials melt flow results necessary.

In general form, it is considered that plasticized starch melt has a shear-thinning behavior, which can be expressed in the power–law equation:

$$\eta s = K\dot{\gamma}^{(n-1)} \tag{1}$$

where ηs is the viscosity, $\dot{\gamma}$ is the shear rate, K is the consistency coefficient, and n is the power-law index (<1). The K value is dependent on thermomechanical treatment, temperature and plasticizer content, which can be expressed as (Tajuddin et al. 2011):

$$K = Ko \exp\left(\frac{E}{RT} - \alpha\text{MC} - \alpha'\text{GC} - \beta\text{SME}\right) \tag{2}$$

where MC is the moisture content, GC is the glycerol content, SME is the specific mechanical energy, T is the temperature, E/R is the reduced flow activation energy, and Ko, α, α', and β are coefficients.

During extrusion processing, the structural modification of starch, which is directly related to the viscosity, depends on the specific mechanical energy, feeding rate, screw speed, screw profile, and barrel pressure in the extruder.

SME is the amount of mechanical energy (work) dissipated as heat inside the material, expressed per unit mass of the material. Specifically, it is the work input from the drive motor into the material being extruded and thus provides a good characterization of the extrusion process, then it is an important process parameter influencing the final product characteristics such as solubility, extrudate density, expansion index, hardness, etc. (Godavarti and Karwe 1997). This energy input into the material is an important parameter because it is related with the product physical and chemical transformations. The heat generated due to viscous dissipation is proportional to the material´s volume. In a corotating twin screw extruder, viscous dissipation of the mechanical energy predominates, especially at low moisture contents, making the extrusion process highly energy efficient and cost-effective (Guerrero et al. 2012).

One of the more straightforward SME calculations is as follows:

$$\mathrm{SME} = \frac{P \times \tau \times \frac{\mathrm{RPM}_{\mathrm{act}}}{\mathrm{RPM}_{\mathrm{rated}}}}{\dot{m}} \tag{3}$$

where P is the motor power, expressed in kW, τ is the difference between the running torque and the torque when the extruder is running empty divided by the maximum allowable torque, $\mathrm{RPM}_{\mathrm{act}}$ is the actual screw RPM, $\mathrm{RPM}_{\mathrm{rated}}$ is the maximum allowable screw RPM, $\dot{m}$ is the mass flow rate of the system (kg/sec).

The output of the calculation will be in kJ/kg (Plattner et al. 2011).

Vergnes y Berzan (2010) founded that starch transformation can be predicted with an error of less than 10 %. They have showed that starch transformation during twin screw extrusion can be accurately predicted using dedicated software in which viscosity and SME are coupled. This may allow the processing conditions or screw profile to be optimized and scale-up problems to be solved.

Mitrus and Moscicki (2014) reported the value of SME in extrusion of starchy components for production of starchy loose-fill foams which can replace expanded polystyrene loose-fill foam products. They prepared blends of different starches mixed with up to a 20 wt% of glycerol; with functional additives such as talc, polyvinyl alcohol, and foaming agent. During the extrusion-cooking process, energy consumption was measured with a wattmeter connected to the extruder and the specific mechanical energy (SME) input was calculated (Janssen et al. 2002). Average value of SME for extrusion-cooking of TPS-based foams was in the order of 2.52×10^5 J kg^{-1}, which is equivalent to 0.07 kW h kg^{-1}. They concluded that the SME value depended on the material composition, temperature, and the screw rotation speed used during processing.

Zhu et al. (2010) extruded blends of high-amylose cornstarch and soy protein concentrate. The increasing SME with increasing screw speed observed in this work is in agreement with previous reports (Jin et al. 1994; Akdogan 1996; Gropper et al. 2002; Schmid et al. 2005; de Mesa et al. 2009).

Besides rheology, there are other aspects to consider in the extrusion of starch which have already been widely studied and are important to emphasize.

In order to produce starch-based materials the starch has to be processed with a plasticizer. Water and glycerol are the most commonly used plasticizers. The use of plasticizers in the extrusion technique is very common because the starch cannot be thermally processed without water since the melting temperature of dry starch is often higher than its decomposition temperature, as it was previously explained.

Plasticizers combinations are reported in literature as well as the use of additives and lubricants, which can be included before entering the extruder. For example, Yan et al. (2012) made mixtures of starch, glycerol (25, 30, 35, 40, 45, 50 wt% of starch), pullulan (5 wt% of starch), stearic acid (1 wt% of starch), and glyceryl monostearate (0.5 wt% of starch) in a mixer at room temperature for 30 min. Blended mixtures were stored in polyethylene bags at room temperature overnight to equilibrate all components and finally after this procedure the extruder was fed with these mixtures.

López et al. (2013) mixed starches powders with glycerol, in different proportions, by hand to achieve plasticizer incorporation and homogeneous mixtures. The mixtures were conditioned for 12 h at ambient temperature before submitting them to the extrusion process.

In a work by Wang et al. (2012), pea starch was fed into the extruder at a rate of 350 g/h using a MT-1 twin screw volumetric feeder. In this case, water was injected into the extruder with a liquid pump at a rate of 85–158 g/h, so that the final mixture had moisture content in the range of 30–40 %.

Mościcki et al. (2012) extruded starch–glycerol mixtures without the addition of water. SEM micrographs of the obtained samples showed many of untreated starchy granules inside its structure, evidencing the presence of ungelatinized zones. Li et al. 2011b compared the results of extruding starch powder and inject water at the extrusion barrel with the ones of extruding a premix of water and starch. They show that, in the first case gelatinization is not homogeneous and instabilities in the torque are generated during extrusion, while premixing leads to a homogeneous material. They also report that without a gases venting step, the extruder cannot work with higher water content than 40 %. Huneault and Li (2012) obtained an extruded homogeneous material feeding the extruder with a slurry composed of 50 wt% starch, 25 wt% water, and 25 wt% glycerol prior. They gelatinized starch during extrusion at 140 °C and then applied two-stage vent. In the last one, residual water was extracted by mechanical pump.

The importance of the plasticizer for the processes following extrusion, such as blowing and injection molding techniques, is exposed in the work of Moscicki et al. (2012). Various blends of potato starch mixed with glycerol were extrusion-cooked to obtain good quality TPS pellets, which were later processed using film blowing and injection molding techniques. It was noticed that with the full range of processing temperatures, independently of the mixtures composition used, the maximal elongation values coincided with increasing injection temperature and glycerin content in the sample. Results of the mechanical properties measurements showed that the extrusion processing parameters, the presence of plasticizer, and the blowing conditions have a critical impact on the film strength and elongation.

3 Starch Nanocomposites

3.1 *Characteristics and Properties*

As it was previously mentioned, the starch materials are strongly limited due their poor mechanical properties and high permeation. In this context, and in order to solve this problem, the incorporation of different fillers such as fibers, flakes, platelets, and particles into starch matrices (García et al. 2009a, 2011; Siqueira et al. 2009; Famá et al. 2010, 2011, 2012), or the starch blending with other polymers, have been the topic of many investigations. It has been well established that the filler size greatly influences physicochemical properties of the final composite. Smaller size fillers have serious advantages over micro-sized particles. For example, when nanofillers are used, the relationship between the area of the interface and the volume of the reinforcement is much higher than in the case of micro or conventional reinforcements. On the other hand, microfillers scatter light and therefore the composites lose transparency, obtaining composites with high opacity. This behavior does not occur in the case of nanofillers. Moreover, the optimal effect is achieved when the light reflection coefficient of the filler is similar to that of the material used as matrix.

In addition, the use of nanofillers improve several mechanical properties, such as Young's modulus, tensile strength, deformations, and toughness (García et al. 2009b; Famá et al. 2011), as well as electrical (Kilbride et al. 2002; Sandler et al. 2003) and thermal properties (Biercuk et al. 2002; Wei et al. 2002). The high aspect ratio ($k = L/d$) and large surface area per unit volume of same nanofiller result in a very good transfer of their properties to the polymer matrix. The effect of the incorporation of nanofillers in the starch retrogradation has also been investigated by some researches who found that the addition of nanoparticles diminishes the retrogradation of the polymer (Angellier et al. 2006; Cao et al. 2007; Ma et al. 2008b).

Interfacial adhesion between filler and matrix plays a vital role in the final properties of polymeric composites. The physicochemical properties of a composite material depend not only on the properties of its components, but also on the bonding between matrix and filler. When nanofillers are used, the ratio between the area of the interface and the volume of the reinforcement is much greater than in the case of micro-sized fillers. Furthermore, in the case of nanometric fillers, a high fraction of the atoms are in the surface resulting in increased reactivity. If good nanofillers dispersion is achieved, the interface region is maximized allowing the conformation of an interfacial percolation network that can improve different properties of the nanocomposite (Qiao and Brinson 2009).

There are different ways to improve the filler–matrix interaction in the interface. The most popular ones between them are:

- Mechanical bonding: Roughness between both surfaces leads to bonding, being better when higher roughness is observed. This type of bonding is ineffective for tensile but effective for shear improvements.
- Electrostatic bonding: This type of bonding occurrs when both surfaces are charged, one positively and the other negatively.
- Chemical bonding: It occurs when the surface of the reinforcement has chemical groups compatible with the chemical groups of the matrix. The strength depends on the number of bonds per unit area.
- Bonding by interdiffusion: In this type of bonding, the surfaces of the reinforcement and of matrix have chains that spread between them. The strength of this bond depends on the number of entanglements between chains and increases with the addition of solvents or plasticizers.

In many cases certain drawbacks such as incompatibility with the polymer matrix due to hydrophilicity differences, tendency to form aggregates during processing, and poor resistance to moisture greatly reduce the potential of the filler to be used as reinforcement in polymer composites. Interfacial coupling agents are often used to improve interfacial properties and control the morphologies of polymeric composites. Coupling agents make the fillers compatible with organic polymers since they are hybrids of organic molecules and minerals; and, as they have reactive functional groups, they are able to generate in situ formation of blocks or grafted copolymers at the interface by hot-melting blending. It has been proven that the addition of these coupling agents generates compatibilization between fillers and matrix, being an effective method for morphology control in a variety of composite systems (Son et al. 2000; Kim et al. 2001; Pagnoulle and Jerome 2001; Wang et al. 2001).

3.2 Processing

Processing nanocomposites results much more complex than neat starch due to the inherent complications related to the nanofillers use. Its high surface/volume ratio makes the electrostatic interactions between them, such as Van der Waals interactions, maximized, generating a large amount of agglomerates which are too difficult to break during processing. Furthermore, most of the employed loads in starch composites are hydrophilic and therefore show strong hydrogen bridge interactions, contributing to the agglutination problem. For that reason, different strategies have been developed in order to incorporate nanofillers to the matrix, with a special focus on a good dispersion achievement.

In the next sections we briefly discuss the processing of nanocomposites by the casting and extrusion techniques.

3.2.1 Casting

Casting is one of the most common techniques for processing lab scale starch composites. Different types of polysaccharide nanofillers such as cellulose from flax, wood, hemp ramie, cassava bagasse, wheat straws, starch from waxy maize, regular maize, and chitin, chitosan, among others; were used to fabricate starch nanocomposites by the solution casting method.

The main drawback associated with this method is that polysaccharide nanofillers tend to agglutinate due to the strong hydrogen bonding interactions between them (especially cellulose nanowhiskers (C-NW) (Cao et al. 2008a; b; Kaushik et al. 2010) and starch nanoparticles (S-NPs) (Angellier et al. 2006; Viguié et al. 2007; García et al. 2009a, 2011).

In general for the dispersion of these fillers in starch solutions, some additional treatments, such as ultrasonication, sonication and homogenization, are required

(Chang et al. 2010a; Kaushick et al. 2010). In some cases special considerations need to be taken into account. Starch nanoparticles (S-NP), for example, must be added at reduced temperature to avoid the gelatinization of their amorphous component and the destruction of their special structure, which affect their reinforcing ability (Xie et al. 2014). A possible attempt to solve this problem was proposed by Ma et al. (2008c), who modified starch nanoparticles with citric acid by dry preparation technique. These modified nanoparticles dispersed in distilled water were added before gelatinization at 90 °C. In this case they reported that S-NP could not be swelled or gelatinized in hot water due the cross-linking induced by the citric acid.

In the case of starch or cellulose nanocrystals obtained by the sulfuric acid treatment, and prepared in aqueous condition, the suspension experiences neither sedimentation nor flocculation, as a consequence of charge repulsion due to the sulfate groups created on their surface. This feature makes it easier to add them into to the starch solution (Angles et al. 2000; Mathew and Dufresne 2002; Angellier et al. 2006; Lu et al. 2006; Viguié et al. 2007; Cao et al. 2008a, b; Chen et al. 2009a, b; García et al. 2009, 2011).

The way in which different nanofillers are added into the matrix varies according to different authors. For example, Viguié et al. (2007) added a nanoparticle suspension of waxy maize starch to gelatinized starch, but they do not specify the suspension preparation protocol.

García et al. (2009), adding the suspension of waxy maize starch nanocrystals in the desired quantities to gelatinized matrix of cassava starch, and then the mixture was stirred for 10 min at 250 rpm and degassed for another 1 h.

Piyada et al. (2013) prepared a rice starch solution with a concentration of 3 % (w/v) in distilled water. They heated the dispersion with continuous stirring until it was completely gelatinized (85 °C for 5 min), and finally cooled it to 45 ± 2 °C. Subsequently, previously lyophilized nanocrystals were dispersed in water, and incorporated to the gelatinized starch suspension.

By contrast, some authors add the suspension together with the starch before gelatinization. In the work by Kaushik et al. (2010), prepared TPS nanocomposites reinforced with cellulose nanofibers (C-NF) extracted from wheat straw. They shear mixed maize starch and glycerol (30 %) for 10 min using a Fluko FA25 homogenizer (10,000 rpm) and prepared a C-NF water dispersion using a bath sonicator for 3 h. The C-NF dispersion was added to the starch–glycerol mixture and the final product was subjected to further shear mixing for 20 min. Finally, the starch + glycerol + C-NF mixture was continuously stirred (at 80–100 rpm) using a mechanical stirrer and heated at 75 ± 3 °C. Noteworthy that these authors mix the starch nanofillers suspension employing high shear.

In the case of Ma et al. (2008c), as already described, added citric acid-modified starch nanoparticles (CA-S-NP) into a water, glycerol, and starch mixture, prior to gelatinization.

Cao et al. (2008a), used starch, glycerol, a cellulose nanocrystallites suspension and distilled water mixed together to obtain a homogeneous dispersion and fabricate plasticized starch/flax cellulose nanocrystals (S/FC-NC) composite films.

The glycerol content was fixed at 36 wt% based on the dry starch matrix. Then this mixture was poured into a flask equipped with a stirrer and heated at 100 °C for 30 min to gelatinize starch. The same procedure is used for the fabrication of the Hemp (*Cannabis sativa*) cellulose nanocrystals and potato starch nanocomposite films S/HC-NC (Cao et al. 2008b).

Summarizing, in the preparation of starch-based nanocomposites by casting technique it is generally necessary take into account how to prevent the nanoparticles clumping.

3.2.2 Extrusion

The problems associated with starch extrusion are increased when starch nanocomposites are processed using hydrophilic fillers. As it was explained before, although these fillers are compatible with starch and glycerol, they tend to clump together obstructing their dispersion in the matrix.

Some authors propose the preparation of a dry starch, glycerol, and nanofiller mixture without water addition (only the starch moisture is used) to be extruded (Galicia-García et al. 2012; Müller et al. 2012; Pelissari et al. 2012); while others, underline the problem of the hydrophilic fillers and propose their incorporation into water suspensions, either by pumping the solution to the molten polymer into a subsequent extrusion stage (Vasquez et al. 2012), or by feeding the extruder with the mixture of all the components with water (Ayadi and Dole 2011; Martinez-Bustos et al. 2011). For example, Mitrus y Moscicki (2014), reported two types of food extruders that were used for processing of prepared blends of starchy components mixed with glycerol added up to 20 % in weight; functional additives like talc, polyvinyl alcohol, and foaming agent (PLASTRONFOAM PDE). The additives were added up to 3 % in weight, while water up to 5 % in weight. The blends were extruded at temperature range 80–170 °C, variable screw rotational speed up to 150 rpm, and die pressure of 12 till 18 MPa.

Hietala et al. (2013), premixed starch, plasticizer, lubricant, and cellulose nanofiber (C-NF) gels with high water contents and then extruded the obtained dispersions. The authors prepared nanocomposite with 0, 5, 10, 15, and 20 wt% of cellulose nanofiber. In order to remove the water excess, an extruder barrel equipped with two atmospheric vents and vacuum ventilation was proposed.

Yurdakul et al. (2013), proposed to mix starch, carbon nanotubes, water and glycerol, gelatinize the obtained suspension and introduce it manually into the extruder.

The extrusion process was performed between 25 and 75 rpm with temperatures between 110 and 140 °C. The obtained material showed a good filler dispersion and improvements in stress and strain at break for filler contents smaller than 1 wt%.

Regardless of the variations in the methodology chosen for the components addition in the extruder, it is important to note that the filler can be dispersed in starch during the gelatinization extrusion process. It is advantageous to use a twin screw extruder with multiple inlet ports so that gelatinization, plasticizer, and filler addition can be separated processing steps within one extrusion cycle. The production of TPS pellets in common plastic extruders is generally not recommended due to the need of a different machine design and processing difficulties.

4 Nanocomposites

4.1 Starch/Starch

Starch nanocrystals or nanoparticles can be prepared principally by four different methodologies, (1) by hydrolysis including acid or enzymatic hydrolysis (nanocrystals), (2) by regeneration using cocrystallization (Kim and Lim 2009), or cross-linking (Tan et al. 2009; Ma et al. 2008), and (3) using gamma radiation (Lamanna et al. 2013) mechanical treatment with a microfluidizer (Le Corre et al. 2010).

It is noteworthy that both amorphous and crystalline particles are obtained in the final reaction of the methods 2 to 4, while method 1 produces preferably nanocrystals.

The mechanical treatment is performed analogously to the production of microfibrillated cellulose. Liu et al. (2011) intended to produce starch nanoparticles by fluidization. They found that crystalline microparticles turned into amorphous nanoparticles with increasing run numbers. The hydrolysis methodology consists basically in removing starch amorphous regions. Attempts to produce starch nanocrystals by enzymatic hydrolysis were reported by Kim et al. (2008), it is believed that the process leads to blocklets rather than nanoparticles. Kim and Lim (2009) reported the preparation of nanocrystals by complex formation between amylose and *n*-butanol, thereafter enzymatic hydrolysis was used to selectively keep crystalline particles. Ma et al. (2008) reported the production of starch nanoparticles by precipitation of gelatinized starch in nonsolvent followed by a cross-linking reaction.

There are other methodologies as used by Lamanna et al. (2013); who obtained starch nanoparticles (S-NP) applying gamma radiation. Another environmental-friendly mechanical method to obtain nanoparticles is extrusion. Eugenius et al. (2000) patented a starch nanoparticles preparing method based on reaction extrusion. Using this technology, commercial product EcoSphere is produced by Ecosynthetix.

In the same way, Song et al. (2011), reported starch nanoparticles prepared by reactive extrusion method. Their results indicate that with the addition of an appropriate cross-linker, starch nanoparticles with an average size of 160 nm could be obtained.

Le Corre et al. (2011), propose another technique against limitations of the acid hydrolyses process for producing starch nanocrystals. Protocol using acid hydrolysis is only for producing small quantities of S-NC and renders a limited yield (10–15 %) after a production time of (5 days). They proposed a continuous extraction technique:

hydrolyzed from wheat starch and filtered using a microfiltration unit equipped with ceramic membranes to assess the cross-flow membrane filtration potential of S-NC suspensions. They show to be an efficient continuous operation for separating S-NC from the bulk suspension whatever the ceramic membrane pore size (0.2–0.8 μm).

Therefore, starch nanoparticles look like a good alternative for developing starch-based nanocomposites with high transparency. This was reported in the study of Gonçalves et al. (2014), with nanocrystals obtained from the seeds of pinhão (Araucaria angustifolia). The greater solubility and reduced turbidity are interesting from a commercial standpoint, indicating that pinhão starch nanoparticles could be useful for development of coating materials or films composites.

However, there are two fundamental problems to solve in order to develop composites using starch nanoparticles: the high agglomeration between them and the degradation possibility due to temperature or shear employed in processing method.

Furthermore, it is well known that the nanoparticles shape plays a very important role in the improvement of barrier properties. Platelet-shaped particles are preferred because they are thought to alter the diffusion path of penetrant molecules and improve the barrier properties of the material. Nanoparticles with this geometry create a sort of winding road hindering and delaying the passage of gases and water.

The nanofillers having these latest features are clays and starch nanocrystals, therefore, both could be fitted for flexible food packaging applications.

The use of starch nanoparticles is receiving a significant amount of attention because of the abundant availability of starch, low cost, renewability, biocompatibility, biodegradability, and nontoxicity (Chakraborty et al. 2005).

Starch nanocrystals has been used in different environmental-friendly polymers such as waterborne polyurethane (Chen et al. 2008) (also called organic solvent free polyurethane), starch (waxy maize, Angellier et al. 2006; Viguié et al. 2007; García et al. 2009a), pullulan (Kristo et al. 2007) (obtained by starch fermentation), PLA (Yu et al. 2008), polyvinyl alcohol (PVA) (Chen et al. 2008), and soy protein isolate (Zhen et al. 2009), but we will focus on its use on starch matrices.

The most common sources of starch to develop the nanocomposites matrix are pea starch, waxy maize starch, cassava starch, normal maize and potato starch. (Chen et al. 2009; García et al. 2009a, b; Jayakody and Hoover 2002; He et al. 2012).

In most publications, authors have opted for a simple casting–evaporation method in the preparation of starch-based nanocomposites. The general method consists in dispersing the nanocrystals in aqueous solution and adds this suspension to the mixture of starch and other components which will be then gelatinized. Other authors prefer to add the suspension after the gelatinized starch. This was already discussed in section casting of nanocomposites.

Other conditions have caught significant attention during investigations, for example, Garcia et al. (2009a) taking care of degassing thermoplastic cassava starch solution before and after the introduction of starch nanoparticles.

The potential use of starch nanocrystals as a reinforcing phase in a polymeric matrix has been evaluated from a mechanical point of view in several publications.

Angellier et al. (2006) showed that the reinforcing effect of starch nanocrystals is more significant in TPS than in other matrices. This is probably a consequence of the strong interactions between the filler and amylopectin chains and a possible crystallization at the filler/matrix interface (Angellier et al. 2006). Besides, the reinforcing effect of S-NC within a TPS matrix was higher than that of other fillers, such as tunicin or sugarcane bagasse cellulose whiskers.

Other authors reported the same reinforcing effect (Viguié et al. 2007; Ma et al. 2008c; García et al. 2009). In particular, Viguié et al. (2007) showed that for higher plasticizer contents (35 wt% sorbitol) the reinforcing relative effect was higher. The Young's modulus and the strength of a film reinforced with 15 wt% starch nanocrystals increased by a factor 7 and 12, respectively, when plasticized with 35 wt% of sorbitol, and only by a factor 2.7 and 4.2, respectively, when plasticized with 25 wt% sorbitol.

Garcia et al. (2009), reported an increase of 380 % of the rubbery storage modulus (at 50 °C), in nanocomposites of cassava starch reinforced with waxy starch nanocrystals.

On the other side, starch nanocrystals' surface can be chemically modified. These modifications consist in transforming the polar hydroxyl group sitting at the surface of starch nanocrystals into moieties capable of enhancing interactions with nonpolar polymers (Labet et al. 2007; García et al. 2012).

Modifications in order to improve starch matrix–starch nanoparticles nanocomposites were also proposed. For example, Ma et al. (2008c), proposed the fabrication and characterization of citric acid-modified starch nanoparticles/plasticized pea starch composites. In dynamic mechanical thermal analysis, the introduction of CA-S-NP could improve the storage modulus and the glass transition temperature of pea starch/CA-S-NP composites. The tensile yield strength and Young's modulus increased from 3.94 to 8.12 MPa and from 49.8 to 125.1 MPa, respectively, when the CA-S-NP contents varied from 0 to 4 wt%.

On the other hand, nanoparticle's moisture resistance was reported by different authors and an important characteristic in many nanocomposites potential applications. The improvements in the water vapor permeability (WVP) of nanocomposites with starch nanoparticles are associated to the introduction of a tortuous pathway for water molecules to pass through, as it was mentioned before. Garcia et al. (2009a) reported a WVP reduction of a 40 % in cassava starch matrix composites reinforced with 2.5 wt% starch nanoparticles respect to the starch matrix. However, in waxy starch matrix composites the permeability to water vapor increased (by 79 %) upon the addition of only 2.5 wt% waxy maize starch nanocrystals. This result is opposite to the cassava starch, where the nanocrystals were well distributed (García et al. 2011).These results are due to the fact that it is considered that starch–SNC is a complex system governed apparently both by the nanocrystals and the plasticizer. García et al. (2011), proposed that that the higher density of –OH groups on the surfaces of S-NPs, which were mainly the crystalline zones of hydrolyzed waxy starch, led to more association of the S-NPs with glycerol molecules. Consequently, more –OH groups of the matrix result available to interact with moisture, and a nanometric fibrillar preferential path for water vapor diffusion was formed, resulting in an increase in the moisture sensitivity.

In composites, starch nanocrystals (S-NC) are candidates of growing interest, for this reason recently has been studied in detail in a recent European Project (FlexPakRenew–FP7/2007-2013—no. 207810).

4.2 Nanocomposites: Starch/Cellulose

Natural cellulose fibers from different biorenewable resources have considerable attraction of research community due to their unique intrinsic properties such as biodegradability, easy availability, environmental friendliness, flexibility, easy processing, and impressive physicomechanical properties (Singha and Thakur 2008a, b, 2010a, b; Thakur et al. 2010a; Thakur and Thakur 2014a, b, c). In addition, considerable efforts are currently being directed toward improving the quality of the interfacial bonding between polymers and fibers by chemical surface modifications (Singha et al. 2009; Thakur et al. 2010b, 2011, 2012).

Moreover, cellulose nanofillers such as nanorod, rod-like cellulose microcrystals, nanowires, and long and straight crystals were investigated.

Since the first announcement cellulose whiskers application as reinforcing phase by Favier et al. in 1995, new nanocomposite materials with original properties were obtained using cellulose nanowhiskers (C-NW) and microfibrillated cellulose (MFC) and led to the development of studies on chitin whiskers.

The C-NW can be obtained from different sources like cassava baggage (Texeira et al. 2009), ramie (Lu et al. 2006), hemp (Cao et al. 2008b), flax (Cao et al. 2008a), peal hull (Chen et al. 2009a, b), jute (Cao et al. 2012), tunicate (Angles et al. 2000; Angles et al. 2001; Mathew et al. 2002; Mathew et al. 2008) (sea animal) or from microcrystalline cellulose (MCC) (Kvien et al. 2007; Chang et al. 2010a). Recent publications also reported cellulose nanocrystals from sweet potato residue and bamboo (Lu et al. 2013; Zhang et al. 2014).

Different methods are presented in literature to obtained cellulose nanowhiskers: chemical hydrolysis, high-pressure homogenization, and enzymatic hydrolysis.

In a classical hydrolyses is obtained cellulose rod-like nanocrystals or nanocrystal systems from hardwood and softwood (Beck-Candanedo et al. 2005); or whiskers (Bondenson et al. 2007) or rod-like nanoparticles (Mesquinfam et al. 2011)

These nanofillers are usually obtained by acid hydrolysis under specified conditions of temperature and processing time, allowing the decrease of the amorphous regions, resulting in a highly a crystalline material. The subsequent procedure to acid hydrolysis consists in a wash by centrifugation, then dialyzed against distilled water and ultrasonicated. The surface grafted sulfate groups, negatively charged, provide that nanocrystals form stable layers at the air–water interface, then the nanofiller is prepared in aqueous condition, with well dispersion and without sedimentation or flocculation. This is an advantage in the preparation of nanobiocomposites, because the suspension can be incorporated directly into the starch solutions. (Angles et al. 2000; Lu et al. 2006; Cao et al. 2008a; Chen et al. 2009; García et al. 2009a, 2011).

Using high-pressure homogenization of wood fibers in water, MFC, composed of liberated semicrystalline microfibrils, is obtained. MFC was first introduced by Turbak et al. (1983), who reported that the fibrils width was between 25 and 100 nm, while the length was longer.

Similarly MFC were obtained applying a high-pressure homogenizer to commercial microcrystal cellulose previously dispersed by a Turrax. The cellulose fibrils were in the range from 28 to 100 nm (Lee et al. 2009).

A novel concept to prepare nanocellulose fibrils has been reported, where enzymatic hydrolysis was used in combination with mechanical shearing and high-pressure homogenization to produce defibrillation of the fiber wall and obtain fibrils with a diameter of around 5–6 nm and aggregates of around 10–20 nm (Paakko et al. 2007).

Recently novel techniques for obtaining of cellulose nanocrystals have appeared in the literature. Cao et al. (2012) showed a controllable method to fabricate cellulose nanowhiskers from jute fibers with a high yield (over 80 %) via a 2,2,6,6-tetramethylpiperidine-1-oxyl radical (TEMPO)/NaBr/NaClO system selective oxidization combined with mechanical homogenization.

Haafiz et al. (2014) showed an isolation method of C-NW from MCC and a detailed characterization of the new filler. MCC was produced from oil palm empty fruit bunch chlorine free pulps. It was swelled and partly separated to whiskers by chemical and ultrasonication treatments using the same method as described by Pereda et al. (2011) based on original procedures described by Oksman et al. (2006).

An exhaustive economic analysis for the production of cellulose nanowhiskers as a coproduct in an ethanol biorefinery and an ASPEN Plus-based process model (http://www.aspentech.com/core/aspen-plus.cfm) was developed to evaluate ethanol production from wheat straw. All the collected data on cellulose nanocrystals in terms of production, characterization, and application suggest that this nanomaterial could be easily extrapolated to bioethanol production (Duran et al. 2011).

The relative degree of crystallinity and the geometrical aspect ratio of obtained cellulose nanocrystals, depend on cellulose origin and the processing conditions, such as time, temperature, and original raw material purity.

The principal characteristics observed with the addition of cellulose nanocrystals to starch-based materials led to improvements in the mechanical properties, thermal properties (represented by an increase in the glass transition temperature, T_g), and moisture resistance.

Several publications reported these effects. In the works of (Cao et al. 2008a), cellulose crystals, prepared by acid hydrolysis of flax fiber (F-CN), were added to a potato starch (PS) matrix, the nanocomposite films exhibited a significant increase in tensile strength and Young's modulus from 3.9 to 11.9 MPa and from 31.9 to 498.2 MPa, respectively, with increasing F-CN content from 0 to 30 wt%.

The resulting nanocomposite films also showed a higher water resistance and the nanocomposite water uptake decreased as the filler content increased. The fillers addition did not change the transition temperature for the glycerol-rich phase, but the T_g value of the starch-rich phase shifted to higher temperatures from 43.3 to 48.8 °C, indicating that F-CN restrict the mobility of starch chains. This can probably be explained by the reinforcement effect from the homogeneously dispersed high-performance F-CN fillers in the PS matrix and the strong hydrogen bonding interaction between F-CN and PS molecules.

The reason for the nanocomposites increased thermal properties is associated to the relocation of the plasticizer(s), water inclusive, from the starch matrix to the cellulose nanocrystals surfaces, which decreases the plasticization effect on the amorphous regions (Angles and Dufresne 2000, 2001). They proposed the formation of the transcrystalline zone, around the cellulose nanofillers in order to explain the results obtained in plasticized waxy maize starch reinforced with tunicin nanowhiskers. This effect is caused by the recrystallization of amylopectin chains assisted by the plasticizer accumulation and the nucleating effect of C-NW.

The improvement in mechanical properties of starch due to the addition of different cellulose nanofillers was reported in different works: Hemp nanocrystals in TPS (Cao et al. 2008b); nanofibrils of wheat starch in regular maize starch (Kaushick et al. 2010); ramie crystallites in commercial industrial-grade wheat starch (Lu et al. 2006), nanofibrils in cassava starch (Texeira et al. 2009), tunicin whiskers in plasticized starch (Angles et al. 2000), between others. The explanation for this effect is ascribed to the formation of a rigid network of nanofillers, the mutual entanglement between the nanofiller and the matrix, and the efficient stress transfer from the matrix to the nanofiller (Kaushik et al. 2009; Siqueira et al. 2009). In works of Mathew and Dufresne (2002) and Mathew et al. (2008), the authors propose an increase in the overall crystallinity of the system resulting from the nucleating effect of the C-NW, this effect can be beneficial for mechanical properties.

Slavutsky and Bertuzzi (2014) recently showed cellulose nanocrystals (C-NC) obtained from sugarcane bagasse. They formulated starch/C-NC composites and their water barrier properties were studied. The measured film solubility, contact angle, and water sorption isotherm indicated that reinforced starch/C-NC films have a lower affinity to water molecules than starch films. The same effect was observed in studies by Savadekar and Mhaske (2012). They studied the effect of C-NC incorporation on TPS matrix and found that the nanofillers addition improved their barrier and mechanical properties.

The main causes for the improved moisture resistance exhibited by the nanocomposites are the less hydrophilic nature of cellulose and the geometrical impedance created by the C-NW, the constraint of the starch swelling due to the presence of the C-NW network, the resistance of the diffusion of water molecules along the nanofiller–matrix interface, and the reduced mobility of the starch chains, resulting from an increase in the T_g or the crystallinity.

4.3 Nanocomposites: Starch/Layer Silicates

Polymer nanocomposites reinforced with layer silicates are very common because the first application of these fillers was in the automotive industry. The packaging industry has focused its attention principally on layered inorganic fillers like clays and silicates due to their low cost, availability, significant augment, and considerably simple processability (Le Corre et al. 2010).

Clay minerals, the micas, chlorite, serpentine, talc, between them can also find within the group of minerals phyllosilicates or layered silicates. They all have different morphology, structure, and texture (Bergaya et al. 2009). The some representative layer silicates used mainly in nanocomposites-based starch are listed in Table 2. It is important to note that the some phyllosilicates do not display a normal layered structure, for example, the sepiolite shows a fibrous structure meanwhile halloysite has spheroidal aggregates morphology (Duquesque et al. 2007; Bergaya et al. 2009).

Many organophilic nanoclays, therefore, have already been studied, and some of their products are already marketed on an industrial scale (Park et al. 2002; Turrit al. 2008). The variation in the fillers dimensions depends on the clay source, particulate silicate and preparation technique, but all these layers have a very high aspect ratio (length/thickness) and surface area (Reedy et al. 2013).

The crystal arrangement in the silicate layers is made up of two tetrahedrally coordinated atoms fused to edge-shared octahedral sheets. These sheets are made up either of magnesium or aluminum hydroxide. Their thickness is about 1 nm and their tangential dimensions range from 300 Å to a few microns. The van der Waals gap between these layers is due to the regular layers stacking; this gap is known as inter-gallery spacing. Each of these layers is negatively charged, and the excess charge is balanced by alkali cations, such as Na^+, Li^+ or Ca^{2+}, that reside in these inter-gallery space between the aluminosilicate layers (Ray and Okamoto 2003; Majdzadeh-Ardakani et al. 2010). To improve their dispersability, clays are often modified with conventional organomodifiers, many of them are surfactants, which are typically the quaternary ammonium salts of long fatty acid chains. These organomodifiers decrease the surface tension of the aluminosilicate particulates, which in turn reduces the mixing endothermic enthalpy.

Table 2 Representative layer silicates used mainly in nanocomposites based starch

Group	Nanofiller	Dimensional type	Matrix starch	Load (%)	References
Kaolinite	Kaolinite	Nanolayer	Cassava	2–6	Mbey et al. (2012)
Kaolinite	Halloysite	Nanotubes	Potato	5	He et al. (2012)
Smectite	Montmorillonite	–	Wheat	3–6	Chivrac et al. (2010)
			Corn	0–5	Aouada et al. (2011)
Sepiolite	Sepiolite	Nanolayer	Potato	–	Du et al. (2013)
Smectite	Hectorite	Nanolayer	Yam	<30	Wilhelm et al. (2005)
Synthetic clay	Somasif	Nanolayer	PLA	–	Borges et al. (2012)
Mica	Laponite	–	Corn	1–5	Aouada et al. (2011)

The most commonly used strategy to increase the interlayer spacing (Lagaly et al. 1986) is the chemical modification of the phyllosilicate surface by cationic exchange using the quaternary ammonium salts of long fatty acid chains. From this technique, different organomodified clays can be obtained.

In nanocomposites-based starch, the most widely studied type of clay is montmorillonite (MMT) or modified montmorillonite, among which the main difference is the nature of the countercation and the cationic exchange capacity (CEC).

In literature, natural Montmorillonite MMT or MMT-Na^+, can also be found under the following names: Cloisite® Na^+, Dellite® LVF, Dellite® HPS, Nanofil® 757, BH Natural. When different counterions are used in order to modify MMT, the trade names used which can be found are: Cloisite® 30B, Cloisite® 10A, Cloisite® 25A, Cloisite® 93A, Cloisite® 20A, Nanomer® I.30E, Nanofil® 784, Nanofil® 804, Nanofil® 948, natural calcium hectorite and Bentone® 109 (all are commercial names registered). The principal counterions used to modify MMT are listed and the synonyms names associated with them are showed in Table 3.

However, some these chemical modifications increase the hydrophobicity of the phyllosilicate, resulting in a reduction of the compatibility with the hydrophilic starch matrix. The surface hydrophobicity of Cloisite MMT nanofillers follows an order: Natural sodium MMT is less hydrophobic than Cloisite® 30B. In turn, Cloisite® 30B is less hydrophobic than Cloisite 10 A, and thereby the order is as follows: 10A < 25A < 93A < 20A < 15A < 6A (Xie et al. 2012).

The dispersion state of a typical phyllosilicate (except sepiolite and halloysite) in a matrix polymer depends on the preparation conditions and the matrix–nanolayer affinity. This effect determines the structure of the resulting composites, which can be either phase separated composites (microcomposites), intercalated nanocomposites, or exfoliated nanocomposites (Alexandre and Dubois 2000).

In nanocomposites, several studies have reported the efficiency of nanoclay incorporation in improving mechanical properties and decreasing water vapor (Müller et al. 2011; Sadegh-Hassani and Nafchi 2014), and oxygen permeability (O_2P) (Tjong 2006; Lu and Mai 2007).

Particularly, starch nanocomposites reinforced by phyllosilicates show generally an increase in tensile strength (σ), Young's modulus (E), storage modulus (measured by dynamic mechanical analysis), thermal stability, moisture resistance, oxygen barrier property (Zeppa et al. 2009), and biodegradation rate (Magalhaes and Andrade 2009). For example, green nanocomposites were prepared by solution induced intercalation method using starch, jute, glutaraldehyde, nanoclay (Montmorillonite K-10) and glycerol. The addition of glutaraldehyde and nanoclay has been found to improve the thermal stability, flame retardancy, dimensional stability, and mechanical strength of the prepared composite. The values for tensile strength changed from 18.62 MPa (0 wt% filler) to 40.63 MPa for 5 wt% of addition filler, and the modulus value changed from 1117 MPa for the matrix to 2344 Mpa for the nanocomposite (Iman and Maji 2012). On the other hand, Sadegh-Hassani et al. (2014), reported bionanocomposite films prepared by the casting method from potato starch with halloysite nanoclay as the reinforcing materials. The composition included potato starch with 40 wt% of a mixture of

Table 3 List of the principal counterions used to modify MMT and the synonyms names associated with them

Counter-cation	Name
Na^+	Natural sodium MMT; MMT-Na^+; Cloisite® Na^+; Dellite® LVF; Dellite® HPS; Nanofil® 757; BH Natural
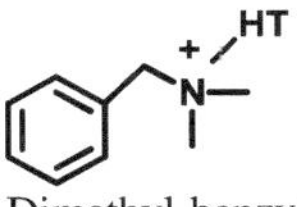 Dimethyl-benzyl-hydrogenated-tallow ammonium	Cloisite® 10A; Bentone® 111; Dellite® 43B
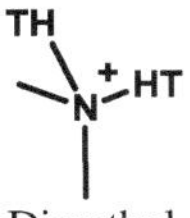 Dimethyl-dihydrogenated tallow ammonium	Cloisite® 20A; Cloisite® 15A; Cloisite® 6A; Dellite® 67G; Dellite® 72T
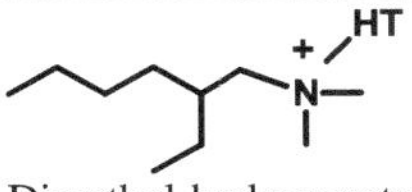 Dimethyl-hydrogenated tallow-2-ethylhexyl ammonium	Cloisite® 25A
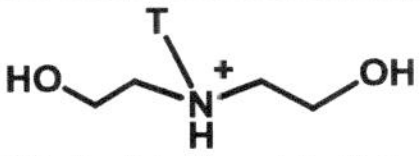 Methyl-tallow-bis-2-hydroxyethyl ammonium	Cloisite® 30B
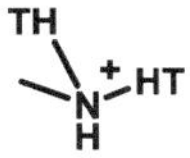 Methyl-dihydrogenated tallow ammonium	Cloisite® 93A
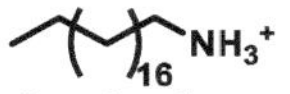 Octadecyl ammonium	Nanomer® I.30E
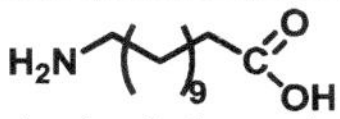 Aminododecanoic acid	Nanofil® 784
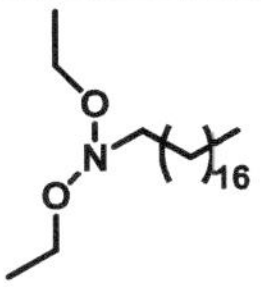 Stearyl-dihydroxyethyl ammonium	Nanofil® 804
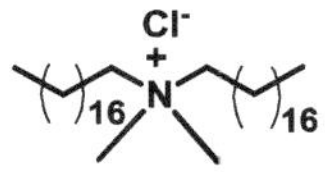 Distearyl dimethyl ammonium chloride	Nanofil® 948
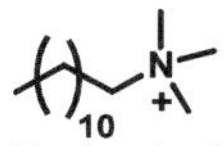 Dodecyl trimethyl ammonium	

(continued)

Table 3 (continued)

Counter-cation	Name
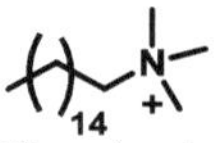 Hexadecyl trimethyl ammonium	
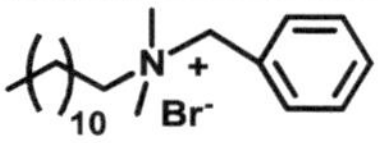 Dodecyl benzyl dimethyl ammonium bromine	
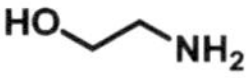 Etanolamine	
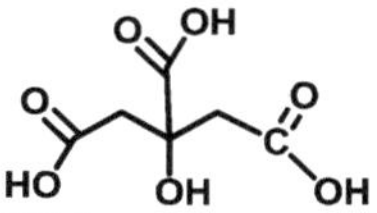 Citric acid	
Ca^{2+}	Natural calcium hectorite; Bentone EA-163

sorbitol/glycerol (weight ratio of 3–1 as plasticizer) with nanoclay (0–5 wt%). The clay nanoparticles incorporation decreased permeability of the material to gaseous molecules and improved the mechanical properties. Tensile strength increased from 7.33 to 9.82 MPa, and elongation at break decreased from 68.0 to 44.0 % with the filler addition. Matsuda et al. (2013) presented biodegradable trays of cassava starch and organically modified montmorillonite (Cloisite® 10A and 30B) using a baking process. The stress at break of the samples was strongly affected by the nanoclay additions for different RH conditions. In addition, independently of the nanoclay type or concentration used, the stress at break was significantly increased. The stress at break value for the control sample was 13.26 MPa while for the nanocomposite at 33 % RH it was 108.18 MPa.

The improvements observed in these properties are due to the good dispersion of silicate layers in the starch matrix, the high aspect (width-to-thickness) ratio and thus accessible exposed surface of the montmorillonite nanofiller, and the strong hydrogen bonding interactions.

4.4 *Nanocomposites: Starch/Antioxidants and/or Antimicrobials*

In food packaging and medical industry, biodegradable thermoplastic materials require the possibility to add antimicrobial and/or antioxidant agents to limit microbial activity.

Antimicrobial agents are additives that may retard microbial growth or cause microbial death, being the main targets of action pathogenic microorganisms producing toxins or causing infections and deteriorative microorganisms whose metabolic end products cause odors and flavors or texture and discoloration problems, in particular in food (Davidson and Zivanovic 2003). In medical science, antimicrobial components contribute to the elimination of infections and stimulate the immune system.

In the case of food, antioxidants are substances with the ability to delay or prevent the development of rancidity and deterioration of sensory attributes related to flavors and aroma; and also function as oxidation inhibitors or retarders. The effectiveness of these additives depends on a number of factors, like: intrinsic factors, such as the composition (lipids, carbohydrates and proteins), pH, water activity and oxide reduction potential; extrinsic factors, such as temperature, storage time, and humidity and atmospheric conditions; processing factors; and microbial factors, such as the type and quantity of microorganisms, resilience microorganisms and cellular composition (Davidson and Taylor 2007).

Different studies have shown that starch films are excellent intermediates for transporting antimicrobials and antioxidants. However, although the incorporation of organic compounds such as essential oils (Tripathi and Dubey 2004), organic acids (Schirmer et al. 2009), enzymes like lysozyme (Appendini and Hotchkiss 1997) and bacteriocins (Han 2005; Gálvez et al. 2007) is effective to improve the quality and safety, and/or to slow the oxidation of a product, typically deplete mechanical resistance and barrier properties (Raybaudi-Massilia et al. 2009). To solve this problem, nanofillers with antimicrobials and/or antioxidants characteristics have been used, to contribute their important antimicrobial properties, without diminishing general properties of the product (Pérez Espitia et al. 2012). The advantage of nanotechnology, which involves nano-sized products, has brought great opportunities for the development of antimicrobial agents with new properties.

The increasing number of studies in the nanocomposites field is probably due to the strong antibacterial activity achieved at low filler concentrations, caused by the high surface area to volume ratio, unique chemical and physical properties (Rai et al. 2009), and stability in extreme conditions (Sawai 2003).

These additives can be classified into two major groups: synthetic and natural nanofillers. The former has been used for many years and they have been generally produced by chemical synthesis. Such classification does not imply that synthetic preservatives are more effective from a microbiological point of view than those of natural origin, or vice versa (Davidson and Taylor 2007; Raybaudi-Massilia et al. 2009).

Natural fillers have low-cost, low density, high specific properties and, also, they provoke a great acceptation between consumers. They are particularly biodegradable and nonabrasive, unlike other reinforcing materials. Besides, they are readily available and their specific properties are comparable to those of other fillers used as reinforcements.

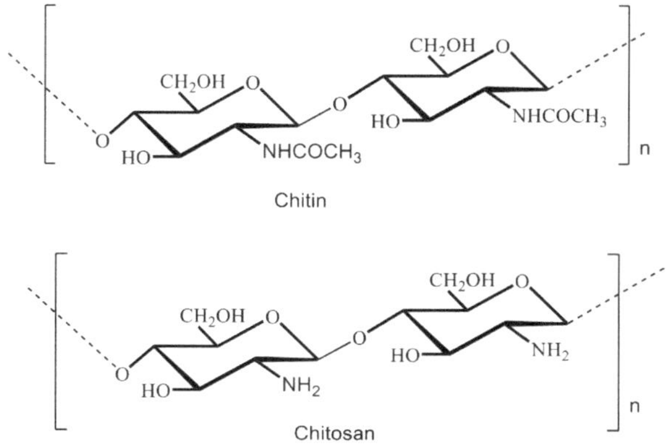

Fig. 7 Chemical structures of chitin and chitosan

Among possible natural nano-sized additives with antimicrobial character, chitosan nanoparticles (CH-NP) have been the focus of discussion of some investigators due to their high compatibility with starch, as both polysaccharides are structurally related (Yumin et al. 1997; Li et al. 2013) (Fig. 7). Chitosan is the main derivative of chitin, which is part of the carapace of crustaceans, mollusks, insects, and other living things. It is nontoxic, biodegradable, biofunctional and biocompatible and has the structure of a linear copolymer of glucosamine and *N*-acetyl glucosamine units linked by β-1,4 glycoside linkages. Chitosan has the advantage of having antibacterial activity (Dutta et al. 2009), which makes it very promising for applications in the previously mentioned industries (Chang et al. 2010b, c).

In order to obtain nanocomposites with potential applications in medicine, agriculture, drug release and edible films packaging, Chang et al. (2010b) fabricated glycerol plasticized starch matrix films reinforced with different concentrations of chitosan nanoparticles. They prepared the nanoparticles by physical crosslinking (electrostatic interaction) between tripolyphosphate and protonised chitosan, instead of chemical crosslinking, obtaining CH-NP from about 50 to 100 nm. For the lower nanofiller contents (0, 2, 4 and 6 wt%) a good degree of dispersion was achieved, and the tensile strength and storage modulus increased in more than four times respect to the matrix. A decrease in the elongation at break was also exhibited, but it was of nearly a third of the matrix value. The similar polysaccharide structures of chitosan and starch and their great interaction were the reasons of that mechanical behavior (Yu et al. 2008). When 8 wt% of CH-NP were used, nanoparticles agglomerations were detected and the tensile strength resulted worse than that of the others composites. The glass transition temperature resulted also sensitive to the chitosan nanoparticles presence. With their addition, the T_g shifted to a higher temperature, indicating that the filler improved the intermolecular interaction in the matrix, reducing the free volume. On the other hand, water vapor permeability of composites decreased, probably due to the introduction of a tortuous path for the water molecules to pass through, in the case of the major concentration (Yu et al. 2008). These nanoparticles also modified the decomposition temperature of the films. The degradation of the composites resulted at higher temperatures respected to that of the matrix. The increment of the thermal stability has been attributed to the better thermal stability of CH-NP, and the better interfacial interaction between the filler and the matrix.

In their work, (Chang et al. 2010b, c) found that the incorporation of chitin nanoparticles uniformly dispersed in a starch matrix at low loading levels (till 5 wt %) led to improvements in mechanical parameters (tensile strength, elastic modulus, and T_g) and water vapor permeability. For higher filler contents, agglomeration occurred, but good interfacial interactions between the nanofiller and the starch could be observed.

Woranucha and Yoksana (2013) conducted a thorough study about the possibility of using chitosan nanoparticles and eugenol-loaded chitosan nanoparticles as antioxidants to obtain active bio-based packaging materials by extrusion. They used a mixture of cassava, rice, and waxy rice flours as base material. The studies of the nanocomposites antimicrobial activity revealed that both chitosan and eugenol-loaded chitosan nanoparticles exhibited high antimicrobial. The incorporation of less than 6 wt% of nanoparticles did not greatly affect the crystallinity of the material. However, some physicochemical properties of these composites were changed by the incorporation of the nanofiller. The degradation and melting temperature of the matrix tended to decrease when CH-NP and eugenol-CH-NP were used, leading to a reduction in thermal stability of composites. The addition of only 3 wt% of this filler did not significantly changed tensile strength and elastic modulus respected to the matrix, but provoked a great reduction on the elongation at break. Although the eugenol has characteristics of plasticizer, which could lead to decrements in tensile strength and modulus and an increment in elongation at break, the incorporation of eugenol-loaded chitosan nanoparticles did not significantly change these mechanical parameters, because of the small amount of eugenol in the final material. In contrast, in composites with CH-NP, the elongation at break behaved as the matrix, but both tensile strength and modulus, decreased considerably, probably due to the particle agglomerations which might cause poor interfacial interaction between CH-NP and matrix as well as induce structural fragility. The incorporation of these nanoparticles caused a reduction of WVP, indicating that this property was enhanced by the incorporation of the nanofillers. Probably, the improvement might be due to the fewer free hydroxyl groups available in the matrix because of H-bond formation with the nanoparticles; and a more tortuous path for water molecules to pass through (Chang et al. 2010; Ghanbarzadeh et al. 2011). In the case of the incorporation of eugenol-CH-NP, WVP resulted lower and O_2P higher than those of starch/C-NP, due to the hydrophobicity of liberated eugenol in the extrusion process.

On the other hand, wood fibers are being used as natural reinforcement of composites in a large number of applications because of their low cost and readily availability. Recently, some researchers have explored the use of bamboo nano-size fibers into polymer materials (Takagi and Takura 2003; Takagi and Ichihara 2004; Ghavami 2005; Liu et al. 2010). Bamboo is a renewable natural bioresource abundant in tropical and subtropical regions of the world (Ghavami 2005). It has several advantages respect to others wood fibers including small environment load, rapid growth, renewability, relatively high strength, and good flexibility (Liu et al. 2010).

Presently bamboo is considered as an important source of fibers. Their fibers are mainly composed by cellulose, hemicelluloses and lignin. There are, in minor concentration, other constituents as protein, fat, pectin, tannins, pigments and ash. These constituents play an important role in the physiological activity of the fibers. The nature of bamboo fibers is one of the most attractive characteristic of this material. This makes bamboo fibers promising alternatives, among the many types of natural fibers, for their use as bio-based composites reinforcements (Jindal 1986; Amada et al. 1997; Kitagawa et al. 2005). Particularly, it is deemed to have one of the most favorable combinations of low density and good mechanical properties (specific strength and stiffness) (Dieu et al. 2004).

Starting from bamboo fibers, a combined nitric acid–potassium chloride treatment and sulfuric acid hydrolysis can be used to obtain bamboo nanocrystals (B-NC). Liu et al. (2010) investigated the effect of B-NC (Fig. 8) incorporation on starch-based materials conformed by casting. The fractured surface of neat starch films was rather smooth, but when the starch matrix contained 1 wt% of the nanofiller, small leaves were pulled out of the matrix surface. The leaves kept increasing on the rugged fracture surface, due to good surface bonding between starch and low-level B-NC. In samples with high concentration of the filler, the interface was smooth with white dots. When filler level increased from 10 to 20 wt %, the composites exhibited increased groups of agglomerates. Micrographs finally detected that the suspended crystals kept their one-dimensional nano-size morphology at low concentrations of the bamboo reinforcement, but at high concentration levels, the crystals congregated into microparticles. With the addition of low filler concentrations (<8 wt%), the humidity content of the nanocomposites greatly

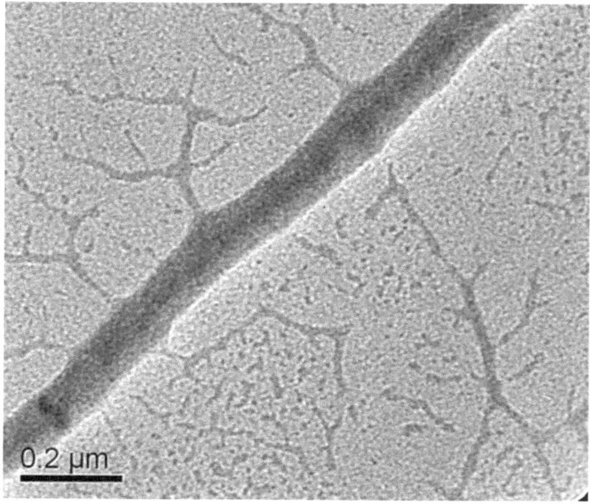

Fig. 8 TEM micrograph of bamboo nano crystals (B-NC). Reproduced with permission from Liu et al. (2010). © 2010, Elsevier Ltd

decreased, however in the case of the materials containing more than 8 wt% of B-NC, the decrease was much slower. Mechanical properties depended on the composites filler's content. Till 8 wt%, tensile strength and Young's modulus increased one order of magnitude, and the elongation at break decreased with increasing B-NC content. When the reinforcement was higher, no improvements were observed for all of these parameters. In that case, the interface adhesion had not been enough to withstand the imposed high stretching forces, resulting in diminished tensile strength. For all the composites, the relaxation transition temperatures shifted toward lower temperatures, what was attributed to the incorporation of a significant degree of crystallinity by the B-NC. The authors concluded that the addition of 8 wt % of bamboo cellulose crystals was quite sufficient to get an important reinforcing effect on a starch-based composite.

Different-sized bamboo fillers were obtained and incorporated into starch matrices. In this context Takagi and Ichihara (2004) investigated the effect of the dimension of short bamboo fibers and their concentration on mechanical properties of a starch–resin material. When bamboo fibers with smaller lengths than 15 mm were used, both tensile and flexural strengths of the composites tended to increase with the increment of the fiber dimensions. The authors also observed that this significant increment in the strengths parameters was experienced for samples with filler contents of up to a 10 wt%, but it did not happen in the cases of composites containing more than a 20 wt% of bamboo fibers.

On the other hand, when Takagi and Takura (2003) analyzed biodegradable composites fabricated with starch and bamboo powder of 500 μm in diameter, they found that the resultant materials had acceptable high bending strength and modulus.

As it was previously explained, many additives studied in order to provide antimicrobial and/or antioxidant properties to polymeric films were detrimental to some basic physical properties of composites. However, in several cases, when these additives were incorporated in powdered form instead of oil, some of them did not cause damage to the characteristics of the final material. One case is the work of Ghasemlou et al. (2013), who studied starch films containing Zataria multiflora Boiss (ZEO) or pennyroyal (MEO). The addition of both ZEO as MEO produced a decrease in water vapor permeability, while it had not lead to significant differences in O_2P, except with the incorporation of the highest concentration of both components. Following to Han et al. (2006), they suggested that the oxygen could have penetrated between the additives particles, facilitating its path for the case of higher concentrations. Another explanation was developed by the authors: in the starch films with ZEO smooth surfaces were observed, however, fine particles and holes distributed homogeneously in the polymer matrix had been found. In the case of the addition of high concentrations of MEO, a thicker microstructure had been seen, probably due to possible agglomerations.

Biocomposites with garlic powder added on glycerol–cassava starch films were studied by Famá et al. (2009b) and Famá et al. (2010). When the structure was evaluated, the authors observed a homogeneous distribution of nano/micro fibers that was attributed to the fact that garlic powder has a high percentage of water soluble components (DTU Food 2009), detecting fibrils in the remainder of the solubilisation process. Due to the high number of hydroxyl groups present in this fiber (Sato et al. 2006; Corzo-Martínez et al. 2007), the number of groups available to be involved in exchange with water increased. Similar values of T_g were reported for both matrix and composites; however, composite presented slightly wider peak. This trend was ascribed to the occurrence of two opposite effects: first, the addition of a filler shifts the T_g to higher temperatures (Wollerdorfer and Bader 1998; Van de Velde and Kiekens 2002) and broadens peaks associated with it, but garlic powder contains chemical moieties that possess hydroxyl and amino groups (Sato et al. 2006; Corzo-Martínez et al. 2007) capable of forming hydrogen bonds, affecting the polymeric network formation, developing a plasticizing action (Chartoff 1981).

The development of new resistant bacteria to existing antibiotics (Singh et al. 2008) led to the search for other types of antibacterial substances that can effectively reduce the harmful effects of microorganisms. In the last years, the advent of nanotechnology made that some investigations focused on the development of inorganic fillers nanostructures. There are lots of inorganic composites that are considered nontoxic and contain mineral elements essential to the human body (Roselli et al. 2003). Most antibacterial inorganic materials are metallic nanoparticles and metal oxide nanoparticles such as silver, copper, titanium oxide, gold and zinc oxide (Sondi and Salopek-Sondi 2004; Cioffi et al. 2005; Chaudhry et al. 2008; Bradley et al. 2011). Besides their promising use in industries such as food or medicine, metal nanoparticles have been attractive because of their unique size-dependent optical (Norman et al. 2002; Ung 2002) and electronic properties (Wessels et al. 2004; Schmid and Simon 2005). In addition, they promise environmental benefit due to the possibility of their applications on nanoscience/ nanotechnology, including new catalysts for environmental improve (Kamat et al. 2002), photovoltaic (Hasobe et al. 2003), thermoelectric materials for cooling without refrigerants (Venkatasubramanian et al. 2001), nanocomposite materials for vehicles (Lloyd and Lave 2003), sensors (Macanás et al. 2006; Muraviev et al. 2007), packaging in food industry (Varaprasad et al. 2010), and biomedical applications (Stodolak et al. 2009).

When an inorganic nanofiller is incorporated in a starch matrix, uniform dispersions and strong interfacial adhesion through hydrogen bonding between the metal and the matrix could be achieved due to the similar chemical structures of the stabilizer and the matrix (Chang et al. 2009; Ma et al. 2009; Yu et al. 2009; Liu et al. 2011). As a result, increments in mechanical properties (Wu et al. 2009; Yun et al. 2011), WVP (Yu et al. 2009; Liu et al. 2011), and UV absorbance (Chang et al. 2009; Ma et al. 2009), can be observed.

Oxide nanoparticles such as zinc oxide are a very innovative alternative for use as antibacterial agents to prevent bacterial growth (Moezzi et al. 2012). Research on zinc oxide (ZnO) as an antimicrobial agent started in mid-twentieth century. However, their real use for this application was late last century (Sawai 2003). Currently, ZnO is *generally recognized as safe* material by the U.S. (FDA 2011). It is an essential micronutrient and has important and critical roles in growth and development in humans and animals (Shi et al. 2008). In addition to its important antimicrobial properties, there is a wide interest in ZnO is for its fundamental characteristic, which lead to potential applications in electronic, structural and biomaterials (Wang et al. 2004). Microelectromechanical devices, sensors, transducers and biomedical applications are a few among the spectrum of applications for ZnO.

Today, the major uses of zinc oxide are in the rubber industry, followed by ceramics (International Zinc Association-Zinc Oxide Information Center 2011), drilling fluids for the oil and gas industry (Lau et al. 1997; Sayyadnejad et al. 2008), LEDs, transparent transistors, solar cells and memory devices (Ozgur et al. 2005; Klingshirn 2007a), pigments (Auer et al. 2005), as an energy-saving coating on windows (Klingshirn 2007b), and as the basis of a transparent conducting oxide for consumer devices, for example in food packaging or medicine (Ellmer et al. 2008; Tankhiwale and Bajpai 2012).

In medicine, ZnO has been in use at least since year 2000 as a constituent of medicinal treatment of boils and carbuncles (Frederickson et al. 2005; Halioua and Ziskind 2005). In vitro tests have shown that zinc oxide nanoparticles (ZnO-NP) have antitumor activity against human colon carcinoma cells (De Berardis et al. 2010). ZnO-NP have been considered to be used for the prevention of infectious diseases through their antimicrobial effects (Zhang et al. 2008; Li et al. 2010; Rajendra et al. 2010).

Yadav et al. (2006) reported that ZnO-NP (Fig. 9) possess higher antibacterial effects on microorganisms *S. aureus* and *Salmonella* that other metal oxide nanoparticles (Jones et al. 2007), and fair activity against *E. coli* and *Bacillus phaeus* (Tam et al. 2008). The antimicrobial activity of these nanoparticles is attributed to several mechanisms, including the release of antimicrobial ions (Kasemets et al. 2009),

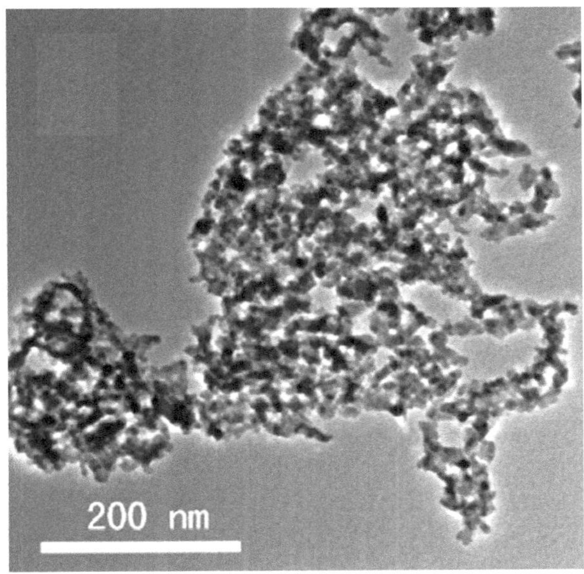

Fig. 9 TEM images of zinc oxide nanoparticles (ZnO-NP). Reproduced with permission from Pantani et al. (2013). © 2013, Elsevier Ltd

interaction of nanoparticles with microorganisms, subsequently damaging the integrity of bacterial cell (Zhang et al. 2008) and the formation of *reactive oxygen species* by the effect of light radiation (Jalal et al. 2010).

In this context, in order to improve the properties of glycerol plasticized starch films, Yu et al. (2009) prepared pea starch-based nanocomposites by casting containing ZnO-NP stabilized by carboxymethylcellulose sodium (CMC). A few agglomerations of the nanofiller with CMC appeared in the composites with higher filler concentrations (more than 5 wt%). The incorporation of the nanofiller at lower concentrations enhanced the pasting viscosity, storage modulus, glass transition temperature and the tensile strength of the nanocomposites. At these filler concentrations, the tensile strength increased more than 100 %, while an important decrement in the elongation at break to nearly half of the matrix value was observed. Besides, the glass transitions shifted to higher temperature with increasing contents of ZnO-NP. The authors explained this behavior in the same way as Chang et al. (2010b) in their investigation of starch/CH-NP nanocomposites, attributing it to the fact that ZnO-NP improved the intermolecular interaction of the matrix, reducing the free volume and increasing the glass transition temperature of composites. WVP of these composites decreased with the increasing of ZnO-NP concentration. This behavior was more marked at very low contents of filler, while for the higher concentrations the decrease of WVP resulted less evident. The addition of these nanoparticles probably introduced a tortuous path for water molecule to pass through (Kristo and Biliaderis 2007), as their water resistance was better than the matrix one. Since ZnO-NP-CMC could achieve a good dispersion degree in the starch matrix at low concentrations, there were few paths for water molecules to pass through. An important conclusion of these authors was that the ZnO-NP protected by CMC could be easily integrated into some relevant systems for pharmaceutical (drug release) and biomedical applications, as well as for agriculture, and packaging fields.

Parallel to the study of Yu et al. (2009), a bionanocomposite based on glycerol plasticized-pea starch containing ZnO-NP stabilized by soluble starch as filler, was developed by Ma et al. (2009). In this work, the authors showed that the incorporation of this nanofiller led to improvements in the pasting viscosity, storage modulus, glass transition temperature and UV absorbance. In the same way of Yu et al. (2009), the authors attributed the results to the interaction between ZnO-NP and starch matrix. Soluble starch played an important role in the stabilization of the filler and in the fabrication of starch/ZnO-NP composites. The strong interaction between the filler and the matrix contributed to the improvement in the bionanocomposite properties.

In the recently years some researches demonstrated the important effect of zinc oxide nanorods (ZnO-NR) incorporating them in a thermoplastic sago starch-based material (Nafchi et al. 2012, 2013; Rahman et al. 2013). In particular, (Nafchi et al. 2012, 2013) investigated physicochemical properties of nanocomposites formed by sago starch and different concentration of ZnO-NR (1–5 wt%). The incorporation of low concentration of this nanofiller significantly increased the viscosity of the solution and decreased water vapor permeability of the composites to less than one third of the matrix one. These behaviors were attributed to the greater water resistance of the filler compared with the matrix, and the introduction of a tortuous pathway for water vapor molecules to pass through (Yu et al. 2009). Solubility, moisture content, and monolayer water content also decreased with the addition of ZnO-NR. According to Tunç and Duman (2010), the increment of the ZnO-NP concentration, results in the formation of more hydrogen bonds between the filler and the matrix components, making that free water molecules cannot interact as strongly with the nanocomposite than with the matrix. Besides, increasing the ZnO-NR content in starch films led to a greater contact angle, indicating greater hydrophobicity of the surface. The authors indicated that contact angle results were consistent with the decrease in water absorption tendency because the surfaces of the composites became more hydrophobic. Finally, these nanofiller did not modify UV transmittance of the films (0 %), and were able to absorb more than 80 % of Near Infrared spectra (Nafchi et al. 2012).

Similar behavior on water absorption capacity, WVP and water solubility of sago starch-based composites containing 0, 1, 3 and 5 wt% of ZnO-NR was reported by Alebooyeh et al. (2012): all these properties decreased by increasing concentration of the nanofiller. In particular, these authors observed that when ZnO-NR content was only 1 wt%, these parameters showed a more marked decrease, especially WVP. Furthermore, the addition of the nanorods demonstrated a decrease in the microbial activity, in particular, the composites showed antimicrobial properties against *E. coli*, which was very important even with the addition of the lowest concentration of ZnO-NR. In other investigation of Nafchi et al. (2013), mechanical, thermal and antibacterial properties of starch/ZnO-NR films were exposed. Significant increments in tensile strength and elastic modulus were observed when the nanorod concentration increased, while elongation at break decreased in all the studied composites. The authors explained that this phenomenon could have been due to two possible reasons. The first is related to the decrease in moisture content with the ZnO-NR addition. Water plays opposite roles in a biopolymer matrix: as plasticizer and as antiplasticizer, decreasing the flexibility of the films (Cheng et al. 2006; Müller et al. 2011). The second reason is related to the interfacial interaction between the nanorods and the starch matrix. In these work, the authors predicted the mechanical parameters of the composites, showing that the experimental values resulted higher than those data predicted, suggesting that there was a significant interaction between the sago starch matrix and the ZnO-NR. Dynamic mechanical properties indicated that the addition of the filler had significantly increased T_g. This behavior resulted consistent with the moisture content reported for this composite (Abdorreza et al. 2011; Nafchi et al. 2012).

Sago starch with ZnO-NR films exhibited excellent antimicrobial activity against *S. aureus*, suggesting these nanocomposites have potential applications as active packaging materials in the pharmaceutical and food industries. Rahman et al. (2013) used concentration of ZnO-NR between zero and 10 wt% to incorporate into a thermoplastic sago starch matrix. Initially, the authors showed no filler particles agglomeration in all film samples indicating an uniform distribution of the nanorods into the starch film. Unusually, FTIR analysis revealed that there was no presence of new functional groups after the incorporation of the ZnO-NR, indicating that the interaction between the filler and the starch was physical or no covalent. By increasing the amount of filler, the authors observed a huge difference in the studied properties, especially on relative dielectric constant and electrical conductivity. When ZnO-NR concentration increased, the electrical conductivity and relative dielectric constant of the composites increased, becoming around 53 and 60 % in the case of the material with highest filler content, respectively. Similar to Nafchi et al. (2012), ultraviolet absorbance showed zero light transmission below 380 nm independently of the filler concentration. The authors concluded that that ZnO-NR can provide UV-shielding properties to TPS materials. When loss tangent with frequency less than 1 Hz was investigated, a decrement in the T_g was observed with the filler addition. In general, tan δ values are expected to increase as concentration of filler increase. The opposite behavior observed in this case was explained taking into account the charge transport through the different chains or interfaces and to some defects as some agglomerations. The authors concluded that the antibacterial mechanism of starch/ZnO-NR composites could preclude the growth of bacteria.

Other metal as silver (Ag) is economical and has important antimicrobial activity properties (Sharma et al. 2009). However, the use of Ag as a reinforcement of thermoplastic materials requires a pretreatment of the metal, due to its natural size and its propensity to form agglomerates in the composite material. Improved properties are generally obtained when small dispersed nanodomains are observed (Wiley et al. 2007). In order to reduce the particle size, some authors synthesized silver nanoparticles (Ag-NP) (Fig. 10). Controllable synthesis of this metal is the

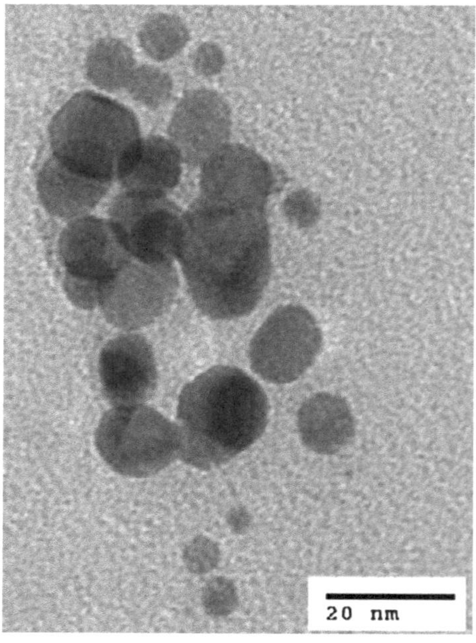

Fig. 10 TEM image of silver nanoparticles (Ag-NP). Taken from Seo et al. (2012). © 2012, Elsevier Ltd

first key challenge to achieve their better applied characteristics (Wiley et al. 2007). Silver nanoparticles can be prepared by physical, biological and chemical methods. Generally, the modification of this metal implies a large number of stabilizers, such as surfactants, proteins, peptides, polymers, oligonucleotides, carbohydrates, plants extracts, and organic solvents (Xie et al. 2007; Sharma et al. 2009). Ag-NP have attracted considerable interest in biological studies because of their easy preparation, good biocompatibility, and relatively large surface area (Wiley et al. 2004; Xie et al. 2007). Additionally, for their potential antibacterial characteristics (Huang et al., 2006), they are used as a sterilizer for removing bacteria from food (Creutz 1981; Davies 1992).

The incorporation of Ag-NP into biodegradable polymers for their potential applications in biotechnology has been a great interest topic in the last years (Huang et al. 2004; Narayanan and El-Sayed 2005; Murugadoss and Chattopadhyay 2008; Sanpui et al. 2008; Rhim et al. 2013; Cheviron et al. 2014). The carbohydrates can act as a reducing and/or stabilizer agent, and also have the possibility to carry Ag-NP with excellent antibacterial activity. The formed nanocomposites result safe, biocompatible, nontoxic, and environmentally friendly (Rhim et al. 2013).

Among all the studies about composites with silver nanoparticles, very few have concerned about the preparation of biodegradable starch–silver nanoparticles nanocomposites for green applications.

Cheviron et al. (2014) prepared environmentally friendly nanocomposite films based on potato starch/glycerol reinforced by colloidal Ag-NP by solution casting. In order to correctly disperse the nanofiller into the biodegradable matrix, the authors used water, glucose and soluble starch as solvent, reducing and stabilizing agent respectively. Two different populations with distinct Ag-NP particle size (diameters: 5 nm and 20–50 nm) were distinguished in their observations. They concluded that the starch presence in the colloidal nanoparticles allows a better dispersion of them in the starch matrix due to their similar chemical structure and the high molar mass polymer chains of potato starch.

Starch-based composites were developed by Khachatryan et al. (2013) using Ag-NP with dialdehyde starch (DAS) in order to reduce and protect agents. Micrographs confirmed formation of spherical Ag-NP of size within 10 and 15 nm, and with crystalline characteristics. TGA curves showed that the incorporation of the filler presented materials with approximately 1 % less water than matrix. It was also possible to note that the decomposition of the materials with the nanoparticles occurred in slightly smaller temperatures than that of DAS.

Yoksan and Chirachanchai (2010) used γ-ray irradiation reduction of silver nitrate in a chitosan solution to obtain Ag-NP with the objective of incorporating them into a starch-based matrix. TEM images showed that after the synthesis, silver nanoparticles presented a spherical shape with size between 20 and 25 nm. γ-ray irradiated chitosan solution containing Ag-NP exhibited an important antimicrobial activity against *E. coli*, *S. aureus* and *B. cereus*. When color was evaluated, matrix resulted transparent and slightly yellowish, while composites were pale brown; and the color difference of the film increased with the concentration of Ag-NP. Physicochemical behavior of these composites was not the most desired compared with that reported in the literature (Khachatryan et al. 2013; Cheviron et al. (2014). In particular, films containing low concentrations of nanoparticles presented not markedly differences in tensile strength and elongation at break respect to the matrix, and a slightly improvement was achieved with the highest filler content, suggesting that the incorporation of Ag-NP could somewhat improve the tensile properties of starch-based films but not with low concentrations of the filler. However, in the case of the elastic modulus, a slight increase was observed for all composites. The addition of Ag-NP slightly increased WVP, probably due to the obstruction of intermolecular hydrogen bond formation between polymer molecules by Ag-NP, causing the incompatibility of the film matrix, adsorption of water vapor at the hydrophilic sites of polysaccharide molecules, and penetration of moisture (Shelma et al. 2008). The best result showed by these authors was the enhancement of O_2P properties of the composites with the addition of the silver nanoparticles.

Some authors have investigated the properties of starch/Ag-NP composites incorporating a biopolymer as part of the matrix or as an additive, for its antimicrobial contribution (Arockianathan et al. 2012; Zepon et al. 2013). The attributes of excellent biocompatibility of biopolymers with versatile biological activities and in some cases antimicrobial activity, have provided great opportunities to improve the properties of nanocomposites, obtaining functional biomaterials of high potential in various fields. In particular, in the investigation of Arockianathan et al. (2012) sago starch-based composite films impregnated with different concentrations of CH-Ag-NP were developed using casting method. FTIR spectrum validated chemical interaction between the polymers. The authors showed that the better composite was the one containing 10 wt% of CH-Ag-NP mix, which increased tensile straight without decreasing elongation at break, and its water absorption capacity after 24 h was lower than the matrix one. The authors used this nanocomposite and the starch matrix as wound dressing materials in experimental wounds of rats, and evaluated the healing pattern, observing faster healing in the cases treated with the material containing CH-Ag-NP compared to the control. They finally suggested that starch/CH-Ag-NP nanocomposite could have possibilities as a dressing material for wound healing applications.

Among the various semiconductor photocatalysts, titanium oxide (TiO_2) is one of the most popular and promising materials because it is stable in various solvents and it is commercially available (Sepone and Pelizzetti 1989; Schiavello 1997). It is also efficient to adsorb light energy. In particular, titanium oxide nanoparticles (TiO_2-NP) have been incorporated in polysaccharide based thermoplastic materials (Miao et al. 2008; Hejri et al. 2012; Nuryetti and Nasikin 2012; Sreekumar et al. 2012; Yun et al. 2012) (Fig. 11). TiO_2-NP have potential remarkable applications as an attractive filler for multifunctional materials, due to their unique properties such as higher stability, long lasting, safe and broad-spectrum antibiosis (Roessler et al. 2002; Cai et al. 2006). Their most striking characteristics are their photocatalytic activities (Kanehira et al. 2008; Wang et al. 2008), related to their microstructure and the powder purification (Weibel et al. 2006; Verran et al. 2007). Common preparation of these types of bionanocomposites included hydrolysis of the metal compounds and dehydration (Montoya et al. 1992), sol–gel (Barringer and Bowen 1982),

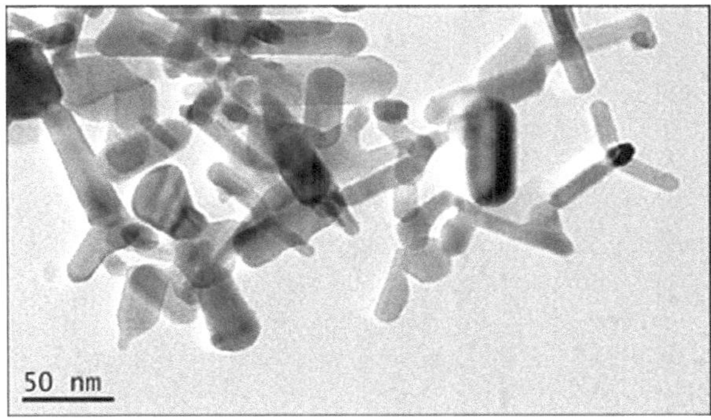

Fig. 11 TEM image of titanium oxide nanoparticles (TiO_2-NP). Reproduced with permission from Miao et al. (2008). © 2008, Elsevier Ltd

a hydrothermal method (Kondo et al. 1994), or modification of commercially available TiO_2.

Two methods are the most frequently used to perform a polymer-based bio-composites containing TiO_2-NP. The filler is directly added into the starch matrix solution (Yun et al. 2011), or directly synthesized in a starch dispersion, which acts not only as the stabilizer but also as the matrix (Liao and Wu 2008).

In plants it has been demonstrate that TiO_2-NP could increase the nitrate reductive enzyme, increase the abilities to absorb and use water and fertilizer, promote antioxidant system, and accelerate germination (Lu et al. 2002; Zheng et al. 2005; Navarro et al. 2008). In particular these effects can be because titanium oxide nanoparticles have high specific surface area.

One example of starch composites formed with TiO_2-NP was the reported by Nuryetti and Nasikin (2012). They prepared tapioca starch matrices containing different concentrations of this filler (1, 3, 5 and 7 wt%) using a melt process, in order to investigated structure, energy band gap, and electrical conductivity of the composites. The energy gap range for the composites with TiO_2-NP resulted very closed in number (3.30–3.22 eV), and similar to the values of the energy gap of semiconductors (Poole and Owens 2003; Sing et al. 2010). UV tests showed that tapioca starch/TiO_2-NP were effective as barriers for UV wavelength range (315–280 nm) (Kim et al. 2000). The incorporation of filler concentrations till 3 wt% resulted in a small increase of the electrical conductivity of the composites, while in the case of 5 wt%, the increment was sharp. However, a decrease in electrical conductivity of the composite with 7 wt% of filler was observed. Finally, taking into account the conductivity behavior, starch/TiO_2-NP presented percolation thresholds when concentrations of the oxide were between 3 and 5 wt%.

5 Conclusions and Future Perspective

In this chapter a review of the most important and recent researches on development and characterization of biodegradable nanocomposites based on starch reinforced by different types of nanofillers were exposed. Particularly, the investigation was focused on the influence of the incorporation of starch, cellulose, layered silicate, and antioxidant and/or antimicrobial nanofillers on the physicochemical properties of the composites.

Some different methods for processing both starch matrix and nanocomposites were discussed. Significant attention was paid on the variables involved in casting and extrusion techniques, being these the most used methods at laboratory and industrial scale processes.

Several physicochemical properties as structure, mechanics and barrier, as well as degradation, antibacterial and antioxidant activity have been presented. There is agreement in the literature about improvements in barrier, thermal and mechanical properties of starch by the addition of nanofillers, only in the cases in which they are well dispersed and are compatible with it. In some cases the methodology for the nanofiller obtaining generates reactive groups that help the filler dispersion in the starch matrix, whereas in other cases it is necessary to functionalize them. In particular in the case of clays, they are usually modified by organic surfactants. Furthermore, the interface between filler and matrix plays a crucial role in the final properties of the composite. Besides, the researches on nanofillers with antibacterial characteristics as chitosan, garlic, bamboo, zinc oxide, silver and titanium oxide, demonstrated that the incorporation of this kind of reinforcement in a starch material serves to retard the growth and the action of pathogenic microorganisms whose produce toxins or cause infections. On the other hand, in some cases the degradation of the composites results at higher temperatures due to the characteristics of the nanofiller.

From the results reported in the different works cited in this chapter, it can be concluded that in order to improve the barrier, mechanical and thermic properties of a starch matrix, the next considerations need to be taken into account: (a) morphological and chemical characteristics of the employed nanofillers; (b) the reactive groups in their surface, as well as their crystalline fraction (this depends on the methodology employed for their obtaining); (c) the variables involved in the biodegradable nanocomposites manufacture.

Taking into the account the serious problems that have been caused by mistreatment ecology due to the increasing use of synthetic materials, necessary awareness must be given to the replacement of these treacherous by materials that are friendly to both the environment and human health. Furthermore, the increasing use of synthetic polymers or plastics as a result of the growing human population and standard of living will result in higher demands on oil production and will contribute to a possible depletion of crude oil before the end of the century. The implement of the bio-based polymers use will reduce the dependency on the plastics derived from the petroleum industry, creating more sustainable alternatives. Nowadays, there are few companies that use starch-based polymer materials in the packaging industry and food products. For example, Biostarch (Singapour), Plantic (Australia), Novamont (Australia), BIOP Biopolymer Technologies (Germany) and Sphere (France) manufacture disposable products as dishwasher tablets, plates, cutlery, glasses, thermoformed trays, film to cover food, packaging, bags for agriculture, etc.

Based on the literature, several nanofillers showed that they can improve the properties of biodegradable starch-based polymers making possible their use in numerous applications.

In particular, nowadays industries with high demand for materials with these nanocomposites' characteristics, such as packaging or biomedicine, do focus on them to be implemented in daily life.

Acknowledgements The authors want to thank the National Scientific and Technical Research Council of Argentina, CONICET (PIP 2013-2015, 11220120100508CO and 11220110100370CO), the University of Buenos Aires (UBACYT 2011-2014, 20020100100350 and 200220100100142, and UBACYT 2012-2015, 20020110200196), and ANPCyT (PICT-2012-1093 and PICT-2012-0717), for their support.

References

Abdorreza MN, Cheng LH, Karim AA (2011) Effects of plasticizers on thermal properties and heat sealability of sago starch films. Food Hydrocolloid 25:56–60

Akdogan H (1996) Pressure, torque, and energy responses of a twin screw extruder at high moisture contents. Food Res Int 29:423–429

Alebooyeh R, Nafchi AM, Jokar M (2012) The effects of ZnO nanorods on the characteristics of sago starch biodegradable films. J Chem Health Risks 2:13–16

Alexandre M, Dubois P (2000) Polymer-layered silicate nanocomposites: preparation, properties and uses of a new class of materials. Mater Sci Eng 28:1–63

Amada S, Ichikawa Y, Munekata T, Nagase Y, Shimuzu H (1997) Fibre texture and mechanical graded structure of bamboo. Compos Part B Eng 28:13–20

Angellier H, Molina-Boisseau S, Dole P, Dufresne A (2006) Thermoplastic starch-waxy maize starch nanocrystals nanocomposites. Biomacromolecules 7:531–539

Angles MN, Dufresne A (2000) Plasticized starch/tunicin whiskers nanocomposites. 1. Structural analysis. Macromolecules 33:8344–8353

Angles MN, Dufresne A (2001) Plasticized starch/tunicin whiskers nanocomposite materials. 2. Mechanical behavior. Macromolecules 34:2921–2931

Aouada FA, Mattoso LHC, Longo E (2011) A simple procedure for the preparation of laponite and thermoplastic starch nanocomposites: structural, mechanical, and thermal characterizations. J Therm Comp Mat 26:109–124

Appendini P, Hotchkiss JH (1997) Immobilization of lysozyme on food contact polymers as potential antimicrobial films. Packag Technol Sci 10:271–279

Arockianathan PM, Sekar S, Kumaran B, Sastry TP (2012) Preparation, characterization and evaluation of biocomposite films containing chitosan and sago starch impregnated with silver nanoparticles. Int J Biol Macromol 50:939–946

Arora A, Padua GW (2010) Review: nanocomposites in food packaging. J Food Sci 75:43–49

Arvanitoyannis I, Billiaderis CG, Ogawa H, Kawasaki N (1998) Biodegradable films made from low-density polyethylene (LDPE), rice starch and potato starch for food packaging applications: part 1. Carbohydr Polym 36:89–104

Atwell WA, Hood LF, Lineback DR, Varriano-Marston E, Zobel HF (1988) The terminology and methodology associated with basic starch phenomena. Cereal Food World 33:306–311

Auer G, Griebler WD, Jahn B (2005) Industrial inorganic pigments. Wiley-VCH Verlag GmbH & Co, KGaA, Weinheim

Avérous L, Boquillon N (2004) Biocomposites based on plasticized starch: thermal and mechanical behaviours. Carbohydr Polym 56:111–122

Ayadi F, Dole P (2011) Stoichiometric interpretation of thermoplastic starch water sorption and relation to mechanical behavior. Carbohydr Polym 84:872–880

Baker R, Baldwin E, Nisperos-Carriedo M (1994) Edible coatings and films to improve food quality. CRC Press, Lancaster, p 392

Barringer EA, Bowen HK (1982) Formation, packing and sintering of monodispersed TiO_2 powders. J Am Ceram Soc 65:199–201

Beck-Candanedo S, Roman M, Gray DG (2005) Effect of reaction conditions on the properties and behavior of wood cellulose nanocrystal suspensions. Biomacromolecules 6:1048–1054

Bergaya F, Jaber M, Lambert JF (2009) Clays and clay minerals as layered nanofillers for (bio) polymers. In: Averous L, Pollet E (eds) Environmental silicate nano-biocomposites (green energy and technology). Springer, London, pp 41–75

Bertan LC, Tanada-Palmu PS, Siani AC, Grosso CRF (2005) Effect of fatty acids and 'Brazilian elemi' on composite films based on gelatin. Food Hydrocolloid 19:73–82

Bertolini C, Souza E, Nelson JE, Huber KC (2003) Composition and reactivity of A- and B-type starch granules of normal, partial waxy, and waxy wheat. Cereal Chem 80:544–549

Biercuk MJ, Llaguno MC, Radosavljevic M, Hyun JK, Fischer JE, Johnson AT (2002) Carbon nanotube composites for thermal management. Appl Phys Lett 80:2767–2769

Bierhalz ACK, da Silva MA, Kieckbusch TG (2012) Natamycin release from alginate/pectin films for food packaging applications. J Food Eng 110:18–25

Borges SV, Dias ML, Pita VJRR, Azuma C, Dias MV (2012) Water vapor permeability and tensile properties of poly(l-lactic acid)/synthetic mica nanocomposites prepared by melt blending. J Plast Film Sheet. doi:10.1177/8756087912463712

Bourtoom T (2009) Edible protein films: properties enhancement. Int Food Res J 16:1–9

Bouwmeester H, Dekkers S, Noordam MY, Hagens WI, Bulder AS, de Heer C (2009) Review of health safety aspects of nanotechnologies in food production. Regul Toxicol Pharmacol 53:52–62

Bouyer E, Mekhloufi G, Rosilio V, Grossiord J-L, Agnely F (2012) Proteins, polysaccharides, and their complexes used as stabilizers for emulsions: alternatives to synthetic surfactants in the pharmaceutical field? Int J Pharma 436:359–378

Bradley EL, Castle L, Chaudhry Q (2011) Applications of nanomaterials in food packaging with a consideration of opportunities for developing countries. Trends Food Sci Technol doi:10.1016/j.tifs.2011.01.002 (in press)

Cai K, Bossert J, Jandt KD (2006) Does the nanometre scale topography of titanium influence protein adsorption and cell proliferation? Colloid Surf B Biointerfaces 49:136–144

Cao X, Chen Y, Chang PR, Huneault MA (2007) Preparation and properties of plasticized starch/multiwalled carbon nanotubes composites. J Appl Polym Sci 106:1431–1437

Cao X, Chen Y, Chang PR, Muir AD, Falk G (2008a) Starch-based nanocomposites reinforced with flax cellulose nanocrystals. Express Polym Lett 2:502–510

Cao X, Chen Y, Chang PR, Stumborg M, Huneault MA (2008b) Green composites reinforced with hemp nanocrystals in plasticized starch. J Appl Polym Sci 109:3804–3810

Cao X, Ding B, Yu J, Al-Deyab S (2012) Cellulose nanowhiskers extracted from TEMPO-oxidized jute fibers. Carbohydr Polym 90:1075–1080

Cioffi N, Torsi L, Ditaranto N, Tantillo G, Ghibelli L, Sabbatini L, Bleve-Zacheo T, D'Alessio M, Zambonin PG, Traversa E (2005) Copper nanoparticle/polymer composites with antifungal and bacteriostatic properties. Chem Mater 17:5255–5262

Copeland L, Blazek J, Salman H, Tang MCM (2009) Form and functionality of starch. Food Hydrocolloid 23:1527–1531

Corzo-Martínez M, Corzo N, Villamiel M (2007) Biological properties of onions and garlic. Trends Food Sci Tech 18:609–625

Creutz C (1981) The complexities of ascorbate as a reducing agent. Inorg Chem 20:4449–4452

Cuq JL, Aymard C, Cheftel C (1977) Effects of hypochlorite treatments on a methionyl peptide. Food Chem 2:309–314

Chang PR, Yu J, Ma X (2009) Fabrication and characterization of Sb_2O_3/carboxymethyl cellulose sodium and the properties of plasticized starch composite films. Macromol Mat Eng 294:762–767

Chang PR, Jian R, Zheng P, Yu J, Ma X (2010a) Preparation and properties of glycerol plasticized-starch (GPS)/cellulose nanoparticle (CN) composites. Carbohydr Polym 79:301–305

Chang PR, Jian R, Yu J, Ma X (2010b) Fabrication and characterisation of chitosan nanoparticles/plasticised-starch composites. Food Chem 120:736–740

Chang PR, Jian R, Yu J, Ma X (2010c) Starch-based composites reinforced with novel chitin nanoparticles. Carbohydr Polym 80:421–426

Chakraborty S, Sahoo B, Teraka I, Miller LM, Gross RA (2005) Enzyme-catalyzed regioselective modification of starch nanoparticles. Macromolecules 38:61–68

Chartoff RP (1981) Thermal characterization of polymeric materials. In: Turi E (ed) Academic Press, San Diego, p 526

Chaudhry Q, Scotter M, Blackburn J, Ross B, Boxall A, Castle L, Aitken R, Watkins R (2008) Applications and implications of nanotechnologies for the food sector. Food Addit Contam A 25:241–258

Chen Y, Cao X, Chang PR, Huneault MA (2008) Comparative study on the films of poly(vinyl alcohol)/pea starch nanocrystals and poly(vinyl alcohol)/native pea starch. Carbohydr Polym 73:8–17

Chen Y, Liu C, Chang PR, Anderson DP, Huneault MA (2009a) Pea starch-based composite films with pea hull fibers and pea hull fiber-derived nanowhiskers. Polym Eng Sci 49:369–378

Chen Y, Liu C, Chang PR, Cao X, Anderson DP (2009b) Bionanocomposites based on pea starch and cellulose nanowhiskers hydrolyzed from pea hull fibre: effect of hydrolysis time. Carbohydr Polym 76:607–615

Cheng LH, Karim AA, Seow CC (2006) Effects of water–glycerol and water–sorbitol interactions on the physical properties of konjac glucomannan films. J Food Sci 71:62–67

Cheviron P, Gouanvé F, Espuche E (2014) Green synthesis of colloid silver nanoparticles and resulting biodegradable starch/silver nanocomposites. Carbohydr Polym 108:291–298

Chivrac CF, Pollet E, Dole P, Avérous L (2010) Starch base nano-biocomposites: plasticizer impact on the montmorillonite exfoliation process. Carbohydr Polym 79:941–947

Davidson PM, Zivanovic S (2003) The use of natural antimicrobials. In: Zeuthen P, Bogh-Sorensen L (eds) Food preservation techniques. CRC Press, Boca Raton, pp 5–30

Davidson PM, Taylor TM (2007) Chemical preservatives and natural antimicrobial compounds. In: Doyle MP, Beuchat LR (eds) Food microbiology: fundamentals and frontiers. ASM Press, Washington, DC, pp 713–746

Davies MB (1992) Reactions of L-ascorbic acid with transition metal complexes. Polyhedron 11:285–321

De Azedero HMC (2009) Nanocomposites for food packaging applications. Food Res Int 42:1240–1253

De Berardis B, Civitelli G, Condello M, Lista P, Pozzi R, Arancia G, Meschini S (2010) Exposure to ZnO nanoparticles induces oxidative stress and cytotoxicity in human colon carcinoma cells. Toxicol Appl Pharm 246:116–127

de Mesa NJE, Sajid A, Singh N, Shi YC, Dogan H, Sang Y (2009) Soy protein fortified expanded extrudates: baseline study using normal corn starch. J Food Eng 90:262–270

Dhakal HN, Zhang Z (2012) Polymer matrix composites: moisture effects and dimensional stability. In: Nicolais L, Borsachiello A (eds) Wiley encyclopedia of composites. Wiley, New York

Dieu T, Liem N, Mai Y, Tung N (2004) Study on fabrication of BMC laminates based on unsaturated polyester resin reinforced by hybrid bamboo/glass fibers. JSME Int J Ser A 47:570–573

DTU Food (2009) Food National Institute Food. Danish Food Composition Databank. (http://www.foodcomp.dk/v7/fcdb_foodnutrlist.asp?CompId=0065)

Du YY, Fang HH, Zheng PW (2013) Porous sepiolite/starch composites: Preparation, structure and absorption properties. Adv Mat Res 1937:634–638

Duquesne E, Moins S, Alexandre M, Dubois P (2007) How can nanohybrids enhance polyester/ sepiolite nanocomposite properties? Macromol Chem Phys 208:2542–2550

Duran N, Lemes AP, Duran M, Freer J, Baeza J (2011) A mini review of cellulose nanocrystals and its potential integration as co-product in bioethanol production. J Chil Chem Soc 56:672–677

Dutta PK, Tripathi S, Mehrotra GK, Dutta J (2009) Perspectives for chitosan based antimicrobial films in food applications. Food Chem 114:1173–1182

Elsabee MZ, Abdou ES (2013) Chitosan based edible films and coatings: a review. Mater Sci Eng C 33:1819–1841

Ellmer K, Klein A, Rech B (2008) Transparent conductive zinc oxide. Springer, Heidelberg

Eugenius GF, Jongboom ROJ, Feil H, Gotlieb KF, Boersma A (2000) Patent WO 2000069916 A1, 20001123

Espitia PJP, Soares NFF, dos Reis Coimbra JS, de Andrade NJ, Cruz RS, Alves Medeiros EA (2012) Zinc oxide nanoparticles: synthesis, antimicrobial activity and food packaging applications. Food Bioprocess Technol 5:1447–1464

European Project (FlexPakRenew–FP7/2007-2013—no. 207810. http://ec.europa.eu/research/ infocentre/converting.cfm

Famá L, Rojas AM, Goyanes S, Gerschenson L (2005) Mechanical properties of tapioca-starch edible films containing sorbates. LWT 38:631–639

Famá L, Flores SK, Gerschenson L, Goyanes S (2006) Physical characterization of cassava starch biofilms with special reference to dynamic mechanical properties at low temperatures. Carbohydr Polym 66:8–15

Famá L, Goyanes S, Gerschenson L (2007) Influence of storage time at room temperature in physicochemical properties of tapioca starch edible films. Carbohydr Polym 70:265–273

Famá L, Gerschenson L, Goyanes S (2009a) Starch-vegetable fiber composites to protect food products. Carbohydr Polym 75:230–235

Famá L, Gerschenson LN, Goyanes S (2009b) Nanocompuestos biodegradables y comestibles: almidón-polvo de ajo. Suplemento de la Revista Latinoamericana de Metalurgia y Materiales S1(3):1235–1240

Famá L, Bittante AMBQ, Sobral PJA, Goyanes S, Gerschenson LN (2010) Garlic powder and wheat bran as fillers: their effect on the physicochemical properties of edible biocomposites. Mat Sci Eng C 30:853–859

Famá LM, Pettarin V, Goyanes S, Bernal CR (2011) Starch based nanocomposites with improved mechanical properties. Carbohydr Polym 83:1226–1231

Famá LM, Gañan P, Bernal CR, Goyanes S (2012) Biodegradable starch nanocomposites with low water vapor permeability and high storage modulus. Carbohydr Polym 87:1989–1993

Famá L, Kumar R (2014) Nanocomposites based on polylactic acid (PLA) reinforced by functionalized carbon nanotubes (CNT). In: Kumar R (ed) Polymer-matrix composites: types, applications and performance. Nova Science Publishers, Inc. USA (in press)

FDA (2011) Part 182-substances generally recognized as safe. Food and drug administration. Washington DC. http://ecfr.gpoaccess.gov/cgi/t/text/text-idx?c0ecfr&sid0786bafc6f6343634fbf79fcdca7061e1&rgn0div5&-

Flores S, Famá L, Rojas AN, Goyanes S, Gerschenson L (2007) Physicochemical properties of tapioca-starch edible films. Influence of gelatinization and drying technique. Food Res Int 4:257–265

Frederickson CJ, Koh JY, Bush AI (2005) The neurobiology of zinc in health and disease. Nat Rev Neurosci 6:449–462

Gálvez A, Abriouel H, López RL, Omar NB (2007) Bacteriocin-based strategies for food biopreservation. Int J Food Microbiol 120:51–70

García MA, Martino MN, Zaritzky NE (1998) Starch-based coatings: effect on refrigerated strawberry (*Fragaria* × *Ananassa*) quality. J Sci Food Agric 76:411–420

García NL, Famá L, Dufresne A, Aranguren A, Goyanes S (2009a) A comparison between the physico-chemical properties of tuber and cereal starches. Food Res Int 42:976–982

García NL, Ribba L, Dufresne A, Aranguren M, Goyanes S (2009b) Physico mechanical properties of biodegradable starch nanocomposites. Macromol Mater Eng 294:169–177

García NL, Ribba L, Dufresne A, Aranguren M, Goyanes S (2011) Effect of glycerol on the morphology of nanocomposites made from thermoplastic starch and starch nanocrystals. Carbohydr Polym 84:203–210

Galicia-García T, Martínez-Bustos F, Jiménez-Arévalo OA, Arencón D, Gámez-Pérez J, Martínez AB (2012) Films of native and modified starch reinforced with fiber: influence of some extrusion variables using response surface methodology. J Appl Polym Sci 126:327–336

Gaspar M, Benko Z, Dogossy G, Reczey K, Czigany T (2005) Reducing water absorption in compostable starch-based plastics. Polym Degrad Stab 90:563–569

Ghanbarzadeh B, Almasi H, Entezami A (2011) Improving the barrier and mechanical properties of corn starch-based edible films: effect of citric acid and carboxymethyl cellulose. Ind Crop Prod 33:229–235

Ghavami K (2005) Bamboo as reinforcement in structural concrete elements. Cement Concr Comp 27:637–649

Ghasemlou M, Aliheidari N, Fahmi R, Shojaee-Aliabadi S, Keshavarz B, Cran MJ, Khaksar R (2013) Physical, mechanical and barrier properties of corn starch films incorporated with plant essential oils. Carbohydr Polym 98:1117–1126

Ghori MU, Alba K, Smith AM, Conway BR, Kontogiorgos V (2014) Okra extracts in pharmaceutical and food applications. Food Hydrocolloid ISSN 0268005X. Available online 26 April 2014 (in press)

Godavarti S, Karwe MV (1997) Determination of specific mechanical energy distribution on a twin-screw extruder. J Agric Eng Res 67:277–287

Godbillot L, Dole P, Joly C, Rogé B, Mathlouthi M (2006) Analysis of water binding in starch plasticized films. Food Chem 96:380–386

González Seligra P, Nuevo F, Lamanna M, Famá L (2013) Covalent grafting of carbon nanotubes to PLA in order to improve compatibility. Compos B Eng 46:61–68

Goyanes S, Aranguren M, García N, Famá L, Ribba L, Dufresne A (2010) International Patent No 20.100.100.044

Gropper M, Moraru CI, Kokini JL (2002) Effect of specific mechanical energy on properties of extruded protein–starch mixtures. Cereal Chem 79:429–433

Guerrero P, Beatty E, Kerry JP, de la Caba K (2012) Extrusion of soy protein with gelatin and sugars at low moisture content. J Food Eng 110:53–59

Gutiérrez TJ, Pérez E, Guzmán R, Tapia MS, Famá L (2014a) Physicochemical and functional properties of native and modified by crosslinking, dark cush-cush yam (*Dioscorea trifida*) and cassava (*Manihot esculenta*) starch. J Polym Biopolym Phys Chem 2:1–5

Gutiérrez TJ, Morales NJ, Tapia MS, Pérez E, Famá L (2014b) Corn starch 80:20 "waxy": regular, "native" and phosphated, as bio-matrixes for edible films. Procedia materials science. Elsevier, New York. ISSN 2211-8128 (in press)

Haafiz MKM, Hassan A, Zakaria Z, Inuwa IM (2014) Isolation and characterization of cellulose nanowhiskers from oil palm biomass microcrystalline cellulose. Carbohydr Polym 103:119–125

Halsall TG, Hirst EL, Jones JKN, Sansome FW (1948) The amylose content of the starch present in the growing potato tuber. Biochem J 43:70–72

Halioua B, Ziskind B (2005) Medicine in the days of the pharaohs. Belknap Press of Harvard University Press. http://www.PalArch.nl, web based Netherlands scientific journal

Han JH (2005) Antimicrobial packaging systems. In: Jung HH (ed) Innovations in food packaging. Academic Press, London, pp 80–107

Han JH, Seo GH, Park IM, Kim GN, Lee DS (2006) Physical and mechanical properties of pea starch edible films containing beeswax emulsions. J Food Sci 71:290–296

Hansen NML, Plackett D (2008) Sustainable films and coatings from hemicelluloses: a review. Biomacromolecules 9:1494–1505

Hasobe T, Imahori H, Fukuzumi S, Kamat PV (2003) Light energy conversion using mixed molecular nanoclusters. Porphyrin and C_{60} cluster films of efficient photocurrent generation. J Phys Chem B 107:12105–12112

He Y, Kong W, Wang W, Liu T, Liu Y, Gong Q, Gao J (2012) Modified natural halloysite/potato starch composite films. Carbohydr Polym 87:2706–2711

He A, Li S, Ma J, Yang Z (2014) Environmental friendly polymer materials for sustainable Development. Int J Polym Sci 2014. Article ID 107028. http://dx.doi.org/10.1155/2014/107028 (in press)

Hejri Z, Ahmadpour A, Seifkordi AA, Zebarjad SM (2012) Role of nano-sized TiO_2 on mechanical and thermal behavior of starch/poly (vinyl alcohol) blend films. Int J Nanosci Nanotechnol 8:215–226

Hietala M, Mathew AP, Oksman K (2013) Bionanocomposites of thermoplastic starch and cellulose nanofibers manufactured using twin-screw extrusion. Eur Polym J 49:950–956

Hoover R, Hughes T, Chung HJ, Liu Q (2010) Composition, molecular structure, properties and modification of pulse starches: a review. Food Res Int 43:399–413

Hotza D (1997) Colagem de Folhas Cerâmicas. Tape Casting Cerâmica 159–166

Huneault MA, Li H (2012) Preparation and properties of extruded thermoplastic starch/polymer blends. J Appl Polym Sci 126:96–108

Huang H, Yuan Q, Yang X (2004) Preparation and characterization of metal–chitosan nanocomposites. Colloid Surf B 39:31–37

Huang Y-F, Lin Y-W, Chang H-T (2006) Growth of various Au-Ag nanocomposites from gold seeds in amino acid solutions. Nanotechnology 17:4885–4894

Iman M, Maji TK (2012) Effect of crosslinker and nanoclay on starch and jute fabric based green nanocomposites. Carbohydr 89:290–297

International Zinc Association-Zinc Oxide Information Center (2011). http://www.znoxide.org/index.html

Jalal R, Goharshadi EK, Abareshi M, Moosavi M, Yousefi A, Nancarrow P (2010) ZnO nanofluids: green synthesis, characterization, and antibacterial activity. Mater Chem Phys 121:198–201

Jane JL (2007) Structure of starch granules. J Appl Glycosci 54:31–36

Janssen LPBM, Moscicki L, Mitrus M (2002) Energy aspects in food extrusion-cooking. Int Agrophys 16:191–195

Jayakody L, Hoover R (2002) The effect of lintnerization on cereal starch granules. Food Res Int 35:665–680

Jiménez A, Fabra MJ, Talens P, Chiralt A (2010) Effect of lipid self-association on the microstructure and physical properties of hydroxypropyl-methylcellulose edible films containing fatty acids. Carbohydr Polym 82:585–593

Jiménez A, Fabra MJ, Talens P, Chiralt A (2013) Phase transitions in starch based films containing fatty acids. Effect on water sorption and mechanical behavior. Food Hydrocolloid 30:408–418

Jin Z, Hsieh F, Huff HE (1994) Extrusion of corn meal with soy fiber, salt, and sugar. Cereal Chem 7:227–234

Jindal UC (1986) Development and testing of bamboo–fibres reinforced plastic composites. J Compos Mater 20:19–29

Jones N, Ray B, Ranjit KT, Manna AC (2007) Antibacterial activity of ZnO nanoparticle suspensions on a broad spectrum of microorganisms. FEMS Microbiol Lett 279:71–76

Kamat PV, Huehn R, Nicolaescu R (2002) A sense and shoot approach for photocatalytic degradation of organic contaminants in water. J Phys Chem B 106:788–794

Kanehira K, Banzai T, Ogino C, Shimizu N, Kubota Y, Sonezaki S (2008) Properties of TiO_2–polyacrylic acid dispersions with potential for molecular recognition. Colloid Surf B 64:10–15

Kasemets K, Ivask A, Dubourguier H-C, Kahru A (2009) Toxicity of nanoparticles of ZnO, CuO and TiO_2 to yeast *Saccharomyces cerevisiae*. Toxicol In Vitro 23:1116–1122

Kaushik A, Singh M, Verma G (2010) Green nanocomposites based on thermoplastic starch and steam exploded cellulose nanofibrils from wheat straw. Carbohydr Polym 82:337–345

Khachatryan K, Khachatryan G, Fiedorowicz M, Para A, Tomasik P (2013) Formation of nanometal particles in the dialdehyde starch matrix. Colloid Surf B 102:578–584

Kilbride BE, Coleman JN, Fournet P, Cadek M, Drury A, Blau WJ (2002) Experimental observation of scaling laws for alternating current and direct current conductivity in polymer-carbon nanotube composite thin films. J Appl Phys 92:4024–4030

Kim E, Jiang ZT, No K (2000) Measurement and calculation of optical band gap of chromium alumunium oxide films. Jpn J Appl Phys 39:4820–4825

Kim H-S, Huber KC (2008) Channels within soft wheat starch A- and B-type granules. J Cereal Sci 48:159–172

Kim J-Y, Lim S-T (2009) Preparation of nano-sized starch particles by complex formation with n-butanol. Carbohydr Polym 76:110–116

Kim J-Y, Park D-J, Lim S-T (2008) Fragmentation of waxy rice starch granules by enzymatic hydrolysis. Cereal Chem 85:182–187

Kim SJ, Shin BS, Hong JL, Cho WJ, Ha CS (2001) Reactive compatibilization of the PBT/EVA blend by maleic anhydride. Polymer 42:4073–4080

Kitagawa K, Ishiaku US, Mizoguchi M, Hamada H (2005) Bamboo-based ecocomposites and their potential applications. In: Amar K, Misra MM, Drzal LT (eds) Natural fibers, biopolymers, and biocomposites. CRC Press, Boca Raton, Chapter 11

Klingshirn C (2007a) ZnO: from basics towards applications. Phys Status Solidi B 244:3027–3073

Klingshirn C (2007b) ZnO: material, physics and applications. Chem Phys Chem 8:782–803

Koch K, Gillgren T, Stading M, Andersson R (2010) Mechanical and structural properties of solution-cast high-amylose maize starch films. Int J Biol Macromol 46:13–19

Kondo M, Shinozaki K, Ooki R, Mizutani N (1994) Crystallization behavior and microstructure of hydrothermally treated monodispersed titanium dioxide particles. J Ceram Soc Jpn 102:742–746

Kristo E, Biliaderis CG (2007) Physical properties of starch nanocrystal-reinforced pullulan films. Carbohydr Polym 68:146–158

Kvien I, Sugiyama J, Votrubec M, Oksman K (2007) Characterization of starch based nanocomposites. J Mater Sci 42:8163–8171

Labet M, Thielemans W, Dufresne A (2007) Polymer grafting onto starch nanocrystals. Biomacromolecules 8:2916–2927

Lagaly G (1986) Interaction of alkylamines with different types of layered compounds. Solid State Ionics 22:43–51

Lamanna M, Morales NJ, García NL, Goyanes S (2013) Development and characterization of starch nanoparticles by gamma radiation: potential application as starch matrix filler. Carbohydr Polym 97:90–97

Larotonda FDS (2007) Biodegradable films and coatings obtained from carrageenan from Mastocarpus stellatus and starch from Quercus suber, PhD Thesis, Universidade do Porto, Portugal, pp 136–140

Lau HC, Hale AH, Bernardi Jr LA (1997) Drilling fluid. Patent H001685, US

Le Corre D, Bras J, Dufresne A (2010) Starch nanoparticles: a review. Biomacromolecules 11:1139–1153

Le Corre D, Bras J, Dufresne A (2011) Ceramic membrane filtration for isolating starch nanocrystals. Carbohydr Polym 86:1565–1570

Lee SY, Chun SJ, Kang IA, Park JY (2009) Preparation of cellulose nanofibrils by high-pressure homogenizer and cellulose-based composite films. J Ind Eng Chem 15:50–55

Lelievre J (1974) Starch gelatinization. J Appl Polym Sci 18:293–296

Lewinski N, Colvin V, Drezek R (2008) Cytotoxicity of nanoparticles. Small 4:26–49

Li XH, Xing YG, Li WL, Jiang YH, Ding YL (2010) Antibacterial and physical properties of poly (vinyl chloride)-based film coated with ZnO nanoparticles. Food Sci Technol Int 16:225–232

Li M, Liu P, Zou W, Yu L, Xie F, Pu H, Liu H, Chen L (2011a) Extrusion processing and characterization of edible starch films with different amylose contents. J Food Eng 106:95–101

Li R, Liu C, Ma J (2011b) Crystallinity in starch plastics: consequences for material properties. Carbohydr Polym 84:631–637

Li L, Sun J, Li X, Zhang Y, Wang Z, Wang C, Dai J (2012) Controllable synthesis of monodispersed silver nanoparticles as standards for quantitative assessment of their cytotoxicity. Biomaterials 33:1714–1721

Li XH, Gao X, Wang Y, Zhang X, Tong Z (2013) Comparison of chitosan/starch composite film properties before and after cross-linking. Int J Biol Macromol 52:275–279

Liao HT, Wu CS (2008) New biodegradable blends prepared from polylactide, titanium tetraisopropylate, and starch. J Appl Polym Sci 108:2280–2289

Liu Q (2005) Understanding starches and their role in foods. In: Cui SW (ed) Food carbohydrates: chemistry, physical properties and applications. CRC Press, Boca Raton, Chapter 7

Liu X, Yu L, Liu H, Chen L, Li L (2008) In situ thermal decomposition of starch with constant moisture in a sealed system. Polym Degrad Stabil 93:260–262

Liu D, Zhong T, Chang PR, Li K, Wu Q (2010) Starch composites reinforced by bamboo cellulosic crystals. Bioresour Technol 101:2529–2536

Liu D, Chang PR, Deng S, Wang C, Zhang B, Tian Y, Huang S, Yao J, Ma X (2011) Fabrication and characterization of zirconium hydroxide-carboxymethyl cellulose sodium/plasticized Trichosanthes Kirilowii starch nanocomposites. Carbohydr Polym 86:1699–1704

Liu Y, Kim H-I (2012) Characterization and antibacterial properties of genipin-crosslinked chitosan/poly(ethylene glycol)/ZnO/Ag nanocomposites. Carbohydr Polym 89:111–116

Lin M-F, Thakur VK, Tan EJ, Lee PS (2011a) Dopant induced hollow $BaTiO_3$ nanostructures for application in high performance capacitors. J Mater Chem 21:16500–16504

Lin M-F, Thakur VK, Tan EJ, Lee PS (2011b) Surface functionalization of $BaTiO_3$ nanoparticles and improved electrical properties of $BaTiO_3$/polyvinylidene fluoride composite. RSC Adv 1:576–578

Lloyd SM, Lave LB (2003) Life cycle economic and environmentals. Environ Sci Technol 37:3458–3466

López O, García MA (2012) Starch films from a novel (Pachyrhizus ahipa) and conventional sources: development and characterization. Mater Sci Eng C 32:1931–1940

López O, Zaritzky N, Grossmann M, García MA (2013) Acetylated and native corn starch blend films produced by blown extrusion. J Food Eng 116:286–329

López O, Garcia MA, Villar MA, Gentili A, Rodriguez MS, Albertengo L (2014) Thermo-compression of biodegradable thermoplastic corn starch films containing chitin and chitosan. LWT 57:106–115

Lu CM, Zhang CYW, Tao MX (2002) Research of the effect of nanometer on germination and growth enhancement of *Gly-cine max* L. and its mechanism. Soybean Sci 21:168–172

Lu Y, Tighzert L, Berzin F, Rondot S (2005) Innovative plasticized starch films modified with waterborne polyurethane from renewable sources. Carbohydr Polym 61:174–182

Lu Y, Weng L, Cao X (2006) Morphological, thermal and mechanical properties of ramie crystallites-reinforced plasticized starch biocomposites. Carbohydr Polym 63:198–204

Lu C, Mai Y-W (2007) Permeability modelling of polymer-layered silicate nanocomposites. Compos Sci Technol 67:2895–2902

Lu H, Gui Y, Zheng L, Liu X (2013) Morphological, crystalline, thermal and physicochemical properties of cellulose nanocrystals obtained from sweet potato residue. Food Res Int 50:112–121

Ma X, Yu JG, Wang N (2008a) Glycerol plasticized-starch/multiwall carbon nanotube composites for electroactive polymers. Compos Sci Technol 68:268–273

Ma X, Chang PR, Yu JG (2008b) Characterizations of glycerol plasticized starch (GPS)/carbon black (CB) membranes prepared by melt extrusion and microwave radiation. Carbohydr Polym 74:895–900

Ma X, Jian R, Chang PR, Yu J (2008c) Fabrication and characterization of citric acid-modified starch nanoparticles/plasticized-starch composites. Biomacromolecules 9:3314–3320

Ma X, Chang PR, Yang J, Yu J (2009) Preparation and properties of glycerol plasticized-pea starch/zinc oxide-starch bionanocomposites. Carbohydr Polym 75:472–478

Macanás J, Farre M, Muñoz M, Alegret S, Muraviev DN (2006) Preparation and characterization of polymer-stabilized metal nanoparticles for sensor applications. Phys Status Solidi A 203:1194–1200

Magalhaes NF, Andrade CT (2009) Thermoplastic corn starch/clay hybrids: effect of clay type and content on physical properties. Carbohydr Polym 75:712–718
Majdzadeh-Ardakani K, Navarchian A, Sadheghi F (2010) Optimization of mechanical properties of thermoplastic starch/clay nanocomposites. Carbohydr Polym 79:547–554
Mali S, Grossmann M, Garcia M, Martino M, Zaritzky N (2005) Mechanical and thermal properties of yam starch films. Food Hydro 19:157–164
Martinez-Bustos F, Viveros-Contreras R, Galicia-Garcia T, Nabeshima EH, Verdalet-Guzman I (2011) Some functional characteristics of extruded blends of fiber from sugarcane bagasse, whey protein concentrate, and corn starch. Ciênc Tecnol Aliment 31:870–878
Martinez-Gutierrez F, Thi EP, Silverman J, de Oliveira CC, Svensson SL, Hoek AV, Sanchez EM, Reiner NE, Gaynor EC, Pryzdial ELG, Conway EM, Orrantia E, Ruiz F, Av-Gay Y, Bach H (2012) Antibacterial activity, inflammatory response, coagulation and cytotoxicity effects of silver nanoparticles. Nanomedicine 8:328–336
Martins JT, Cerqueira MA, Vicente AA (2012) Influence of α-tocopherol on physicochemical properties of chitosan-based films. Food Hydrocolloid 27:220–227
Mathew AP, Dufresne A (2002) Morphological investigation of nanocomposites from sorbitol plasticized starch and tunicin whiskers. Biomacromolecules 3:609–617
Matsuda DKM, Verceheze AES, Carvalho GM, Yamashita F, Mali S (2013) Baked foams of cassava starch and organically modified nanoclays. Ind Crops Prod 44:705–711
Mbey JA, Hoppe S, Thomas F (2012) Cassava starch–kaolinite composite film. Effect of clay content and clay modification on film properties. Carbohydr Polym 88:213–222
Meskinfam M, Sadjadi MAS, Jazdarreh H, Zare K (2011) Biocompatibility evaluation of nano hydroxyapatite-starch biocomposites. J Biomed Nanotechnol 7:455–459
Miao Z, Ding K, Wu T, Liu Z, Han B, An G, Miao S, Yang G (2008) Fabrication of 3D-networks of native starch and their application to produce porous inorganic oxide networks through a supercritical route. Microporous Mesoporous Mat 111:104–109
Mitrus M, Moscicki L (2014) Extrusion-cooking of starch protective loose-fillfoams. Chem Eng Res Des 9:778–783
Moad G (2011) Chemical modification of starch by reactive extrusion. Progr Polym Sci 36:218–237
Moezzi A, McDonagh AM, Cortie MB (2012) Zinc oxide particles: synthesis, properties and applications. Chem Eng J 185–186:1–22
Montoya IA, Viveros T, Dominguez JM, Canales LA, Shifter I (1992) On the effects of the sol-gel synthesis parameters on textural characteristics of TiO_2. Catal Lett 15:207–217
Moongngarm A (2013) Chemical compositions and resistant starch content in starchy foods. Am J Agr Biol Sci 8:107–113
Mościcki L, van Zuilichem DJ (2011) Extrusion-cooking and related technique. In: Moscicki L (ed) Extrusion-cooking techniques: applications, theory and sustainability. Wiley-VCH Verlag GmbH & Co. KGaA, Weinheim
Mościcki L, Mitrus M, Wojtowicz A, Oniszczuk T, Rejak A, Janssen L (2012) Application of extrusion-cooking for processing of thermoplastic starch (TPS). Food Res Int 47:291–329
Mościcki L, Mitrus M, Wojtowicz A, Oniszczuk T, Rejak A (2013) Extrusion-cooking of starch, advances in agrophysical research. In: Grundas S (ed) Tech doi: 10.5772/52323. http://www.intechopen.com/books/advances-in-agrophysical-research/extrusion-cooking-of-starch
Müller C, Laurindo J, Yamashita F (2011) Effect of nanoclay incorporation method on mechanical and water vapor barrier properties of starch-based films. Ind Crop Prod 33:605–610
Müller C, Yamashita F, Laurindo J (2008) Evaluation of the effects of glycerol and sorbitol concentration and water activity on the water barrier properties of cassava starch films through a solubility approach. Carbohydr Polym 72:82–87
Müller C, Laurindo J, Yamashita F (2009) Effect of cellulose fibers addition on the mechanical properties and water vapor barrier of starch-based films. Food Hydrocolloid 23:1328–1333
Müller C, Laurindo J, Yamashita F (2012) Composites of thermoplastic starch and nanoclays produced by extrusion and thermopressing. Carbohydr Polym 89:504–510

Muraviev DN, Macanás J, Esplandiu MJ, Farre M, Muñoz M, Alegret S (2007) Simple route for intermatrix synthesis of polymer stabilized core-shell metal nanoparticles for sensor applications. Phys Status Solidi A 204:1686–1692
Murillo-Martínez MM, Pedroza-Islas R, Lobato-Calleros C, Martinez-Ferez A, Vernon-Carter EJ (2011) Designing W-1/O/W-2 double emulsions stabilized by protein-polysaccharide complexes for producing edible films: rheological, mechanical and water vapour properties. Food Hydrocolloid 25:577–585
Murugadoss A, Chattopadhyay A (2008) A 'green' chitosan–silver nanoparti-cles composite as a heterogeneous as well as micro-heterogeneous catalyst. Nanotechnol 19:015603/1-015603/9
Myllärinen P, Partanen R, Sepalla J, Forsell P (2002) Effect of glicerol on behavior of amylose and amylopectin films. Carbohydr Polym 50:355–361
Nafchi AM, Alias AK, Mahmud S, Robal M (2012) Antimicrobial, rheological, and physicochemical properties of sago starch films filled with nanorod-rich zinc oxide. J Food Eng 113:511–519
Nafchi AM, Nassiri R, Sheibani S, Ariffin F, Karim AA (2013) Preparation and characterization of bionanocomposite films filled with nanorod-rich zinc oxide. Carbohydr Polym 96:233–239
Narayanan R, El-Sayed MA (2005) Catalysis with transition metal nanopariclesin colloidal solution: nanoparticle shape dependence and stability. J Phys Chem B 109:12663–12676
Navarro E, Baun A, Behra R, Hartmann NB, Filser J, Miao A, Quigg A, Santschi PH, Sigg L (2008) Environmental behavior and ecotoxicity of engineered nanoparticles to algae, plants, and fungi. Ecotoxicology 17:372–386
Norajit K, Kim MK, Ryu GH (2010) Comparative studies on the characterization and antioxidant properties of biodegradable alginate films containing ginseng extract. J Food Eng 98:377–384
Norman TJ, Grant CD, Magana D, Zhang JZ, Liu J, Cao D, Bridges F, Buuren AV (2002) Near infrared optical absorption of gold nanoparticle aggregates. J Phys Chem B 106:7005–7012
Nuryetti HH, Nasikin M (2012) Structure, energy band gap and electrical conductivity of Tapioca/metal oxide composite. J Eng Chem 6:911–919
Oksman K, Mathew AP, Bondeson D, Kvien I (2006) Manufacturing process of cellulose whiskers/polylactic acid nanocomposites. Comp Sci Tech 66:2776–2784
Oliveira de Moraes J, Scheibe AS, Sereno A, Laurindo JB (2013) Scale-up of the production of cassava starch based films using tape-casting. J Food Eng 119:800–808
Olsson E, Hedenqvist M, Johansson C, Järnström L (2013) Influence of citric acid and curing on moisture sorption, diffusion and permeability of starch films. Carbohydr Polym 94:765–772
Orliac O, Rouilly A, Silvestre F, Rigal L (2003) Effects of various plasticizers on the mechanical properties, water resistance and aging of thermo-moulded films made from sunflower proteins. Ind Crop Prod 18:91–100
Ozgur U, Ya IA, Liu C, Teke A, Reshchikov MA, Dogan S, Avrutin V, Cho SJ, Morkoc H (2005) A comprehensive review of ZnO materials and devices. J Appl Phys 98:041301
Pääkkö M, Ankerfors M, Kosonen H, Nykänen A, Ahola S, Osterberg M, Ruokolainen J, Laine J, Larsson PT, Ikkala O, Lindström T (2007) Enzymatic hydrolysis combined with mechanical shearing and high-pressure homogenization for nanoscale cellulose fibrils and strong gels. Biomacromolecules 8:1934–1941
Pagnoulle C, Jerome R (2001) Particle-in-particle morphology for the dispersed phase formed in reactive compatibilization of SAN/EPDM blends. Polymer 42:1893–1906
Pantani R, Gorrasi G, Vigliotta G, Murariu M, Dubois P (2013) PLA-ZnO nanocomposite films: water vapor barrier properties and specific end-use characteristics. Eur Polym J 49:3471–3482
Pastor C, Sánchez-González L, Chiralt A, Cháfer M, González-Martínez C (2013) Physical and antioxidant properties of chitosan and methylcellulose based films containing resveratrol. Food Hydrocolloid 30:272–280
Pelissari FM, Yamashita F, Garcia MA, Martino MN, Zaritzky NE, Grossmann MVE (2012) Constrained mixture design applied to the development of cassava starch–chitosan blown films. J Food Eng 108:262–267
Pereda M, Amica G, Rácz I, Marcovich NE (2011) Structure and properties of nanocomposite films based on sodium caseinate and nanocellulose fibers. J Food Eng 103:76–83

Pérez E, Segovia X, Tapia MA, Schroeder M (2012) Native and cross-linked modified Dioscorea trifida (cush-cush yam) starches as bio-matrices for edible films. J Cell Plast 48:545–556

Petersson M, Stading M (2005) Water vapor permeability and mechanical properties of mixed starch-monoglyceride films and effect of film forming conditions. Food Hydrocolloid 19:123–132

Piyada K, Waranyou S, Thawien W (2013) Mechanical, thermal and structural properties of rice starch films reinforced with rice starch nanocrystals. Int Food Res J 20:439–449

Plattner BS, Wenger L, Rokey GJ (2011) Extruded, highly cooked, non-sticky starch products. Patent US 20110086150 A1. Application number US 12/829,948

Polesi LF, Sarmiento SBS, dos Anjos CBP (2011) Composition and characterization of pea and chickpea starches. Braz J Food Technol 14:74–81

Poole CP, Owens FJ (2003) Introduction to nanotechnology. Wiley, Chichester

Prajapati VD, Jani GK, Moradiya NG, Randeria NP (2013) Pharmaceutical applications of various natural gums, mucilages and their modified forms. Carbohydr Polym 92:1685–1699

Ptaszek A, Lukasiewicz M, Bednarz S (2013) Environmental friendly polysaccharide modification —rheological properties of oxidized starches water systems. Starch Stärke 65:134–145

Qiao R, Brinson LC (2009) Simulation of interphase percolation and gradients in polymer nanocomposites. Compos Sci Technol 69:491–499

Rahman MAA, Mahmud S, Alias AK, Nor AFM (2013) Effect of nanorod zinc oxide on electrical and optical properties of starch-based polymer nanocomposites. J Phys Sci 24:17–28

Rai M, Yadav A, Gade A (2009) Silver nanoparticles as a new generation of antimicrobials. Biotechnol Adv 27:76–83

Rajendra R, Balakumar C, Ahammed H, Jayakumar S, Vaideki K, Rajesh E (2010) Use of zinc oxide nano particles for production of antimicrobial textiles. Int J Eng Sci Technol 2:202–208

Ratnayake WS, Jackson DS, Steve LT (2008) Starch gelatinization. Adv Food Nutr Res 55:221–268

Ray S, Okamoto M (2003) Polymer/layered silicate nanocomposites: a review from preparation to processing. Prog Polym Sci 28:1539–1641

Raybaudi-Massilia RM, Mosqueda-Melgar J, Soliva-Fortuny R, Martin-Belloso O (2009) Control of pathogenic and spoilage microorganisms in fresh-cut fruits and fruit juices by traditional and alternative natural antimicrobial. Comp Rev Food Sci F 8:157–180

Reddy MM, Vivekanandhan S, Misra M, Bhatia SK, Mohanty AK (2013) Biobased plastics and bionanocomposites: current status and future opportunities. Prog Polym Sci 38:1653–1689

Rhim JW, Wang LF, Hong SL (2013) Preparation and characterization of agar/silver nanoparticles composite films with antimicrobial activity. Food Hydrocolloid 33:327–335

Richard ME, Twiname ER (2000) Tape casting. In: American Ceramics Society, USA, Theory and Practice, p 293

Rindlav-Westling A, Stading M, Gatenholm P (1998) Structure, barrier and mechanical properties of amylose and amylopectin films. Carbohydr Polym 36:217–224

Roessler S, Zimmermann R, Scharnweber D, Werner C, Worch H (2002) Characterization of oxide layers on Ti6Al4V and titanium by streaming potential and streaming current measurements. Colloid Surf B 26:387–395

Rojas-Graü MA, Avena-Bustillos RJ, Olsen C, Friedman M, Henika OR, Martín-Belloso O, Pan Z, McHugh TH (2007) Effects of plant essential oils and oils compounds on mechanical, barrier and antimicrobial properties of alginate–apple puree edible films. J Food Eng 81:634–641

Romero-Bastida CA, Bello-Pérez LA, García MA, Martino MN, Solorza-Feria J, Zarintzky NE (2005) Physicochemical and microstructural characterization of films prepared by thermal and cold gelatinization from non-conventional sources of starches. Carbohydr Polym 60:235–244

Roselli M, Finamore A, Garaguso I, Britti MS, Mengheri E (2003) Zinc oxide protects cultured enterocytes from the damage induced by Escherichia coli. J Nutr 133:4077–4082

Rosin PM, Lajolo FM, Menezes EW (2002) Measurement and characterization of dietary starches. J Food Compos Anal 15:367–377

Rubilar JF, Cruz RMS, Silva HD, Vicente AA, Khmelinskii I, Vieira MC (2013) Physico-mechanical properties of chitosan films with carvacrol and grape seed extract. J Food Eng 115:466–474

Russell PL (1987) Gelatinisation of starches of different amylose/amylopectin content. A study by differential scanning calorimetry. J Cereal Sci 6:133–145

Sadegh-Hassani F, Nafchi AM (2014) Preparation and characterization of bionanocomposite films based on potato starch/halloysite nanoclay. Int J Biol Macromol 67:446–458

Salman H, Blazek J, Lopez-Rubio A, Gilbert EP, Hanley T, Copeland L (2009) Structure-function relationships in A and B granules from wheat starches of similar amylose content. Carbohydr Polym 75:420–427

Sandhu KS, Singh N (2007) Some properties of corn starches II: physicochemical, gelatinization, retrogradation, pasting and gel textural properties. Food Chem 101:1499–1507

Sandler JKW, Kirk JE, Kinloch IA, Shaffer MSP, Windle AH (2003) Ultra-low electrical percolation threshold in carbon-nanotube-epoxy composites. Polymer 44:5893–5899

Sandstedt RM (1961) The function of starch in the baking of bread. Baker Dig 35:36–44

Sanpui P, Murugadoss A, Prasad PVD, Ghosh SS, Chattopadhyay A (2008) The antibacterial properties of a novel chitosan–Ag-nanoparticle composite. Int J Food Microbiol 124:142–146

Santiago-Silva P, Soares NFF, Nóbrega JE, Júnior MAW, Barbosa KBF, Volp ACP, Zerdas ERMA, Würlitzer NJ (2009) Antimicrobial efficiency of film incorporated with pediocin (ALTA®2351) on preservation of sliced ham. Food Control 20:85–89

Sato E, Kohno M, Hamano H, Niwano Y (2006) Increased antioxidative potency of garlic by spontaneous short-term fermentation. Plant Foods Hum Nutr 61:157–160

Sayanjali S, Ghanbarzadeh B, Ghiassifar S (2011) Evaluation of antimicrobial and physical properties of edible film based on carboxymethyl cellulose containing potassium sorbate on some mycotoxigenic Aspergillus species in fresh pistachios. LWT 44:1133–1138

Sayyadnejad MA, Ghaffarian HR, Saeidi M (2008) Removal of hydrogen sulfide by zinc oxide nanoparticles in drilling fluid. Int J Environ Sci Technol 5:565–569

Savadekar NR, Mhaske ST (2012) Synthesis of nano cellulose fibers and effect on thermoplastics starch based films. Carbohydr Polym 89:146–151

Sawai J (2003) Quantitative evaluation of antibacterial activities of metallic oxide powders (ZnO, MgO and CaO) by conductimetric assay. J Microbiol Method 54:177–182

Schiavello M (1997) Heterogeneous photocatalysis. Wiley, New York

Schirmer BC, Heiberg R, Eie T, Moretro T, Maugesten T, Carlehog M, Langsrud S (2009) A novel packaging method with a dissolving CO_2 headspace combined with organic acids prolongs the shelf life of fresh salmon. Int J Food Microbiol 133:154–160

Schmid G, Simon U (2005) Gold nanoparticles: assembly and electrical properties in 1–3 dimensions. Chem Commun 6:697–710

Schmid AH, Dolan KD, Ng PKW (2005) Effect of extruding wheat flour at lower temperatures on physical attributes of extrudates and on thiamin loss when using carbon dioxide gas as a puffing agent. Cereal Chem 82:305–313

Seo SY, Lee GH, Lee SG, Jung SY, Lim JO, Choi JH (2012) Alginate-based composite sponge containing silver nanoparticles synthesized in situ. Carbohydr Polym 90:109–115

Sepone N, Pelizzetti E (1989) Photocatalysis: fundamentals and applications. Wiley, New York

Shanks R, Kong I (2012) Thermoplastic starch. In: El-Sonbati A (ed) Thermoplastic elastomers. http://www.intechopen.com/books/thermoplasticelastomers/thermoplastic-starch

Sharma VK, Yngard RA, Lin Y (2009) Silver nanoparticles: green synthesis and their antimicrobial activities. Adv Colloid Interfac 145:83–96

Shelma R, Paul W, Sharma CP (2008) Chitin nanofibre reinforced thin chitosan films for wound healing application. Trends Biomater Artif Organs 22:107–115

Shi L, Zhou J, Gunasekaran S (2008) Low temperature fabrication of ZnO-whey protein isolate nanocomposite. Mater Lett 62:4383–4385

Siqueira G, Bras J, Dufresne A (2009) Cellulosic bionanocomposites: a review of preparation and properties of nanocomposites. Biomacromolecules 10:425–432

Sing P, Rhee HW, Tomar SK, Nagarale RK (2010) Ternary semiconductor nanoparticles embedded in PEO-polymer electrolyte matrix. J Thermoplast Compos 23:227–237

Singh M, Singh S, Prasad S, Gambhir IS (2008) Nanotechnology in medicine and antibacterial effect of silver nanoparticles. Digest J Nanomater Biostruct 3:115–122

Singha AS, Thakur VK (2008a) Saccaharum cilliare fiber reinforced polymer composites. Eur J Chem 5:782–791

Singha AS, Thakur VK (2008b) Effect of fibre loading on urea-formaldehyde matrix based green composites. Iran Polym J 17:861–873

Singha AS, Thakur VK (2009a) Fabrication and characterization of H. sabdariffa fiber-reinforced green polymer composites. Polym Plast Technol Eng 48:482–487

Singha AS, Thakur VK (2009b) Physical, chemical and mechanical properties of Hibiscus sabdariffa fiber/polymer composite. Int J Polym Mater 58:217–228

Singha AS, Thakur VK (2009c) Grewia optiva fiber reinforced novel, low cost polymer composites. J Chem 6:71–76

Singha AS, Thakur VK (2009d) Fabrication and characterization of S. cilliare fibre reinforced polymer composites. Bull Mater Sci 32:49–58

Singha AS, Thakur VK (2009e) Synthesis, characterisation and analysis of Hibiscus sabdariffa fibre reinforced polymer matrix based composites. Polym Polym Compos 17:189–194

Singha AS, Thakur VK, Mehtac IK, Shama A, Khanna AJ, Rana RK, Rana AK (2009) Surface-modified Hibiscus sabdariffa fibers: physicochemical, thermal, and morphological properties evaluation. Int J Polym Anal Charact 14:695–711

Singha AS, Thakur VK (2010a) Mechanical, morphological, and thermal characterization of compression-molded polymer. Biocomposites 15:87–97

Singha AS, Thakur VK (2010b) Synthesis, characterization and study of pine needles reinforced polymer matrix based composites. J Reinf Plast Compos 29:700–709

Slavutsky AM, Bertuzzi MA (2014) Water barrier properties of starch films reinforced with cellulose nanocrystals obtained from sugarcane bagasse. Carbohydr Polym 110:53–61

Son Y, Ahn KH, Char K (2000) Effect of processing conditions and reactive compatibilizer on the morphology of injection molded modified poly(phenylene oxide)/polyamide-6 blends. Polym Eng Sci 40:1385–1394

Sondi I, Salopek-Sondi B (2004) Silver nanoparticles as antimicrobial agent: a case study of *E. coli* as a model for gram-negative bacteria. J Colloid Interf Sci 275:177–182

Song D, Thio YS, Deng Y (2011) Starch nanoparticle formation via reactive extrusion and related mechanism study. Carbohydr Polym 85:208–214

Sorrentino A, Gorrasi G, Vittoria V (2007) Potential perspectives of bionanocomposites for food packaging applications. Trends Food Sci Technol 18:84–95

Souza AC, Goto GEO, Mainardi JA, Coelho ACV, Tadini CC (2013) Cassava starch composite films incorporated with cinnamon essential oil: antimicrobial activity, microstructure, mechanical and barrier properties. LWT 54:346–352

Sreekumar PA, Al-Harthi MA, De SK (2012) Reinforcement of starch/polyvinyl alcohol blend using nano-titanium dioxide. J Compos Mater 46:3181–3187

Stodolak E, Paluszkiewicz C, Bogun M, Blazewicz M (2009) Nanocomposite fibers for medical applications. J Mol Struct 924–926:208–213

Sullivan JW, Johnson JA (1964) Measurement of starch gelatinization by enzyme susceptibility. Cereal Chem 41:73–77

Suvakanta D, Narsimha MP, Pulak D, Joshabir C, Biswajit D (2014) Optimization and characterization of purified polysaccharide from Musa sapientum L as a pharmaceutical excipient. Food Chem 149:76–83

Svegmark K, Helmersson K, Nilsson G, Nilsson P-O, Andersson R, Svensson E (2002) Comparison of potato amylopectin starches and potato starches: influence of year and variety. Carbohydr Polym 47:331–340

Tajuddin S, Xie F, Nicholson TM, Liu P, Halley PJ (2011) Rheological properties of thermoplastic starch studied by multipass rheometer. Carbohydr Polym 83:914–919

Takagi H, Takura R (2003) The manufacture and mechanical properties of composite boards made from starch-based biodegradable plastic and bamboo powder. J Soc Mater Sci 4:357–361

Takagi H, Ichihara Y (2004) Effect of fiber length on mechanical properties of "green" composites using a starch-based resin and short bamboo fibers. JSME Int J Ser A 47:551–555

Takeda C, Takeda Y, Hizukuri S (1989) Structure of amylomaize. Cereal Chem 66:22–25

Talja RA, Helén H, Roos YH, Jouppila K (2007) Effect of various polyols and polyol contents on physical and mechanical properties of potato starch-based films. Carbohydr Polym 67:288–297

Tam KH, Djurisic AB, Chan CMN, Xi YY, Tse CW, Leung YH, Chan WK, Leung FCC (2008) Antibacterial activity of ZnO nanorods prepared by a hydrothermal method. Thin Solid Films 516:6167–6174

Tankhiwale R, Bajpai SK (2012) Preparation, characterization and antibacterial applications of ZnO-nanoparticles coated polyethylene films for food packaging. Colloid Surf B 90:16–20

Teixeira EdM, Pasquini D, Curvelo AAS, Corradini E, Belgacem MN, Dufresne A (2009) Cassava bagasse cellulose nanofibrils reinforced thermoplastic cassava starch. Carbohydr Polym 78:422–431

Thakur VK, Singha AS, Mehta IK (2010a) Renewable resource-based green polymer composites: analysis and characterization. Int J Polym Anal Charact 15:137–146

Thakur VK, Singha AS, Kaur I, Nagarajarao RP, Liping Y (2010b) Silane functionalization of saccaharum cilliare fibers: Thermal, morphological, and physicochemical study. Int J Polym Anal Charact 15:397–414

Thakur VK, Singha AS, Misra N (2011) Graft copolymerization of methyl methacrylate onto cellulosic biofibers. J Appl Polym Sci 122:532–544

Thakur VK, Singha AS, Thakur MK (2012a) In-air graft copolymerization of ethyl acrylate onto natural cellulosic polymers. Int J Polym Anal Charact 17:48–60

Thakur VK, Singha AS, Thakur MK (2012b) Graft Copolymerization of Methyl Acrylate onto Cellulosic Biofibers: Synthesis, Characterization and Applications. J Polym Environ 20:164–174

Thakur VK, Singha AS, Thakur MK (2012c) Biopolymers based green composites: mechanical, thermal and physico-chemical characterization. J Polym Environ 20:412–421

Thakur VK, Singha AS, Thakur MK (2012d) Modification of natural biomass by graft copolymerization. Int J Polym Anal Charact 17:547–555

Thakur VK, Singha AS, Thakur MK (2012e) Green composites from natural fibers: mechanical and chemical aging properties. Int J Polym Anal Charact 17:401–407

Thakur VK, Thakur MK (2014a) Recent trends in hydrogels based on psyllium polysaccharide: a review. J Clean Prod 82:1–15

Thakur VK, Thakur MK (2014b) Processing and characterization of natural cellulose fibers/ thermoset polymer composites. Carbohydr Polym 109:102–117

Thakur VK, Thakur MK (2014c) Processing and characterization of natural cellulose fibers/ thermoset polymer composites. Carbohydr Polym 109:102–117

Thakur VK, Thakur MK, Raghavan P, Kessler MR (2014a) Progress in green polymer composites from lignin for multifunctional applications: A Review. ACS Sustainable Chem Eng 2:1072–1092

Thakur VK, Thakur MK, Gupta RK (2014b) Review: Raw natural fiber–based polymer composites. Int J Polym Anal Charact 19:256–271

Thakur VK, Thunga M, Madbouly SA, Kessler MR (2014c) PMMA-g-SOY as a sustainable novel dielectric material. RSC Adv 4:18240–18249

Thakur VK, Grewell D, Thunga M, Kessler MR (2014d) Novel Composites from Eco-Friendly Soy Flour/SBS Triblock Copolymer. Macromol Mater Eng 299:953–958

Thakur VK, Vennerberg D, Madbouly SA, Kessler MR (2014e) Bio-inspired green surface functionalization of PMMA for multifunctional capacitors. RSC Adv 4:6677–6684

Tharanathan RN (2003) Biodegradable films and composite coatings: Past, present and future. Trends Food Sci Technol 14:71–78

Thunwall M, Kuthanová V, Boldizar A, Rigdahl M (2008) Film blowing of thermoplastic starch. Carbohydr Polym 71:583–590

Tjong SC (2006) Structural and mechanical properties of polymer nanocomposites. Mater Sci Eng R Rep 53:73–97

Torres-Castro A, González González VA, Navarro MG, González EG (2011) Síntesis de nanocompósitos de plata con almidón. Ingenierías XIV No 50

Tripathi P, Dubey NK (2004) Exploitation of natural products as an alternative strategy to control postharvest fungal rotting of fruit and vegetables. Postharvest Biol Technol 32:235–245

Tunç S, Duman O (2010) Preparation and characterization of biodegradable methyl cellulose/montmorillonite nanocomposite films. Appl Clay Sci 48:414–424

Turbak AF, Snyder FW, Sandberg KR (1983) Microfibrillated cellulose, a new cellulose product: properties, uses, and commercial potential. J Appl Polym Sci Appl Polym Symp 37:815–827

Ung T, Liz-Marzan LM, Mulvaney P (2002) Gold nanoparticle. Thin Films Colloid Surf A 202:119–126

Van der Maarel MJEC, Van der Veen B, Uitdehaag JCM, Leemhuis H, Dijkhuizen L (2002) Properties and applications of starch-converting enzymes of the alpha-amylase family. J Biotechnol 94:137–155

Varaprasad K, Mohan IM, Ravindra S, Reddy NN, Vimala K, Monika K, Sreedhar B, Raju KM (2010) Hydrogel–silver nanoparticle composites: a new generation of antimicrobials. J Appl Polym Sci 115:1199–1207

Vasanthan T, Bergthaller W, Driedger D, Yeung J, Sporns P (1999) Starch from Alberta potatoes: wet-isolation and some physicochemical properties. Food Res Int 32:355–365

Van de Velde K, Kiekens P (2002) Biopolymers: overview of several properties and consequences on their applications. Polym Test 21:4433–4442

Vazquez A, Cyras VP, Alvarez VA, Morán JI (2012) Starch/clay nano-biocomposites. In: Averous L, Pollet E (eds) Environmental silicate nano-biocomposites. Springer, London

Venkatasubramanian R, Siivola E, Colpitts T, O'Quinn B (2001) Thin-film thermoelectric devices with high room-temperature figures of merit. Nature 413:597–602

Vergnes B, Berzin F (2010) Predicting starch transformation in twin screw extrusion. Society of Plastic Engineers. http://www.4spepro.org/view.php?article=002986-2010-06-22

Verran J, Sandoval G, Allen NS, Edge M, Stratton J (2007) Variables affecting the antibacterial properties of nano and pigmentary titania particles in suspension. Dyes Pigments 73:298–304

Vicentini N, Sobral P, Cereda M (2002) The influence of the thickness on the functional properties of cassava starch edible films. Plant Biopolym Sci Food and Non-Food Appl 291–300

Vieira MGA, da Silva MA, dos Santos LO, Beppu MM (2011) Natural-based plasticizers and biopolymer films: a review. Eur Polym J 47:254–263

Viguié J, Molina-Boisseau S, Dufresne A (2007) Processing and characterization of waxy maize starch films plasticized by sorbitol and reinforced with starch nanocrystals. Macromol Biosci 7:1206–1216

Wang H, Niu J, Long X, He Y (2008) Sonophotocatalytic degradation of methyl orange by nano-sized Ag/TiO_2 particles in aqueous solutions. Ultrason Sonochem 15:386–392

Wang H, Sun XZ, Seib P (2001) Trengthening blends of poly(lactic acid) and starch with methylenediphenyl diisocyanate. J Appl Polym Sci 82:1761–1767

Wang ZL, Kong XY, Ding Y, Gao P, Hughes WL, Yang R, Zhang Y (2004) Semiconducting and piezoelectric oxide nanostructures induced by polar surfaces. Adv Funct Mater 14:943–956

Wang N, Maximiuk L, Toews R (2012) Pea starch noodles: effect of processing variables on characteristics and optimisation of twin-screw extrusion process. Food Chem 133:742–753

Wang J, Cheng F, Zhu P (2014) Structure and properties of urea-plasticized starch films with different urea contents. Carbohydr Polym 101:1109–1115

Wei C, Srivastava D, Cho K (2002) Thermal expansion and diffusion coefficients of carbon nanotube-polymer composites. Nano Lett 2:647–650

Weibel A, Bouchet R, Knauth P (2006) Electrical properties and defect chemistry of anatase (TiO_2). Solid State Ionics 177:229–236

Wessels JM, Nothofer H, Ford WE, von Wrochem F, Scholz F, Vossmeyer T, Schroedter A, Weller H, Yasuda A (2004) Optical and electrical properties of three-dimensional interlinked gold nanoparticle assemblies. J Am Chem Soc 126:3349–3356

Wiley B, Herricks T, Sun Y, Xia Y (2004) Polyol synthesis of silver nanoparticles: Use of chloride and oxygen to promote the formation of single-crystal, truncated cubes and tetrahedrons. Nano Lett 4:1733–1739

Wiley B, Sun Y, Xia Y (2007) Synthesis of silver nanostructures with controlled shapes and properties. Acc Chem Res 40:1067–1076

Wilhelm H-M, Sierakowski M-R, Reicher F, Wypych F, Souza GP (2005) Dynamic rheological properties of Yam starch/hectorite composite gels. Polym Int 54:814–822

Wollerdorfer M, Bader H (1998) Influence of natural fibres on the mechanical properties of biodegradable polymers. Ind Crop Prod 8:105–112

Wong M, Paramsothy M, Xu XJ, Ren Y, Li S, Liao K (2003) Physical interactions at carbon nanotube-polymer interfaces. Polymer 44:7757–7764

Wonisch A, Polfer P, Kraft T, Dellert A, Heunisch A, Roosen A (2011) A comprehensive simulation scheme for tape casting: from flow behavior to anisotropy development. J Am Ceram Soc 94:2053–2060

Woranucha S, Yoksana R (2013) Eugenol-loaded chitosan nanoparticles: II. Application in bio-based plastics for active packaging. Carbohydr Polym 96:586–592

Wu M, Wang M, Ge M (2009) Investigation into the performance and mechanism of SiO_2 nanoparticles and starch composite films. J Text I 100:254–259

Xie J, Lee JY, Wang DIC, Ting IP (2007) Silver nanoplates: from biological to biomimetic synthesis. ACS Nano 1:429–439

Xie F, Yu L, Liu H, Dean K (2006) Effect of compatibilizer distribution on thermal and rheological properties of gelatinized starch/biodegradable polyesters blends. Int Polym Proc 21:379–385

Xie F, Halley PJ, Averous L (2011a) Bio-nanocomposites based on starch. In: Mittal V (ed) Nanocomposites with biodegradable polymers: synthesis, properties and future perspectives. Oxford University Press, Oxford, pp 234–260

Xie Y, Chang PR, Wang S, Yu J, Ma X (2011b) Preparation and properties of halloysite nanotubes/ plasticized Dioscorea opposite Thunb starch composites. Carbohydr Polym 83:186–191

Xie F, Halley PJ, Averous L (2012) Rheology to understand and optimize processibility, structures and properties of starch polymeric materials. Prog Polym Sci 37:595–623

Xie F, Liu P, Yu L (2014) Processing of plasticized starch-based materials: state of the art and perspectives. In: Halley P, Avérous L (eds) Starch polymers. From genetic engineering to green applications, 1st edn. Elsevier, Amsterdam, pp 257–289

Yadav A, Prasad V, Kathe AA, Raj S, Yadav D, Sundarmoorthy C, Vigneshvaran N (2006) Functional finishing in cotton fabrics using zinc oxide nanoparticles. Bull Mater Sci 29:641–645

Yan Q, Hou H, Guo P, Dong H (2012) Effects of extrusion and glycerol content on properties of oxidized and acetylated corn starch-based films. Carbohydr Polym 87:707–712

Yoksan R, Chirachanchai S (2010) Silver nanoparticle-loaded chitosan–starch based films: Fabrication and evaluation of tensile, barrier and antimicrobial properties. Mater Sci Eng C 30:891–897

You S, Stevenson SG, Izydorczyk MS, Preston KR (2002) Separation and characterization of barley starch polymers by a flow field-flow fractionation technique in combination with multiangle light scattering and differential refractive index detection. Cereal Chem J 79:624–630

Yu J, Ai F, Dufresne A, Gao S, Huang J, Chang PR (2008a) Structure and mechanical properties of poly(lactic acid) filled with (starch nanocrystal)-graft-poly(ε-caprolactone). Macromol Mater Eng 293:763–770

Yu J, Wang N, Ma XF (2008b) Fabrication and characterization of poly(lactic acid)/acetyl tributyl citrate/carbon black as conductive polymer composites. Biomacromolecules 9:1050–1057

Yu J, Yang J, Liu B, Ma X (2009) Preparation and characterization of glycerol plasticized-pea starch/ ZnO-carboxymethylcellulose sodium nanocomposites. Bioresour Technol 100:2832–2841

Yurdakul H, Durukan O, Seyhan AT, Celebi H, Oksuzoglu M, Turan S (2013) Microstructural characterization of corn starch-based porous thermoplastic composites filled with multiwalled carbon nanotubes. J App Polym Sci 127:812–820

Yumin D, Zuyong X, Rong L (1997) Blend films of chitosan/starch. Wuhan Univ J Nat Sci 2: 220–224

Yun YH, Hwang KJ, Wee YJ, Yoon SD (2011) Synthesis, physical properties, and characterization of starch-based blend films by adding nano-sized TiO_2/poly(methyl metacrylate-co-acrylamide). J Appl Polym Sci 120:1850–1858

Yun YH, Youn YN, Yoon SD, Lee JU (2012) Preparation and physical properties of starch-based nanocomposite films with the addition of titanium oxide nanoparticles. J Ceram Process Res 13:59–64

Zavareze E, Guerra Días A (2011) Impact of heat–moisture treatment and annealing in starches: a review. Carbohydr Polym 83:317–328

Zepon KM, Vieira LF, Soldi V, Salmoria GV, Kanis LA (2013) Influence of process parameters on microstructure and mechanical properties of starch-cellulose acetate/silver sulfadiazine matrices prepared by melt extrusion. Polym Test 32:1123–1127

Zeppa C, Gouanvé Espuche E (2009) Effect of a plasticizer on the structure of biodegradable starch/clay nanocomposites: thermal, water-sorption, and oxygen-barrier properties. J App Polym Sci 112:2044–2056

Zhang L, Ding Y, Povey M, York D (2008) ZnO nanofluids: a potential antibacterial agent. Prog Nat Sci 18:939–944

Zhang PP, Tong DS, Lin CX, Yang HM, Zhong ZK, Yu WH, Wang H, Zhou CH (2014) Effects of acid treatments on bamboo cellulose nanocrystals. Asia-Pacific J Chem doi:10.1002/apj.1812 (in press)

Zhao L, Wang H, Huo K, Cui L, Zhang W, Ni H, Zhang Y, Wu Z, Chu PK (2011) Antibacterial nano-structured titania coating incorporated with silver nanoparticles. Biomaterials 32:5706–5716

Zheng H, Ai F, Chang PR, Huang J, Dufresne A (2009) Structure and properties of starch nanocrystal-reinforced so y protein plastics. Polym Compos 30:474–480

Zheng L, Hong F, Lu S, Liu C (2005) Effect of nano-TiO_2 on strength of naturally aged seeds and growth of spinach. Biol Trace Elem Res 105:83–91

Zhu L, Shukri R, de Mesa-Stonestreet NJ, Alavi S, Dogan H, Shi Y (2010) Mechanical and microstructural properties of soy protein: high amylose corn starch extrudates in relation to physiochemical changes of starch during extrusion. J Food Eng 100:232–238

Zobel HF (1994) Starch granule structure. In: Alexander RJ, Zobel HF (eds) Developments in carbohydrate chemistry. The American Association of Cereal Chemists, St Paul Minnesota, pp 1–36

Zuraida A, Yusliza Y, Anuar H, Mohd Khairul Muhaimin R (2012) The effect of water and citric acid on sago starch bio-plastics. Int Food Res J 19(2):715–719

Nanocomposites of Polyhydroxyalkanoates Reinforced with Carbon Nanotubes: Chemical and Biological Properties

A.P. Lemes, T.L.A. Montanheiro, F.R. Passador and N. Durán

Abstract The polyhydroxyalkanoates (PHA) is one of the most investigated polymers in the development of eco-friendly nanocomposites. Biotechnology is used for their production and the mechanisms of their biodegradation make them very interesting polymers to replace conventional polymers in applications where the biodegradability is a desirable characteristic. PHA applications include medical field (suture fasteners, staples, screws, valves, orthopedic pins, etc.) besides agriculture and packaging sectors. The introduction of nanofillers in the polyhydroxyalkanoates matrixes is one of the ways used in an attempt to improve their properties or to reach new properties. With this goal, PHA/carbon nanotubes (CNT) nanocomposites have been quite studied. The remarkable properties shown by carbon nanotubes such as high Young's modulus, high thermal stability and electrical conductivity, and their low chemical reactivity is the key to achieve excellent properties from PHA nanocomposites, and to maintaining the matrix biodegradability. CNT cause changes in PHA characteristics that can affect the biodegradation rate as crystallinity degree, porosity, surface roughness, and hydrophilicity of polymer matrix. Many researches have shown the effects and advances caused by CNT filler in the mechanical resistance, crystallinity degree, thermal properties, and other important characteristics of PHA nanocomposites. However, these works have disregarded the study of the biodegradation of PHA/CNT nanocomposites, what is essential to define the application field of final product.

Keywords Polyhydroxyalkanoates · Carbon nanotubes · Nanocomposites · Biodegradable polymers

A.P. Lemes · T.L.A. Montanheiro · F.R. Passador
Institute of Science and Technology, Universidade Federal de São Paulo, SP, Brazil

A.P. Lemes · N. Durán (✉)
Institute of Chemistry, Universidade Estadual de Campinas, SP, Brazil
e-mail: duran@iqm.unicamp.br

N. Durán
Laboratory on Nanostructures Synthesis and Biological Interactions (NanoBioss), Universidade Estadual de Campinas, SP, Brazil

V.K. Thakur and M.K. Thakur (eds.), *Eco-friendly Polymer Nanocomposites*, Advanced Structured Materials 75, DOI 10.1007/978-81-322-2470-9_3

1 Introduction

Polyhydroxyalkanoates (PHA) are a family of intracellular biopolymers synthesized by many bacteria as intracellular carbon and energy storage granules (Jendrossek and Handrick 2002). Several applications have been proposed for PHA polymers in the fields of medicine, agriculture, and packaging. For the PHA family, the most widely used material is the polyhydroxybutyrate (PHB) and its copolymers with hydroxyvalerate units (PHBV). Compared with other biodegradable plastics from chemically synthesized polymers or starch-based biodegradable plastics, PHA are the only 100 % biodegradable and biosynthetic polymers, completely keeping the sustainable development of energy and environment (Shan et al. 2011; Yu et al. 2010).

Biodegradable polymers have strong commercial potential for bioplastics. However most of the biodegradable polymers-based materials exhibit low mechanical properties, low heat distortion temperature, relatively high gas permeability, low melt viscosity, and low thermal and electrical conductivity. These deficiencies narrow their applications (Sanchez-Garcia et al. 2010). One way to improve these properties is the introduction of small amount of nanofillers as nanoclay, talc, carbon nanofibers, and carbon nanotubes. In recent years, polymer nanocomposites materials prepared using different kinds of nanomaterials have generated great interests in the polymer industry as a type of novel material because of their superior properties such as high heat-deflection temperature, gas barrier performance, dimensional stability, enhanced mechanical properties optical clarity, and flame retardancy when compared with the pure polymer or composites having conventional fillers (Durmus et al. 2007).

There are researches showing property improvements in biodegradable materials with the addition of carbon nanotubes (CNT). An advantage of the CNT/polymer nanocomposites is the possibility to exploit the many advantages provided by carbon: hardness, thermal stability, and enhanced electrical conductivity. Unlike traditional polymer composites containing microscale fillers, the incorporation of nanoscale CNT into a polymer system results in very short distance between the fillers, and thus the properties of composites can be largely modified even at an extremely low content of filler (Ma et al. 2010). In this chapter, we present the main concepts involving the family of PHA and the effects in the crystallinity degree, mechanical properties, electron conductivity, and biodegradability of the PHA polymer matrix.

2 Polyhydroxyalkanoates

2.1 Brief History

The existence of PHA in bacteria was first reported in 1920s by Lemoigne (Koller et al. 2011). In a soil bacillus, the anaerobic degradation of an unknown material resulted in excretion of 3-hydroxybutyrate acid. Lemoigne identified this unknown material as a hydroxyl acid homopolyester, 3-hydroxybutyrate, or poly(3-hydroxybutyrate) (PHB), describing it experimentally as a storage material (Braunegg et al. 1998; El-Hadi et al. 2002; Lenz and Marchessault 2005; Sudesh et al. 2000). Many years later, PHA were found in activated sludge, and it took nearly a decade to report the existence of PHA with 11 varieties of linear and branched repeating units of four to eight carbon atoms (Keshavarz and Roy 2010).

However, for a long time, the interest in PHA was restricted to their physiological significance, as a substance of microbial origin. Few considerations were made regarding the possibility of using these polymers for humanity, as a feedstock for manufacturing products (Lenz and Marchessault 2005).

In the first half of the 1960s, potential application of PHA had already been recognized, as shown in that time patents related to production methods by fermentation and extraction of produced biomass. However, manufacturing plastics with petrochemical origin was easy and less expensive, and there was not strong public interest in environmental issues, the industrial use of natural plastics was still unfeasible (Braunegg et al. 1998).

In mid-1970, due to the petroleum crisis, the production of plastics from renewable resources became economically attractive (Lenz and Marchessault 2005). The price of an oil barrel increased at very high values, and the same occurred to all petroleum products. So, at that time, an extensive search of materials that could replace synthetic polymers took place. Many polymers were proposed and investigated regarding to its biodegradability and its possibility of industrial application, such as cellulose, starch, polycaprolactone, poly(lactic acid), and PHA. Among these polymers, PHA are of particular interest due to their biodegradability, biocompatibility, and mainly because of their similarity to conventional thermoplastics (Zinn et al. 2001).

2.2 Chemical Composition and Properties of Polyhydroxyalkanoates

Polyhydroxyalkanoates (PHA) are a family of biodegradable polyesters which are naturally synthesized by microorganisms from several carbon substrates, and exist inside the cell as granules, as carbon or energy storage (Zribi-Maaloul et al. 2013). A broad variety of procariotic organisms store PHA from 30 to 80 % of their cellular dry weight (Bordes et al. 2009) and their production occurs by fermentation under restricted feeding conditions (Braunegg et al. 1998; Reddy et al. 2013; Sudesh et al. 2000).

The granules storage occurs specifically when nutrients such as nitrogen and phosphorus are available in limit concentrations, while there is excess of carbon sources (Khanna and Srivastava 2005).

The industrial production of PHA is already a reality, but the utilization of pure microbial cultures leads to high production costs, making the final product significantly expensive compared to conventional thermoplastics. For this reason, efforts are made to create PHA through economical production strategies, using mixed microbial cultures (Fradinho et al. 2013).

PHA have a wide range of industrial applications due to their properties as biocompatibility, biodegradability, and low cytotoxicity. They have been considered as effective substitutes for petroleum-based polymers (Zribi-Maaloul et al. 2013), but their use on industrial scale is still limited mainly by their thermal instability near melting temperature (Aoyagi et al. 2002).

The main polymer of PHA family is the poly(hydroxybutyrate), PHB, but there is a variety of copolymers such as poly(hydroxybutyrate-co-hydroxyvalerate), PHBV, poly(hydroxybutyrate-co-hydroxyhexanoate), PHBHx, poly(hydroxybutyrate-co-hydroxyoctanoate), PHBO, and poly(hydroxybutyrate-co-hydroxyoctadecanoate) PHBOd (Bordes et al. 2009). The general structural formula of PHA is shown in the Fig. 1.

PHBV is a straight-chain semicrystalline copolymer with smaller amount of hydroxyvalerate (HV) units and higher proportion of hydroxubutyrate (HB) units. The great interest in this copolymer is due to the fact that it shows better mechanical and physical properties compared to the homopolymer PHB (Choi and Park 2004). PHBV has lower melting point, lower crystallinity degree, greater ductility, flexibility, ease of molding, and larger processing window compared to PHB (Anderson and Dawes 1990; Braunegg et al. 1998; Chen et al. 2002; Fei et al. 2004; Ferreira et al. 2002; Liu et al. 2002).

The physical and mechanical properties of PHBV are directly related to the amount of HV units. The increase of HV units can improve the flexibility and impact resistance of the material (Hu et al.2004; Shang et al. 2011). Thus, its properties can be modified, changing the copolymer composition by variations in the feeding supply to the bacteria (Ferreira et al. 2002), which allows to obtain a wide range of properties in the material, according to the content of HV units in the polymer (Braunegg et al. 1998).

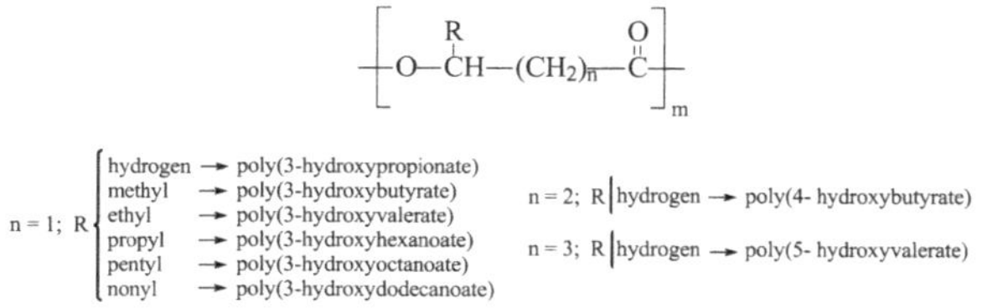

$$\left[-O-\overset{R}{\overset{|}{C}}H-(CH_2)_n-\overset{O}{\overset{\|}{C}}- \right]_m$$

n = 1; R
hydrogen → poly(3-hydroxypropionate)
methyl → poly(3-hydroxybutyrate)
ethyl → poly(3-hydroxyvalerate)
propyl → poly(3-hydroxyhexanoate)
pentyl → poly(3-hydroxyoctanoate)
nonyl → poly(3-hydroxydodecanoate)

n = 2; R | hydrogen → poly(4- hydroxybutyrate)

n = 3; R | hydrogen → poly(5- hydroxyvalerate)

Fig. 1 Chemical structure of polyhydroxyalkanoates

PHBV has physical and mechanical properties that can be compared to some thermoplastics, as polyethylene, polypropylene, poly(lactic acid), and poly(butylene succinate) (El-Hadi et al. 2002; Gursel et al. 1998; Hu et al. 2004; Luo and Netravali 2003; Reddy et al. 2013; Shang et al. 2011).

Degradation rate of this copolymer is slower than other biodegradable polymers such as poly(lactic acid) and poly(lactic-co-glycolic acid), therefore suitable for use in implants (Hu et al. 2004).

Despite showing improved properties compared to PHB, especially with regard to reducing the fragility (Dufresne et al. 2003), the applicability of PHBV is also limited compared to conventional polymers due to its high crystallinity degree, low impact strength, low elongation, narrow processing window (Han et al. 2004; Liu et al. 2002), relatively high water vapor permeability (Yu et al. 2014), and relatively high cost (Avella et al. 2002).

PHA generally decreases in molecular weight at temperatures around 170 °C, which is proportional to exposure time, and hence the low stability of PHBV. Therefore, the molecular weight reduction begins soon after their melting, which explains their narrow processing window. The mechanism of thermal degradation of PHB, for example, follows a random breakage of ester groups in accordance with elimination of a β hydrogen (Fig. 2) (Avella et al. 2000; Gonzalez et al. 2005).

Due to their biodegradability and biocompatibility, PHBV and other PHA exhibit great potential for application in packaging (Zribi-Maaloul et al. 2013), to produce bioabsorbable scaffolds (Ke et al. 2014), membranes for controlled drug delivery, sutures, wound dressings, artificial heart valves, etc. (Ruiz et al. 2011) and in the agricultural sector, such as reforestation tubes, toiletries, and other sectors as toy and automotive industries.

The crystallization rate of PHBV compared with most polymers used in industrial processes is very slow, due to the low nucleation that is a result of its high purity and stereoregularity and the slow nuclear growth rate, resulting from the HV units inclusion in the PHB crystalline cell (Reinsch and Kelley 1996).

Due to the limitations found in PHA processing, the development of polymeric nanocomposites is an alternative to improve the materials performance (Bordes et al. 2009). Reinforcing agents of several sizes and shapes have been used in

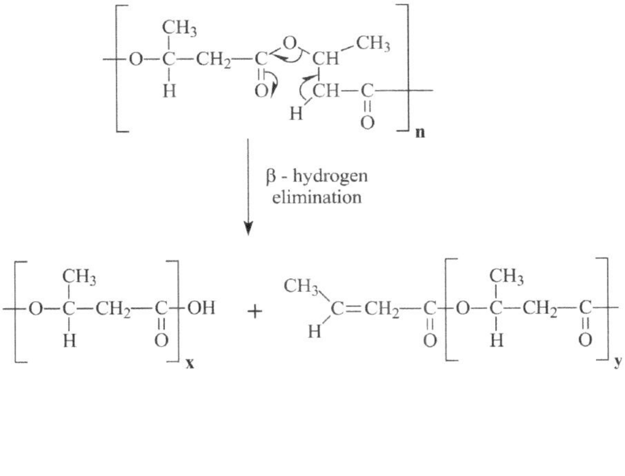

Fig. 2 PHB thermal degradation mechanism

$$\left[-O-\underset{H}{\overset{CH_3}{C}}-CH_2-\underset{\underset{O}{\|}}{C}-O-\underset{\underset{H}{CH}-\underset{\underset{O}{\|}}{C}-}{CH}-CH_3 \right]_n$$

β - hydrogen elimination

$$\left[-O-\underset{H}{\overset{CH_3}{C}}-CH_2-\underset{\underset{O}{\|}}{C}- \right]_x OH + \underset{H}{\overset{CH_3}{}}C=CH_2-\underset{\underset{O}{\|}}{C}-\left[-O-\underset{H}{\overset{CH_3}{C}}-CH_2-\underset{\underset{O}{\|}}{C}- \right]_y$$

polymeric composites; however, nanosized fillers are preferred for many applications, due to the larger surface area, need of lower concentrations to obtain reinforcing effect, and ability to improve mechanical properties, processing, thermal stability, among other properties (Ten et al. 2010).

Several studies are being conducted in order to improve PHBV mechanical properties and/or decrease its cost and enabling its industrial application. Among many studies it is possible to find addition of plasticizers (Choi and Park 2004), production of blends, including blends with natural polymers such as polycaprolactone, poly(L-lactic acid), starch, and cellulose (El-Hadi et al. 2002; Ferreira et al. 2002) and also with synthetic polymers such as polyvynyl phenol (Fei et al. 2004), oxidized polyethylene, and polypropylene (Avella et al. 2002), blends with natural rubber (Han et al. 2004), among others.

Another way to improve the polymer properties is the incorporation of fillers in the polymer matrix, which often contributes, in the case of thermoplastics, to obtain better price and an increase in thermal and mechanical properties (Avella et al. 2000; Dufresne et al. 2003; Lo and Yu 2002).

Among the fillers most used in the production of green polymer composites are the lignin and cellulose. These fillers and their derivatives have been large studied in the reinforcement of synthetic and natural polymers. In the review work of Thakur et al. (2014), recent advances in multifunctional applications of composites from lignin are elucidated in different kinds of matrixes. The authors highlight the changes and improvement of several properties obtained in thermoplastic, thermosetting, rubber, and biodegradable matrixes composites by the addition of lignin.

In the case of cellulose, the reinforcement provided by cellulose fibers in the properties of thermosetting composites is recently discussed in the work of Thakur and Thakur (2014), including polyester, epoxy, phenolic and vinyl ester matrixes. The main properties that justify the utilization of cellulose fibers and their composites were considerable are toughness, flexibility, easy processing, recyclability, and eco-friendliness.

Cellullose in the form of natural fibers (Luo and Netravali 2003), nanofibers (Srithep et al. 2013), and nanowhiskers or nanocrystals (Jiang et al. 2008a, b; Ten et al. 2013; Yu et al. 2011) have been combined to PHBV matrix to obtain environmental friendly composites or nanocomposites. It was found that the addition of nanofibrillated cellulose and nanowhiskers acts as effective PHBV nucleation agent (Srithep et al. 2013; Ten et al. 2010; Yu et al. 2011) and increase its mechanical properties (Srithep et al. 2013; Ten et al. 2010).

Several nanocomposites of PHA (PHB and PHBV) and nanoclay were mentioned on Bordes review (2009). It was possible to obtain PHB/nanoclay and PHBV/nanoclay intercalated structures. The well-dispersed nanoclay also act as nucleating agents on PHBV matrix (Choi et al. 2003; Wang et al. 2005a), besides acting as reinforcing agents (Choi et al. 2003). Thiré et al. (2011) produced PHBV/attapulgite nanocomposites and unlike observed by Choi et al. (2003), they verified a reduction in the crystallinity degree of PHBV with addition of attapulgite, that is a fibrous clay mineral.

It is also possible to find PHBV composites with oak wood flour (Srubar et al. 2012) and bamboo pulp fiber (Jiang et al. 2008a, b). The addition of oak wood flour increased PHBV matrix tensile modulus, but reduced the crystallinity degree and induced embrittlement in the composites (Srubar et al. 2012). Bamboo pulp fiber, added to PHBV matrix resulted in increased crystallization ability and mechanical properties of the polymer matrix (Jiang et al. 2008a, b).

Other important nanofiller used to produce nanocomposites with PHA, which will have a special attention in this chapter, is the carbon nanotube. Carbon nanotubes have been used to produce nanocomposites for several applications, due to their unique and extraordinary properties.

3 Carbon Nanotubes (CNT)

Carbon nanotubes are one-dimensional carbon materials with high aspect ratio (greater than 1000) and excellent mechanical, electrical, and thermal properties when compared to other carbon materials, such as graphite and fullerene. CNT is one of the most promising filler for nanocomposites and have generated great interests in the polymer industry due the technical applications in electrical conductivity, thermal conductivity, and improvements in mechanical properties (Choi et al. 2014).

The chemical bonding of CNT is composed entirely of sp^2 carbon–carbon bonds. The four valence electrons in carbon, when shared equally create isotropically strong diamond. However, when only three valence electrons are shared covalently between neighbors in a plane and the fourth valence electron is allowed to be delocalized among all atoms, the resulting material is graphite. The sp^2 carbon–carbon bonds build a layered structure with strong in-plane bonds and weak out-of-plane bonding of the van der Waals type. Therefore, the graphite sheets have the ability to sliding along the plane (Ajayan 1999). This bonding structure provides CNT extremely high mechanical and thermal properties.

Graphene consists of a single sheet of graphite, while the carbon nanotubes are graphitic sheets concentrically rolled with nanoscale diameters. Typically, the end of the CNT cylinders is closed with a structure similar to half of a fullerene (Hussain 2006).

CNT exist in a wide variety of forms and few of them are *single-walled carbon nanotubes—SWCNT* (Ma et al. 2010), *multiwalled carbon nanotubes—MWCNT*, and *double walled carbon nanotubes—DWCNT* (Stobinski et al. 2010; Sahoo et al. 2010). The aspect ratios of the types of carbon nanotubes are large since their lengths are in the range of several micrometers. Among them, two are the most widely used:

- *Single-walled carbon nanotubes* (*SWCNT*): this is close to an ideal fullerene fiber; consisting of a single layer of a graphite crystal that is rolled up into a seamless cylinder, with one atom thick (Dresselhaus et al. 2005). This category of carbon nanotube possess good uniformity in diameter (thereabout 1–2 nm) (Ajayan 1999);

- *Multiwalled carbon nanotubes* (*MWCNT*): they are made of concentric cylinders of graphite layers placed around a common central hollow, with a spacing between the layers close to that of the interlayer distance in graphite (0.34 nm) (Lai et al. 2004; Roy et al. 2012).

Both varieties of nanotubes can be regarded as aggregates of nanotubes units, the MWCNT consists of concentric assembly and SWCNT ropes of close packed nanotube units (Ajayan 1999).

According to the rolling angle of the graphene sheet, CNT have three chiralities: armchair, zigzag, and chiral one. The chirality of nanotubes has significant impact on their electronic properties (Ma et al. 2010).

Carbon nanotubes can be produced using various methods. Three of these methods are currently the most widely used in the preparation of carbon nanotubes: electric-arc discharge, laser ablation (high-temperature methods), and chemical vapor deposition—CVD (low-temperature method). Electric-arc discharge and laser ablation methods are based on the condensation of carbon generated by the carbon′ s evaporation from a solid precursor, typically high-purity graphite. The average temperature in these processes is extremely high of the order of 4000 K (Arras et al. 2013; Luo et al. 2013; Uchino et al. 2005). The CVD method is based on the decomposition of precursor gases or vapors containing carbon atoms, typically a hydrocarbon of a metal catalyst, and occurs at a temperature below 1000 K (Uchino et al. 2005).

The growth mechanism of nanotubes is not well understood. Generally, in a synthesis it is obtained by a mixture of metallic and semiconducting nanotubes (Coleman et al. 2006a, b).

Since the discovery of multiwalled carbon nanotubes in 1991, this material has been the focus of extensive global research to explore the remarkable physical and chemical properties of this new form of carbon. This kind of carbon nanotubes has unique properties and is promising for the development of new material systems (Roy et al. 2012; Stobinski et al. 2010).

A large number of investigations have reported remarkable physical and mechanical properties of this new form of carbon, such as high elastic modulus, tensile strength, and fracture toughness (Luo et al. 2013). The carbon nanotubes have an average value of Young's modulus in the TPa range (Mylvaganam and Zhang 2007) and an average tension strength in the range of 10–50 GPa, while for steel these values are 0.2 TPa and 0.25 GPa, respectively (Luo et al. 2008; Tang et al. 2003). The method used to obtain the carbon nanotubes can produce significant changes in the properties.

Measurements of the thermal conductivity of a single CNT obtained values of about 3000 W m^{-1} K^{-1} for MWCNT (Kim et al. 2001) and above 2000 W m^{-1} K^{-1} for SWCNT (Yu et al. 2005).

The MWCNT have diameters ranging from 2 to 100 nm and feature similar to a metal electrical conductivity. Already SWCNT have diameter of 0.4–2 nm and their electrical properties are defined by their chirality and diameter (Gooding 2005). In the metallic state, the electrical conductivity of carbon nanotubes is estimated in billions of A cm^{-2}, due to the few defects in its structure, and that could spread the electrons. The electrical conductivity of copper wire is limited to one million A cm^{-2}, since this current causes the fusion of cooper (Poole and Owens 2003).

Due to their unique properties, the CNT have been one of the most investigated nanofiller in the development of polymer nanocomposites. The addition of CNT as nanofiller to improve and/or to obtain new properties in polymeric matrix composed by synthetic and natural polymers is the objective of several researches (Anandhan and Bandyopadhyay 2011; Gaharwar et al. 2011; Liu and Choi 2012; Mylvaganam and Zhang 2007; Reddy et al. 2013; Roy et al. 2012; Sahoo et al. 2010).

The polymer/CNT nanocomposites can be used for some other applications, for which conductivity requirements are not as severe. Among these applications may be mentioned electromagnetic interference shielding, electrostatic dissipation, touch panels, display panels, which need to be transparent in addition to presence certain conductivity. The nanocomposites may be also used to dissipate heat or as flame retardant, and in the biological field, as part of sensors (Grady 2011).

Carbon nanotubes have the ability to absorb electromagnetic radiation and convert the energy into heat; so if embedded in a polymer, they may be used as photothermal actuators, due to the polymer's high thermal expansion coefficient (Yang et al. 2008).

However, there are some important factors in the production of polymer/CNT nanocomposites that must be considered to obtaining success in the development of these nanocomposites. The factors include the interfacial adhesion between the polymer and the CNT, which is essential to permit a good transfer of tension between matrix and filler; and the method to produce the nanocomposites, since it is important to obtain an efficient distribution and dispersion of CNT inside the matrix.

The improvement of mechanical, thermal, electrical, and other properties in polymer/CNT nanocomposites occurs when the CNT are well-dispersed and adhered in the polymeric matrix, which are the major challenges in the production of polymer/CNT nanocomposites.

The possible methods to produce polymer/CNT nanocomposites and the functionalization ways to modify the surface of CNT and, consequently, to improve their dispersion and adhesion in the polymer matrix will be addressed in the next section.

3.1 Functionalization of CNT in the Production of Polymer/CNT Nanocomposites

The carbon nanotubes have a very stable surface structure and hard surface interaction; so when the carbon nanotubes are used as nanofiller in polymer matrices, this high stability becomes a problem in the interaction between matrix and filler.

Another difficulty in the use of carbon nanotubes is their strong tendency to form agglomerates (Yu et al. 2014), due to the small size and high surface area, which makes the carbon nanotubes strongly bound together by van der Waals attractive forces, forming bundles and large agglomerates of CNT (Spitalsky et al. 2010). Therefore, in many cases, it is necessary to modify the CNT surface for an effective use of them in polymer/nanocomposites.

The functionalization of CNT includes the modification of CNT surface by the introduction of functional groups or the attachment of molecules. This can be used to prevent the CNT agglomeration, due the decrease of van der Waals attractive force, which leads to better dispersion and stability of CNT in the polymer matrix (Hirsch 2002).

In addition, the functionalization can be used to promote a better interaction between nanofiller and matrix or even the chemical bond between CNT and polymer.

Despite the functionalization be a useful tool in the production of polymer/CNT nanocomposites, the control of the functionalization reactions of the hydrophobic and chemically inert surface of the carbon nanotubes remains a major challenge for their use (Likodimos et al. 2014).

The functionalized carbon nanotubes can present different mechanical and electrical properties when compared to the unfunctionalized carbon nanotubes. These chemically modified structures can be used to facilitate the interaction of the nanotubes with organic and biological molecules and with other chemical groups as toxic molecules, drugs, and even viruses and bacteria (Filho and Fagan 2007).

There are several methods of functionalization of the carbon nanotubes which can be separated in covalent and noncovalent functionalization (Hirsch 2002).

The noncovalent functionalization method includes modifications of the carbon nanotubes using surfactants agents, involvement or absorption of polymers through noncovalent interaction between the carbon nanotubes and the functional groups presents in the polymers matrices. The advantage of this method of functionalization is the preservation of the integrity of the carbon nanotubes while keeping the mechanical and electrical properties. On the other hand, the main disadvantage is that the interaction forces between the molecules used to modify the surface and the carbon nanotubes may be too weak and this modification cannot be effective (Han and Fina 2011).

An example of noncovalent functionalization is the use of anionic surfactants such as sodium dodecyl sulfate (SDS) and sodium dodecyl benzene sulfonate (NaDDBS) to reduce aggregation of carbon nanotubes in water (Sahoo et al. 2010).

In the case of covalent functionalization, the local stresses due to misalignment of π orbitals of carbon atoms with sp^2 hybridization make the carbon nanotubes more reactive than a graphene sheet, facilitating covalent bonding of chemical species. However, even when extensive damage to the carbon nanotubes structure are avoided, a notable disadvantage of this process of functionalization is the breakdown of the conjunction of the carbon nanotubes in with the conversion of carbon with hybridization sp^2–sp^3 (Han and Fina 2011).

The reaction of covalent functionalization of carbon nanotubes is always preceded by an oxidation reaction for obtaining of carbonyl and carboxyl groups, which can be created during the oxidation by oxygen, air, concentrated sulfuric or nitric acid, aqueous hydrogen peroxide and acid mixture. The oxidative processes induce the opening of the ends and the formation of holes in the walls of the carbon nanotubes (Sahoo et al. 2010; Tasis et al. 2006).

The presence of carboxylic groups on the surface of the carbon nanotube is more convenient than other groups once a variety of chemical reactions can be carried out with this functional group, for example the silanization reactions, graft of polymers, esterification, thiolation, alkylation, and reaction with biomolecules (Ma et al. 2010). According to Yu et al. (2014), if the carbon nanotubes functionalization is performed with the same polymer used as matrix, one can obtain better compatibility by formation of a continuous interface between polymer matrix and filler, which can improve the interfacial adhesion that is responsible for good properties in nanocomposites. Byrne et al. (2008) reported increase of 25 % in Young's modulus and 50 % in tensile strength in polystyrene (PS)/carbon nanotubes nanocomposites with 0.25 wt% of carbon nanotube functionalized with butyl radicals. The functionalization reactions were carried out using *n*-butyllithium in tetrahydrofuran (THF). Her and Lai (2013) studied the effect of addition of MWCNT functionalized with carboxylic groups (COOH) in epoxy resins. There was an increase in interfacial adhesion between the epoxy resin and nanofiller with increase of stiffness of the nanocomposite. The functionalization process also improved the dispersion of the carbon nanotubes in solvents. The effect of the functionalization of MWCNT through oxidation reactions using a mixture of nitric and sulfuric acids was also reported by Sahoo et al. (2007). The MWCNT were functionalized with COOH groups and subsequently chemically bonded to the polyurethane (PU). The authors reported a significant increase in thermal stability of the nanocomposites when compared to neat PU, and an increase in tensile strength of 90 % for the nanocomposite with 5 wt% of carbon nanotubes.

High concentrations (>5 wt%) of functionalized carbon nanotubes with carboxylic groups with 1.23 wt% of COOH groups were dispersed in poly(dimethylsiloxane) (PDMS) and the effect of functionalization in the surface of the carbon nanotubes was verified by Liu and Choi (2012). CNT and CNT-COOH were dispersed with PDMS in chloroform, sonicated and subjected to mechanical stirring. The resulting solution was well-dispersed, but after 4 h the carbon nanotubes were already agglomerated while the CNT–COOH remained dispersed.

Damian et al. (2010) used amine groups to functionalization of carbon nanotubes. Epoxy/carbon nanotubes were prepared and the thermomechanical properties were evaluated. It was observed a reduction on the thermal degradation of the nanocomposite probably due to the low degree of crosslinking obtained in the presence of carbon nanotubes.

After functionalization, functional groups are formed on the CNT surface, and these groups may be positively or negatively charged. The presence of same sign electrostatic charges on CNT surface makes them repel one another, maintaining the material dispersed in a colloidal suspension.

Fig. 3 Aqueous dispersion of carbon nanotubes 2 hours after sonication in an ultrasonic bath, (*1*) pristine MWCNT; (*2*) MWCNT functionalized with carboxylic groups; (*3*) MWCNT functionalized with hydroxyl groups

Our research groups functionalized MWCNT with carboxylic and hydroxyl groups. Pristine and functionalized MWCNT were dispersed in distilled water in ultrasonic bath and photographed after sonication in time intervals. The photo in Fig. 3 confirms that functionalization improves CNT dispersion, since after 2 h the functionalized MWCNT remain dispersed in water while most of pristine MWCNT formed agglomerates in the bottom of flask.

3.2 Methods for Producing Polymer/CNT Nanocomposite

Among the principal methods for producing nanocomposites, three are the most adopted: solution blending, melt blending, and in situ polymerization (Lee et al. 2005).

Solution blending method consists of mixing the nanoparticles with desired polymer in a suitable solvent. Usually, this method follows three steps: (1) dispersing the nanoparticles in a suitable solvent or a polymer solution; (2) mixing the nanoparticles with the polymer solution at room temperature or under heating; and (3) obtaining the nanocomposite by precipitation or by solvent casting. This method is the mostly used for nanocomposites preparation and its main advantage is in the way of stirring the nanoparticles in the solvent, and the possibility of working on a small scale. Stirring may be promoted by magnetic bar, mechanical mixer, or ultrasonication, and compared to other methods it results in a better dispersion of the nanoparticles (Coleman et al. 2006a, b; Moniruzzaman and Winey 2006).

Ultrasonication is an important method in the preparation of nanostructured materials (Bang and Suslick 2010), and it may be mild when performed on an ultrasound bath, or severe when performed in an ultrasonic processor (ultrasonic probe). The main effect of ultrasound is the cavitation, which consists of a cycle of formation, growth, and collapse of micrometric bubbles during sonication (Korn et al. 2003; Zhang et al. 2009).

So, the breakdown of the carbon nanotubes agglomerates is caused by the collapse of microbubbles in the interstices of the aggregates and by collisions between aggregates at high speeds and by liquid microjets that reach the surface of aggregates and bundles of carbon nanotubes.

Both ultrasonic bath and ultrasonic processor show the same mechanisms of nanoparticles dissociation, but their power and frequency of the ultrasonic waves are different.

Ultrasonication can be used to produce metastable suspensions of nanoparticles, or to blend these nanoparticles with the desired polymer in different solvents. However, the intensity and time of sonication must be first analyzed, since longer periods of sonication may damage the nanoparticles (Moniruzzaman and Winey 2006).

If there is a good interaction between solvent and nanoparticle, certainly the clumps will be disrupted and a temporarily stable suspension of nanoparticles may be obtained; however, the best method of solvent casting must be studied, once the nanoparticles tend to reaggregate during this period (Lee et al. 2005).

Although solution blending method consists in the most efficient nanoparticles dispersion technique in the polymeric matrix, it is mostly not compatible with current large-scale industrial processes.

Thus, melt blending method of nanocomposites preparation is a common alternative, and is particularly useful for thermoplastics processing. This method is based on high temperatures and shear to disperse nanoparticles on the polymer matrix, and has advantages as its speed and simplicity, besides being the most compatible with currently industrial processes, among which, extrusion is the most practiced (Coleman et al. 2006a, b; Esawi et al. 2010; Lee et al. 2005; Moniruzzaman and Winey 2006).

The disadvantage of this method is less efficiency in dispersing the nanoparticles when compared to the method of solution blending. To improve nanoparticles dispersion on the polymer matrix, it is usually necessary a high shear rate, which may damage the nanoparticle, reducing its reinforcement action and other properties (Lee et al. 2006).

In situ polymerization method has been widely explored, and it consists of dispersing nanoparticles on monomers, followed by their polymerization. This technique is an important alternative, when the polymer is not soluble in most solvents, or even thermally unstable, making it impracticable to produce the nanocomposite through solution blending or melt blending. In general, in situ polymerization may be used to prepare almost all polymeric nanocomposite containing nanoparticles. Nanoparticles can be covalently or noncovalently bonded to the polymer matrix. Noncovalent interaction between polymer and nanoparticle may involve physical adsorption or van der Waals forces (Coleman et al. 2006a, b; Moniruzzaman and Winey 2006; Mylvaganam and Zhang 2007).

4 Polyhydroxyalkanoates/Carbon Nanotubes Nanocomposites

The PHA/CNT nanocomposites exhibit interesting mechanical, thermal, and electrical properties. To attain these properties, several parameters are important such as the type of processing, the type of carbon nanotube, and solvent besides the type of PHA used as polymer matrix.

Some recent studies have showed that nanocomposites of PHB/CNT are suitable to develop advanced electronic devices, being integrated in a multilayer film with photoresponsive properties (Valentini et al. 2014). It was also shown that transparent bionanocomposites made of PHBV and functionalized MWCNT with PHBV chains have great potential for applications in food and beverage packaging (Yu et al. 2014).

There are few reports in the literature about the use of poly(hydroxybutyrate), PHB, as polymeric matrix in CNT nanocomposites. Xu and Qiu (2009) produced nanocomposites of PHB with carboxyl functionalized multiwalled carbon nanotubes (MWCNT–COOH) through a solution mixing method using chloroform as solvent. The contents of MWCNT–COOH in the polymer matrix were 0.5, 1 and 2 wt%. For the preparation of the nanocomposite, the mixture was sonicated and poured into a dish to allow the solvent to evaporate at room temperature. The morphological characterization showed homogeneous distribution of MWCNT–COOH in the PHB matrix for all MWCNT–COOH contents, and the authors attributed the good dispersion of the carbon nanotubes due the presence of carboxyl groups on MWCNT surface, which prevent aggregation of the CNT in the polymeric matrix. Another important observation was the increase of PHB crystallization due to nucleation effect of modified MWCNT.

Valentini et al. (2014) showed that PHB/CNT nanocomposites can be used as advanced electronic devices. The authors studied the effect of the type of carbon nanotube in several properties of the nanocomposites. In order to develop new electronic devices, solvent casting was used to prepare thin films of PHB/SWCNT and PHB/MWCNT with 0.25 wt% of CNT, using chloroform as solvent and glycerol as plasticizer. The addition of both types of CNT improved the storage modulus of both nanocomposites; however, a more significant improvement in mechanical properties was verified in PHB/SWCNT, it was attributed to a larger surface area of SWCNT when compared to MWCNT and a partial aggregation of MWNT bundles observed in the nanocomposites. Electrical resistance was measured using a two-probe method, and the results showed an ohmic behavior. For PHB/SWCNT nanocomposite, the electrical resistance value is about 2×10^8 Ohm and for PHB/MWCNT nanocomposite the value is about 3×10^7 Ohm.

Among the PHAs, poly(3-hydroxybutyrate–co–3-hydroxyvalerate), PHBV, was mostly used to produce nanocomposites with carbon nanotubes. Lai et al. (2004) produced nanocomposites of PHBV (4.5 mol% of HV) and MWCNT. PHBV/MWCNT nanocomposites with 2 wt% of MWCNT were prepared using trichloromethane as solvent. The mixture was sonicated, then the solvent was allowed to evaporate, and the residue was pressed in a hot press at 180 °C to form samples with 1 mm in thickness. They showed that MWCNT were uniformly dispersed in the PHBV matrix and that the diameter of the nanotubes was four or five times broader than that of neat MWCNT. One explanation for this is because PHBV form a covering layer on the nanotubes. Furthermore, the addition of carbon nanotubes increased the crystallization degree of the PHBV in the nanocomposite, increasing the crystal nucleation and acting as nucleating agent in the polymer matrix.

Similar results were obtained by Shan et al. (2011) that prepared PHBV/MWCNT-COOH nanocomposites using PHBV with 2.57 mol% of HV and carboxyl MWCNT (with 2.56 wt% of –COOH). On this study, PHBV was used to produce nanocomposites with 0.5, 1, 3, 5, and 7 wt% of MWCNT and the nanocomposites were prepared by melt processing using traditional machines to produce composite materials (extruder and injection molding machine). It was showed that MWCNT–COOH increases the crystallinity degree and crystallite sizes of PHBV, acting as heterogeneous nucleation agent, without influencing the basic PHBV crystalline structure. Furthermore, the incorporation of only 0.5 wt% of MWCNT–COOH elevated substantially the crystallization temperature.

Vidhate et al. (2012) prepared PHBV/MWCNT nanocomposites using PHBV with 18 mol% of hydroxyvalerate by melt blending in a counter rotating mixer and samples were prepared by compression molding. The nanocomposites were produced with 0, 1, 5, and 10 wt% of MWCNT. The process used for the preparation of the nanocomposites promoted excellent dispersion of the carbon nanotubes in the polymer matrix especially when compared to the nanocomposites produced by solution. The addition of MWCNT promoted the crystallization of PHBV, increasing its crystallization temperature. When MWCNT are well-dispersed in the polymer matrix, there are an increased number of sites available for nucleation, and the crystallization rate is enhanced. The authors reported that the Young´s modulus increased for all MWCNT contents, and the tensile strength only increased for the nanocomposites with 1 and 5 wt% of MWCNT. Compression test results showed an increase in the compressive strength with increase of carbon nanotubes contents. The modulus values only increased for 1 and 5 wt% of MWCNT, while strain at break was reduced for 1 and 5 wt% of carbon nanotubes. The nanocomposite with 10 wt% of MWCNT showed a very high improvement in strain at break value under compressive load, while under tensile loading the same nanocomposite had a poor performance. Electrical resistivity was obtained by a two-probe method and the results showed lower volume resistivity with increasing MWCNT content. It could also be interfered that electrical percolation threshold occurred between 2 and 5 wt% of MWCNT.

Nanocomposites of PHBV (21 mol% of HV) and acidified MWCNT-COOH were prepared by solution blending followed by solvent casting at 70° C by Ma et al. (2012). Chloroform was used as solvent and the solution was subjected to ultrasonication. Nanocomposites were produced with 0, 0.5, 1, and 1.5 wt% of MWCNT-COOH and the acidification of MWCNT was made using a mixture of concentrated sulfuric and nitric acids. Adding MWCNT-COOH to PHBV implied in a growth of crystallinity in the nanocomposites, increase in crystallization temperature, and changes the dynamics and kinetics of its nonisothermal crystallization. Mechanical tests were performed and it was observed the increase of elastic modulus, tensile strength, and fracture elongation for contents of MWCNT-COOH until 1 wt%. On the other hand, the addition of higher contents of carbon nanotubes decreased the mechanical properties. Lower contents of MWCNT-COOH with uniform distribution imply in a strong interface between the carbon nanotubes and the polymer matrix and this contributes to improve the mechanical properties.

Yu et al. (2013) prepared PHBV/MWCNT nanocomposites using MWCNT previously treated in a mixture of concentrated sulfuric and nitric acids (3:1 v/v) to obtain –COOH functionalized MWCNT. First, the authors prepared a PHBV/MWCNT-COOH (3 wt%) nanocomposite by physically mixed in DMF under ultrasonic irradiation at room temperature. After PHBV chains were modified by transesterification reaction with ethylene glycol to produce hydroxyl groups in the ends of the chains, PHBV-OH. This way, PHBV-OH chains were grafted onto MWCNT-COOH surface through esterification reaction between carboxylic groups of MWCNT-COOH and hydroxyl groups of PHBV-OH, producing PHBV-*g*-MWCNT-COOH.

In the PHBV/MWCNT–COOH nanocomposites, it was observed that MWCNT–COOH were well-dispersed into the polymer matrix, with no aggregation. On the other hand, PHBV–*g*–MWCNT showed a core-shell structure, with MWCNT–COOH as the core and a PHBV layer as the shell. Polarized optical microscopy was also used to investigate the crystals of neat PHBV and nanocomposites at an isothermal crystallization temperature at 45 °C. The addition of MWCNT–COOH to PHBV reduced the size of the spherulites, comparing to neat PHBV, but increased the nucleation density of the crystals, confirming the heterogeneous nucleating effect of carbon nanotubes. For PHBV–*g*–MWCNT–COOH, broken and imperfect spherulites was observed, due to covalent attachment between PHBV and MWCNT–COOH, which restricted conformational and structural changes of PHBV chains. The authors showed that MWCNT–COOH act as nucleation agent to PHBV matrix, enhancing nucleating rate, facilitating the crystallization, and improve the perfection of the crystals. For PHBV–*g*–MWCNT, the crystallinity degree was reduced due to the limited segmental mobility of PHBV grafted onto MWCNT–COOH surface, leading to the formation of imperfect crystals. The thermal stability of PHBV–*g*–MWCNT–COOH was higher than that of PHBV/MWCNT–COOH and that of PHBV, as verified by thermogravimetric analysis (TGA). The maximum decomposition temperature of the PHBV–*g*–MWCNT increased about 14.4 °C compared with PHBV/MWCNT and about 23.7 °C compared with that of the original PHBV. This result indicates that the covalent attachment of PHBV–OH in the MWCNT–COOH can be processed in a larger processing window compared to PHBV/MWCNT and original PHBV.

In a posterior study, Yu et al. (2014) produced nanocomposites of PHBV and PHBV–*g*–MWCNT with contents from 1 to 10 wt% of PHBV–*g*–MWCNT. The functionalized MWCNT was produced following the procedure previously described (Yu et al. 2013). In this case, the nanocomposites were prepared by solution blending using chloroform as solvent. The morphology shows well-dispersed PHBV–*g*–MWCNT in the polymer matrix with carbon nanotubes smaller than 7 wt%. For 10 wt% content of carbon nanotubes, it was observed the presence of slight agglomerates. For nanocomposites with about 7 wt% MWCNT makes the crystallization easily, leading to the formation of crystals with great perfection. On the other hand, for higher nanofiller content the entanglement between polymer chain grafted to PHBV–*g*–MWCNT surface and PHBV from the matrix suppressed crystallization process, resulting in imperfect crystals and reducing the crystallinity degree. Mechanical properties of the nanocomposites were substantially enhanced with PHBV–*g*–MWCNT with lower content (<7 wt%). The tensile strength was greatly improved (88 %) as well as and Young´s modulus (172 %), while the elongation at break was reduced (3.3 %). For nanofiller contents higher than 7 wt% the mechanical properties were reduced, due to PHBV–*g*–MWCNT agglomeration.

Other reports in the literature show a different efficiencies on the crystallinity degree of the PHBV with the addition of carbon nanotubes. Sanchez-Garcia et al. (2010) produced solution-cast films of PHBV with MWCNT using chloroform as solvent. PHBV with 12 mol% of HV was used to produce films with 1, 3, 5, and 10 wt% of MWCNT. The MWCNT solutions in chloroform were mixed in a homogenizer and then stirred with the polymer. The authors observed that the addition of MWCNT increases the glass transition temperature (T_g) of the nanocomposites, compared to neat PHBV. The increasing values of T_g was higher for nanocomposites with lower contents of carbon nanotubes and this fact was correlated with the better dispersion of MWCNT. The crystallinity degree for the nanocomposites decreased with the addition of carbon nanotubes that hindered the crystallization process of PHBV. On the other hand, the electrical conductivity of nanocomposites increases with the increase of MWCNT. This effect is more pronounced with filler contents higher than 5 %, due to percolation threshold for electrical conductivity.

Lemes et al. (2008) verified that the functionalization of MWCNT with carboxylic groups (MWCNT–COOH) affected the dispersion of CNT in the matrix in PHBV nanocomposites, even in extremely low MWCNT–COOH concentration (0.06 %). SEM micrographs of fractured surface of PHBV/MWCNT and PHBV/MWCNT–COOH showed that the diameter of regions of CNT agglomerates were larger in PHBV/MWCNT than in the PHBV/MWCNT–COOH nanocomposites. Changes in the thermal behavior of PHBV matrix and a nucleating effect promoted by CNT were observed in both nanocomposites.

Other methods of processing can be used to produce PHA/CNT nanocomposites. Chan et al. (2010) prepared randomly oriented and aligned fibers from PHBV (8 mol% of HV) embedded with MWCNT by electrospinning. A PHBV solution was prepared for electrospinning, dissolving the purified polymer in a mixture of chloroform and 1, 2-dichloroethane (2/3; v/v). The PHBV/MWCNT solution was prepared in a mixture of solvent of chloroform and 1, 2-dichloroethane (2/3; v/v) at 17 wt% via ultrasonification, with MWCNT content changing from 0.2 to 0.8 wt%. Morphological observation revealed MWCNT aligned to the fiber axis and protruding from the surface and a single electrospun fiber with 0.2 wt% of MWCNT showed electrical conductivity of 2.07 cm^{-1}. Yun et al. (2010) used a spray drying process to produce single-wall carbon nanotube (SWCNT)/polyhydroxyalcanoates composites. Particles of SWCNT/poly(3-hydroxybutyrate) (PHB) and SWCNT/poly(3-hydroxyoctanoate) (PHO) were produced. The polymers were dissolved in chloroform with concentration of 5 % (w/v), then carbon nanotubes were also dispersed in chloroform with concentration of 3 % (w/v) using an ultrasonic liquid processor. The solutions of PHB/SWCNT and PHO/SWCNT were then mixed together with different contents relative to PHA polymers and sonicated and then dried using a spray drying equipment. The particles produced by spray drying methods were observed through morphological techniques. Scanning electron microscopy (SEM) images showed that particles of neat PHA had well-defined spherical morphology, and particles with SWCNT had rosette shapes. Transmission electron microscopy (TEM) images of the particles show that SWCNT were well-dispersed in the polymer matrix of PHA. The authors affirm that spray drying technique is environmentally safe and cost effective, and that the powders produced by spray drying with well-dispersed SWCNT in the polymer matrix may be good precursors for further processing.

Thus, according to works reported in the literature about PHA/CNT nanocomposites, we can conclude that the introduction of CNT in the PHA matrix can change and improve the nanocomposite properties. Thermal behavior, electrical conductivity, mechanical properties, and mainly the crystallization of PHA were the most affected properties by the addition CNT. Obviously, the properties of PHA/CNT depends on several factors, such as the characteristics of PHA matrix and CNT, amount of CNT in the nanocomposite, method for producing nanocomposite, and functionalization of CNT. The choice and control of all these factors should be made through the choice of properties to be reached.

However, an important property of PHA/CNT nanocomposites that has been disregarded consists in the biodegradability of these nanocomposites. The study of biodegradability is essential to define the application field of final product, and for this reason this subject will have a special attention in this chapter.

5 Biodegradation of Polyhydoxyalkanoates

As the PHA consist in a family of polyesters, they can be depolymerized by hydrolysis reactions, at which the molecules of water cause the scission of ester bonds, producing carboxylic acid and alcohol groups (Fig. 4).

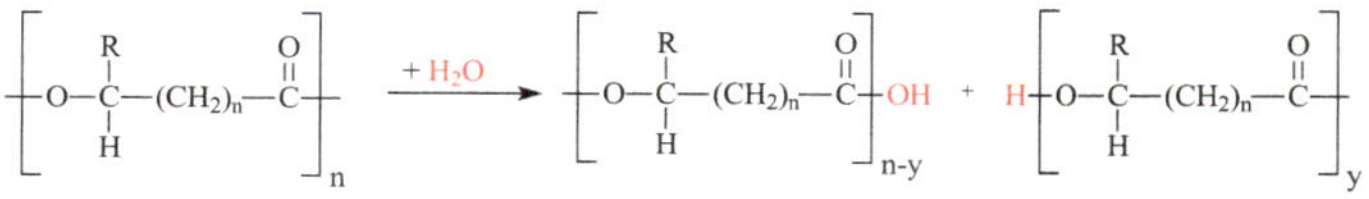

$$\left[-O-\underset{H}{\overset{R}{C}}-(CH_2)_n-\overset{O}{\overset{\|}{C}}-\right]_n \xrightarrow{+H_2O} \left[-O-\underset{H}{\overset{R}{C}}-(CH_2)_n-\overset{O}{\overset{\|}{C}}-\right]_{n-y}OH + H\left[-O-\underset{H}{\overset{R}{C}}-(CH_2)_n-\overset{O}{\overset{\|}{C}}-\right]_y$$

Fig. 4 Hydrolysis reaction of polyhydroxyalkanoates

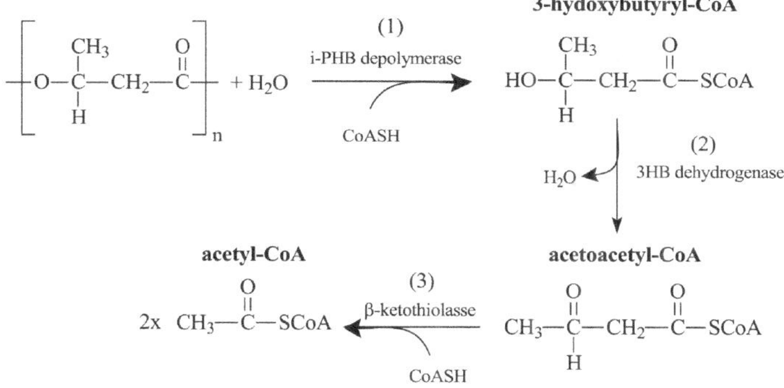

Fig. 5 Mechanism of intracellular biodegradation of PHB

The biodegradation of PHA can occur inside the cell, by microorganisms that produce PHA, or outside the cell by microorganisms that do not produce PHA, but can also use it as carbon and energy sources. However, in both cases the hydrolysis is catalyzed by specific enzymes, intracellular, or extracellular PHA depolymerases.

For nonproducing PHA microorganisms, the high molecular weight of PHA chains prevents their penetration through cell wall. This way, the scission of PHA chain in water soluble monomers and oligomers is necessary in order to enable their absorption by cells and then their metabolization in the cytoplasm.

Eggers and Steinbüchel (2013) state in their work that the degradation process of extracellular PHB depolymerases is well understood; however the reaction mechanism of the intracellular PHB depolymerases is still in progress of elucidation.

Take for example the intracellular degradation of PHB, a simplified mechanism can be observed in the scheme shown in the Fig. 5. This mechanism begins by the hydrolysis of chain polymer catalyzed by a i-PHB depolymerase leads to (*R*)-3HB-CoA (1), after a 3HB dehydrogenase converts the (*R*)-3HB-CoA to acetoacetyl-CoA (2) and a 3-ketoacyl-CoA thiolase converts the acetoacetyl-CoA to acetyl-CoA (3). After that, under aerobic conditions, the acetyl-CoA enters in the citric acid cycle and is oxidized to CO_2 (Eggers and Steinbüchel 2013; Jendrossek and Handrick 2002; Lenz and Marchessault 2005; Philip et al. 2007; Senior and Dawes 1973).

As the PHA are insoluble in water, their enzymatic degradation consists of heterogeneous reactions that involve two steps. In the first step, the enzyme adsorbs in the PHA substrate by binding dominions; and in the second step, the catalytic domain realize the PHA hydrolysis (Mukai et al. 1993). According to Yamashita et al. (2003), the enzyme adsorption is extremely important to degradation rate of PHB.

PHA can be biodegraded by many bacteria as *Alcaligenes faecalis* T1, *Pseudomonas lemoignei, Pseudomonas fluorescens* GK13, *Streptomyces exfoliates* K10, *Comamonas testosterone, Comamonas acidovorants,* etc., and many fungi as *Aspergillus penicilloides, Aspergillus fumigatos, Penicillium funiculosum, Penicillium daleae, Paecilomyces marquandii, Candida guilliermondii,* etc. (Bhatt et al. 2011).

The knowledge about the mechanisms of biodegradation reactions and the factors that control the rate of biodegradation is essential for the development of materials that biodegrade in a desirable period according to their application and environment where they will be disposed.

Many researches underline the influence of crystallinity of PHA samples in the velocity of biodegradation (Abe et al. 2000; Iwata et al. 1997; Maiti et al. 2007; Nobes et al. 1996; Numata et al. 2006, 2009). The hydrolysis reactions occur faster in the polymer chains that compound the amorphous regions than in the polymer chains that compound the crystalline regions (Spyros et al. 1997). For this reason, the biodegradation process can be accompanied by an increase in the crystallinity degree of PHA samples due to consume of amorphous fraction. So, several works that aim the study of PHA biodegradation include in the characterization essays, techniques to determine the crystallinity degree of biodegraded samples, as X-ray diffraction (XRD) (Bonartsev et al. 2012; Gonçalves and Franchetti 2013), and differential scanning calorimetry (DSC) (Madbouly et al. 2014).

The study of biodegradation of monocrystals of PHA reveals that enzymatic attack occurs preferentially by edges than the surface of crystalline lamella. This way, the hydrolysis of polymer chains in the edges of lamella result in its section, forming strips whose tips become thinner along biodegradation. This process is carry on by action of the PHA depolymerase and the lamellar crystals is gradually converted into fibril-like crystals in advanced stages of biodegradation (Iwata et al. 1997; Nobes et al. 1996, 1998; Numata et al. 2009; Numata et al.2006; Spyros et al. 1997). The characterization techniques used to study the biodegradation of monocrystals of PHA include transmission electron microscopy (TEM) and atomic force microscopy (AFM).

In addition to crystallinity degree, other characteristics of samples affect the rate of biodegradation as chemical composition of PHA (monomer units) (Abe and Doi 1999; Feng et al. 2004) dimensions (thickness, wide and length) and porosity, since they determine the superficial area of sample and molar weight (Bonartsev et al. 2012). The rate of biodegradation is still affect by factors related to environment conditions as temperature, humidity, pH, type and concentration of microorganisms, in soil composition.

The biodegradation of PHA has been verified in several environments, including different kinds of soil: garden soil, paddy field soil, farm soil, infertile garden soil, river bank soil (Wang et al. 2005b), sea water (Akmal et al. 2003; Deroiné et al. 2014; Volova et al. 2010; Wang et al. 2005b); aerobic activated sludge (Akmal et al. 2003; Wang et al. 2005b); lake water (Akmal et al. 2003), river water (Wang et al. 2005b), waste landfill model reactors undergo aerobic and anaerobic conditions (Ishigaki et al. 2004).

Lotto et al. 2004 evaluated the effect of temperature in the biodegradation of PHB in soil compost at 24 and 46 °C. The higher temperature raised the biodegradation rate of PHB samples, probably due to a direct action on mechanism of the hydrolysis or an indirect action by enhancing the activity of thermophylic microorganisms, as stated by authors. It is well known that the decrease of temperature reduces and can inhibit the bacterial activity decreasing the biodegradation rate, as verified by Deroiné et al. (2014).

An example of the effect of environmental conditions in the biodegradation of PHA can be observed in the work of Boyandin et al. (2012). In this work, the biodegradation tests were realized in the root zone of coniferous and deciduous tree during two field seasons: 1st July to October 2007 and 2nd June to September 2010. The main difference between the two filed seasons was the temperature of the winter that preceded the season. In the second season, the temperature of winter was 13.5 °C lower than the first season. The authors stated that the biodegradation rate was affect by chemical composition of PHA, temperature, humidity, and microbial soil component. The biodegradation was verified by the decrease of molar weight of studied polymers and the increase of crystallinity degree during biodegradation indicated that the polymer chains in the amorphous phase were preferentially consumed.

The biodegradability of PHB and PHBV was studied in tropical marine water in the South of China Sea by Volova et al. (2010). In this case, the chemical composition of PHA did not affect the biodegradation, since PHB and PHBV samples showed similar biodegradation rate. The biodegradation was rather influenced by the shape of samples, showing the important effect of superficial area in the biodegradation rate. The crystallinity degree of both PHA remained unchanged, indicating that the amorphous and the crystalline phases were equally disintegrated.

This way, while in some works, some variables affect the biodegradation rate of PHA samples, in other works it does not occur, which leads us to believe that some variables can be more important in some kind of environments than in other ones.

The biodegradation of PHA over time is verified by the visual analysis of biodegraded samples (photos), scanning electron microscopy (SEM), atomic force microscopy (AFM), optical microscopy, average molar mass, mechanical resistance, infrared spectroscopy, etc. The micrographs of biodegraded samples obtained by SEM and AFM reveal in details the changes caused by the attack of microorganisms.

The most usual technique to quantify the biodegradation is the monitoring of mass loss. However, some times the removal of biofilm formed on the surface of samples subjected to the test of biodegradation to determine their mass loss is difficult due to increasing of sample fragility. For this reason, the error usually increases over the biodegradation time.

The biodegradation rate can also be used to quantify the respirometric evaluations (Gonçalves and Franchetti 2013). In this case, the biodegradation rate is not measured by mass loss but by the amount of CO_2 produced or the oxygen demand originated by microbial activity. This technique eliminates the aforementioned errors about the measurements of mass loss.

The biodegradation tests performed in field show the advantage to obtain data that represent the real conditions where the polymers will be biodegraded. However, the laboratorial tests show many advantages, as the facility to work in a small scale, shorter time of test, and better reproducibility and control of conditions.

As the rate of biodegradation is affected by environmental conditions and the properties of PHA samples mentioned previously it is expected that the addition of nanofiller in biodegradable polymers causes some changes in their biodegradability, since the nanofillers can alter the properties of PHA samples.

The study about time and mechanisms of biodegradation of PHA nanocomposites has been disregarded in many researches nevertheless be essential to allow their applications.

Few works have been investigated the effects of nanofillers in the biodegradation of PHA nanocomposites. For example, the influence of organically modified montmorillonite (MMT) and dimethyl dialkyl ammonium (MAE) in the biodegradation of PHB was first investigated by Maiti et al. (2003). In the first 4 weeks, the biodegradation rate of PHB/MMT nanocomposite was higher than pure PHB and PHB/MAE. The authors suggested that this can occur due to the presence of Al Lewis acid sites in the octahedral layers of montmorillonite and/or the higher amount of hydroxyl groups on the surface of MMT, which could catalyze the hydrolysis of ester bonds. After 4 weeks, the biodegradation rate of pure PHB was higher than both PHB nanocomposites. This result was attributed to the better barrier properties of nanocomposites due to presence of nanoclays, which could impede the diffusion of oxygen and moisture through the sample, decreasing the biodegradation rate. However, in a later study, it was verified that the biodegradation rate of PHB/MMT was higher than pure PHB during all time of test (8 weeks) at room temperature (Maiti et al. 2007).

The biodegradation rate of PHB/MMT and PHB/MAE nanocomposites at room temperature and at 60 °C was also evaluated by Maiti et al. (2007). The results revealed a decrease of 2–3 times of biodegradation rate at 60 °C for all samples. The decrease in the biodegradation rate was attributed to two factors: a reduction of concentration of microorganisms at 60 °C and/or an increase of crystallinity degree. This is because as higher the temperature of essay above Tg of matrix ($\sim$16 °C) higher is the mobility of polymer chains in the amorphous phase, and consequently higher the possibility of their organization in a crystalline dominions.

Other important result discussed in the work of Maiti et al. (2007) was the influence of MMT in the spherulite size and, consequently, in the biodegradation rate. The nanoclays enhanced nucleation, leading to smaller crystallite size. The decrease in the spherulite size implies the increase of interspherulitic area, which is composed of amorphous phase. This way, as the amorphous phase is more affected by hydrolysis reactions the biodegradation rate of PHB/MMT is higher than pure PHB.

On the other hand, Wang et al. (2005a) verified a decrease of biodegradation rate of PHBV/MMT nanocomposites in function of increasing of MMT amount, although the smaller size of PHBV spherulites in the PHBV/MMT nanocomposites. The authors suggest that the MMT act as physical barrier that protect the polymer chains and impede the diffusion of water through the sample. So, the lower biodegradability of PHBV/MMT than pure PHBV is occasioned by the lower availability of water to hydrolysis reactions.

Han et al. (2012) studied the biodegradation of P(3HB–co–4HB)/silica nanocomposites by enzymatic degradation in phosphate buffer (pH 7.4) containing lipase from *Pseudomonas mendocina*. The measurement of mass loss showed that the introduction of silica nanoparticles increased the enzymatic degradation rate in function of silica nanoparticles amount. The higher enzymatic degradation of P (3HB–co–4HB)/silica nanocomposites was attributed to the increase of hydrophilicity due to hydroxyl groups on the surface of silica nanoparticles. The higher hydrophilicity improves the interactions between polymer and enzyme and, consequently, accelerates the enzymatic degradation.

In a similar way, the presence of carbon nanotubes in PHA/CNT nanocomposites can cause changes in the biodegradation of PHA matrix. The effect of CNT in biodegradation of a similar polyester was investigated by Zeng et al. (2006). In this work, chains of poly(ε-caprolactone), PCL, were grafted onto surfaces of multiwalled carbon nanotubes (MWCNT) functionalized with hydroxyl group, by in situ ring-opening polymerization. The enzymatic degradation of MWCNT–*g*–PCL was evaluated in a phosphate buffer solution, using pseudomonas (PS) lipase as the bioactive enzyme. The degradation was monitored by the decrease of diameter of MWCNT–*g*–PCL due to consumption of PCL grafted onto MWCNT by hydrolysis reaction. The degradation rate of PCL in MWCNT–*g*–PCL was the same to pure PCL, indicating that its biodegradability was preserved.

In the development of PHA/CNT nanocomposites, the changes caused by CNT in the PHA matrix have been wide investigated. As previously mentioned in this chapter, the introduction of CNT implies in changes in the mechanical properties, thermal decomposition, and morphology especially the crystallinity of PHA matrix, which is one of the most affected properties. However, although the crystallinity has an important role in the biodegradation rate of PHA samples, the influence of CNT in the biodegradation of PHA matrix has not been investigated yet. There are other properties that can be affected by CNT in PHA/CNT nanocomposites, such as roughness, hydrophilicity, porosity, etc., which can also affect the biodegradation rate of PHA.

Preliminary works realized in our research group have shown that the biodegradability of PHBV/CNT nanocomposites with 1 % (w/w) of CNT is preserved. The addition of CNT did not prevent the biodegradation of PHA matrix in soil. SEM micrographs showed that originally the surfaces of PHBV/CNT samples are completely smooth, but after the biodegradation in soil, the surface becomes rough and irregular. The biodegradation of these samples results still in the exposition of CNT on the surface of the PHBV/CNT nanocomposites, as can be verified in Fig. 6.

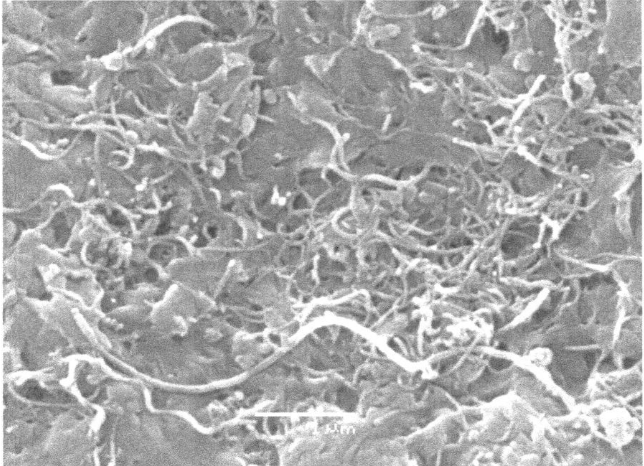

Fig. 6 SEM micrograph of PHBV/CNT 1 % (w/w) surface after 20 days of biodegradation in soil

Apparently, the biodegradation rate of PHBV/CNT nanocomposites was slower than neat PHBV, but more studies are underway (Silva et al. 2013).

The study about the influence of CNT in the biodegradation rate and mechanisms of PHA/CNT nanocomposites is essential to enable the application of this nanocomposite. The knowledge about the environment conditions, interaction with different microorganisms, soil compositions, among others, in this biodegradation process will allow the development of PHA/CNT nanocomposites with requested chemical, physical, and biodegradable profile for specific applications.

6 Conclusions and Future Perspective

The improvement in the properties of PHA matrices obtained by the addition of CNT as reinforcement filler shows that PHA/CNT nanocomposites have a great field of applications in medical, agriculture, packaging, and other sectors. The renewable source and the biodegradability of PHA make them an interesting polymer to replace conventional polymers in applications where the biodegradability is a desirable characteristic. The functionalization of carbon nanotubes has been an important tool in the development of PHA/CNT nanocomposite, since it can promote better adhesion and dispersion of CNT in PHA matrix. However, few kinds of functionalization have already been investigated, so it is a great field to be explored as well as the methods used in the PHA/CNT production. Although the study of PHA/CNT biodegradation is essential to determine their application, the researches about it are very scarce. Our preliminary studies show that PHA/CNT nanocomposites remain biodegradable and apparently only the biodegradation rate can be affect by the presence of CNT. This indicates that CNT can be used in future to reach excellent properties in PHA nanocomposites, maintaining matrix biodegradability.

References

Abe H, Doi Y (1999) Structural effects on enzymatic degradabilities for poly [(R)-3-hydroxybutyric acid] and its copolymers. Int J Biol Macromol 25:185–192

Abe H, Kikkawa Y, Iwata T, Aoki H, Akehata T, Doi Y (2000) Microscopic visualization on crystalline morphologies of thin films for poly [(R)-3-hydroxybutyric acid] and its copolymer. Polymer 41:867–874

Ajayan PM (1999) Nanotubes from carbon. Chem Rev 99:1787–1800

Akmal D, Azizan MN, Majid MIA (2003) Biodegradation of microbial polyesters P(3HB) and P (3HB–co–3HV) under the tropical climate environment. Polym Degrad Stab 80:513–518

Anandhan S, Bandyopadhyay S (2011) Polymer nanocomposites: from synthesis to applications. In: Nanocomposites and polymers with analytical methods, pp 3–28

Anderson AJ, Dawes EA (1990) Occurrence, metabolism, metabolic role, and industrial uses of bacterial polyhydroxyalkanoates. Microbiol Rev 54:450–472

Aoyagi Y, Yamashita K, Doi Y (2002) Thermal degradation of poly [(R)-3-hydroxybutyrate], poly [e-caprolactone], and poly [(S)-lactide]. Polym Degrad Stab 76:53–59

Arras MML, Schillai C, Keller TF, Schulze R, Jandt KD (2013) Alignment of multi-wall carbon nanotubes by disentanglement in ultra-thin melt-drawn polymer films. Carbon 60:366–378

Avella M, Errico ME, Rimedio R, Sadocco P (2002) Preparation of biodegradable polyesters/high-amylose-starch composites by reactive blending and their characterization. J Appl Polym Sci 83:1432–1442

Avella M, Martuscelli E, Raimo M (2000) Properties of blends and composites based on poly (3-hydroxy) butyrate (PHB) and poly(3-hydroxybutyrate-hydroxyvalerate) (PHBV) copolymers. J Mater Sci 35:523–545

Bang JH, Suslick KS (2010) Applications of ultrasound to the synthesis of nanostructured materials. Adv Mater 22:1039–1059

Bhatt R, Patel K, Trivedi U (2011) Biodegradation of poly(3-hydroxyalkanoates). In: A handbook of applied biopolymer technology: synthesis, degradation and applications, pp 311–331

Bonartsev A, Boskhomdziev A, Voinova V, Makhina T, Myshkina V, Yakovlev S, Iordanskii A (2012) Degradationof poly(3-hydroxybutyrate) and its derivatives: characterization and kinect behavior. Chem Chem Technol 6:385–392

Bordes P, Pollet E, Averous L (2009) Nano-biocomposites: biodegradable polyester/nanoclay systems. Prog Polym Sci 34:125–155

Boyandin AN, Rudnev VP, Ivonin VN, Prudnikova SV, Korobikhina KI, Filipenko ML, Volova TG, Sinskey AJ (2012) Biodegradation of polyhydroxyalkanoate films in natural environments. Macromol Symp 320:38–42

Braunegg G, Lefebvre G, Genser KF (1998) Polyhydroxyalkanoates, biopolyesters from renewable resources: Physiological and engineering aspects. J Biotechnol 65:127–161

Byrne MT, McNamee WP, Gun'ko YK (2008) Chemical functionalization of carbon nanotubes for the mechanical reinforcement of polystyrene composites. Nanotechnology 19:1–9

Chan KHK, Wong SY, Tiju WC, Li X, Kotaki M, He CB (2010) Morphologies and electrical properties of electrospun poly [(R)-3-hydroxybutyrate-*co*-(R)-3-hydroxyvalerate]/multiwalled carbon nanotubes fibers. J Appl Polym Sci 116:1030–1035

Charles PP, Owens FJ (2003) Introduction to nanotechnology. Wiley

Chen Y, Yang G, Chen Q (2002) Solid-state NMR study on the structure and mobility of the noncrystalline region of poly(3-hydroxybutyrate) and poly(3-hydroxybutyrate–co–3-hydroxyvalerate). Polymer 43:2095–2099

Choi EY, Roh SC, Kim CK (2014) Noncovalent functionalization of multi-walled carbon nanotubes with pyrene-linked nylon66 for high performance nylon66/multi-walled carbon nanotube composites. Carbon 72:160–168

Choi JS, Park WH (2004) Effect of biodegradable plasticizers on thermal and mechanical properties of poly(3-hydroxybutyrate). Polym Test 23:455–460

Choi WM, Kim TW, Park OO, Chang YK, Lee JW (2003) Preparation and characterization of Poly(hydroxybutyrate-co-hydroxyvalerate)–organoclay nanocomposites. J Appl Polym Sci 90:525–529

Coleman JN, KhanU Blau WJ, Gun'ko YK (2006a) Small but strong: a review of the mechanical properties of carbon nanotube–polymer composites. Carbon 44:1624–1652

Coleman JN, Khan U, Gun'ko YK (2006b) Mechanical reinforcement of polymers using carbon nanotubes. Adv Mater 18:689–706

Damian C, Andreea M, Iovu H (2010) Ethylenediamine functionalization effect on the thermo-mechanical properties of epoxy nanocomposites reinforced with multiwall carbon nanotubes. U.P.B. Sci Bull 72:163–174

Deroiné M, Le Duigou A, Corre YM, Le Gac PY, Davies P, César G, Bruzaud S (2014) Seawater accelerated ageing of poly(3-hydroxybutyrate–co–3-hydroxyvalerate). Polym Degrad Stab 105:237–247

Dresselhaus MS, Dresselhaus G, Saito R, Jorio A (2005) Raman spectroscopy of carbon nanotubes. Phys Rep 409:47–99

Dufresne A, Dupeyre D, Paillet M (2003) Lignocellulosic flour-reinforced poly(hydroxybutyrate–co–valerate) composites. J Appl Polym Sci 87:1302–1315

Durmus A, Kasgoz A, Macosko CW (2007) Linear low density polyethylene (LLDPE)/clay nanocomposites. Part I: structural characterization and quantifying clay dispersion by melt rheology. Polymer 48:4492–4502

Eggers J, Steinbüchel A (2013) Poly(3-hydroxybutyrate) degradation in Ralstonia eutropha H16 is mediated stereoselectively to (S)-3-hydroxybutyryl coenzyme A (CoA) via crotonyl-CoA. J Bacteriol 195:3213–3223

El-Hadi A, Schnabel R, Straube E, Müller G, Henning S (2002) Correlation between degree of crystallinity, morphology, glass temperature, mechanical properties and biodegradation of poly (3-hydroxyalkanoate) PHAs and their blends. Polym Test 21:665–674

Esawi AMK, Salem HG, Hussein HM, Ramadan AR (2010) Effect of processing technique on the dispersion of carbon nanotubes within polypropylene carbon nanotube-composites and its effect on their mechanical properties. Polym Compos 31:772–780

Fei B, Chen C, Chen S, Peng S, Zhuang Y, An Y (2004) Crosslinking of poly [(3-hydroxybutyrate)–co–(3-hydroxyvalerate)] using dicumyl peroxide as initiator. Polym Int 53:937–943

Feng L, Wang Y, Inagawa Y, Kasuya K, Saito T, Doi Y, Inoue Y (2004) Enzymatic degradation behavior of comonomer compositionally fractionated bacterial poly(3-hydroxybutyrate–co–3-hydroxyvalerate)s by poly(3-hydroxyalkanoate) depolymerases isolated from Ralstonia pickettii T1 and Acidovorax sp. TP4. Polym Degrad Stab 84:95–104

Ferreira BMP, Zavaglia CAC, Duek EAR (2002) Films of PLLA/PHBV: thermal, morphological, and mechanical characterization. J Appl Polym Sci 86:2898–2906

Filho A, Fagan S (2007) Funcionalização de nanotubos de carbono. Quim Nova 30:1695–1703

Fradinho JC, Domingos JMB, Carvalho G, Oehmen A, Reis MAM (2013) Polyhydroxyalkanoates production by a mixed photosynthetic consortium of bacteria and algae. Bioresour Technol 132:146–153

Gaharwar AK, Schexnailder PJ, Schmidt G (2011) Nanocomposite polymer biomaterials for tissue repair of bone and cartilage: amaterial science perspective. In: Nanobiomaterials handbook. CRC Press, pp 1–28

Gonçalves SPC, Franchetti SMM (2013) Respirometric evaluation of the biodegradability of films of PE/PHBV blends. Int J Mater Sci 3:54–60

Gonzalez A, Irusta L, Fernández-Berridi MJ, Iriarte M, Iruin JJ (2005) Application of pyrolysis/gas chromatography/Fourier transform infrared spectroscopy and TGA techniques in the study of thermal degradation of poly(3-hydroxybutyrate). Polym Degrad Stab 87:347–354

Gooding JJ (2005) Nanostructuring electrodes with carbon nanotubes: a review on electrochemistry and applications for sensing. Electrochim Acta 50:3049–3060

Grady BP (2011) Carbon nanotube—polymer composites. Wiley, New Jersey

Gursel I, Balcik C, Arica Y, Akkus O, Akkas N, Hasirci V (1998) Synthesis and mechanical properties of interpenetrating networks of polyhydroxybutyrate–co–hydroxyvalerate and polyhydroxyethyl methacrylate. Biomaterials 19:1137–1143

Han CC, Ismail J, Kammer HW (2004) Melt reaction in blends of poly(3-hydroxybutyrate–co–3-hydroxyvalerate) and epoxidized natural rubber. Polym Degrad Stab 85:947–955

Han L, Han C, Cao W, Wang X, Bian J, Dong L (2012) Preparation and characterization of biodegradable poly(3-hydroxybutyrate–co–4-hydroxybutyrate)/silica nanocomposites. Polym Eng Sci 52:250–258

Han Z, Fina A (2011) Thermal conductivity of carbon nanotubes and their polymer nanocomposites: a review. Prog Polym Sci 36:914–944

Her SC, Lai CY (2013) Dynamic behavior of nanocomposites reinforced with multi-walled carbon nanotubes (MWCNTs). Materials 6:2274–2284

Hirsch A (2002) Functionalization of single-walled carbon nanotubes. Angew Chem 41:1853–1859

Hu SG, Jou CH, Yang MC (2004) Biocompatibility and antibacterial activity of chitosan and collagen immobilized poly(3-hydroxybutyric acid–co–3-hydroxyvaleric acid). Carbohydr Polym 58:173–179

Hussain F (2006) Review article: polymer-matrix nanocomposites, processing, manufacturing and application: an Overview. J Compos Mater 40:1511–1575

Ishigaki T, Sugano W, Nakanishi A, Tateda M, Ike M, Fujita M (2004) The degradability of biodegradable plastics in aerobic and anaerobic waste landfill model reactors. Chemosphere 54:225–233

Iwata T, Doi Y, Tanaka T, Akehata T, Shiromo M, Teramachi S (1997) Enzymatic degradation and adsorption on poly [(R) -3-hydroxybutyrate] single crystals with two types of extracellular PHB depolymerases from comamonas acidovorans YM1609 and alcaligenes faecalis T1. Macromolecules 30:5290–5296

Jendrossek D, Handrick R (2002) Microbial degradation of polyhydroxyalkanoates. Annu Rev Microbiol 56:403–432

Jiang L, Huang J, Qian J, Chen F, Zhang J, Wolcott MP, Zhu Y (2008a) Study of poly(3-hydroxybutyrate–co–3-hydroxyvalerate) (PHBV)/bamboo pulp fiber composites: effects of nucleation agent and compatibilizer. J Polym Environ 16:83–93

Jiang L, Morelius E, Zhang J, Wolcott M, Holbery J (2008b) Study of the poly(3-hydroxybutyrate–co–3-hydroxyvalerate)/cellulose nanowhisker composites prepared by solution casting and melt processing. J Compos Mater 42:2629–2645

Ke Y, Wu G, Wang Y (2014) PHBV/PAM scaffolds with local oriented structure through UV polymerization for tissue engineering. BioMed Res Int 2014:1–9

Keshavarz T, Roy I (2010) Polyhydroxyalkanoates: bioplastics with a green agenda. Curr Opin Microbiol 13:321–326

Khanna S, Srivastava AK (2005) Statistical media optimization studies for growth and PHB production by Ralstonia eutropha. Process Biochem 40:2173–2182

Kim P, Shi L, Majumdar E, McEuen PL (2001) Thermal transport measurements of individual multiwalled nanotubes. Phys Rev Lett 87:1–4

Koller M, Gasser I, Schmid F, Berg G (2011) Linking ecology with economy: insights into polyhydroxyalkanoate-producing microorganisms. Eng Life Sci 11:222–237

Korn M, Andrade MVAS, Borges SS (2003) Procedimentos analíticos assistidos por ultra-som. Analytica 3:34–39

Lai M, Li J, Yang J, Liu J, Tong X, Cheng H (2004) The morphology and thermal properties of multi-walled carbon nanotube and poly(hydroxybutyrate–co–hydroxyvalerate) composite. Polym Int 53:1479–1484

Lee JH, Kim SK, Kim NH (2006) Effects of the addition of multi-walled carbon nanotubes on the positive temperature coefficient characteristics of carbon-black-filled high-density polyethylene nanocomposites. Scripta Mater 55:1119–1122

Lee LJ, Zeng C, Cao X, Han X, Shen J, Xu G (2005) Polymer nanocomposite foams. Compos Sci Technol 65:2344–2363

Lemes AP, Marcato PD, Ferreira OP, Alves OL, Duran N (2008) Nanotechnology and applications. In: Nanocomposites of poly(3-Hydroxybutyrate–co–3-Hydroxyvalerate) reinforced with carbon nanotubes and oxidized carbon nanotubes, pp 615–085

Lenz RW, Marchessault RH (2005) Bacterial polyesters: biosynthesis, biodegradable plastics and biotechnology. Biomacromolecules 6:1–8

Likodimos V, Steriotis TA, Papageorgiou SK, Romanos GE, Marques RRN, Rocha RP, Faria JL, Pereira MFR, Figueiredo JL, Silva AMT, Falaras P (2014) Controlled surface functionalization of multiwall carbon nanotubes by HNO3 hydrothermal oxidation. Carbon 69:311–326

Liu CX, Choi JW (2012) Improved dispersion of carbon nanotubes in polymers at high concentrations. Nanomaterials 2:329–347

Liu WJ, Yang L, Wang Z, Dong LS, Liu JJ (2002) Effect of nucleating agents on the crystallization of poly(3-hydroxybutyrate–co–3-hydroxyvalerate). J Appl Polym Sci 86:2145–2152

Lo WH, Yu J (2002) Effects of the energy dissipation rate and surface erosion on the biodegradation of poly(hydroxybutyrate–co–hydroxyvalerate) and its blends with synthetic polymers in an aquatic medium. J Appl Polym Sci 83:1036–1045

Lotto NT, Calil MR, Guedes CGF, Rosa DS (2004) The effect of temperature on the biodegradation test. Materials Science and Engineering C 24:659-662

Luo JT, Wen H, Wu WF, Chou CP (2008) Mechanical research of carbon nanotubes/PMMA composite films. Polym Compos 29:1285–1290

Luo M, Li Y, Jin S, Sang S, Zhao L, Wang Q, Li Y (2013) Microstructure and mechanical properties of multi-walled carbon nanotubes containing Al2O3–C refractories with addition of polycarbosilane. Ceram Int 39:4831–4838

Luo S, Netravali A (2003) A study of physical and mechanical properties of poly(hydroxybutyrate–co–hydroxyvalerate) during composting. Polym Degrad Stab 80:59–66

Ma PC, Siddiqui NA, Marom G, Kim JK (2010) Dispersion and functionalization of carbon nanotubes for polymer-based nanocomposites: a review. Compos A Appl Sci Manuf 41:1345–1367

Ma Y, Zheng Y, Wei G, Song W, Hu T, Yang H, Xue R (2012) Processing, structure, and properties of multiwalled carbon nanotube/poly(hydroxybutyrate-co-valerate) biopolymer nanocomposites. J Appl Polym Sci 125:E620–E629

Madbouly SA, Schrader JA, Srinivasan G, Liu K, McCabe KG, Grewell D, Graves WL, Kessler MR (2014) Biodegradation behavior of bacterial-based polyhydroxyalkanoate (PHA) and DDGS composites. Green Chem 16:1911–1920

Maiti P, Batt CA, Giannelis EP (2007) New biodegradable polyhydroxybutyrate/layered silicate nanocomposites. Biomacromolecules 8:3393–3400

Maiti P, Batt CA, Giannelis EP (2003) Biodegradable polyester/layered silicate nanocomposites. Mater Res Soc Symp Proc 740(I5):3

Moniruzzaman M, Winey KI (2006) Polymer nanocomposites containing carbon nanotubes. Macromolecules 39:5194–5205

Mukai K, Yamada K, Doi Y (1993) Kinetics and mechanism of heterogeneous hydrolysis of poly [(R)-3-hydroxybutyrate] film by PHA depolymerases. Int J Biol Macromol 15:361–366

Mylvaganam K, Zhang LC (2007) Fabrication and application of polymer composites comprising carbon nanotubes. Recent Pat Nanotechnol 1:59–65

Nobes GAR, Marchessault RH, Briese BH, Jendrossek D (1998) Microscopic visualization of the enzymatic degradation of poly(3HB–co–3HV) and poly(3HV) single crystals by PHA depolymerases from Pseudomonas lemoignei. J Environ Polym Degrad 6:99–107

Nobes GAR, Marchessault RH, Chanzy H, Briese BH, Jendrossek D (1996) Splintering of poly (3-hydroxybutyrate) single crystals by PHB-depolymerase a from Pseudomonas lemoignei. Macromolecules 29:8330–8333

Numata K, Abe H, Iwata T (2009) Biodegradability of poly(hydroxyalkanoate) materials. Materials 2:1104–1126

Numata K, Kikkawa Y, Tsuge T, Iwata T, Doi Y, Abe H (2006) Adsorption of biopolyester depolymerase on silicon wafer and poly [(R)-3-hydroxybutyric acid] single crystal revealed by real-time AFM. Macromol Biosci 6:41–50

Philip S, Keshavarz T, Roy I (2007) Polyhydroxyalkanoates : biodegradable polymers with a range of applications. J Chem Technol Biotechnol 82:233–247

Reddy MM, Vivekanandhan S, Misra M, Bhatia SK, Mohanty AK (2013) Biobased plastics and bionanocomposites: current status and future opportunities. Prog Polym Sci 38:1653–1689

Reinsch VE, Kelley SS (1996) Crystallization of poly(hydroxybutyrate–co–hydroxyvalerate) in wood fiber-reinforced composites. J Appl Polym Sci 64:1785–1796

Roy N, Sengupta R, Bhowmick AK (2012) Modifications of carbon for polymer composites and nanocomposites. Prog Polym Sci 37:781–819

Ruiz I, Hermida ÉB, Baldessari A (2011) Fabrication and characterization of porous PHBV scaffolds for tissue engineering. J Phys: Conf Ser 332:1–10

Sahoo NG, Jung YC, So HH, Cho JW (2007) Synthesis of polyurethane nanocomposites of functionalized carbon nanotubes by in-situ polymerization Methods. J Korean Physycal Soc 51:S1–S6

Sahoo NG, Rana S, Cho JW, Li L, Chan SH (2010) Polymer nanocomposites based on functionalized carbon nanotubes. Prog Polym Sci 35:837–867

Sanchez-Garcia MD, Lagaron JM, Hoa SV (2010) Effect of addition of carbon nanofibers and carbon nanotubes on properties of thermoplastic biopolymers. Compos Sci Technol 70:1095–1105

Senior PJ, Dawes EA (1973) The regulation of poly-beta-hydroxybutyrate metabolism in Azotobacter beijerinckii. Biochem J 134:225–238

Shan GF, Gong X, Chen WP, Chen L, Zhu MF (2011) Effect of multi-walled carbon nanotubes on crystallization behavior of poly(3-hydroxybutyrate–co–3-hydroxyvalerate). Colloid Polym Sci 289:1005–1014

Shang L, Fei Q, Zhang YH, Wang XZ, Fan DD, Chang HN (2011) Thermal properties and biodegradability studies of poly(3-hydroxybutyrate–co–3-hydroxyvalerate). J Polym Environ 20:23–28

Silva AP, Montanheiro TLA, Duran N, Lemes AP (2013) Characterization of PHBV/carbon nanotubes nanocomposites after biodegradation in soil. In: XII Brazilian MRS Meeting 6JS2

Spitalsky Z, Tasis D, Papagelis K, Galiotis C (2010) Carbon nanotube–polymer composites: chemistry, processing, mechanical and electrical properties. Prog Polym Sci 35:357–401

Spyros A, Kimmich R, Briese BH, Jendrossek D (1997) ^{1}H NMR imaging study of enzymatic degradation in poly(3-hydroxybutyrate) and poly(3-hydroxybutyrate–co–3-hydroxyvalerate). Evidence for preferential degradation of the amorphous phase by PHB depolymerase B from pseudomonas lemoignei. Macromolecules 30:8218–8225

Srithep Y, Ellingham T, Peng J, Sabo R, Clemons C, Turng L, Pilla S (2013) Melt compounding of poly(3-hydroxybutyrate–co–3-hydroxyvalerate)/nanofibrillated cellulose nanocomposites. Polym Degrad Stab 98:1439–1449

Srubar WV, Pilla S, Wright ZC, Ryan CA, Greene JP, Frank CW, Billington SL (2012) Mechanisms and impact of fiber–matrix compatibilization techniques on the material characterization of PHBV/oak wood flour engineered biobased composites. Compos Sci Technol 72:708–715

Stobinski L, Lesiak B, Kövér L, Tóth J, Biniak S, Trykowski G, Judek J (2010) Multiwall carbon nanotubes purification and oxidation by nitric acid studied by the FTIR and electron spectroscopy methods. J Alloy Compd 501:77–84

Sudesh K, Abe H, Doi Y (2000) Synthesis, structure and properties of polyhydroxyalkanoates: biological polyesters. Prog Polym Sci 25:1503–1555

Tang W, Santare MH, Advani SG (2003) Melt processing and mechanical property characterization of multi-walled carbon nanotube/high density polyethylene (MWNT/HDPE) composite films. Carbon 41:2779–2785

Tasis D, Tagmatarchis N, Bianco A, Prato M (2006) Chemistry of carbon nanotubes. Chem Rev 106:1105–1136

Ten E, Jiang L, Wolcott MP (2013) Preparation and properties of aligned poly(3-hydroxybutyrate–co–3-hydroxyvalerate)/cellulose nanowhiskers composites. Carbohydr Polym 92:206–213

Ten E, Turtle J, Bahr D, Jiang L, Wolcott M (2010) Thermal and mechanical properties of poly(3-hydroxybutyrate–co–3-hydroxyvalerate)/cellulose nanowhiskers composites. Polymer 51:2652–2660

Thakur VK and Thakur MK (2014) Processing and characterization of natural cellulose fibers/thermoset polymer composites. Carbohydr Polym 109:102–117

Thakur VK, Thakur MK, Raghavan P, Kessler MR (2014) Progress in green polymer composites from lignin for multifunctional applications: a review. ACS Sustain Chem Eng 2:1072–1092

Thiré RMDSM, Arruda LC, Barreto LS (2011) Morphology and thermal properties of poly(3-hydroxybutyrate–co–3-hydroxyvalerate)/attapulgite nanocomposites. Mater Res 14:340–344

Uchino T, Bourdakos KN, Groot CH, Ashburn P, Kiziroglou ME, Dilliway GD, Smith DC (2005) Metal catalyst-free low-temperature carbon nanotube growth on SiGe islands. Appl Phys Lett 86:233110

Valentini L, Fabbri P, Messori M, Esposti MD, Bon SB (2014) Multilayer films composed of conductive poly(3-hydroxybutyrate)/carbon nanotubes bionanocomposites and a photoresponsive conducting polymer. J Polym Sci, Part B: Polym Phys 52:596–602

Vidhate S, Innocentini-mei L, Souza NAD (2012) Mechanical and electrical multifunctional poly (3-hydroxybutyrate–co–3-hydroxyvalerate)—multiwall carbon nanotube nanocomposites. Polym Eng Sci 52:1367–1374

Volova TG, Boyandin AN, Vasiliev AD, Karpov VA, Prudnikova SV, Mishukova OV, Boyarskikh UA, Filipenko ML, Rudnev VP, Xuânf BB, Dungf VV, Gitelson II (2010) Biodegradation of polyhydroxyalkanoates (PHAs) in tropical coastal waters and identification of PHA-degrading bacteria. Polym Degrad Stab 95:2350–2359

Wang S, Song C, Chen G, Guo T, Liu J, Zhang B, Takeuchi S (2005a) Characteristics and biodegradation properties of poly(3-hydroxybutyrate–co–3-hydroxyvalerate)/organophilic montmorillonite (PHBV/OMMT) nanocomposite. Polym Degrad Stab 87:69–76

Wang S, Song C, Mizuno W, Sano M, Maki M, Yang C, Zhang B, Takeuchi S (2005b) Estimation on biodegradability of poly(3-hydroxybutyrate–co–3-hydroxyvalerate) (PHB/V) and numbers of aerobic PHB/V degrading microorganisms in different natural environments. J Polym Environ 13:39–45

Xu C, Qiu Z (2009) Nonisothermal melt crystallization and subsequent melting behavior of biodegradable poly(hydroxybutyrate)/ multiwalled carbon nanotubes nanocomposites. J Polym Sci, Part B: Polym Phys 47:2238–2246

Yamashita K, Funato T, Suzuki Y, Teramachi S, Doi Y (2003) Characteristic interactions between poly(hydroxybutyrate) depolymerase and poly [(R)-3-hydroxybutyrate] film studied by a quartz crystal microbalance. Macromol Biosci 3:694–702

Yang L, Setyowati K, Li A, Gong S, Chen J (2008) Reversible infrared actuation of carbon nanotube–liquid crystalline elastomer nanocomposites. Adv Mater 20:2271–2275

Yu C, Shi L, Yao Z, Li D, Majumdar A (2005) Thermal conductance and thermopower of an individual single-wall carbon nanotube. Nano Lett 5:1842–1846

Yu H, Qin Z, Zhou Z (2011) Cellulose nanocrystals as green fillers to improve crystallization and hydrophilic property of poly(3-hydroxybutyrate–co–3-hydroxyvalerate). Prog Nat Sci: Mater Int 21:478–484

Yu HY, Qin ZY, Sun B, Yang XG, Yao JM (2014) Reinforcement of transparent poly(3-hydroxybutyrate–co–3-hydroxyvalerate) by incorporation of functionalized carbon nanotubes as a novel bionanocomposite for food packaging. Compos Sci Technol 94:96–104

Yu HY, Yao JM, Qin ZY, Liu L, Yang XG (2013) Comparison of covalent and noncovalent interactions of carbon nanotubes on the crystallization behavior and thermal properties of poly (3-hydroxybutyrate–co–3-hydroxyvalerate). J Appl Polym Sci 130:4299–4307

Yu W, Lan CH, Wang SJ, Fang PF, Sun YM (2010) Influence of zinc oxide nanoparticles on the crystallization behavior of electrospun poly(3-hydroxybutyrate–co–3-hydroxyvalerate) nanofibers. Polymer 51:2403–2409

Yun SI, Lo V, Noorman J, Davis J, Russel RA, Holden PJ, Gadd GE (2010) Morphology of composite particles of single wall carbon nanotubes/biodegradable polyhydroxyalkanoates prepared by spray drying. Polym Bull 64:99–106

Zeng H, Gao C, Yan D (2006) Poly(ε-caprolactone)-functionalized carbon nanotubes and their biodegradation properties. Adv Funct Mater 16:812–818

Zhang K, Park BJ, Fang FF, Choi HJ (2009) Sonochemical preparation of polymer nanocomposites. Molecules 14:2095–2110

Zinn M, Witholt B, Egli T (2001) Occurrence, synthesis and medical application of bacterial polyhydroxyalkanoate. Adv Drug Deliv Rev 53:5–21

Zribi-Maaloul E, Trabelsi I, Elleuch L, Chouayekh H, Ben Salah R (2013) Purification and characterization of two polyhydroxyalcanoates from Bacillus cereus. Int J Biol Macromol 61:82–88

Biodegradable Polymer/Clay Nanocomposites

Leandro Ludueña, Juan Morán and Vera Alvarez

Abstract This chapter is a comprehensive source for biodegradable polymer/clay nanocomposite research. The work focuses on several strategies to obtain well-dispersed nanocomposites with improved mechanical properties. Different strategies for optimization of biodegradable polymer-clay nanocomposites were identified and treated herein. The two most important strategies are based on the chemical modification of the clay and on the processing conditions. The combination of these two factors greatly affects the resulting structure of the nano composite, being always the goal to obtain a good dispersion of the reinforcement within the polymeric matrix. Challenges in processing conditions have been analyzed and the prospective for future research has been addressed.

Keywords Nanocomposites · Clay · Performance · Polymeric matrix · Chemical modification · Processing conditions

1 Introduction

Because of waste accumulation at the end of the life cycle of traditional polymer products, the development of environmental friendly, degradable, polymeric materials has attracted extensive interest (Bordes et al. 2009; Singha and Thakur 2010; Thakur et al. 2010; Thakur et al. 2014a, b, c; Thakur and Thakur 2014a, b, c). Nevertheless, the properties of such kind of polymers are lower than that of traditional ones.

It is known that performance of polymeric matrices can be greatly enhanced by dispersion of nanometer-size particles. Such materials are called nanocomposites and have the interesting characteristic that the mechanical properties (Ludueña et al.

L. Ludueña (✉) · J. Morán · V. Alvarez
Composite Materials Group (CoMP), INTEMA, National University of Mar del Plata, Solís 7575 Mar del Plata, B7608FDQ Buenos Aires, Argentina
e-mail: alvarezvera@fi.mdp.edu.ar

V.K. Thakur and M.K. Thakur (eds.), *Eco-friendly Polymer Nanocomposites*,
Advanced Structured Materials 75, DOI 10.1007/978-81-322-2470-9_4

2007; Viville et al. 2003); the barrier properties (Becker et al. 2004; Gorrasi et al. 2003; Messersmith and Giannelis 1995); the thermal properties (Kojima et al. 1993; Leszczyńska et al. 2007), and some others such as flammability (Chiang and Ma 2004; Gilman et al. 2000) and water adsorption (Gorrasi et al. 2003; Pérez et al. 2008) can be greatly enhanced with addition of a small amount of filler (usually less than 10 wt%).

One kind of nanometer-size reinforcements are layered silicates or clays. Among these reinforcements, the use of montmorillonites and bentonites is interesting because added to their environmental and economic importance (Zampori et al. 2008), their natural abundance and their mechanical and chemical resistance makes them very useful as polymeric material reinforcements. This type of clay is characterized by a moderate negative charge (known as cation exchange capacity, CEC, and expressed in meq/100 gr) (Mandalia and Bergaya 2006; He et al. 2006).

A key point is the dispersion of the layered silicate particles in the matrix. In order to achieve better properties it is necessary to obtain a totally exfoliated structure (this means that the silicate layers are completely and uniformly dispersed in a continuous polymeric matrix) (Roelofs and Berben 2006) but the tendency of the particles to agglomerate has been difficult to overcome. There are different ways of increasing or optimizing the dispersion of the clay inside a biodegradable matrix, but the most popular ones are related to the chemical modification of one of the components and optimizing the processing technique and parameters.

Another kind of nanometer-size reinforcements studied during the past decades are nanocellulose fibers. The attention has been initially driven to the use of agro-based fibers (natural fibers) as an alternative to synthetic inorganic fibers and has lately been directed toward the use of pure cellulose and nanocellulose fibers as reinforcement for both synthetic and bio-polymers. Based on a number of desirable characteristics of nanocellulose, they are currently the focus of much scientific research, as evidenced by the increasing number of scientific articles and reviews published in the past years (Siqueira et al. 2010; Hubbe et al. 2008; Eichhorn et al. 2010; Habibi et al. 2010; Siró and Plackett 2010; Klemm et al. 2011). The most interesting properties of nanocellulose include renewability, abundance, and low cost of the raw material, large surface-to-volume ratio, high strength and stiffness, very low coefficient of thermal expansion, low weight, low density, high aspect ratio, and biodegradability (Charreau et al. 2012). Applications of nanocellulose particles are vast including, reinforcement of composite materials, moistening masks for cosmetic applications, filtration media, thickening agents, rheology modifiers, adsorbents, paper reinforcement, synthesis of polymers with liquid crystalline behavior for electronic applications, delivery of drugs, flavor carriers, suspension stabilizer, optically transparent films, among others (Noorani et al. 2007; Jalal Uddin et al. 2011). Moreover, in the past years an outstanding number of uses of nanocellulose in biomedical applications have been reported, including not only tissue engineering scaffolds but also wound coverage, regeneration of damaged organs, blood vessel replacement, and more (Charreau et al. 2012; Svensson et al. 2005; Czaja et al. 2006).

The aim of this chapter is to review the strategies used to improve the dispersion of layered silicate particles inside biodegradable matrix in order to produce final products with optimized properties.

2 Biodegradable Polymers

Biodegradable polymers or biopolymers are a specific type of polymer in which at least one of the steps in the degradation process is through metabolism of naturally occurring organisms (Rhim et al. 2013). This process results in the production of gases (CO_2, N_2), water, biomass, and inorganic salts (Bastioli 2011; Avérous and Pollet 2012). These polymers are naturally occurring and also synthetically made, and generally consist of ester, amide, and ether functional groups. Their properties and breakdown mechanism are determined by their exact structure. However, in order to attain complete biodegradability, proper handling of residual plastics needs to take place; that is, after they have been used, the plastics should be disposed under certain conditions that enable their biological decomposition (i.e., composting).

In general, biodegradable polymers can be grouped into two large groups based on their structure and synthesis (Fig. 1). One of these groups is natural or agro-polymers, which are derived from biomass (Bastioli 2011; Avérous and Pollet 2012). The other consists of biopolyesters, usually derived from microorganisms or synthetically made from naturally or synthetic monomers. Natural biopolymers include polysaccharides, like starch found in potatoes or wood, and proteins, such as animal-based whey or planet derived gluten, gelatine, collagen, and casein (Bastioli 2011; Avérous and Pollet 2012). Polysaccharides consist of glycosidic bonds, which take a hemiacetal of a saccharide and bind it to alcohol via loss of water. Proteins are made from amino acids, which contain various functional groups, which form peptidic bonds through condensation reactions. On the other hand, biodegradable polymers can be manufactured from natural or synthetic monomers or synthesized by microorganisms. Examples of biopolyesters include polyhydroxybutyrate and polylactic acid (Bastioli 2011; Avérous and Pollet 2012). A recent breakthrough in this class of bioplastics is the development of technology to synthesize polymers like polyethylene, polypropylene, and nylon from biological resources.

Among the most important biodegradable polymers, we can specially mention thermoplastic starch (TPS), polylactide (PLA), polycaprolactone (PCL), and polyhydroxybutyrate and polyhydroxyalkanoate (PHA) due to their promising properties.

Thermoplastic starch (TPS) is an attractive source for the development of biodegradable plastics. It is one of the lowest cost biodegradable materials currently available in the global market (Mathew and Dufresne 2002; Shen et al. 2009). It can be found in the form of discrete semicrystalline particles, whose size, shape, morphology, and composition depend on the botanical origin (corn, potato, maize,

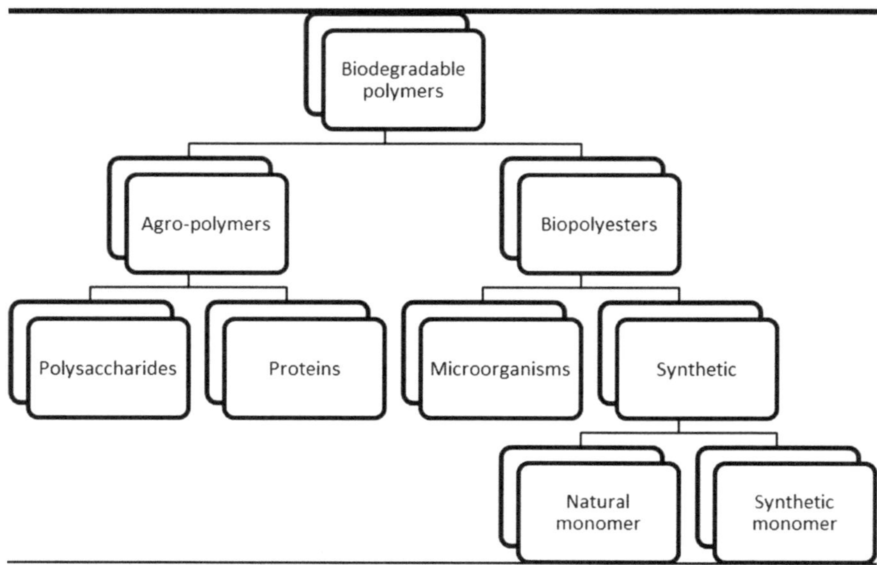

Fig. 1 Biodegradable polymers organization based on structure and occurrence (adapted from Averous and Pollet 2012)

cassava, etc.). After gelatinization (starch granules disruption), TPS shows similar properties and processing conditions to those of polyolefins. Starch is composed mainly of two glucopyranose homopolymers: amylose and amylopectin. While it is possible to produce plastics from native starch, they are not suitable for use as packaging material due to their poor mechanical properties and high moisture susceptibility (Slavutsky et al. 2014).

The simplest of the family of polyhydroxyalkanoate (PHA) biopolymers is poly-R-3-hydroxybutyrate or PHB. This polymer was first discovered in 1925 by Lemoigne and was initially described as a lipid inclusion in the bacterium Bacillus megaterium. The technical challenges for PHA are its narrow processing window and high brittleness. To overcome those problems, it is usual to make the copolymer with valerate resulting in PHBV that exhibit reduced crystallinity.

Polylactide or poly (lactic acid), otherwise known as PLA, is biodegradable thermoplastic polyester that is manufactured by biotechnological processes from renewable resources (e.g., corn). Although other sources of biomass can be used, corn has the advantage of providing the required high-purity lactic acid. The use of alternative starting materials (e.g., woody biomass) is being pursued in order to reduce process costs; however, the number of steps involved in deriving pure lactic acid from such raw materials means that their use remains much less cost-effective at present.

Poly(e-caprolactone) or PCL is an oil-derived biodegradable and semicrystalline polyester. PCL has good water, oil, solvent, and chlorine resistance, a low melting point and low viscosity, and is easily processed using conventional melt blending technologies (Nair and Laurencin 2007; Gross and Kalra 2002). PCL has low tensile strength (approximately 23 MPa) but an extremely high elongation at break [>700 % (Ludueña et al. 2007)].

As described previously, the inherent difficult processing operations and relatively poor mechanical and barrier properties have limited the industrial use of biodegradable polymers. One of the possible routes for overcoming biopolymers' inherent shortcomings is through the application of nanocomposites technology.

3 Layered Silicates

The general structure, with a small number of exceptions, consists of sheets arranged in structural layers. Each individual layer is formed by 2, 3, or 4 sheets which are formed either by tetrahedral, T, $[SiO_4]_4$ or by octahedral, O, $[AlO_3(OH)_3]_6$. All the clay minerals are composed either of TO or of TOT layers. Especially, the smectites (between one of the most famous is the montmorillonite) are the main choice in the preparation of polymer-based nanocomposites. Their crystal lattice consists of two-dimensional layers, where a central O sheet of alumina or magnesia is fused to two external silica T sheets so that the oxygen ions of the octahedral sheet do also belong to the T sheets (Fig. 2). The thickness of a layer is around 1 nm, whereas the lateral dimensions are in the range from 300 Å to several microns or even larger depending on the particular silicate, the source of the clay and the method of preparation. The layers are organized in the form of stacks with a regular van der Waals gap between them denominated interlayer or gallery. As the forces that hold the stacks together are relatively weak, the intercalation of small molecules between the layers is easy. So, added to their low cost, the rich intercalation chemistry of these clays allows them to be chemically modified and to make them compatible with polymeric matrices.

4 Dispersion of the Clay Inside the Matrix

Two types of polymer/clay nanocomposite structures (Fig. 3) can be obtained, namely intercalated, where polymer chains intercalate between the layers resulting in finite expansion and exfoliated nanocomposites where silicate layers are completely delaminated in the polymer matrix (Pavlidou and Papaspyrides 2008; Sinha Ray and Okamoto 2003).

As we have previously explained, there are two main options to improve the dispersion and distribution of the clay inside a polymeric matrix: the chemical modification of one component and the variation of the processing technique and parameters (Fig. 4).

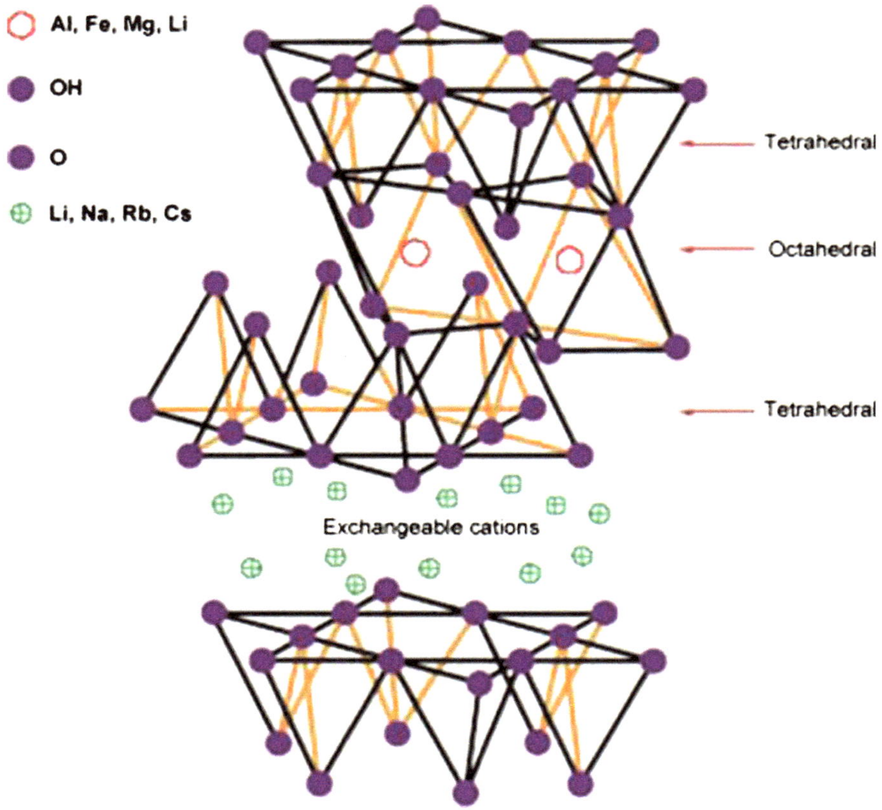

Al, Fe, Mg, Li
OH
O
Li, Na, Rb, Cs
Tetrahedral
Octahedral
Tetrahedral
Exchangeable cations

Fig. 2 Structure of a 2:1 layered silicate

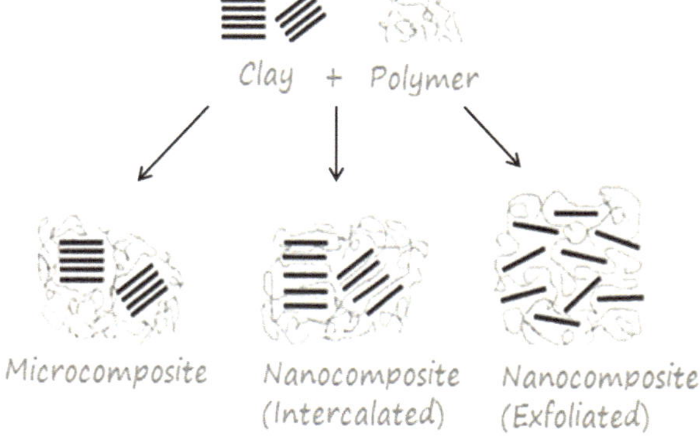

Fig. 3 Schematic illustration the morphologies of polymer/clay composites

Clay + Polymer
Microcomposite
Nanocomposite (Intercalated)
Nanocomposite (Exfoliated)

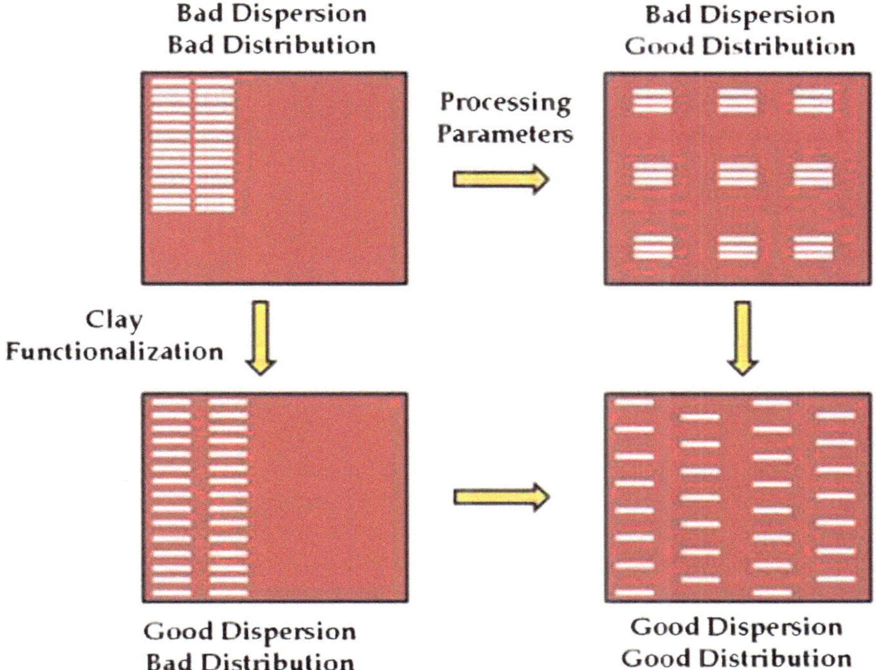

Fig. 4 Importance of clay functionalization and processing on the dispersion and distribution of clay inside a polymeric matrix

5 Chemical Treatments of Clays

Most polymers are hydrophobic in nature and, so that, the hydrophilic cations from the interlaminar spaces of clay should be exchanged by organic cations in order to make them compatible with polymeric matrices (Xi et al. 2004). Hence, the synthesis of organoclays is based on the possible reactions between the clay minerals and the organic compounds. There are several types of chemical treatments that can be grouped as follows (Fig. 5).

5.1 Purification

Commercial clay materials are unrefined clays and they typically contain different impurities (carbonates, cristobalite, feldspars, quartz, organic matter, iron hydroxides, among others (Bergaya et al. 2012). So that, the purification of clays by means of sedimentation isolates of the smectite portion is commonly required and drives to

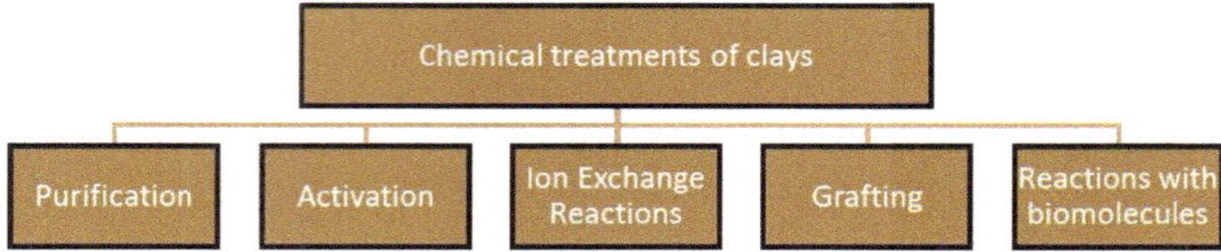

Fig. 5 Chemical treatments of clays

highly pure montmorillonite and, as a consequence, a material with improved properties (Leite et al. 2009).

5.2 Activation

The activation treatment is commonly used to improve the absorb-ability of clay. It has been studied for several years since the catalytic properties of clays are dependent on the chemical composition, structure, and pretreatment conditions of clays (Auer and Hofmann 1993; Ravichandran et al.1996). The most common treatment consists in the use of sulfuric acid (H_2SO_4), which produces an enlargement of the specific surface area but retains the ability to absorb organic molecules and layered morphology (Steudel et al. 2009). Both dissolution of the octahedral cations and edge attack also took place during this treatment.

5.3 Ion Exchange Reactions

This is a simple treatment that allows the exchange of any of the interlaminar cations by a desirable cation (Lee and Tiwari 2012). The technique consists in the exchange of interlaminar cations of the clay by quaternary cations from a salt (Fig. 6). The surfactant goes into the interlaminar spaces and could replace the initial inorganic cations. In this case, electrostatic force is the principal linkage between the modifier and the clay platelets (Lagaly 1981). This treatment produces changes from hydrophilic to hydrophobic character of clay and an important increase in interlaminar spacing (Kornmann 2001; Le Pluart et al. 2002; LEE et al. 2005; He et al. 2005; Edwards et al. 2005). In addition the incorporated cations could provide functional groups that eventually can react with the polymer matrix

Fig. 6 Schematic representation of ionic exchange treatment

or initiate a polymerization reaction, improving the strength of the interface between the inorganic platelet and the polymer matrix (Xi et al. 2007).

The quantity of organic salts to be used in the modification is always related to its CEC (cation exchange capacity), which can be determined by several methods and which is defined as the number of cationic charges retained by a fixed mass of the clay mineral sample, generally either 1 or 100 g, so it is commonly expressed as (meq/g) or (meq/100 g). Other relevant parameters of this kind of treatments are the time and temperature used during the exchange process and the surfactant chain length.

5.4 Grafting

The most common grafting process is silylation, which consists of covalently bonding the silane molecules to the clay platelets. The covalent bond is produced from a condensation reaction of the reactive hydroxyl groups of the clay, located at the edges of the platelets and structural defects, with the hydroxyl groups generated from the hydrolysis of an alkoxysilane (Katsarava and Vygodskii 1992; LeBaron 1999; Letoffe et al. 2004; Rong et al. 2006; Kiliaris and Papaspyrides 2010). As montmotillonites and bentonites are swelling clays, the silanes can be easily intercalated into the interlaminar space, silylating both the internal and external surfaces, depending on the process conditions (He et al. 2005; Shen et al. 2007; Shen et al. 2008).

This method gives covalent bond between the added organic component and the clay minerals, producing immobilization of the organic moieties in the silylated products (da Fonseca and Airoldi 2000; LeBaron 1999; Takahashi and Kuroda 2011), and avoiding their leaching into the reaction solutions.

5.5 Reactions with Biomolecules

Clays have been modified with different kinds of biomolecules such as proteins, enzymes, amino acids, or peptides. One of the possibilities is the use of amino acids which are present at very low toxicity and various possible structures in comparison to chemically synthetic modifiers (Parbhakar et al. 2007; Katti et al. 2005); in this case, the reaction consists of an ion exchange process of the sodium cations of the protonated amino acid clay.

It is essential to point out that these clays are really attractive as reinforcing materials in the preparation of bionanocomposites, because they are not harmful to human health and they may have curative properties and can also be employed in pharmaceutical formulations (Bergaya et al. 2012).

6 Processing of Biodegradable Clay Nanocomposites

Polymer layered silicate nanocomposites have been studied for nearly 50 years, but few references deal with the importance of how the organoclay was processed together with the plastic of choice. Many articles have been focused on the importance of the chemistry used to modify the surface of the clay, usually montmorillonite or kaolinite, but without including the important role of processing. As clearly stated by Ojijo et al. and Ludueña et al. (Ojijo and Sinha Ray 2013; Ludueña et al. 2011a, b), the different processing techniques significantly affect the level of clay intercalation and/or exfoliation reached and the final disposition of clay platelets within the polymeric matrix. The final properties of the composite are strongly improved as a result of the proper interactions between the nanoparticles and the biopolymer. Three thermodynamically feasible nanocomposite structures can be obtained: intercalated, flocculated, and exfoliated. The main objective during nanocomposite fabrication is to obtain an exfoliated disposition.

A compromise condition between processing technique and regime is reached, where the need for well-dispersed particles needs to be balanced with the need to ensure the structural integrity of the polymer and matrix (i.e., degradation). Clay bionanocomposites are usually prepared by three main processing techniques (Ojijo and Sinha Ray 2013):

(a) Solvent casting: a method where a solvent system is used to pre-swell the clays and then, the intercalation of polymer chains within the clay galleries is facilitated. This method is only feasible when the polymer is soluble in the same solvent used for clay swelling.
(b) In situ polymerization: a method where clay nanoparticles are pre-mixed with the monomer solution. After the monomer solution has swelled the clay platelets, the polymerization process is started by any of the normal methods (light, heat, etc.).
(c) Melt intercalation: a method where molten polymer molecules are mixed with the clay nanoparticles, usually under the application of simultaneous high shear rates. Probably, this method is one of the most appealing from an industrial perspective, since it can be readily adapted to current industrial processes such as extrusion or injection molding.

The selection of the appropriate processing technique and conditions strongly depend on the type and hydrophilicity/hydrophobicity of the polymer, the type of clay being used, and also on the clay compatibilization degree. An excellent, and probably the most exhaustive, review on different processing techniques for the production of bionanocomposites has been prepared by Ojijo and Sinha Ray (2013). Table 1 shows a selection of the information presented in this contribution. The information has been reordered and presented in such a way that the final structure of the clay platelets achieved after processing is clearly accessible for each type of polymer. Only the three aforementioned processing techniques together with the most important biopolymers were included, although Ojijo and Sinha Ray

(2013) cited additional publications reporting the preparation of bionanocomposites by other methods and included several other biopolymers such as proteins and cellulose derivatives.

The first polymer reviewed in Table 1 is starch. As mentioned in the previous section, starch can be obtained from different sources in nature and, in order to be a suitable polymer for use as a plastic, it needs to undergo a gelatinization process. This process requires very high amounts of hydrophilic solvents (water, glycerol, etc.). The preparation of starch-based clay nanocomposites can be done simultaneously with the gelatinization process (solvent casting-like process) or after gelatinization where starch is referred as thermoplastic starch (TPS) and can be melt-compounded in an intensive mixer or extruder. Many researchers have studied the preparation of starch/clay nanocomposites. Best results were obtained by using unmodified or slightly modified sodium MMT or fluoromica (Pandey and Singh 2005; Schlemmer et al. 2010; Qiao et al. 2005; Dean et al. 2007) Good results have been achieved with either solvent casting or melt intercalation techniques using water or glycerol as plasticizer. The use of different plasticizers and/or modified clays (C30B, C25A, hectorite, etc.) generally resulted in fully or partially intercalated structures at best (Cyras et al. 2008; Mondragón et al. 2008; Pandey and Singh 2005; Chung et al. 2010; Schlemmer et al. 2010; Chiou et al. 2007; Dean et al. 2007; Park et al. 2003; Chen and Evans 2005; Park et al. 2002) or in the formation of microcomposites in many cases (Wilhelm et al. 2003; Kampeerapappun et al. 2007; Zeppa et al. 2009; Chaudhary 2008; Chen and Evans 2005; Chiou et al. 2007). The need for solutions in this regard has led many researchers to investigate the feasibility of blending starch with other biodegradable polymers in order to obtain nanocomposites. This topic has been of great interest and some results are presented herein. Starch-based blends containing different amounts of clay can be successfully prepared by solvent casting in presence of polyvinyl alcohol (PVA). In this case, a good dispersion of clay within the matrix and an intercalated structure is obtained by using unmodified sodium montmorillonite (Ali et al. 2011). Intercalated structures are also obtained by melt blending and compounding of TPS with PCL, PLA, PBSA and PVA and most types of compatibilized clays (Ali et al. 2011; Pérez et al. 2007; Lee et al. 2008; Lee and Hanna 2009; Lee and Hanna 2008; Bocchini et al. 2010; Majdzadeh-Ardakani and Nazari 2010). However, there are a few cases where the melt intercalation process resulted into well-exfoliated structures as reported by Perez et al. (Pérez et al. 2007) who worked with starch/PCL/C30B, Majdzadeh-Ardakani and Nazari (2010) and Dean et al. (2008) who prepared starch/PVA/MMT-Na+ nanocomposites, and Bocchini et al. (2010) who worked with starch/PBSA/C30B.

Several researchers have focused their attention on the preparation of polylactide (PLA) nanocomposites. PLA is one type of hydrophobic aliphatic polyester, and as such, it is not water soluble. Therefore, solvent casting processes are carried out in nonpolar solvents such as chloroform or dichloromethane. In order to achieve intercalated structures, the use of modified clays is mandatory (McLauchlin and Thomas 2009; Jaffar Al-Mulla 2011; Rhim et al. 2009; Chang and An 2002; Krikorian and Pochan 2003). Best results, i.e., partial or complete exfoliation, are

Table 1 Processing techniques for clay bionanocomposites

Biopolymer	Processing technique	Structure	System	Clay	References
Starch	Solvent casting	Microcomposite	Starch/glicerol	MMT-Mac-gel, Hectorite-Ca++	Wilhelm et al. (2003); Kampeerapappun et al. (2007)
			Starch/glycerol/urea	MMT-Na+, C30B	Zeppa et al. (2009)
			Starch/ethanolamine	MMT-Na+, C30B	Zeppa et al. (2009)
		Intercalated	Starch/glicerol	MMT-Na+	Cyras et al. (2008); Mondragón et al. (2008); Pandey and Singh (2005); Chung et al. (2010)
			Starch/pequi oil	MMT(K10)	Schlemmer et al. 2010)
		Exfoliated	Starch/glicerol	MMT-Na+	Pandey and Singh (2005)
			Starch/pequi oil	MMT(K10)	Schlemmer et al. (2010)
	Melt intercalation	Microcomposite	Starch/wáter	C25A	Chaudhary (2008)
			Starch/glicerol	MMT-Na+, Bentone109	Chen and Evans (2005); Chiou et al. (2007)
		Intercalated	Starch/wáter	MMT-Na+, FHT-Na+	Dean et al. (2007)
			Starch/glicerol	MMT-Na+, C30B, C10A, C6A, Hectorite	Chen and Evans (2005); Park et al. (2002, 2003); Chiou et al. (2007)
		Exfoliated	Starch/wáter	Fluoromica-Na+	Dean et al. (2007)
			Starch/glicerol	OMMT	Qiao et al. (2005)
Starch-based blends	Solvent casting	Intercalated	Starch/PVA	MMT-Na+	Ali et al. (2011)
	Melt intercalation	Intercalated	Starch/PCL	MMT-Na+	Pérez et al. (2007)
			Starch/PLA	MMT-Na+, C10A, C25A, C93A, C15A, C30B	Lee et al. (2008); Lee and Hanna (2008, 2009)
			Starch/PBSA	MMT-Na+	Bocchini et al. (2010)
			Starch/PVA	C30B, CMMT	Majdzadeh-Ardakani and Nazari (2010)
		Exfoliated	Starch/PCL	C30B	Pérez et al. (2007)
			Starch/PVA	MMT-Na+	Dean et al. (2008; Majdzadeh-Ardakani and Nazari 2010)
			Starch/PBSA	C30B	Bocchini et al. (2010)

(continued)

Table 1 (continued)

Biopolymer	Processing technique	Structure	System	Clay	References
PLA	Solvent casting	Microcomposite	PLA/Chloroform	MMT-Na+	Jaffar Al-Mulla (2011)
		Intercalated	PLA/Chloroform	MMT-CAB, C30B, C20A, MMT-FA, MMT-FHA	McLauchlin and Thomas (2009); Jaffar Al-Mulla (2011); Rhim et al. (2009)
			PLA/DMAc	MMT-C16, Mica-C16	Chang and An (2002)
			PLA/dichloromethane	C25A, C15A	Krikorian and Pochan (2003)
		Exfoliated	PLA/Chloroform	MMT-CDFA	Jaffar Al-Mulla (2011)
			PLA/dichloromethane	C30B, C25A, m-MMT-CTAB	Krikorian and Pochan (2003; Wu and Wu (2006)
	Melt intercalation	Microcomposite	PLA	Smectite-Ph, Smectite-C8	Sinha Ray et al. (2002)
		Intercalated	PLA	MMT-C18, MMT-3C18, SFM-O, SAP-O, MMT-C16, Mica-C16, Smectice-C12, C15A, C93A, C30B, C25S, C15S, C20A	Sinha Ray et al. (2002, 2003a, b); Ray and Okamoto (2003); Sinha Ray et al. (2002); Chen and Yoon (2005); Krishnamachari et al. (2009); Pluta et al. (2006)
			PLA-o-PCL	MMT-C18	Sinha Ray et al. (2002)
			PLA/diglycerine tetra-acetate	MMT-c18	Shibata et al. (2006)
			PLA/acetyltriethyl citrate ester	C25A, C15A	Thellen et al. (2005)
			PLA/o-PEG	MMT-poly(ethylene glycol) stearylamine	Shibata et al. (2006)
			PLA/PEG(5–20 %)	MMT-Can+, C30B, C25A, C20A	Paul et al. (2003)
		Exfoliated	PLA	SFM-O, TFC, OOMT	Sinha Ray et al. (2003a, b); Chen and Yoon (2005); Feijoo et al. (2005)
	In-situ polymerization	Intercalated	PLA/Tin(II) octoate	C30A	Pluta et al. (2006)
			PLLA/triethylaluminum	MMT-Na+, C25A	Pluta et al. (2006)
		Exfoliated	PLA/Tin(II) octoate	C30B	Pluta et al. (2006)
			PLLA/triethylaluminum	C30B	Pluta et al. (2006)

(continued)

Table 1 (continued)

Biopolymer	Processing technique	Structure	System	Clay	References
PHA	Solvent casting	Microcomposite	PHB/HV-chloroform	Organo-Mica	Sanchez-Garcia and Lagaron (2010)
		Intercalated	PHB/HV-chloroform	C20A, C25A, C15A	Lim et al. (2003); D'Amico et al. (2012); Cabedo et al. (2009)
			PHB/chloroform	MMT-Na+	D'Amico et al. (2012)
	Melt intercalation	Microcomposite	PHB	MMT-Na+, Kaolinite	Bordes et al. (2008; Botana et al. (2010); Sanchez-Garcia and Lagaron (2010)
		Intercalated	PHB	MEE, MAE, C30B, Organo-Kaolinite, C20A	Maiti et al. (2007); Bordes et al. (2008); Botana et al. (2010); Sanchez-Garcia and Lagaron (2010)
			PHB/HV	C30B, C20A, OOMT	Bordes et al. (2008); Mook Choi et al. (2003); Cabedo et al. (2009)
			PHB/PCL	C15A, OOMT	Wang et al. (2005a, b; Chen et al. 2004); Chen et al. (2002); Bruzaud and Bourmaud (2007)
PCL	Solvent casting	Microcomposite	PCL/chloroform	OMMT	Jimenez et al. (1997)
		Intercalated	PCL/DMF	MMT-Na+	Wu et al. (2009)
			PCL/dichloroformethane	C30B	Ludueña (2007)
			PCL/DMF/dichloroformethane	C30B	Ludueña (2007)
		Exfoliated	PCL/dichloroformethane	C30B	Ludueña (2007)
			PCL/DMF/dichloroformethane	C30B	Ludueña (2007)
	Melt intercalation	Microcomposite	PCL	MMT-Na+, MMT-COOH	Pantoustier et al. (2002); Gain et al. (2005)
		Intercalated	PCL	C10A, C30B, C25A, C93A, OMMT	Lepoittevin et al. (2002a, b); Gain et al. (2005); Di et al. (2003); Dean et al. (2009); Messersmith and Giannelis (1995); Ludueña et al (2007, 2011a, b); Okada et al. (2003)

(continued)

Table 1 (continued)

Biopolymer	Processing technique	Structure	System	Clay	References
		Exfoliated	PCL	C20A, C30B, C25A	Ludueña et al (2007); Ludueña et al. (2011a, b); Okada et al. (2003)
	In situ polymerization	Intercalated	PCL	MMT-Na+, Fluorohectorite-Cr + 3	Gorrasi et al. (2003); Messersmith and Giannelis (1995)
			PCL/dibutylin(IV) dimethoxide	C25A, MMT-COOH, Laponite, Sapionite	Pantoustier et al. (2002); Hrobarikova et al. (2004)
			PCL/tin(II)octoate	MMT-Na+, MMT-COOH, C25A, MMT-OH/Me(<50 %)	Pantoustier et al. (2002); Lepoittevin et al. (2002a, b); Liao et al. (2007); Gorrasi et al. (2004)
			PCL/triethylaluminum	MMT-OH	Gain et al. (2005)
		Exfoliated	PCL	MMT-COOH12	Messersmith and Giannelis (1995)
			PCL/dibutylin(IV) dimethoxide	C30B	Pantoustier et al. (2002; Hrobarikova et al. (2004)
			PCL/tin(II)octoate	C30B, MMT-OH/Me(>50 %)	Hrobarikova et al. (2004); Pantoustier et al. (2002); Liao et al. (2007); Tasdelen (2011)
			PCL/triethylaluminum	C30B	Gain et al. (2005)
PBS	Solvent casting	Microcomposite			
		Intercalated	PBS	MMT-CPC	Shih et al. (2007, 2008)
		Exfoliated	PBS	MMT-CPC/dicumyl peroxide, MMT-CTAB	Shih et al. (2007, 2008)
	Melt intercalation	Microcomposite	PBS	MMT-Na+	Someya et al. (2004)
		Intercalated	PBS	MMT-C18, MMT-LEA, MMT-DA, MMT-ODA	Ray et al. (2002); Sinha Ray et al. (2003); Someya et al. (2004)
		Exfoliated	PBS	SAP-C16	Ray et al. (2002)
	In situ polymerization	Intercalated	PBS/titanium(VI) isobutoxide	C10A, C30B	Hwang et al. (2009)
		Exfoliated	PBS/titanium(VI) isobutoxide	C30B-M (1.6 diisocuanaotohexane)*	Hwang et al. (2009)

only achieved when highly modified clays are used in the presence of organic solvents like dichloromethane. As a first approach, organo-modified synthetic fluoro-micas (OSFMs) resulted in better degrees of exfoliation than organo-modified montmorillonites (OMMTs) (Chang and An 2002). Krikorian and Pochan were the first who obtained exfoliated PLA/MMT nanocomposites by solvent casting. They used different commercially available clays and obtained totally exfoliated structures with Cloisite 30B (Southern Clay Products, USA) which was demonstrated by the total absence of the d001 peak in X-ray spectra (Krikorian and Pochan 2003). Partially exfoliated structures were also obtained with C25B. Later, other researchers also reported the preparation of exfoliated PLA-OMMT clays via solvent casting in chloroform and dichloromethane (Jaffar Al-Mulla 2011; Wu and Wu 2006).

The preparation of PLA/clay nanocomposites by in situ polymerization has seldom been reported by different researchers. Although it produces good results, it requires the use of organometallic catalysts which reduces its suitability for industrial applications (Ojijo and Sinha Ray 2013). Pluta et al. (2006) obtained intercalated, partially exfoliated and exfoliated nanocomposites by in situ polymerization of PLA with MMT, C25A, C30A and C30B. Exfoliated nanocomposites were obtained when C30B clay was used.

Melt intercalation has also been explored by several authors and is the preferred method to prepare PLA/clay nanocomposites (Ojijo and Sinha Ray 2013). However, the usual result is to obtain partially or completely intercalated structures (Sinha Ray et al. 2002, 2003a, b; Ray and Okamoto 2003; Sinha Ray et al. 2002; Chen and Yoon 2005; Krishnamachari et al. 2009; Pluta et al. 2006). Different plasticizers and compatibilizers have been used, finding little to no exfoliation when montmorillonite clays are involved (Sinha Ray et al. 2002; Shibata et al. 2006; Thellen et al. 2005; Paul et al. 2003). A good exfoliation degree has been reported by Chen et al., who used organo-modified synthetic fluoro-micas (SFMs) with dual functionalization (Chen and Yoon 2005). Partial exfoliation has been achieved with some OMMTs (Sinha Ray et al. 2003a, b; Feijoo et al. 2005). Although melt intercalation is the preferred method, its direct application is most often inefficient to produce exfoliated structures. For this reason the preparation of masterbatches by in situ polymerization followed by melt processing has been an alternative route (Ojijo and Sinha Ray 2013).

Due to its attractive properties, the preparation of polycaprolactone (PCL)—clay nanocomposites has been extensively investigated by different researchers. The three above-mentioned techniques were used with different results. Jimenez et al. (1997) obtained microcomposites by a casting process of PCL dissolved in chloroform and mixed with OMMT particles. Wu et al. (2009) reported the obtention of partially intercalated structures by solvent casting of PCL/MMT-Na+ in dimethylformamide. Ludueña and co-workers (2007) investigated the effect of two types of solvent used for preparing PLA/C30B nanocomposites and the time used for swelling the clays under ultrasonic agitation. Their most interesting finding is that the main factor determining clay intercalation/exfoliation is swelling time. They obtained both intercalated and exfoliated structures (Ludueña et al. 2007).

Processing PCL/clay nanocomposites by melt intercalation techniques led in most cases to well-intercalated or exfoliated structures. Only a few researchers have reported the formation of microcomposites by melt intercalation of PCL and unmodified sodium montmorillonite or COOH-modified montmorillonite (Pantoustier et al. 2002; Gain et al. 2005). When using organo-modified clays, the usual result is well-dispersed nanocomposites with either intercalated (Lepoittevin et al. 2002a, b; Gain et al. 2005; Di et al. 2003; Dean et al. 2009; Messersmith and Giannelis 1995; Ludueña et al. 2007; Ludueña et al. 2011a, b; Okada et al. 2003) or exfoliated structures (Ludueña et al. 2007; Ludueña et al. 2011a, b; Okada et al. 2003). Processing conditions remain as the most important factor for the achievement of exfoliated structures.

In situ polymerization techniques have shown to be very effective to obtain well-intercalated and/or exfoliated nanocomposites since it has been found that the ring-opening polymerization reaction of lactone monomers can be catalyzed by layered silicates, as reported by Messersmith and Giannelis (1995). Complex reaction mechanisms involved resulted in several researchers studying the in situ polymerization of PCL in the presence of clays. Hydroxyl groups attached to the inner silicate surfaces resulted in a single PCL chain and, consequently, a lower number average molecular weight. Further information on this important topic can be found in the mentioned researchers' publications (Messersmith and Giannelis 1993, 1995; Pantoustier et al. 2002; Gorrasi et al. 2004; Lepoittevin et al. 2002a, b). Exfoliated structures were obtained especially when C30B clay was used in the presence of particular catalysers (Messersmith and Giannelis 1995; Pantoustier et al. 2002; Hrobarikova et al. 2004; Liao et al. 2007; Tasdelen 2011; Gain et al. 2005). When using other commercially available clays or in-house organo-modified clays, regardless of the catalyzer involved, intercalated nanocomposites were obtained (Pantoustier et al. 2002; Lepoittevin et al. 2002a, b; Liao et al. 2007; Gorrasi et al. 2004; Messersmith and Giannelis 1995; Hrobarikova et al. 2004; Gain et al. 2005).

Finally, within the group of polyesters we can find the polybutylensuccinate or PBS. The preparation of PBS/clay nanocomposites has been investigated by a number of researchers. The formation of microcomposites has been reported by Someya et al. (2004) when processing PBS and MMT-Na+ by melt intercalation. On the other hand, many researchers have reported obtaining well-intercalated structures when processing PBS with organo-modified clays such as C10A, C30B, MMT-LEA by either solvent casting, melt intercalation, or in situ polymerization (Ray et al. 2002, 2003; Someya et al. 2004; Shih et al. 2007, 2008; Hwang et al. 2009). Some authors have also reported the preparation of completely exfoliated PBS/clay nanocomposites by solvent casting (Shih et al. 2007, 2008), melt intercalation (Ray et al. 2002), and in situ polymerization (Rong et al. 2002) as well.

As mentioned earlier, the family of bacterial-synthesized polyhydroxyesters so-called polyhyroxyalkanoates (PHAs), poly(3-hydroxybutyrate) or PHB is the most important one and represents an interesting alternative to synthetic polymers. Another interesting variant is the copolymer poly(3-hydroxybutyrate-co-3-hydroxyvalerate) or PHB/HV. Due to its high hydrophobicity, the preparation of nanocomposites by solvent casting requires the use of organic solvents. Sanchez-García and Lagaron (2010) have reported the preparation of PHB/HV and organo-modified mica microcomposites by solution casting in chloroform. Intercalated structures have been achieved by dispersing modified montmorillonite clays such as MMT-Na+, C15A, C20A, and C25A in chloroform by many authors (Lim et al. 2003; D'Amico et al. 2012).

PHAs, and especially PHB and PHB/HV, present a very narrow melt processing window due to the small difference between melting and degradation temperatures. Thus, melt processing is somehow limited because polymer degradation occurs rapidly at temperatures slightly above melting point. However, the use of melt intercalation for the preparation of PHB-based clay nanocomposites has been reported by several researchers, showing great difficulty to produce clay intercalation. No intercalation and particle aggregation has been reported by different authors when processing pure PHB and unmodified clays (montmorillonite and kaolinite), which was conducted for formation of microcomposites (Bordes et al. 2008; Botana et al. 2010; Sanchez-Garcia and Lagaron 2010). The best results have been obtained when processing PHB, PHB/HV, and PHB/PCL together with organo-modified clays such as C20A, C30B, etc., as reported by several researchers (Maiti et al. 2007; Bordes et al. 2008; Botana et al. 2010; Sanchez-Garcia and Lagaron 2010; Mook Choi et al. 2003; Feijoo et al. 2005; Chen et al. 2002; Wang et al. 2005a, b; Chen et al. 2004; Bruzaud and Bourmaud 2007).

Other important biodegradable polymers that have been investigated include cellulose, chitosan, and protein-based. In the case of cellulose, some authors reported the preparation of cellulose acetate (CA) nanocomposites, obtaining intercalated structures for solution casting of CA with MMT-Na+ and slightly modified clays (Romero et al. 2009; Rodríguez et al. 2012), intercalated nanocomposites when melt compounding CA with C30B (Park et al. 2006; Wibowo et al. 2006; Park et al. 2004b) and exfoliated structures when TEC "green" plasticizers were used (Park et al. 2004a, b). In the case of chitosan-based biopolymers, several authors claim to have obtained exfoliated structures for different types of clays by solution casting in water or diluted acids (Wang et al. 2005a, b; Xu 2006; Hong et al. 2011). In some other cases, the obtention of microcomposites (Rhim et al. 2006; Xu et al. 2006) or partially intercalated structures has been reported (Oguzlu and Tihminlioglu 2010; Zhang et al. 2009; Wang et al. 2010).

The previous revision on the different types of structures obtained during processing of diverse biopolymer-based clay nanocomposites shows that an important amount of research is needed for each matrix in order to obtain an adequate set of conditions that produces exfoliated-clay nanocomposites. The most important factors are: (a) processing method; (b) processing parameters; matrix/clay compatibilization.

In this regard, the most appealing processing technique for industrial applications appears to be the melt intercalation process, since it is continuous and easily adaptable to the existing processing facilities (twin screw extruders). However, scarce information can be found in the literature regarding the impact of the high shear melt intercalation extreme conditions on the delicate organo-modifiers used for compatibilizing the different clays with hydrophobic polymers. Some authors

(VanderHart et al 2001a, b; Alexandre and Dubois 2000; Xie et al. 2001, 2002; Luduena et al. 2013) demonstrated that clay organo-modifiers can be degraded during the melt blending process. Therefore, not only the clay/polymer compatibility but also the processing stability of the clay organo-modifier should be the key to obtain well-dispersed polymer/organo-modified clay nanocomposites by melt blending. Ludueña et al. (2011a, b) studied the effect of natural and organo-modified montmorillonite on the final performance of PCL-based nanocomposites prepared by melt intercalation on a twin screw extruder. They found that Cloisite 20A (C20A) was the organo-modified clay with the best balance between processing stability and chemical compatibility with the PCL. Thus, PCL/C20A nanocomposites showed the best clay dispersion degree and hence, the best mechanical performance. Results shown by Homminga et al. (2005) demonstrate that shear forces in the melt-preparation of PCL/layered-mineral nanocomposites facilitate the breakup of large-sized agglomerates, but further exfoliation of the mineral layers is determined by the chemical compatibility between the polymer matrix and the mineral layers rather than by shear forces. Therefore, when the polymer/clay system is not optimally compatibilized, stronger shear forces induced by extrusion parameters such as low temperature, high screw speed, and long residence time can improve the final clay dispersion degree inside the nanocomposite but the nanocomposite may remain thermodynamically unstable and so, a second melting process can produce the partial reagglomeration of the clay platelets (Vaia and Giannelis 1997). An important conclusion presented by Ludueña et al. (2011a, b) is that once the optimal chemical compatibility and processing stability of the polymer/clay system prepared by melt blending is obtained, changing the extrusion parameters will not further improve the morphology and final properties of the nanocomposites. Other methods besides the improvement of the clay dispersion degree can further increase the efficiency of clays as reinforcement of biodegradable polymeric matrices such as the orientation of clay platelets due to fluid flow as stated by Weon and Sue (2005).

7 Conclusions

An increased awareness of the need to protect the environment has changed the way companies approach the material selection process for designing their products. Both governmental regulations and consumer pressure forced the incorporation of environmental considerations in the design process. Regarding this, the substitution of synthetic nonbiodegradable polymers is an attractive solution to the increasing amount of plastic waste that remains un-recycled every year. Several biopolymers derived from biomass or synthesized by microorganisms appear as potential substitutes. However, most biodegradable polymers show many shortcomings like low mechanical properties, high hydrophilicity, low-dimensional stability, poor barrier properties, among others. In order to be able to readily use these polymers for industrial applications such as packaging, there is an urgent need to overcome these problems. The preparation of biodegradable nanocomposites based on natural reinforcements appears to be the perfect solution. Nanoclays offer a great potential for overcoming each and every limitation of biodegradable polymers. They offer high mechanical properties, provide dimensional stability and increased barrier properties, and are also environmentally friendly. However, it was shown that processing biodegradable polymer-clay nanocomposites is a challenging task. Clays come densely packed and need to be separated into individual clay platelets (intercalation/exfoliation) in order to really improve polymer's properties. Several ways have been proposed in order to obtain intercalated and/or exfoliated structures. It was shown that both processing conditions and clay compatibilization are equally effective and important to achieve the goal. When these strategies are correctly applied, the obtained nanocomposites exhibit improved thermal, mechanical, and barrier properties. It was also shown that the key factors in controlling the final behavior of nanocomposites are the final clay content and clay dispersion degree, which can be increased by selecting an optimum balance between the polymer/clay chemical compatibility and the thermomechanical stability of the clay organo-modifier. Future research on the topic should be aimed at studying the optimization of chemical compatibilization of clays together with a set of processing parameters that ensure no degradation of organo-modifiers yielding nanocomposites with improved properties.

Acknowledgments Authors acknowledge CONICET, ANPCyT and UNMdP for the financial support of the present work.

References

Alexandre M, Dubois P (2000) Polymer-layered silicate nanocomposites: preparation, properties and uses of a new class of materials. Mater Sci Eng, R 28(1–2):1–63

Ali SS, Tang X, Alavi S, Faubion J (2011) Structure and physical properties of starch/poly vinyl alcohol/sodium montmorillonite nanocomposite films. J Agric Food Chem 59(23):12384–12395

Auer H, Hofmann H (1993) Pillared clays: characterization of acidity and catalytic properties and comparison with some zeolites. Appl Catal A 97(1):23–38

Avérous L, Pollet E (2012) Environmental silicate nano-biocomposites. Green En Tech 50

Bastioli C (2011) Handbook of biodegradable polymers

Becker O, Varley RJ, Simon GP (2004) Thermal stability and water uptake of high performance epoxy layered silicate nanocomposites. Eur Polym J 40(1):187–195

Bergaya F, Jaber M, Lambert JF (2012) Clays and clay minerals as layered nanofillers for (bio) polymers. Green En Tech 50:41–75

Bocchini S, Battegazzore D, Frache A (2010) Poly (butylensuccinate co-adipate)-thermoplastic starch nanocomposite blends. Carbohydr Polym 82(3):802–808

Bordes P, Pollet E, Averous L (2009) Nano-biocomposites: biodegradable polyester/nanoclay systems. Prog Polym Sci 34(2):125–155

Bordes P, Pollet E, Bourbigot S, Avérous L (2008) Structure and properties of PHA/clay nano-biocomposites prepared by melt intercalation. Macromol Chem Phys 209(14):1473–1484

Botana A, Mollo M, Eisenberg P, Torres Sanchez RM (2010) Effect of modified montmorillonite on biodegradable PHB nanocomposites. Appl Clay Sci 47(3–4):263–270

Bruzaud S, Bourmaud A (2007) Thermal degradation and (nano)mechanical behavior of layered silicate reinforced poly(3-hydroxybutyrate-co-3-hydroxyvalerate) nanocomposites. Polym Test 26(5):652–659

Cabedo L, Plackett D, Giménez E, Lagarón JM (2009) Studying the degradation of polyhydroxybutyrate—co—valerate during processing with clay-based nanofillers. J Appl Polym Sci 112(6):3669–3676

Chang J-H, An YU (2002) Nanocomposites of polyurethane with various organoclays: thermomechanical properties, morphology, and gas permeability. J Polym Sci, Part B: Polym Phys 40(7):670–677

Charreau H, Foresti ML, Vazquez A (2012) Nanocellulose patents trends: a comprehensive review on patents on cellulose nanocrystals, microfibrillated and bacterial cellulose. Recent Pat Nanotechnol 26:26–26

Chaudhary DS (2008) Understanding amylose crystallinity in starch–clay nanocomposites. J Polym Sci, Part B: Polym Phys 46(10):979–987

Chen B, Evans JRGG (2005) Thermoplastic starch-clay nanocomposites and their characteristics. Carbohydr Polym 61(4):455–463

Chen GX, Hao GJ, Guo TY, Song MD, Zhang BH (2002) Structure and mechanical properties of poly(3-hydroxybutyrate-co-3-hydroxyvalerate) (PHBV)/clay nanocomposites. J Mater Sci Lett 21(20):1587–1589

Chen GX, Hao GJ, Guo TY, Song MD, Zhang BH (2004) Crystallization kinetics of poly (3-hydroxybutyrate-co-3-hydroxyvalerate)/clay nanocomposites. J Appl Polym Sci 93(2):655–661

Chen G-X, Yoon J-S (2005) Clay functionalization and organization for delamination of the silicate tactoids in poly(L-lactide) matrix. Macromol Rapid Commun 26(11):899–904

Chiang C-L, Ma C-CM (2004) Synthesis, characterization, thermal properties and flame retardance of novel phenolic resin/silica nanocomposites. Polym Degrad Stab 83(2):207–214

Chiou B-S, Wood D, Yee E, Imam SH, Glenn GM, Orts WJ (2007) Extruded starch–nanoclay nanocomposites: Effects of glycerol and nanoclay concentration. Polym Eng Sci 47(11):1898–1904

Chung Y-L, Ansari S, Estevez L, Hayrapetyan S, Giannelis EP, Lai H-M (2010) Preparation and properties of biodegradable starch–clay nanocomposites. Carbohydr Polym 79(2):391–396

Cyras VP, Manfredi LB, Ton-That M-T, Vázquez A (2008) Physical and mechanical properties of thermoplastic starch/montmorillonite nanocomposite films. Carbohydr Polym 73(1):55–63

Czaja WK, Young DJ, Kawecki M, Brown RM (2006) The future prospects of microbial cellulose in biomedical applications. Biomacromolecules 8(1):1–12

da Fonseca MG, Airoldi C (2000) New layered inorganic–organic nanocomposites containing n-propylmercapto copper phyllosilicates. J Mater Chem 10(6):1457–1463

D'Amico DA, Manfredi LB, Cyras VP (2012) Relationship between thermal properties, morphology, and crystallinity of nanocomposites based on polyhydroxybutyrate. J Appl Polym Sci 123(1):200–208

Dean K, Yu L, Wu DY (2007) Preparation and characterization of melt-extruded thermoplastic starch/clay nanocomposites. Compos Sci Technol 67(3–4):413–421

Dean KM, Do MD, Petinakis E, Yu L (2008) Key interactions in biodegradable thermoplastic starch/poly(vinyl alcohol)/montmorillonite micro- and nanocomposites. Compos Sci Technol 68(6):1453–1462

Dean KM, Pas SJ, Yu L, Ammala A, Hill AJ, Wu DY (2009) Formation of highly oriented biodegradable polybutylene succinate adipate nanocomposites: effects of cation structures on morphology, free volume, and properties. J Appl Polym Sci 113(6):3716–3724

Di Y, Iannace S, Di Maio E, Nicolais L (2003) Nanocomposites by melt intercalation based on polycaprolactone and organoclay. J Polym Sci, Part B: Polym Phys 41(7):670–678

Edwards G, Halley P, Kerven G, Martin D (2005) Thermal stability analysis of organo-silicates, using solid phase microextraction techniques. Thermochim Acta 429(1):13–18

Eichhorn S, Dufresne A, Aranguren M, Marcovich N, Capadona J, Rowan S, Weder C, Thielemans W, Roman M, Renneckar S, Gindl W, Veigel S, Keckes J, Yano H, Abe K, Nogi M, Nakagaito A, Mangalam A, Simonsen J, Benight A, Bismarck A, Berglund L, Peijs T (2010) Review: current international research into cellulose nanofibres and nanocomposites. J Mater Sci 45(1):1–33

Feijoo JL, Cabedo L, Giménez E, Lagaron JM, Saura JJ (2005) Development of amorphous PLA-montmorillonite nanocomposites. J Mater Sci 40(7):1785–1788

Gain O, Espuche E, Pollet E, Alexandre M, Dubois P (2005) Gas barrier properties of poly(ε-caprolactone)/clay nanocomposites: influence of the morphology and polymer/clay interactions. J Polym Sci, Part B: Polym Phys 43(2):205–214

Gilman JW, Jackson CL, Morgan AB, Harris R, Manias E, Giannelis EP, Wuthenow M, Hilton D, Phillips SH (2000) Flammability properties of polymer—Layered-Silicate Nanocomposites. Polypropylene and polystyrene nanocomposites. Chem Mater 12(7):1866–1873

Gorrasi G, Tortora M, Vittoria V, Pollet E, Alexandre M, Dubois P (2004) Physical properties of poly(ε-caprolactone) layered silicate nanocomposites prepared by controlled grafting polymerization. J Polym Sci, Part B: Polym Phys 42(8):1466–1475

Gorrasi G, Tortora M, Vittoria V, Pollet E, Lepoittevin B, Alexandre M, Dubois P (2003) Vapor barrier properties of polycaprolactone montmorillonite nanocomposites: effect of clay dispersion. Polymers 44(8):2271–2279

Gross RA, Kalra B (2002) Biodegradable polymers for the environment. Science 297(5582):803–807

Habibi Y, Lucia LA, Rojas OJ (2010) Cellulose nanocrystals: chemistry, self-assembly, and applications. Chem Rev 110(6):3479–3500

He H, Ding Z, Zhu J, Yuan P, Xi Y, Yang D, Frost RL (2005) Thermal characterization of surfactant-modified montmorillonites. Clays Clay Miner 53(3):287–293

He H, Duchet J, Galy J, Gérard J-F (2006) Influence of cationic surfactant removal on the thermal stability of organoclays. J Colloid Interface Sci 295(1):202–208

Homminga D, Goderis B, Hoffman S, Reynaers H, Groeninckx G (2005) Influence of shear flow on the preparation of polymer layered silicate nanocomposites. Polymers 46(23):9941–9954

Hong SI, Lee JH, Bae HJ, Koo SY, Lee HS, Choi JH, Kim DH, Park S-H, Park HJ (2011) Effect of shear rate on structural, mechanical, and barrier properties of chitosan/montmorillonite nanocomposite film. J Appl Polym Sci 119(5):2742–2749

Hrobarikova J, Robert JL, Calberg C, Jérôme R, Grandjean J (2004) Solid-state NMR study of intercalated species in poly(ε-caprolactone)/clay nanocomposites. Langmuir 20(22):9828–9833

Hubbe MA, Rojas OJ, Lucia LA, Sain M (2008) Cellulosic nanocomposites: a review. Bioresources 3(3):929–980

Hwang SY, Yoo ES, Im SS (2009) Effect of the urethane group on treated clay surfaces for high-performance poly(butylene succinate)/montmorillonite nanocomposites. Polym Degrad Stab 94(12):2163–2169

Jaffar Al-Mulla EA (2011) Preparation of new polymer nanocomposites based on poly(lactic acid)/fatty nitrogen compounds modified clay by a solution casting process. Fibers Polym 12(4):444–450

Jalal Uddin A, Araki J, Gotoh Y (2011) Toward "strong" green nanocomposites: polyvinyl alcohol reinforced with extremely oriented cellulose whiskers. Biomacromolecules 12(3):617–624

Jimenez G, Ogata N, Kawai H, Ogihara T (1997) Structure and thermal/mechanical properties of poly (ε-caprolactone)-clay blend. J Appl Polym Sci 64(11):2211–2220

Kampeerapappun P, Aht-ong D, Pentrakoon D, Srikulkit K (2007) Preparation of cassava starch/montmorillonite composite film. Carbohydr Polym 67(2):155–163

Katsarava RD, Vygodskii YS (1992) Silylation in polymer chemistry. Russ Chem Rev 61(6):629–650

Katti DR, Ghosh P, Schmidt S and Katti KS (2005) Mechanical properties of the sodium montmorillonite interlayer intercalated with amino acids. Biomacromolecules 6(6):3276–3282

Kiliaris P, Papaspyrides CDD (2010) Polymer/layered silicate (clay) nanocomposites: an overview of flame retardancy. Prog Polym Sci 35(7):902–958

Klemm D, Kramer F, Moritz S, Lindström T, Ankerfors M, Gray D, Dorris A (2011) Nanocelluloses: a new family of nature-based materials. Angew Chem Int Ed 50(24):5438–5466

Kojima Y, Usuki A, Kawasumi M, Okada A, Fukushima Y, Kurauchi T, Kamigaito O (1993) Mechanical properties of nylon 6-clay hybrid. J Mater Res 8(05):1185–1189

Kornmann X (2001) Synthesis of epoxy–clay nanocomposites. Influence of the nature of the curing agent on structure. Polymers 42(10):4493–4499

Krikorian V, Pochan DJ (2003) Poly (L-lactic acid)/layered silicate nanocomposite : fabrication, characterization, and properties. Chem Mater 15(13):4317–4324

Krishnamachari P, Zhang J, Lou J, Yan J, Uitenham L (2009) Biodegradable poly(lactic acid)/clay nanocomposites by melt intercalation: a study of morphological, thermal, and mechanical properties. Int J Polym Anal Charact 14(4):336–350

Lagaly G (1981) Characterization of clays by organic compounds. Clay Miner 16(1):1–21

Le Pluart L, Duchet J, Sautereau H, Gérard JF (2002) Surface modifications of montmorillonite for tailored interfaces in nanocomposites. J Adhes 78(7):645–662

LeBaron P (1999) Polymer-layered silicate nanocomposites: an overview. Appl Clay Sci 15(1–2):11–29

Lee S, Cho W, Hahn P, Lee M, Lee Y, Kim K (2005) Microstructural changes of reference montmorillonites by cationic surfactants. Appl Clay Sci 30(3–4):174–180

Lee SM, Tiwari D (2012) Organo and inorgano-organo-modified clays in the remediation of aqueous solutions: an overview. Appl Clay Sci 59–60:84–102

Lee SY, Chen H, Hanna MA (2008) Preparation and characterization of tapioca starch–poly(lactic acid) nanocomposite foams by melt intercalation based on clay type. Ind Crops Prod 28(1):95–106

Lee SY, Hanna MA (2009) Tapioca starch-poly(lactic acid)-Cloisite 30B nanocomposite foams. Polym Compos 30(5):665–672

Lee S-Y, Hanna MA (2008) Preparation and characterization of tapioca starch-poly(lactic acid)-Cloisite NA+ nanocomposite foams. J Appl Polym Sci 110(4):2337–2344

Leite IF, Soares APS, Carvalho LH, Raposo CMO, Malta OML, Silva SML (2009) Characterization of pristine and purified organobentonites. J Therm Anal Calorim 100 (2):563–569

Lepoittevin B, Pantoustier N, Alexandre M, Calberg C, Jerome R, Dubois P (2002a) Layered silicate/polyester nanohybrids by controlled ring-opening polymerization. Macromol Symp 183(1):95–102

Lepoittevin B, Pantoustier N, Devalckenaere M, Alexandre M, Kubies D, Calberg C, Jérôme R, Dubois P (2002b) Poly(ε-caprolactone)/clay nanocomposites by in-Situ intercalative polymerization catalyzed by dibutyltin dimethoxide. Macromolecules 35(22):8385–8390

Leszczyńska A, Njuguna J, Pielichowski K, Banerjee JR (2007) Polymer/montmorillonite nanocomposites with improved thermal properties. Thermochim Acta 453(2):75–96

Letoffe J-M, Putaux J-L, David L, Bourgeat-Lami E (2004) Aqueous dispersions of silane-functionalized laponite clay platelets. a first step toward the elaboration of water-based polymer/clay nanocomposites. Langmuir 20(5):1564–1571

Liao L, Zhang C, Gong S (2007) Preparation of poly(ε-caprolactone)/clay nanocomposites by microwave-assisted in situ ring-opening polymerization. Macromol Rapid Commun 28 (10):1148–1154

Lim ST, Hyun YH, Lee CH, Choi HJ (2003) Preparation and characterization of microbial biodegradable poly(3-hydroxybutyrate)/organoclay nanocomposite. J Mater Sci Lett 22 (4):299–302

Luduena L, Kenny J, Vazquez A, Alvarez V (2013) Effect of extrusion conditions and post-extrusion techniques on the morphology and thermal/mechanical properties of polycaprolactone/clay nanocomposites. J Compos Mater 48(17):2059–2070

Ludueña LN, Va Alvarez, Vazquez A (2007) Processing and microstructure of PCL/clay nanocomposites. Mater Sci Eng, A 460–461:121–129

Ludueña LN, Vazquez A, Kenny JM, Alvarez VA (2011a) The effect of processing technique on the performance of PCL/clay nanocomposite. In: Frontiers in polymer science, second international symposium, Lyon, France

Ludueña LNN, Kenny JMM, Vázquez A, Alvarez VA (2011b) Effect of clay organic modifier on the final performance of PCL/clay nanocomposites. Mater Sci Eng, A 529(0):215–223

Maiti P, Batt CA, Giannelis EP (2007) New biodegradable polyhydroxybutyrate/layered silicate nanocomposites. Biomacromolecules 8(11):3393–3400

Majdzadeh-Ardakani K, Nazari B (2010) Improving the mechanical properties of thermoplastic starch/poly(vinyl alcohol)/clay nanocomposites. Compos Sci Technol 70(10):1557–1563

Mandalia T, Bergaya F (2006) Organo clay mineral–melted polyolefin nanocomposites effect of surfactant/CEC ratio. J Phys Chem Solids 67(4):836–845

Mathew AP, Dufresne A (2002) Plasticized waxy maize starch: effect of polyols and relative humidity on material properties. Biomacromolecules 3(5):1101–1108

McLauchlin AR, Thomas NL (2009) Preparation and thermal characterisation of poly(lactic acid) nanocomposites prepared from organoclays based on an amphoteric surfactant. Polym Degrad Stab 94(5):868–872

Messersmith PB, Giannelis EP (1993) Polymer-layered silicate nanocomposites: in situ intercalative polymerization of e-caprolactane in layered silicates. Chem Mater 5(8):1064–1066

Messersmith PB, Giannelis EP (1995) Synthesis and barrier properties of poly(ε-caprolactone)-layered silicate nanocomposites. J Polym Sci, Part A: Polym Chem 33(7):1047–1057

Mondragón M, Mancilla JE, Rodríguez-González FJ (2008) Nanocomposites from plasticized high-amylopectin, normal and high-amylose maize starches. Polym Eng Sci 48(7):1261–1267

Mook Choi W, Wan Kim T, Ok Park O, Keun Chang Y, Woo Lee J (2003) Preparation and characterization of poly(hydroxybutyrate-co-hydroxyvalerate)-organoclay nanocomposites. J Appl Polym Sci 90(2):525–529

Nair LS, Laurencin CT (2007) Biodegradable polymers as biomaterials. Prog Polym Sci 32(8–9):762–798

Noorani S, Simonsen J, Atre S (2007) Nano-enabled microtechnology: polysulfone nanocomposites incorporating cellulose nanocrystals. Cellulose 14(6):577–584

Oguzlu H, Tihminlioglu F (2010) Preparation and barrier properties of chitosan-layered silicate nanocomposite films. Macromol Symp 298(1):91–98

Ojijo V, Sinha Ray S (2013) Processing strategies in bionanocomposites. Prog Polym Sci 38(10–11):1543–1589

Okada K, Mitsunaga T, Nagase Y (2003) Properties and particles dispersion of biodegradable resin/clay nanocomposites. Korea-Aust Rheol J 15(1):43–50

Pandey JK, Singh RP (2005) Green nanocomposites from renewable resources: effect of plasticizer on the structure and material properties of clay-filled starch. Starch/Stärke 57(1):8–15

Pantoustier N, Lepoittevin B, Alexandre M, Dubois P, Kubies D, Calberg Cd, Jérôme R (2002) Biodegradable polyester layered silicate nanocomposites based on poly(ε-caprolactone). Polym Eng Sci 42(9):1928–1937

Parbhakar A, Cuadros J, Sephton MA, Dubbin W, Coles BJ, Weiss D (2007) Adsorption of l-lysine on montmorillonite. Colloids Surf A 307(1–3):142–149

Park H-M, Lee W-K, Park C-Y, Cho W-J, Ha C-S (2003) Environmentally friendly polymer hybrids part I mechanical, thermal, and barrier properties of thermoplastic starch/clay nanocomposites. J Mater Sci 38(5):909–915

Park H-M, Li X, Jin C-Z, Park C-Y, Cho W-J, Ha C-S (2002) Preparation and properties of biodegradable thermoplastic starch/clay hybrids. Macromol Mater Eng 287(8):553–558

Park H-M, Liang X, Mohanty AK, Misra M, Drzal LT (2004a) Effect of compatibilizer on nanostructure of the biodegradable cellulose acetate/organoclay nanocomposites. Macromolecules 37(24):9076–9082

Park H-M, Misra M, Drzal LT, Mohanty AK (2004b) "Green" nanocomposites from cellulose acetate bioplastic and clay: effect of eco-friendly triethyl citrate plasticizer. Biomacromolecules 5(6):2281–2288

Park H-M, Mohanty AK, Drzal LT, Lee E, Mielewski DF, Misra M (2006) Effect of sequential mixing and compounding conditions on cellulose acetate/layered silicate nanocomposites. J Polym Environ 14(1):27–35

Paul M-A, Alexandre M, Degée P, Henrist C, Rulmont A, Dubois P (2003) New nanocomposite materials based on plasticized poly(l-lactide) and organo-modified montmorillonites: thermal and morphological study. Polymers 44(2):443–450

Pavlidou S, Papaspyrides CD (2008) A review on polymer–layered silicate nanocomposites. Prog Polym Sci 33(12):1119–1198

Pérez CJ, Alvarez VA, Mondragón I, Vázquez A (2007) Mechanical properties of layered silicate/starch polycaprolactone blend nanocomposites. Polym Int 56(5):686–693

Pérez CJ, Alvarez VA, Mondragón I, Vázquez A (2008) Water uptake behavior of layered silicate/starch–polycaprolactone blend nanocomposites. Polym Int 57(2):247–253

Pluta M, Paul M-A, Alexandre M, Dubois P (2006) Plasticized polylactide/clay nanocomposites. II. The effect of aging on structure and properties in relation to the filler content and the nature of its organo-modification. J Polym Sci, Part B: Polym Phys 44(2):299–311

Qiao X, Jiang W, Sun K (2005) Reinforced thermoplastic acetylated starch with layered silicates. Starch/Stärke 57(12):581–586

Ravichandran J, Lakshmanan CM, Sivasankar B (1996) Acid activated montmorillonite and vermiculite clays as dehydration and cracking catalysts. React Kinet Catal Lett 59(2):301–308

Ray SS, Okamoto K, Maiti P, Okamotoa M (2002) New poly(butylene succinate)/layered silicate nanocomposites: preparation and mechanical properties. J Nanosci Nanotechnol 2(2):171–176

Ray SS, Okamoto M (2003) Biodegradable polylactide and its nanocomposites: opening a new dimension for plastics and composites. Macromol Rapid Commun 24(14):815–840

Rhim J-W, Hong S-I, Ha C-S (2009) Tensile, water vapor barrier and antimicrobial properties of PLA/nanoclay composite films. LWT Food Sci Technol 42(2):612–617

Rhim J-W, Hong S-I, Park H-M, Ng PKW (2006) Preparation and characterization of chitosan-based nanocomposite films with antimicrobial activity. J Agric Food Chem 54 (16):5814–5822

Rhim J-W, Park H-M, Ha C-S (2013) Bio-nanocomposites for food packaging applications. Prog Polym Sci 38(10–11):1629–1652

Rodríguez FJ, Galotto MJ, Guarda A, Bruna JE (2012) Modification of cellulose acetate films using nanofillers based on organoclays. J Food Eng 110(2):262–268

Roelofs JCAA, Berben PH (2006) Preparation and performance of synthetic organoclays. Appl Clay Sci 33(1):13–20

Romero RB, Leite CAP, Gonçalves MdC (2009) The effect of the solvent on the morphology of cellulose acetate/montmorillonite nanocomposites. Polymers 50(1):161–170

Rong MZ, Zhang MQ, Liu Y, Zhang ZW, Yang GC, Zeng HM (2002) Effect of stitching on in-plane and interlaminar properties of sisal/epoxy laminates. J Compos Mater 36(12):1505–1526

Rong MZ, Zhang MQ, Ruan WH (2006) Surface modification of nanoscale fillers for improving properties of polymer nanocomposites: a review. Mater Sci Technol 22(7):787–796

Sanchez-Garcia MD, Lagaron JM (2010) Novel clay-based nanobiocomposites of biopolyesters with synergistic barrier to UV light, gas, and vapour. J Appl Polym Sci 118(1):188–199

Schlemmer D, Angélica RS, Sales MJA (2010) Morphological and thermomechanical characterization of thermoplastic starch/montmorillonite nanocomposites. Compos Struct 92(9):2066–2070

Shen L, Haufe J, Patel MK, Excellence EB, European Polysacharide Network of Excellence (2009) Product overview and market projection of emerging bio-based plastics. Utrecht, Netherlands, Group Science, Technology and Society (STS); Copernicus Institute for Sustainable Development and innovation, Utrecht University

Shen W, He H, Zhu J, Yuan P, Frost RL (2007) Grafting of montmorillonite with different functional silanes via two different reaction systems. J Colloid Interface Sci 313(1):268–273

Shen W, He H, Zhu J, Yuan P, Ma Y, Liang X (2008) Preparation and characterization of 3-aminopropyltriethoxysilane grafted montmorillonite and acid-activated montmorillonite. Chin Sci Bull 54(2):265–271

Shibata M, Someya Y, Orihara M, Miyoshi M (2006) Thermal and mechanical properties of plasticized poly(L-lactide) nanocomposites with organo-modified montmorillonites. J Appl Polym Sci 99(5):2594–2602

Shih YF, Wang TY, Jeng RJ, Wu JY, Teng CC (2007) Biodegradable nanocomposites based on poly(butylene succinate)/organoclay. J Polym Environ 15(2):151–158

Shih YF, Wang TY, Jeng RJ, Wu JY, Wuu DS (2008) Cross-linked and uncross-linked biodegradable nanocomposites. I. Nonisothermal crystallization kinetics and gas permeability. J Appl Polym Sci 110(2):1068–1079

Singha AS, Thakur VK (2010) Mechanical, morphological, and thermal characterization of compression-molded polymer biocomposites. Int J Polym Anal Charact 15(2):87–97

Sinha Ray S, Maiti P, Okamoto M, Yamada K, Ueda K (2002) New polylactide/layered silicate nanocomposites. 1. preparation, characterization, and properties. Macromolecules 35(8):3104–3110

Sinha Ray S, Okamoto K and Okamoto M (2003a) Structure—property relationship in biodegradable poly(butylene succinate)/layered silicate nanocomposites. Macromolecules 36 (7):2355–2367

Sinha Ray S, Yamada K, Okamoto M, Fujimoto Y, Ogami A, Ueda K (2003b) New polylactide/layered silicate nanocomposites. 5. Designing of materials with desired properties. Polymers 44(21):6633–6646

Sinha Ray S, Okamoto M (2003) Polymer/layered silicate nanocomposites: a review from preparation to processing. Prog Polym Sci 28(11):1539–1641

Siqueira G, Bras J, Dufresne A (2010) Cellulosic bionanocomposites: a review of preparation. Properties and applications. Polymers 2(4):728–765

Siró I, Plackett D (2010) Microfibrillated cellulose and new nanocomposite materials: a review. Cellulose 17(3):459–494

Slavutsky AM, Bertuzzi MA, Armada M, García MG, Ochoa NA (2014) Preparation and characterization of montmorillonite/brea gum nanocomposites films. Food Hydrocolloids 35 (Complete):270–278

Someya Y, Nakazato T, Teramoto N, Shibata M (2004) Thermal and mechanical properties of poly(butylene succinate) nanocomposites with various organo-modified montmorillonites. J Appl Polym Sci 91(3):1463–1475

Steudel A, Batenburg LF, Fischer HR, Weidler PG, Emmerich K (2009) Alteration of swelling clay minerals by acid activation. Appl Clay Sci 44:105–115

Svensson A, Nicklasson E, Harrah T, Panilaitis B, Kaplan DL, Brittberg M, Gatenholm P (2005) Bacterial cellulose as a potential scaffold for tissue engineering of cartilage. Biomaterials 26 (4):419–431

Takahashi N, Kuroda K (2011) Materials design of layered silicates through covalent modification of interlayer surfaces. J Mater Chem 21(38):14336–14336

Tasdelen MA (2011) Poly(ε-caprolactone)/clay nanocomposites via "click" chemistry. Eur Polym J 47(5):937–941

Thakur VK, Singha AS, Mehta IK (2010) Renewable resource-based green polymer composites: analysis and characterization. Int J Polym Anal Charact 15(3):137–146

Thakur VK, Thakur MK (2014a) Processing and characterization of natural cellulose fibers/thermoset polymer composites. Carbohydr Polym 109:102–117

Thakur VK, Thakur MK (2014b) Recent trends in hydrogels based on psyllium polysaccharide: a review. J Cleaner Prod 82:1–15

Thakur VK, Thakur MK (2014c) Recent advances in graft copolymerization and applications of chitosan: a review. ACS Sustain Chem Eng 2:2637–2652

Thakur VK, Thakur MK, Gupta RK (2014a) Review: raw natural fiber-based polymer composites. Int J Polym Anal Charact 19:256–271

Thakur VK, Vennerberg D, Kessler MR (2014b) Green aqueous surface modification of polypropylene for novel polymer nanocomposites. ACS Appl Mater Interfaces 6:9349–9356

Thakur VK, Thakur MK, Raghavan P, Kessler MR (2014c) Progress in green polymer composites from lignin for multifunctional applications: a review. ACS Sustain Chem Eng 2(5):1072–1092

Thellen C, Orroth C, Froio D, Ziegler D, Lucciarini J, Farrell R, D'Souza NA, Ratto JA (2005) Influence of montmorillonite layered silicate on plasticized poly(l-lactide) blown films. Polym 46(25):11716–11727

RA Vaia, Giannelis EP (1997) Lattice model of polymer melt intercalation in organically-modified layered silicates. Macromolecules 30:7990–7999

VanderHart DL, Asano A, Gilman JW (2001a) NMR measurements related to clay-dispersion quality and organic-modifier stability in nylon-6/clay nanocomposites. Macromolecules 34 (12):3819–3822

VanderHart DL, Asano A, Gilman JW (2001b) Solid-state NMR investigation of paramagnetic nylon-6 clay nanocomposites. 2. Measurement of clay dispersion, crystal stratification, and stability of organic modifiers. Chem Mater 13:3796–3809

Viville P, Lazzaroni R, Pollet E, Alexandre M, Dubois P, Borcia G, Pireaux J-J (2003) Surface characterization of poly(ε-caprolactone)-based nanocomposites. Langmuir 19(22):9425–9433

Wang S, Song C, Chen G, Guo T, Liu J, Zhang B, Takeuchi S (2005a) Characteristics and biodegradation properties of poly(3-hydroxybutyrate-co-3-hydroxyvalerate)/organophilic montmorillonite (PHBV/OMMT) nanocomposite. Polym Degrad Stab 87(1):69–76

Wang SF, Shen L, Tong YJ, Chen L, Phang IY, Lim PQ and Liu TX (2005b) Biopolymer chitosan/montmorillonite nanocomposites: preparation and characterization. Polym Degrad Stab 90(1):123–131

Wang X, Liu B, Ren J, Liu C, Wang X, Wu J, Sun R (2010) Preparation and characterization of new quaternized carboxymethyl chitosan/rectorite nanocomposite. Compos Sci Technol 70 (7):1161–1167

Weon JI, Sue HJ (2005) Effects of clay orientation and aspect ratio on mechanical behavior of nylon-6 nanocomposite. Polym 46:6325–6334

Wibowo AC, Misra M, Park H-M, Drzal LT, Schalek R, Mohanty AK (2006) Biodegradable nanocomposites from cellulose acetate: mechanical, morphological, and thermal properties. Compos A 37(9):1428–1433

Wilhelm HMM, Sierakowski MRR, Souza GPP, Wypych F (2003) Starch films reinforced with mineral clay. Carbohydr Polym 52(2):101–110

Wu T, Xie T, Yang G (2009) Preparation and characterization of poly(ε-caprolactone)/Na+-MMT nanocomposites. Appl Clay Sci 45(3):105–110

Wu T-M, Wu C-Y (2006) Biodegradable poly(lactic acid)/chitosan-modified montmorillonite nanocomposites: preparation and characterization. Polym Degrad Stab 91(9):2198–2204

Xi Y, Ding Z, He H, Frost RL (2004) Structure of organoclays—an X-ray diffraction and thermogravimetric analysis study. J Colloid Interface Sci 277:116–120

Xi Y, Frost RL, He H (2007) Modification of the surfaces of Wyoming montmorillonite by the cationic surfactants alkyl trimethyl, dialkyl dimethyl, and trialkyl methyl ammonium bromides. J Colloid Interface Sci 305(1):150–158

Xie W, Gao Z, Liu K, Pan WP, Vaia R, Hunter D, Singh A (2001) Thermal characterization of organically modified montmorillonite. Thermochim Acta 367–368:339–350

Xie W, Xie R, Pan W-P, Hunter D, Koene B, Tan L-S, Vaia R (2002) Thermal stability of quaternary phosphonium modified montmorillonites. Chem Mater 14(11):4837–4845

Xu Y, Ren X, Hanna MA (2006) Chitosan/clay nanocomposite film preparation and characterization. J Appl Polym Sci 99(4):1684–1691

Zampori L, Gallo Stampino P, Dotelli G, Botta D, Natali Sora I, Setti M (2008) Interlayer expansion of dimethyl ditallowylammonium montmorillonite as a function of 2-chloroaniline adsorption. Appl Clay Sci 41(3–4):149–157

Zeppa C, Gouanvé F, Espuche E (2009) Effect of a plasticizer on the structure of biodegradable starch/clay nanocomposites: thermal, water-sorption, and oxygen-barrier properties. J Appl Polym Sci 112(4):2044–2056

Zhang K, Xu J, Wang KY, Cheng L, Wang J, Liu B (2009) Preparation and characterization of chitosan nanocomposites with vermiculite of different modification. Polym Degrad Stab 94 (12):2121–2127

Static and Dynamic Mechanical Analysis of Coir Fiber/Montmorillonite Nanoclay-Filled Novolac/Epoxy Hybrid Nanocomposites

Sudhir Kumar Saw

Abstract A simple method for the preparation of novel nanocomposites involving different weight ratios of blends of epoxy novolac (ENR) and diglycidyl ether of bisphenol A (DGEBA) resin, natural coir fiber, and organically modified montmorillonite (OMMT) nanoclay is described. It was found that on blending ENR with DGEBA, the storage modulus at room temperature are enhanced by about 100 % or more in the case of 50 and 65 % ENR-containing matrices; whereas, the enhancement in the case of 20 and 35 % ENR-containing matrices is only 50 % that of the pure matrix. It was also observed that the tan δ peak heights of the composite containing 50 and 65 % ENR are closer to that of 35 % ENR-containing composite. The plausible explanation is made on the basis of experimental findings of static and dynamic mechanical analysis (DMA) of ENR and DGEBA resin blends. X-ray diffraction (XRD) studies and Fourier transform infrared (FTIR) spectroscopy was used to obtain information on the modification of nanoclay and also thermal stability of various nanocomposites were determined from thermo gravimetric analyses. DMA showed that the modification of the clay strongly influences the stiffness and glass transition temperature (T_g) of the nanocomposites. It is possible to manufacture coir composites with increased stiffness without sacrificing their ductility.

Keywords Coir fibers · Epoxy resin · Composite materials · Mechanical properties · Fiber-matrix interaction · Synergistic effect

1 Introduction

Epoxy resins have been widely used as engineering plastics because of their high-performance characteristics, such as good mechanical, thermal, and electrical properties. To use epoxy resins properly, it is important to understand their

S.K. Saw (✉)
Central Instrumentation Facility, Birla Institute of Technology,
Mesra, Ranchi 835215, India
e-mail: sudhirsaw@yahoo.co.in

V.K. Thakur and M.K. Thakur (eds.), *Eco-friendly Polymer Nanocomposites*,
Advanced Structured Materials 75, DOI 10.1007/978-81-322-2470-9_5

structure–property relationship. A considerable amount of research work has already been done in this field on the relationship between the properties of the cured epoxy resins and the structures of the epoxy resin compounds; however, these studies were mostly made on the basis of bisphenol A-type resins using various kinds of hardeners (Lee and Lichtenhan 1999).

Recently, because high-performance epoxy resin is strict in its requirements, polyfunctional epoxy has been offered in practical fields. In particular, epoxy novolac resin (ENR) is largely used as electronic encapsulation material because of its well-known thermal resistance properties. Because the structures of ENR exert a significant influence on the properties of the cured resins, it is necessary to understand their structure–property relationship. Despite various advantages, epoxy needs modifications to overcome some crucial disadvantages like limited solubility in polar solvents, higher cost of bisphenol A-based epoxy and maximum service temperature of only about 100 °C.

From the early stage of epoxide resin technology research, constant effort has been devoted to obtain resin systems that maintain their properties at higher temperature. One approach was to increase the crosslink density in the cured polymer network, either by use of polyfunctional resins having average epoxide content greater than 2 or by using curing agents of higher functionality. The ENR mostly used in practice has 3.6 epoxide groups per molecule ($n = 1.6$ in general formula). Attention must be given that no unreacted phenolic groups remain in the resin because this would limit storage stability and cause volatiles to be evolved during cure. Depending on the cure temperature and the choice of the amine curing agents, strengths as high as 3000 psi at room temperature and over 1000 psi at the range of 55–150 °C were reported in the literature (Datta et al. 2002). On blending ENR with DGEBA, the essential thermal stability of the ENR resin is preserved, while the crosslinking possibility of the epoxide group can be concomitantly utilized, as seen from the molecular structure of the polyglycidyl ether. Incorporation of more aromatic rings in the chain enhances the thermoresistant profile of the system. Thus, such materials partake of the nature of both components and show greatly improved high temperature performance compared with that of a straight epoxy (Lee and Neville 1967).

Lignocellulosic materials have played a major role in human life (Thakur et al. 2013a, b, c, d, e). Lignocellulosic fibers found to have extensive applications in building, civil, and automotive engineering fields (Thakur and Thakur 2014). In case of synthetic fiber-based composites, despite the usefulness in service, these are difficult to be recycled after designed service life. However, natural fiber-based composites are environment friendly to a large extent (Thakur et al. 2014a). Though hydrophilic character of natural fibers would leads to composites with weak interface but surface modification of natural fibers are aimed at improving the adhesion between fibers and matrix (Singha and Thakur 2008a, b, c, d). In surface modifications, either hydroxyl groups get activated or new moieties are added that can effectively interlock with the matrix. Mercerization (Singha et al. 2009), acrylation, acetylation, silane treatment (Singha et al. 2009), and peroxide treatment with various coupling agents, graft copolymerization (Thakur et al. 2011, 2012a, 2013a, b, c, d, e), plasma treatments, and other pretreatments of natural fibers have achieved various levels of success in improving fiber strength, fiber fitness, and fiber-matrix adhesion in natural fiber reinforced composites (Singha and Thakur 2010a, b, c). Simple chemical treatments can be applied to the fibers to change surface tension and polarity through modification of fiber surface. Over the last 30 years composite materials, plastics and ceramics have been the dominant emerging materials. The volume and number of applications of composite materials have grown steadily, penetrating and conquering new markets relentlessly (Singha and Thakur 2009a, b). Modern composite materials constitute a significant proportion of the engineered materials market ranging from everyday products to sophisticated applications.

Natural fibers are environmentally friendly, biodegradable, and non-polluting substance. The cellulosic materials have same properties of strength as plastics. My research work is definitely an effort toward better qualities of composites which are filled with cellulose material and do not disturb the ecology of the biosphere. Among the various types of natural biomasses, coconut trees are abundantly found in India. The coir fiber and pith materials are a major cause of fire during summer season when plants shed these fruits. The burning of this biomass causes not only environmental pollution but also destroy the other flora and fauna. The literature review has revealed that a very little work has been done on utilizing these fibers as reinforcing material in the polymer composites. Though India is full of natural biomass yet this precious wealth of nature is not exploited for better end products.

Researchers have reported that the introduction of inorganic materials into the lignocellulose polymer matrix may significantly improve the performance of a composite materials (Blanchard and Blanchet 2011; Mai and Militz 2004) and have a positive effect on the compression strength of lignocellulose/plastic composites. Chen et al. (2004) investigated the effect of temperature on the mechanical and surface properties of calcium carbonate-filled bamboo fibers and polypropylene composites and discovered that the treatments improved compatibility between fibers and polypropylene matrix. Consequently, the tensile strength and modulus of the composites were increased. Ma et al. (2012a) synthesized wood powder/$CaCO_3$ composites via hydrothermal route using dewaxed wood powder. Moreover, the types of lignocellulose were found to play an important role in the microstructure and morphologies of the lignocellulose/$CaCO_3$ composites (Ma et al. 2012b). Recently, they described a microwave-assisted method for the preparation of $CaCO_3$ particles-filled wood powder nanocomposites as well as high value-added applications of wood powder utilized in biomedical applications (Ma et al. 2014). Li et al. (2011a, b) also investigated a combination of cellulose fibers and silver nanocomposites by microwave-assisted method and discussed in details their thermal stability and antimicrobial property. Moreover, Jia et al. (2011, 2012) synthesized cellulose/calcium silicate and cellulose/F-substituted hydroxyapatite nanocomposites via microwave heating, respectively.

Polymer nanocomposites have attracted significant attention recently owing to their unique chemical and physical properties (Thakur et al. 2012b, 2014c). The nanofillers usually used for fabrication of the nanocomposites include nanoclays, nano-ceramic, boron nitride nanotubes (BNNTs), carbon nanotubes (CNTs), and graphene nanosheets (Lin et al. 2011a, b; Thakur et al. 2012b, 2014c). Among the various nanofillers, nanoclays montmorillonite have been explored to improve the mechanical, thermal, and electrical properties of several polymers. The improvement in thermomechanical properties was observed by the introduction of modified montmorillonite into the blends of laboratory prepared ENR and DGEBA resin. For characterization of polymeric materials such as homopolymers, copolymers, terpolymers, blends, and composites and their evaluation for consideration in stress and safety sensitive applications, dynamic mechanical analysis (DMA) at a selected fixed frequency over a range of temperature has advanced as an increasingly useful technique (Ghosh et al. 2003; Otaigbe 1991; Wingard and Beatty 1990). Dynamic mechanical tests, in general, give more information than other tests about a composite material. Dynamic tests, over a wide range of temperature and frequency, are especially sensitive to all kinds of transitions and relaxation process of matrix resin and also to the morphology of the composites. Such studies enable the determination of the temperature dependencies of the dynamic modulus, stress relaxation, mechanical loss, and damping phenomena.

In this study, modified montmorillonite and natural coir fiber was added to epoxy resin blends to produce hybrid nanocomposites and emphasized the effects of the use of different weight proportions of ENR. The lignocellulosic coir fiber, having been given no chemical treatments and bearing many alcoholic groups, limited (acidic) carboxylic groups, reducing (aldehydic) groups, and phenolic groups, provides ample scope for chemical anchorage of epoxide group segments on it through reaction with some of these groups present at the fiber surface. The cellulosic, hemicellulosic, and lignin constituents in the coir fiber, which itself is viewed as a composite material, thus become integral parts of the amine-cured ENR/DGEBA epoxy networks. The interaction between the filler and the matrix was studied by observing thermal and mechanical properties of nanocomposites. The performance testing confirmed that the improvements in the strength, stiffness, and thermal stability of hybrid nanocomposites were influenced by the addition of ENR. Blending of resin was better able to enhance the strength, stiffness, and thermal stability of composites but did not effectively enhance the T_g and elongation of nanocomposites. The thermal analysis result showed improved thermal stability of the hybrid nanocomposites.

2 Experimental Details

2.1 Materials Used

The coir fiber (diameter = 200–240 μm; density = 1.2–1.4 g/cm^3 and micro-fibrillar angle = 30°–39°) was obtained from Central Coir Research Institute, Coir Board, Kerala, India. The coconut husk was cut down into small pieces of approximately 10 mm long before making composites. ENR was prepared in the laboratory. Phenol and formaldehyde (37 % w/w) were obtained from E. Merck, India, and used as received. Reagent grade of acetone and toluene were obtained from E. Merck, India. Epichlorohydrin and sodium hydroxide were procured from S.D. Fine Chemicals,

India. DGEBA-based epoxy resin (Diglysidyl ether of bisphenol A, EPG-180) and room temperature aromatic polyamine (Cyclohexyl amine, CA)-based curative EH-400, are purchased from SIP industries Ltd., India. The unmodified montmorillonite nanoclay (Na-montmorillonite or MMT) with a 92.5 mequiv./100 g clay cationic exchange capacity (CEC) and the modifier agents, 1-Octadecyl amine was purchased from Sigma Aldrich, USA. The modified montmorillonite (OMMT) was prepared by ion-exchange reaction using octadecyl ammonium chloride according to reported method (Datta et al. 2002). The modified montmorillonite (OMMT) and coir fiber (CF) was used as fillers in the fabrication of nanocomposites.

2.2 Preparation of Epoxy Novolac Resin

The ENR was prepared by two step reactions. The first step was involved in the formation of novolac resin through condensation of phenol and formaldehyde under oxalic acid catalyzed with the molar ratio of phenol to formaldehyde was 1:0.8. In the second step, the epoxidation of previously prepared novolac resin was carried out by reacting novolac with excess of epichlorohydrin with 1:8 molar ratio under 40 % sodium hydroxide catalyst (Saw et al. 2011b, 2012). The schematics of synthesis are shown in Fig. 1.

2.3 Preparation of Organically Modified Montmorillonite (OMMT)

The method used for clay modification consists in cation exchange reactions, silane grafting, and adsorption of polar polymers. There are many factors, such as charge density of layered silicate, CEC of clay, curing condition (curing agent concentration, curing temperature, curing degree, resin-hardener ratio), and swelling time of clay that exhibit some important influence on the synthesis and final structures of nanocomposites (Garea et al. 2010).

Na-montmorillonite (MMT) was purified by dispersion of crude clay into deionized water and separation of non-colloidal impurities. To obtain cation exchange process, the purified MMT was swollen in deionized water for 24 h agitation at room temperature and a certain quantity of 1-octadecylamine was added. The system was maintained at around 68 °C for about 4 h and then filter and repeatedly washed with deionized water. The product was then dried, crushed, and sieved with 325-mesh to obtain organically modified montorillonite (OMMT). The main objective for the modification of nanoclay montmorillonite consists in organophilization of clay in order to improve the compatibility with epoxy resin blends. Otherwise, the polymer and the clay will form a separate phase system with a low interface among the various components.

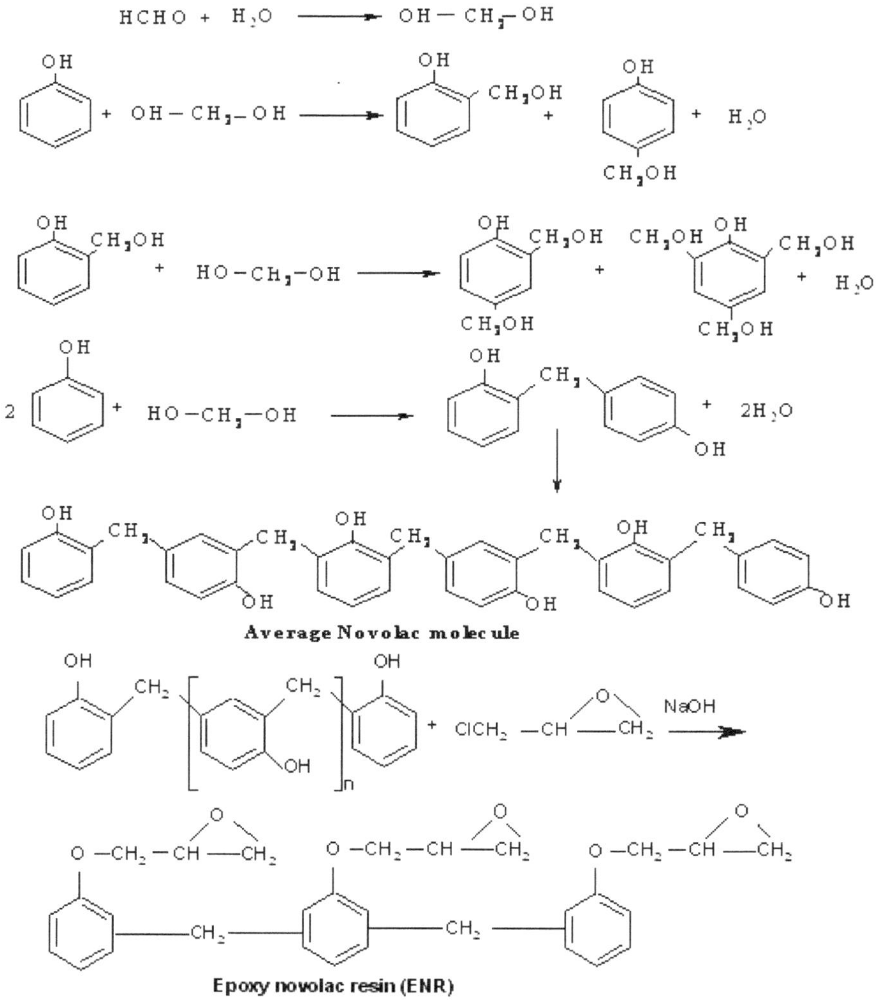

Average Novolac molecule

Epoxy novolac resin (ENR)

Fig. 1 The schematics of synthesis of **a** novolac resin **b** epoxy novolac resin

2.4 Fabrication of Nanocomposite Materials

Novel nanocomposites were fabricated from epoxy resin blends, polyamine curing agents, modified montmorillonite, and coir fibers using a simple method by hand lay-up technique followed by compression molding. A weighed amount of resin blends (blend of ENR and DGEBA) dissolved in similar weight proportions of acetone reinforced with two fillers; 30 wt% coir fiber and 2 wt% modified montmorillonite is admixed at a relevant room temperature with rapid mechanical stirring for 30 min and then a curative, cyclohexyl amine is added and to keep continue the stirring for 10 min. Acetone is used as the distribution medium for the resin to be adhered to the reinforcements. Two forms of epoxy resin were used, one is ENR, which is prepared in the laboratory, and the other is general-purpose bisphenol A-based epoxy resin (DGEBA) purchased from market. The weight mixing ratio of ENR to DGEBA resins were 0:100, 20:80, 35:65, 50:50, 65:35, and 100:00. The total hardener (curative) content was 20 wt% in all formulated composites. First, all nanocomposites were allowed to cure for 24 h at room temperature (30 °C) and it is then compression molded at 105 °C of temperature and 5 kg/cm^2 of pressure. Constant pressure and temperature was given to all composites. The pre-pressure had to be done before the hot-press in order to make the raw materials maintain a certain shape. The hot-press temperature was then set to 105 °C for 30 min. The thickness gauges were adopted to ensure the 3 mm thickness of the samples. Finally, postcuring of compression molded nanocomposites is done in an oven at 100 °C for 4 h. Test pieces of required dimensions were cut out from the composite sheets for static and dynamic mechanical properties. Composites of neat ENR and neat DGEBA (With fillers and without fillers) were also prepared for comparison the various properties.

3 Characterization of Nanoclay and Nanocomposites

3.1 FTIR Spectroscopy

A Fourier transform infrared spectrophotometer (FTIR, model-IR Prestige-21, Schimadzu Corporation, Japan) was used for the structural determination of functional groups and compounds. The FTIR spectra of modified and unmodified clay in the solid state using potassium bromide (KBr) as a reference material were recorded by diffuse reflectance spectroscopy (DRS technique). This technique consists of preparation of sample by mixing about 0.5 mg of powder sample with 50 mg of high-purity infrared-grade KBr powder (Aldrich). The KBr was previously oven dried to reduce the interference of water. The spectra were recorded in the wave number range of 4000–400 cm^{-1} with a resolution of 4 cm^{-1} and 20 scans were carried for each sample in transmittance mode.

3.2 X-ray Diffraction Analysis

X-ray powder diffraction studies were performed under ambient condition on a Rigaku miniflex, Japan make X-ray diffractometer using Ni-filtered CuKα radiation (λ = 1.542 Å) and a scintillation counter as a detector at 40 kV and 30 mA on rotation between 1° and 20° at 2θ scale at a step size of 0.02 and scanning speed of 2° per minute. Corundum was the reference sample used to verify and calibrate the instrument. The finely powdered sample was evenly distributed in the cavity of the sample holder with the help of a glass slide. The glass slide was carefully removed without disturbing the surface of the sample. The randomly oriented powdered sample with uniform surface was exposed to X-rays from all possible planes.

3.3 Thermogravimetric Analysis (TGA)

Dynamic TGA of various formulated composites were executed in the DTG-60 (Schimadzu, Japan) from 30 to 800 °C in platinum pan at constant heating rate of 10 °C/min under dynamic nitrogen atmosphere (Flow rate—30 ml/min) to find out the decomposition, thermal stability, and % char residue of composites. For each scan 8–10 mg of composite sample was taken.

3.4 Mechanical Analysis

Tensile testing of the samples was carried out at 25 ± 0.5 °C, using a rectangular shape test specimen in a Universal Testing Machine (model: UTM 3366, Instron, UK). The specification of the test specimen is as follows: gauge length, 60 mm; width, 10 mm; and thickness, 3 mm with end tabs, and loaded with serrated jaw sedge grips. A strain rate of 1 mm/min was used throughout the investigation. In all cases, six specimens were tested, and average values are reported.

3.5 Dynamic Mechanical Analysis

A dynamic mechanical thermal analyzer (model: Q_{800}, TA Instruments, USA) was used for the evaluation of dynamic moduli and mechanical damping (tan δ). Rectangular test specimens having dimensions of 60 mm × 10 mm × 3 mm were used for the dynamic mechanical experiments. Three point-bending modes were used. The temperature range over which properties were measured was 30–200 °C (303–473 K) at a linear heating rate of 5 °C/min using a fixed frequency (5 Hz) in nitrogen atmosphere (N_2).

4 Results and Discussion

4.1 FTIR Analysis of Montmorillonite Nanoclay

The clay organophilization step was performed through the cationic exchange reaction between protonated adducts and clay and was investigated from FTIR spectra as shown in Fig. 2. The analysis of FTIR spectra (Fig. 2a) shows that there

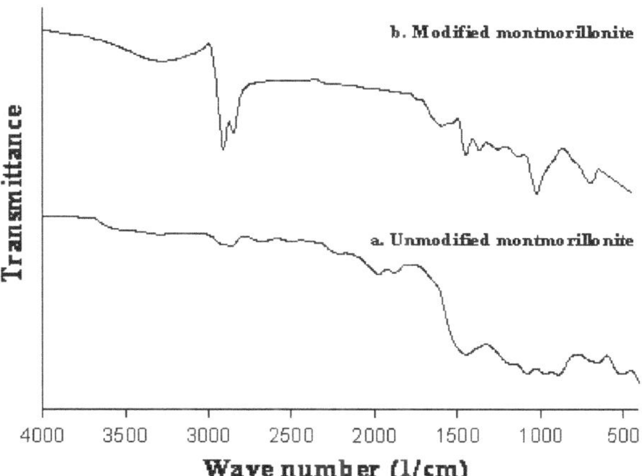

Fig. 2 The join visible FTIR spectra of **a** unmodified and **b** modified montmorillonite nanoclay

is a strong absorption band of siloxane bond, Si–O–Si around 1190, 1075, 880, and 640 cm^{-1} in unmodified montmorillonite. The silane alkoxide is able to condense to form silicate structure. The broadband centered near 3400 cm^{-1} is due to –OH stretching band for interlayer water. The bands at 3620 and 3690 cm^{-1} are due to –OH band stretch for Al–OH and Si–OH. The shoulders and broadness of the structural –OH band are mainly due to contributions of several –OH groups occurring in the clay.

It is also found from the FTIR spectra that the peak intensity and broadness of –OH (hydroxyl) stretching bond of unmodified montmorillonite at 3300–3700 cm^{-1} is reduced or almost disappears after modification. The appearance of new peaks at 2922 and 2874 cm^{-1}, which correspond to –CH stretching vibration and its corresponding –CH in-plane bending vibration peak at around 1452 cm^{-1} from CH_2 groups. The overlaid absorption peak in the region of 1640 cm^{-1} in the FTIR spectra of modified montmorillonite is attributed to –OH bending mode in water (adsorbed water). The spectra of modified montmorillonite indicate vibrational bands of organic modifier without causing any distortion of structure of clay (Garea et al. 2010).

4.2 X-ray Diffraction (XRD) Analysis of Montmorillonite Nanofiller

The most important argument for the silicate layers intercalation by protonated adducts was revealed by XRD analysis, which gives the value of the basal distance between silicate layers as shown in Fig. 3. From Fig. 3, one may observe that the basal distance of modified montmorillonite is always higher than of unmodified clay. The modification directed that all the new protonated adducts were successively intercalated between the silicate layers during the cationic exchange process.

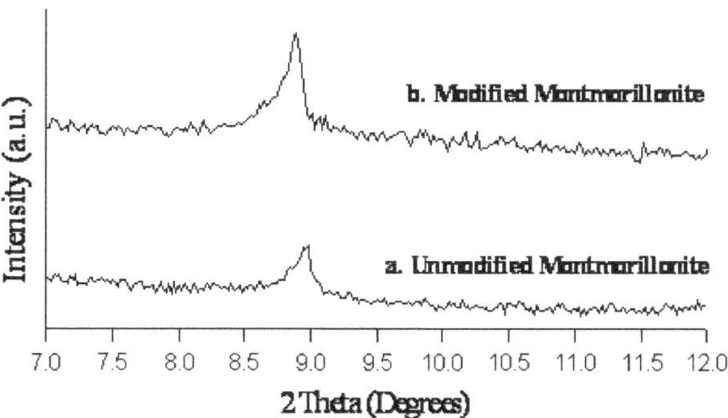

Fig. 3 The overlaying XRD peaks of **a** unmodified and **b** modified montmorillonite nanoclay

The intercalation of clay exhibits high basal distances (18–25 Å) at $2\theta = 8.8$ due to long hydrophobic (polymeric) chains, which include 14–18 carbon atoms, while the basal distance of unmodified montmorillonite is only 11.2 Å at $2\theta = 8.7$ (Basara et al. 2005).

4.3 Thermo Gravimetric Analysis of Various Formulated Composites

The thermal stability of the ENR/DGEBA-based epoxy hybrid nanocomposites in nitrogen atmosphere was evaluated by thermo gravimetric analysis (TGA) and compared with that pure ENR and pure DGEBA-based unhybridized nanocomposites. The TGA curves obtained for pure DGEBA composites (without filler), pure ENR composites (without filler), pure DGEBA composites (with filler, 2 % OMMT and 30 % coir fiber), pure ENR composites (with filler, 2 % OMMT and 30 % coir fiber), and blends of ENR/DGEBA-based epoxy hybrid nanocomposites are presented in Fig. 4. It was observed that the thermal degradation of all the samples has taken place within the programmed temperature range of 30–800 °C. The onset temperature of degradation (T_0), Temperature of maximum rate of mass loss (T_{max}), and final decomposition temperature (T_f) were noted in Table 1. It can be seen from Fig. 4 and Table 1 that the thermal decomposition of the hybrid nanocomposites started at comparatively higher temperature than that of the unhybridized nanocomposites. The onset temperature of degradation in composites of pure DGEBA and pure ENR containing no filler (0 wt% OMMT and 0 wt% coir filler) is started at 196 and 221 °C, respectively, and nearly 100 % decomposition occurred at 600 and 650 ° C, respectively. The onset temperature of degradation is shifted at higher temperature at 223 and 236 °C in composites of pure DGEBA and pure ENR-containing filler of 2 wt% OMMT and 30 wt% coir filler, respectively, and nearly 100 % decomposition occurred at 750 and 800 °C, respectively. This may be due to heat-deflection properties of coir fiber and addition of nanoclay improves the properties. The hybrid nanocomposites (composites of ENR/DGEBA blends) started loosing weight in the

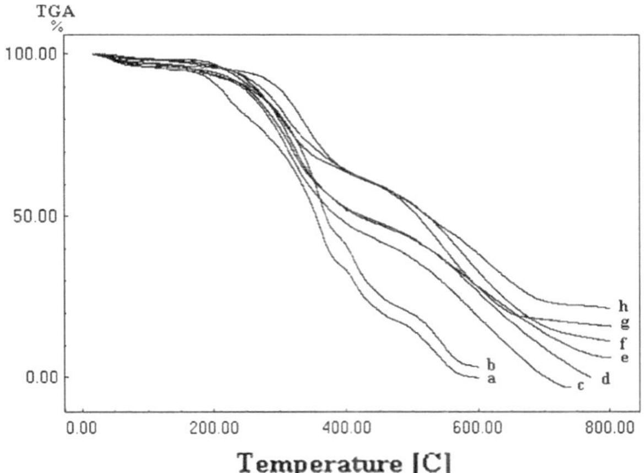

Fig. 4 TGA thermogram of pure resins and its formulated hybrid nanocomposites (*a* Pure DGEBA composite (Without filler); *b* Pure ENR composite (Without filler); *c* Pure DGEBA composite (With filler); *d* Pure ENR composite (With filler); *e* ENR:DGEBA (20:80) hybrid nanocomposite; *f* ENR:DGEBA (35:65) hybrid nanocomposite; *g* ENR:DGEBA (50:50) hybrid nanocomposite; *h* ENR:DGEBA (65:35) hybrid nanocomposite)

Table 1 TGA data obtained for pure resins and its hybrid nanocomposites

Sample	T_o (°C)[a]	T_{max} (°C)[b]	T_f (°C)[c]	% Residue
Pure DGEBA composite (without filler)	196.0	312.1	600.00	0.00
Pure ENR composite (without filler)	221.2	327.3	650.00	0.00
Pure DGEBA composite (with filler)	223.4	371.9	750.00	0.00
Pure ENR composite (with filler)	235.9	393.8	800.00	0.00
ENR:DGEBA (20:80) hybrid nanocomposite	247.7	440.5	800.00	10.35
ENR:DGEBA (35:65) hybrid nanocomposite	261.6	451.2	800.00	16.96
ENR:DGEBA (50:50) hybrid nanocomposite	294.7	485.7	800.00	28.19
ENR:DGEBA (65:35) hybrid nanocomposite	274.4	461.6	800.00	22.78

[a]Onset degradation temperature
[b]Temperature of maximum rate of mass loss
[c]Final degradation temperature

range of 248–295 °C and a certain quantity of charred residue of carbonaceous products was left. This indicates that the blending of resins enhanced the thermal stability of the composites. The addition of ENR in nanocomposites enhanced the crosslink density of the resin blends, which resist some thermal energy. The increase of loading of ENR content increased the crosslink density up to a certain limit. The 50:50 ratio of ENR/DGEBA epoxy blends produced experimentally excellent properties due to better adhesion and ordered structure of nanocomposites. At higher loading of ENR (65 %), there is intense disruption of the ordered structure of nanocomposites and not favorable for good dispersion.

4.4 Mechanical Properties Testing

The mechanical properties can also give indirect information about interfacial behavior in the composite system, because the interaction between the components has a great effect on the mechanical properties of the composites. The interphase region found between the reinforcement and bulk polymer in polymeric composites is a complicated chemical and mechanical system. The results of mechanical properties are tabulated in Table 2. The properties of hybrid nanocomposites were strongly influenced by the incorporation of ENR. The blending of ENR with DGEBA influenced the improvement in the strength and stiffness of hybrid nanocomposites. The bonding performance of ENR was better than that of DGEBA. When ENR content was 50 %, the composites show relatively excellent properties.

All the nanocomposites were prepared with similar content of 2 wt% of OMMT and 30 wt% of coir fiber as filler to improve the properties. Composites without filler also prepared and compared its properties with the nanocomposites of pure ENR and pure DGEBA containing 2 wt% of OMMT and 30 wt% of coir fiber as filler. The formulated composites showed about 500 % differences in mechanical properties when reinforced with 2 % OMMT and 30 % coir fiber. These results represent major progress in understanding the relationship between the chemistry and the nanomechanical properties in the interphase region of polymeric composites.

There are two reasons to explain the above phenomenon. One is that the polymer chains of ENR and DGEBA are tied on the surface of the OMMT and coir fiber through hydrogen bonds, which improves crosslink density of ENR/DGEBA resin blends. The other may be that the catalytic effect of OMMT accelerates cure rate of ENR resin in greater extent compared to that of DGEBA resin.

Pure ENR and pure DGEBA nanocomposites reinforced with 2 wt% OMMT and 30 wt% coir fiber have some lower value of tensile strength compared to that of ENR/DGEBA epoxy blends-based nanocomposites. This is due to aggregation of large number of OMMT layers in pure ENR and pure DGEBA nanocomposites. Pure ENR nanocomposites have higher tensile strength than pure DGEBA nanocomposites due to good mechanical interlocking and bonding behaviors.

From Table 2, it has been observed that the tensile strain value of composites having no filler is lower than all other nanocomposites due to brittleness of polymeric

Table 2 Mechanical properties of pure resins and its hybrid nanocomposites

Sample	Tensile strength (MPa)	Tensile strain (%)
Pure DGEBA composite (without filler)	6.5	1.4
Pure ENR composite (without filler)	7.4	2.2
Pure DGEBA composite (with filler)	31.3	13.5
Pure ENR composite (with filler)	37.2	14.2
ENR:DGEBA (20:80) hybrid composite	46.7	17.4
ENR:DGEBA (35:65) hybrid composite	52.8	18.7
ENR:DGEBA (50:50) hybrid composite	65.9	19.8
ENR:DGEBA (65:35) hybrid composite	61.1	17.9

materials. The major problems with polymers in engineering applications are the low strength and stiffness when compared with metals and fiber or particle reinforced composites. To offset these deficiencies, reinforced particles or fibers may be added to the resin. The addition of filler improves the strength and strain in composites. The coir (*Cocos nucifera*) is an important lognocellulosic hard and stiff fiber obtained from coconut trees. It has a low tensile strength material among other natural fiber such as jute bagasse, banana, and so on. It possesses some unique properties, which makes it very attractive. It has high lignin content (46 %) and elongation at break value (33 %). It shows high failure strain value of 40–45 % between fiber and polymer. This is the reason to explain the phenomena of comparable tensile strain value among various nanocomposites. Pure ENR and pure DGEBA nanocomposites reinforced with 2 wt% OMMT and 30 wt% coir fiber have some lower value of tensile strain compared to that of ENR/DGEBA epoxy blends-based nanocomposites. Among the blends of epoxy-based nanocomposites, ENR/DGEBA (50:50)-based nanocomposites showed highest value of tensile strain due to high degree of strain compatibility between fiber and matrices (Saw et al. 2011a, 2014).

4.5 Dynamic Mechanical Analysis

Dynamic storage modulus (E') is the most important property to assess the load-bearing capability of a composite material. The ratio of the loss modulus (E'') to be storage modulus (E') is known as a mechanical loss factor (tan δ), which quantifies the measure of balance between the elastic phase and the viscous phase in a polymeric structure. This can relate to impact properties of a material. Generally, the tan δ peak (at low frequency) is at a temperature 10–20 °C above the T_g as measured by dilatometer or differential thermal analysis (DTA). The temperature of maximum loss modulus E'' is very close to T_g.

4.5.1 Storage Modulus (E′)

Figure 5 shows variations in E' values with temperature (30–200 °C). The plots clearly reveal that storage modulus of ENR-based nanocomposites shows 50 % enhancement compared to that of DGEBA-based epoxy nanocomposite, both reinforced with 2 wt% OMMT and 30 wt% coir fiber. It is interesting to note that on blending ENR with DGEBA-based epoxy, the storage modulii at room temperature are enhanced by about 100 % or more in the case of 50 and 65 % ENR–containing matrices; whereas, the enhancement in the case of 20 and 35 wt% ENR-containing matrices is only 50 % of the pure matrix. This observation is as expected because the crosslink density of ENR is much higher compared to that of pure DGEBA; thus with increasing temperature, the slippage is greater in the pure DGEBA compared to that in the pure ENR, resulting in higher storage modulus (E').

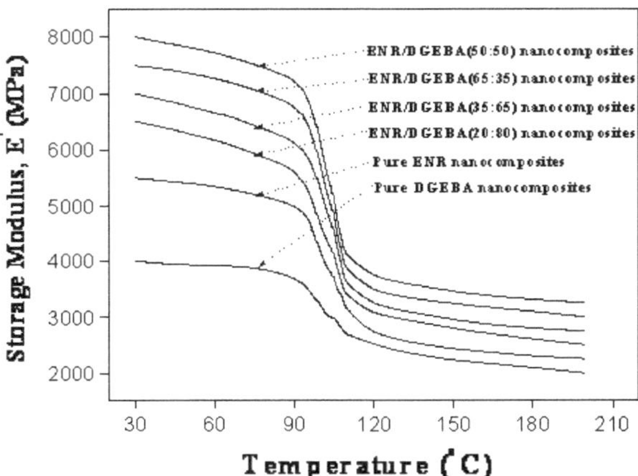

Fig. 5 Variation of storage modulus (E') with temperature for different percentages of ENR-DGEBA epoxy nanocomposites

4.5.2 Loss Modulus (E'')

Figure 6 shows trends of variation of E'' for the different nanocomposite laminates with variation of temperature. The maximum heat dissipation occurs at the temperature where E'' is at a maximum, indicating the T_g of the system. It can be seen that loss modulus peak temperature of DGEBA-based OMMT/coir composite is lower (72 °C) compared to that of ENR-based OMMT/coir composite (81 °C).

Again, by blending ENR and DGEBA-based epoxy, the values of loss modulii peaks are shifted toward higher temperatures. It is interesting to note that the nanocomppsites with 20, 35, and 50 % ENR have T_g of 88, 97, and 99 °C, respectively, whereas 65 % ENR-based nanocomposite has a lower T_g (93 °C).

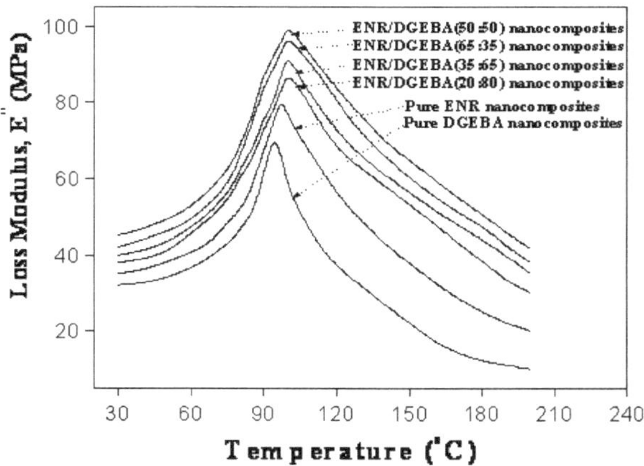

Fig. 6 Variation of loss modulus (E'') with temperature for different percentages of ENR-DGEBA epoxy nanocomposites

It is evident from this observations that ENR offers more rigid networks after curing, its fiber–matrix interaction is strong compared to that of DGEBA-based nanocomposite that has lower stress at maximum load and also a lower T_g compared to that of 20 and 35 % ENR-based composites. The possible explanation may be the synergistic effect of good fiber–matrix interaction attributed to DGEBA and higher extent of crosslinking to the presence of ENR (Otaigbe 1991; Wingard and Beatty 1990).

4.5.3 Mechanical Loss Factor (Tan δ)

Figure 7 shows tan δ versus temperature plots for different nanocomposites. The incorporation of stiff fibers reduces the tan δ peak height by restricting the movement of polymer chains. It is also seen that the tan δ peak heights of coir composites with 20, 35, 50, and 65 % ENR are reduced compared to that of pure DGEBA-based OMMT/coir nanocomposite and pure ENR-based OMMT/coir nanocomposite. The tan δ peak height of ENR-based OMMT/coir nanocomposite is smaller than that of DGEBA-based OMMT/coir nanocomposite, which is probably the result of the higher crosslinking density of the ENR matrix. It is observed in many cases that the improvement of stiffness markedly reduces the ductility (Ghosh et al. 2003). However, through blending ENR with DGEBA-based epoxy resin it is possible to manufacture coir composites with increased stiffness without sacrificing their ductility. It may be seen from the observed value of tensile strain tabulated in Table 2. Thus it is observed that temperatures corresponding to the maximum tan δ of 50 and 65 % ENR-based coir nanocomposites are closer to that of 20 and 35 % ENR-based coir nanocomposite.

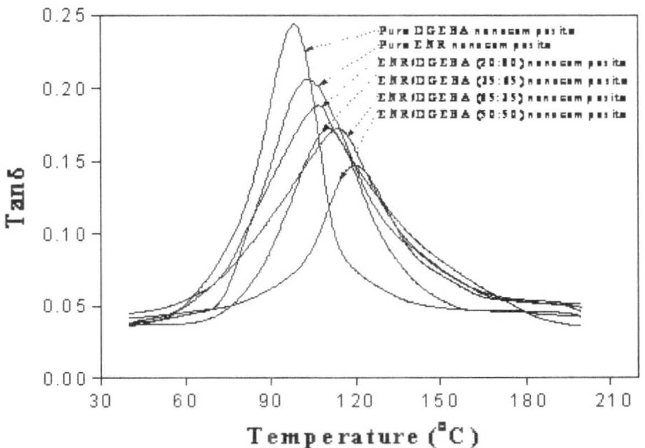

Fig. 7 Variation of tan δ with temperature for different percentages of ENR-DGEBA epoxy nanocomposites

5 Conclusions and Future Perspectives

It is concluded from this study that OMMT/coir fiber and ENR/DGEBA epoxy blends exhibit synergistic effect on the phase structure and thermomechanical properties of ENR/DGEBA/OMMT/coir fiber hybrid nanocomposites. The work reported in this paper attempts to shed light on the critical interphase region. TGA data demonstrates that the exfoliated degree of OMMT and adhesion property of coir fiber in ENR/DGEBA/OMMT/coir fiber nanocomposites is greater than that in pure DGEBA/OMMT/coir fiber and pure ENR/OMMT/coir fiber nanocomposite. Mechanical and DMA shows that ENR/DGEBA/OMMT/coir fiber nanocomposites present a better interface network structure than pure ENR/OMMT/coir fiber and pure DGEBA/OMMT/coir fiber nanocomposites. The special network structure obviously improves dynamic mechanical and thermal properties of ENR/DGEBA/OMMT/coir fiber nanocomposites.

The proximity of the stiff fiber and preferential adsorption of readily diffusible constituents (usually the low molecular weight curative) on the fiber surface may impose a relatively high crosslink density, and hence lead to the development of a more than expected stiffness level at the fiber-matrix interface. At the same time, this may lead to some softening of the matrix in the zone next to the interface because of the notable depletion of the amine curative. The differential opposing effects of matrix stiffening and matrix softening as above in different unitary and hybrid composite laminates may also partly explain the odd observed effects in their dynamic mechanical thermal properties. The observed odd effects about the dynamic properties of blends of ENR/DGEBA epoxy-OMMT/coir fiber-based nanocomposites detailed here should be viewed in the light of the differences in the uncertain nature and degrees of interactions at the fiber–matrix interfaces designed in different odd combinations with respect to selection of matrix blend systems. Subtle variations in reinforcing fiber selection, hand lay-up design, or sequence conditions may lead to odd variations in adhesion zones, buildup of uncertain presence of voids, and odd thermal effects ascribed to the difference in the expansion coefficient of the main phases. These features may combine to cause odd differences in the dynamic properties of the composites as highlighted.

The present work can be further extended to study the aspects of such hybrid nanocomposites. In practice, these nanocomposites are designed to perform in different static and dynamic conditions. The dielectric and corrosion behaviors of nanocomposites are need to be investigated for finding their potential applications as dielectric materials.

Acknowledgments I would like to thank Central Instrumental Facility (CIF) of Birla Institute of Technology, Mesra, Ranchi, India for providing equipment access.

References

Basara C, Yilmazer U, Bayram G (2005) Synthesis and characterization of epoxy based nanocomposites. J Appl Polym Sci 98:1081–1086

Blanchard V, Blanchet P (2011) Color stability for wood products during use: effects of inorganic nanoparticles. BioResour 6:1219–1229

Chen HT, Gao J, Wang G, Shi SQ, Zhang SB, Cai LP (2004) Effect on temperature on calcium carbonate deposition in situ on bamboo fiber and polymer interfaces. Wood Fiber Sci 46:1–12

Datta C, Basu D, Banerjee AN (2002) Mechanical and dynamical mechanical properties of jute fibers-Novolac-Epoxy composite laminates. J App Polym Sci 85:2800–2807

Garea SA, Nicolescu A, Deleanu C, Iovu H (2010) New nanocomposites based on epoxy resins reinforced with modified montmorillonite. Int J Polym Anal Charact 15:497–508

Ghosh P, Mitra PS, Banerjee AN (2003) Photopolymerization of methyl methacrylate with the use of bromine as photoinitiator. J Polym Sci Polym Chem Ed 11(8):2021–2030

Jia N, Li SM, Ma MG, Sun RC, Zhu L (2011) Green microwave-assisted synthesis of cellulose/calcium silicate nanocomposites in ionic liquids and recycled ionic liquids. Carbohydr Res 34:2970–2974

Jia N, Li SM, Ma MG, Sun RC (2012) Rapid microwave-assisted fabrication of cellulose/F-substituted hydroxyapatite nanocomposites using green ionic liquids as additives. Matter lett 68:44–46

Lee A, Lichtenhan JD (1999) Thermal and viscoelastic property of epoxy-clay and hybrid inorganic-organic nanocomposites. J Appl Polym Sci 73:1993–2001

Lee H, Neville K (1967) Handbook of Epoxy Resins. McGraw-Hill, New York

Li SM, Jia N, Ma MG, Zhang Z, Liu QH, Sun RC (2011a) Cellulose-silver nanocomposites: microwave-assisted synthesis, characterization, their thermal stability, and antimicrobial property. Carbohydr Polym 86:441–447

Li WY, Sun N, Stoner B, Jiang XY, Lu XM, Rogers RD (2011b) Rapid dissolution of lignocellulosic biomass in ionic liquids using temperatures above the glass transition of lignin. Green Chem 13:2038–2047

Lin M-F, Thakur VK, Tan EJ, Lee PS (2011a) Dopant induced hollow $BaTiO_3$ nanostructures for application in high performance capacitors. J Mater Chem 21:16500–16504

Lin M-F, Thakur VK, Tan EJ, Lee PS (2011b) Surface functionalization of $BaTiO_3$ nanoparticles and improved electrical properties of BaTiO3/polyvinylidene fluoride composite. RSC Adv 1:576–578

Ma MG, Fu LH, Li SM, Zhang XM, Sun RC, Dai YD (2012a) Hydrothermal synthesis and characterization of wood powder/$CaCO_3$ composites. Carbohydr Polym 88:1470–1475

Ma MG, Fu LH, Sun RC, Jia N (2012b) Compare study on the cellulose/$CaCO_3$ composites via microwave-assisted method using different cellulose types. Carbohydr Polym 90:309–315

Ma MG, Fu LH, Yao K, Tian CH (2014) Microwave-assisted synthesis and characterization of $CaCO_3$ particles-filled wood powder nanocomposites. BioRes 9:3909–3918

Mai C, Militz H (2004) Modification of wood with silicon compounds, inorganic silicon compounds and sol-gel systems: a review. Wood Sci Technol 37:339–348

Otaigbe JU (1991) Dynamic mechanical response of a thermoplastic sheet molding compound-glass fiber composite. Polym Eng Sci 31:104–109

Saw SK, Sarkhel G, Choudhury A (2011a) Surface modification of coir fiber involving oxidation of lignins followed by reaction with furfuryl alcohol: characterization and stability. App Surf Sci 257:3763–3769

Saw SK, Sarkhel G, Choudhury A (2011b) Dynamic mechanical analysis of randomly oriented short bagasse/coir hybrid fibre reinforced epoxy novolac composites. Fibers Polym 12:506–513

Saw SK, Sarkhel G, Choudhury A (2012) Preparation and characterization of chemically modified jute-coir hybrid fiber reinforced epoxy novolac composites. J Appl Polym Sci 125:3038–3049

Saw SK, Akhtar K, Yadav N, Singh AK (2014) Hybrid composites made from jute/coir fibers: Water absorption, thickness swelling, density, morphology and mechanical properties. J Nat Fibers 11:39–53

Singha AS, Thakur VK (2008a) Fabrication of hibiscus sabdariffa fiber reinforced polymer composites. Iran Polym J 17(7):782–791

Singha AS, Thakur VK (2008b) Effect of fiber loading on urea formaldehyde matrix based green composites. Iran Polym J 17(11):861–873

Singha AS, Thakur VK (2008c) Fabrication and study of lignocellulosic hibiscus sabdariffa fiber reinforced polymer composites. Bioresources 3:1173–1186

Singha AS, Thakur VK (2008d) Synthesis and characterization of pine needles reinforced RF matrix based biocomposites. E-J Chem 5:1055–1062

Singha AS, Thakur VK (2009a) Chemical resistance, mechanical and physical properties of biofiber based polymer composites. Polym Plast Technol Eng 48(7):736–744

Singha AS, Thakur VK (2009b) Fabrication and characterization of S. cilliare fiber reinforced polymer composites. Bull Mater Sci 32(1):49–58

Singha AS, Thakur VK (2010a) Synthesis, characterization and study of pine needles reinforced polymer matrix based composites. J Reinf Plast Compos 29(5):700–709

Singha AS, Thakur VK (2010b) Synthesis and characterization of short Grewia optiva fiber based polymer composites. Polym Compos 31:459–470

Singha AS, Thakur VK (2010c) Renewable resources based green polymer composites: analysis and characterization. 15:127–214

Singha AS, Thakur VK, Mehta IK, Sharma A, Khanna AJ, Rana RK, Rana AS (2009) Surface-modified hibiscus sabdariffa fibers: physicochemical, thermal, and morphological properties evaluation. Int J Polym Anal Charact 14(8):695–711

Thakur VK, Thakur MK (2014) Processing and characterization of natural cellulose fibers/thermoset polymer composites. Carbohydr Polym 109:102–117

Thakur VK, Singha AS, Misra BN (2011) Graft copolymerisation of methyl methacrylate onto cellulosic biofibers. J Appl Polym Sci 122(1):532–544

Thakur VK, Singha AS, Thakur MK (2012a) Graft copolymerisation of methyl acrylate onto cellulosic biofibers: synthesis, characterization and applications. J Polym Environ 20:164–172

Thakur VK, Yan J, Lin M-F et al (2012b) Novel polymer nanocomposites from bioinspired green aqueous functionalization of BNNTs. Polym Chem 3:962–969

Thakur VK, Thakur MK, Gupta RK (2013a) Graft copolymers from cellulose: synthesis, characterization and evaluation. Carbohydr Polym 97:18–25

Thakur VK, Singha AS, Thakur MK (2013b) Fabrication and physico-chemical properties of high-performance pine needles/green polymer composites. Int J Polym Mater Polym Biomater 62:226–230

Thakur VK, Singha AS, Thakur MK (2013c) Ecofriendly biocomposites from natural fibers: mechanical and weathering study. Int J Polym Anal Charact 18:64–72

Thakur VK, Singha AS, Thakur MK (2013d) Synthesis of natural cellulose-based graft copolymers using methyl methacrylate as an efficient monomer. Adv Polym Technol 32: E741–E748

Thakur VK, Thakur MK, Gupta RK (2013e) Development of functionalized cellulosic biopolymers by graft copolymerization. Int J Biol Macromol 62:44–51

Thakur VK, Thakur MK, Raghavan P, Kessler MR (2014a) Progress in green polymer composites from lignin for multifunctional applications: a review. ACS Sustain Chem Eng 2(5):1072–1092

Thakur VK, Thakur MK, Gupta RK (2014b) Review: raw natural fiber–based polymer composites. Int J Polym Anal Charact 19(3):256–271

Thakur VK, Vennerberg D, Kessler MR (2014c) Green aqueous surface modification of polypropylene for novel polymer Nanocomposites. ACS Appl Mater Interfaces 6:9349–9356

Wingard CD, Beatty CL (1990) Crosslinking of an epoxy with a mixed amine as a function of stoichiometry. II Final properties via dynamic mechanical spectroscopy. J Appl Polym Sci 41:2539–2544

Multifunctionalized Carbon Nanotubes Polymer Composites: Properties and Applications

Nurhidayatullaili Muhd Julkapli, Samira Bagheri and S.M. Sapuan

Abstract Carbon nanotubes (CNTs) is a rigid rod-like nanoscale material produced from carbon in powder, liquid, or gel form via acid or chemical hydrolysis. Due to its unique and exceptional renewability, biodegradability, mechanical, physicochemical properties, and abundance, the incorporation associated with a small quantity of CNTs to polymeric matrices enhance the mechanical and thermal resistance, and also stability of the latter by several orders of magnitude. Moreover, NCC-derived carbon materials are of no serious threat to the environment, providing further impetus for the development and applications of this green and renewable biomaterial for lightweight and degradable composites. Surface functionalization of CNTs remains the focus of CNTs research in tailoring its properties for dispersion in hydrophilic and hydrophobic media. Through functionalization, the attachment of appropriate chemical functionalities between conjugated sp^2 of CNTs and polymeric matrix is established. It is thus of utmost importance that the tools and protocols for imaging CNTs in a complex matrix and quantify its reinforcement, antimicrobial, stability, hydrophilicity, and biodegradability are be developed.

Keywords CNTs · Composites · Polymer · Functionalization and applications

1 Introduction

1.1 Polymeric Nanocomposites: Advantages and Limitation

Polymer composites are made up of a polymeric matrix with some physically distinct distributed phases called reinforcements, or fillers (Richard and Giannelis

N.M. Julkapli · S. Bagheri
Nanotechnology & Catalysis Research Centre (NANOCAT), IPS Building, University Malaya, 50603 Kuala Lumpur, Malaysia

S.M. Sapuan (✉)
Department of Mechanical and Manufacturing Engineering, Universiti Putra Malaysia, 43400 UPM Serdang, Selangor, Malaysia
e-mail: sapuan@upm.edu.my

V.K. Thakur and M.K. Thakur (eds.), *Eco-friendly Polymer Nanocomposites*, Advanced Structured Materials 75, DOI 10.1007/978-81-322-2470-9_6

2001; Zheng-Ming et al. 2003; Paul et al. 2007; Rohan and Darrin 2007; Xiaofeng et al. 2011). The reinforcing fillers combined with the polymeric matrix result in preferred qualities, such as high stiffness, strength, flame redundancy, scratch/wear resistance, toughness, thermal/electrical conductivity, electromagnetic shielding, coefficient thermal expansion, wear, and damping resistances (Zheng-Ming et al. 2003). The polymeric nanocomposites consist of a polymer with nanoparticles or nanofillers dispersed in its matrix (Paul et al. 2007; Thakur et al. 2012). It is expected that the transition from micro to nanoparticles increase the surface area-to-volume ratio (Thakur et al. 2014a, b). This in turn results in a prominent increment of the behavior of the atoms on the surface of the particles. It affects the properties of the particles when they react with other particles (Yuan-Qing et al. 2008). Due to the higher specific surface area of nanoparticles, the interaction with other particles within the mixture became more intense (Dubois and Alexandre 2006). This consequently results in positive properties, such as high temperature capability, resistance against corrosion, noise damping, low in cost/manufacturer, ductile, high specific stiffness and strength, high thermal conductivity, and low coefficient of thermal expansion (Fig. 1). Another advantage of polymer nanocomposites is that it could be fabricated via rapid and precise manufacturing methods, such as injection molding, compression molding, vacuum bag molding, contact molding, and resin

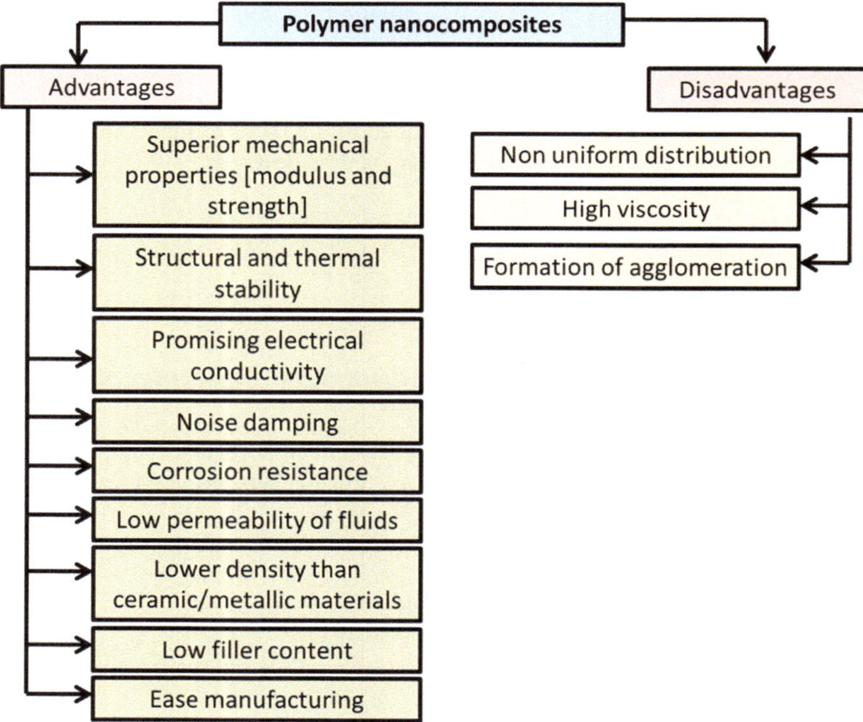

Fig. 1 Advantages and disadvantages of polymeric nanocomposites

transfer molding (Peter and Richard 2002; Hua and Brinson 2005). Therefore, polymeric nanocomposites are posited as appropriate options in overcoming the inherent restrictions of microcomposite and monolithic, while posing preparation challenges related to the control of elemental composition and stoichiometry in the nanocluster phase.

In the development of polymer nanocomposites, there are several challenges and limitations. For example, polymeric nanocomposites require controllable mixing/compounding, stabilization of the dispersion, and orientation of the dispersed phase. Despite the fact that the modulus of polymeric nanocomposites increases with the increasing nanofiller content, toughness, and thus it impacts the strength which decreases when the materials became more brittle (Kiliaris and Papaspyrides 2010; Xiao-Lin et al. 2004). The viscosities of polymeric nanocomposites also increase with the nanofiller content, which render manufacturing difficult (Adams and Charles 2001). Furthermore, a highly viscous flow of polymer melts induced large forces or cause short shots during extrusion and injection molding. In other words, the effect of nanofiller on the polymer properties differs from predicted using the thermodynamic studies for reduced particle size filler (Gary and Dimitris 2008). Studies and modeling using continuum mechanics revealed that the enhanced properties of nanocomposites strongly depend on particular features of the nanofiller system, particularly its content, aspect ratio, and the ratio of filler mechanical properties to those of the matrix. Furthermore, uniform dispersion of fillers (micro/nano) particles/fibers within the polymer matrix is limited due to the formation of agglomerates (Singha et al. 2009a, b; Yuan-Qing et al. 2008; Thakur et al. 2012). Agglomeration induced defects that limit the mechanical performance of the polymeric composite materials (Gary and Dimitris 2008; Thakur et al. 2014a, b, c).

2 CNTs: General View

Carbon nanotubes (CNTs) were discovered in 1991, and since then many studies were dedicated to it and its related nanomaterials due to its superior electronic, chemical, and mechanical properties (Li et al. 1996; Micheal et al. 2002; Min-Feng et al. 2000a, b; Philip et al. 2000). The general structure of CNTs is depicted as a rolled up sheet of a planar-hexagonal arrangement of carbon atoms dispersed in a honeycomb lattice (Micheal et al. 2002). There are two major categories of CNTs; single-walled (SWCNTs) and multiwalled (MWCNTs).

2.1 CNTs: Properties

CNTs exhibited unique mechanical, thermal, and field emission properties and electrical conductivity (Min-Feng et al. 2000a, b) (Table 1). It is claimed that CNTs have an elastic modulus that is higher than carbon fibers, and is five times stronger

Table 1 List on properties and characterization of CNTs

Properties	Testing/analysis	Results	References
Stiffness	Observation the amplitude of thermal vibration inside the transmission electron microscopy	1.8–1.25 TPa	Min-Feng et al. (2000a, b)
Tensile strength	A stress–strain measurement utilizing a nanostressing stage operating in the scanning electron microscope	Outer shell of MWCNTs (11–63 GPa)	Min-Feng et al. (2000b) and Demczyk et al. (2002)
		Fracture strains (12 %)	
		Modulus (270–950 GPa)	
		Strength is 10–100 times more than the strongest steel	
Elastic modulus	A stress–stains curve	1 TPa	Treacy et al. (1996) and Jian (1997)
Thermal/ electrical stability	High-temperature differential scanning calorimetric analysis	2800 °C	Savas et al. (2000) and Hone et al. (1999)
		Electrical conductivity twice higher than diamond and 1000 times higher than Cu wire	

than carbon fibers. Its strength is determined by the number of defects, bundles of SWCNTs, and interlayer interactions within MWCNTs (Philip et al. 2000). The structural defects, together with twists or bends, considerably influence the mechanical strength of the CNT. It is shown that the CNTs absorb near-infrared light at wavelengths that are optically transparent to native tissues (Kenji et al. 2000). This allows selective drug delivery that is capable of sufficient heating and killing the target cell. In addition to the mechanical behavior, CNTs also possess very high intrinsic electrical conductivity. The electrical conductivity of CNTs is in the range of 107–108 S m^{-1}, which is comparable to metals (Li et al. 1996). The room temperature conductivity of metallic SWCNTs was found to be 105–106 S m^{-1}, and for CNTs, which is a semiconductor, it is about 10 S m^{-1}. Its electrical conductivity assisted in imparting conductivity in remarkably insulating materials (Min-Feng et al. 2000a, b). Certain theoretical studies on the electronic properties of SWCNTs pointed out that CNTs shells depend on which helicity could be metallic or semiconducting (Tang et al. 2001). This was analyzed due to the weak control on generation; more than 30 % of SWCNTs formed are metallic, while the rest are semiconductors. The axial thermal conductivity of individual, perfect CNTs were showed to be as high as 3300 W m^{-1} K^{-1} (Brigitte et al. 2000). Due to these factors and also their superior electrical and thermal properties, lots of consideration have been dedicated to the use of CNTs as reinforcement in polymeric composite systems (Li et al. 1996).

2.2 CNTs: Synthesis Process

One of the most intriguing problems in the synthesis of CNTs is to understand its microscopic growth mechanism and determine ways of controlling it (Journet et al. 1995). Currently, experimental techniques have been developed, and CNTs could be produced with various methods and very flexible environments (from higher than 3000 °C of arc discharge, to laser ablation to as low as 500 °C of chemical vapor deposition methods to control the growth of CNTs (Table 2). The optimum pressure of the arc-discharge method of up to 500 Torr resulted in more than 75 % conversion of CNTs in large quantities (Hutchisona et al. 2001). In addition to the aforementioned main generation methods, there are other parameters to these routes, such as ball milling, cold water, SiC decomposition, graphene scrolling, and flame synthesis (Hwa-Jeong et al. 2005; Zhu et al. 2005; Pierard et al. 2001; Li et al. 1999). Due to its unique quasi one-dimensional structures, CNTs have different chirality, diameters, and layers, which were in turn brought about from different growth conditions and behaviors (Scott et al. 2001). For example, hollow tubes of CNTs, with a C range between 2 and 50 nm in diameter, are produced by a mixture of benzene and H2 decomposition using arc-discharge apparatus at low pressures of argon (100 Torr) (Journet et al. 1995). Furthermore, the addition of catalyst to the synthesis of CNTs plays an important role in its nucleation and sustained growth (Stig et al. 2004; Chris et al. 2000). A noteworthy aspect is the emergence on non-metal catalyst; these might well replace metallic catalyst system due to their potential of yielding high-purity samples, and compatibility with silicon technology (Hwa-Jeong et al. 2005).

Meanwhile, CNTs with a fully interconnected two-dimensional ring network has been grown by low-temperature chemical vapor deposition prepared from nano-channel network template in porous anodic alumina (Stig et al. 2004; Chris et al. 2000). In this case, CNTs strictly grow in a both-tip mechanism; with the ends open and growing forward in both directions by the incorporation of C clusters (Fig. 2). Meanwhile, MWCNTs have 2–50 walls or concentric tubes prepared by the deposition of carbon evaporation from the anode for condensation on the cathode (Chris et al. 2000).

Therefore, different growth mechanisms were proposed to explain the underlying initiating process and dynamical growth, which focuses on the metal catalyst-assisted growth. In this case, the precipitation of diffusion of C atoms at the catalyst's surface is believed to precipitate the continued growth of CNTs (Pierard et al. 2001; Li et al. 1999). However, these proposed mechanisms are still highly controversial, due to the lack of experimental proofs and inability of explaining the growth behavior.

In the last few years, great advances have been made regarding SWCNTs separation, based on metallicity. Positive developments were made in controlling and optimizing the generation of CNTs, as well as its separation and purification via chirality and metallicity (Hongjie et al. 2002; Rodney et al. 2002). Therefore, SWCNTs were formed as a small amount of metal particle placed on a dimple

Table 2 The synthesis methods on production of CNTs

Synthesis methods	Description	Advantages	Disadvantages	References
Arc discharge	The arc ignited between two electrodes of graphite in an H_2 gas	Mass production of CNTs and fullerenes	Multimorphology shoots productions	Hutchisona et al. (2001)
	The arcing evaporates the C and while it cools and condenses that some of the product forms as filamentous C on the cathode	Production of both SWCNTs and MWCNTs	Require several purification steps	
			High temperature process	
Laser ablation	Direction of intense pulse of laser light on a C surface in a stream of He gas	Combination of a metal catalyst in the C target results in the formation of SWCNTs with a tiny diameter distribution and high	Not suitable for mass production	Scott et al. (2001)
	The evaporated material condenses to yield fullerenes	SWCNTs yield and diameter distribution could be varied by controlling the process parameters		
Chemical vapor deposition (CVD)	The CVD process in that volatile precursors utilized to provide a C feed source to a catalyst particle or pore at elevated temperature around 350–1000 ° C, heated flow of CO, pressure between 1 and 10 atm	An extensive method which also shows multivariable process can adjust in a several manner like plasma enhanced CVD, thermochemical CVD, aerogel supported, high pressure CO disproportionation, alcohol catalytic CVD, aerosol assisted CVD, and hybrid laser assisted thermal CVD	Modification process parameters needed to control SWCNTs diameter distribution and yield	Stig et al. (2004) and Chris et al. (2000)
		Mass production		

cathode and a mixture of argon and methane atmosphere utilized during arc discharge (Hutchisona et al. 2001). For example, 2 at.% Co-containing anode utilized in the arc-discharge apparatus under the atmosphere results in an 80 % selectivity of

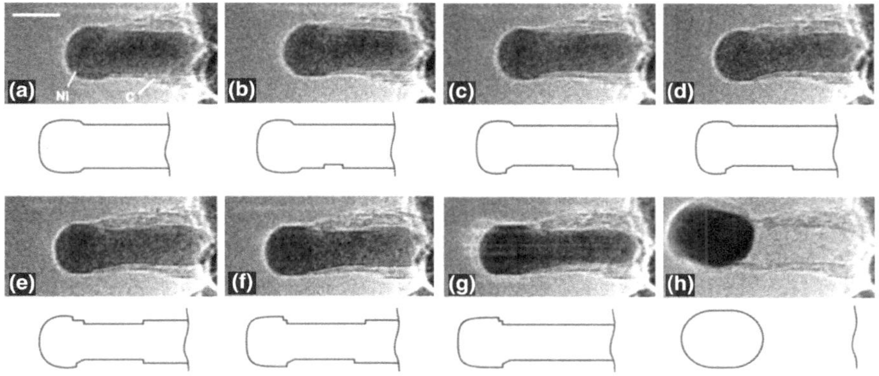

Fig. 2 In situ HRTEM image sequence of the growing CNTs (Scale bar = 5 nm). Images (**a–h**) show one cycle in the elongation/contraction process (Stig et al. 2004)

SWCNTs. Then, the synthesis of SWCNTs via evaporation of a hot (1200 °C) transition metal containing C target by the laser ablation technique is followed by the condensation on a cold finger, yielding up to 80 % purity at 50 kg day^{-1} (Shigeo et al. 2002; Flahaut et al. 2000).

2.3 CNTs: Applications

The CNTs have exposed completely new paths intended for establishing novel functional materials. Thus, some applications seek to exploit CNTs with respect to different fields (Table 3). The combination of CNTs with some macromolecules improves the conductivity of the material, representing one avenue of application (Brigitte et al. 2000; Demczyk et al. 2002). The large environmental window and the electrochemical stability draw essential values onto CNTs (Jian 1997). Additionally, the extension of the functional techniques for two-dimensional carbon, including grapheme, is a growing area in semiconductor applications.

Table 3 List on applications of CNTs corresponding to its characteristics

Applications of CNTs	Characteristics of CNTs
Structural applications	High tensile strength fibers
	Fire resistance properties
Artificial muscles	Electromagnetic properties
Loudspeakers	Parallel carbon sheets
Air/water filtration	Electroacoustic potential properties
	Fast oscillators
Electronic devices	Semiconductor properties
	Field emission properties

3 CNTs: In Polymeric Composites

Since surface characteristics influence its proinflammatory effect, embedding CNTs in polymeric materials modifies the surface environment. This, in turn, could modify its toxicity, thus representing a useful strategy in reducing adverse health effects of industrially produced CNTs (Joseph et al. 2005; Andrews and Weisenberger 2004; Huisheng 2008; Peng et al. 2008). Furthermore, there are great challenges and opportunities expected for the CNTs as nanoscopic reinforcement in polymer matrices (Andrews and Weisenberger 2004). These opportunities include CNTs with a small number of defects per unit length possessing 27,500 times higher specific surface area per gram according to the equivalent volume fraction of typical carbon fiber, and a high aspect ratio, mostly exhibiting great tensile, thermal, and electrical properties (Huisheng 2008).

In addition to the economic advantages caused by combining expensive CNTs and cheap polymer, it is also possible where a synergy presents itself between the CNTs and polymeric materials (Breuer and Uttandaraman 2004). This, in turn, brought about the simple rule of mixture, which fully utilize CNTs properties in producing a composite system with promising properties.

Furthermore, due to their hollow nature, CNTs can be opened and filled with various materials such as biological molecules, which in turn generate technological opportunities (Chenyang et al. 2003; Petra et al. 2002; Myounggu et al. 2008). This combination addresses the challenges in producing homogeneous dispersion and strong interfacial interactions, improving surface grafting/functionalization. To tailor and optimize the properties of CNT-filled polymer composites, it is necessary to disperse the CNTs homogeneously with the sustenance of strong interaction and adhesion of composite components via several proposed methods (Table 4). Finally, nanoreinforcements using biodegradable polymers possess a substantial possibility of the structure of eco-friendly green materials regarding future applications (Joseph et al. 2005).

3.1 CNTs: In Synthetic Polymeric Composites

3.1.1 CNTs: In Thermoset Polymeric Composites

Thermoset resin is a petrochemical in a viscous state or soft solid, which changes irreversibly straight into an infusible, insoluble polymer system via curing (Pickering et al. 2000; Wim and Richard 2004; Makki et al. 2005; Torresa et al. 2000). The curing process of thermosets could be induced via radiation or heat. The actual curing procedure converts the resin into a rubber or plastic through cross-linking (Wim and Richard 2004). Adding energy and catalysts results in the

Table 4 Mainly used methods with commercial viability in preparation of CNTs-filled polymer composites

Preparation methods	Advantages	Disadvantages	References
Solution mixing	Simplest and most widely used methods	Compatibility issue between functionlized CNTs and polymer matrix	Zhaoxia et al. (2001) and Zdenko et al. (2010)
	CNTS and polymer mixed with a suitable solvent, evaporated in control conditions	Agglomeration of CNTs takes place after evaporation process	
	Acceptable for wide range of polymer		
	Obtain a good dispersion with ultrasonic agitation methods		
Melt processing	Acceptable for polymer with solution mixing approach problem	Less efficient than solution mixing due to the high viscosity of thermoplastic polymer	Wenzhong et al. (2003) and Haggenmuellera et al. (2000)
	Methods involved melting of the polymer to form viscous liquid followed by blending with CNTs	Hindrance in achieving uniform dispersion of CNTs	
	Dispersion of CNTs improved by shear mixing		
In situ polymerization	CNTs dispersion into the monomer matrix in the presence or absence of solvent which followed by standard methods of polymerization	Limited number of polymer used	Seung et al. (2003) and Fenga et al. (2003)
	Enables the grafting of polymer molecules on CNTs		
	Better dispersion coefficients		
	Better interactions between CNTs and polymeric matrix		
	Process deal with insoluble and thermally unstable polymer		

molecular chains being able to react at chemically active sites, linked into a rigid three-dimensional structure. The cross-linking procedure forms a molecule with a much larger molecular weight, leading to a material with a much higher melting point. Throughout the reaction, the molecular weight increased to a point so that the melting point exceeds the ambient temperature, and the material forms a solid material (Toressa et al. 2000). The CNT-filled thermoset polymer composites have been fabricated and studied with different kinds of thermosets, such as epoxides, polyester, and polyimide resin. Most of the mechanical, conductivity, and thermal stability of the thermoset resin increases with the addition of low content CNTs.

CNTs: In Epoxides Polymeric Composites

The reinforcement of CNTs into epoxy improved mechanical properties, such as strength, stiffness, and durability (Huang et al. 2014a, b, c; Battisti et al. 2014a, b; De Borbon et al. 2014). Furthermore, MWCNTs with epoxy coatings increase the adhesion strength of the matrix, and exhibit hydrophobicity, low water intake, and corrosion resistance, flame retardant, and antioxidation properties (Koreyam et al. 2014; Grabowski et al. 2014). CNT-filled epoxy composites comprised most of studies utilizing the in situ polymerization method, where the CNTs were first dispersed in the resin and cured with a hardener (Korayem et al. 2014). Meanwhile, there are some studies that prepare epoxy composites by this method, utilizing carboxylated end-cap SWCNTs and an esterification reaction to fabricate composites with enhanced tensile modulus (Guo et al. 2014). It is worthy pointing out that as polymerization moves along, the viscosity is enhanced (Gardea and Lagoudas 2014; Wang et al. 2014a, b, c). Thus, the addition of CNTs into epoxy resin increases the mechanical properties, but only to a point. For example, monotonic increased with hardness was observed to up to a factor of 3.5 by loading of 2 wt% of SWNT into the epoxy matrix (Wang et al. 2014a, b, c; Bal and Saha 2014). Furthermore, the measured fracture energy increased from 133 to 223 J m^{-2} with the addition of 0.5 wt% of CNTs. In order to transfer the superior properties of the epoxy matrix, the functionalization of as-prepared CNTs is crucial for realizing proper dispersion and strong interfacial bonding (Fig. 3) (Kuzhir et al. 2013; Rajendra et al. 2013; Li et al. 2013a, b, c, d, e; Florian et al. 2003).

CNT's electrical conductivity-filled epoxy nanocomposites, with less than 0.5 wt % of CNTs, were improved by several orders of magnitude (Jiang et al. 2013a, b; Russ et al. 2013; Prolongo et al. 2013, He et al. 2013; Safdari and Al-Haik 2013). Furthermore, thermal conductivity of epoxy matrix at room temperature increased by 300 % on 3 wt% SWCNTs loading, and an additional increase of 10 % once they are magnetically aligned (Li et al. 2013a, b, c, d, e). Similar observations were found on epoxy nanocomposites with 1 wt% raw laser oven SWCNTs, and recorded a 125 % thermal conductivity enhancement. In addition, the CNT's alignment plays a key role in improving the transport properties of CNT-filled epoxy composites (He et al. 2013). Compared to its non-aligned counterpart, 10 % increment in thermal conductivity was recorded with aligned MWCNTs. Moreover, the ultra-low electrical percolation threshold of the 0.0025 wt% in aligned CNT-filled epoxy composites were correspondingly recorded (Russ et al. 2013). This affect the aspect ratio of CNT-filled polymer composite vis-à-vis the electrical shielding properties. Some studies focused on the electrical conductivity properties of CNT-filled epoxy nanocomposites with respect to the aspect ratio and percolation threshold of CNTs. It is found that there is an eight times decrease in the threshold concentration in MWCNT-filled epoxy composites as its length increased from 1 to 50 μm (He et al. 2013). Meanwhile, the minimum percolation threshold concentration of MWCNT-filled epoxy was recorded at 0.0021 wt% of MWCNTs (Safdari and Al-Haik 2013). Furthermore, there are some reports on the effect of surface functionalization of CNTs toward the electrical conductivity of the

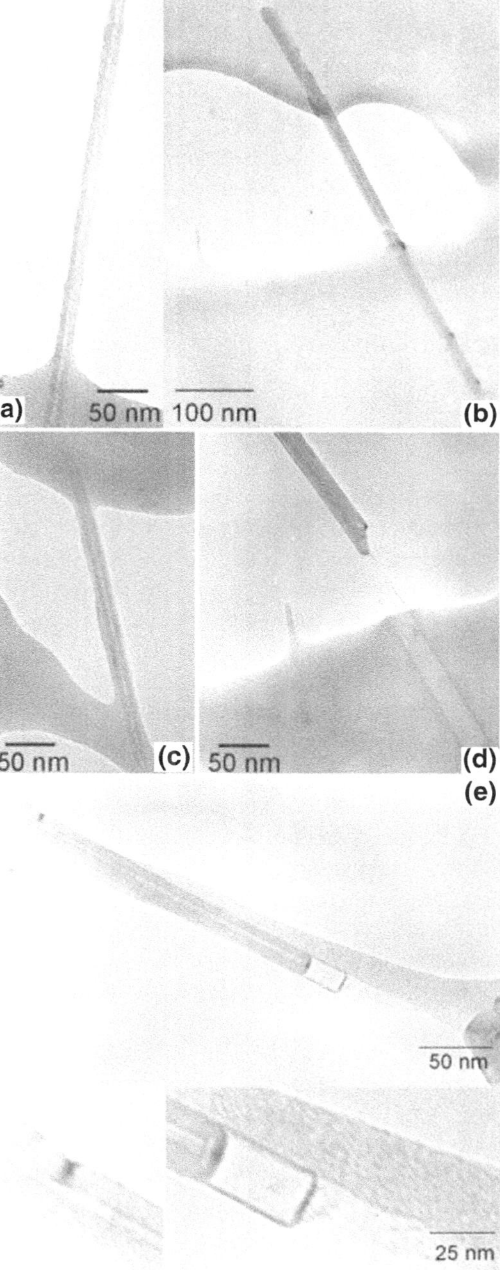

Fig. 3 TEM images of functionalized CNTs. Epoxides covers the surface of the CNTs which indicates an improved interaction (**a**). CNTs improve the fracture toughness by bridging pores and microcracks in the epoxies (**b** and **c**). Telescopic pull-outs substantiate the evidence of improved interactions (**d** and **e**) (Florian et al. 2013)

nanocomposites. It is found that for nanocomposites, the incorporation of octadecylated and acid functionalized MWCNTs in the epoxy resin, reducing the electrical conductivity (Abu et al. 2006; Jeena et al. 2010). Therefore, it is very important to improve the modification reagent or condition of CNTs to minimize the degradation of electrical properties.

A noteworthy enhancement in the mechanical and electrical conductivity of CNT-filled epoxy composites lead to the development of conductive materials for electronics, automotive shielding, electrostatic dissipation, conductive coating, multilayer printed circuits, and electromagnetic inference.

CNTs: In Polyester Polymeric Composites

The CNT-filled unsaturated polyester composites with styrene have extensive usage in industrial applications included structural (Jung and Park 2013), automotive, (Seyhana et al. 2007a, b, c) aerospace (Liang et al. 2009), and others. Conventionally, the composites were fabricated through three-roll ill and sonication technique, which fabricated CNTs with and without NH_2 functional groups and polyester. The CNT-filled polyester suspension demonstrated a shear loss behavior, while the polyester resin combination behaves in the manner of a Newtonian fluid. Improvements within the character of the rheology of the CNTs/polyester suspension checked like a function of the level of energy, introduced via ultrasonic horn mixing and associated with microscopic observations.

However, the reported improvement on mechanical and thermal properties of CNT-filled polyester composites is considerably lower than the expectation due to difficult alignment of CNTs, weak dispersion, and poor interface between polyester matrix and CNTs, which are usually associated with geometrical properties of CNTs, polyester properties, and fabrication methods (Liao et al. 2011; Hossain et al. 2011; Agnihotri and Kar 2007). To overcome these obstacles, various efforts included ultrasonication, surface treatment, shear mixing, bi-tri-axial rolling, extrusion, and combination process, all of which were designed to properly accomplish excellent dispersion of CNTs in polyester (Hossain et al. 2011). Furthermore, several methods were suggested for managing CNT alignment in polyester by utilizing shear, elongation, and melt processing, as well as magnetic field or electrical spinning (Matthew and Virginia 2009; Qiao et al. 2006). Furthermore, the self-polymerization and styrene evaporation at high temperatures are main issues that need to be accounted for whenever a thermoset polyester resin was blended together with CNTs by utilizing the three-roll milling and sonication technique. It is surmized that the three-roll milling technique is more suitable for dispersing CNTs in polyester resin blends compared to other techniques such as direct mixing and sonication (Matthew and Virginia 2009). Another study prepared CNTs/polyester composites by shear mixing with no solvents. In this case, additional energetic mixing of the condition generated greater dispersion at both the nanoscopic and microscopic levels. The results demonstrate that the dispersion depends on the high shear conditions on the structure and nature of nanofilaments

(Seyhana et al. 2007a, b, c). The most effective dispersion is realized by demonstrating the minimum percolation threshold, which did not correspond to the most energetic mixing conditions. Moreover, lower nanofilament concentrations resulted in a much better dispersion, which demonstrates superior mechanical performance. With regards to electrical resistivity properties, the quality of the CNTs dispersion within the polyester matrix was studied using optical microscopy (Cao et al. 2003). The results showed that polyester matrix is suitable for the preparation of electrically conductive thermosetting nanocomposites at low CNT concentrations.

Moreover, surface functionalization of CNTs influenced the final properties of the composites. Thus, they are focused on enhancing the CNTs/polyester master batches without styrene through various kinds of functional groups to obtain the desired mechanical properties and microstructure of composites (Seyhana et al. 2009; Hilmi et al. 2010; Esteban et al. 2013; Ziyan et al. 2014).

CNTs: In Polyimides Polymeric Composites

Polyimides are broadly utilized in the manufacture of aircraft assemblies, packaging materials, microelectronic devices, interlayer dielectrics, and circuit boards (Cui et al. 2013; Ko et al. 2014; Jiang et al. 2014a, b). This is due to its special structure, flexibility, good dielectric properties, great glass transition temperature, excellent thermal stability, radiation resistance, and thermal and mechanical characteristics (Wu et al. 2013a, b, c). The electrostatic charge is accumulated on the surface of polyimides, due to its insulating nature, which causes local heating, and consequently leads to premature material degradation. Therefore, the promising mechanical strength, thermal stability, and surface resistivity of Polyimide could be realized by the addition of CNTs as filler (Jia et al. 2012). Polyimide/CNTs composites can be prepared using various fabrication techniques, such as polymerization, wet casting, and efficient solution.

It is suggested that in situ polymerization is one the most suitable fabrication technique of CNT-filled polyimides composite, which also results in the introduction of certain level of electrical conductivity despite lower loadings of CNTs (Wang et al. 2014a, b, c). There are some reports on the synthesis of SWCNT-reinforced polyamide composites via the sonication of in situ polymerization of diamine and dianhydride (Chen et al. 2011; Schlea et al. 2012). Other studies have fabricated CNT-filled polyimides composites by in situ polymerization, utilizing 4,4′-oxydianilline, MWCNTs, and pyromellitics dianhydride, continued with casting, evaporation, and also thermal imidization (Xiaowen et al. 2006; Hyang et al. 2007). The incorporation of 3 wt% MWCNTs improved the mechanical features of polyimide due to the presence of a robust interfacial interaction between the CNTs and polymer matrix (Hyang et al. 2007). It is also pointed out that the tensile strength increased from 102 MPa for neat polyimide, to 134 MPa for the 6.98 wt% MWCNTs/polymides composites (Xiaowen et al. 2006). Furthermore, pretreatment of CNTs in solvent released enough CNTs, which resulted in the percolation of solvents into the network. This, in turn, forms a large amount of entanglements between CNTs and the polymer molecular chains (Zhang 2011; Pei et al. 2011). CNT-filled with polyamide could be fabricated via the wet casting technique (Tang et al. 2010a, b; Lu et al. 2011; Ribeiro et al. 2012a, b). In this case, both CNTs and polyimide matrices are soluble in certain organic solvents; allow close mixing of solutions, and subsequent fabrication of the composites (Lu et al. 2011). Moreover, in situ polymerization, with the dispersion of the CNTs, leads to a composite with good electrical, mechanical, optical, and thermal properties (Tang et al. 2010a, b). Furthermore, an effectual solution process (Lu et al. 2011) could prepare the polyimides/MWCNTs nanocomposites. Through this method, the MWCNTs were well dispersed, and their structures remained similar in the final resulting nanocomposites.

The electrical conductivity of polyimides is improved by more than 11 orders of magnitudes to 10^{-4} S cm^{-1} at the percolation threshold by the addition of 0.15 % vol CNTs (Tsai et al. 2010). Moreover, the nanocomposites containing 10 wt% of MWCNTs resulted in the dielectric constant reaching 60, which are about 17 times of 3.5 for pure polyimide (Thuau et al. 2009). The electrical resistivity of the nanocomposites surface was reduced from 1.28 × 1015 ohm cm^{-2} for neat polyimide, to 7.5 × 106 ohm cm^{-2} by the addition of 6.98 wt% of MWCNTs (Myung et al. 2010; Sun et al. 2008). The frequency behavior of specific admittance of 0.05 vol% of CNTs-filled polyimides composites determined that its conductivity properties follow a percolation-like power law, with a comparatively low percolation threshold (Tzeng et al. 2008; Guo et al. 2009; Zha et al. 2013). The measurement of the current-voltage demonstrated that the composites displayed a non-ohmic behavior, representing a quantum tunneling conduction procedure (Itoh et al. 2008). Thus, it is concluded that the conductivity of the composites results from the formation of conducting pathways to the polyimides by CNTs (Bong et al. 2006; Yang et al. 2007; Kim et al. 2007; Li and Bai 2011). Therefore, based on the concentration of CNTs, it is possible to modify the conductivity of the composite (Shigeta et al. 2006; Ogasawara et al. 2004).

3.1.2 CNTs: In Thermoplastic Polymeric Composites

The CNT-filled thermoplastic composites have been effectively introduced into an extensive range of applications formerly owned by thermoset composites (Ortengren 2000; Kanagaraj 2010; Nie and Fisher 2013; Zaminpayma 2014; Pang et al. 2014; Kulathunga and Ang 2014). Generally, thermoplastic possess high toughness, larger impact resistance, and ease of shaping and recycling compared to thermoset. However, the use of thermoplastic as a matrix of CNTs composites is traditionally limited due to impregnation difficulties and high temperatures (Nie and Fisher 2013). The processing methods included Fulcrum thermoplastic composites technology, comingled thermoplastic fabrics, powder/sheath fibres bundles, wet processing method, direct reinforcement fabrication technology powder pre-impregnation, filament winding, and film stacking (Panamoottil et al. 2013;

Gao et al. 2013; Huang et al. 2013). The processing of CNT-filled thermoplastic leads to different mechanisms of residual stress formation, especially crystallization shrinkage in semicrystalline thermoplastics (Gao et al. 2013). This great impact resistance and large volume production potential make CNT-filled thermoplastic composites attractive as structural materials in ground and rail vehicles, mass transit, aircraft, and military structures. They have outstanding potential to preserve the integrity in case of impact, due to their catastrophic failure resistance (Huang et al. 2013).

CNTs: In Polyacrylic/Polymethylacrylic Polymeric Composites

Studies using melt-processed CNTs-filled polymethylmethacrylate polymer composites have been quite limited. The particular inclination of CNTs to form agglomerates may be minimized by the suitable application of shear throughout melt mixing (Chen and Lin 2010; Nie et al. 2012). Therefore, some studies applied a combined solvent casting and melt processing to produce polymethylmethacrylate-containing SWCNTs. They press little pieces of cast films between warm plates, and subsequently breaking the resulting film all over again into little pieces, and repeated the process many times (Nie et al. 2012). The particular film acquired by this melt processing technique had more homogenous CNTs distribution than the cast film, and led to superior mechanical properties. Other studies used a miniature mixer-molder to produce small quantities (approximately 0.4 g) of well-dispersed mixtures of the CNTs in polymethylmethacrylate. The well-dispersed mixture was then compressed into thin films for the purpose of investigating the dynamic mechanical properties, with a significant improvement in storage modulus (Vicente et al. 2009). It has also been recorded that CNT-filled polyacrylic acid composite film generated by the electrophoretic deposition technique in polyacrylic acid solution is utilized as electrodes for capacitive deionization (Antolin-Ceron et al. 2008). In this case, polyacrylic acid serves as the matrix to incorporate CNTs and cation-exchange polymer. The unit cell, according to the CNT-filled polyacrylic acid composite film electrode demonstrated an 83 % NaCl removal, with excellent regeneration ability, meaning that it is 51 % higher than the cell based on pure CNTs electrodes (Chen and Lin 2010).

Furthermore, binary CNT-filled polyacrylic composite system was introduced in the belief that a miscible polyacrylic blends attract host materials where CNTs could be inserted, since this kind of mixtures has a degree of mixing down to the molecular level (Nie et al. 2012). For example, CNTs contain composite materials films, which were obtained after evaporating the solvent used to prepare solutions of the four types of binary polymer blends of poly[ethylene-co-(acrylic acid)]. The evidence of H-bond formation was verified for the composite materials (Antolin-Ceron et al. 2008). The Young's moduli and crystallinity of the CNTs-filled poly [ethylene-co-(acrylic acid)] composites were improved compared to single polyacrylic.

CNTs: In Polyethylene Polymeric Composites

Currently, polyethylene, as high-strength matrix composites, is widely used for protective systems, due to its flexibility, high Young's Modulus, good impact resistance, and lightness (Pollanen et al. 2011; Raja et al. 2013a, b; Maizatulnisa et al. 2013; Sedláková et al. 2014a, b). It is reported that CNT-filled polyethylene composites enhance dynamic stab-resistance compared with plain polyethylene (Haznedar et al. 2013; Ma et al. 2014a, b). During mechanical testing, polyethylene maintains the position and orientation of the CNTs and distributed the load due to the impact among CNT fillers (Kanagaraj et al. 2011; Sulong and Park 2011; Yesil and Bayram 2011). In this case, the weak CNT-to-polyethylene adhesion is required to allow the composite to undergo maximum deformation. Polyethylene also protects the CNTs from environmental factors, such as decreased impact resistance under conditions of high humidity and the decrease of mechanical behaviour due to the photocatalytic degradation caused by ultraviolet radiations (Kim et al. 2010; Mehta et al. 2011; Sulong et al. 2011).

Certain studies focused on the impact of CNT's diameter and temperature on the interaction energy of CNT-filled polyethylene composites; and at low temperatures, a large radius CNT displays the toughest interaction energy with the polyethylene matrix (Hida et al. 2012; Ibrahim et al. 2012; Xie et al. 2013; Hao et al. 2013). Additionally, the studies also indicated a direct relationship between interaction energy and mechanical properties, which render CNT-filled polyethylene a promising candidate for ultra-strong lightweight materials. Meanwhile, some reports focused on the temperature-dependent electrical behaviour of MWCNTs/high density polyethylene (HDPE) composites prepared by solution precipitation. The electrical intensity for MWNT/HDPE composites can reach 104 by 5.4 wt% loading of MWCNTs (Ibrahim et al. 2012). Furthermore, the addition of neat CNTs improved the gas permeability properties of the polyethylene composites (Xie et al. 2013). For example, in the composite membranes, organic vapour are much more permeable than permanent gases, permeability of hexane and toluene is higher about two orders of magnitude than permanent gas permeability (Mhlanga et al. 2013a, b; Zhao et al. 2013; Li et al. 2013a, b, c, d, e). The CNT-filled polyethylene membranes results offer perspectives for vapor/gas separation applications.

CNTs: In Polypropylene Polymeric Composites

Surface properties of CNTs induce chemical interactions between CNTs and polypropylene, which in turn improve the mechanical behaviour of the composites (Girei et al. 2012; Pascual et al. 2012; Kim et al. 2013a, b, c). With that in mind, the interface between CNTs and polypropylene was simulated using contact elements. It is recorded that the length of CNTs significantly affects the reinforcement phenomenon of the polypropylene composites (Sulong et al. 2013; Wu et al. 2013a, b, c). Indeed, to increase the surface properties of CNT-filled polypropylene composites, some studies focused on the surface functionalization of CNTs. For

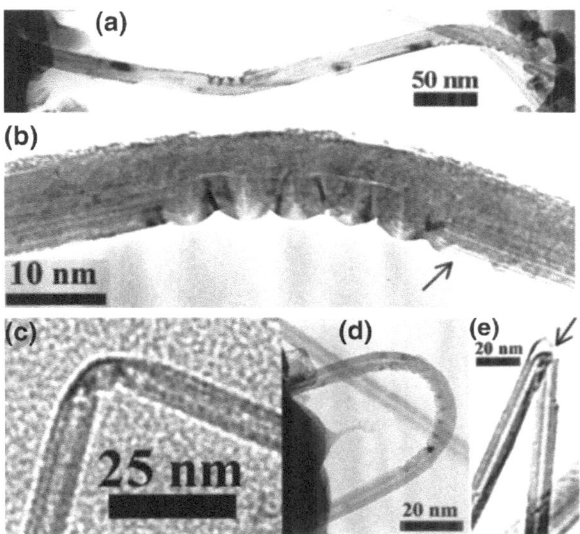

Fig. 4 TEM images of nanotube–polymer composites which show the buckled CNTs. The ends of the nanotubes, embedded in the polymer matrix. **a** Buckled nanotube bridging a micro-crack in the composite. **b** A close up of a buckled region which indicates the narrowing of the inner diameter and the *arrow* shows the change the inter-shell spacing. **c** CNTs with thin walls where single buckles were typical. **d** A buckled nanotube with 18 % tensile strain in the outer wall. **e** A fractured CNT (Bower et al. 1999)

example, butyllithium, which is functionalized by MWCNTs, developed in a manner that can covalently bond to chlorinated polypropylene (Yazdani-Pedram et al. 2013). By adding 0.6 vol% MWCNTs, the modulus improved by three orders of magnitude, and both toughness and tensile strength were enhanced by 4 times (from 27 to 108 J g^{-1}), and 3.8 times (from 13 to 49 MPa), respectively (Zhou et al. 2012). The micrograph on the break surface showed that while CNTs have been pulled out from the polypropylene matrix, its outer wall remained in the polypropylene matrix (Fig. 4). Moreover, the polypropylene's percolation threshold reinforced CNT composites being prepared by diluting a master batch with different kinds of polypropylene, varying from 1.1 to 2.0 vol%. Only poor van der Waals forces are present between the different concentric tubes of MWCNTs; whereas, the outer tubes are covalently attached to the polypropylene matrix (Georgiev et al. 2011; Pötschke et al. 2011; Ma et al. 2014a, b).

Furthermore, better CNT-filled polypropylene composite system was obtained by ultrasonic treatment, demonstrating superior storage modulus, viscosity, electrical, and mechanical properties (Yang et al. 2013; Zhang et al. 2014; Huang et al. 2014a, b, c). The lower fractal dimension of CNT and higher backbone fractal dimension result in comparatively better dispersions (Zhong et al. 2014).

CNTs: In Polystyrene Polymeric Composites

The MWCNT-filled polystyrene nanocomposites were prepared by solution evaporation method after sonication. By adding 1 wt% of MWCNTs to polystyrene, the elastic modulus and break stress increased by 36–42 and 25 %, respectively. The verification of the external load transfers to nanotubes was efficiently achieved by tensile tests and in situ transmission electron microscopy, showing that nucleation of cracks takes place at low-density area of CNTs, and after that, propagates along the poor CNT-polystyrene interfaces or relatively low CNT density regions (Kumar Sachdev et al. 2013; Tang et al. 2014). When the crack dimension exceeds 800 nm, CNTs start to break and/or even remove itself from the polystyrene matrix. Increased CNT concentrations result in a significant decrease in both tensile strength and elongation at break. Furthermore, with excess content of CNTs, decrease in fluidity with increasing CNT loading becomes an impediment to the formation of a uniform microstructure (Suemori et al. 2013). The super hydrophobic aligned layer of polystyrene nanotubes layer showed strong adhesion to water (Tang et al. 2014). This, in turn, disclosed the fact that aligned CNTs structure could not only improve hydrophobicity, but also give rise to a high adhesion force. The CNTs incorporated into the polystyrene matrix are applicable to the tire industry (Kumar Sachdev et al. 2013).

CNTs: In Polyvinyl Chloride Polymeric Composites

The effective application of CNTs in polyvinyl chloride is based on the improvement of electrical conductivity and mechanical properties, and its capability of dispersing homogeneously in the polyvinyl chloride matrix (Suemori et al. 2013). However, homogenous dispersion of CNTs is difficult due to the van der Waals interactions between the CNTS, consequently leading to the formation of agglomerations (Song et al. 2013). In this respect, the melt-mixing method is the preferred method of fabricating CNT-filled polyvinyl chloride (Farsheh et al. 2011; Aljaafari et al. 2012). The MWCNT-filled polyvinyl chloride mixed matrix membrane is suitable for gas separation applications, as well as an indicator electrode in potentiometric titrations (Abu-Abdeen 2012). In addition, it was determined that the dispersion of CNTs and morphology changes from CNT breakages are closely related to the electrical conductivity of the composites (Suemori et al. 2013). Therefore, a large morphological change in CNTs occurs at a specific processing time, and a significant decrease in the electrical conductivity of polyvinyl chloride was reinforced by CNT composites (Song et al. 2013). For example, a meaningful increase of electrical and mechanical properties was observed in the composites with about 1–2 wt% CNT contents sintered at 200 °C after being milled for 20 min (Mamunya et al. 2008).

The actual concentration dependence of the thermophysical and electrical behaviour of composites depends on polyvinyl chloride filled with MWCNTs discovered that the great anisotropy of the MWCNTs and the actual presence associated with segregated structure of MWCNTs within the polyvinyl chloride permitted the attainment of very low value of the electrical percolation threshold of 0.00047 (Zhou et al. 2010). The improvement of thermal conductivity in CNTs volume content was attained following a minimum value. Thus, the addition of CNTs influenced the heat flow through the composite (Aljaafari et al. 2012). The experimental values obtained for poly(vinyl chloride)/CNT composites were utilized to estimate the thermal conductivity of the CNT fillers (Suemori et al. 2013).

Furthermore, the addition of CNT affected the thermal properties of polyvinyl chloride. The suspension polyvinyl chloride and the MWCNTs within the concentration range of 0.01 and 0.05 wt% resulted in a lower glass transition temperatures, and an obvious relationship between the frequency, CNT content, and the glass transition temperature was determined (Sterzynski et al. 2010; Jin and Matuana 2010). By increasing the charging frequency, the glass transition temperature improved by about 3 °C via frequencies $f = 1$ Hz and $f = 10$ Hz, and 9 °C by $f = 1000$ Hz, respectively (Sterzynski et al. 2010). The maximum glass transition temperature was realized when the CNT concentrations are at 0.01–0.02 wt%. This might be due to the multiple response of CNTs distribution on the temperature-dependent chain mobility of polyvinyl chloride (Jin and Matuana 2010).

3.1.3 CNTs: In Elastomer Polymeric Composites

The introduction of CNTs to the collection of possible fillers provides new opportunities to tailor the behavior of elastomers through blending with comparatively small volume fractions of CNTs. These kinds of enhancements rely on good alignment and dispersion of the CNTs and excellent bonding of composite components (Sementsov et al. 2010; Raja et al. 2013a, b). Issues with bonding and alignment might be good for optimally improving the composites behavior which could possibly be detrimental to improving elastomeric mechanical properties (Singha and Thakur 2008a, b, c, d; Shi et al. 2013). Generally, the applications of elastomeric need the significant deformation extensibility and resilience of the elastomer. After incorporation of CNTs, as highly rigid fillers in elastomers, it typically needs to improve the stiffness of overall large-strain deformation behavior of composites (Cadambi and Ghassemieh 2012). Besides, this approach is likewise maintaining the key features of large strain-to-break behavior as well as large strain resilient of composites. Additionally, if stiffness improvement mainly consequences from unbending of the waviness of CNTs as opposed to axial straining of the CNTs, depends on good bonding and shear lag load transfer from the elastomer to the CNTs, the stiffness improvement will not be lost with large strains (Le et al. 2014).

CNTs: In Polyisoprene Polymeric Composites

Currently, the polyisoprene vulcanized offers many attributes of great interest at a technological perspective, included damping, mechanical, age and heat resistance, dynamic fatigue resistance, compression set, low temperature flexibility, electrical and swelling resistance properties (Galimberti et al. 2013). The addition of CNTs as filler achieved the level and range of properties in polyisoprene to offer a suitable amount of reinforcement such as tear resistance, tensile strength, and abrasion resistance. To have a high degree of reinforcement, the quantity of CNTs filler loading has elevated significantly which is difficult to improve these types of attributes in order to same optimal level (Yu et al. 2012). In advance, vulcanization of CNTs-filled polyisoprene composites transforms predominantly the polyisoprene into elastic or hard Ebonite-like state. This procedure is termed as curing or "cross-linking." It involved the association of macromolecules through the reactive sites. In addition, in irradiated CNTs-filled polyisoprene composites by powerful radiation, H_2 atoms of the chain, chiefly groups of methylene proportional to double bonds are ejected and radical sites are formed and combined into C–C cross-links. However, the radiation cross-linking efficiency of polyisoprene is insignificant, because of the loose packing of polyisoprene molecules with the cis structure and the groups of methyl (Yu et al. 2013a, b).

Polyisoprene is known to form carbon–carbon cross-links under pressure at controllable process parameters. The results of the cross-linking and inclusion of CNTs into polyisoprene studied by in situ thermal conductivity and tensile test revealed that Polyisoprene reinforced MWCNTs composite showed an increment in stiffness with growing MWCNTs content, retained stiffness to large strains, but with the increase in MWCNTs content the failure strains decreased (Yu et al. 2013a, b).

CNTs: In Polybutadiene Polymeric Composites

Elastomer polybutadiene copolymers composed of different ratios of styrene and butadiene, influencing strongly their macroscopic properties. The polybutadiene composites using CNTs show higher enhancement in the electrical properties that can interrelate within the framework of percolation theory (Zhou et al. 2005). The electrical percolation for polybutadiene often observed in the greater CNTs content around between 2 and 14 wt% (Speltini et al. 2013). The application range of polybutadiene once suitably reinforced with CNTs can extend to a variety of products such as sensors/actuators, materials with electromagnetic shielding properties, vapor and infrared sensor, and capacitors.

In advanced approach, the CNT is incorporated to a 50:50 blend of styrene–butadiene rubber and butadiene rubber solution (Das et al. 2008; Mari and Schaller 2009; Yu et al. 2011). The predispersed CNTs in ethanol is formed and after that the CNT-alcohol suspension is mixed with the polybutadiene at elevated temperature. CNTs-filled polybutadiene nanocomposites prepared by a technique which show meaningfully improved physical behavior already at very low concentrations of the CNTs (Mari and Schaller 2009). The particular high ratio of the CNTs enabled the formation of a conductive percolating network in the composites at concentrations lower than 2 wt%. By the presence of CNTs, as opposed to the electrical conduction properties, the thermal conductivity of the composites not influenced meaningfully. The dynamic mechanical analysis designates that the CNTs incorporation affects the glass transition (T_g) behavior of polybutadiene by reducing the height of the tan δ peak significantly (Yu et al. 2011). As mentioned earlier T_g the storage modulus has been improved after incorporating a small amount of CNTs (Das et al. 2008).

CNTs: In Nitrile Rubber Polymeric Composites

The nitrile butadiene rubber is a random copolymer of acrylonitrile and butadiene. Melt mixing of nitrile rubber with CNTs is described using a two-step process; internal mixer and two-roll mild which fond a powerful dependency of the surface resistivity of the composites on processing parameters (Perez et al. 2009; Likozar 2010; Boonbumrung et al. 2013). The CNTs-filled nitrile rubber composites were prepared by blending in a two-roll mill. The CNTs dispersion in nitrile rubber regularly began with treatment of CNTs in organic solvent ultrasonically; included toluene and ethanol followed by adding of the ethanol dispersion to the nitrile rubber compounds. In this case, it should be consider that beside the homogeneous dispersion of the CNTs in the nitrile rubber matrix, the vulcanization might have a significant effect on the final composite properties (Likozar 2010). For the filler network developed by the CNTs above, the percolation threshold will probably be interpenetrating the network of cross-linked nitrile rubber (Perez et al. 2009) (Figure 5).

Undoubtedly, it is represented as a good potential for the conceptualization of CNTs-filled nitrile rubber for many reasons included, nitrile rubber degradation process occurred meanwhile the melt mixing procedure results in the formation of free-radicals on chains of polymer (Fang et al. 2011). This, in turn, increases the

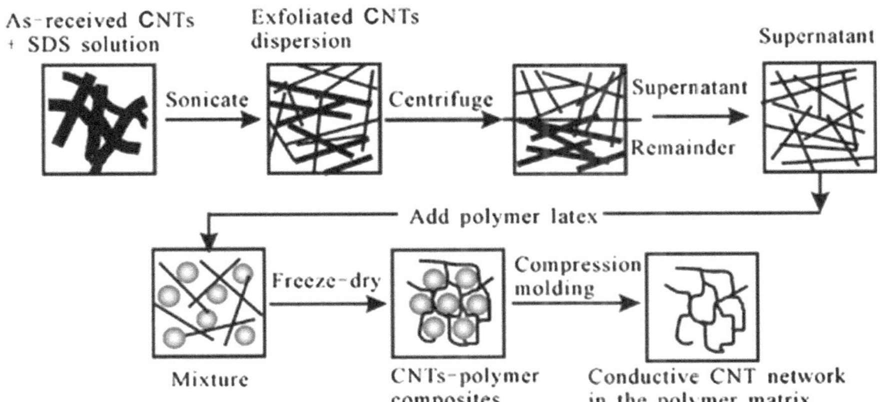

Fig. 5 Schematic description of CNT/polymer composites, which prepare by using nitrile rubber technology

affinity between acrylonitrile and CNTs components, which consequently give no important effect of poisoning by CNTs on vulcanization procedure. It is highlighted that, degradation process of polymer in terms of thermo-oxidative and/or thermo-mechanical happens during the melt blending results in the covalent grafting nitrile runner on the surface of CNTs (Verge et al. 2010). For example, the stiffness of the nitrile rubber matrix increases the use of CNTs as filler that has a large specific surface area. This is due to the large surface area to a more developed CNTs–CNTs networks, which in consequence generate strong hysteresis under dynamic operating conditions (Yue et al. 2006). The stiffness also imparted by a certain mass of CNTs and clearly observed once the CNTs aggregates. This effect attributed to the hydrodynamic effect, which is an analog to the effects of CNTs on the viscosity of the nitrile rubber (Perez et al. 2009).

CNTs: In Silicon Rubber Polymeric Composites

Silicone rubber-based material owns great mechanical elasticity as it certainly has 100 % tensile strain without showing any structural failure. The study on silicone rubber elastomers filled with SWCNTs shows a remarkable improvement in preliminary stiffness with small fractions of SWCNTs (Liu et al. 2013a, b, c). Though the improvement in stiffness is lost after just 10–20 % strain where the tangent stiffness of the nanocomposites returns to that of the parent elastomer because of debonding of the CNTs from the silicon matrix; the tensile strain-to-break found to decline meaningfully with growth in volume fraction of the CNTs (Li et al. 2011; Zhang et al. 2011). Furthermore, introduction of infra red light assisted the actuating mechanism of the silicon rubber (Tarawneh and Ahmad 2012). The actuating aspect of the mechanical properties contributed to the resilient and reversible behavior required for a superior candidate of improving the mechanical behavior of silicon rubber by means of incorporation of small amount of CNTs (Tarawneh and Ahmad 2012).

CNTs: In Polyurethane Polymeric Composites

Elastomer polyurethanes are multiblock copolymers keeping the common replicate unit structure $(A_mB_n)_p$ (Liu et al. 2013a, b, c; Gupta et al. 2013; Yu et al. 2013a, b; Gu et al. 2014). As a result of modifications within the individual block features, including the chemical identity and molecular weight; polyurethanes fabricate to be soft or hard (Raja et al. 2014; Jiang et al. 2014a, b). The CNTs-filled polyurethane composites fabricated either via melt mixing, dispersion of CNTs in the solvent, and the dissolution of the polyurethane in the same solvent, followed by solvent evaporation or the reaction of the monomers or pre-polyurethane in the presence of dispersed CNTs (Gupta et al. 2013; Liu et al. 2013a, b, c). The former has not been tried; most probably due to weakly melting material has less tendency to disperse CNTs, since the latter technique is only industrially practical (Fonseca et al. 2013;

Zheng et al. 2013). The latter technique accustomed to generate MWCNTs-filled polyurethane composites. In this attempt, isophorone diisocyanate in an organic solvent reacted with poly(tetramethylene oxide, after that this mixture emulsified in H_2O, and dispersed CNTs added. The ethylenediamine just added as a chain extender to react with the terminal groups of isocyanate (Gurunathan et al. 2013; Loos et al. 2013a, b; Tijing et al. 2013). As an indicative of an excellent CNTs dispersion, the percolation threshold is extremely low, approximately 0.1 wt% (Loos et al. 2013a, b). In a comparable approach, the CNTs dispersed in the liquid soft segment, and after that in a single step the reaction is completed (Yu and Li 2012; Yan et al. 2012). This mixture is added at one time to the methylene diisocyanate, isocyanate, and chain extender, 1,4-butanediol (Wu et al. 2012; Raja et al. 2011).

3.2 CNTs: In Biopolymer System

Different types of biopolymers-based materials have been used in a number of applications either as the polymer matrix or as the reinforcement due to their enormous advantages (Thakur and Thakur 2014a, b, c; Thakur et al. 2014c, d, e, f). The biopolymer or biodegradable plastics are polymeric materials which degrade in one-step through metabolism of the organism occuring naturally (Parvinzadeh et al. 2013; Singha and Thakur 2008a, b, c, d). In suitable temperature, O_2 availability, and moisture, biodegradation of polymer is induced into disintegration or fragmentation with no toxicity. Generally, biopolymer is divided into three main categories as listed in Fig. 6.

However, most of the biopolymers-based materials show relatively weak barrier and mechanical behavior, which presently limit their industrial utilization for the targeted applications (Zhang et al. 2012; Yang et al. 2014; Thakur et al. 2010a, b). In particular, low-heat distortion temperature, brittleness, high vapor and gas permeability, weak resistance to protected processing operation have powerfully limit its applications (Singha and Thakur 2009a, b, c, d, e; Alimohammadi et al. 2013). Thus, biopolymers have been filled with CNTs nanoparticles for improving their required properties, whereas retaining the biodegradability in a reasonably economic ways.

The incorporation of CNTs into the biopolymer system is achieved predominantly by adsorption and/or chemical binding. An ideal method should assist the interactions of the biopolymer toward CNTs within its environment.

3.2.1 CNTs: In Cellulose Polymeric Composites

The CNTs-filled cellulose nanocomposites prepared with various fabrication techniques included phase inversion, vacuum filtration, and flash freezing (Li et al. 2013a, b, c, d, e; Qi et al. 2013a, b, c; El Badawi et al. 2014). The CNTs-filled

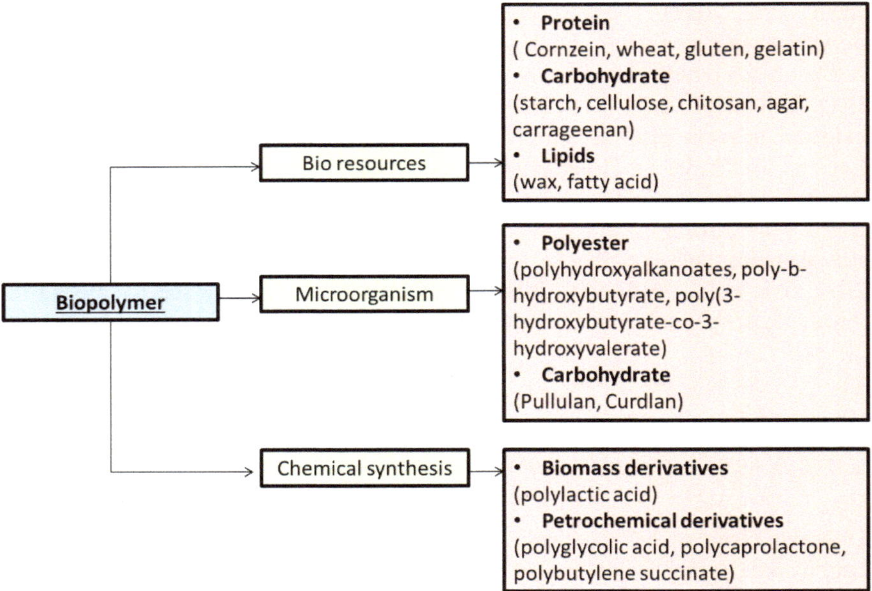

Fig. 6 Categories of biopolymer based on the origin of raw materials and their manufacturing process

cellulose nanocomposites membrane are prepared by phase inversion using acetone as solvent and 20 wt% deionized water as nonsolvent. It revealed that permeation rate found to improve by 54 % with a minimal decrease in salt retention (6 %) for the membrane with only 0.01 wt% of CNTs (Callone et al. 2008). Further addition of CNTs caused a reduction in permeation rate, which attributed to the decreased porosity and surface area (Nadagouda and Varma 2008). Another fabrication of CNTs-filled cellulose composites is done by the flash freezing and lyophilization process using wet-gel precursors. The fabricated composites exhibited both a nanostructured solid network with specific surface area between 140 and 160 $m^2 g^{-1}$ and nanoporous network (Fugetsu et al. 2008). The Young's modulus of the composites tuned to reach 90 MPa with conductivity about 2.3×10^{-4} to 2.2×10^{-2} S cm^{-1} (Loos and Manas-Zloczower 2013). In consequence, composite materials consisting of CNTs combined with cellulose paper have developed, and found that the composite is able of shielding electromagnetic interference over the examined range of 15–40 GHz, mainly in the range of 30–40 GHz, with absorption as the critical shielding mechanism (Wang et al. 2012; Won et al. 2013). It is also found in other studies that both normal flexible paper and conducting CNTs demonstrated in the composite systems with a controllable volume resistivity within a range of 1.35–540 Ohm cm (Tanaka et al. 2013). It is also documented that the composites are physically strong and yet highly flexible (Wang et al. 2012). Approximately 10 wt% CNT is needed to attain composite paper with 20-dB far-field EMI SE. In another case, the composite electrodes generated by CNTs vacuum filtration, followed by rebuilding of cellulose dissolved 1-ethyl-3-methylimidazolium acetate, which is an ionic liquid, for the oxidation process of glucose oxide (Qi et al. 2013a, b, c; Kim et al. 2013a, b, c). The result shows that direct electron transfer between glucose oxide and composite electrodes is achieved. It is also found that the glucose oxide immobilized on the composite electrodes retained catalytic oxidation of the glucose (Qi et al. 2013a, b, c). A similar result is obtained using bacterial cellulose as a matrix for the CNTs filler; it revealed that ultra-strong, transparent, and highly conductive CNTs-filled bacterial cellulose is obtained with a good biocompatibility for direct electron transfer to glucose oxide (Kim et al. 2013a, b, c). The electrical conductive properties of CNTs-filled cellulose also used as a water sensor. The composites demonstrated high sensitivity and fast response with an electrical resistance change of 5500–500 % with CNTs loading up to 2–10 wt% (Koga et al. 2013). Thus, CNTs/cellulose composite systems have high potential to use in H_2O/CH_2OH solution fraction process. In another study, a good alignment and dispersion of MWCNTs in cellulose is attained by dissolution in an ionic liquid and subsequent grinding and spinning (Lin et al. 2011a, b, c; Peng et al. 2013). This simple technique of preparing regenerated-cellulose/MWCNTs composite fibers can result in the carbon fibers production from a renewable resource (Peng et al. 2013).

3.2.2 CNTs: In Chitosan Polymeric Composites

Chitosan is the only cationic biopolymer that has the solution sensitivity of positive charged NH_2 groups in its molecular chains (Yu et al. 2014). Thus, it possesses beneficial properties included biodegradability, biocompatibility, and adsorption capacity (Shawky et al. 2012; Nitayaphat and Jintakosol 2014; Popuri et al. 2014). It has many vital biological applications in immunity, tissue engineering, catalyst support, permeable membranes, biological carrier, and drug delivery. In addition, besides excellent electrical and mechanical properties, CNTs have described to be biocompatible with chitosan matrix (Popuri et al. 2014).

When compared to chitosan, the composites composed of 2 wt% MWCNTs show more doubled Young's modulus and tensile strength (Shin et al. 2006). The micrograph analysis shows that the produced composites have a three-dimensional network with lamellar structure and macrospores (Spinks et al. 2006). This makes CNTs-filled chitosan composites as suitable candidates for the well-defined microchannel porous structure, biodegradable and biocompatible support for culture growth (Wu and Yan 2013). It is recorded that the composites have a promising adsorption properties (Salehi et al. 2012). Thus, the composite systems could offer exclusive properties as a composite in removal of heavy metal ions and treatment of wastewater (Zheng et al. 2008; Lu et al. 2009). For example, the maximum adsorption capacity 0.393 mg g^{-1} of silver ions (Ag^+) adsorbed only 0.01 wt% of CNTs in chitosan composites. It is also found that the maximum monolayer adsorption of copper ions (Cu^{2+}) of CNTs-filled chitosan composites recorded at 454.55 mg g^{-1} (Zheng et al. 2008). Modulated release of dexamethasone from CNTs-filled chitosan recorded to be faster than unfilled chitosan film. In this case, the produced composites have very low self-standing ability and density that makes it having a very worthy penetrability and process ability (Babaei and Babazadeh 2011).

The CNTs-filled chitosan composites did not cause significant cytotoxic effects on the tissue culture plate. Thus, manipulation of CNTs/chitosan composites gave a positive signal for scaffold and living cell applications (Sahithi et al. 2010; Takahashi et al. 2009). However, at high densities, the CNTs in chitosan composites might exert inhibitory effects by inducing apoptosis (Takahashi et al. 2009).

3.2.3 CNTs: In Collagen Polymeric Composites

Incorporation of CNTs into the collagen matrix leads to a considerable improvement of mechanical behavior, thermal stability, and infrared emissivity (Roy et al. 2010; Ribeiro et al. 2012a, b; Mao et al. 2014). The choice of CNTs for reinforcement of collagen has motivated by two considerations. First, according the viscoelastic and calorimetric analysis, collagen is thermodynamically immiscible, thus it is a promising matrix to wrap the CNTs surface (Roy et al. 2010). Second, the probable worthy adhesion between collagen and CNTs combined with the self-assembly capabilities of collagen can result in the alignment of CNTs in the polymer matrix, improving the mechanical behavior at low loading level (PourAkbar Saffar et al. 2009). Besides, CNTs are striking for being used in filler-reinforced composite materials because of their high aspect ratio, combined with good electrical properties. Thus, this material possesses potential applications in some fields such as biomedicine, biosensor, medical devices, tissue engineering, and substrates for electrical stimulation of cells, transducers, and infrared camouflage (Mao et al. 2014). Most probably, the composites materials including the collagen matrix with implanted CNT are prepared by blending solubilized and polymerization (Cho and Borgens 2010).

It is shown that the mixture of SWCNTs with collagen supports smooth muscle cell growth; with mouse fibroblast has effectively grown on CNTs (Lee et al. 2010). Furthermore, constructs containing 201 wt% CNTs demonstrated delayed gel compaction, relative to lower concentrations that compacted at the same rate as pure collagen control (Boccaccini and Gerhardt 2010). Under the micrograph analysis, the collagen/CNTs composites formed rigid fibril bundles, which polarized the growth and differentiation of human embryonic stem cell.

The conductivity of collagen increased uniformly with increasing CNTs content from 0.8 to 4.0 wt% and displayed modest frequency dependence, suggesting that the electrical percolation threshold had not been reached in the CNTs-filled collagen composites (Cho and Borgens 2010). Furthermore, there is some report on the improvement of the mechanical behavior of SWCNTs-filled collagen composites. It is found that a dramatic toughness (700 %), Young modulus (260 %), tensile strength (300 %) could expect with the classical rule of mixture between CNTs and collagen matrix (Chahine et al. 2008).

4 Functionalized CNTs: In Polymeric System

In principle, the outer wall of pristine CNTs conceived as chemically inert. However, this is not always desirable for applications in polymeric composite systems. Indeed, the surface energy of CNTs significantly different form that of the polymer matrices thus makes CNTs may not have chemical affinity to organic matrices (Song et al. 2010; Ntim et al. 2011). Therefore, the dispersion of CNTs into matrices becomes the biggest obstacle in practice. In addition, the seamless surface of CNTs cannot provide physical interaction within the interface of CNTs and polymer matrices (Abe et al. 2011a, b; Kotchey et al. 2013). The nature of the dispersion problem for CNTs is different from other conventional filler (Song et al. 2010). This is due to its small diameter in nanometer scale with high aspect ratio and thus possessing large surface area (Vijay et al. 2011; Abe et al. 2011a, b). A typical molecular dynamic stimulation theoretically attributed the aggregation of CNTs to the solvation interaction causes the H atoms of H_2O molecules point to the surface of CNTs. This leads to greater interaction of H_2O molecules around CNT surface than in the bulk H_2O. The orientated H_2O molecules give rise to the energy of those molecules around CNTs and force CNTs aggregate into bundles to minimize the system energy rise (Lei and Ju 2010). Indeed, the commercialized CNTs supplied in the form of heavily entangled bundles, resulting in inherent difficulties in dispersion. For this reason, further modifications on properties of CNTs in a controlled manner through several functionalization routes have thought to make the CNTs chemically active. For example, amine functionalized CNTs is completely dispersed in the polymer matrix in comparison to unmodified CNTs. The functionalization can mean in lattice doping, intercalation, molecule/particle adsorption, encapsulation, or even other nonexplored modifications (Lei and Ju 2010; Prajapati et al. 2011).

4.1 CNTs: Covalent Functionalization

One of the major obstacles in the processing of CNTs is their inherent poor solubility in organic and aqueous solvents. It is thought that the formation of covalent links significantly multiplies the solubility of CNTs in a variety of solvents at the same time assures the structural integrity of the CNTs skeleton. This, consecutively, modifies the intrinsic physical properties and polydispersity of the CNTs caused by the modification of the sp^2 C skeleton. Direct covalent sidewall functionalization is coupled with a change of hybridization from sp^2 to sp^3 and a simultaneous loss of the p-conjugation system (Wang et al. 2010; Lee et al. 2013). The end caps of CNTs consist of highly curved fullerene-like hemispheres, which are hence highly reactive, when compared with the sidewalls (Wang et al. 2010). Such modification of CNTs together with their low reactivity impedes the chemical functionalization and the characterization of the corresponding reaction products with high-chemical reactivity (Lee et al. 2013). Further, this covalent bond considerably develops the interfacial contact between matrix and filler that enables a stress (Yi et al. 2010). It is supposed that the solubility of CNTs is enhanced with modification and fine-tune on physical properties of CNTs. The modification is supposed to improve the compatibility between CNTs and the foreign matrix and makes available the direct grafting with little or no structural damage to CNTs available (Lee et al. 2013).

Overall, covalent functionalization of CNTs has diverse mechanical and electrical attributes caused by the intervention of the attached moieties and the modification of the structural p-network (Wang et al. 2010). This structural alteration occurred at the termini of the tubes and/or at the sidewalls. Moreover, the direct sidewall functionalization associated with rehybrization of one or more sp^2 C atc of C network into a sp^3 configuration and concurrent loss of conjugation (Lee et al. 2013).

4.1.1 CNTs: Carboxylation Functionalization

The conventional covalent functionalization strategy of CNTs, most commonly initiated through the carboxylation procedure by chemical acid oxidation treatment including HNO_3, H_2SO_4 or a combination of them. Powerful oxidation agents such as $KMnO_4$, ozone, reactive plasma tend to open the CNTs tubes, and consequently create oxygenated functional groups like COOH, COH, OH, and ester, which function to bind various types of chemical moieties onto the ends and defect location of CNTs. These functional groups have rich chemistry and the CNTs can be used as originators for further chemical reactions, such as silanation and polymer grafting (Battigelli et al. 2013a, b; Liu et al. 2014a, b, c, d, e). For instance, the oxidation of MWCNTs with HNO_3/H_2O_2 and HNO_3/H_2SO_4 leaded into some COOH groups on CNTs, which improved their stability in H_2O at room temperature for over 100 days (de Lannoy et al. 2013a, b; Frohlich et al. 2013; Li et al. 2013a, b, c, d, e). Consequently, the water-stable CNTs easily embedded in water-soluble polymer contained poly(vinyl alcohol), providing CNTs-filled polymer composites the homogeneous dispersion of CNTs. Oxidized CNTs well show an exceptional stability in other solvents including caprolactam, which is applied in the production of polyamide (Frohlich et al. 2013). Study on carboxylation of CNTs has shown a considerable enhancement in interfacial bonding between CNTs and polymer matrices, which consecutively triggered stronger CNTs-polymer interaction, leading to improvement of Young's modulus and mechanical strength (Hashimi et al. 2012; Shi et al. 2009).

Conversely, dramatic amounts of induced defects throughout functionalization hamper the intrinsic mobility of carriers along CNTs, which is not desirable in any case (Naeimi et al. 2009; Liu et al. 2011; Zhong et al. 2011). The carboxylation technique not only functionalizes the CNTs exterior with COOH groups, but also leaves behind unfavorable structures, thus hampering their potential for practical purposes (Zhong et al. 2011). This in turn compromises the mechanical properties of CNTs. Moreover; the concentrated acids or strong oxidants often used for CNTs functionalization are environmental unfriendly (Liu et al. 2011).

4.1.2 CNTs: Amidation Functionalization

Polymer molecules can further graft on the surface of CNTs in the presence of NH_2 functional groups. This grafting method carried out either by grafting from or grafting to technique (Table 5).

Conversely, at some point in the functionalization reaction, chiefly along with the damaging ultrasonic process, a large number of defects unavoidably formed on the CNTs sidewalls (Abe et al. 2011a, b; Singh et al. 2012; Ng and Manickam 2013; Li et al. 2013a, b, c, d, e; Jiang et al. 2013a, b). In some circumstances, CNTs fragmented into smaller chunks and altered the C hybridization from sp^2 to sp^3 (Singh et al. 2012; Ng and Manickam 2013). These detrimental effects bring about severe degradation in mechanical properties of CNTs besides disruption of π electron system in CNTs (Li et al. 2013a, b, c, d, e). The disruption of H electrons is disadvantageous to transport properties of CNTs caused by the defect sites scattered electrons and photons that are responsible for electrical and thermal conductions of CNTs (Ng and Manickam 2013).

4.1.3 CNTs: Halogenation Functionalization

The fluorination of CNTs becomes prevalent for early investigation of the covalent functionalization due to the fact that CNTs sidewalls are expected to be inert (Karousis et al. 2010; Li et al. 2012). The fluorinated CNTs have C-F bonds that are easily broken than those in alkyl fluorides, and therefore providing substitution sites

Table 5 Polymer grafting methods of amidation functionalized CNTs

Items	Grafting from technique	Grafting to technique	References
Synthesis methods	The initial immobilization of NH_2 initiators onto the CNTs surface, followed by in-situ polymerization with the formation of polymer molecules attached to CNTs	Attachment of already functionalized polymer molecules to the functionalized CNTs surface via appropriate chemical reactions	Chen and Hseih (2010)
Advantages	High grafting density	High grafting selectivity	Jiang et al. (2010)
		Commercially available polymers containing reactive groups can be utilized	
Disadvantages	Process needs a strict control of the amounts of initiator and substrate	Low product density	Coto et al. (2011)
Polymer matrix	Poly(methyl methacrylate)	Poly(ethylene glycol)	Jain et al. (2011) and Mases et al. (2011)
	Poly(*n*-butyl methacrylate)	Poly epoxy-polyamidoamine	

for additional functionalization, successfully replacements of the fluorine atoms by NH_2, CH_3, and OH groups have been achieved (Xu et al. 2013).

4.1.4 CNTs: Acylation Functionalization

The acylation of CNTs is a hopeful strategy to not only advance its dispersion, but also offer a method for creating microscopic interlinks (Heidari et al. 2013; Saidi 2013; Ye et al. 2011). Overall, acylation of CNTs improves the reactivity, enhances the specificity, and provides an avenue for additional chemical modification of CNTs; Considerable achievements have been made improving various functionalities of CNTs-filled polymer composites, generally not possible for each of the components independently (Saidi 2013). The approach is conceptualized based on CNTs chemistry via direct Friedel-Craft acylation technique, which has higher operational simplicity (Ye et al. 2011). This is not only a mild and a substitute path to functionalize CNTs, this approach also has previously shown to be less-detrimental and/or nondestructive reaction form for the proficient dispersion and functionalization of CNTs (Heidari et al. 2013). Consequently, CNT damage from severe chemical treatments including oxidation and sonication can avoid largely. Hence, greatest improvement in properties can be expected from enhanced dispersion stability on top of a chemical affinity with matrices (Saidi 2013).

4.2 CNTs: Noncovalent Functionalization

The suggested application of CNTs in polymeric composite systems has reduced because of their functional insolubility in aqueous and organic solvents (Chen et al. 2013a, b; Wu et al. 2013a, b, c; Yan et al. 2014; Battisti et al. 2014a, b). Because of their high polarizability and flat surface, CNTs, specifically SWCNTs, produced bundles and ropes characteristics (Wu et al. 2013a, b, c). Hence, numerous CNTs line up in parallel to each other through a high van der Waals attraction (0.5 eV mm^{-1}) (Yan et al. 2014). Besides, CNTs obtained as mixtures that demonstrate different chiralities, diameter, and length, in which non-CNTs carbon and metal catalyst represent in the ultimate CNTs product (Battisti et al. 2014a, b). A number of these limitations can be conquered by controlling defect and sidewall functionalization of CNTs. Yet the most prominent effect on this functionalization is that the natural conductivity of the CNTs is destroyed (Werengowska-Ciecwierz et al. 2014). A substitute approach for maintaining the inherent electronic and mechanical characteristics of CNTs is based on the noncovalent or super-molecular alteration of CNTs (Liu et al. 2014a, b, c, d, e). Such interactions, chiefly involve hydrophobic, van der Waals, and electrostatic forces, and necessitate the physical adsorption of suitable molecules onto the sidewalls of the CNTs (Wu et al. 2013a, b, c; Yan et al. 2014). Noncovalent functionalization is attained by polymer wrapping, adsorption of surfactants or small aromatic molecules, and interaction with porphyrins or biomolecules (Li et al. 2014; Sedláková et al. 2014a, b). Moreover, the major benefit of noncovalent functionalization is that it does not break the conjugated system of CNTs sidewalls, and as a result, it does not influence the final structural properties of the matter. The noncovalent functionalization of CNTs can do much to maintain their preferred properties, while enhancing the solubility rather remarkably (Werengowska-Ciecwierz et al. 2014). The noncovalent entities interact with the sidewalls of CNTs via π–π stacking interactions, and consequently opening up the track for the noncovalent functionalization of CNTs (Yan et al. 2014).

4.2.1 Oxidized Functionalization CNTs

At the primitive stage, nearly all noncovalent functionalization of CNTs focused on sorting out and dispersing them by chemical oxidation in acidic media, where the acid not only breaks up any residual metal catalyst but also eliminates the CNT caps, leaving behind COOH residue (Yin et al. 2014; Jerez et al. 2014). The oxidized CNTs are easily dispersible in a variety of NH-R organic solvents, under the impact of an ultrasonic force field (Lertrojanachusit et al. 2013; Parveen et al. 2013). In a following work, treating MWCNTs by sonication in H_2O caused the implementation of O-containing functionalities (OH, C–O–C and COOH) and no considerable harm to the basic CNTs structure (Vanyorek et al. 2014; Parodi et al. 2014). The production of functional groups is reflected in the withdrawal of $-CH_n$ groups existing on the pristine CNTs and the presence of H bonding between the CNTs and the aqueous medium (Wang et al. 2014a, b, c; Sato et al. 2013). Previously, soluble and oxidized SWCNTs arranged by supramolecular attachment of functionalized organic crown ethers (2-aminomethyl-18-crown-6). The obtained CNTs yielded concentrations of dissolved products in H_2O and CH_2OH. The composition of produced CNTs reflected in noncovalent, zwitterionic chemical interaction involving COOH groups and NH_2 moieties (Masinga et al. 2013).

The advanced oxidized functionalization of CNTs required a vast ultrasonic treatment in a mixture of concentrated HNO_3 and H_2SO_4 (Kim et al. 2013a, b, c). Such extreme conditions bring about the opening of the CNTs caps in addition to the formation of holes in the sidewalls (Vanyorek et a. 2014). This is persisted by an oxidative etching along the CNTs wall with the simultaneous discharge of CO_2. The ultimate products are CNTs with fragment length of 100–300 nm, whose ends and sidewalls decorated with an elevated density of diverse O_2 containing groups (Wang et al. 2014a, b, c).

4.2.2 Small Molecules Functionalized CNTs

The interaction between CNTs and a series of small molecules involving cyclohexane, cyclohexene, cyclohexadiene, and benzene is studied in gas phase and confirmed that p–p interactions are essential for the adsorption on CNTs (Nxumalo et al. 2013). It is discovered that the CNTs and small molecule interactions in this series regulated by coupling of the p-electrons of the molecules in the electronic p-system of the CNTs (Liu et al. 2013a, b, c; Mhlanga et al. 2013a, b). Undeniably, the coupling of p-electrons involving CNTs and aromatic molecules is noticed as an effective way to solubilize individual CNTs, which consecutively controls electronic properties (Song et al. 2012).

In addition, the solubility of CNTs with biological elements is definitely more appropriate by introducing and incorporating tiny biomolecules (Pang et al. 2010). The biomolecules for noncovalent functionalization of CNTs involve simple saccharides, enzyme, protein, DNA, and others. A range of biomaterials consisting of *n*-decyl-*β*-Dmaltoside, *Υ*-cyclodextrin, *n*-cyclodextrin, chitosan, pullulan, and phospholipid-dextran have been employed for noncovalent functionalization of CNTs (Braga et al. 2014; Lu et al. 2014; Ahmad et al. 2013). They have assisted the process for the reason that such biomolecules have nearly no light adsorption in UV-Vis wavelength region, in order to that the CNTs polymeric composites can be characterized by photochemical and are mostly biocompatible and appropriate for many medicinal purposes (Xu et al. 2010; Bai et al. 2010; Krause et al. 2010).

4.2.3 Derivatives Functionalized CNTs

The adsorption of multiple derivatives, replaced with groups with different electronic properties and volume size onto the sidewalls of cut SWCNTs, expected to make better p–p bond interactions controlling the adsorption process (Table 6). This procedure along with an electron donor–acceptor charge transfer interacts between the aromatic adsorbents and the SWCNTs sidewall, which technicality causes a considerable change in the electrical dipole moment along its primary axis. This shift alters the local electrostatic potential in the CNTs, modifies its conductance with elevated threshold voltage current flows (Tan et al. 2011; Martin et al. 2009). Functionalization of CNTs by means of derivatives with positive or negative charge like nitrogenated bases, alkyl ammonium ion, through p–p interaction carried on by the assembling of the energy/electron donor molecules complementary electrostatics, axial coordination or crown ether-alkyl ammonium ion interactions, in order (Martin et al. 2009). This brought astable donor–acceptor system with maximum preservation of the mechanical and electronic characteristics of CNTs. To the best our knowledge, self-assembly via ammonium ion-crown ether derivatives is held as one of the most potent methods as it proposes a high level of directionality with binding energies up to 50–200 kJ mol^{-1} (Feng and Chen 2006).

4.2.4 Polymer Functionalized CNTs

Polymers, particularly conjugated polymers, have proved to serve as exceptional wrapping materials for the noncovalent functionalization of CNTs due to π–π stacking and van der Waals interactions between the conjugated polymer chains

Table 6 List of derivatives functionalized CNTs with its advantages and potential applications

Derivatives	Advantages	Potential applications	References
Chromophore	Reversible and repeatable conductance change over a long period of time	Integrated nano photo-detector	Sanip et al. (2009)
Ammonium amphiphiles	Excellent solubility in H_2O	Semiconductor	Matsuoka et al. (2014)
	Transparent solution of CNTs		
	High single nanotube chiral index		
N-succinimidyl-1-pyrenebutanoate	Nucleophilically be substituted by primary/secondary NH_2	Protein detector/sensor	Ghasemi et al. (2014)
	Allow immobilization of the biopolymer on the CNTs surface		
Glycodendrimers	Mitigation of toxicity	Active coating materials	Bandaru and Voelcker (2012)
N-acetyl-D-glucosamine	Interface biocompatibility with living cells	Donor acceptor hybrid	Sevilla et al. (2014)
	Detect the dynamic secretion of biomolecules		
	Able to transform sunlight into electrical/chemical energy		
Pyrene-tetrathiafulvalene	Formation of flexible and medium length chains	Template for immobilization of electroactive unit on CNTs surface	Ehli et al. (2008)
Pyrene-pyropheophorbide	Favor a facile interaction with CNTs surface		

having aromatic rings and surface of CNTs within. The physical adsorption of polymer on the CNTs surface reduced the surface tension of CNTs that successfully averted the aggregation of CNTs (Adhikari et al. 2014). The success of this method relied deeply on the properties of polymer and medium chemistry. There are two types of polymers used, nonionic and cationic polymer (Table 7). In addition to achievable enhancement in the mechanical and electrical properties of polymers, the functionalization with CNTs regarded as a useful approach for integrating CNTs into polymer-based devices (Liu et al. 2014a, b, c, d, e; Chehata et al. 2014). For noncovalently functionalized CNTs with polymers, quite a lot of strategies have been taken on and involved physical mixing in solution, in situ polymerization of monomers in the presence of CNTs, surfactant-assisted processing of composites and chemical functionalization (Roy et al. 2014). For instance, polymers such as poly(m-phenylene-co-2,5-dioctoxy-p-phenylenevinylene) were employed to wrap around CNTs in organic solvent contained $CHCl_3$. Polymers which hold a polar side chain, including poly(vinyl pyrrolidone) or poly(styrene silfonate) gave stable solutions of SWCNTs-filled polymer complexes in H_2O medium (Ten et al. 2014;

Table 7 List of polymer functionalized CNTs with different types and potential applications

Polymer	Types	Advantages	References
Polymeric amphiphiles	Polyvinylpyrrolidone	A stable composites materials	Han et al. (2014)
	Polystyrene sulfonate	High glass transition temperature	
		Elastic modulus 30 and higher in relative to native sample	
		Water soluble	
	Polydiallyldimethylammonium chloride	Increase the hydrophilicity of CNTs	Lee and Cui (2011)
		Positively charged polyelectrolytes	
		Formation of strong interface via electrostatic interactions	
	Polystyrene sulfonate + Polydiallyldimethylammonium chloride	Systematic and molecularly controlled organization of CNTs	Huaming et al. (2005)
	N-ethyl-4-vinylpyridinum bromide-co-4-vinylpyridine	Stable CNTs formation	Vladimir et al. (2005)
		Efficiency of macromolecular dispersion	
		Promising electronic interactions in CNTs bundles	
		Formation of uniform coating (1.0–1.5 thick)	
	Polyethylene glycol	Good solubility in organic solvents obtained by covalent/ionic attachment of long chain aliphatic NH_2 onto COOH groups	Nozomi et al. (2007)
Biopolymer	Polypeptide	Able to fold around the graphitic surface of CNTs	Davide et al. (2003)
		Good dispersion in aqueous solutions by noncovalent interactions	
		The size and morphology of coated CNTs can control by peptide–peptide interactions	
		Highly ordered structure	

(continued)

Table 7 (continued)

Polymer	Types	Advantages	References
	Single-chain lipid	Polar part of the lipids could participate in the selective immobilization of histidine-tagged protein through metal ion chelates	Cyrille et al. (2003)
		The lipid membrane found to maintain its fluidity and mobility of lipid molecules	
	Oligonucleotides	Effective in dispersing CNTs in H_2O	Davide et al. (2004)
		A stable solution	
		Slow structure re-arrangement	
		Aligned parallel to the CNTs surface with a high degree of orientation order	

He et al. 2014). While surfactants may be effective in the solubilization of CNTs, they proven permeable plasma membranes (Primo et al. 2013; Mallakpour and Zadehnazari 2013; Rath et al. 2013). They are toxic for biological purpose, hence imperfect for biomedical applications (Chen et al. 2013a, b; Fisher et al. 2013; de Lannoy et al. 2013a, b; Amirilargani et al. 2013).

To defeat the imperfections, biopolymer functionalized CNTs have persistently studied. CNTs liquid crystal phase creation and selective chiral SWCNTs enrichment assisted by biopolymer uncovered that biopolymer is a promising agent of high quality on surface functionalization of CNTs (Battigelli et al. 2013a, b; Albuerne et al. 2013; Loos et al. 2013a, b; Wei et al. 2005). Now widely available, large-scale production and low-price polysaccharides including chitosan, gelling gum, hydraulic acid, and others have realized to be easier and commercially acceptable; therefore making a high-concentration CNTs in a single dispersion becomes more convenient. For instance, physical purification of CNTs by chitosan functionalization has been endorsed to be easy processing and also is efficient. Besides, the CNT-H_2O interface direction, the ordered organization of lipid derivatives onto CNTs by supra molecular self-assembly by biopolymers on the CNT surface has noticed to be of tremendous effect on CNTs dispersion (Hordy et al. 2013). Gum Arabia, the primeval biopolymer dispersant presented to stabilize SWCNTs (Yulong et al. 2006; Rajdip et al. 2002; Nadia et al. 2005). The dispersion can concentrate into suspension of SWCNTs concentration as much as 150 mg mL^{-1}; the most favorable concentration of SWCNTs (Rajdip et al. 2002). The hyaluronic acid functionalized CNTs at high concentration of 10 mg mL^{-1} demonstrated anisotropic birefringence phenomenon, representing the liquid crystal part of biopolymer functionalized CNTs. Aligning CNTs throughout a liquid crystal phase of CNTs by polysaccharide has improved as well.

5 CNTs/Polymer: Applications

5.1 Structural Applications

The structural characteristics of CNTs-filled polymer composites are very imperative in automotive, aerospace, paint, protectors, and other (Hyang et al. 2007). As mechanical properties play a crucial role in structural purposes, load transfer from the polymer matrix to CNTs filler grows to be essential. Load transfer between polymer matrix and CNTs is subject to the interfacial shear stress between the composite components (Kumar Sachdev et al. 2013). To make the reinforcement efficient, it is required that the CNTs must be adequately long and the interface between CNTs and polymer matrix is strong (Xiaowen et al. 2006). Since CNTs has some surface defects, including changeable diameter and bend/twist as a result of nonhexagonal defects, along CNTs, mechanical interlocking do play a role in CNTs to polymer interface (Hyang et al. 2007). The outstanding mechanical characteristics of CNTs are proposed that incorporation of very little amounts of CNTs into the polymer matrix initiated structural materials application with considerably higher strength and modulus. For instance, the addition of 1 wt% of MWCNTs in the polystyrene by solution evaporation procedure brought about 36–42 and almost 25 % enhancement in tensile modulus and tensile strength, in order (Suemori et al. 2013). Whereas, the improvement in indentation resistance was recorded up to 3.5 times by supplementing 2 wt% SWCNTs in epoxy resin (Shi et al. 2009; Huaming et al. 2005). Some studies observed a major enhancement in modulus and hardness (1.8 times and 1.6 times) with the integration of 1 wt% MWCNTs in polyvinyl alcohol (Song et al. 2013). Homogeneous dispersion and alignment of CNTs had a considerable result in mechanical properties of CNTs-filled polymer composites particularly in the structural applications (Kanagaraj et al. 2011). It is reported that by enhancing the dispersion of CNTs through the in situ polymerization, great mechanical strength of CNTs-filled polymer composites could employ to make some high-end sporting goods as well as tennis rackets, baseball bat, and consequently delivering excellent performance (Hida et al. 2012).

The mechanical property study of CNTs-filled polymer is also described by morphology studies (Girei et al. 2012). For example, pullout process proposes that effective load transfer arises from the polymer matrix to the outer layer of CNTs, caused by the sturdy covalent bonding within its interfacial region (Shi et al. 2009). This observation anticipated that the efficiency of property improvement relies on the form of CNTs, processing technique, and compatibility between CNTs and polymeric matrix (Pascual et al. 2012). Additionally, the modulus and strength of composites mainly traded for high fracture toughness. In contrast with traditional polymer composites containing micron-scale fillers, the integration of nanoscale CNTs into a polymer system causes the very tiny distance between the fillers; these characteristics of composites can largely modified even at an exceedingly low content of filler (Li et al. 2011). Even though chemical functionalization of CNTs enhanced the compatibility between CNTs and polymer, which consecutively improved the mechanical properties, but it has a worsening effect on the other properties such as electrical and thermal conductivity (Raja et al. 2014).

5.2 Medical Application

The stable dispersed CNTs by biopolymer set up into biomedical purposes as well as tissue engineering and drug delivery system (Yu and Li 2012). For the bioactivity of biopolymer, their composites with CNTs offer exceptional sensing performance (Liu et al. 2014a, b, c, d, e). The biomimetic actuation founded on CNTs-filled biopolymer devices have as well initially proved to be of large and fast actuation displacement under low voltage electrical stimulation (Chahine et al. 2008).

The CNTs, and in particular SWCNTs, with surface area as high as 2600 $m^2\ g^{-1}$ is very appropriate for acting as a drug carrier for biomedical purposes. For instance, CNTs has presented as a template for hosting bioactive peptides to the immune system (Davide et al. 2003). In this case, B cell epitope of the foot and mouth disease virus covalently adhered to the NH_2 groups functionalized CNTs (Serrano et al. 2014). This, sequentially, increases the formation of peptides around CNT adopting the appropriate secondary structure because recognized by specific monoclonal and polyclonal antibodies. The function of CNTs as vaccine delivery further developed via the interaction with the complement system (Battigelli et al. 2013a, b). It is found that pristine CNT activates the complement following both classical and alternative ways to make selective adsorption of some of its protein, which consequently enhances antibody response leading to immunization with peptide-CNT conjugates (Davide et al. 2003, 2004). It also recorded that CNTs encouraged delivery of DNA or any bioactive mediator to cells. As CNTs surface functionalized to attach either electrostatically or covalently to DNA and RNA, the residual unfunctionalized and hydrophobic segment of CNTs attracted to the hydrophobic areas of the cells (Rath et al. 2013). Besides, CNTs-filled chitosan biopolymer offers localized delivery of therapeutic agents initiated by external sources (Battigelli et al. 2013a, b).

5.3 Sensor Applications

The CNTs-filled polymer composites used as an implantable sensor that is capable of transmitting information extracorporeally. Such a sensor made real-time date related to the physiological relevant parameters such as pH, O_2 concentration, and glucose level available. In addition, the good biocompatibility with high electrical and electrochemical sensitivity assisted implantable biosensor applications (Qi et al. 2013a, b, c). The early research found that CNTs-filled polymer composites are able to detect serum proteins, as well as disease markers, autoantibodies, and antibodies (Grabowski et al. 2014).

An important composite biosensor is derived from CNTs-filled chitosan. The CNTs-filled chitosan composites have discovered to be an excellent biocompatibility for neutral cell growth (Spinks et al. 2006). Their suspension coated on glassy carbon electrode is capable of detecting 90 % of NaDH in less than 5 s. The stability and sensitivity of CNTs-filled chitosan composites as a biosensor allowed interference-free determination of glucose in physiological matrix (Gopalan et al. 2009). A composite of MWCNTs-filled chitosan composites employed as a matrix for capturing lactate dehydrogenase into a glassy carbon electrode to produce amperometric biosensor (Mao et al. 2014). Moreover, CNTs-filled chitosan-lactate dehydrogenase composite film demonstrats the abilities to boost the current responses, to reduce the electro-oxidation potential of β-nicotinamide adenine dinucleotide and to thwart the electrode surface fouling. It found that, the efficient biosensor for this kind of system has the sensitivity of lactate up to 0.0083 AM^{-1} cm^{-2} with response time of 3 s (Shawky et al. 2012; Popuri et al. 2014). The entrapment of acetychlolinesterase on CNTs-filled chitosan biosensor recorded that the inhibition of organophosphorous insecticide to the enzymatic activity of acetylcholinesterase, using triazophos as a model of compounds is relative to its concentrations (Zhang et al. 2012). The acetylcholinesterase could regenerate using pralidoxime iodide within 8 min. Therefore, the CNTs-filled chitosan biosensor has outstanding characteristics and performance, such as high precision and reproducibility, suitable stability and accuracy, quick response, and low detection threshold (Spinks et al. 2006). It has a potential function in the characterization of enzyme inhibitors and detection of toxic compounds against enzyme (Gopalan et al. 2009).

5.4 Semiconductor Applications

The CNTs composites have anticipated as a potential replacement for Cu interconnects in future technologies because of its high mechanical stability, high thermal conductivity, large current ability and compatibility with present-day silicon technologies (Peter and Richard 2002). For instance, the improvement in electrical conductivity and insulating of CNTs in the polymer matrix to a high extent has accomplished with a very small loading (0.021 wt%) of CNTs. The current through CNTs is either sublinear or superlinear with voltage, in the same way as many other metallic and semiconducting nanowires/nanotubes (Gardea and Lagoudas 2014). The remarkable fact is the large current carrying capability of CNTs composites as compared to Cu as well as to superconductors and its rise with increasing diameter (Bal and Saha 2014). For example, SWCNTs composites are capable of carrying a current in the microampere random while 100 nm diameter MWCNTs composites revealed to transmit up to mili-ampere current. In reality, the current carrying capability of MWCNTs is much greater than SWCNTs, attributable to its larger conduction at outer shell (Abu et al. 2006). The boost in electrical conductivity of polymer material with CNT addition is the greatest benefit of production of CNT-filled polymer composites (Jiang et al. 2014a, b).

The electrical conductivity of CNTs-filled polymer composites is subject to on many features including type of CNYs, aspect ratio, surface functionalization and CNT content (Xiaowen et al. 2006). For example, the electrical conductivity of nanocomposites rises with increasing CNTs loading up until a significant filler concentration, where a dramatic boost in conductivity recorded. This critical CNTs concentration is called as electrical percolation threshold concentration. At this stage, CNTs particle shapes three-dimensional conductive networks inside a polymer matrix, therefore electron tunnels from one filler to another are created and result in a high resistance presented by insulating polymer matrix. Since the creation of percolating networks is associated with both intrinsic conductivity and aspect ratio of CNTs particles, the CNTs-filled polymer composites have exhibited very low percolation threshold. This is due to the high conductivity and the aspect ratio of CNTs (Ma et al. 2014a, b). Consequently, the percolation threshold concentration and nanocomposites conductivity determined by polymer type, synthesis technique, aspect ratio of CNTs, the extrication of CNTs agglomerates, uniform spatial distribution of single CNTs and the level of alignment (Aljaafari et al. 2012). Another considerable factor, which influenced the electrical conductivity of nanocomposites is the chemical functionalization of CNTs. This is because of the interference with the extended π-conjugation of CNTs and thus decreases the electrical conductivity of isolated CNTs (Qi et al. 2013a, b, c).

As a result, these composite systems could be employed to protect electromagnetic interference and as electrostatic discharge components (Song et al. 2013). As a major progress in electrical conductivity monitored at very low CNT loading, this composite system found application as lightweight, economical, and highly effective shielding materials (Ramoa et al. 2013). Because of very great aspect ratio and impressive electrical properties of CNTs, some reports have revealed that ultra-low electrical percolation limit was observed with merely 0.0025 wt% in aligned CNTs-filled epoxy composites (Zeng et al. 2013). As a result, CNTs-filled polymer composites are in increasing demand in various application area such as transparent conductive coatings, electrostatic dissipation, electrostatic painting, and electromagnetic interference protecting applications (Qi et al. 2013a, b, c; Matsuoka et al. 2014).

5.5 Thermal Conductor Application

The thermal characteristics of polymer matrix as well altered by CNT addition including increment on glass transition, melting and thermal decomposition temperatures caused by hindered chain, and segmental mobility of the polymers (Reddy and Ramu 2008). Besides, the CNTs also influenced the crystallization speed and percentage of crystallinity by performing as nucleating agents in CNTs-filled polymer system (Table 8). Integration of 1 wt% surfactant in the role of a wetting

Table 8 Shielding properties of different kinds of CNTs-filled polymer composites

CNTs-filled polymer composite systems	Shielding properties (dB)	References
7.5 vol% MWCNTs-filled polypropylene	35	Lopez Manchadoa et al. (2006)
7 wt% MWCNTs-filled polystyrene	20	Guoxing et al. (2010)
20 wt% MWCNTs polyurethane	17	Hsu-Chiang et al. (2006)
40 wt% Polymethyl methacrylic	27	Fangming et al. (2003)

agent enhanced the glass transition temperature of CNTs-filled polymer composites up to 25 to 40 °C (Hone et al. 1999). Additionally, the thermal decomposition temperature of polypropylene in N_2 improves by 12 °C on 2 vol% CNTs loading. These observations imply that shielding characteristics of the composites determined by numerous factors including fabrication method and purification of CNTs.

It is recorded that enhancement in both properties boosts its mechanical and processing properties as well (Maizatulnisa et al. 2013). Besides, the microwave-absorbing capacity of CNTs could utilize to heat temporary accommodation structures (Raja et al. 2013a, b).

Owing to the excellent thermal conductivity of CNTs, integration of CNTs drastically enhances the thermal transport properties, which makes possible its usage as printed circuit boards, connectors, thermal interface equipments, heat sinks, and other high performance thermal management system (Ibrahim et al. 2012; Xie et al. 2013).

6 Conclusion

The study on CNT-filled polymer nanocomposite surmized that CNTs are capable of altering the properties of polymer matrices. The great challenge in realizing the full ability of CNTs is to accomplish homogenous dispersion of CNTs with the intention that the maximum filler surface area is accessible for load-transfer between the composite constituents. The functionalization of CNTs offers a suitable route to develop dispersion and compatibility without negatively affecting the properties of the resulting composite. Three chief processing techniques of CNT-filled polymer composites involve solution, melting, and in situ polymerization. Solution blending yields high-quality composite, but melt compounding is much simpler, and offers alternatives to large-scale production. The greatest improvement in mechanical properties of CNTs-filled polymer composites is detected in the case of in situ polymerization, which forms a covalent bond between CNTs and the polymer.

References

Abe S, Nakayama K, Hayashi D, Akasaka T, Uo M, Watari F, Takada T (2011a) Development of a novel transparent substrate coated by carbon nanotubes through covalent bonding. Phys Procedia 14:147–151

Abe S, Nakayama K, Kobayashi H, Kiba T, Akasaka T, Sato S-I, Uo M, Watari F, Takada T (2011b) Versatile surface modification by carbon nanotubes through an amide-bond formation. Nano Biomed 3(1):208–216

Abu BS, Joohyuk P, Naesung L, Jeungchoon G (2006) Wear behavior of functionalized multi-walled carbon nanotube reinforced epoxy matrix composites. J Compos Mater 40(21):1947–1960

Abu-Abdeen M (2012) Investigation of the rheological, dynamic mechanical, and tensile properties of single-walled carbon nanotubes reinforced poly(vinyl chloride). J Appl Polym Sci 124(4):3192–3199

Adams T, Charles AW (2001) Photo-oxidation of polymeric-inorganic nanocomposites: chemical, thermal stability and fire retardancy investigations. Polym Degrad Stab 74:33–37

Adhikari PD, Jeon S, Cha M-J, Jung DS, Kim Y, Park C-Y (2014) Immobilization of carbon nanotubes on functionalized graphene film grown by chemical vapor deposition and characterization of the hybrid material. Sci Technol Adv Mater 15(1):015007

Agnihotri P, Kar KK (2007) Hybrid nanocomposites of carbon nanotubes (CNTs) grown on carbon fiber in polyester matrix with improved thermomechanical properties. In: Proceedings of the annual technical conference—ANTEC, vol 4, pp 2191–2195

Ahmad AL, Jawad ZA, Low SC, Sharif Zein SH (2013) The functionalization of beta-cyclodextrins on multi walled carbon nanotubes: effects of the dispersant and non aqueous media. Curr Nanosci 9(1):93–102

Albuerne J, Zenkel C, Munirasu S (2013) Functionalization and polymerization on the CNT surfaces. Curr Org Chem 17(17):1867–1879

Alimohammadi F, Parvinzadeh Gashti M, Shamei A (2013) Functional cellulose fibers via polycarboxylic acid/carbon nanotube composite coating. J Coat Technol Res 10(1):123–132

Aljaafari A, Abu-Abdeen M, Aljaafari M (2012) Mechanical and electrical properties of poly(vinyl chloride) loaded with carbon nanotubes and carbon nanopowder. J Thermoplast Compos Mater 25(6):679–699

Amirilargani M, Ghadimi A, Tofighy MA, Mohammadi T (2013) Effects of poly(allylamine hydrochloride) as a new functionalization agent for preparation of poly vinyl alcohol/multiwalled carbon nanotubes membranes. J Membr Sci 447:315–324

Andrews R, Weisenberger MC (2004) Carbon nanotube polymer composites. Curr Opin Solid State Mater Sci 8(1):31–37

Antolín-Cerón VH, Gómez-Salazar S, Soto V, Ávalos-Borja M, Nuño-Donlucas SM (2008) Polymer nanocomposites containing carbon nanotubes and miscible polymer blends based on poly[ethylene-co-(acrylic acid)]. J Appl Polym Sci 108(3):1462–1472

Babaei A, Babazadeh M (2011) Multi-walled carbon nanotubes/chitosan polymer composite modified glassy carbon electrode for sensitive simultaneous determination of levodopa and morphine. Anal Methods 3(10):2400–2405

Bai Y, Xu GY, Sun HY, Hao AY, Mao HZ, Dong SL, Shi XF, Xin X, Ao MQ, Pang JY, Yang XD (2010) Effect of substituted group of β-cyclodextrin derivatives on the dispersing of carbon nanotubes. J Dispers Sci Technol 31(3):353–358

Bal S, Saha S (2014) Assesment of electrical and mechanical properties of carbon nanomaterial doped polymer composites. In: Proceedings of the 2014 international conference on advances in energy conversion technologies—intelligent energy management: technologies and challenges, ICAECT 2014, art no 6757069, pp 99–103

Bandaru NM, Voelcker NH (2012) Glycoconjugate-functionalized carbon nanotubes in biomedicine. J Mater Chem 22(18):8748–8758

Battigelli A, Ménard-Moyon C, Da Ros T, Prato M, Bianco A (2013a) Endowing carbon nanotubes with biological and biomedical properties by chemical modifications. Adv Drug Deliv Rev 65(15):1899–1920

Battigelli A, Ménard-Moyon C, Da Ros T, Prato M, Bianco A (2013b) Endowing carbon nanotubes with biological and biomedical properties by chemical modifications. Adv Drug Deliv Rev 65(15):1899–1920

Battisti A, Esqué-de los Ojos D, Ghisleni R, Brunner AJ (2014a) Single fiber push-out characterization of interfacial properties of hierarchical CNT-carbon fiber composites prepared by electrophoretic deposition. Compos Sci Technol 95:121–127

Battisti A, Esqué-de los Ojos D, Ghisleni R, Brunner AJ (2014b) Single fiber push-out characterization of interfacial properties of hierarchical CNT-carbon fiber composites prepared by electrophoretic deposition. Compos Sci Technol 95:121–127

Boccaccini AR, Gerhardt L-C (2010) Carbon nanotube composite scaffolds and coatings for tissue engineering applications. Key Eng Mater 441:31–52

Bong SK, Sang HB, Park Y-H, Kim J-H (2006) Polyimide/carbon nanotubes composite films: a potential for FPCB. In: Proceedings of the 2006 international conference on nanoscience and nanotechnology, ICONN, art no 4143419, pp 407–410

Boonbumrung A, Sae-Oui P, Sirisinha C (2013) Dispersion enhancement of multi-walled carbon nanotube (MWCNT) in nitrile rubber (NBR). Adv Mater Res 747:59–62

Bower C, Rosen R, Jin L, Han J, Zhou O (1999) Deformation of carbon nanotubes in nanotube–polymer composites. Appl Phys Lett 74:3317–3320

Braga SS, Marques J, Heister E, Diogo CV, Oliveira PJ, Paz FAA, Santos TM, Marques MPM (2014) Carriers for metal complexes on tumour cells: the effect of cyclodextrins vs CNTs on the model guest phenanthroline-5,6-dione trithiacyclononane ruthenium(II) chloride. BioMetals (article in press)

Breuer O, Uttandaraman S (2004) Big returns from small fibers: a review of polymer/carbon nanotube composites. Polym Compos 25(6):630–645

Brigitte V, Alain P, Claude C, Cédric S, René P, Catherine J, Patrick B, Philippe P (2000) Macroscopic fibers and ribbons of oriented carbon nanotubes. Science 290(5495):1331–1334

Cadambi RM, Ghassemieh E (2012) Optimized process for the inclusion of carbon nanotubes in elastomers with improved thermal and mechanical properties. J Appl Polym Sci 124(6):4993–5001

Callone E, Fletcher JM, Carturan G, Raj R (2008) A low-cost method for producing high-performance nanocomposite thin-films made from silica and CNTs on cellulose substrates. J Mater Sci 43(14):4862–4869

Cao M-S, Gao Z-J, Zhu J (2003) Research on microwave absorbability towards CNTs/polyester composites. J Mater Eng 1(2):34–37

Chahine NO, Collette NM, Thompson H, Loots GG (2008) Biocompatibility of carbon nanotubes for cartilage tissue engineering. In: Technical proceedings of the 2008 NSTI nanotechnology conference and trade show, NSTI-nanotech. Nanotechnology 1:125–128

Chehata N, Ltaief A, Bkakri R, Bouazizi A, Beyou E (2014) Conducting polymer functionalized multi-walled carbon nanotubes nanocomposites: optical properties and morphological characteristics. Mater Lett 121:227–230

Chen J, Hsieh K (2010) Polyacrylamide grafted on multi-walled carbon nanotubes for open-tubular capillary electrochromatography: comparison with silica hydride and polyacrylate phase matrices. Electrophoresis 31(23–24):3937–3948

Chen J-L, Lin Y-C (2010) The role of methacrylate polymerized as porous-layered and nanoparticle-bound phases for open-tubular capillary electrochromatography: substitution of a charged monomer for a bulk monomer. Electrophoresis 31(23–24):3949–3958

Chen D, Wang R, Tjiu WW, Liu T (2011) High performance polyimide composite films prepared by homogeneity reinforcement of electrospun nanofibers. Compos Sci Technol 71(13):1556–1562

Chen C, Zhang J, Peng F, Su D (2013a) Efficient functionalization of multi-walled carbon nanotubes by nitrogen dioxide. Mater Res Bull 48(9):3218–3222

Chen J, Tong H, Gao Z, Zhu J, Zhang X, Liang Y (2013b) Preparation of polyaniline covalently grafted carbon nanotubes supported Pt catalysts and its electrocatalytic performance for methanol. Acta Chim Sinica 71(12):1647–1655

Chenyang L, Jun Z, Jiasong H, Guohua H (2003) Gelation in carbon nanotube/polymer composites. Polymer 44(24):7529–7532

Cho Y, Borgens RB (2010) The effect of an electrically conductive carbon nanotube/collagen composite on neurite outgrowth of PC12 cells. J of Biomed Mater Res Part A 95A(2):510–517

Chris B, Otto Z, Wei ZDJW, Sungho J (2000) Nucleation and growth of carbon nanotubes by microwave plasma chemical vapor deposition. Appl Phys Lett 77:2767–2770

Coto B, Antia I, Blanco M, Martinez-De-Arenaza I, Meaurio E, Barriga J, Sarasua J-R (2011) Molecular dynamics study of the influence of functionalization on the elastic properties of single and multiwall carbon nanotubes. Comput Mater Sci 50(12):3417–3424

Cui C, Qian W, Zhao M, Ding F, Jia X, Wei F (2013) High strength composites using interlocking carbon nanotubes in a polyimide matrix. Carbon 60:102–108

Cyrille R, Fabrice B, Patrick S, Thomas WE, Charles M (2003) Supramolecular self-assembly of lipid derivatives on carbon nanotubes. Science 300(5620):775–778

Das A, Stöckelhuber KW, Jurk R, Saphiannikova M, Fritzsche J, Lorenz H, Klüppel M, Heinrich G (2008) Modified and unmodified multiwalled carbon nanotubes in high performance solution-styrene-butadiene and butadiene rubber blends. Polymer 49(24):5276–5283

Davide P, Charalambos DP, Johan H, Fred B, Ed K, Jean-Paul B, Sylviane M, Maurizio P, Alberto B (2003) Immunization with peptide-functionalized carbon nanotubes enhances virus-specific neutralizing antibody responses. Chem Biol 10(10):961–966

Davide PD, Ravi S, McC David, Mathieu E, Jean-Paul B, Maurizio P, Kostas K, Alberto B (2004) Functionalized carbon nanotubes for plasmid DNA gene delivery. Angew Chem 116 (39):5354–5358

De Borbón F, Ambrosini D, Curadelli O (2014) Damping response of composites beams with carbon nanotubes. Compos B Eng 60:106–110

de Lannoy C-F, Soyer E, Wiesner MR (2013a) Optimizing carbon nanotube-reinforced polysulfone ultrafiltration membranes through carboxylic acid functionalization. J Membr Sci 447:395–402

de Lannoy C-F, Soyer E, Wiesner MR (2013b) Optimizing carbon nanotube-reinforced polysulfone ultrafiltration membranes through carboxylic acid functionalization. J Membr Sci 447:395–402

Demczyk BG, Wang YM, Cumings J, Hetman M, Han W, Zettl A, Ritchie RO (2002) Direct mechanical measurement of the tensile strength and elastic modulus of multiwalled carbon nanotubes. Mater Sci Eng A 334:173–178

Dubois P, Alexandre M (2006) Performant clay/carbon nanotube polymer nanocomposites. Adv Eng Mater 8(3):147–154

Ehli C, Guldi DM, Ángeles Herranz M, Martín N, Campidelli S, Prato M (2008) Pyrene-tetrathiafulvalene supramolecular assembly with different types of carbon nanotubes. J Mater Chem 18(13):1498–1503

El Badawi N, Ramadan AR, Esawi AMK, El-Morsi M (2014) Novel carbon nanotube-cellulose acetate nanocomposite membranes for water filtration applications. Desalination 344:79–85

Esteban EU-B, Matthew JK, Virginia AD (2013) Dispersion and rheology of multiwalled carbon nanotubes in unsaturated polyester resin. Macromolecules 46(4):1642–1650

Fang L, Xue Y, Lin H, Shuai C (2011) Friction properties of carbon nanotubes reinforced nitrile composites under water lubricated condition. Adv Mater Res 284–286:611–614

Fangming D, John EF, Karen IW (2003) Coagulation method for preparing single-walled carbon nanotube/poly(methyl methacrylate) composites and their modulus, electrical conductivity, and thermal stability. J Polym Sci Part B: Polym Phys 41(24):3333–3338

Farsheh AT, Talaeipour M, Hemmasi AH, Khademieslam H, Ghasemi I (2011) Investigation on the mechanical and morphological properties of foamed nanocomposites based on wood flour/PVC/multi-walled carbon nanotube. BioResources 6(1):841–852

Feng L, Chen Z (2006) Light-emitting conjugated molecule containing 1,3,4-oxadiazole, carbazole and naphthalene units. Spectrochim Acta Part A: Mol Biomol Spectrosc 63(1):15–20

Fenga W, Baib XD, Liana YQ, Liang J, Wanga XG, Yoshinoc K (2003) Well-aligned polyaniline/carbon-nanotube composite films grown by in-situ aniline polymerization. Carbon 41(8):1551–1557

Fisher RA, Watt MR, Jud Ready W (2013) Functionalized carbon nanotube supercapacitor electrodes: a review on pseudocapacitive materials. ECS J Solid State Sci Technol 2(10): M3170–M3177

Flahaut E, Peigney A, Laurent C, Rousset A (2000) Synthesis of single-walled carbon nanotube–Co–MgO composite powders and extraction of the nanotubes. J Mater Chem 10:249–252

Florian HG, Jacek N, Zbigniew R, Karl S (2003) Surface modified multi-walled carbon nanotubes in CNT/epoxy-composites. Chem Phys Lett 370(5):820–824

Fonseca MA, Abreu B, Gonçalves FAMM, Ferreira AGM, Moreira RAS, Oliveira MSA (2013) Shape memory polyurethanes reinforced with carbon nanotubes. Compos Struct 99:105–111

Fröhlich E, Meindl C, Höfler A, Leitinger G, Roblegg E (2013) Combination of small size and carboxyl functionalisation causes cytotoxicity of short carbon nanotubes. Nanotoxicology 7 (7):1211–1224

Fugetsu B, Sano E, Sunada M, Sambongi Y, Shibuya T, Wang X, Hiraki T (2008) Electrical conductivity and electromagnetic interference shielding efficiency of carbon nanotube/cellulose composite paper. Carbon 46(9):1256–1258

Galimberti M, Coombs M, Riccio P, Riccò T, Passera S, Pandini S, Conzatti L, Ravasio A, Tritto I (2013) The role of CNTs in promoting hybrid filler networking and synergism with carbon black in the mechanical behavior of filled polyisoprene. Macromol Mater Eng 298(2):241–251

Gao J-F, Huang H-D, Yan D-X, Ren P-G, Zeng X-B, Li Z-M (2013) Resistivity relaxation of anisotropic conductive polymer composites. J Macromol Sci Part B Phys 52(6):788–796

Gardea F, Lagoudas DC (2014) Characterization of electrical and thermal properties of carbon nanotube/epoxy composites. Compos B Eng 56:611–620

Gary DS, Dimitris CL (2008) A micromechanics model for the thermal conductivity of nanotube-polymer nanocomposites. J Appl Mech 75(4):1–10

Georgiev G, McIntyre MB, Judith R, Gombos EA, Cebe P (2011) Interplay between the crystal and liquid crystalline ordering of iPP and carbon nanotube composites under melt-shear. Mater Res Soc Symp Proc 1308:1–6

Ghasemi S, Karami H, Khanezar H (2014) Hydrothermal synthesis of lead dioxide/multiwall carbon nanotube nanocomposite and its application in removal of some organic water pollutant. J Mater Sci 49(3):1014–1024

Girei SA, Thomas SP, Atieh MA, Mezghani K, De SK, Bandyopadhyay S, Al-Juhani A (2012) Effect of -COOH functionalized carbon nanotubes on mechanical, dynamic mechanical and thermal properties of polypropylene nanocomposites. J Thermoplast Compos Mater 25(3):333–350

Gopalan AI, Lee K-P, Ragupathy D (2009) Development of a stable cholesterol biosensor based on multi-walled carbon nanotubes-gold nanoparticles composite covered with a layer of chitosan-room-temperature ionic liquid network. Biosens Bioelectron 24(7):2211–2217

Grabowski K, Zbyrad P, Uhl T (2014) Development of the strain sensors based on CNT/epoxy using screen printing. Key Eng Mater 588:84–90

Gu S-Y, Liu L-L, Yan B-B (2014) Effects of ionic solvent-free carbon nanotube nanofluid on the properties of polyurethane thermoplastic elastomer. J Polym Res 21(2), art no 356

Guo Q-H, Zhou X-P, Wang S-Q, Fu H-W, Li Y-H, Hou H-Q (2009) Heat-resistant polyimide electrical conductive composites. Polym Mater Sci Eng 25(2):52–58

Guo Y, Zhu G, Qu P, Jia Y (2014) Toughness enhancement by aligned multi-walled carbon nanotubes pullout from polymer matrix of multi-scale composites. Key Eng Mater 575–576:160–165

Guoxing S, Guangming C, Zhengping L, Ming C (2010) Preparation, crystallization, electrical conductivity and thermal stability of syndiotactic polystyrene/carbon nanotube composites. Carbon 48(5):1434–1440

Gupta KK, Abbas SM, Srivastava A, Nasim M, Saxena AK, Abhyankar A (2013) Microwave interactive properties of cotton fabrics coated with carbon nanotubes/polyurethane composite. Indian J Fibre Text Res 38(4):357–365

Gurunathan T, Rao CRK, Narayan R, Raju KVSN (2013) Polyurethane conductive blends and composites: synthesis and applications perspective. J Mater Sci 48(1):67–80

Haggenmuellera R, Gommansb HH, Rinzlerb AG, Fischera JE, Winey KI (2000) Aligned single-wall carbon nanotubes in composites by melt processing methods. Chem Phys Lett 330 (3):219–225

Han T, Qu L, Luo Z, Wu X, Zhang D (2014) Enhancement of hydroxyl radical generation of a solid state photo-Fenton reagent based on magnetite/carboxylate-rich carbon composites by embedding carbon nanotubes as electron transfer channels. New J Chem 38(3):942–948

Hao XY, Chien A-T, Hua XY, Lu J, Liu YD (2013) Dispersion of pristine CNTs in UHMWPE solution to prepare CNT/UHMWPE composite fibre. Mater Res Innov 17(Suppl 1):123–125

Hashmi S, Ghavaminejad A, Obiweluozor FO, Vatankhah-Varnoosfaderani M, Stadler FJ (2012) Supramolecular interaction controlled diffusion mechanism and improved mechanical behavior of hybrid hydrogel systems of zwitterions and cnt. Macromolecules 45(24):9804–9815

Haznedar G, Cravanzola S, Zanetti M, Scarano D, Zecchina A, Cesano F (2013) Graphite nanoplatelets and carbon nanotubes based polyethylene composites: electrical conductivity and morphology. Mater Chem Phys 143(1):47–52

He Z, Zhang X, Chen M, Li M, Gu Y, Zhang Z, Li Q (2013) Effect of the filler structure of carbon nanomaterials on the electrical, thermal, and rheological properties of epoxy composites. J Appl Polym Sci 129(6):3366–3372

He Q, Yuan T, Yan X, Ding D, Wang Q, Luo Z, Shen TD, Wei S, Cao D, Guo Z (2014) Flame-retardant polypropylene/multiwall carbon nanotube nanocomposites: effects of surface functionalization and surfactant molecular weight. Macromol Chem Phys 215(4):327–340

Heidari A, Beheshty MH, Rahimi H (2013) Functionalization of multi-walled carbon nanotubes via direct friedel-crafts acylation in an optimized PPA/P2O5 medium. Fuller Nanotubes Carbon Nanostruct 21(6):516–524

Hida S, Shiga T, Maruyama S, Elliott JA, Shiomi J (2012) Influence of thermal boundary resistance and interfacial phonon scattering on heat conduction of carbon nanotube/polymer composites. Trans Jpn Soc Mech Eng Part B 78(787):634–643

Hilmi Y, Seyhana TA, Servet T, Metin T, Wolfgang B, Karl S (2010) Electric field effects on CNTs/vinyl ester suspensions and the resulting electrical and thermal composite properties. Compos Sci Technol 70(14):2102–2110

Hone J, Whitney M, Piskoti C, Zettl A (1999) Thermal conductivity of single-walled carbon nanotubes. Phys Rev 59:R2514–R2520

Hongjie D (2002) Carbon nanotubes: synthesis, integration, and properties. Acc Chem Res 35 (12):1035–1044

Hordy N, Coulombe S, Meunier J-L (2013) Plasma functionalization of carbon nanotubes for the synthesis of stable aqueous nanofluids and poly(vinyl alcohol) nanocomposites. Plasma Process Polym 10(2):110–118

Hossain ME, Hossain MK, Hosur MV, Jeelani S(2011) Investigation of carbon nanofibers (CNFs) effects on the flexural and thermal behavior of E-glass/polyester composites. In: Proceedings of the ASME international mechanical engineering congress and exposition (IMECE), vol 12, pp 135–143

Hsu-Chiang K, Chen-Chi MM, Wei-Ping C, Siu-Ming Y, Hsin-Ho W, Tzong-Ming L (2006) Synthesis, thermal, mechanical and rheological properties of multiwall carbon nanotube/waterborne polyurethane nanocomposite. Compos Sci Technol 65(11):1703–1710

Hua L, Brinson LC (2005) A hybrid numerical-analytical method for modeling the viscoelastic properties of polymer nanocomposites. J Appl Mech 73(5):758–768

Huaming L, Fuyong C, Andy MD, Alex A (2005) Functionalization of single-walled carbon nanotubes with well-defined polystyrene by "click" coupling. J Am Chem Soc 127(41):14518–14524

Huang J-Q, Zhang Q, Zhang S-M, Liu X-F, Zhu W, Qian W-Z, Wei F (2013) Aligned sulfur-coated carbon nanotubes with a polyethylene glycol barrier at one end for use as a high efficiency sulfur cathode. Carbon 58:99–106

Huang G, Wang S, Song P, Wu C, Chen S, Wang X (2014a) Combination effect of carbon nanotubes with graphene on intumescent flame-retardant polypropylene nanocomposites. Compos A Appl Sci Manuf 59:18–25

Huang J, Gao M, Pan T, Zhang Y, Lin Y (2014b) Effective thermal conductivity of epoxy matrix filled with poly(ethyleneimine) functionalized carbon nanotubes. Compos Sci Technol 95:16–20

Huang J, Gao M, Pan T, Zhang Y, Lin Y (2014c) Effective thermal conductivity of epoxy matrix filled with poly(ethyleneimine) functionalized carbon nanotubes. Compos Sci Technol 95:16–20

Huisheng P (2008) Aligned carbon nanotube/polymer composite films with robust flexibility, high transparency, and excellent conductivity. J Am Chem Soc 130(1):42–43

Hutchisona JL, Kiselevb NA, Krinichnayac AEP, Krestininc AV, Loutfyd RO, Morawsky AP, Muradyan VE, Obraztsova ED, Sloan J, Terekhov SV, Zakharov DN (2001) Double-walled carbon nanotubes fabricated by a hydrogen arc discharge method. Carbon 39:761–770

Hwa-Jeong L, Se-Jin O, Ja-Young C, Jin WK, Jungwan H, Loon-Seng T, Jong-Beom B (2005) In situ synthesis of poly(ethylene terephthalate) (PET) in ethylene glycol containing terephthalic acid and functionalized multiwalled carbon nanotubes (MWNTs) as an approach to MWNT/PET nanocomposites. Chem Mater 17(20):5057–5064

Hyang HS, Jae WC, Nanda GS (2007) Effect of carbon nanotubes on mechanical and electrical properties of polyimide/carbon nanotubes nanocomposites. Eur Polym J 43(9):3750–3756

Ibrahim S, Yasin SMM, Nee NM, Ahmad R, Johan MR (2012) Conductivity, thermal and infrared studies on plasticized polymer electrolytes with carbon nanotubes as filler. J Non-Cryst Solids 358(2):210–216

Itoh E, Kato Y, Sano Y, Miyairi K (2008) Field emission from conducting polymer/single-walled carbon nanotube composite prepared by AC coupled electrophoresis. Jpn J Appl Phys 47(4 PART 1):2016–2020

Jain S, Thakare VS, Das M, Godugu C, Jain AK, Mathur R, Chuttani K, Mishra AK (2011) Toxicity of multiwalled carbon nanotubes with end defects critically depends on their functionalization density. Chem Res Toxicol 24(11):2028–2039

Jeena JK, Narasimha Murthy HNM, Rai KS, Krishna M, Sreejith M (2010) Effect of amine functionalization of CNF on electrical, thermal, and mechanical properties of epoxy/CNF composites. Polym Bull 65(8):849–861

Jerez J, Isaguirre AC, Bazán C, Martinez LD, Cerutti S (2014) Determination of scandium in acid mine drainage by ICP-OES with flow injection on-line preconcentration using oxidized multiwalled carbon nanotubes. Talanta 124:89–94

Jia X, Zhang Q, Zhao M-Q, Xu G-H, Huang J-Q, Qian W, Lu Y, Wei F (2012) Dramatic enhancements in toughness of polyimide nanocomposite via long-CNT-induced long-range creep. J Mater Chem 22(14):7050–7056

Jian PL (1997) Elastic properties of carbon nanotubes and nanoropes. Phys Rev Lett 79:1297–1300

Jiang F, Wang Y, Hu X, Shao N, Na N, Delanghe JR, Ouyang J (2010) Carbon nanotubes-assisted polyacrylamide gel electrophoresis for enhanced separation of human serum proteins and application in liverish diagnosis. J Sep Sci 33(21):3393–3399

Jiang J, Xiao H, Li H (2013a) Electrical resistivity and piezoresistivity of Ni-CNT filled epoxy-based composites. In: Proceedings of SPIE—the international society for optical engineering, vol 8689, art no 86890W

Jiang Y, Zhang H, Wang Y, Chen M, Ye S, Hou Z, Ren L (2013b) Modulation of apoptotic pathways of macrophages by surface-functionalized multi-walled carbon nanotubes. PLoS ONE 8(6), art no e65756

Jiang L, Zhang C, Liu M, Yang Z, Tjiu WW, Liu T (2014a) Simultaneous reinforcement and toughening of polyurethane composites with carbon nanotube/halloysite nanotube hybrid. Compos Sci Technol 91:98–103

Jiang Q, Wang X, Zhu Y, Hui D, Qiu Y (2014b) Mechanical, electrical and thermal properties of aligned carbon nanotube/polyimide composites. Compos B Eng 56:408–412

Jin S, Matuana LM (2010) Wood/plastic composites co-extruded with multi-walled carbon nanotube-filled rigid poly(vinyl chloride).cap layer. Polym Int 59(5):648–657

Joseph W, Jinhua D, Travis Y (2005) Carbon nanotube—conducting-polymer composite nanowires. Lagmuir 21(1):9–12

Journet C, Maser WK, Bernier P, Loiseau A, Chapelle ML, Lefrant S, Deniard P, Lee R, Fischer JE (1995) Large-scale production of single-walled carbon nanotubes by the electric-arc technique. Nature 388:756–758

Jung Y-T, Park Y-B (2013) Carbon-nanotube-based structural health monitoring for wind turbine applications. In: Proceedings of the international SAMPE technical conference, pp 2902–2909

Kanagaraj S (2010) Polyethylene nanotube nanocomposites, polymer nanotube nanocomposites: synthesis, properties, and applications, pp 113–139

Kanagaraj S, Fonseca A, Guedes RM, Oliveira MSA, Simoes JAO (2011) Thermo-mechanical behaviour of ultrahigh molecular weight polyethylene-carbon nanotubes composites under different cooling techniques. Defect Diffus Forum 312–315:331–340

Karousis N, Tagmatarchis N, Tasis D (2010) Current progress on the chemical modification of carbon nanotubes. Chem Rev 110(9):5366–5397

Kenji H, Don NF, Kohei M, Tatsunori N, Motoo Y, Sumio I (2000) Water-assisted highly efficient synthesis of impurity-free single-walled carbon nanotubes. Science 306(5700):1362–1364

Kiliaris P, Papaspyrides CD (2010) Polymer/layered silicate (clay) nanocomposites: an overview of flame retardancy. Prog Polym Sci 35:902–958

Kim BS, Bae SH, Park Y-H, Kim J-H (2007) Preparation and characterization of polyimide/ carbon-nanotube composites. Macromol Res 15(4):357–362

Kim B-J, Byun J-H, Park S-J (2010) Effects of graphenes/CNTs co-reinforcement on electrical and mechanical properties of HDPE matrix nanocomposites. Bull Korean Chem Soc 31(8):2261–2264

Kim H-K, Roh KC, Kang K, Kim K-B (2013a) Synthesis of nano-Li4Ti5O12 decorated on non-oxidized carbon nanotubes with enhanced rate capability for lithium-ion batteries. RSC Adv 3 (34):14267–14272

Kim M-S, Yan J, Kang K-M, Joo K-H, Kang Y-J, Ahn S-H (2013b) Soundproofing ability and mechanical properties of polypropylene/exfoliated graphite nanoplatelet/carbon nanotube (PP/ xGnP/CNT) composite. Int J Precis Eng Manuf 14(6):1087–1092

Kim Y-H, Park S, Won K, Kim HJ, Lee SH (2013c) Bacterial cellulose-carbon nanotube composite as a biocompatible electrode for the direct electron transfer of glucose oxidase. J Chem Technol Biotechnol 88(6):1067–1070

Ko H-H, Cheng Y-Y, Dai C-A (2014) Silane modified multiwall carbon nanotubes/polyimide composites prepared using in-situ polymerization. Nanosci Nanotechnol Lett 6(3):190–196

Koga H, Saito T, Kitaoka T, Nogi M, Suganuma K, Isogai A (2013) Transparent, conductive, and printable composites consisting of TEMPO-oxidized nanocellulose and carbon nanotube. Biomacromolecules 14(4):1160–1165

Korayem AH, Barati MR, Simon GP, Zhao XL, Duan WH (2014) Reinforcing brittle and ductile epoxy matrices using carbon nanotubes masterbatch. Compos A Appl Sci Manuf 61:126–133

Kotchey GP, Zhao Y, Kagan VE, Star A (2013) Peroxidase-mediated biodegradation of carbon nanotubes in vitro and in vivo. Adv Drug Deliv Rev 65(15):1921–1932

Krause RWM, Mamba BB, Dlamini LN, Durbach SH (2010) Fe-Ni nanoparticles supported on carbon nanotube-co-cyclodextrin polyurethanes for the removal of trichloroethylene in water. J Nanopart Res 12(2):449–456

Kulathunga DDTK, Ang KK (2014) Modeling and simulation of buckling of embedded carbon nanotubes. Comput Mater Sci 81:233–238

Kumar Sachdev V, Bhattacharya S, Patel K, Kumar Sharma S, Chand Mehra N, Pal Tandon R (2013) Electrical and EMI shielding characterization of multiwalled carbon nanotube/ polystyrene composite. J Appl Polym Sci

Kuzhir P, Paddubskaya A, Plyushch A, Volynets N, Maksimenko S, MacUtkevic J, Kranauskaite I, Banys J, Ivanov E, Kotsilkova R, Celzard A, Fierro V, Zicans J, Ivanova T, Merijs Meri R, Bochkov I, Cataldo A, Micciulla F, Bellucci S, Lambin P (2013) Epoxy composites filled with high surface area-carbon fillers: optimization of electromagnetic shielding, electrical, mechanical, and thermal properties. J Appl Phys 114(16), art no 164304

Le HH, Sriharish MN, Henning S, Klehm J, Menzel M, Frank W, Wießner S, Das A, Stöckelhuber K-W, Heinrich G, Radusch H-J (2014) Dispersion and distribution of carbon nanotubes in ternary rubber blends. Compos Sci Technol 90:180–186

Lee D, Cui T (2011) Suspended carbon nanotube nanocomposite beams with a high mechanical strength via layer-by-layer nano-self-assembly. Nanotechnology 22(16), art no 165601

Lee M, Ku SH, Ryu J, Park CB (2010) Mussel-inspired functionalization of carbon nanotubes for hydroxyapatite mineralization. J Mater Chem 20(40):8848–8853

Lee G-H, Min K-M, Kim D-W (2013) Synthesis and Li electroactivity of MnS/carbon nanotube composites. J Korean Ceram Soc 50(6):539–544

Lei J, Ju H (2010) Nanotubes in biosensing. Wiley Interdiscip Rev: Nanomed Nanobiotechnol 2 (5):496–509

Lertrojanachusit N, Pornsunthorntawee O, Kitiyanan B, Chavadej J, Chavadej S (2013) Separation and purification of carbon nanotubes using froth flotation with three sequential pretreatment steps of catalyst oxidation, catalyst removal, and silica dissolution. Asia-Pac J Chem Eng 8 (6):830–842

Li J, Bai T (2011) The effect of CNT modification on the mechanical properties of polyimide composites with and without MoS2. Mech Compos Mater 1–6

Li WZ, Xie SS, Qian LX, Chang BH, Zou BS, Zhou WY, Zhao RA, Wang G (1996) Large-scale synthesis of aligned carbon nanotubes. Science 274(5293):1701–1703

Li YB, Wei BQ, Liang J, Yu Q, Wu DH (1999) Transformation of carbon nanotubes to nanoparticles by ball milling process. Carbon 37:493–497

Li Z, Chen H, Zhu Z, Zhang Y (2011) Study on thermally conductive ESBR vulcanizates. Polym Bull 67(6):1091–1104

Li Y, Chen YF, Feng YY, Zhao SL, Lü P, Yuan XY, Feng W (2012) Progress of synthesizing methods and properties of fluorinated carbon nanotubes. Sci China Technol Sci 53(5):1225–1233

Li R, Wang X, Ji Z, Sun B, Zhang H, Chang CH, Lin S, Meng H, Liao Y-P, Wang M, Li Z, Hwang AA, Song T-B, Xu R, Yang Y, Zink JI, Nel AE, Xia T (2013a) Surface charge and cellular processing of covalently functionalized multiwall carbon nanotubes determine pulmonary toxicity. ACS Nano 7(3):2352–2368

Li R, Wang X, Ji Z, Sun B, Zhang H, Chang CH, Lin S, Meng H, Liao Y-P, Wang M, Li Z, Hwang AA, Song T-B, Xu R, Yang Y, Zink JI, Nel AE, Xia T (2013b) Surface charge and cellular processing of covalently functionalized multiwall carbon nanotubes determine pulmonary toxicity. ACS Nano 7(3):2352–2368

Li S, Guo ZP, Wang CY, Wallace GG, Liu HK (2013c) Flexible cellulose based polypyrrole-multiwalled carbon nanotube films for bio-compatible zinc batteries activated by simulated body fluids. J Mater Chem A 1(45):14300–14305

Li W, He D, Bai J (2013d) The influence of nano/micro hybrid structure on the mechanical and self-sensing properties of carbon nanotube-microparticle reinforced epoxy matrix composite. Compos A Appl Sci Manuf 54:28–36

Li X, Pignatello JJ, Wang Y, Xing B (2013e) New insight into adsorption mechanism of ionizable compounds on carbon nanotubes. Environ Sci Technol 47(15):8334–8341

Li J, Chen C, Zhang S, Ren X, Tan X, Wang X (2014) Critical evaluation of adsorption-desorption hysteresis of heavy metal ions from carbon nanotubes: Influence of wall number and surface functionalization. Chem Asian J 9(4):1144–1151

Liang F, Tang Y, Gou J, Gu C, Song G (2009) Multifunctional nanocomposites with high damping performance for aerospace structures. Mech Solids Struct Fluids 11:1–10

Liao L, Wang X, Fang P, Liew KM, Pan C (2011) Interface enhancement of glass fiber reinforced vinyl ester composites with flame-synthesized carbon nanotubes and its enhancing mechanism. ACS Appl Mater Interfaces 3(2):534–538

Likozar B (2010) Diffusion of ionic liquids into elastomer/carbon nanotubes composites and tensile mechanical properties of resulting materials. Scientia Iranica 17(1 F):35–42

Lin X-P, Guan P, Hu X-L, Tang Y-M (2011a) Preparation of carbon nanotube composites in ionic liquids. Mod Chem Indus 31(9):14–18

Lin M-F, Thakur VK, Tan EJ, Lee PS (2011b) Surface functionalization of BaTiO3 nanoparticles and improved electrical properties of BaTiO3/polyvinylidene fluoride composite. RSC Adv 1:576–578

Lin M-F, Thakur VK, Tan EJ, Lee PS (2011c) Dopant induced hollow BaTiO3 nanostructures for application in high performance capacitors. J Mater Chem 21:16500–16504

Liu Z, Zhao L, Chen M, Yu J (2011) Effect of carboxylate multi-walled carbon nanotubes on the performance of thermoplastic starch nanocomposites. Carbohydr Polym 83(2):447–451

Liu J, Liu R, Jiang J, Liu X (2013a) Design and synthesis of water-soluble photosensitive α-cyclodextrin and its application in dispersing carbon nanotubes. J Appl Polym Sci 130 (4):2588–2593

Liu M, Zhang C, Tjiu WW, Yang Z, Wang W, Liu T (2013b) One-step hybridization of graphene nanoribbons with carbon nanotubes and its strong-yet-ductile thermoplastic polyurethane composites. Polym (UK) 54(12):3124–3130

Liu X-M, Gao F, Cai W-T, Liu P, Miao W, Huang Y (2013c) Analysis of pressure-resistance calculating model of carbon nanotubes/carbon black/silicone rubber composite material. J Funct Mater 44(5):669–672

Liu C-C, Sadhasivam S, Savitha S, Lin F-H (2014a) Fabrication of multiwalled carbon nanotubes-magnetite nanocomposite as an effective ultra-sensing platform for the early screening of nasopharyngeal carcinoma by luminescence immunoassay. Talanta 122:195–200

Liu L, Zhang H, Zhou Y (2014b) Quasi-static mechanical response and corresponding analytical model of laminates incorporating with nanoweb interlayers. Compos Struct 111(1):436–445

Liu T, Yang F, Li Y, Ren L, Zhang L, Xu K, Wang X, Xu C, Gao J (2014c) Plasma synthesis of carbon nanotube-gold nanohybrids: efficient catalysts for green oxidation of silanes in water. J Mater Chem A 2(1):245–250

Liu X-L, Lu H-J, Xing L-Y (2014d) Morphology and microwave absorption of carbon nanotube/bismaleimide foams. J Appl Polym Sci 131(9), art no 40233

Liu Z, Dong X, Song L, Zhang H, Liu L, Zhu D, Song C, Leng X (2014e) Carboxylation of multiwalled carbon nanotube enhanced its biocompatibility with L02 cells through decreased activation of mitochondrial apoptotic pathway. J Biomed Mater Res Part A 102(3):665–673

Loos MR, Manas-Zloczower I (2013) Micromechanical models for carbon nanotube and cellulose nanowhisker reinforced composites. Polym Eng Sci 53(4):882–887

Loos MR, Nahorny J, Fontana LC (2013a) Plasma modification of carbon nanotubes. Curr Org Chem 17(17):1880–1893

Loos MR, Yang J, Feke DL, Manas-Zloczower I, Unal S, Younes U (2013b) Enhancement of fatigue life of polyurethane composites containing carbon nanotubes. Compos B Eng 44 (1):740–744

López Manchadoa MA, Valentinib L, Biagiottib J, Kennyb JM (2006) Thermal and mechanical properties of single-walled carbon nanotubes–polypropylene composites prepared by melt processing. Carbon 43(7):1499–1505

Lu Y-M, Gong Q-M, Liang J (2009) Preparation of carbon nanotubes/activated carbon composite microspheres and their application to adsorption of VB12. Acta Phys Chim Sin 25(8):1697–17202

Lu W, Ding G, Wang Y, Deng M, Zhou Z (2011) Nanomanufacturing of multi-walled carbon nanotubes enhanced electrochemical electrode through surface micromachining. Int J Nanomanuf 7(2):104–115

Lu D, Lin S, Wang L, Li T, Wang C, Zhang Y (2014) Sensitive detection of luteolin based on poly (diallyldimethylammonium chloride)-functionalized graphene-carbon nanotubes hybrid/β-cyclodextrin composite film. J Solid State Electrochem 18(1):269–278

Ma Y, Wu D, Liu Y, Li X, Qiao H, Yu Z-Z (2014a) Electrically conductive and super-tough polypropylene/carbon nanotube nanocomposites prepared by melt compounding. Compos B Eng 56:384–391

Ma Y, Wu D, Liu Y, Li X, Qiao H, Yu Z-Z (2014b) Electrically conductive and super-tough polypropylene/carbon nanotube nanocomposites prepared by melt compounding. Compos B Eng 56:384–391

Maizatulnisa O, Tan KH, Mohd Yusof H, Halisanni K, Ruzaidi G, Nazarudin Z, Paridah T, Noor Azlina H (2013) Effects of multi-walled carbon nanotubes (MWCNTS) on the mechanical and thermal properties of plasticized polylactic acid nanocomposites. Adv Mater Res 812:181–186

Makki A, Sami B, Mohamed NB, Alain D, Alessandro G (2005) Modification of cellulose fibers with functionalized silanes: effect of the fiber treatment on the mechanical performances of cellulose–thermoset composites. J Appl Polym Sci 98(3):974–984

Mallakpour S, Zadehnazari A (2013) Functionalization of multiwalled carbon nanotubes with S-valine amino acid and its reinforcement on amino acid-containing poly(amide-imide) bionanocomposites. High Perform Polym 25(8):966–979

Mamunya Ye, Boudenne A, Lebovka N, Ibos L, Candau Y, Lisunova M (2008) Electrical and thermophysical behaviour of PVC-MWCNT nanocomposites. Compos Sci Technol 68 (9):1981–1988

Mao H, Kawazoe N, Chen G (2014) Cellular uptake of single-walled carbon nanotubes in 3D extracellular matrix-mimetic composite collagen hydrogels. J Nanosci Nanotechnol 14 (3):2487–2492

Mari D, Schaller R (2009) Mechanical spectroscopy in carbon nanotube reinforced ABS. Mater Sci Eng A 521–522:255–258

Martín R, Jiménez LB, Álvaro M, Scaiano JC, Garcia H (2009) Photoinduced formation and characterization of electron-hole pairs in azaxanthylium-derivatized short single-walled carbon nanotubes. Chem Eur J 15(35):8751–8759

Mases M, Noë M, Mercier G, Dossot M, Vigolo B, Mamane V, Fort Y, Soldatov AV, McRae E (2011) Effects on Raman spectra of functionalisation of single walled carbon nanotubes by nitric acid. Phys Status Solidi (B) Basic Res 248(11):2552–2555

Masinga SP, Nxumalo EN, Mamba BB, Mhlanga SD (2013) Microwave-induced synthesis of β-cyclodextrin/N-doped carbon nanotube polyurethane nanocomposites for water purification. Phys Chem Earth

Matsuoka M, Tatami J, Wakihara T (2014) Control of dispersion and agglomeration of CNTs for their networking—mechanical and electrical properties of CNT/alumina composites. Ceram Trans 243:117–120

Matthew JK, Virginia AD (2009) Viscoelasticity and shear stability of single-walled carbon nanotube/unsaturated polyester resin dispersions. Macromolecules 42(17):6624–6632

Mehta H, Wangshul H, Kanagaraj S (2011) Studies on mechanical properties of carbon nanotubes/ high-density polyethylene nanocomposites by small punch technique. Int J Nanosci 10(1–2):247–251

Mhlanga SD, Masinga SP, Bambo MF, Mamba BB, Nxumalo EN (2013a) A facile procedure to synthesize a three-component β-cyclodextrin polyurethane nanocomposite matrix containing Ag decorated N-CNTs for water treatment. Nanosci Nanotechnol Lett 5(3):341–348

Mhlanga SD, Phao N, Nxumalo EN, Marnba BB (2013b) Synthesis and study of novel nanostructured membranes incorporating N-doped CNTs for water treatment. In: Technical proceedings of the 2013 NSTI nanotechnology conference and expo, NSTI-nanotech 2013, vol 3, pp 665–668

Michael JO'C, Sergei MB, Chad BH, Valerie CM, Michael SS, Erik HH, Kristy LR, Peter JB, William HN, Carter K, Jianpeng M, Robert HH, Bruce RW, Richard ES (2002) Band gap fluorescence from individual single-walled carbon nanotubes. Science 297(5581):593–596

Min-Feng Y, Bradley SF, Sivaram A, Rodney SR (2000a) Tensile loading of ropes of single wall carbon nanotubes and their mechanical properties. Phys Rev Lett 84:5552–5555

Min-Feng Y, Oleg L, Mark JD, Katerina M, Thomas FK, Rodney SR (2000b) Strength and breaking mechanism of multiwalled carbon nanotubes under tensile load. Science 287 (5453):637–640

Myounggu P, Hyonny K, Jeffrey PY (2008) Strain-dependent electrical resistance of multi-walled carbon nanotube/polymer composite films. Nanotechnology 19(5):1–10

Myung JL, Cheol-Jin L, Singh VR, Kum-Pyo Y, Nam-Ki M (2010) Humidity sensing characteristics of plasma functionalized multiwall carbon nanotube-polyimide composite films. Proc IEEE Sens 430–433, art no 4716470

Nadagouda MN, Varma RS (2008) Noble metal decoration and alignment of carbon nanotubes in carboxymethyl cellulose. Macromol Rapid Commun 29(2):155–159

Nadia G, Joachim L, Cor EK (2005) Strategies for dispersing carbon nanotubes in highly viscous polymer. J Mater Chem 15:2349–2352

Naeimi H, Mohajeri A, Moradi L, Rashidi AM (2009) Efficient and facile one pot carboxylation of multiwalled carbon nanotubes by using oxidation with ozone under mild conditions. Appl Surf Sci 256(3):631–635

Ng CM, Manickam S (2013) Improved functionalization and recovery of carboxylated carbon nanotubes using the acoustic cavitation approach. Chem Phys Lett 557:97–101

Nie M, Fisher FT (2013) A nano-hybrid shish kebab approach to modifying the interface in carbon nanotube—semicrystalline polymer nanocomposites. In: Proceedings of the 28th annual technical conference of the American society for composites 2013, ASC 2013, vol 2, pp 1804–1814

Nie C, Pan L, Liu Y, Li H, Chen T, Lu T, Sun Z (2012) Electrophoretic deposition of carbon nanotubes-polyacrylic acid composite film electrode for capacitive deionization. Electrochim Acta 66:106–109

Nitayaphat W, Jintakosol T (2014) Removal of silver (I) from aqueous solutions by chitosan/carbon nanotube nanocomposite beads. Adv Mater Res 893:166–169

Nozomi N-R, Sarunya B, Xiaoming S, Kevin W, Hongjie D (2007) Noncovalent functionalization of carbon nanotubes by fluorescein—polyethylene glycol: supramolecular conjugates with ph-dependent absorbance and fluorescence. J Am Chem Soc 129(9):2448–2449

Ntim SA, Sae-Khow O, Witzmann FA, Mitra S (2011) Effects of polymer wrapping and covalent functionalization on the stability of MWCNT in aqueous dispersions. J Colloid Interface Sci 355(2):383–388

Nxumalo EN, Msomi PF, Mhlanga SD, Mamba BB (2013) Production of N-doped carbon nanotubes using α- and β-cyclodextrins: the effect of solubility. Mater Lett 100:66–69

Ogasawara T, Ishida Y, Ishikawa T, Yokota R (2004) Characterization of multi-walled carbon nanotube/phenylethynyl terminated polyimide composites. Compos A Appl Sci Manuf 35 (1):67–74

Ortengren U (2000) On composite resin materials. Degradation, erosion and possible adverse effects in dentists. Swed Dent J Suppl 141:1–61

Panamoottil SM, Pötschke P, Lin RJT, Bhattacharyya D, Fakirov S (2013) Conductivity of microfibrillar polymer-polymer composites with CNT-loaded microfibrils or compatibilizer: a comparative study. Express Polym Lett 7(7):607–620

Pang J, Xu G, Bai Y, Yuan S, He F, Wang Y, Sun H, Hao A (2010) Molecular dynamics simulations of the interactions between β-cyclodextrin derivatives and single-walled carbon nanotubes. Comput Mater Sci 50(2):283–290

Pang H, Bao Y, Yang S-G, Chen C, Zhang W-Q, Chen J, Ji X, Lei J (2014) Preparation and properties of carbon nanotube/binary-polymer composites with a double-segregated structure. J Appl Polym Sci 131(2), art no 39789

Parodi B, Londonio A, Polla G, Savio M, Smichowski P (2014) On-line flow injection solid phase extraction using oxidised carbon nanotubes as the substrate for cold vapour-atomic absorption determination of Hg(ii) in different kinds of water. J Anal At Spectrom 29(5):880–885

Parveen S, Husain S, Kumar A, Ali J, Husain M, Harsh M, Husain M (2013) Enhanced field emission properties of carbon nanotube based field emitters by dynamic oxidation. Curr Nanosci 9(5):619–623
Parvinzadeh Gashti M, Almasian A (2013) UV radiation induced flame retardant cellulose fiber by using polyvinylphosphonic acid/carbon nanotube composite coating. Compos B Eng 45 (1):282–289
Pascual J, Peris F, Boronat T, Fenollar O, Balart R (2012) Study of the effects of multi-walled carbon nanotubes on mechanical performance and thermal stability of polypropylene. Polym Eng Sci 52(4):733–740
Paul P, Amit KK, Ellen MA, Anthony MW, Bong SS, Jiadi X, Himabindu N, Benjamin GP, Joerg L, Ayyalusamy R, Nicholas AK (2007) Ultrastrong and stiff layered polymer nanocomposites. Science 318(5847):80–83
Pei L, Abbott J, Zufelt K, Davis A, Zappe M, Decker K, Liddiard S, Vanfleet R, Linford MR, Davis R (2011) Processing of thin carbon nanotube-polyimide composite membranes. Nanosci Nanotechnol Lett 3(4):451–457
Peng CM, Ben ZT, Jang-Kyo K (2008) Effect of CNT decoration with silver nanoparticles on electrical conductivity of CNT-polymer composites. Carbon 46(11):1497–1505
Peng R, Wang Y, Tang W, Yang Y, Xie X (2013) Progress in imidazolium ionic liquids assisted fabrication of carbon nanotube and graphene polymer composites. Polymers 5(2):847–872
Perez LD, Zuluaga MA, Kyu T, Mark JE, Lopez BL (2009) Preparation, characterization, and physical properties of multiwall carbon nanotube/elastomer composites. Polym Eng Sci 49 (5):866–874
Peter KHH, Richard HF (2002) π-electronic and electrical transport properties of conjugated polymer nanocomposites: poly(p-phenylenevinylene) with homogeneously dispersed silica nanoparticles. J Chem Phys 116:6782–6790
Petra P, Fornes TD, Paul DR (2002) Rheological behavior of multiwalled carbon nanotube/ polycarbonate composites. Polymer 43(11):3247–3255
Philip GC, Keith B, Masa I, Zettl A (2000) Extreme oxygen sensitivity of electronic properties of carbon nanotubes. Science 287(5459):1801–1804
Pickering SJ, Kelly RM, Kennerley JR, Rudd CD, Fenwick NJ (2000) A fluidised-bed process for the recovery of glass fibres from scrap thermoset composites. Compos Sci Technol 60(4):509–523
Pierard N, Fonseca A, Konya Z, Willems I, Van Tendeluo G, Nagy JB (2001) Production of short carbon nanotubes with open tips of ball milling. Chem Phys Lett 335:1–8
Pöllänen M, Pirinen S, Suvanto M, Pakkanen TT (2011) Influence of carbon nanotube-polymeric compatibilizer masterbatches on morphological, thermal, mechanical, and tribological properties of polyethylene. Compos Sci Technol 71(10):1353–1360
Popuri SR, Frederick R, Chang C-Y, Fang S-S, Wang C-C, Lee L-C (2014) Removal of copper (II) ions from aqueous solutions onto chitosan/carbon nanotubes composite sorbent. Desalin Water Treat 52(4–6):691–701
Pötschke P, Villmow T, Pegel S, John A, Rentenberger R (2011) Melt mixed polymer-MWCNT composites for liquid sensing applications. Mater Res Soc Symp Proc 1410:31–42
PourAkbar Saffar K, JamilPour N, Rouhi G, Raeisi Najafi A, Arshi AR, Sudak L (2009) Fracture toughness of carbon nanotube reinforced artificial bone tissue. In: Proceedings of the 12th international conference on fracture 2009, ICF-12, vol 8, pp 5895–5902
Prajapati VK, Awasthi K, Gautam S, Yadav TP, Rai M, Srivastava ON, Sundar S (2011) Targeted killing of leishmania donovani in vivo and in vitro with amphotericin β attached to functionalized carbon nanotubes. J Antimicrob Chemother 66(4):874–879, art no dkr002
Primo EN, Gutierrez FA, Luque GL, Dalmasso PR, Gasnier A, Jalit Y, Moreno M, Bracamonte MV, Rubio ME, Pedano ML, Rodríguez MC, Ferreyra NF, Rubianes MD, Bollo S, Rivas GA (2013) Comparative study of the electrochemical behavior and analytical applications of (bio) sensing platforms based on the use of multi-walled carbon nanotubes dispersed in different polymers. Anal Chim Acta 805:19–35
Prolongo SG, Del Rosario G, Ureña A (2013) Coupled thermal-electrical analysis of carbon nanotube/epoxy composites. Polym Eng Sci

Qi H, Liu J, Gao S, Mäder E (2013a) Multifunctional films composed of carbon nanotubes and cellulose regenerated from alkaline-urea solution. J Mater Chem A 1(6):2161–2168

Qi H, Mäder E, Liu J (2013b) Unique water sensors based on carbon nanotube-cellulose composites. Sens Actuators B: Chem 185:225–230

Qi H, Mäder E, Liu J (2013c) Electrically conductive aerogels composed of cellulose and carbon nanotubes. J Mater Chem A 1(34):9714–9720

Qiao YJ, Cao M, Zhang L (2006) Investigation on potential microwave absorbability of polyester-composites filled with carbon nanotubes. In: Proceedings of 1st IEEE international conference on nano micro engineered and molecular systems, IEEE-NEMS, art no 4135190, pp 1331–1334

Raja M, Shanmugharaj AM, Ryu SH, Subha J (2011) Influence of metal nanoparticle decorated CNTs on polyurethane based electro active shape memory nanocomposite actuators. Mater Chem Phys 129(3):925–931

Raja M, Ryu SH, Shanmugharaj AM (2013a) Thermal, mechanical and electroactive shape memory properties of polyurethane (PU)/poly(lactic acid) (PLA)/CNT nanocomposites. Eur Polym J 49(11):3492–3500

Raja M, Ryu SH, Shanmugharaj AM (2013b) Thermal, mechanical and electroactive shape memory properties of polyurethane (PU)/poly(lactic acid) (PLA)/CNT nanocomposites. Eur Polym J 49(11):3492–3500

Raja M, Ryu SH, Shanmugharaj AM (2014) Influence of surface modified multiwalled carbon nanotubes on the mechanical and electroactive shape memory properties of polyurethane (PU)/poly(vinylidene diflouride) (PVDF) composites. Colloids Surf A 450(1):59–66

Rajdip B, Einat N-R, Oren R, Rachel Y-R (2002) Stabilization of individual carbon nanotubes in aqueous solutions. Nano Lett 2(1):25–28

Rajendran Royan NR, Sulong AB, Sahari J, Suherman H (2013) Effect of acid- and ultraviolet/ozonolysis-treated MWCNTs on the electrical and mechanical properties of epoxy nanocomposites as bipolar plate applications. J Nanomater, art no 717459

Ramôa SD, Barra GM, Oliveira RV, De Oliveira MG, Cossa M, Soares BG (2013) Electrical, rheological and electromagnetic interference shielding properties of thermoplastic polyurethane/carbon nanotube composites. Polym Int 62(10):1477–1484

Rath D, Chahataray R, Nayak PL (2013) Synthesis and characterization of conducting polymers multi walled carbon nanotube-chitosan composites coupled with poly(metachloroaniline). Middle East J Sci Res 18(5):635–641

Reddy CC, Ramu TS (2008) Polymer nanocomposites as insulation for HV DC cables—investigations on the thermal breakdown. IEEE Trans Dielectr Electr Insul 15(1):221–229

Ribeiro R, Banda S, Ounaies Z, Ucisik H, Usta M, Liang H (2012a) A tribological and biomimetic study of PI-CNT composites for cartilage replacement. J Mater Sci 47(2):649–658

Ribeiro R, Banda S, Ounaies Z, Ucisik H, Usta M, Liang H (2012b) A tribological and biomimetic study of PI-CNT composites for cartilage replacement. J Mater Sci 47(2):649–658

Richard A AV, Giannelis Emmanuel P (2001) Polymer nanocomposites: status and opportunities. MRS Bull 26(5):394–401

Rodney A, David J, Dali Q, Terry R (2002) Multiwall carbon nanotubes: synthesis and application. Acc Chem Res 35(12):1008–1017

Rohan AH, Darrin JP (2007) Polymer nanocomposites for biomedical applications. MRS Bull 32 (04):354–358

Roy D, Bhattacharyya S, Rachamim A, Plati A, Saboungi M-L (2010) Measurement of interfacial shear strength in single wall carbon nanotubes reinforced composite using Raman spectroscopy. J Appl Phys 107(4), art no 043501

Roy S, Das T, Ming Y, Chen X, Yue CY, Hu X (2014) Specific functionalization and polymer grafting on multiwalled carbon nanotubes to fabricate advanced nylon 12 composites. J Mater Chem A 2(11):3961–3970

Russ M, Rahatekar SS, Koziol K, Farmer B, Peng H-X (2013) Length-dependent electrical and thermal properties of carbon nanotube-loaded epoxy nanocomposites. Compos Sci Technol 81:42–47

Safdari M, Al-Haik MS (2013) Synergistic electrical and thermal transport properties of hybrid polymeric nanocomposites based on carbon nanotubes and graphite nanoplatelets. Carbon 64:111–121

Sahithi K, Swetha M, Ramasamy K, Srinivasan N, Selvamurugan N (2010) Polymeric composites containing carbon nanotubes for bone tissue engineering. Int J Biol Macromol 46(3):281–283

Saidi WA (2013) Functionalization of single-wall zigzag carbon nanotubes by carboxyl groups: clustering effect. J Phys Chem C 117(19):9864–9871

Salehi E, Madaeni SS, Rajabi L, Vatanpour V, Derakhshan AA, Zinadini S, Ghorabi S, Ahmadi Monfared H (2012) Novel chitosan/poly(vinyl) alcohol thin adsorptive membranes modified with amino functionalized multi-walled carbon nanotubes for Cu(II) removal from water: preparation, characterization, adsorption kinetics and thermodynamics. Sep Purif Technol 89:309–319

Sanip SM, Ismail AF, Aziz M, Soga T (2009) Functionalized carbon nanotubes for mixed matrix membrane. IEICE Trans Electron E92-C(12):1427–1431

Sato Y, Yokoyama A, Nodasaka Y, Kohgo T, Motomiya K, Matsumoto H, Nakazawa E, Numata T, Zhang M, Yudasaka M, Hara H, Araki R, Tsukamoto O, Saito H, Kamino T, Watari F, Tohji K (2013) Long-term biopersistence of tangled oxidized carbon nanotubes inside and outside macrophages in rat subcutaneous tissue. Sci Rep 3, art no 2516

Savas B, Young-Kyun K, David T (2000) Unusually high thermal conductivity of carbon nanotubes. Phys Rev Lett 84:4613–4620

Schlea MR, Meree CE, Gerhardt RA, Mintz EA, Shofner ML (2012) Network behavior of thermosetting polyimide/multiwalled carbon nanotube composites. Polymer 53(4):1020–1027

Scott CD, Arepalli S, Nikolaev P, Smalley RE (2001) Growth mechanisms for single-wall carbon nanotubes in a laser-ablation process. Appl Phys A 72(5):573–580

Sedláková Z, Clarizia G, Bernardo P, Jansen JC, Slobodian P, Svoboda P, Kárászová M, Friess K, Izak P (2014a) Carbon nanotube- and carbon fiber-reinforcement of ethylene-octene copolymer membranes for gas and vapor separation. Membranes 4(1):20–39

Sedláková Z, Clarizia G, Bernardo P, Jansen JC, Slobodian P, Svoboda P, Kárászová M, Friess K, Izak P (2014b) Carbon nanotube- and carbon fiber-reinforcement of ethylene-octene copolymer membranes for gas and vapor separation. Membranes 4(1):20–39

Sementsov YI, Prikhod'Ko GP, Melezhik AV, Aleksyeyeva TA, Kartel MT (2010) Physicochemical properties and biocompatibility of polymer/carbon nanotubes composites. Nanomater Supramol Struct Phys Chem Appl 347–368

Serrano MC, Gutiérrez MC, del Monte F (2014) Role of polymers in the design of 3D carbon nanotube-based scaffolds for biomedical applications. Progr Polym Sci

Seung JP, Min SC, Sung TL, Hyoung JC, Myung SJ (2003) Synthesis and dispersion characteristics of multi-walled carbon nanotube composites with poly(methyl methacrylate) prepared by in-situ bulk polymerization. Macromol Rapid Commun 24(18):1070–1073

Sevilla M, Yu L, Zhao L, Ania CO, Titiricic M-M (2014) Surface modification of CNTs with N-doped carbon: an effective way of enhancing their performance in supercapacitors. ACS Sustain Chem Eng 2(4):1049–1055

Seyhana AT, Florian HG, Metin T, Karl S (2007a) Critical aspects related to processing of carbon nanotube/unsaturated thermoset polyester nanocomposites. Eur Polym J 43(2):374–379

Seyhana AT, Gojnyb FH, Metin T, Karl S (2007b) Critical aspects related to processing of carbon nanotube/unsaturated thermoset polyester nanocomposites. Eur Polym J 43(2):374–379

Seyhana AT, Gojnyb FH, Tanoğlua T, Schulteb K (2007c) Rheological and dynamic-mechanical behavior of carbon nanotube/vinyl ester–polyester suspensions and their nanocomposites. Eur Polym J 43(7):2836–2847

Seyhana AT, Metin T, Karl S (2009) Tensile mechanical behavior and fracture toughness of MWCNT and DWCNT modified vinyl-ester/polyester hybrid nanocomposites produced by 3-roll milling. Mater Sci Eng A 523(1–2):85–92

Shawky HA, El-Aassar AHM, Abo-Zeid DE (2012) Chitosan/carbon nanotube composite beads: preparation, characterization, and cost evaluation for mercury removal from wastewater of some industrial cities in Egypt. J Appl Polym Sci 125(SUPPL. 1):E93–E101

Shi L, Li G, Sui G, Yang X (2009) Preparation and mechanical properties of epoxy resin reinforced with Jeffamines-grafted carbon nanotubes. Adv Mater Res 79–82:553–556

Shi Y-Y, Zhang W-B, Yang J-H, Huang T, Zhang N, Wang Y, Yuan G-P, Zhang C-L (2013) Super toughening of the poly(L-lactide)/thermoplastic polyurethane blends by carbon nanotubes. RSC Adv 3(48):26271–26282

Shigeo M, Ryosuke K, Yuhei M, Shohei C, Masamichi K (2002) Low-temperature synthesis of high-purity single-walled carbon nanotubes from alcohol. Chem Phys Lett 360:229–234

Shigeta M, Komatsu M, Nakashima N (2006) Individual solubilization of carbon nanotubes using polyimides. Polym Preprints Jpn 55(1):1591–1593

Shin SR, Park SJ, Yoon SG, Lee CK, Shin KM, Gu BK, Shin MK, Kim MS, Kim YJ, Kim SJ (2006) Anomalous pH actuation of a chitosan/SWNT microfiber hydrogel with improved mechanical property. Mater Res Soc Symp Proc 915:11–16

Singh RP, Das M, Thakare V, Jain S (2012) Functionalization density dependent toxicity of oxidized multiwalled carbon nanotubes in a murine macrophage cell line. Chem Res Toxicol 25(10):2127–2137

Singha AS, Thakur VK (2008a) Saccaharum cilliare fiber reinforced polymer composites. E-J Chem 5:782–791

Singha AS, Thakur VK (2008b) Effect of fibre loading on urea-formaldehyde matrix based green composites. Iran Polym J 17:861–873

Singha AS, Thakur VK (2008c) Synthesis and characterization of pine needles reinforced RF matrix based biocomposites. E-J Chem 5:1055–1062

Singha AS, Thakur VK (2008d) Synthesis and characterization of grewia optiva fiber-reinforced PF-based composites. Int J Polym Mater 57:1059–1074

Singha AS, Thakur VK (2009a) Fabrication and characterization of S. cilliare fibre reinforced polymer composites. Bull Mater Sci 32:49–58

Singha AS, Thakur VK (2009b) Grewia optiva fiber reinforced novel, low cost polymer composites. J Chem 6:71–76

Singha AS, Thakur VK (2009c) Physical, chemical and mechanical properties of Hibiscus sabdariffa fiber/polymer composite. Int J Polym Mater 58:217–228

Singha AS, Thakur VK (2009d) Synthesis, characterization and analysis of Hibiscus sabdariffa fibre reinforced polymer matrix based composites. Polym Polym Compos 17:189–194

Singha AS, Thakur VK (2009e) Fabrication and characterization of H. sabdariffa fiber-reinforced green polymer composites. Polym-Plast Technol Eng 48:482–487

Singha AS, Thakur VK, Mehta IK et al (2009a) Surface-modified Hibiscus sabdariffa fibers: physicochemical, thermal, and morphological properties evaluation. Int J Polym Anal Charact 14:695–711

Singha AS, Thakur VK, Mishra BN (2009b) Study of grewia optiva fiber reinforced urea-formaldehyde composites. J Polym Mater 26:81–90

Song D, Liu Y, Leng J (2010) EMI shielding performance study of tri-layer nano stealth composites. In: Proceedings of SPIE—the international society for optical engineering, vol 7644, art no 764423

Song X-J, Yang F, Wang X, Xuan H (2012) Preparation of β-cyclodextrin-modified multi-walled CNTs and its application in capturing β-naphthol from wastewater. Micro Nano Lett 7(9):892–895

Song BJ, Ahn JW, Cho KK, Roh JS, Lee DY, Yang YS, Lee JB, Hwang DY, Kim HS (2013) Electrical and mechanical properties as a processing condition in polyvinylchloride multi walled carbon nanotube composites. J Nanosci Nanotechnol 13(11):7723–7727

Speltini A, Merli D, Dondi D, Milanese C, Galinetto P, Bozzetti C, Profumo A (2013) Radiation-induced grafting of carbon nanotubes on HPLC silica microspheres: theoretical and practical aspects. Analyst 138(13):3778–3785

Spinks GM, Shin SR, Wallace GG, Whitten PG, Kim SI, Kim SJ (2006) Mechanical properties of chitosan/CNT microfibers obtained with improved dispersion. Sens Actuators B: Chem 115 (2):678–684

Sterzyński T, Tomaszewska J, Piszczek K, Skórczewska K (2010) The influence of carbon nanotubes on the PVC glass transition temperature. Compos Sci Technol 70(6):966–969

Stig H, Carlos Lo´ pez-C, Jens S, Poul LH, Bjerne SC, Jens RR-N, Frank A-P, Jens KN (2004) Atomic-scale imaging of carbon nanofibre growth. Nature 427:1–10

Suemori K, Hoshino S, Kamata T (2013) Flexible and lightweight thermoelectric generators composed of carbon nanotube-polystyrene composites printed on film substrate. Appl Phys Lett 103(15), art no 153902

Sulong AB, Park J (2011) Alignment of multi-walled carbon nanotubes in a polyethylene matrix by extrusion shear flow: mechanical properties enhancement. J Compos Mater 45(8):931–941

Sulong AB, Park J, Azhari CH, Jusoff K (2011) Process optimization of melt spinning and mechanical strength enhancement of functionalized multi-walled carbon nanotubes reinforcing polyethylene fibers. Compos B Eng 42(1):11–17

Sulong AB, Ramli MI, Hau SL, Sahari J, Muhamad N, Suherman H (2013) Rheological and mechanical properties of carbon nanotube/Graphite/SS316L/ polypropylene nanocomposite for a conductive polymer composite. Compos B Eng 50:54–61

Sun KJ, Wincheski RA, Park C (2008) Magnetic property measurements on single wall carbon nanotube polyimide composites. J Appl Phys 103(2), art no 023908

Takahashi K, Shizume R, Uchida K, Yajima H (2009) Improved blood biocompatibility of composite film of chitosan/carbon nanotubes complex. J Biorheol 23(1):64–71

Tan A, Yildirimer L, Rajadas J, De La Peña H, Pastorin G, Seifalian A (2011) Quantum dots and carbon nanotubes in oncology: a review on emerging theranostic applications in nanomedicine. Nanomedicine 6(6):1101–1114

Tanaka T, Sano E, Imai M, Akiyama K (2013) Low-frequency noise in carbon-nanotube/cellulose composite paper. Jpn J Appl Phys 49(5 Part 1):0551011–0551013

Tang ZK, Lingyun Z, Wang N, Zhang XX, Wen GH, Li GD, Wang JN, Chan CT, Ping S (2001) Superconductivity in 4 Angstrom single-walled carbon nanotubes. Science 292(5526):2462–2465

Tang Q-Y, Chan Y-C, Wong N-B, Cheungc R (2010a) Surfactant-assisted processing of polyimide/multiwall carbon nanotube nanocomposites formicroelectronics applications. Polym Int 59(9):1240–1245

Tang Q-Y, Chen J, Chan YC, Chung CY (2010b) Effect of carbon nanotubes and their dispersion on thermal curing of polyimide precursors. Polym Degrad Stab 95(9):1672–1678

Tang L-C, Wang X, Gong L-X, Peng K, Zhao L, Chen Q, Wu L-B, Jiang J-X, Lai G-Q (2014) Creep and recovery of polystyrene composites filled with graphene additives. Compos Sci Technol 91:63–70

Tarawneh MA, Ahmad SH (2012) Improved mechanical properties of thermoplastic natural rubber (TPNR) composite using a hybrid SiC-CNTs nano-filler. In: Proceedings of the 8th Asian-Australasian conference on composite materials 2012, ACCM 2012—composites: enabling tomorrow's industry today, vol 2, pp 1279–1284

Ten E, Ling C, Wang Y, Srivastava A, Dempere LA, Vermerris W (2014) Lignin nanotubes as vehicles for gene delivery into human cells. Biomacromolecules 15(1):327–338

Thuau D, Koutsos V, Cheung R (2009) Electrical and mechanical properties of carbon nanotube-polyimide composites. J Vac Sci Technol B: Microelectron Nanometer Struct 27(6):3139–3144

Thakur VK, Thakur MK (2014a) Processing and characterization of natural cellulose fibers/ thermoset polymer composites. Carbohydr Polym 109:102–117

Thakur VK, Thakur MK (2014b) Recent trends in hydrogels based on psyllium polysaccharide: a review. J Clean Prod 82:1–15

Thakur VK, Thakur MK (2014c) Recent advances in graft copolymerization and applications of chitosan: a review. ACS Sustain Chem Eng 2:2637–2652

Thakur VK, Singha AS, Kaur I et al (2010a) Silane functionalization of Saccaharum cilliare fibers: thermal, morphological, and physicochemical study. Int J Polym Anal Charact 15:397–414

Thakur VK, Singha AS, Mehta IK (2010b) Renewable resource-based green polymer composites: analysis and characterization. Int J Polym Anal Charact 15(3):137–146

Thakur VK, Yan J, Lin M-F et al (2012) Novel polymer nanocomposites from bioinspired green aqueous functionalization of BNNTs. Polym Chem 3:962–969

Thakur VK, Vennerberg D, Kessler MR (2014a) Green aqueous surface modification of polypropylene for novel polymer nanocomposites. ACS Appl Mater Interfaces 6:9349–9356

Thakur VK, Thakur MK, Raghavan P, Kessler MR (2014b) Progress in green polymer composites from lignin for multifunctional applications: a review. ACS Sustain Chem Eng 2:1072–1092

Thakur VK, Thakur MK, Gupta RK (2014c) Review: raw natural fiber-based polymer composites. Int J Polym Anal Charact 19:256–271

Thakur VK, Vennerberg D, Madbouly SA, Kessler MR (2014d) Bio-inspired green surface functionalization of PMMA for multifunctional capacitors. RSC Adv 4:6677–6684

Thakur VK, Thunga M, Madbouly SA, Kessler MR (2014e) PMMA-g-SOY as a sustainable novel dielectric material. RSC Adv 4:18240–18249

Thakur VK, Grewell D, Thunga M, Kessler MR (2014f) Novel composites from eco-friendly soy flour/sbs triblock copolymer. Macromol Mater Eng 299:953–958

Tijing LD, Park C-H, Kang S-J, Amarjargal A, Kim T-H, Pant HR, Kim HJ, Lee DH, Kim CS (2013) Improved mechanical properties of solution-cast silicone film reinforced with electrospun polyurethane nanofiber containing carbon nanotubes. Appl Surf Sci 264:453–458

Torresa A, Marcoa ID, Caballeroa BM, Laresgoitia MF, Legarretaa JA, Cabreroa MA, Gonzáleza A, Chomóna MJ, Gondrab K (2000) Recycling by pyrolysis of thermoset composites: characteristics of the liquid and gaseous fuels obtained. Fuel 79(8):897–902

Treacy MJ, Ebbesen TW, Gibson JM (1996) Exceptionally high Young's modulus observed for individual carbon nanotubes. Nature 381:678–680

Tsai J-L, Tzeng S-H, Chiu Y-T (2010) Characterizing elastic properties of carbon nanotubes/polyimide nanocomposites using multi-scale simulation. Compos B Eng 41(1):106–115

Tzeng S-H, Tsai J-L, Chiu Y-T (2008) Characterizing the elastic properties of carbon nanotubes/polyimide nanocomposites. In: Progress of composites 2008 in Asia and Australasia—proceedings of the 6th Asian-Australasian conference on composite materials, ACCM 2008, pp 80–83

Vanyorek L, Meszaros R, Barany S (2014) Surface and electrosurface characterization of surface-oxidized multi-walled N-doped carbon nanotubes. Colloids Surf A 448(1):140–146

Verge P, Peeterbroeck S, Bonnaud L, Dubois P (2010) Investigation on the dispersion of carbon nanotubes in nitrile butadiene rubber: role of polymer-to-filler grafting reaction. Compos Sci Technol 70(10):1453–1459

Vicente A, Ortiz AJ, Bravo LA (2009) Microleakage beneath brackets bonded with flowable materials: effect of thermocycling. Eur J Orthod 31(4):390–396

Vijay KT, Amar SS, Bhupendra NM (2011) Graft copolymerization of methyl methacrylate onto cellulosic biofibers. J Appl Polym Sci 122(1):532–554

Vladimir AS, Muhammed KG, Alexander AY, Anna AR, Kai S, Arif AM, James PW, Nicholas AK (2005) Aqueous dispersions of single-wall and multiwall carbon nanotubes with designed amphiphilic polycations. J Am Chem Soc 127(10):3463–3472

Wang Z, Yu X, Pan B, Xing B (2010) Norfloxacin sorption and its thermodynamics on surface-modified carbon nanotubes. Environ Sci Technol 44(3):978–984

Wang J, Li L, Wong CL, Madhavi S (2012) Flexible single-walled carbon nanotube/polycellulose papers for lithium-ion batteries. Nanotechnology 23(49), art no 495401

Wang H, Chang L, Yang X, Yuan L, Ye L, Zhu Y, Harris AT, Minett AI, Trimby P, Friedrich K (2014a) Anisotropy in tribological performances of long aligned carbon nanotubes/polymer composites. Carbon 67:38–47

Wang J, Jiao Q, Li H, Zhao Y, Guo B (2014b) In situ preparation of polyimide/amino-functionalized carbon nanotube composites and their properties. Polym Compos

Wang P-C, Liao Y-C, Liu L-H, Lai Y-L, Lin Y-C, Hsu Y-J (2014c) Upgrading non-oxidized carbon nanotubes by thermally decomposed hydrazine. Appl Surf Sci

Wei W, Sébastien W, Giorgia P, Monica B, Cédric K, Jean-Paul B, Renato G, Maurizio P, Alberto B (2005) Targeted delivery of amphotericin B to cells by using functionalized carbon nanotubes. Angew Chem Int Ed 44(39):6358–6362

Wenzhong T, Michael HS, Suresh GA (2003) Melt processing and mechanical property characterization of multi-walled carbon nanotube/high density polyethylene (MWNT/HDPE) composite films. Carbon 41(14):2779–2785

Werengowska-Ciećwierz K, Wiśniewski M, Terzyk AP, Gurtowska N, Olkowska J, Kloskowski T, Drewa TA, Kiełkowska U, Druzyński S (2014) Nanotube-mediated efficiency of cisplatin anticancer therapy. Carbon 70:46–58

Wim T, Richard PW (2004) Butyrated kraft lignin as compatibilizing agent for natural fiber reinforced thermoset composites. Compos A Appl Sci Manuf 35(3):327–338

Won K, Kim Y-H, An S, Lee HJ, Park S, Choi Y-K, Kim JH, Hwang H-I, Kim HJ, Kim H, Lee SH (2013) Glucose oxidase/cellulose-carbon nanotube composite paper as a biocompatible bioelectrode for biofuel cells. Appl Biochem Biotechnol 171(5):1194–1202

Wu T, Yan J (2013) Porous CNTs/chitosan composite with lamellar structure prepared by ice-templating. In: Proceedings of SPIE—the international society for optical engineering, vol 8923, art no 89233A

Wu Z, Wang H, Zheng K, Xue M, Cui P, Tian X (2012) Incorporating strong polarity minerals of tourmaline with carbon nanotubes to improve the electrical and electromagnetic interference shielding properties. J Phys Chem C 116(23):12814–12818

Wu H, Wang K, Meng Y, Lu K, Wei Z (2013a) An organic cathode material based on a polyimide/ CNT nanocomposite for lithium ion batteries. J Mater Chem A 1(21):6366–6372

Wu W, Jiang W, Zhang W, Lin D, Yang K (2013b) Influence of functional groups on desorption of organic compounds from carbon nanotubes into water: insight into desorption hysteresis. Environ Sci Technol 47(15):8373–8382

Wu Z, Wang H, Tian X, Ding X, Xue M, Zhou H, Zheng K (2013c) Mechanical and flame-retardant properties of styrene-ethylene-butylene-styrene/carbon nanotube composites containing bisphenol A bis(diphenyl phosphate). Compos Sci Technol 82:8–14

Xiaofeng L, Wanjin Z, Ce W, Ten-Chin W, Yen W (2011) One-dimensional conducting polymer nanocomposites: synthesis, properties and applications. Prog Polym Sci 36(5):671–712

Xiao-Lin X, Qing-Xi L, Robert K-YL, Xing-Ping Z, Qing-Xin Z, Zhong-Zhen Y, Yiu-Wing M (2004) Rheological and mechanical properties of PVC/CaCO3 nanocomposites prepared by in situ polymerization. Polymer 45:6665–6673

Xiaowen J, Yuezhen B, Masaru M (2006) Electrical and mechanical properties of polyimide–carbon nanotubes composites fabricated by in situ polymerization. Polymer 46(18):7418–7424

Xie F, Liang H, Ren XJ, Yifa C (2013) Isothermal crystallization of PET/PTT-CNTs composites. Adv Mater Res 750–752:191–194

Xu H, Zhang X, Zhan J (2010) Determination of pentachiorophenol at carbon nanotubes modified electrode incorporated with β-cyclodextrin. J Nanosci Nanotechnol 10(11):7654–7657

Xu Y, Chen L, Wang X (2013) Synthesis of modified carbon nanotube-supported Pd and the catalytic performance for hydrodehalogenation of aryl halides. Can J Chem 91(5):307–314

Yan D, Xu L, Chen C, Tang J, Ji X, Li Z (2012) Enhanced mechanical and thermal properties of rigid polyurethane foam composites containing graphene nanosheets and carbon nanotubes. Polym Int 61(7):1107–1114

Yan W, Yan L, Duan J, Jing C (2014) Sorption of organophosphate esters by carbon nanotubes. J Hazard Mater 273:53–60

Yang Z, Chen X, Pu Y, Zhou L, Chen C, Li W, Xu L, Yi B, Wang Y (2007) Facile approach to obtain individual-nanotube dispersion at high loading in carbon nanotubes/polyimide composites. Polym Adv Technol 18(6):458–462

Yang J-J, Fang H-G, Zhang Y-Q, Shi W-T, Chen P, Wang Z-G (2013) Studies on influences of viscoelastic properties on melt extrusion of carbon nanotubes-filled isotactic polypropylene composites. Acta Polym Sin 10:1325–1333

Yang S, Han C, Wang X, Nagatsu M (2014) Characteristics of cesium ion sorption from aqueous solution on bentonite- and carbon nanotube-based composites. J Hazard Mater 274:46–52

Yazdani-Pedram M, Menzel C, Toro P, Quijada R, May-Pat A, Avilés F (2013) Mechanical and thermal properties of multiwalled carbon nanotube/polypropylene composites using itaconic acid as compatibilizer and coupling agent. Macromol Res 21(2):153–160

Ye S, Zhang H, Wang Y, Jiao F, Lin C, Zhang Q (2011) Carboxylated single-walled carbon nanotubes induce an inflammatory response in human primary monocytes through oxidative stress and NF-jB activation. J Nanopart Res 13(9):4239–4252

Yesil S, Bayram G (2011) Effect of carbon nanotube purification on the electrical and mechanical properties of poly(ethylene terephthalate) composites with carbon nanotubes in low concentration. J Appl Polym Sci 119(6):3360–3371

Yi C, Qi S, Zhang D, Yang M (2010) Covalent conjugation of multi-walled carbon nanotubes with proteins. Methods Mol Biol (Clifton, NJ) 625:9–17

Yin G, Zheng Z, Wang H, Du Q, Zhang H (2014) Strings of polymer microspheres stabilized by oxidized carbon nanotubes. J Colloid Interface Sci 426:137–144

Yu G, Li X (2012) Fabrication and bio-tribological properties of medical polyurethane/carbon nanotubes composites. Energy Educ Sci Technol Part A: Energy Sci Res 30(1):475–480

Yu J, Tonpheng B, Gröbner G, Andersson O (2011) Microstructural and property changes in high pressure treated carbon nanotube/polybutadiene composites. J Mater Chem 21(35):13672–13682

Yu J, Tonpheng B, Gröbner G, Andersson O (2012) A MWCNT/polyisoprene composite reinforced by an effective load transfer reflected in the extent of polymer coating. Macromolecules 45(6):2841–2849

Yu J, Yao M, Gröbner G, Sundqvist B, Tonpheng B, Liu B, Andersson O (2013a) Buckminsterfullerene: a strong, covalently bonded, reinforcing filler and reversible cross-linker in the form of clusters in a polymer. ACS Macro Lett 2(6):511–517

Yu X, Lv Z, Tong Y, Yang R (2013b) The effect of inorganic particles on microphase separation of polyether polyurethanes. Polym Mater Sci Eng 29(7):101–105

Yu J-G, Zhao X-H, Yang H, Chen X-H, Yang Q, Yu L-Y, Jiang J-H, Chen X-Q (2014) Aqueous adsorption and removal of organic contaminants by carbon nanotubes. Sci Total Environ 482–483(1):241–251

Yuan-Qing L, Shao-Yun F, Yang Y, Yiu-Wing M (2008) Facile synthesis of highly transparent polymer nanocomposites by introduction of core-shell structured nanoparticles. Chem Mater 20(8):2637–2643

Yue D, Liu Y, Shen Z, Zhang L (2006) Study on preparation and properties of carbon nanotubes/rubber composites. J Mater Sci 41(8):2541–2544

Yulong D, Hajar A, Dongsheng W, Richard AW (2006) Heat transfer of aqueous suspensions of carbon nanotubes (CNT nanofluids). Int J Heat Mass Transf 49(2):240–250

Zaminpayma E (2014) Molecular dynamics simulation of mechanical properties and interaction energy of polythiophene/polyethylene/poly(p-phenylenevinylene) and CNTs composites. Polym Compos

Zdenko S, Dimitrios T, Konstantinos P, Costas G (2010) Carbon nanotube–polymer composites: chemistry, processing, mechanical and electrical properties. Prog Polym Sci 35(3):357–401

Zeng Y, Liu P-F, Zhao L, Du J-H, Liu C (2013) High electrical sensitivity of polyurethane carbon nanotube composites to tensile strain. New Carbon Mater 28(2):88–93

Zha J-W, Sun F, Dang Z-M (2013) Fabrication and properties of high performance polyimide nanofibrous films by electrospinning. In: Proceedings of IEEE international conference on solid dielectrics, ICSD, art no 6619776, pp 923–926

Zhang JG (2011) The effect of carbon fibers and carbon nanotubes on the mechanical properties of polyimide composites. Mech Compos Mater 47(4):447–450

Zhang Y, Huang Y, Liu C, Ge Y (2011) Research of carbon nanotubes/silicon rubber pressure-sensitive composites for artificial skin, Huazhong Keji Daxue Xuebao (Ziran Kexue Ban). J Huazhong Univ Sci Technol (Nat Sci Edn) 39(SUPPL. 2):306–315

Zhang M, Dai W, Yan M, Ge S, Yu J, Song X, Xu W (2012) Ultrasensitive electrochemiluminescence immunosensor using PtAg@carbon nanocrystals composites as labels and carbon nanotubes-chitosan/gold nanoparticles as enhancer. Analyst 137(9):2112–2118

Zhang B, Chakoli AN, Zang W, Tian Y, Zhang K, Chen C, Li Y (2014) A comparative study on effect of aromatic polyimide chain conformation on reinforcement of carbon nanotube/polyimide nanocomposites. J Appl Polym Sci 131(13), art no 40479

Zhao H, Li H, Yu H, Chang H, Quan X, Chen S (2013) CNTs-TiO2/Al2O3 composite membrane with a photocatalytic function: fabrication and energetic performance in water treatment. Sep Purif Technol 116:360–365

Zhaoxia J, Pramodab KP, Guoqin X, Suat HG (2001) Dynamic mechanical behavior of melt-processed multi-walled carbon nanotube/poly(methyl methacrylate) composites. Chem Phys Lett 337(1):43–47

Zheng W, Chen YQ, Zheng YF (2008) Adsorption and electrochemistry of hemoglobin on Chi-carbon nanotubes composite film. Appl Surf Sci 255(2):571–573

Zheng Z, Wang Z, Feng Q, Zhang F, Du Y, Wang C (2013) Preparation of surface-silvered graphene-CNTs/polyimide hybrid films: processing, morphology and properties. Mater Chem Phys 138(1):350–357

Zheng-Ming H, Zhang Y-Z, Kotakic M, Ramakrishna S (2003) A review on polymer nanofibers by electrospinning and their applications in nanocomposites. Compos Sci Technol 63 (15):2223–2253

Zhong J, Xie T, Deng J, Sun X, Pan X, Bao X, Wu Z (2011) Direct observation and spectroscopy of nanoscaled carboxylated carbonaceous fragments coated on carbon nanotubes. Chem Commun 47(29):8373–8375

Zhong J, Isayev AI, Huang K (2014) Influence of ultrasonic treatment in PP/CNT composites using masterbatch dilution method. Polymer (UK) 55(7):1745–1755

Zhou J, Bai H-P, Li F-S, Cui P, Chen L-Y, Jiang W (2005) Preparation of carbon nanotubes/HTPB and research of catalytic performance. J Solid Rocket Technol 28(3):195–197

Zhou Z, Xu Y, Hojamberdiev M, Liu W, Wang J (2010) Enhanced cycling performance of silicon/disordered carbon/carbon nanotubes composite for lithium ion batteries. J Alloy Compd 507 (1):309–311

Zhou K, Gu S-Y, Zhang Y-H, Ren J (2012) Effect of dispersion on rheological and mechanical properties of polypropylene/carbon nanotubes nanocomposites. Polym Eng Sci 52(7):1485–1494

Zhu Y, Ang TP, Keith C, John AM, Narayan SH, Masao T (2005) Substituted carborane-appended water-soluble single-wall carbon nanotubes: new approach to boron neutron capture therapy drug delivery. J Am Chem Soc 127(27):9875–9880

Metallic Nanocomposites: Bacterial-Based Ecologically Benign Biofabrication and Optimization Studies

Kannan Badri Narayanan, Anil K. Suresh and Natarajan Sakthivel

Abstract Metallic nanocomposites are gaining considerable attention and are widely being implemented in several biomedical and engineering applications due to their potent physicochemical properties. To ease wide application of nanoparticles, research is focused on novel and better synthesis strategies. This brief chapter details on the biofabrication of diverse forms of metallic nanoparticles using various bacterial systems, and the cellular impact, illustrated using suitable examples. Demonstration on the biosynthesis of silver nanoparticles using the cell-free extract of *P. plecoglossicida* is presented. This chapter will also describe the influence of various physicocultural parameters such as the growth medium, concentration of precursor salt; pH and temperature on the biotransformation, so as to attain desirable morphological and surface characteristics of nanoparticles. Overall, this chapter aims to discuss the recent progress in relation to bacterial-based biosynthesis so as to have a better understanding on their safe use for various biomedical and engineering applications.

Keywords Bacteria · Nanoparticles · Silver · Biosynthesis · Physicocultural parameters

K.B. Narayanan · A.K. Suresh (✉) · N. Sakthivel (✉)
Department of Biotechnology, School of Life Sciences,
Pondicherry University, Kalapet 605014, Puducherry, India
e-mail: anil.dbt@pondiuni.edu.in

N. Sakthivel
e-mail: puns2005@gmail.com

K.B. Narayanan
School of Biotechnology, Graduate School of Biochemistry and Research
Institute of Protein Sensor, Yeungnam University, Gyeongsan 7122749, South Korea

V.K. Thakur and M.K. Thakur (eds.), *Eco-friendly Polymer Nanocomposites*,
Advanced Structured Materials 75, DOI 10.1007/978-81-322-2470-9_7

1 Introduction

Noble metallic nanocomposites are gaining recognition to pursuit in various fields of nanosciences and technology. Particularly, monometallic and bimetallic nanocomposites that exhibit different physical and chemical properties are widely used in catalysis, biotechnology, biomedicine, bioengineering, surface-enhanced spectroscopy, textile engineering, nonlinear optical materials, waste management, water treatment, optoelectronics, nonlinear optical materials, etc. (Karthikeyan 2012; Loganathan and Karthikeyan 2013). For instance, silver/palladium core–shell nanocomposites have been reported to have sensing and Raman Scattering applications (Sathiyadevi et al. 2014). Several reports exist on the use of chemical and physical methods for the synthesis of various forms of metallic nanocomposites. However, most of these methods are cumbersome, ecologically unfriendly and might involve the use of toxic and combustible solvents and precursors. And often, the produced particles are not biocompatible, are insoluble in aqueous suspensions, tend to self-aggregate losing their stability and also poses environmental and health risks exhibiting different levels of toxicity to living organisms (Borm and Berube 2008; Li et al. 2008). Considering the environmental risks of the as-synthesized nanoparticles and the artifacts due to the reagents used, eco-friendly and green chemistry approaches are appreciated to avoid possible environmental contamination (Albrecht et al. 2006). Also, the use of nanoparticles should be effectively engineered depending on its application to guarantee the safe use of nanoparticles. In this respect, embedment of nanoparticles into organic or inorganic matrices minimizes its mobility and its interaction with the environment. Thus, the use of nanocomposites in an effective way by increasing their stability, and considering the safety of nanoparticles is of paramount concern. An important advantage of nanocomposites is the accessibility of substrates to the functional metallic nanoparticles. Although, these nanoparticles are immobilized into solid matrices, depending on the nature of nanophase and the matrix, it is classified as biologically-derived, biologically-inspired, and polymer–metal nanocomposites (Kim and Van der Bruggen 2010).

Biologically-derived and bioinspired metallic nanocomposites are garnering considerable attention and are widely being implemented in several biomedical and engineering applications due to their potent physicochemical properties (Laurent et al. 2008; Lin et al. 2011a, b; Thakur et al. 2012a, b). Some of the natural biopolymers such as flax, hemp, coir, jute, kenaf, Grewia optiva, sisal, ramie, *Hibiscus sabdariffa* (Singha and Thakur 2008a; Singha et al. 2009; Thakur et al. 2011), pine needles (Thakur et al. 2010a, b, 2012b; Singha and Thakur 2010b), and *Saccaharum cilliare* (Singha and Thakur 2008b, 2010a; Thakur et al. 2010a, b) have also been used as natural entities along with polymers/resins for the production of low-cost eco-friendly materials, and with properties such as acceptable specific strength, low density, high toughness, good thermal properties (Thakur and Thakur 2014a, b, c; Thakur et al. 2014a, b, c). In particular, bacterial-based microbial biosynthesis is emerging as a novel technique for the production of diverse forms of nanoparticles, as the technique is economic, nonlaborious, and most importantly eco-friendly. The synthesis of microbially-matrixed metal nanoparticles is a new composite material and this chapter is an attempt to describe the biosynthesis of diverse forms of metallic nanoparticles using various bacterial systems, and the cellular impact using suitable examples. Another important issue related to microbial-based synthesis is the attainment of required morphological and surface characteristics of nanoparticles. This chapter will also illustrate the influence of various physicocultural parameters such as the growth medium, concentration of precursor salt; pH and temperature on the biotransformation. Overall, this chapter aims to discuss the recent progress in relation to bacterial-based biosynthesis so as to have a better understanding on their safe use for various biomedical and engineering applications.

2 Eco-friendly Biofabrication of Metallic Nanocomposites

Biologically-matrixed nanomaterial containing metal nanoparticles were used as catalysts in various chemical reactions (Sharma et al. 2007). This will help to retain the nanoparticles for continuous usage in bioreactors. In Alaskar placer, *Pedomicrobium*-like budding bacteria that were initially reported in iron and manganese oxide deposition process have now been found to accumulate gold (Mann 1992). In an other study, *Bacillus subtilis* 168 was found to reduce Au(III) ions to Au(0) inside the cell wall (Beveridge and Murray 1980; Southam and Beveridge 1994). Similarly, heterotrophic sulfate-reducing bacterial (SRB) enrichment from a gold mine was used to destabilize gold(I)–thiosulfate complex to elemental gold in the bacterial envelope (Lengke and Southam 2006). Iron-reducing bacterium, *Geobacter ferrireducens*, gold was intracellularly precipitated in periplasmic space (Kashefi et al. 2001). Similarly, in anaerobic conditions, iron(III)-reducing *Shewanella algae* reduced Au^{+3} ions in periplasmic space and on bacterial surfaces (Konishi et al. 2007a, b). As in the case of Gram-negative bacteria that produce membrane vesicles like outer membrane proteins, lipopolysaccharides and phospholipids to protect themselves against toxic agents, *Plectonema boryanum* UTEX485, a filamentous cyanobacterium precipitated gold nanoparticles in abiotic and cyanobacterial systems (Lengke et al. 2006a). Recently, bionanocomposites of face centered cubic (fcc) structured silver nanoaprticles were produced using the fungus, *Cylindrocladium floridanum* (Narayanan and Sakthivel 2011b). As can be made out from the TEM image (Fig. 1a), the Ag nanoparticles were neatly embedded on the outer surface of the fungal cell wall. Histogram analysis of the particles revealed them to be within a range of 5–35 nm with an average size distribution of 15 ± 5 nm (Fig. 1b).

Engineered *Escherichia coli* DH5α also mediated the bioreduction of chloroauric acid to Au(0) nanoparticles (Lengke et al. 2006b). These cell-bound nanoparticles have been reported for promising applications in realizing the direct electrochemistry of hemoglobin and other proteins (Du et al. 2007). Similarly, the bioreduction of trivalent aurum was also reported in photosynthetic bacterium, *Rhodobacter capsulatus*, which showed biosorption capacity of 92.43 mg $HAuCl_4$/g dry weight in

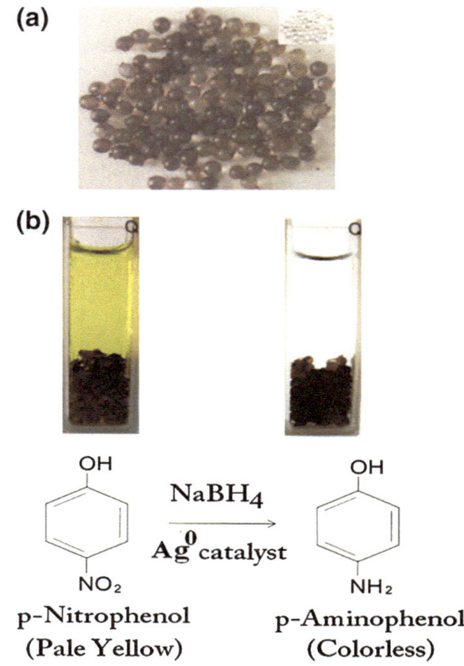

Fig. 1 **a** The image of borosilicate glass beads coated with bionanocomposite-containing silver nanoparticles. **b** Shows the catalytic conversion of p-Nitrophenol [*pale yellow color*, before the reaction (cuvette on the *left*)] to p-Aminophenol [*colorless*, after the reaction (cuvette on the *right*)] using the silver bionanocomposites embedded in borosilicate glass beads (**a**) (Reproduced from Narayanan and Sakthivel (2011b) with permission from Elsevier publishers)

the logarithmic phase of its growth. In some microbes carotenoids and NADPH-dependent enzymes embedded in plasma membrane were found to be involved in the biosorption and bioreduction of metal ions to respective nanoparticle forms on the plasma membrane (Feng et al. 2007). Recently, Correa-Llanten et al. (2013) reported the synthesis of intracellular gold nanoparticles using a thermophilic bacterium *Geobacillus* sp. strain ID17 from Antartica. Biosorbent *Phoma* sp.3.2883 was also suited for preparing silver nanoparticles (Chen et al. 2003) and *Phoma* PT35 was able to selectively accumulate silver (Pighi et al. 1989). Fungal biomass of *Verticillium* sp. when exposed to aqueous silver nitrate solution accumulated silver nanoparticles on the cell surface (Senapati et al. 2004; Mukherjee et al. 2001b). Likewise, *Aspergillus flavus* also accumulated silver nanoparticles on the surface of its cell wall when incubated with silver nitrate solution (Vigneshwaran et al. 2007).

In an other investigation, a rhizosopheric strain *Stenotrophomonas maltophilia* SELTE02, isolated from the rhizospheric soil of selenium hyperaccumulator legume plant *Astragalus bisulcatus* showed promising transformation of selenite (SeO_3^{-2}) to elemental selenium (Se^0) and accumulating selenium granules either in the cell cytoplasm or in the extracellular space (Gregorio et al. 2005). A facultative purple non-sulfur anaerobic bacterium, *Enterobacter cloacea* SLD1a-1, *Rhodospirullum rubrum* in oxic and anoxic condition, and *Desulfovibrio desulfuricans* have also been found to bioreduce selenite to selenium inside the cell (Tomei et al. 1995; Losi and Frankenberger 1997; Kessi et al. 1999). Simultaneously, *E. coli* was also able to deposit elemental selenium both within its periplasmic space and the cytoplasm (Gerrard et al. 1974; Silverberg et al. 1976), *Psueomonas stutzeri* also aerobically reduced selenite to elemental selenium (Lortie et al. 1992). Yadav et al. (2008) showed that rhizospheric seleniferous soil strain *P. aeruginosa* SNT1 biosynthesized nanostructured selenium by biotransforming selenium oxyanions to spherical amorphous allotrophic elemental red selenium intracellularly. The resting cells of *S. algae* when incubated anaerobically in H_2PtCl_6 solution at room temperature and neutral pH, it reduced $PtCl_6^{-2}$ ions to metallic platinum in the presence of lactate as the electron donor and deposited in the periplasmic space between inner and outer membrane of *S. algae* (Konishi et al. 2007a, b). Yet another iron-reducing bacterium, *S. oneidensis* MR-1 reduced Pd(II) to Pd(0) nanoparticles in the presence of formate, and accumulated the particles on the cell wall and within the periplasmic space. This cell-matrixed (nano-)bioPd was used as biocatalyst in the dechlorination of polychlorinated biphenyl 21 (2,3,4-chloro biphenyl) (De Windt et al. 2005). In a recent application point of view, silver bionanocomposite produced using fungus, *C. floridanum* was successfully used as a biocatalyst for the degradation of 4-nitrophenol to 4-aminophenol (Narayanan and Sakthivel 2011b) (Fig. 1).

Magnetosomes comprise both crystallite and noncrystallite magnetic crystals. A microaerophilic bacterium *Aquaspirillum magnetotacticum*, isolated from sediments, produced crystals of magnetite (Fe_3O_4) particles (Mann et al. 1984). Marine magnetotactic bacterium MV-1 isolated from sulfide-rich sediments of an estuarine salt marsh anaerobically bioreduced nitrous oxide and ferric quinate to iron-rich magnetosomes. Another magnetotactic bacterium isolated from brackish and marine sulfide-rich water and sediments deposited intracellular single crystals of ferromagnetic iron sulfide; greigite (Fe_3S_4), which was also found to be associated with nonmagnetic mineral, iron pyrite (FeS_2) and were aligned in chains that contained approximately ten nanoparticles. Magnetotactic bacterium, *Magnetospirullum magnetotacticum* aligned magnetite nanoparticles (Fe_3O_4) intracellularly in the form of a chain parallel to the longitudinal axis of the cell (Kornig et al. 2014). The authors noted that the radionuclides initially in the soluble form somehow became insoluble through microbial activity. Gram-positive sulfate-reducing microbe, *Desulfosporosinus* sp., isolated from sediments reduced hexavalent uranium U(VI) to tetravalent uranium U (IV) in turn precipitated as uraninite (UO_2) crystals (Suzuki et al. 2002).

Metal semiconductor nanocrystals have applications in various fields of research particularly in biomedical fields. Cunningham and Lundie (1993) demonstrated the precipitation of bright yellow CdS crystals on the cell surfaces of *Clostridium thermoaceticum* by an enzyme, cysteine desulfhydrase activity. Likewise, Smith et al. (1998) produced CdS nanoparticles on the cell surface of another microorganism *Klebsiella pneumoniae*. Interestingly, these "Bio-CdS" possessed the optical and photoactive traits analogous to the chemically synthesized CdS nanoparticles. The use of eukaryotic microorganisms for the biological synthesis of metal nanoparticles was also demonstrated (Mukherjee et al. 2001a). For example, *Verticillium* sp. synthesized gold nanoparticles on the surface and on the cytoplasmic membrane of fungal mycelium when exposed to auric chloride. In addition, intracellular synthesis of gold nanoparticles was produced by *Verticillium* sp. (Mukherjee et al. 2001a),

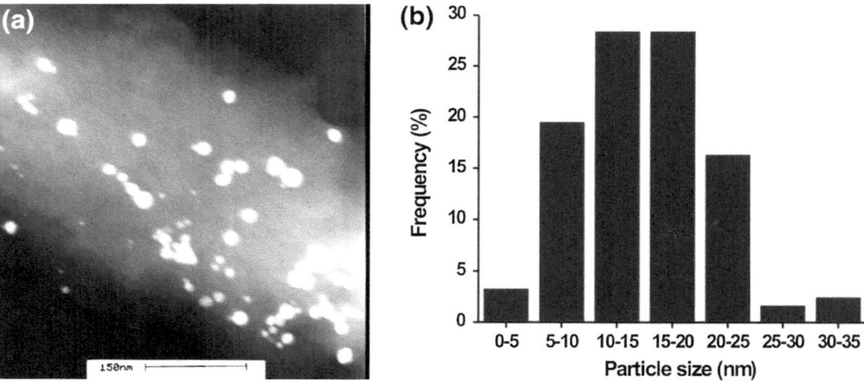

Fig. 2 **a** Transmission electron micrograph image of the Ag nanoparticles formed on the surface of the fungus, *C. floridanum*. Ag particles assembled on the cell wall surface can be clearly seen. **b** Particle size histogram distribution obtained from the TEM image by counting ~100 particles so as to determine particles diameter (Reproduced from Narayanan and Sakthivel (2011a) with permission from Elsevier publishers)

Trichothecium sp. (Ahmad et al. 2005), and V. *luteoalbum* (Gericke and Pinches 2006a, b). Similarly, Narayanan and Sakthivel (2011a) produced gold bionanocomposite using filamentous fungus *Cylindrocladium floridanum* and used it as heterogeneous catalyst in the degradation of pollutant 4-nitrophenol (Fig. 2). Alkalotolerant actinomycete *Rhodococcus* sp. was also demonstrated in the intracellular accumulation of gold nanoparticles with the aid of reductases secreted on the cell wall (Ahmad et al. 2003). Yeast, *Candida glubrata* produced intracellular monodispersed spherical shaped peptide-functionalized CdS quantum crystallites of 20 Å by forming metal–thiolate complex with phytochelatins (Dameron et al. 1989). Kowshik et al. (2002) for the first time reported the synthesis of PbS nanocrystallites exhibiting quantum semiconductor properties by *Torulopsis* sp. which was intracellularly produced and matrixed in the vacuoles when incubated with Pb^{2+} ions. These nanoparticles were used to biofabricate diode heterojunction with poly (p-phenylenevinylene).

In addition, baker's yeast, *S. cerevisiae* was also reported to biosorb and reduces Au^{+3} to elemental gold in the peptidoglycan layer of the cell wall in situ by the aldehyde group present in reducing sugars (Lin et al. 2005). Marine yeast, *Yarrowia lipolytica* NCIM 3589, also synthesized gold nanoparticles associated with cell wall. The reduction of gold ions occurred in pH-dependent manner (Agnihotri et al. 2009). Yet again, *S. cerevisiae* was also found to produce spherical-shaped antimony oxide (Sb_2O_3) nanoparticles at room temperature exhibiting semiconductor properties. This was presumably due to the radial tautomerization of membrane-bound quinines or by membrane-bound or cytosolic pH-dependent oxidoreductases (Jha et al. 2009). Intracellular biosynthesis and removal of copper nanoparticles by dead biomass of the yeast *Rhodotorula mucilaginosa* isolated from the wastewater of a mine (Salvadori et al. 2014) was also demonstrated. Interestingly, tobacco mosaic virus (TMV) was used as biotemplate for the synthesis of iron oxides by oxidative hydrolysis, co-crystallization of CdS and PbS, and the synthesis of SiO_2 by sol–gel condensation. It happened with the help of external groups of glutamate and aspartate on the external surface of the virus (Shenton et al. 1999). Thus, inorganic–organic nanotube composites from template mineralization of TMV. Self-assembled viral capsids of genetically engineered viruses were exploited as biological templates for the assembly of quantum dot nanowires.

2.1 Bacterial-Based Biofabrication of Silver Nanoparticles

As aforementioned, similar to biologically-derived and bioinspired metallic nanocomposites, ceramic–metal nanocomposites (cermets) which shows interesting optical and electrical properties such as spectrally selective light absorption and tunable electrical resistivity can be prepared by biological methods. In the last decade, Joerger et al. (2000) reported a new microbial-based production method of cermet consisting of bioorganic carbon matrix with embedded silver nanoparticles synthesized by *Pseudomonas stutzeri* AG259.

Silver-based crystals such as equilateral, triangles, and hexagons with particle sizes were synthesized and accumulated in the periplasmic space of a silver mine bacterium (Klaus et al. 1999). Among several metal nanoparticles used, silver nanoparticles have special advantages because of its antimicrobial and catalytic property. Biosorption and bioreduction of Ag(I) on cell surface was also reported in *Lactobacillus* sp. A09 (Fu et al. 2000). Even the dried cells of *Corynebacterium* sp. SH09 produced silver nanoparticles at 60 °C on the cell wall in the size range of 10–15 nm with diamine silver complex $[Ag(NH_3)_2]^+$ (Zhang et al. 2005). Normally, bacteria overcome silver toxicity with the help of small periplasmic silver-binding proteins, which bind silver at the cell surface, and the efflux pumps propels the incoming metals and protects the cytoplasm from toxicity (Li et al. 1997; Gupta and Silver 1998). Therefore, it is believed that the organic matrix contains silver-binding proteins that provide amino acid moieties, which serve as nucleation sites for the formation of silver nanoparticles (Naik et al. 2002). An airborne bacterium, *Bacillus* sp. isolated from the atmosphere was also found to reduce Ag^+ ions into Ag^0 that accumulated metallic silver nanoparticles of 5–15 nm in diameter in the periplasmic space of the cell (Pugazhenthiran et al. 2009). Nair and Pradeep (2002) found that *Lactobacillus* sp. in buttermilk produced microscopic gold, silver, and gold–silver alloy crystals of well-defined morphology within the cell with no disturbance in its viability. Recently, a marine bacterium, *Idiomarina* sp. PR58-8 was reported on the intracellular synthesis of silver nanoparticles (Seshadri et al. 2012). Haloarchaeal isolate *Halococcus salifodinae* BK intracellularly synthesized spherical silver nanoparticles (Srivastava et al. 2013).

2.1.1 Bacterial-Based Biofabrication Methodology

Bacterium, *P. plecoglossicida* was maintained on Luria–Bertani agar (LBA) petri dishes [casein enzymatic hydrolysate 1 % (w/v), yeast extract 0.5 % (w/v), sodium chloride 1 % (w/v), and agar 2 % (w/v)] at 25 °C. A single bacterial colony from an overnight LBA plate served as inoculum for 100 ml of LB broth [casein enzymatic hydrolysate 1 % (w/v), yeast extract 0.5 % (w/v), and sodium chloride 1 % (w/v)] in a 500 ml Erlenmeyer flask, followed by incubation at 25 °C on a rotary shaker (200 rpm) for 24 h. The cells at the stationary phase were collected by centrifugation (5000 × g, 25 °C, 20 min) and washed with distilled water under sterile conditions. In a 500 ml Erlenmeyer flask, ~3–5 g wet bacterial biomass was suspended in 100 ml of distilled water for 2 days, after which the cells were separated using centrifugation (5000 × g, 25 °C, 20 min) and to the supernatant containing the cell extract silver nitrate was added at an effective concentration of 1 mM, and incubated further. Control reactions included 1 mM $AgNO_3$ added to uninoculated LB medium. Synthesis of silver nanoparticles was monitored by visually and by UV-Vis absorption spectroscopy (200–700 nm). After completion of the reaction process (18–24 h) the particles were collected from the reaction mixture upon filtration using sterile 0.2 μm syringe filter followed by sedimentation.

2.1.2 Characterization of the Silver Nanoparticles

The first indication for the formation of silver nanoparticles is the change in its color from colorless to brown (Fig. 3a) that is due to the size and shape dependent surface plasmon resonance (SPR). Metallic nanoparticles in particular possess this

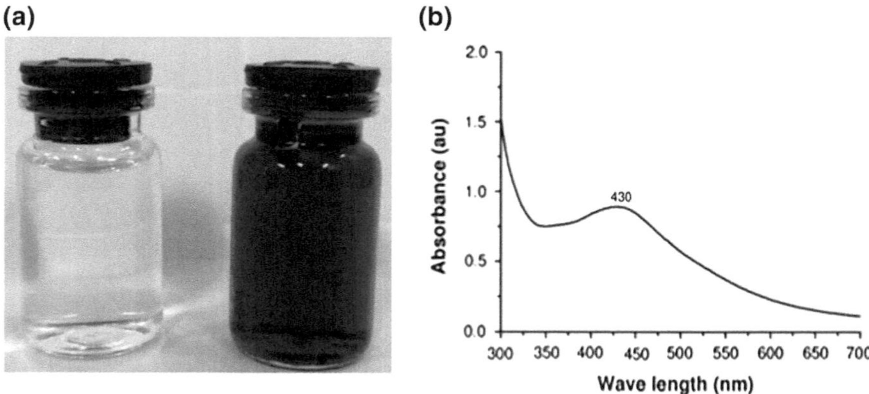

Fig. 3 **a** Image of the glass vial containing $AgNO_3$ solution before (glass vial on the *left*) and after (glass vial on the *right*) the formation of silver nanoparticles by the *P. plecoglossicida* cell extract. **b** UV-Vis spectra of the silver nanoparticles with intense plasma peak at 430 nm

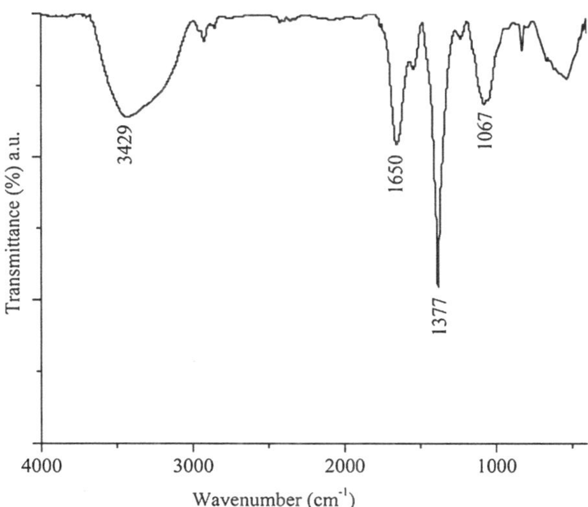

Fig. 4 Fourier transform infrared spectra of the silver nanoparticles produced using the cell extract of *P. plecoglossicida*. Significant vibration bands are labeled

SPR property, for example, gold, silver, copper, tin, lead, mercury, cadmium, indium including the alkali metals. The SPR varies with the nanoparticle form and thereby every nanocomposite tends to have a characteristic color, for example, gold; ruby red, silver; brown, cadmium; yellow, magnetite; black; platinum, brownish red. UV-Visible spectra in the present study were recorded on a Spectra Max 190 plate reader (Molecular Devices, Sunnyvale, USA) which was operated at a resolution of 1 nm. The SPR for silver nanoparticles, of sizes $\sim 24 \pm 2.5$ nm, showed an intense band centered at 430 nm, which is due to the excitation of SPR in the metal nanoparticles, as can be seen in Fig. 3b, and suggests the presence of silver nanoparticles.

To determine the nature stabilizing agent surrounding the Ag nanoparticles produced by the *P. plecoglossicida* cell free extract, Fourier transform infrared spectroscopy measurement was performed. FTIR analysis of the dried samples in KBr pellet was performed on a Thermo Nicolet model 6700 spectrophotometer in a diffuse transmittance mode at a resolution of 4 cm^{-1}. The FTIR analysis revealed the presence of strong bands centered at 1067, 1377, 1650, 2425, and 3429 cm^{-1} (Fig. 4). The vibrational band at 1067 cm^{-1} corresponds to alcoholic and carbonyl group. Small peak at 1377 cm^{-1} corresponds to amide III functional group. Band at 1650 cm^{-1} is due to the presence of carbonyl (–C–O–C– or –C–O) stretch and –N–H– stretch in amide linkage, clearly suggesting the presence of protein or a peptide on the surface that likely appears to be acting as a capping molecule. The vibrational peak at 2425 cm^{-1} indicates the presence of C–H group in $C–CH_3$ compounds. Peak at 3429 cm^{-1} is the characteristic of the hydroxyl functional (–OH) group in alcohols and phenolic compounds.

Another important characterization technique used to assess the purity, structure, and crystallinity of the nanoparticles is the X-ray diffraction (XRD). XRD analysis

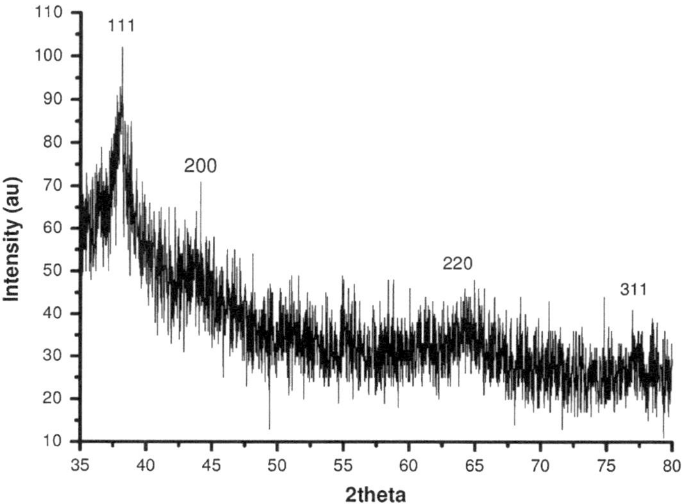

Fig. 5 X-ray diffraction analysis of the silver nanoparticle powder

of dried silver nanoparticle powder was performed on a PANalytical X'pert PRO X-ray diffractometer (Netherlands). The pattern was recorded by Cu Kα1 radiation with λ of 1.5406 Å and nickel monochromatic filtering the wave at tube voltage of 40 kV and tube current of 30 mA. The scanning was done in the region of 2θ from 30° to 80° at a speed of 0.02°/min and the time constant was 2 s. XRD of Ag nanoparticle powder showed intense Bragg peaks at (111), (200), (220), and (311) in the 2θ range of 35θ–70θ (Fig. 5) and agree with the values that are reported for silver nanocrystals, thereby confirming the purity and crystallinity of the biogenic-Ag nanoparticles.

To evaluate the size and shape distributions of the Ag nanoparticles transmission electron microscopy (TEM) was performed. Samples for the TEM measurements were prepared by drop coating the nanoparticles sample on carbon-coated copper TEM grids followed by drying the grids for a few hours. TEM measurements were performed on a TEM (JEOL model 3010) with an accelerating voltage of about 200 keV with wavelength (λ) of 0.0251 Å. TEM images of the Ag nanoparticles revealed polydispersed particles as shown in Fig. 6a. A particle size histogram plot (plotted using Image J software) from the TEM image showed the size distribution of silver nanoparticles to be in range between ~10 and 45 nm with the largest number of particles being 24 ± 6 nm (Fig. 6b). Closer inspection of the particle morphology, at higher magnifications and different locations of the grid showed spherical single well-dispersed nanoparticles as well as aggregates. An estimate of the size of the particles was also made from the line broadening of the (112) reflection pattern using Debye Scherrer's formula ($D = 0.94\ \lambda/b\ \cos\ \theta$), and are in good agreement with the nanoparticles size estimated by TEM analysis.

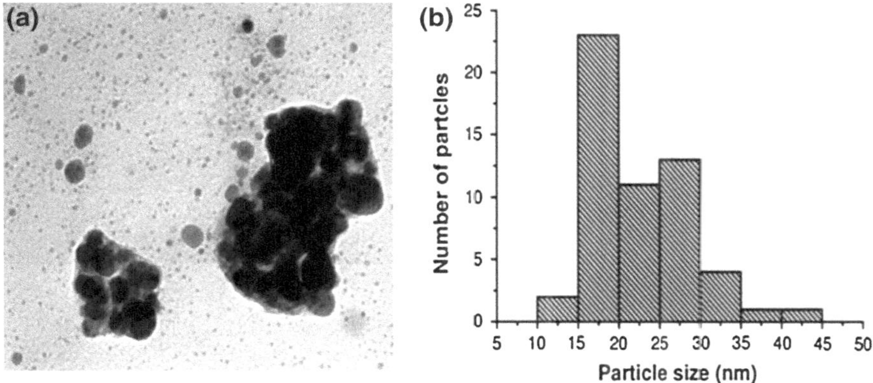

Fig. 6 **a** TEM image of the silver nanoparticles. **b** Histogram particle size distribution measurements made from the TEM image by counting ~100 particles in order to obtain average particle diameter

3 Optimization of the Various Major Physicocultural Parameters

3.1 Influence of Various Growth Mediums

To assess the influence of various growth mediums, the bacteria was maintained and cultured as mentioned above except that various compositions of growth mediums including Luria–Bertani (casein enzymatic hydrolysate 1 % w/v, yeast extract 0.5 % w/v and sodium chloride 1 % w/v), malt glucose yeast extract peptone (malt extract 0.3 % w/v, glucose 1 % w/v, yeast extract 0.3 % w/v and peptone 0.5 % w/v), nutrient broth (peptic digest of animal tissue 0.5 % w/v, beef extract 0.15 % w/v, yeast extract 0.15 % w/v and sodium chloride 0.5 % w/v), tryptone soy broth (pancreatic digest of casein 1.7 % w/v, papaic digest of soybean meal 0.3 % w/v, dextrose 0.25 % w/v, dibasic potassium phosphate 0.25 % w/v and sodium chloride 0.5 % w/v), and King's B broth (protease peptone 2 % w/v, glycerol 1.5 % w/v, dipotassium hydrogen phosphate 0.5 % w/v and magnesium sulphate 0.15 % w/v) were used separately.

3.2 Influence of Various Concentrations of Precursor Salt

To evaluate the influence of precursor salt concentrations, the bacteria was maintained and cultured as mentioned above, upon which the unaltered cell extract was treated with various increasing doses; 50, 250, 500, 750, and 1000 μM of silver nitrate separately.

3.3 Influence of Various Hydrogen Ion Concentrations (pH)

To determine the influence various hydrogen ion concentration (pH), the bacteria was maintained and cultured as mentioned above, after which the cell-free extract was attained with various desired pHs of 3, 6, 7, 8, 9 and 11 separately, followed by treatment with 1000 μM of silver nitrate.

3.4 Influence of Various Temperatures

To assess the influence of various temperatures, the bacteria was maintained and cultured as mentioned above, upon which the unaltered cell extract after adding 1000 μM silver nitrate was incubated at various temperatures including 25, 35, 45, 55, 75, 100 °C separately. Percent theoretical yield of the Ag nanoparticles was determined by taking 1 ml aliquot of the silver nanoparticles suspension was dried overnight at 60 °C, followed by weighing and quantifying the mass of the Ag nanoparticles relative to the Ag ions in the initial reaction mixture.

3.5 Optimization of Physicocultural Conditions for the Biological Synthesis of Silver Nanoparticles

As described above, when we assessed the influence of various physic cultural parameters on the *P. plecoglossicida* cell extract-based synthesis of silver nanoparticles. It was noted that among various growth mediums that were used in this study; LB, NB, TSB and KB, the formation of silver nanoparticles was observed only in LB growth medium, clearly demonstrating the influence of growth medium on the production of nanoparticles itself. We observed no significant effect concentration of salt on the nanoparticle, with respect to formation or size and shape distributions; as expected, the amount of silver nanoparticles produced increased with increase in salt concentration. Likewise, pH also had a drastic influence on the formation of silver nanoparticles, lower and higher pH are used in our studies, at lower pH, we observed that the nanoparticles were precipitated, whereas at higher pH did not had much influence. However, based on our experimentation neutral pH of 7.0 can be considered as the right pH for the synthesis of nanoparticles. Temperature has a drastic influence on the yield of the nanoparticles. The amount of silver nanoparticle produced was more at higher temperature (100 °C). The optimal conditions determined for biological synthesis of silver nanoparticles using *Pseudomonas plecoglossicida* TND35 are given in Table 1.

Table 1 Optimal physicocultural conditions for the biological synthesis of silver nanoparticles *Pseudomonas plecoglossicida*

Physicocultural parameter	Optimal conditions
pH	7.0
Temperature	100 °C
Growth medium	Luria–Bertani broth
Precursor salt concentration	1 mM

4 Physical Characterizations

UV-Vis absorbance spectroscopy measurements were performed on a Spectra Max 190 plate reader (Molecular devices, Sunnyvale, USA) operated at a resolution of 1 nm. Fourier transform infrared (FTIR) analysis of the samples dried in KBr pellet was performed on a Nicolet Model 6700 spectrophotometer in a diffuse transmittance mode at 4 cm^{-1} resolution. XRD of dried silver nanoparticle samples drop-coated on glass substrate was performed on a PANalytical X'pert PRO X-ray diffractometer (Netherlands), operated at 40 kV and at a current of 40 mA. TEM measurements for the samples prepared on carbon-coated copper grids were performed on a HR-TEM (JEOL-Model 3010) operated at an accelerating voltage of 200 keV.

5 Conclusion and Future Perspective

In summary, bacteria have been successfully implemented for the production, nucleation, and assembly of diverse forms of metal, metal oxide and metal sulfide nanocomposites, and are considered green. We briefly described the various biologically-matrixed metallic nanoparticles produced using bacteria. These bioorganic-matrixed inorganic metal nanoparticles composites produced by microbes are advantageous in the fields of catalysis, biotechnology/biomedicine, environmental science and technology. The design and optimization of eco-friendly bionanocomposite material can be made by selection of the biological entity and improving the culture conditions, viz concentration of metal ions, temperature, pH, and duration of the reaction.

Acknowledgments The authors thank the Department of Science and Technology (DST), Government of India for financial support through funds for improvement of science and technology infrastructure in various universities and higher educational institutions (FIST) Programme coordinated by Prof. N. Sakthivel

References

Agnihotri M, Joshi S, Kumar AR, Zinjarde S, Kulkarni S (2009) Biosynthesis of gold nanoparticles by the tropical marine yeast *Yarrowia lipolytica* NCIM 3589. Mater Lett 63:1231–1234

Ahmad A, Senapati S, Kumar R, Khan MI, Ramani R, Srinivas V, Sastry M (2003) Intracellular synthesis of gold nanoparticles by a novel alkalotolerant actinomycete, *Rhodococcus* species. Nanotechnology 14a:824–828

Ahmad A, Senapati S, Khan MI, Kumar R, Sastry M (2005) Extra-/intracellular biosynthesis of gold nanoparticles by an alkalotolerant fungus, *Trichothecium* sp. J Biomed Nanotechnol 1:47–53

Albrecht MA, Evans CW, Raston CL (2006) Green chemistry and the health implications of nanoparticles. Green Chem 8:417–432

Beveridge TJ, Murray RGE (1980) Sites of metal deposition in the cell wall of *Bacillus subtilis*. J Bacteriol 141:876–887

Borm P, Berube D (2008) A tale of opportunities, uncertainties, and risks. NanoToday 3:56–59

Chen JC, Lin ZH, Ma XX (2003) Evidence of the production of silver nanoparticles via pretreatment of *Phoma* sp.3.2883 with silver nitrate. Lett Appl Microbiol 37:105–108

Correa-Llanten DN, Munoz-Ibacache SA, Castro ME, Munoz PA, Blamey JM (2013) Gold nanoparticles synthesized by *Geobacillus* sp. strain ID17 a thermophilic bacterium isolated from Deception island, Antarctica. Microb Cell Fact 12:75

Cunningham DP, Lundie LL (1993) Precipitation of cadmium by *Clostridium thermoaceticum*. Appl Environ Microbiol 59:7–14

Dameron CT, Reese RN, Mehra RK, Kortan AR, Carroll PJ, Steigerwald ML, Brus LE, Winge DR (1989) Biosynthesis of cadmium sulphide quantum semiconductor crystallites. Nature 338:596–597

De Windt D, Aelterman P, Verstraete W (2005) Bioreductive deposition of palladium (0) nanoparticles on *Shewanella oneidensis* with catalytic activity towards reductive dechlorination of polychlorinated biphenyls. Environ Microbiol 7:314–325

Du L, Jiang H, Xiaohua H, Wang E (2007) Biosynthesis of gold nanoparticles assisted by *Escherichia coli* DH5α and its application on direct electrochemistry of hemoglobin. Electrochem Commun 9:1165–1170

Feng Y, Yu Y, Wang Y, Lin X (2007) Biosorption and bioreduction of trivalent aurum by photosynthetic bacteria *Rhodobacter capsulatus*. Curr Microbiol 55:402–408

Fu JK, Liu YY, Gu PY, Tang D, Lin Z, Yao BX, Weng SZ (2000) Spectroscopic characterization on the biosorption and bioreduction of Ag(I) by *Lactobacillus* sp. A09. Acta Phys Chim Sin 16:779–782

Gericke M, Pinches A (2006a) Biological synthesis of metal nanoparticles. Hydrometallurgy 83:132–140

Gericke M, Pinches A (2006b) Microbial production of gold nanoparticles. Gold Bull 39:22–28

Gerrard TL, Telford JN, Williams HH (1974) Detection of selenium deposits in *Escherichia coli* by electron microscopy. J Bacteriol 119:1057–1060

Gregorio SD, Lampis S, Vallini G (2005) Selenite precipitation by a rhizospheric strain of *Stenotrophomonas* sp. isolated from the root system of *Astragalus bisulcatus*: a biotechnological perspective. Environ Int 31:233–241

Gupta A, Silver S (1998) Silver as a biocide: will resistance become a problem? Nat Biotechnol 16:888

Jha AK, Prasad K, Prasad K (2009) A green low-cost biosynthesis of Sb_2O_3 nanoparticles. Biochem Eng J 43:303–306

Joerger R, Klaus T, Granqvist CG (2000) Biologically produced silver-carbon composite materials for optically functional thin-film coatings. Adv Mater 12:407–409

Karthikeyan B (2012) Optical studies on thermally surface plasmon tuned Au, Ag and Au: Ag nanocomposite polymer films. Spectrochim Acta A Mol Biomol Spectrosc 96:456–460

Kashefi K, Tor JM, Nevin KP, Lovley DR (2001) Reductive precipitation of gold by dissimilatory Fe(III)-reducing bacteria and archaea. Appl Environ Microbiol 67:3275–3279

Kessi J, Ramuz M, Wehrli E, Spycher M, Bachofen R (1999) Reduction of selenite and detoxification of elemental selenium by the phototrophic bacterium *Rhodospirillum rubrum*. Appl Environ Microbiol 65:4734–4740

Kim J, Van der Bruggen B (2010) The use of nanoparticles in polymeric and ceramic membrane structures: review of manufacturing procedures and performance improvement for water treatment. Environ Poll 158:2335–2349

Klaus T, Joerger R, Olsson E, Granqvist CG (1999) Silver-based crystalline nanoparticles, microbially fabricated. Proc Natl Acad Sci USA 96:13611–13614

Konishi Y, Ohno K, Saitoh N, Nomura T, Nagamine S, Hishida H, Takahashi Y, Uruga T (2007a) Bioreductive deposition of platinum nanoparticles on the bacterium *Shewanella algae*. J Biotechnol 128:648–653

Konishi Y, Tsukiyama T, Tachimi T, Saitoh N, Nomura T, Nagamine S (2007b) Microbial deposition of gold nanoparticles by the metal-reducing bacterium *Shewanella algae*. Electrochim Acta 53:186–192

Kornig A, Dong J, Bennet M, Widdrat M, Andert J, Muller FD, Schuler D, Klumpp S, Faivre D (2014) Probing the mechanical properties of magnetosome chains in living magnetotactic bacteria. Nano Lett 14:4653–4659

Kowshik M, Deshmukh N, Vogel W, Urban J, Kulkarni SK, Paknikar KM (2002) Microbial synthesis of semiconductor CdS nanoparticles, their characterization, and their use in the fabrication of an ideal diode. Biotechnol Bioeng 78:583–588

Laurent S, Forge D, Port M, Roch A, Robic C, Vander Elst L, Muller RN (2008) Magnetic iron oxide nanoparticles: synthesis, stabilization, vectorization, physicochemical characterizations, and biological applications. Chem Rev 108:2064–2110

Lengke M, Southam G (2006) Bioaccumulation of gold by sulfate-reducing bacteria cultured in the presence of gold(I)-thiosulfate complex. Geochim Cosmochim Acta 70:3646–3661

Lengke M, Ravel B, Fleet ME, Wanger G, Gordon RA, Southam G (2006a) Mechanisms of gold bioaccumulation by filamentous cyanobacteria from gold(III)-chloride complex. Environ Sci Technol 40:6304–6309

Lengke M, Fleet ME, Southam G (2006b) Morophology of gold nanoparticles synthesized by filamentous cyanobacteria from gold(I)-thiosulfate and gold(III)-chloride complexes. Langmuir 22:2780–2787

Li XZ, Nikaido H, Williams KE (1997) Silver-resistant mutants of Escherichia coli display active efflux of Ag^+ and are deficient in porins. J Bacteriol 179:6127–6132

Li Q, Mahendra S, Lyon DY, Brunet L, Liga MV, Li D, Alvarez PJJ (2008) Antimicrobial nanomaterials for water disinfection and microbial control: potential applications and implications. Water Res 42:4591–4602

Lin Z, Wu J, Xue R, Yang Y (2005) Spectroscopic characterization of Au^{3+} biosorption by waste biomass of *Saccharomyces cerevisiae*. Spectrochim Acta Part A 61:761–765

Lin M-F, Thakur VK, Tan EJ, Lee PS (2011a) Dopant induced hollow $BaTiO_3$ nanostructures for application in high performance capacitors. J Mater Chem 21:16500–16504

Lin M-F, Thakur VK, Tan EJ, Lee PS (2011b) Surface functionalization of $BaTiO_3$ nanoparticles and improved electrical properties of $BaTiO_3$/polyvinylidene fluoride composite. RSC Adv 1:576–578

Loganathan B, Karthikeyan B (2013) Au core Pd/Pt shell in trimetallic Au/Pd/Pt colloidal nanocomposites: physicochemical characterization study. Colloids Surf A Physicochem Eng Asp 436:944–952

Lortie L, Gould WD, Rajan S, McCready RGL, Cheng KJ (1992) Reduction of selenite and selenite to elemental selenium by a *Pseudomonas stutzeri* isolate. Appl Environ Microbiol 58:4042–4044

Losi ME, Frankenberger WT Jr (1997) Reduction of selenium oxyanions by *Enterobacter cloacae* SLD1a-1: isolation and growth of the bacterium and its expulsion of selenium particles. Appl Environ Microbiol 63:3079–3084

Mann S (1992) Bacteria and the midas touch. Nature 357:358–360

Mann S, Frankel RB, Blakemore RP (1984) Structure, morphology and crystal growth of bacterial magnetite. Nature 310:405–407

Mukherjee P, Ahmad A, Mandal D, Senapati S, Sainkar SR, Khan MI, Ramani R, Parischa R, Ajayakumar PV, Alam M, Sastry M, Kumar R (2001a) Bioreduction of $AuCl4^{-}$ ions by the fungus, *Verticillium* sp. and surface trapping of the gold nanoparticles formed. Angew Chem Int Ed 40:3585–3588

Mukherjee P, Ahmad A, Mandal D, Senapati S, Sainkar SR, Khan MI, Parishcha R, Ajaykumar PV, Alam M, Kumar R, Sastry M (2001b) Fungus-mediated synthesis of silver nanoparticles and their immobilization in the mycelial matrix: a novel biological approach to nanoparticle synthesis. Nano Lett 1:515–519

Naik RR, Stringer SJ, Agarwal G, Jones SE, Stone MO (2002) Biomimetic synthesis and patterning of silver nanoparticles. Nat Mater 1:169–172

Nair B, Pradeep T (2002) Coalescence of nanoclusters and formation of submicron crystallites assisted by *Lactobacillus* strains. Cryst Growth Des 2:293–298

Narayanan KB, Sakthivel N (2011a) Synthesis and characterization of nano-gold composite using *Cylindrocladium floridanum* and its heterogeneous catalysis in the degradation of 4-nitrophenol. J Hazard Mater 189:519–525

Narayanan KB, Sakthivel N (2011b) Heterogeneous catalytic reduction of anthropogenic pollutant, 4-nitrophenol by silver-bionanocomposite using *Cylindrocladium floridanum*. Bioresour Technol 102:10737–10740

Pighi L, Pumpel T, Schinner F (1989) Selective accumulation of silver by fungi. Biotechnol Lett 11:275–280

Pugazhenthiran N, Anandan S, Kathiravan G, Prakash NKU, Crawford S, Ashokkumar M (2009) Microbial synthesis of silver nanoparticles by *Bacillus* sp. J Nanopart Res 11:1811–1815

Salvadori MR, Ando RA, Oller do Nascimento CA, Correa B (2014) Intracellular biosynthesis and removal of copper nanoparticles by dead biomass of yeast isolated from the wastewater of a mine in the Brazilian Amazonia. PLoS One 9:e87968

Sathiyadevi G, Loganathan B, Karthikeyan B (2014) Solvent-mediated eco-friendly synthesis and characterization of monodispersed bimetallic Ag/Pd nanocomposites for sensing and Raman scattering applications. J Nanosci 762453

Senapati S, Mandal D, Ahmad A, Khan MI, Sastry M, Kumar R (2004) Fungus mediated synthesis of silver nanoparticles: a novel biological approach. Ind J Phys 78A:101–105

Seshadri S, Prakash A, Kowshik M (2012) Biosynthesis of silver nanoparticles by marine bacterium, *Idiomarina* sp. PR58-8. Bull Mater Sci 35:1201–1205

Sharma NC, Sahi SV, Nath S, Parsons JG, Gardea-Torresdey JL, Pal T (2007) Synthesis of plant-mediated gold nanoparticles and catalytic role of biomatrix-embedded nanomaterials. Environ Sci Technol 41:5137–5142

Shenton W, Douglas T, Young M, Stubbs G, Mann S (1999) Inorganic-organic nanotube composites from template mineralization of tobacco mosaic virus. Adv Mater 11:253–256

Silverberg RA, Wong PTS, Chau YK (1976) Localization of selenium in bacterial cells using TEM and energy dispersive X-ray analysis. Arch Microbiol 107:1–6

Singha AS, Thakur VK (2008a) Effect of fibre loading on urea-formaldehyde matrix based green composites. Iran Polym J 17:861–873

Singha AS, Thakur VK (2008b) *Saccaharum cilliare* fiber reinforced polymer composites. E-J Chem 5:782–791

Singha AS, Thakur VK (2010a) Mechanical, morphological, and thermal characterization of compression-molded polymer biocomposites. Int J Polym Anal Character 15:87–97

Singha AS, Thakur VK (2010b) Synthesis, characterization and study of pine needles reinforced polymer matrix based composites. J Reinf Plast Compos 29:700–709

Singha AS, Thakur VK, Mehta IK, Sharma A, Khanna AJ, Rana RK, Rana AK (2009) Surface-modified *Hibiscus sabdariffa* fibers: physicochemical, thermal, and morphological properties evaluation. Int J Polym Anal Character 14:695–711

Smith PR, Holmes JD, Richardson DJ, Russell DA, Sodeau JR (1998) Photophysical and photochemical characterisation of bacterial semiconductor cadmium sulfide particles. J Chem Soc Faraday Trans 94:1235–1241

Southam G, Beveridge TJ (1994) The in vitro formation of placer gold by bacteria. Geochim Cosmochim Acta 58:4527–4530

Srivastava P, Braganca J, Ramanan S, Kowshik M (2013) Synthesis of silver nanoparticles using haloarchaeal isolate *Halococcus salifodinae* BK_3. Extremophiles 17:821–831

Suzuki Y, Kelly SD, Kemner KM, Banfield JF (2002) Radionuclide contamination: nanometer-size products of uranium bioreduction. Nature 419:134

Thakur VK, Thakur MK (2014a) Recent trends in hydrogels based on psyllium polysaccharide: a review. J Clean Prod 82:1–15

Thakur VK, Thakur MK (2014b) Processing and characterization of natural cellulose fibers/ thermoset polymer composites. Carbohydr Polym 109:102–117

Thakur VK, Thakur MK (2014c) Recent advances in graft copolymerization and applications of chitosan: a review. ACS Sustain Chem Eng 2:2637–2652

Thakur VK, Singha AS, Mehta IK (2010a) Renewable resource-based green polymer composites: analysis and characterization. Int J Polym Anal Character 15:137–146

Thakur VK, Singha AS, Kaur I, Nagarajarao RP, Liping Y (2010b) Silane functionalization of *Saccaharum cilliare* fibers: thermal, morphological, and physicochemical study. Int J Polym Anal Character 15:397–414

Thakur VK, Singha AS, Misra BN (2011) Graft copolymerization of methyl methacrylate onto cellulosic biofibers. J Appl Polym Sci 122:532–544

Thakur VK, Yan J, Lin M-F et al (2012a) Novel polymer nanocomposites from bioinspired green aqueous functionalization of BNNTs. Polym Chem 3:962–969

Thakur VK, Singha AS, Thakur MK (2012b) In-air graft copolymerization of ethyl acrylate onto natural cellulosic polymers. Int J Polym Anal Character 17:48–60

Thakur VK, Vennerberg D, Kessler MR (2014a) Green aqueous surface modification of polypropylene for novel polymer nanocomposites. Acs Appl Mater Interfaces 6:9349–9356

Thakur VK, Thakur MK, Raghavan P, Kessler MR (2014b) Progress in green polymer composites from lignin for multifunctional applications: a review. ACS Sustainable Chem Eng 2:1072–1092

Thakur VK, Thakur MK, Gupta RK (2014c) Review: raw natural fiber-based polymer composites. Int J Polym Anal Charact 19:256–271. doi:10.1080/1023666X.2014.880016

\Tomei FA, Barton LL, Lemanski CL, Zocco TG, Fink NH, Sillerud LO (1995) Transformation of selenite and selenite to elemental selenium by *Desulfovibrio desulfuricans*. J Ind Microbiol 14:329–336

Vigneshwaran N, Ashtaputre NM, Varadarajan PV, Nachane RP, Paralikar KM, Balasubramanya RH (2007) Biological synthesis of silver nanoparticles using the fungus *Aspergillus flavus*. Mater Lett 61:1413–1418

Yadav V, Sharma N, Prakash R, Raina KK, Bharadwaj LM, Prakash NT (2008) Generation of selenium containing nano-structures by soil bacterium, *Pseudomonas aeruginosa*. Biotechnology 7:299–304

Zhang H, Li Q, Lu Y, Sun D, Lin X, Deng X (2005) Biosorption and bioreduction of diamine silver complex by *Corynebacterium*. J Chem Technol Biotechnol 80:285–290

Bio-based Wood Polymer Nanocomposites: A Sustainable High-Performance Material for Future

Ankita Hazarika, Prasanta Baishya and Tarun K. Maji

Abstract Numerous studies are underway on the preparation and applications of petroleum-based polymer nanocomposites. The depletion of world oil pool, non-biodegradability, and raising cost of petroleum-based materials are some of the disadvantages allied with these polymers-based products. The utilization of renewable materials has attracted researchers because of its easy availability and low cost. They can potentially remove the harmful effects of petroleum-based materials and thus show a greener path in the fields of application of composites. The biocomposites developed by using renewable polymers such as furfuryl alcohol, poly(lactic acid), gluten, starch, soy flour, etc., and naturally available fibers have been gaining considerable attention because of their environment-friendly nature. Wood is a biologically derived biodegradable raw material which requires minimum processing energy. Wood polymer composites (WPC) have tremendous advantageous properties and it rapidly improves the mechanical, physical, chemical as well as other properties of the composite suitable for different outdoor and indoor applications. The properties of the WPC can be improved to the desired level through the application of nanotechnology, cross-linking agents, flame retardants, grafting, etc. Nano-based wood polymer composite provides versatile advantages in their properties compared to the conventional WPC. Flame retardants obtained from renewable resource such as the gum of the plant *Moringa oleifera* can efficiently improve the flame retardancy along with other properties of the composites. This chapter discusses the various properties of renewable polymer-based wood polymer nanocomposites as a potential, sustainable, green composite to attain durability without using harmful chemicals.

Keywords Wood · Bioresins · Bionanocomposites · Properties

A. Hazarika · P. Baishya · T.K. Maji (✉)
Department of Chemical Sciences, Tezpur University, Assam 784028, India
e-mail: tkm@tezu.ernet.in

V.K. Thakur and M.K. Thakur (eds.), *Eco-friendly Polymer Nanocomposites*,
Advanced Structured Materials 75, DOI 10.1007/978-81-322-2470-9_8

1 Introduction

Polymer nanocomposite (PNC) is a promising multidisciplinary research activity in the field of material research that might expand the utilization of polymers for various industrial applications (Hussain et al. 2006; Paul and Robeson 2008; Leszczy´nska et al. 2007). Generally, the desirable properties including low gas permeability, good transparency, high thermostability, high mechanical strength, light weight, high chemical resistance, etc., required for many advanced applications, are not found in case of commercial polymers. The service of polymers has been extended in different fields as PNC enhance the properties of polymer to obtain a product with essentially new set of properties.

The raise in the environmental awareness has led governments to make more stringent regulations. This assists in evolution of simpler chemical processes or innovative designed product for future generations by the chemical industries that should create least environmental impact. An interest in naturally available renewable materials has been developed due to the global environmental concern. Therefore, there has been attempt to develop high-performance new materials at reasonable expense in recent years. Wood is one of the renewable resources with an outstanding strength-to-weight ratio. It is one of the preferred construction materials because products manufactured from wood require much less energy compared to those produced from competitive materials like concrete, plastic, or steel. It has an extensive sort of applications which include construction of materials, paper, pulp, and as source for energy (Giudice and Pereyra 2007; Wegner et al. 2010). Woods are of two types—hard- and softwoods. The softwoods remain as a biowaste and are mostly used for fuel purposes because of their poor properties, whereas the hardwoods are used for construction applications. These soft woods can be designed as a value-added product by forming wood polymer composite (WPC) (Hetzer and Kee 2008; Ashori 2008). Wood polymer composites prepared from a variety of renewable matrix and their properties are included in this chapter.

The most important disadvantage associated with the synthetic composite materials is that they are derived from nonrenewable resources. However, other disadvantageous properties of these materials include nonbiodegradability, high cost, etc. Thus, natural fibers like silk, wood, keratin biofibers obtained from chicken feather, etc., are acquiring more priority over the materials obtained from nonrenewable sources. The keratin biofiber can be found in hair and feathers. The advantages of keratin biofibers obtained from feathers are its biodegradable, eco-friendly, nonabrasive properties. They possess a very high mechanical strength as they are insoluble in organic solvents. Moreover, they are hydrophobic in nature, have low density, and low cost. Thus, keratin fibers obtained from chicken feather can be suitable for using as a high structural reinforcement material in polymer composites (Martinez-Hernandez and Velasco-Santos 2012). The keratin biofiber extracted from chicken feathers were used as short-fiber reinforcement for a poly (methyl methacrylate) matrix and can significantly improve the dynamic mechanical and thermal properties of the composites (Martínez-Hernández et al. 2007).

The natural fiber like wood possesses certain advantages like low cost, low density, light weight, high specific properties, corrosion protective, and the most significant is its biodegradability and nonabrasiveness (Raj et al. 1989).

The various properties of wood like appearance; pulp quality; strength properties, resistance to penetration by water and chemicals, decay, etc., are influenced by the chemical and anatomic composition of wood (Panshin and de Zeeuw 1980; Haygreen and Bowyer 1982). The cellular structure of wood consists of cellulose, lignin, hemicelluloses, and minor amounts (5–10 %) of extraneous materials. The quantity and distribution of these constituents of wood lead to variations in the characteristics of wood and the nonconformity in cellular structure causes wood hard or soft, bulky or light and stiff, rigid or flexile. The major constituent of the wood cell wall is cellulose, normally 40–50 % by mass of the dry wood. The cellulose is a polymer of glucose residues attached by 1, 4-β-glucosidic bonds. Hemicellulose is a combination of nebulous branched-chain polysaccharides containing a few hundred sugar residues. Lignin is very difficult to separate in a natural state and is an insoluble, amorphous organic polymer. Chemically, lignin is a methoxy-substituted propylphenol moiety which is bonded asymmetrically by ether and carbon–carbon linkages (Rowell et al. 1997). It consists of 18–30 % by weight of the dry wood. It is mainly centralized in the layered cell wall and compound middle lamella of wood. The rigidness in the structure of the cell wall of wood occurs due to the presence of lignin and imparts a woody, rigid structure to the cell walls. It differentiates the fibrous plant of minor lignin content from wood. There is presence of abundant free hydroxyl groups in the structure of cellulose that can form hydrogen bonds with the moisture present in the atmosphere without any difficulty (Xie et al. 2011). The swelling and shrinking of wood is the consequence of bond formation of wood with the water molecules depending upon its moisture content. Besides, the properties of wood are destructed and deteriorated by biotic agents which include decay, mold, fungi, bacteria insects, etc., and abiotic agents which include sun wind, water, and certain chemicals. The sapwood of wood has very low durability. Since wood is an organic compound, it is a nutritional product of the degrading fungi and assist in its decay. The hydrophilic –OH group of wood participates in chemical bonding with the polymers on formation of wood polymer composites and thus gets converted to hydrophobic groups (Hill et al. 1998). This prevents wood from shrinking and swelling and the bacteria and fungi can no longer recognize wood as their food source within their service life (Li et al. 2013).

The use of synthetic polymer-based composites has led to disposable problems because of their nonbiodegradable nature. Naturally available water-soluble polymer can be used in the different application as a substitute of the synthetic polymers due to scarcity of petroleum resources and environmental awareness. Water is the best solvent among all the green solvents because it is nonpolluting, inexpensive, and renewable.

There are several limitations associated with the conventional composites, although they have higher mechanical and other properties. To overcome the limitations, the concept of nanocomposite arises. Nano-dimensional particles-based WPC are reported to improve fire retardancy, moisture barrier, UV protection, etc.,

(Hazarika and Maji 2013a, b, 2014a, b, c, d). Dimensional stability, UV protection, and weather resistance are very significant properties for wood-based materials used in exterior applications. Bio-based wood polymer nanocomposites will have superior properties and can be regarded as high-performance material for the future generation.

2 Wood Polymer Nanocomposites

Nanobiocomposites obtained by incorporating nanofillers to biopolymers are extremely promising product because they provide better properties with conservation of the material biodegradability, avoiding eco-toxicity (Xie et al. 2013). WPC is a paradigm of a structured multiphase product. It is the amalgamation of wood with polymer that can be either polymer filled inside the void spaces of solid pieces of wood or polymer reinforced with wood fiber. Wood polymer nanocomposites (WPNC) are a novel class of wood products with significantly enhanced physical, biological, mechanical, and chemical properties (Deka and Maji 2011; Md. Islama et al. 2012). The impregnation technique is one of the preferred methods for the preparation of WPNC. Composites prepared through this technique can attain the desirable properties of wood completely (Devi and Maji 2011). In this method, the empty cell lumens of wood are occupied by the appropriate monomer or prepolymer, cross-linking agents, and nanoparticles under the influence of vacuum or high pressure, which is then polymerized inside the pores of wood. The wood samples are placed in the impregnation chamber at room temperature (30 °C) and vacuum is applied to eliminate air or water present inside the cavity of cell wall of wood. The prepolymeric mixture is poured to the impregnation chamber to completely immerse the samples. The samples are kept in the chamber for few hours (4–6 h) after attaining atmospheric pressure. The excess chemicals are wiped out from the surface of the samples after taking out of the chamber. They are wrapped in aluminum foil and cured in an oven at 90 °C for 24 h. The samples are further dried at 105 °C for another 24 h. To remove the homopolymers, the samples are Soxlet extracted for 24 h. All the desirable properties of wood like water repellence, mechanical properties, and thermal stability can be achieved by the resultant wood polymer composite. Besides this, the wood fibers have also been used as reinforcing agents with the polymer for developing wood polymer composite (WPC). The wood fibers are added as reinforcing filler in the polymer matrix, pressed and molded in the presence of high temperature and pressure. The addition of various additives like coupling agents, plasticizer, flame retardant along with nanoparticles assist to obtain tailor-made the finished product as per end-use application (Deka and Maji 2012; Adeosun et al. 2012). Initially, the collected wood is chopped into small strips and washed with 1 % soap solution, followed by washing with 1 % NaOH solution and ultimately with cold water. It is dried at 105 ± 5 °C, ground in a mixer, and sieved at different mesh sizes. The polymer slurry is prepared by using suitable resin, plasticizer, coupling agents, nanofillers, etc., in a mechanical stirrer. The dried wood flour is added to the polymer solution

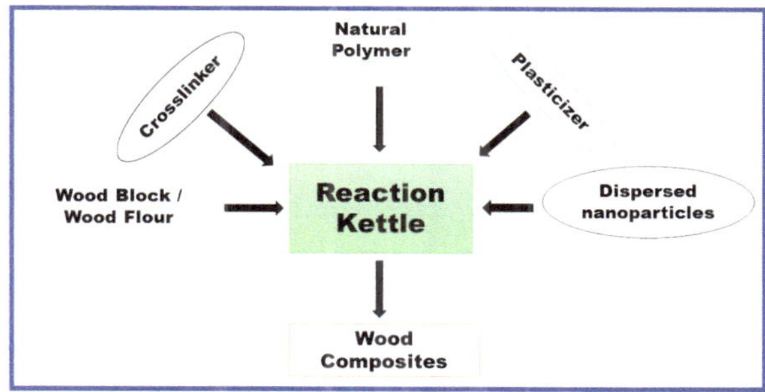

Fig. 1 Methodology for the preparation of wood polymer nanocomposites

and stirred for specified time period. It is oven dried and ground. The wood polymer nanocomposite in powder form is put into the compression molding press under pressure and temperature for certain period of time. It is then cooled at room temperature to obtain WPNC sheets (Deka and Maji 2013). The chemicals used in the formation of WPNC must be chosen appropriately. The prepared WPNC should not discharge any poisonous substances throughout service condition and can be recyclable or easily discarded at the termination of service life. Figure 1 shows the schematic diagram for the preparation of wood polymer nanocomposites.

3 Bio-based Polymers

Cellulosic polymers occupy a vital position in extensive variety of applications such as fabrics, apparel, food, molded goods, varnishes, and plastics. Among the different cellulosic polymers, starch is one of the most commercially significant class of bio-based materials and is being used for decades (Pandey et al. 2005; Yang et al. 2007; John and Thomas 2008).

Wood polymer nanocomposites are generally prepared using thermoplastics or thermosetting resin. The constant increase in the price of crude oil, problem of rising large-scale waste, and the exhaustion of fossil fuels have provoked attention in utilization of natural sources. A 'greener' solution to these issues lies in the utilization of renewable resources. The biopolymers obtained from renewable agricultural feedstock are easily available (Scott 2000; Johnson et al. 2003). Even though the biopolymers are derived from renewable resources, the extent of biodegradability of these polymers depends on its chemical structure. The higher the degree of cross-linking the lesser will be its biodegradability. These green polymers provide perceptible advantages over usual polymers regarding effluence of toxic gases during their service condition, energy consumption while synthesis and waste production. Biodegradability, high compatibility with other polymers, low melting temperatures are some of the attractive characteristics of the biopolymers. But the main drawback of the biopolymers is lack of mechanical properties which limits their applications. If the drawback is removed, it will be feasible to make wood-based green nanocomposites with adequate mechanical properties, durability, and enhanced biodegradability. The bio-based composites are the eco-products that are obtained from natural sources with recycling capabilities (Mohanty et al. 2002).

3.1 Polyfurfuryl Alcohol

Furfuryl alcohol (Fig. 2) is a clear light yellow colored liquid when pure, but it turns into yellowish brown color on long-standing. Furfuryl alcohol has been used in the formation of wood polymer composite for a long time (Baysal et al. 2004; Lande et al. 2010). It can be impregnated into the cells of wood owing to its low molecular weight. After impregnation, it is polymerized inside the wood cell wall and binded with the wood through radiation, catalysts, heat, or other reactants. The use of vegetable biomass as a substitute of petroleum-based chemicals is one of the sustainable solutions for an eco-friendly environment. Monomers such as furfuryl alcohol are particularly derived from hemicellulose. It is synthesized industrially through the catalytic reduction of furfural that in turn is derived from corncob and sugarcane bagasse. During the process of impregnation, furfuryl alcohol is generally transformed into furanic resin prepolymers. Modification of wood through furfurylation represents an exquisite eco-friendly method. The difficulty of using zinc chloride as catalyst in the process of furfurylation is that it can cause degradation of cellulose and hence affect the long-term strength properties of the resultant wood polymer composites. Generally, cyclic carboxylic system is found to be better than the zinc chloride system in the furfurylation process. Among the cyclic carboxylic system, maleic anhydride is the most appropriate one for furfurylation (Schneider 1995). With the increase in the weight percent gain, the water repellence of the wood polymer composites impregnated with bio-derived furfuryl alcohol has increased (Lande et al. 2004b). The degradation of furfurylated wood does not release any volatile organic compounds and therefore it is considered as an environmentally suitable product (Lande et al. 2004a).

3.2 Starch

Starch is one of the most fascinating and potential sources for biodegradable polymers among the wide class of sustainable polymers due to its low cost,

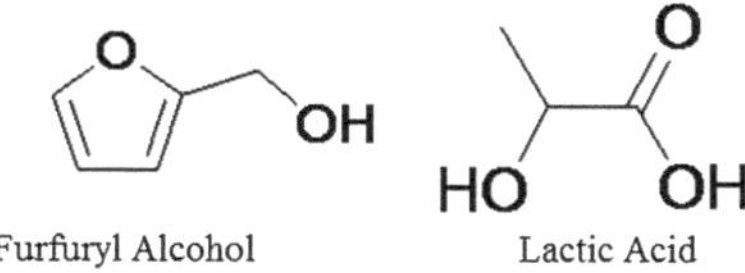

Fig. 2 Structure of furfuryl alcohol and lactic acid

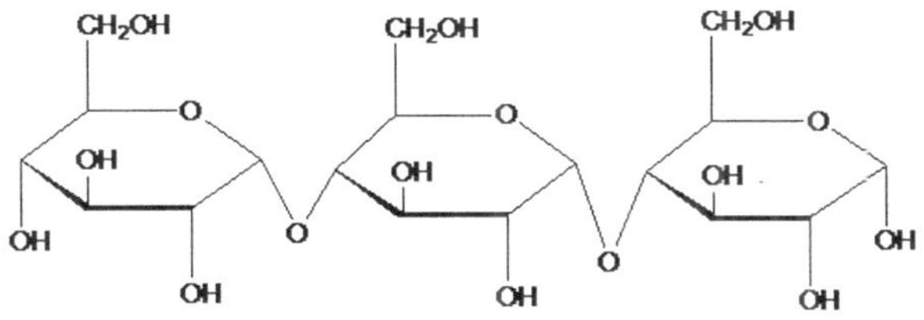

Fig. 3 Structure of amylose which contain 20–30 % structure of starch

Fig. 4 Structure of amylopectin which contains 70–80 % structure

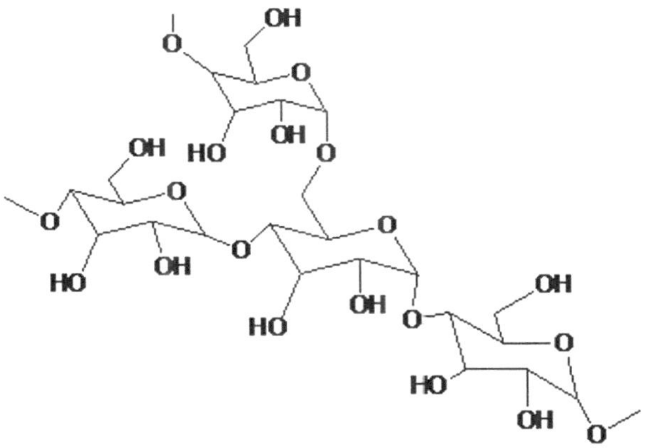

biodegradability, easy availability, and simplicity in the method of its chemical modifications (Mathew and Dufresne 2002; Mohanty et al. 2000). Pure starch is a white, scentless, and tasteless powder that is insoluble in cold water or alcohol but becomes soluble in water when heated under shearing condition. It is the main polysaccharide which is produced by most of the green plants and composed of a large number of glucose units attached through glycosidic bonds. The linear and helical amylose and the branched amylopectin are the two main constituent polysaccharides of starch. The amount of amylose in starch varies from 20 to 30 % (Fig. 3) and that of amylopectin (Fig. 4) ranges from 70 to 80 % by weight depending on the plant. Amylopectin is a much larger molecule than amylose. Starch is a semicrystalline granule due to the presence of amylopectin and the granules vary in size depending upon its source of production. The starch granules obtained from rice (about 2 μm) is much smaller in size than the granules obtained from potato (up to 100 μm).

Plasticized starch has been gaining substantial attention and is used as a substitute for synthetic polymers where long-term durability is not required and degradation is a benefit (Tábi and Kovács 2007). Plasticized starch is also called 'thermoplastic starch.' It is obtained by disintegration of starch in water followed by plasticization with water and plasticizer (e.g., glycerol) using thermomechanical energy in a constant extrusion process. However, plasticized starch-based composites possess many drawbacks like water sensitivity, poor mechanical properties, etc., in comparison with conventional synthetic thermoplastics-based composites (Cao et al. 2008a, b; Santayanon and Wootthikanokkhan 2003). Moreover, the properties of plasticized starch attain stability after several weeks. These plasticized starch are coupled with other compounds to remove the disadvantages associated with their properties. These drawbacks can be overcome by means of different physical or chemical means, including the chemical modification, (Cao et al. 2005) graft copolymerization, (Suda et al. 2002) blending with other synthetic polymers (Cao et al. 2008a, b), and incorporating fillers such as clay (Chen and Evans 2005) and nanocrystalline cellulose.

3.3 Polylactic Acid

Poly(lactic acid) (PLA) is one of the most common and interesting biodegradable polymers that belongs to the class of aliphatic polyesters usually made from α-hydroxyacids. PLA (Fig. 2) is derived from agricultural resources such as corn, sugarcane, etc., to produce articles that are employed in the biocompatible medical device market and industrial packaging field. The fermentation of corn dextrose yields lactic acid as a by-product which is the main source for production of PLA (Vink et al. 2003). PLA is notable for its extremely biodegradable and biocompatible behavior. PLA is able to substitute petrochemical-based polymers and therefore it has acquired substantial industrial cognizance for its utility in commodity resins. It is a high-strength thermoplastic polymer which has superior mechanical properties over pure polypropylene (PP). The tensile strength and modulus of neat PLA is found to be 62 MPa and 2.7 GPa, respectively, whereas neat PP has tensile strength and modulus of 36 MPa and 1.2 GPa, respectively (Huda et al. 2005). The glass-transition temperature (T_g) and the melting temperature (T_m) of PLA are 54 and 172 °C, respectively, so it is capable of processing in blow molding, injection molding, and film forming to produce film, molded parts, and fibers (Hartmann 1998; Oksman et al. 2003). However, PLA is considered brittle for many marketable applications, though it has mechanical properties appropriate for industrialized plastic appliances (Oksman et al. 2003). The problem of brittleness and poor process capability of stiff polymers can be solved by incorporation of other materials into it. The high-molecular-weight poly(lactic acid) has similar properties with that of polystyrene and it is a rigid, colorless, lustrous, thermoplastic polymer. Poly(lactic acid) is used for single-use packaging applications because it has a rational shelf life and, when properly discarded, it will hydrolyze to natural, nontoxic products. Thus, the problem of disposal of the large amount of plastic packaging can be solved by using PLA industrially. The biodegradable green composites based on wood and PLA present attractive properties of the composites and lesser cost than competitive materials. Wood polymer composites treated with PLA have satisfactory mechanical properties comparable to that of conventional thermoplastic-based wood polymer composites (Huda et al. 2006).

3.4 Soy flour

Soy flour is a mixture of soy protein and its related carbohydrate and is accessible in the form of by-product from the soybean industry. Proteins are widely available in nature in different forms from plants such as corn, soy, wheat gluten, etc., and from animals such as collagen, gelatine, etc. Biopolymers useful for various applications are derived from these proteins (Sakurada et al. 1962). Soy protein is a reactive, globular protein and is soluble in water in contrast to that of synthetic polymers which are nonreactive, planar or helical structured, and insoluble in water (Lavoine et al. 2012a, b; Siro and Plackett 2010). It has several advantages over the petroleum-based synthetic polymers and is one of the most inexpensive biopolymers. The various advantages include the purification procedure for soy protein which is an eco-friendly sustainable process. The raw material is relatively less costly and easily available because soybean is an annual crop and available worldwide. Soy proteins can be employed as resins because they can form viscous and ductile polymers. It has been reported that biocomposites based on soy flour utilized for various structural applications such as automotive parts, housing, and truck showed comparable properties to that of structural woods. Thus, these bio-based composites are appropriate for structural applications (Wool et al. 2002). For such purposes, the resin is cured at high temperature. The curing of the resin is possible at room temperature through vacuum-assisted resin transfer molding method thereby diminishing the cost and hence ecological risks.

3.5 Polyhydroxyalkanoates

Polyhydroxyalkanoate (PHA) is another important eco-friendly biopolymer that is synthesized by a wide variety of microorganisms as carbon and energy storage under nutrient-limiting conditions in the existence of excess carbon source (Khanna and Srivastava 2007). The polyhydroxyalkanoates family includes an extensive range of materials, from rigid and brittle to flexible and elastomeric materials (Misra et al. 2006). The processability of PHB is one of its main problems as it is of high stiff and brittle nature. The processing of PHB, under different conditions of temperature and rotor speeds in an internal mixer, will render its crystallinity to increase. Therefore, to improve the processing ability of this highly crystalline but brittle polymer, plasticizers like PVA and stearic acid are employed, as like with starch. The biodegradability and biocompatibility (with livingcells) are the main characteristics of this class of polymer. Polyhydroxyalkanoate has been successfully applied to manufacture nonwoven fibrous materials, films, in biomedical and pharmaceutical applications, including drug delivery systems, implant materials, transplantology, tissue engineering, and nontoxic surgical sutures (Valappil et al. 2006).

3.6 Epoxidised Vegetable Oil

Vegetable oil is one of the most inexpensive and inexhaustible natural resource obtainable in huge quantities. It has its inherent biodegradability as well as low toxicity and the reactive sites in the oil are the double bonds present in these vegetable oils. They can be functionalized by epoxidation and the epoxidized vegetable oils show outstanding properties as it is an inexpensive, renewable materials suitable for various industrial applications. Vegetable oils are lipids derived from plants and are widely extracted from oil seeds (e.g., linseed, sunflower, hemp), although some oils may be derived from other plant sources such as pulp (e.g., avocado, palm, and olive). Chemically, the constituent of vegetable oil is triglyceride molecules consisting of a glycerol molecule bonded to three fatty acid chains. Figures 5 and 6 show the structure of a triglyceride molecule and oleic acid, linoleic acid as well as linolenic acid, respectively. A fatty acid is a carboxylic acid bonded to a long unbranched aliphatic carbon chain. The length of the majority of common fatty acids found in vegetable oils usually varies between 14 and 22 carbons atoms.

$H_2C—O—C(=O)—(CH)_{16}—CH_3$

$HC—O—C(=O)—(CH_2)_7CH{=}CH—(CH_2)_7CH_3$

$CH_2—O—C(=O)—(CH_2)_7CH{=}CH(CH_2)_4CH_3$

Fig. 5 Structure of a typical triglyceride molecule

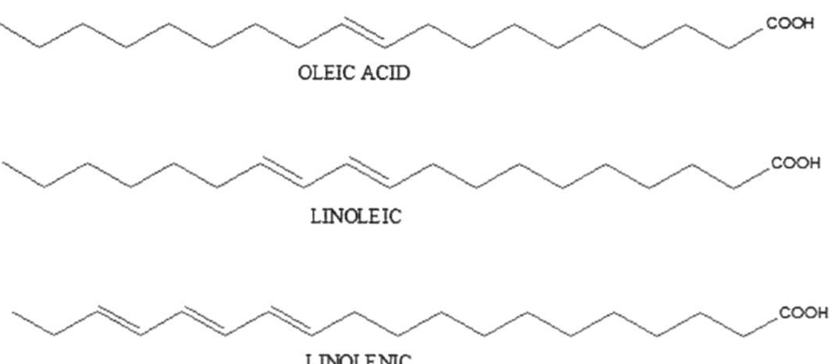

Fig. 6 Structure of oleic, linoleic, and lilonenic acid

The presence of the C=C in unsaturated fatty acids is employed as the reaction sites in the formation of cross-linking. The networked structure can be achieved by a functional reaction of the C=C to enable use in high-molecular-weight products. The properties of the epoxidized resin-based composites are reliant on the extent of cross-linking reaction with the materials. The higher the cross-link density, the better is the mechanical and thermal properties.

4 Modification of Natural Polymers: Grafting

There are considerable choices to produce pioneering polymeric materials by using polymers derived from natural sources. Manufacturing products based on renewable resources that have least environmental impact and reduced reliance on fossil fuels are the motivating forces from environmental as well as economic panorama. It is necessary to improve the properties of a polymer to meet the requirements for its end-use applications. There are various ways to alter the properties of polymers, viz., grafting, blending, and curing (Bhattacharya and Misra 2004). Different techniques have been used for graft copolymerization of various monomers on polymeric backbones which include grafting by coupling chemistries, free radical polymerisation, controlled radical polymerisation, etc. (Roy et al. 2009; Tizzotti et al. 2010; Malmstrom and Carlmark 2012). 'Grafting' is a technique in which monomers become bonded covalently (modified) onto the polymer chain. An active site is formed at a point on the polymer chain other than its end. Graft polymerization occurs on exposure of this active site to a monomer. New hybrid products with controlled properties are formed on grafting specific synthetic polymer chains to different renewable polymers including starch, nanocrystalline cellulose, polysaccharides, cellulose, and chitosan. The naturally available polymers are least water resistant and do not possess mechanical properties of sufficient magnitude for structural applications. Therefore, synthetic polymers grafted biocomposites offer a new means to achieve innovative products with controlled properties and structure. Grafting is considered an important method to modify chemical as well as physical properties of the polymers. Novel renewable-based products have been designed through ring-opening polymerization, controlled (or living) radical polymerization, and thiol-ene or click chemistry presenting interestingly new opportunities. Generally, graft copolymers are produced through radical polymerization. In most of the cases, chain transfer reactions proceed through the removal of hydrogen atoms. Natural polymer graft copolymers have been increasingly significant due to their prospective utilization in industrial application. Thus, grafting is an influential technique for implanting considerable modification on the properties of natural polymers by using various vinyl and other monomers (Athawale and Rathi 1997).

5 Cross-linking Agents

Cross-linking involves formation of three-dimensional arrangement of associated molecules by physicochemical bond formation linking molecular chains. Cross-linking agents are used to treat two semi-compatible or incompatible materials to get the best benefits of their properties and are used in small quantities. They are usually bifunctional compounds where one functionality is bonded with one component (say polymer) and other functionality with another (say wood) (Yanga et al. 2007). The polymers remain in the form of a three-dimensional network structure inside the wood structure or within the composite and hence assist in interaction between the polymers and wood. The resultant wood polymer composites will show improved mechanical properties dimensional stability, weather resistance, etc. Wood polymer composite developed by treatment with biopolymers have many disadvantages associated with their properties. The biopolymers have minimum water resistance capacity, lacking in terms of their mechanical properties, chemical resistance, etc. Addition of small amount of cross-linker enhances the properties of the composites as they form network structure inside the composites thereby stiffening the composites. The cross-linked structure of the composites improved the tortuous pathway for transmission of water as well as chemicals and as a consequence swelling decreased. Hence water resistance and chemical resistance of the composites can be achieved through the use of cross-linking agents. Various cross-linking agents such as *n*-methylol compounds which include dimethyloldihydroxy-ethyleneurea (DMDHEU), *n*-methylol acrylamide (NMA), 2-hydroxyethyl methacrylate (HEMA) are essential multifunctional cross-linking agents that improve the resistance of wood to weathering. DMDHEU can cross-link with wood cell wall and dimensionally stabilize the composites. A remarkable improvement in water resistance and mechanical properties has been found on treatment of WPC with DMDHEU (Hazarika and Maji 2012). NMA can also provide water resistance and enhance strength properties of the resultant composites (Hazarika and Maji 2013a, b).

6 Flame Retardants

Resistance to fire is one of the most desirable properties of the WPC. The fire resistant properties are essential particularly for application in furniture industry and residential construction (windows, roof tiles, decking). The study in the area of flame resistance of WPC is very important with regard to its safety provisions (Sain et al. 2004). The volatile substances evolved from wood during heating react with oxygen vigorously. The hydrocarbons contained in wood ignite indirectly during combustion. Thus wood is vulnerable to burning and ignition.

There has been an untiring attempt to improve flame retardancy of wood to increase its service. Additives that are incorporated into WPC to enhance its flame resistance are called flame-retarding agents. The objective of these flame-retarding agents is to slow down the propagation of fire or circumvent the source of a fire. The main perception of flame retardancy in the condensed phase lies in the development of a protecting barrier (Jimenez et al. 2006). Thus the underlying material is shielded from heat transfer through barrier formation (Weil et al. 1999). Flame retardants are either applied to pretreated wood or blended with the impregnating monomers prior to impregnation into wood. The use of halogenated flame retardant like organic brominated compounds can enhance the flame retardancy of WPC. However, these flame retardants yield carbon dioxide and smoke because of their ineffective combustion. The other commonly used flame retardants are various silicates, organoantimony compounds, borates compounds, and organophosphorus (Watanabe et al. 2009; Li and He 2004; Baysal 2002). Most of them while ignition release noxious gases and fumes that are extremely detrimental to health and thus contaminate the environment. The service life of the product may deteriorate due to leaching of the flame retarding agents to the surface of the product. The leaching problem of the flame retardants can be receded by using polymeric flame retarding agent because they are of high molecular weight and hence enhance the service life of the product. The utilization of polymeric flame retardant derived from natural resources is advantageous from ecological viewpoint. The gum obtained from the plant *Moringa olifera* is used as flame retardant. The flammability and biodegradability of starch-based biodegradable film modified with gum derived from *Moringa olifera* has been found to improve. (Jana et al. 2000). Wood polymer composites treated with the gum polymer showed a considerable improvement in flame retardancy (Hazarika and Maji 2013a, b; Hazarika and Maji 2014a, b, c, d). The gum polymer prohibited combustion because of its phosphorus content (Ghosh and Maiti 1998). The mechanism of fire control is supposed to be liberation of oxides of phosphorus during the process of ignition which displaces the oxygen present on the surface of the material, thus prohibiting combustion.

7 Different Nano Reinforcing Agents

The progress in the field of polymer nanocomposites (polymer composites with nano-sized reinforcement), appears as a multidisciplinary research activity. Different industries have been benefitted through the wide application of polymers nanocomposite (Sengupta et al. 2007; Hambir et al. 2002). Wood polymer nanocomposites are prepared through the interesting technique of nanotechnology. Enhanced properties of bio-based wood polymer nanocomposites can be achieved when nanofillers are incorporated along with the polymer and cross-linker. The nanoparticles should have one of its dimensions lower than 100 nm and the nano-sized reinforcement elements are dispersed in the polymer matrix during the formation of the nanocomposites. This nanoreinforcement offers nanocomposites with exceptional and excellent properties that are not found in conventional wood polymer composites. The various nanoparticles include nanoclay, metal oxide nanoparticles, carbon nanotubes.

7.1 Montmorillonite

It typically belongs to 2:1 phyllosilicates and their structure of the crystal lattice is composed of two-dimensional layers in which a middle octahedral sheet of alumina or magnesia is merged to two exterior silica tetrahedron by the tip in order that the oxygen ions of the octahedral sheet also belong to the tetrahedral sheets. The lateral dimension may differ from 300 Å to several microns, whereas the layer width is about 1 nm. The interlayer space of clay can be extended under precise experimental conditions, and the polymer chains can enter within these gallery layers of clay. The thermal and mechanical properties of clay treated wood/PLA nanocomposites were found to improve significantly (Meng et al. 2011a, b). Besides, addition of clay also improves biodegradability, dynamic mechanical properties, water resistance, chemical resistance of the nanocomposites (Hazarika et al. 2014; Hazarika and Maji 2012).

7.2 Metal Oxide Nanoparticles

The metal oxide nanoparticles can display exclusive chemical and physical properties because of their small size and a high density of boundary surface sites. Nanoparticle should have a little surface free energy so as to exhibit mechanical stability (Fernández-Garcia and Rodriguez 2007). The metal oxide nanoparticles used as nanofillers in wood polymer nanocomposites include ZnO, TiO_2, SiO_2, etc., ZnO nanoparticles are generally used as photostabilizer. In addition to its UV-stabilizing effect, nano ZnO can also enhance thermal stability of polymer nanocomposites. In polymer nanocomposite, SiO_2 nanopowder is also one of the extensively used nanofiller. SiO_2 improves the mechanical as well as thermal properties of the nanocomposite. TiO_2 has been well documented as the most outstanding photocatalyst for the decomposition of many organic pollutants in water and air (Hoffmann et al. 1995). The UV resistance along with other physical properties of the wood polymer nanocomposites improves substantially on treatment with TiO_2 (Hazarika and Maji 2013a, b).

7.3 Carbon Nanotubes (CNT)

CNT are cylindrical nanostructure (Fig. 7) having length starting from hundreds of nanometers to micron and even millimeters and radius in nanometer (Saito et al. 1998). The excellent mechanical and physical properties of CNTs based nanocomposites has motivated researchers in developing high-performance wood polymer nanocomposites based on carbon nanotubes (Hazarika and Maji 2014a, b, c, d). The

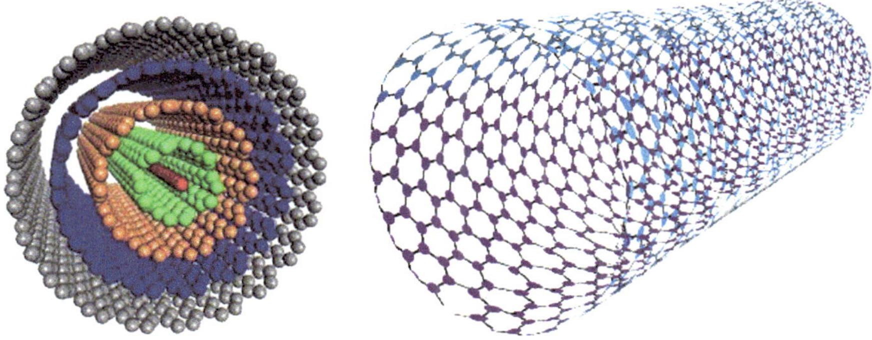

Fig. 7 Structure of CNT

various forms of CNT include, single-wall and multi-wall, networks or isolated. Starch-based MWCNT composites leads to decreased water sensitivity and improved mechanical properties (Cao et al. 2007).

7.4 Nanocellulose

Apart from the above nanoparticles, nanocelluse has been the topic of a broad range of research as reinforcing agents in nanocomposites because of their nanoscale dimension, renewability, availability, light weight, low cost, and most importantly they have minimum environmental impact and have little effect on animal/human health (Lavoine et al. 2012a, b). They offer significant properties of cellulose including its extensive ability of chemical modification, very high aspect ratio leading to the formation of versatile semicrystalline fibers which is the unique characteristic of nano materials as reinforcing agents. There is a presence of strong and complex network of hydrogen bonds which are stabilized by the ordered regions of chain packages of cellulose (Habibi et al. 2010) that resembles nanocrystalline rods. Based on their preparative methods and structure, there are two main types of nanocellulose: (i) nanocrystalline and (ii) microfibrillated cellulose.

(i) Nanocrystalline cellulose: Nanocrystalline cellulose, which are extremely crystalline and rigid nanoparticles, are also called cellulose nanowhiskers or cellulose nanocrystals. This can be prepared from native fibers through acid hydrolysis. A new class of bio-based products with a broad range of applications including automotive industry, construction material, etc., have been developed by using nanocrystals as reinforcing agents. Addition of small amount of nanocrystal can increase the strength, stiffness, and resistance of the material to stress threefold its original strength. Thus incorporation of nanocrystal makes the nanocomposites an interesting high-performance material. It is also a promising green substitute for carbon nanotubes as reinforcing agents in polymers nanocomposites and concrete. Nanocrystal reinforced nanocomposites is used in a variety of applications such as biodegradable plastic bags, textiles, wound dressings, etc.

(ii) Microfibrillated cellulose (MFC): The constituent of MFC is nano-sized cellulose fibril having high aspect ratio. The fibrils are extracted from wood pulp through high temperature, high-pressure, and high-velocity impact which can be employed in polymer nanocomposites of high mechanical capacity (Nakagaito and Yano 2004). The strength properties of these nanocomposites are very high and the Young's modulus is found to be approximately 20 GPa. Thus, MFC-based nanocomposites that are derived from wood pulp is a promising class of substance with outstandingly high mechanical performance. The Young's modulus of the cellulose crystal is about 134 GPa, therefore MFC nanofibers are estimated to provide high stiffness to the resultant nanocomposites (Sakurada et al. 1962).

However, another type of nanocellulose is known as bacterial cellulose. Specific bacteria mainly Gluconacetobacter strains secrets these cellulose nanofibers extracellularly (Klemm et al. 2009; Siro and Plackett 2010). These bacterial celluloses have exceptional mechanical and physical properties due to its special fibrillar nanostructure. Its properties include high strength, high porosity, high crystallinity (up to 84–89 %, Czaja et al. 2004), and high elastic modulus (Guhados et al. 2005). Currently, bacterial cellulose is the topic of research in several fields of applications, reinforcement in nanocomposites (Juntaro et al. 2008; Nogi and Yano 2008), biomedical applications and fuel cell membranes (Evans et al. 2003).

8 Properties of Wood Polymer Nanocomposites

8.1 Dimensional Stability

The moisture content of wood depends on the atmospheric circumstance, temperature, and the relative humidity of the surrounding air. Wood is hydrophilic in nature that can soak up moisture in humid surroundings and elude it in dry surroundings. Studies reveal that thermoplastic starch-based wood composites showed lower water uptake capacity on formation of the composites (Agnantopoulou et al. 2012). After the formation of WPC, water absorption in the composites differs depending upon the content of nanoparticles and cross-linking agents. Addition of cross-linking agent forms a networked structure in the composites which hindered the diffusion of water molecules through the cross-linked structure and as a result the water absorption capacity decreases. The water molecule first saturates the porous cell wall, lumen and tubular structure of the wood fibers, and then fills up the void spaces of wood (Das et al. 2000). As the addition of nanoparticles to the composites occupies the cell wall and the lumen, the transmission of water molecules into the composites by the capillary action is prohibited. Thus the

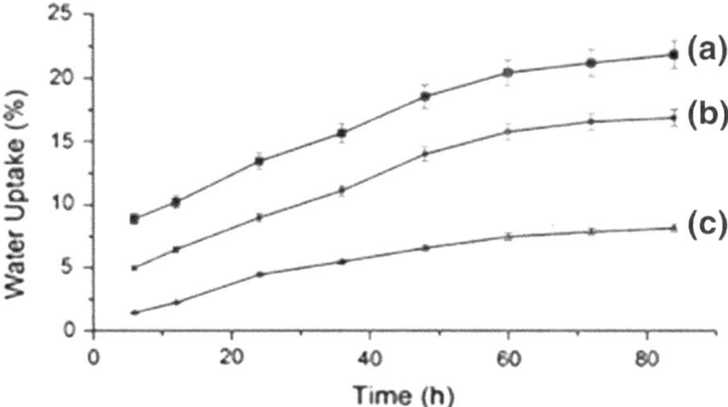

Fig. 8 Water absorption of *a* wood/MMA-*g*-starch/glutaraldehyde *b* wood/MMA-*g*-starch/*n*-methylol acrylamide *c* wood/MMA-*g*-starch/DMDHEU

nanoparticles fill up the voids and the lumens of the fibers of the composite and diminish the available sites for water absorption. Wood treated with starch-grafted MMA and DMDHEU has reduced water absorption capacity (Baishya and Maji 2014) as shown in Fig. 8.

8.2 Mechanical Properties

The various parameters like fiber aspect ratio, volume fraction of the fibers, its orientation, stress transfer at the interface and fiber matrix adhesion influence the properties of natural fiber reinforced composites. The mechanical properties of the composites greatly depend on the influence of various treatments of fibers, fiber content, and the use of cross-linking agents (Garcia et al. 1995; Vollenberg and Heiken 1989).

The polymer and fiber properties are equally significant in enhancing the mechanical performance of the composites. The properties of fibers in a composite affect its modulus, while the matrix properties are more responsive to its tensile strength. A strong interfacial interaction, low stress concentration, and fiber orientation are some of the significant criteria to enhance the tensile strength of the composites. Tensile modulus is improved by fiber wetting in the matrix phase, high-fiber aspect ratio, and fiber orientation. The aspect ratio is one of the most vital characteristics in determining the fracture properties of the composites. An optimum interfacial adhesion is essential for good impact strength of the composites. The impact strength is also sensitive to the fiber pullout, degree of adhesion, and a mechanism to absorb energy.

The mechanical properties of wood-fiber-reinforced poly(lactic acid) composites were similar to conventional composites based on thermoplastic polypropylene. Wood-fiber-reinforced poly(lactic acid) composites had appreciably superior mechanical properties than that of the virgin resin. The use of nano reinforcing

Table 1 Mechanical properties of starch-based wood composites with different cross-linker

Sample	Flexural properties		Tensile strength (MPa)	Hardness (Shore D)
	Strength (MPa)	Modulus (MPa)		
W/MMA-*g*-starch/GA	30.8	2152	13.6	64.1 (±1)
W/MMA-*g*-starch/N	42.7	4369	14.8	66.3(±0.5)
W/MMA-*g*-starch/D	44.7	5011	15.6	72.0 (±1)

agents in the composites improved the mechanical properties of the composites remarkably (Liu et al. 2013; Meng et al. 2011a, b). Addition of clay imposes a restriction in the mobility of the polymer chains that are entrapped in between the layers of the clay thereby stiffening of the composites.

The wood composites prepared from starch with water as a solvent and DMDHEU as cross-linker showed good mechanical properties. The results of mechanical properties are shown in Table 1 (Baishya and Maji 2014).

8.3 Chemical Resistance

The study of the chemical resistance properties of WPC is very important to improve the desirable properties of WPC for their target application and hence service life. As WPCs are mostly used for outdoor applications and they may come in regular contact to the external atmosphere or aqueous media such as acid, alkali, water and various solvents, so it is necessary to enhance their chemical resistance.

Even though there are several literatures on wood polymer composites, yet little information are available on the topic of the chemical resistance properties of these composites. The strength properties and esthetic appearance are deteriorated by chemicals. The hemicelluloses present in wood have low resistance to acid (Sjostrom 1993). Thus, wood polymer composites in which wood contains higher percentage of hemicellulose show substantially lower chemical resistance.

Wood is extremely resistant to mild chemicals and as such it provides a significant advantage over many alternative construction materials. The various alternative construction materials, like concrete and steel are prone to corrosion. WPC can effectively substitute these materials as it is resistant to acidic salt solutions, mild acids, and corrosive agents. The chemical resistance properties enhance further in case of wood polymer nanocomposites (Deka et al. 2012). The nanoparticles offer a tortuous path for diffusion of chemicals throughout the composites thereby improving its chemical resistance.

8.4 Biodegradation Properties

Generally, WPC are prepared by using synthetic thermoplastic monomers. The moisture uptake is retarded by plastic component and hence it slows down the commencement of degradation process. These plastics are obtained from nonrenewable sources and are not biodegradable when disposed after their service life. They remain as nonbiodegradable waste and as a result they are not considered environment-friendly. Pure wood is more vulnerable to fungal attack, decay and bacterial degradation, and thus have a reduced service life. The benefits of WPC are that it can be used as an alternative to pure wood for various outdoor applications such as railing, decking, etc. They provide weather resistance and involve low maintenance. But the main disadvantage is the nondegradable plastic constituent present in the composite (Raberg and Hafren 2008). Therefore, replacement of these synthetic polymers by polymers obtained from renewable resources in WPC will remove the problem associated with their disposal at the end of their service life. Thus bio-based wood polymer composites can effectively be employed in their target applications and have a sufficient period of service life without causing any threat to the environment.

In contrast to the other constructional material such as metal, alloys and concrete, wood polymer nanocomposites are more eco-friendly due to the biodegradation nature of wood. The bacterial growth occurs in the composites because of prevailing cellulolytic as well as pectinolytic activity of bacteria. Apart from cellulose and pectin, one of the main constituents of wood is lignin which is degraded by *Bacillus* sp. (El-Hanafy et al. 2008). Addition of nanoclay into WPNC can enhance the biodegradation process of wood polymer nanocomposites (WPNC). The WPC shows enhanced biodegradability because of the presence of clay which acted as catalyst in the biodegradation process (Hazarika and Maji 2014a, b, c, d; Karak 2006).

9 Applications

WPNC products can be effectively employed in those fields of applications where WPC commodities were in use previously. WPNC has similar functions with that of solid wood but it requires lower maintenance, and has much lesser mass to strength ratio and enhanced service life. It is an exceptional composite material, experiencing worldwide high expansion rates and constituted of wood, polymer, and other additives in various proportions. WPNC can be utilized suitably for both indoor and outdoor applications. WPC are produced commercially from the mid 1969s using the radiation process. But very few bio-based wood polymer nanocomposites have been developed, with majority of their technologies remaining in the research and development stages. These eco-friendly composites endow with the designers for new substitute to meet the challenging requirements. WPNC products can be useful in electronics, constructions, automotive, etc. The bio-based WPNC may be used as an alternative to steel and fiberglass and thus can be employed as a replacement for the automotive parts which is the most important market recognized for the utilization of WPCs (Ashori 2008). These composite materials have been efficiently used in many areas of applications including furniture industry, measurement engineering building industry, automotive industry, flooring such as solid plank flooring, laminated flooring, and fillets for parquet flooring. The main advantages of parquet flooring are its abrasion resistance and hardness which is useful in traffic commercial installations. Though it has high cost, its ease of maintenance and the long service life have justified the high price of parquet flooring over conventional flooring. They can be successfully utilized for making various sports equipment such as baseball bats, hockey sticks, golf club heads, etc., and musical instruments such as finger boards of stringed instruments, wind instruments, mouthpieces of flutes and trumpets, etc.

10 Conclusion and Future Prospect

The lower grade wood that remain unutilized and exists as a biowaste can be fabricated into valuable product by indigenous technical knowledge. Biopolymer-based wood polymer nanocomposites is of significant interest due to its renewable nature, low relative density, biodegradability, low cost, ease of processing advantages, and high specific strength. The development of WPNC by means of different polymers and nano reinforcing agents results in enhancement of properties of WPNC. A notable enhancement in properties of the nanocomposites is indicated by the laboratory study, yet a methodical study has to be performed for extensive industrial application of the product. The furfurylated wood has already been commercialized for numerous applications like roof decking, furniture, parquet flooring, ship decking, and playground equipment. The production of furfurylated wood from Lithuania is 500 m^3/year and that from Norway is 5000 m^3/year. These bio-based WPNC will find much more application in various areas in the near future.

The final properties of the nanocomposites are determined by interfacial interaction between the wood fibers, polymers, nanoparticles, and other additives. Research on preparation of WPCs by using techniques like gamma radiation, electron beam (EB), or radio frequency (RF) to polymerize the monomer(s) within the composite is nearly instantaneous. The inclusion of pigments into impregnation solutions like supercritical fluid, e.g., SC–CO_2 as the medium of impregnation may be studied extensively to investigate diversified value-added uses for modified wood, particularly for flooring and other value-added applications.

The performances of these bio-based nanocomposites are judged by their properties. One of the efficient ways to improve the properties of the final composites is the addition of nanoparticles. Attempt should be made to prepare bio-based WPNC by using nanomaterials obtained from lignocellulosic materials such as cellulose nanowhiskers and nanofibres to enhance their properties. The flame retardant obtained from the *Moringa oleifera* has remarked influence on the thermal stability and flame retardancy of the composites and moreover it is environment friendly. The use of vegetable oil-based resin along with natural additives including hardener may be tried to get fully biodegradable wood polymer composites.

Therefore, attempts should also be made to obtain cross-linking agents, reinforcing agents, resins from various natural sources to develop bio-based WPNC with enhanced properties or develop new technology suitable for the future ecological goals. The preparation of biodegradable polymer-based wood composites with new processing technique and improved properties related to physical, mechanical, chemical properties are some of the aspects that need a lot of consideration. However, the use of these biocomposites is restricted in many fields because of the high cost of the some of the biopolymers. Development of new manufacturing technology, raw materials from which the polymers are derived may decrease the cost of the bioresins for production of composites. Thus, investigation in the area of bio-based WPNC may lead to new perspective to meet with the global challenges and maybe utilized as substitute to conventional composites that may broaden its horizon of applications.

References

Adeosun SO, Lawal GI, Balogun SA, Akpan EI (2012) Review of green polymer nanocomposites. J Miner Mater Charact Eng 11:385–416

Agnantopoulou E, Tserki V, Marras S, Philippou J, Panayiotou C (2012) Development of biodegradable composites based on wood waste flour and thermoplastic starch. J Appl Polym Sci 126:E272–E280

Ashori A (2008) Wood–plastic composites as promising green-composites for automotive industries. Bioresour Technol 99:4661–4667

Athawale VD, Rathi SC (1997) Synthesis and characterization of starch–poly(methacrylic acid) graft copolymers. J Appl Polym Sci 66:1399–1403

Baishya P, Maji TK (2014) Studies on Effects of different crosslinkers on the properties of starch based wood composites doi:10.1021/sc5002325

Baysal E (2002) Determination of oxygen index levels and thermal analysis of scots pine (Pinussylvestris L.) impregnated with melamine formaldehyde-boron combinations. J Fire Sci 20:373–389

Baysal E, Ozaki SK, Yalinkilic MK (2004) Dimensional stabilization of wood treated with furfuryl alcohol catalysed by Borates. Wood Sci Technol 38:405–415

Bhattacharya A, Misra BN (2004) Grafting: a versatile means to modify polymers techniques, factors and applications. Prog Polym Sci 29:767–814

Cao X, Chang PR, Huneault MA (2008a) Preparation and properties of plasticized starch modified with poly caprolactone based waterborne polyurethane. Carbohydr Polym 71:119–125

Cao X, Chen Y, Chang PR, Huneault MA (2007) Preparation and properties of plasticized starch/ multi walled carbon nanotubes composites. J Appl Polym Sci 106:1431–1437

Cao X, Chen Y, Chang PR, Muir AD, Falk G (2008b) Starch-based nanocomposites reinforced with flax cellulose nanocrystals. express Polym Lett 2:502–510

Cao X, Wang Y, Zhang L (2005) Effects of ethyl and benzyl groups on the miscibility and properties of castor oil-based polyurethane/starch derivative semi-interpenetrating polymer networks. Macromol Biosci 5:863–871

Chen B, Evans JRG (2005) Thermoplastic starch-clay nanocomposites and their characteristics. Carbohydr Polym 61:455–463

Czaja W, Romanovicz D, Brown RM (2004) Structural investigations of microbial cellulose produces in sattionary and agitated culture. Cellulose 11:403–411

Das S, Saha AK, Choudhury PK, Basak R, Mitra BC, Todd T, Lang S, Rowel RM (2000) Effect of steam pretreatment of jute fiber on dimensional stability of jute composite. J Appl Polym Sci 76:1652–1661

Deka BK, Maji TK (2011) Effect of TiO_2 and nanoclay on the properties of wood polymer nanocomposite. Compos Part A 42:2117–2125

Deka BK, Maji TK (2012) Effect of nanoclay and ZnO on the physical and chemical properties of wood polymer nanocomposite. J Appl Polym Sci 124:2919–2929

Deka BK, Maji TK (2013) Effect of SiO_2 and nanoclay on the properties of wood polymer nanocomposite. Polym Bull 70:403–417

Deka BK, Mandal M, Maji TK (2012) Effect of nanoparticles on flammability, UV resistance, biodegradability, and chemical resistance of wood polymer nanocomposite. Ind Eng Chem Res 51:11881–11891

Devi RR, Maji TK (2011) Preparation and characterization of wood/styrene-acrylonitrile co-polymer/mmt nanocomposite. J Appl Polym Sci 122:2099–2109

El-Hanafy AA, Elsalam HA, Hafez EE, Borg EL (2008) Molecular characterization of two native Egyptian ligninolytic bacterial strains. J Appl Sci Res 4:1291–1296

Evans BR, O'Neill HM, Malyvanh VP, Lee I, Woodward J (2003) Palladium bacterial cellulose membranes for fuel cells. Biosens Bioelectron 18:917–923

Fernández-Garcia M, Rodriguez JA (2007) Metal oxide nanoparticles, nanomaterials: inorganic and bioinorganic perspectives doi: 10.1002/9781119951438.eibc0331. (Encyclopedia of Inorganic and Bioinorganic Chemistry)

Garcia ZF, Martinez E, Castillo AA, Castano VM (1995) Numerical analysis of the experimental mechanical properties in polyester resins reinforced with natural fibers. J Reinf Plast Compos 14:641–649

Ghosh SN, Maiti S (1998) Adhesive performance, flammability evaluation and biodegradation study of plant polymer blends. Eur Polym J 34:849–854

Giudice CA, Pereyra AM (2007) Fire resistance of wood impregnated with soluble alkaline silicates. Res Lett 2007:1–4

Guhados G, Wan WK, Hutter JL (2005) Measurement of the elastic modulus of single cellulose fibers using atomic force microscopy. Langmuir 21:6642–6646

Habibi Y, Lucia LA, Rojas OJ (2010) Cellulose nanocrystals: chemistry, self-assembly, and applications. Chem Rev 110:3479–3500

Hambir S, Bulakh N, Jog JP (2002) Polypropylene/clay nanocomposites: effect of compatibilizer on the thermal, crystallization and dynamic mechanical behavior. Polym Eng Sci 42:1800–1807

Hartmann MH (1998) Biopolymers from renewable resources. In: Kaplan DL (ed) Springer, Berlin, Chapter 15, pp 367–411

Haygreen JG, Bowyer JL (1982) Forest products and wood science: an Introduction, 1st edn. Iowa State University Press, Ames, Iowa

Hazarika A, Maji TK (2014a) Properties of softwood polymer composites impregnated with nanoparticles and melamine formaldehyde furfuryl alcohol copolymer. Polym Eng Sci 54:1019–1029

Hazarika A, Maji TK (2012) Effect of different crosslinkers on properties of melamine formaldehyde-furfuryl alcohol copolymer/montmorillonite impregnated softwood (*Ficus hispida*). Polym Eng Sci 53:1394–1404

Hazarika A, Maji TK (2013a) Study on the properties of wood polymer nanocomposites based on melamine formaldehyde-furfuryl alcohol copolymer and modified clay. J Wood Chem Technol 33:103–124

Hazarika A, Maji TK (2013b) Synergistic effect of nano-TiO_2 and nanoclay on the ultraviolet degradation and physical properties of wood polymer nanocomposites. Ind Eng Chem Res 52:13536–13546

Hazarika A, Maji TK (2014b) Properties of softwood polymer composites impregnated with nanoparticles and melamine formaldehyde furfuryl alcohol copolymer. Polym Eng Sci 54:1019–1029

Hazarika A, Maji TK (2014c) Strain sensing behavior and dynamic mechanical properties of carbon nanotubes/nanoclay reinforced wood polymer nanocomposite. Chem Eng J 247:33–41
Hazarika A, Maji TK (2014d) Thermal decomposition kinetics, flammability, and mechanical property study of wood polymer nanocomposite. J Therm Anal Calorim 115:1679–1691
Hazarika A, Mandal M, Maji TK (2014) Dynamic mechanical analysis, biodegradability and thermal stability of wood polymer nanocomposites. Compos Part B 60:568–576
Hetzer M, Kee D (2008) Wood/polymer/nanoclay composites, environmentally friendly sustainable technology: a review. Chem Eng Res Des 86:1083–1093
Hill CAS, Abdul KHPS, Hale MD (1998) A study of the potential of acetylation to improve the properties of plant fibres. Ind Crops Prod 8:53–63
Hoffmann MR, Martin ST, Choi WY, Bahnemann W (1995) Environmental application of semiconductor photocatalysis. Chem Rev 95:69–96
Huda MS, Drzal LT, Misra M, Mohanty AK (2006) Wood-fiber-reinforced poly(lactic acid) composites: evaluation of the physicomechanical and morphological properties. J Appl Polym Sci 102:4856–4869
Huda MS, Mohanty AK, Misra M, Drzal LT, Schut EJ (2005) Green composites from recycled cellulose and poly (lactic acid): physico-mechanical and morphological properties evaluation. Mater Sci 40:4221–4229
Hussain F, Hojjati M, Okamoto M, Gorga RE (2006) Review article: polymer-matrix nanocomposites, processing, manufacturing, and application: an overview. J Compos Mater 40:1511–1575
Jana T, Roy BC, Maiti S (2000) Biodegradable film modification of the biodegradable film for fire retardancy. Polym Degrad Stab 69:79–82
Jimenez M, Duquesne S, Bourbigot S (2006) Intumescent fire protective coating: toward a better understanding of their mechanism of action. Thermochim Acta 449:16–26
John MJ, Thomas S (2008) Biofibres and biocomposites. Carbohydr Polym 71:343–364
Johnson MR, Tucker N, Barnes S (2003) Impact performance of miscanthus/ novamont mater bi biocomposites. Polym Test 22:209–215
Juntaro J, Pommet M, Kalinka G, Mantalaris A, Shaffer MSP, Bismarck A (2008) Creating hierarchical structures in renewable composites by attaching bacterial cellulose onto sisal fibers. Adv Mater 20:3122–3126
Karak N (2006) Polymer (epoxy) clay nanocomposites. J Polym Mater 23:1–20
Khanna S, Srivastava AK (2007) Production of poly (3-hydroxybutyric-co-3-hydroxyvaleric acid) having a high hydroxyvalerate content with valeric acid feeding. J Ind Microbiol Biot 34:457–461
Klemm D, Schumann D, Kramer F, Hesler N, Koth D, Sultanova B (2009) Nanocellulose materials—different cellulose, different functionality. Macromol Symp 280:60–71
Lande S, Høibø OA, Larnøy E (2010) Variation in treatability of Scots pine (Pinussylvestris) by the chemical modification agent furfuryl alcohol dissolved in water. Wood Sci Technol 44:105–118
Lande S, Westin M, Schneider M (2004a) Chemistry and ecotoxicology of furfurylated wood. Scand J For Res 19:14–21
Lande S, Westin M, Schneider M (2004b) Properties of furfurylated wood. Scand J For Res 19:22–30
Lande S, Westin M, Schneider MH (2003) Development of modified wood products based on furan chemistry. Mol Cryst Liq Cryst 484:367–378
Lavoine N, Desloges I, Dufresne A, Bras J (2012a) Microfibrillated cellulose—its barrier properties and applications in cellulosic materials: a review. Carbohydr Polym 90:735–764
Lavoine N, Desloges I, Dufresne A, Bras J (2012b) Microfibrillated cellulose—its barrier properties and applications in cellulosic materials: a review. Carbohydr Polym 90:735–764
Leszczy´nska A, Njuguna J, Pielichowski K, Banerjee JR (2007) Polymer/montmorillonitenanocomposites with improved thermal properties. Part II. Thermal stability of montmorillonite nanocomposites based on different polymeric matrixes. Thermochim Acta 454:1–22

Li B, He JM (2004) Investigation of mechanical property, flame retardancy and thermal degradation of LLDPE–wood-fibre composites. Polym Degrad Stab 83:241–246

Li Y, Liu Z, Dong X, Fu Y, Liu Y (2013) Comparison of decay resistance of wood and wood-polymer composite prepared by in-situ polymerization of monomers. Int Biodeter Biodegr 84:401–406

Liang F, Wang Y, Sun XS (1999) Green composites using cross-linked soy flour and flax yarns. J Polym Eng 19:383–393

Liu R, Cao J, Luo S, Wang X (2013) Effects of two types of clay on physical and mechanical properties of poly(lactic acid)/wood flour composites at various wood flour contents. J Appl Polym Sci 127:2566–2573

Malmstrom E, Carlmark A (2012) Controlled grafting of cellulose fibres—an outlook beyond paper and cardboard. Polym Chem 3:727–733

Martinez-Hernandez AL, Velasco-Santos C (2012) Keratin fibers from chicken feathers: structure and advances in polymer composites. Nova Publishers, New York, pp 149-211

Martínez-Hernández AL, Velasco-Santos C, de-Icaza M, Castaño VM (2007) Dynamical–mechanical and thermal analysis of polymeric composites reinforced with keratin biofibers from chicken feathers. Compos Part B Eng 38:405–410

Mathew AP, Dufresne A (2002) Morphological investigation of nanocomposites from sorbitol plasticized starch and tunicin whiskers. Biomacromolecules 3:609–617

Md. Islama S, Hamdana S, Talibb ZA, Ahmeda AS, Md. Rahmana R (2012) Tropical wood polymer nanocomposite (WPNC): The impact of nanoclay on dynamic mechanical thermal properties. Compos Sci Technol 72:1995–2001

Meng QK, Hetzer M, De Kee D (2011a) PLA/clay/wood nanocomposites: nanoclay effects on mechanical and thermal properties. J Compos Mater 45:1145–1158

Meng QK, Hetzer M, Kee DD (2011b) PLA/clay/wood nanocomposites: nanoclay effects on mechanical and thermal properties. J Compos Mater 45:1145–1158

Misra SK, Valappil SP, Roy I, Boccaccini AR (2006) Polyhydroxyalkanoate (PHA)/inorganic phase composites for tissue engineering applications. Biomacromolecules 7:2249–2258

Mohanty AK, Misra M, Drzal LT (2002) Sustainable bio-composites from renewable resources: opportunities and challenges in the green materials world. J Polym Environ 10:19–26

Mohanty AK, Misra M, Hinrichsen G (2000) Biofibres, biodegradable polymers and biocomposites: an overview. Macromol Mater Eng 276–277:1–24

Nakagaito AN, Yano H (2004) The effect of morphological changes from pulp fiber towards nano-scale fibrillated cellulose on the mechanical properties of high-strength plant fiber based composites. Appl Phys A 78:547–552

Nogi M, Yano H (2008) Transparent nanocomposites based on cellulose produced by bacteria offer potential innovation in the electronics device industry. Adv Mater 20:1849–1852

Oksman K, Skrifvars M, Selin JF (2003) Natural fibres as reinforcement in polylactic acid (PLA) composites. Compos Sci Technol 63:1317–1324

Pandey JK, Kumar AP, Misra M, Mohanty AK, Drzal LT, Singh RP (2005) Recent advances in biodegradable nanocomposites. J Nanosci Nanotechnol 5:497–526

Panshin AJ, de Zeeuw C (1980) Textbook of wood technology: structure, identification, uses, and properties of the commercial woods of the United States, 4th edn. McGraw Hill Inc., New York

Paul DR, Robeson LM (2008) Polymer nanotechnology: nanocomposites. Polymer 49:3187–3204

Raberg U, Hafren J (2008) Biodegradation and appearance of plastic treated solid wood. Int Biodeterior Biodegrad 62:210–213

Raj RG, Kokta BV, Maldas D, Daneault C (1989) Use of wood fibers in thermoplastics. VII the effect of coupling agents in polyethylene-wood fiber composites. J Appl Polym Sci 37:1089–1103

Rowell RM, Young RA, Rowell JK (1997) Paper and composites from agro-based resources. CRC Lewis Publishers, Boca Raton FL

Roy D, Semsarilar M, Guthrie JT, Perrier S (2009) Cellulose modification by polymer grafting: a review. Chem Soc Rev 38:2046–2064

Sain M, Park HS, Suhara F, law S (2004) Flame retardant and mechanical properties of natural fibre–PP composites containing magnesium hydroxide. Polym Degrad Stab 83:363–364
Saito R, Dresselhaus G, Dresselhaus MS (1998) Physical properties of carbon nanotubes. Imperial College Press, London
Sakurada I, Nukushina Y, Ito T (1962) Experimental determination of the elastic modulus of crystalline regions in oriented polymers. J Polym Sci 57:651–659
Santayanon R, Wootthikanokkhan J (2003) Modification of cassava starch by using propionic anhydride and properties of the starch-blended polyester polyurethane. Carbohydr Polym 51:17–24
Schneider MH (1995) New cell wall and cell lumen wood polymer composites. Wood Sci Technol 29:135–158
Scott G (2000) Green- polymers. Polym Degrad Stab 68:1–7
Sengupta R, Chakraborty S, Bandyopadhyay S, Dasgupta S, Mukhopadhyay R, Auddy K, Deuri AS (2007) A short review on rubber/clay nanocomposites with emphasis on mechanical properties. Polym Eng Sci 47:1956
Siro I, Plackett D (2010) Microfibrillated cellulose and new nanocomposite materials: a review. Cellulose 17:459–494
Sjostrom E (1993) Wood chemistry: fundamentals and applications, 2nd edn. Academic Press, New York
Agustin MB, Ahmmad B, Leon ERPD, Buenaobra JL, Salazar JR, Hirose F (2013) Starch-based biocomposite films reinforced with cellulose nanocrystals from garlic stalks. Polym Compos 34:1325–1332
Suda K, Kanlaya M, Manit S (2002) Synthesis and property characterization of cassava starch grafted poly[acrylamide-co-(maleic acid)] superabsorbent via-γ irradiation. Polymer 43:3915–3924
Tábi T, Kovács JG (2007) Examination of injection molded thermoplastic maize starch. Express Polym Lett 1:804–809
Tizzotti M, Charlot A, Fleury E, Stenzel M, Bernard J (2010) Modification of polysaccharides through controlled/living radical polymerization grafting-towards the generation of high performance hybrids. Macromol Rapid Commun 31:1751–1772
Valappil SP, Misra SK, Boccaccini AR, Roy I (2006) Expert Rev Med Devices 3:853–868
Vink ETH, Rabago KR, Glassner DA, Gruber PR (2003) Applications of life cycle assessment to nature works polylactide (PLA) production. Polym Degrad Stab 80:403–419
Vollenberg PHT, Heiken D (1989) Particle size dependence of the Young's modulus of filled polymers: 1 Preliminary experiments. Polymer 30:1656–1662
Watanabe M, Sakurai M, Maeda M (2009) Preparation of ammonium polyphosphate and its application to flame retardant. Phosphorus Res Bull 23:35–44
Wegner T, Skog KE, Ince PJ, Michler CJ (2010) Uses and desirable properties of wood in the 21st Century. J Forest 108:165–173
Weil ED, Levchik SV, Ravey M, Zhu W (1999) A Survey of recent progress in phosphorus-based flame retardants and some mode of action studies. Phosphorus, sulfur, Silicon Relat Elem 144:17–20
Wool RP, Khot SN, Lascala JJ, Bunker SP, Lu J, Thielemans W (2002) Affordable composites and plastics from renewable resources Part II: Manufacture of composites. Advancing sustainability through green chemistry and engineering. ACS Symp Ser 823:205–224
Xie F, Pollet E, Halleya PJ, Avérous L (2013) Starch-based nano-biocomposites. Prog Polym Sci 38:1590–1628
Xie Y, Hill CAS, Xiao Z, Mai C, Militz H (2011) Dynamic water vapor sorption properties of wood treated with glutaraldehyde. Wood Sci Technol 45:49–61
Yang KK, Wang XL, Wang YZ (2007) Progress in nanocomposite of biodegradable polymer. J Ind Eng Chem 13:485–500
Yanga HS, Kimb HJ, Parkc HJ, Leed BJ, Hwang TS (2007) Effect of compatibilizing agents on rice-husk flour reinforced polypropylene composites. Compos Struct 77:45–55

Water Soluble Polymer-Based Nanocomposites Containing Cellulose Nanocrystals

Johnsy George, S.N. Sabapathi and Siddaramaiah

Abstract Among the eco-friendly polymers, water soluble polymers are increasingly gaining importance to industry and academia, as they are easy to process, low cost, easily available, and more environmentally friendly than any other polymers. Water soluble polymers are widely used as stabilizers, thickeners, drug delivery materials, protective colloids, dispersants, flocculants, materials for oil recovery, etc. However, replacing nondegradable and nonrenewable plastic materials with these water soluble polymers for several applications remains as a big challenge. Several water soluble polymers, like those derived from naturally occurring proteins, polysaccharides, etc., and those obtained from synthetic methods are not having sufficient properties to replace the existing non-degradable plastic materials for most of the applications. Incorporation of nanomaterials into polymer matrices enhances the mechanical properties like tensile strength, modulus, stiffness, and impact strength significantly. Also other physical properties like barrier, optical, thermal resistance, nonflammability, etc., can also be improved by the introduction of nanomaterials. It is believed that the advances in polymer nanocomposite field will revolutionize the design, development, and performance of water soluble polymer-based materials, which ultimately have negligible adverse impact on the environment. Nanotechnology could be able to play an important role in solving this problem with the development of water soluble nanocomposite materials, which holds the key to future advances in the field of eco-friendly packaging systems. Several nanomaterials have been investigated for reinforcing water soluble polymers; however, rod-shaped cellulose nanocrystals (CNs) having high aspect ratios are found to be a promising nanomaterial for these types of applications. This chapter deals with the development and characterization of water soluble polymer-based nanocomposites containing cellulose nanocrystals and their applications.

J. George (✉) · S.N. Sabapathi
Food Engineering and Packaging Division, Defence Food Research Laboratory, Siddartha Nagar, Mysore 570 011, India
e-mail: g.johnsy@gmail.com

Siddaramaiah
Department of Polymer Science and Technology, Sri Jayachamarajendra College of Engineering, Mysore 570 006, India

V.K. Thakur and M.K. Thakur (eds.), *Eco-friendly Polymer Nanocomposites*, Advanced Structured Materials 75, DOI 10.1007/978-81-322-2470-9_9

Keywords Water soluble polymers · Polymer nanocomposites · Cellulose nanocrystals

1 Water Soluble Polymers

Water soluble polymers cover a wide range of polymer families of natural or synthetic origin. A water soluble polymer is defined as a polymer that can be dissolved in water, with or without the assistance of cosolvents and neutralizing agents, to form transparent solutions. Economic and environmental aspects are contributing to the growing interest in water soluble polymers, due to their biodegradability, low toxicity, low manufacturing costs, lower disposal costs, and renewability. Water soluble polymers have a wide range of industrial applications in food industries, pharmaceuticals, paints, textiles, paper, adhesives, coatings, etc. Water soluble polymers can be classified into two main types: naturally occurring polymers and synthetic polymers.

1.1 Naturally Occurring Water Soluble Polymers

As the name indicates, these polymers are obtained from nature. Polysaccharides and protein-based polymers were dominant among the naturally occurring water soluble polymers.

1.1.1 Polysaccharide-Based Water Soluble Polymers

Polysaccharides are long chain carbohydrate molecules in which several monosaccharide units are joined together by glycosidic linkages (Thakur and Thakur 2014a, b, c). Polysaccharides have a general formula of $C_x(H_2O)_y$ where, x is usually a large number between 200 and 2500. Under the broad heading of naturally occurring polysaccharide-based water soluble polymers, different polymers like carrageenan, alginate, etc., are having wide range of applications. Water soluble polysaccharides can also be produced by physical or chemical modification of natural polymers. Common varieties of such water soluble polymers are chemically modified cellulose and starch. Many derivatives of these polymers can be synthesized by chemical routes such as substitution, oxidation, cross-linking, or partial hydrolysis.

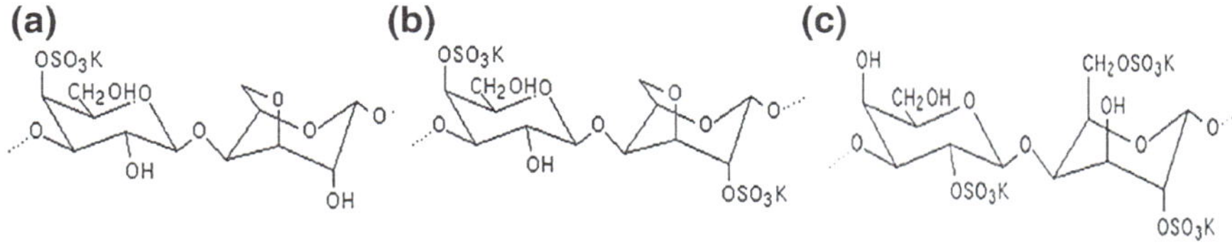

Fig. 1 Chemical structure of **a** kappa, **b** iota, and **c** lambda *Carrageenan*

Carrageenan

Carrageenan is a naturally occurring viscous and gel-forming polysaccharide, which is obtained from certain species of red seaweeds of the class of *Rhodophyceae*. This polymer is a sulfated polygalactan with 15–40 % of ester-sulfate content and an average molecular weight above 100 kDa. Its polymer chain is formed by alternate units of *D*-galactose and 3, 6-anhydrogalactose joined by α-1, 3 and β-1, 4-glycosidic linkage (Necas and Bartosikova 2013). Different forms of carrageenan such as kappa, iota, and lambda are available (Fig. 1). All have similar *D*-galactose backbones, but they differ in solubility, extent of branching, degree of sulfation, cation binding, and gelling ability. λ-carrageenan is the least branched and the least gel forming and hence it is readily soluble in cold water, compared to other forms. Carrageenans are large, highly flexible molecules that can form helical structures, which give them the ability to form a variety of gels at room temperature. They are widely used in food industries as thickening and stabilizing agents. They are also used as polymer matrices for the delivery of drugs, proteins, etc., and for tissue regeneration applications.

Alginate

Alginate is a polysaccharide widely obtained from the cell walls of certain brown seaweeds. Alginate absorbs water quickly and is capable of absorbing 200–300 times of its own weight to form a viscous gel. Alginate is a linear anionic copolymer with homopolymeric blocks of (1-4)-linked β-*D*-mannuronate (M) and α-*L*-guluronate (G) residues (Fig. 2), respectively, covalently linked together in

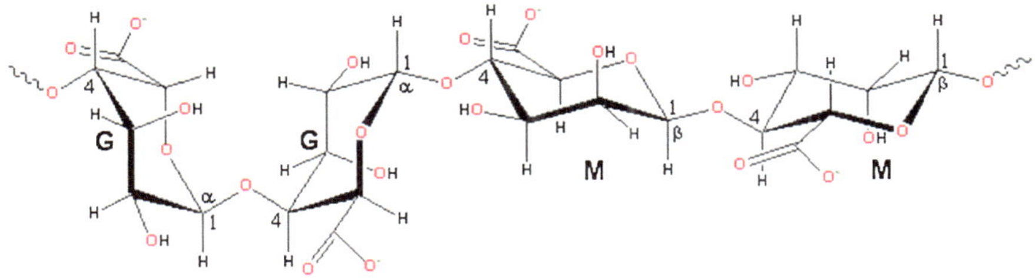

Fig. 2 Chemical structure of alginate

different sequences or blocks (Matsumoto et al. 1992). Alginate is having a large number of free hydroxyl and carboxyl groups in the chain backbone. This makes it an ideal candidate for chemical functionalization. By functionalizing the available hydroxyl and carboxyl groups, the properties of alginates such as solubility, hydrophobicity, physicochemical, and biological characteristics can be altered (Yang et al. 2011). Alginate is used as an additive in different dehydrated products as they can absorb water quickly. It is also used for waterproofing and fireproofing fabrics, as a gelling agent, for thickening drinks, ice cream, and cosmetics. It is the only polysaccharide that naturally contains carboxyl groups in each constituent residue and possesses various functionalities for applications as functional materials. Sodium salt of alginic acid, which is commonly known as sodium alginate is the most widely used water soluble polymer for the synthesis of nanocomposites. Its biocompatibility, flexibility, and the ability to entrap biomolecules make it an ideal polymer for drug and protein delivery systems.

Derivatives of Cellulose

Cellulose is the most abundant organic polymer which is made up of D-anhydroglucopyranose units linked together with β-(1-4) glycosidic bonds (Singha and Thakur 2009a, b, c, d, e, f). Cellulose is an important component of all natural fibers and is the most abundant natural polymer available on the earth (Thakur et al. 2010a, b, 2014a, b). Cellulose is abundant in natural fibers such as cotton, sisal, pine, flax, jute, kenaf, coir, hibiscus sabdariffa, henequen, Grewia optiva, etc. (Thakur et al. 2014b; Singha and Thakur 2010a, b, c). Naturally available cellulose is not soluble in water due to its dense hydrogen bonding network, however, by chemical functionalization, water soluble cellulose can be synthesized. The hydroxyl groups of cellulose can be partially or fully replaced with ethers and esters by employing substitution reactions. Typical modifications like etherification and esterification of the hydroxyl groups of cellulose make them water soluble (Fig. 3). Cellulose ethers can be prepared by treating alkali cellulose with various alkyl or aryl halides (Cohen et al. 1953). Purification is accomplished by washing the reaction product with hot water. The esterification reactions of cellulose involve a typical equilibrium reaction in which an alcohol and acid react to form ester and water. Cellulose is esterified with

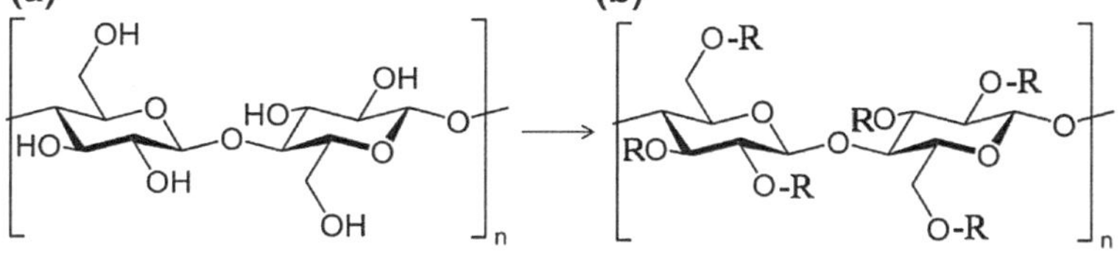

Fig. 3 The chemical structure of **a** cellulose and **b** chemically modified cellulose, where, R can be $-CH_3$, $-COOCH_3$, $-COCH_2CH_2CH_3$, $-CH_2CH(OH)CH_3$, $-CH_2COOH$ or $-CH_2CH_3$

certain acids such as acetic acid, nitric acid, sulfuric acid, phosphoric acid, etc., (Mischnick and Momcilovic 2010). The most important organic ester is cellulose acetate, which is prepared by the reaction of acetic anhydride with cellulose in the presence of sulfuric acid. Other commonly used water soluble cellulose derivatives are methyl cellulose (MC), hydroxyethyl cellulose (HEC), hydroxypropyl cellulose (HPC), hydroxypropyl methyl cellulose (HPMC), and carboxymethyl cellulose (CMC).

Methyl Cellulose

MC is a chemical compound derived from cellulose by heating, cellulose with a solution of sodium hydroxide and methyl chloride. Due to the substitution reaction, the hydroxyl groups of cellulose are replaced by methoxide ($-OCH_3$) groups (Fig. 4a). MC is obtained in the form of white powder and dissolves in cold water, forming a clear viscous solution or gel. Methylcellulose (MC) has excellent film-forming properties and efficient oxygen and lipid barrier properties (Debeaufort et al. 1995). It is used as a thickener and emulsifier in various foods and cosmetic products, and also as a treatment of constipation. Like cellulose, it is not digestible, nontoxic, and not an allergen.

Hydroxyethyl Cellulose

HEC is manufactured by reacting purified cellulose with sodium hydroxide to produce swollen alkali cellulose. By treating alkali cellulose with ethylene oxide, a series of HEC ethers can be produced. In this reaction, the hydrogen atoms in the hydroxyl groups of cellulose are replaced by hydroxyethyl groups (Fig. 4b), which confer water solubility to the product when the molar substitution reaches a certain value (Gorgieva and Kokol 2011). HEC is a nonionic, water soluble polymer that can thicken, emulsify, and form films in nature. It is readily soluble in hot or cold water and can be used to prepare solutions with a wide range of viscosities. HEC has many industrial applications like paints, textile finishing, thickening cement mortar, and paper making.

Hydroxypropyl Cellulose

HPC is synthesized by chemical modification of cellulose with an etherifying agent, propylene oxide that results in the introduction of hydroxypropyl side chains onto

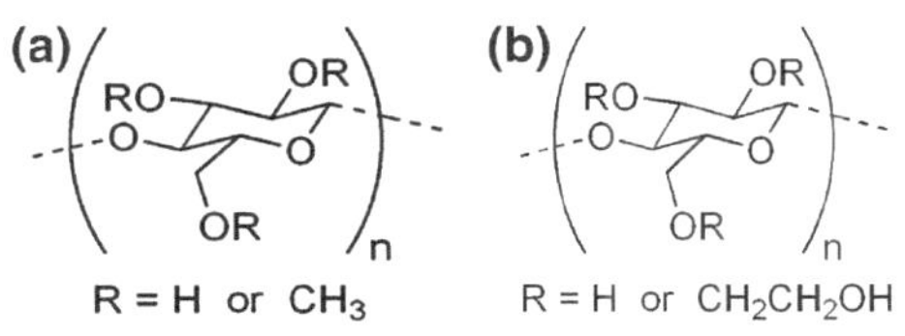

(a) R = H or CH_3 (b) R = H or CH_2CH_2OH

Fig. 4 Chemical structure of **a** methyl cellulose and **b** hydroxyethyl cellulose

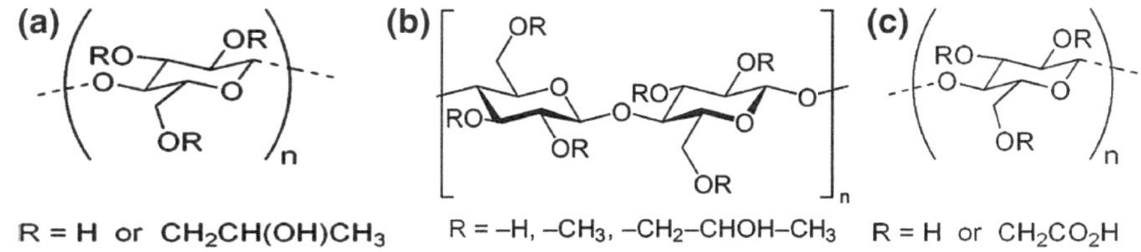

(a) R = H or $CH_2CH(OH)CH_3$ (b) R = –H, $–CH_3$, $–CH_2–CHOH–CH_3$ (c) R = H or CH_2CO_2H

Fig. 5 Chemical structure of **a** hydroxypropyl cellulose, **b** hydroxypropyl methyl cellulose and **c** carboxymethyl cellulose

the polymeric backbone of cellulose. In substitution reactions by etherification, the cellulose is first converted to alkali cellulose followed by a reaction with propylene oxide which results in the formation of HPC (Fig. 5a) through bimolecular substitution reaction (Schagerlf et al. 2006). HPC is widely used in industrial applications like suspension polymerization, as a protective colloid, binder for ceramics, films, etc. HPC cellulose films possess excellent flexibility, lack of tackiness, and barrier to oil and fat. HPC can also be used to modify other coating resins such as shellac, ethylcellulose, CMC, and starches. Generally, it improves flexibility and toughness, reduces water resistance, and tendency of the film to crack.

Hydroxy Propyl Methyl Cellulose

Hydroxy propyl methyl cellulose (HPMC) is a water soluble, viscous 1, 4-glucose polymer that is produced by the addition of methyl and hydroxylpropyl groups to the cellulose backbone (Fig. 5b). This modification leads to a polymer with a high surface activity and unique hydration-dehydration solution dependent on temperature (Sarker et al. 1995). HPMC is obtained in the form of powder and can be formed into granules. The compound forms colloids when dissolved in water and also have superior foam stabilization ability. HPMC is used as a food additive, an emulsifier, thickening and suspending agent, and an alternative to animal gelatin.

Carboxy Methyl Cellulose

CMC is a cellulose derivative with carboxymethyl groups attached to some of the hydroxyl groups of the cellulose backbone. It is synthesized by the alkali-catalyzed reaction of cellulose with chloroacetic acid (Fig. 5c). The polar carboxyl groups make the cellulose chemically reactive and water soluble. The functional properties of CMC depend on the degree of substitution of the cellulose structure, as well as the chain length of the cellulose backbone structure and the degree of clustering of the carboxymethyl substituents. CMC is a film-forming polymer and resistant to oils, fats, and organic solvents. CMC also possesses a good degree of adhesiveness, suspending characteristics, and retention of water. CMC is used in different applications like biomedical, cosmetics, pharmaceutical, food etc., because of its thickening, absorbing, stabilizing, and film-forming properties (Barbucci et al. 2000).

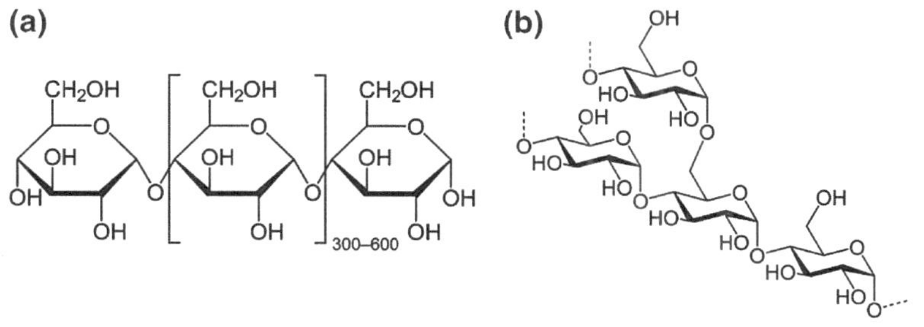

Fig. 6 Chemical structure of **a** amylose and **b** amylopectin

Derivatives of Starch

Starch is the most abundant organic compound in nature after cellulose. It is an amorphous polymeric carbohydrate consisting of anhydrous glucose units linked primarily through α-D-(1–4) glycosidic bonds. Starch is generally extracted from food products like potatoes, wheat, maize, rice, cassava, etc. Starch is composed of two types of alphaglucan, like the linear and helical, amylose, and branched amylopectin (Fig. 6), which represent approximately 98–99 % of the dry weight (Tester et al. 2004). Amylose and amylopectin differ significantly in their properties and functionality. Amylose has a high tendency to retrograde and produce tough gels and strong films while amylopectin when dispersed in water produces soft gels and weak films (Perez and Bertoft 2010). Depending on their origin, starch generally contains 20–35 % amylose and 65–80 % amylopectin by weight. The industrial utilization of native starches is limited because of inherent imperfect nature, such as water insolubility and their tendency to easily retrograde.

Pure starch is a white, tasteless, and odorless powder, which is insoluble in water. However, when heated in the presence of water, their intermolecular interactions get affected, leading to an irreversible transition, which makes them soluble in water. Cold water soluble starch can also be prepared by physical, enzymatic, or chemical treatment of native starch. Physical modification of starch improves its water solubility and reduces its particle size. The physical methods involve treating native starch granules under different temperature/moisture combinations, pressure, shear, and irradiation. Mechanical attrition to change the physical size of starch granules also comes under the category of physical modification. Chemical modification of starch is carried out by introducing different functional groups into the starch molecules, which alters the physicochemical properties of starch. Such modification makes changes in the proximate compositions, gelatinization, retrogradation, and pasting characteristics. Chemical modification is intended to facilitate intra- and intermolecular bonds at random locations in the starch granule for their stabilization. Chemical modification is generally achieved through acetylation, cationization, oxidation, acid hydrolysis, and cross-linking reactions (Ashogbon and Akintayo 2014).

Modified or unmodified starch is one of the most used biopolymer for making biodegradable or edible packaging film because of its low cost, availability, and film-forming ability. Starch-based films are transparent, flavorless, tasteless, and colorless. Due to their hydrophilic nature, the films produced from starch possess poor gas and water vapor barrier properties These films also have poor mechanical strength and hence several modifications and additives are required to improve the barrier and mechanical properties. This limits their potential to be used as a raw material for the development of food packaging materials to a large extent.

Apart from the polymers described above, there are many other polysaccharide-based water soluble polymers such as galactomannan gums, guar gums, locust bean gums, pectin, bacterial polysaccharides such as xanthan, scleroglucan, dextrin, etc., polysaccharides derived from plant sources such as Dendrobium officinale, Psyllium, etc., which can also be used for many applications (Thakur and Thakur 2014b).

1.1.2 Protein-Based Water Soluble Polymers

Proteins are naturally occurring polymers of α-amino acids joined together by peptide bonds. They are long chain molecules typically folded into globular or fibrous form. These materials are the building blocks of animal tissues while some of them are formed from vegetable sources also. Most of the protein-based polymers are water soluble even at lower temperatures. Out of several proteins, gelatin is the most important polymer in this class used for nanocomposite preparation.

Gelatin

Gelatin is a soluble protein compound obtained by partial hydrolysis of collagen, which is the most abundant, naturally occurring animal protein found in the flesh and connective tissues of animals. Collagen is composed of a triple helix, which generally consists of two identical chains and an additional chain that differs slightly in its chemical composition (Gomez-Guillen et al. 2011). Heat and chemical treatment can cleave the hydrogen and covalent bonds of collagen to destabilize the triple helix, resulting in helix-to-coil transition, and conversion into soluble gelatine. Depending on the pretreatment procedure, two types of gelatin can be produced, known commercially as type-A and type-B gelatin. Gelatin is not a single chemical entity, but a mixture of fractions composed entirely of amino acids joined by peptide linkages to form polymers varying in molecular mass from 15,000 to 400,000 (Fig. 7). Gelatin is a slightly yellow, transparent, brittle protein, which is widely used in foods, drugs, and photographic film applications (Kolodziejska et al. 2004). Gelatin has been extensively used in the manufacturing of hard and soft capsules in pharmaceutical industry, as an ingredient to improve the elasticity, consistency, and stability of foods in food industry, etc. Gelatin is also used as a gelling agent in cooking and also used for the clarification of juices.

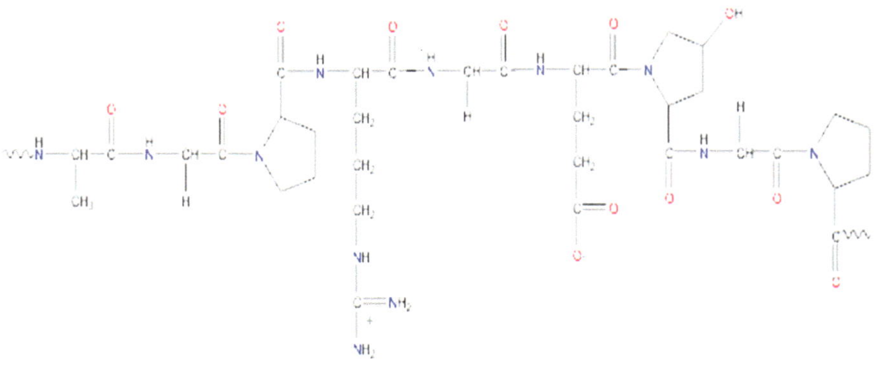

Fig. 7 Chemical structure of gelatin

1.2 Synthetic Water Soluble Polymers

Synthetic water soluble polymers are polymeric materials that dissolve, disperse, or swell in water, which are man-made polymers. They are also capable of modifying the physical properties of aqueous systems in the form of gellation, thickening, or emulsification/stabilization. These polymers usually have hydrophilic groups attached to their chain backbone. Commonly used synthetic water soluble polymers in various applications are poly vinyl alcohol (PVA), poly vinyl pyrrolidone (PVP), poly acrylic acid (PAA), poly acrylamide (PAm), etc. Among these polymers, PVA is widely used for making films suitable for packaging applications.

1.2.1 Poly vinyl Alcohol

PVA is a linear synthetic water soluble polymer with a pendant hydroxyl group (Fig. 8). Since the monomer, vinyl alcohol is not stable; PVA is generally produced by the hydrolysis of polyvinyl acetate. The synthesis process is based on the partial replacement of ester groups in the vinyl acetate with hydroxyl groups and is done in the presence of anhydrous sodium methylate or aqueous sodium hydroxide. Commercial PVA grades are available with a high degree of hydrolysis.

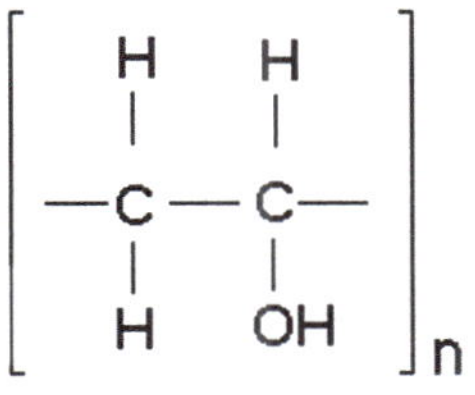

Fig. 8 Chemical structure of polyvinyl alcohol

The amount of hydroxylation determines the physical, chemical, and mechanical properties of the PVA. PVA grades with a high degree of hydrolysis exhibit low solubility in water. PVA has a melting point of 230 °C for the fully hydrolyzed grades and a thermal decomposition temperature above 250 °C. PVA is atactic; however, it exhibits crystallinity as the hydroxyl groups are small enough to fit into the lattice without disrupting it. PVA films cast from water solutions result in a transparent, tough, tear-resistant and puncture-resistant films. It provides a unique combination of water solubility, high barrier to oxygen, and resistance to oil, grease, and solvents.

PVA is biocompatible, easy to process, hydrophilic, and has good physical and chemical properties. PVA exhibits excellent barrier properties against oils and fats, aromas, and small gas molecules [48]. In addition to its biocompatibility, PVA is also an environment-friendly polymer as it can be biodegraded into CO_2 and water by the action of microorganisms. PVA is widely used in pharmaceutical and ophthalmic formulations, food, adhesive, textile, paper and packaging industries. PVA is used as an industrial polymer mainly because of its low environmental impact due to its aqueous solubility, high chemical resistance, and biodegradability (Baker et al. 2012).

1.2.2 Polyacrylic Acid

PAA is another biodegradable water soluble polymer obtained by the polymerization of acrylic acid monomer with numerous industrial applications. PAA is well known for its water absorbing capability and is often known as superabsorbent. PAA can absorb large quantity of water due to the presence of carboxylic acid groups on the chain backbone (Fig. 9a), which strongly associate with water molecules. Carboxylic groups are readily ioniziable and sensitive to the effects of pH and ionic strength. Thus, the equilibrium swelling of PAA is often affected by the pH of the solution in which they are swelling. The unique property of PAA is due to that it exists as a liquid at pH 5 and as a gel at pH 7. PAA is also used in water treatment as a metal ion scavenger and in the treatment of metal surfaces prior to coating. Poly (acrylic acid) is also a pH-sensitive synthetic polymer used in the area of the site-specific drug delivery to specific regions of the gastrointestinal tract (Yang 2008). PAA and its derivatives have been shown to have good mucoadhesive properties and is used in various mucoadhesive drug delivery systems.

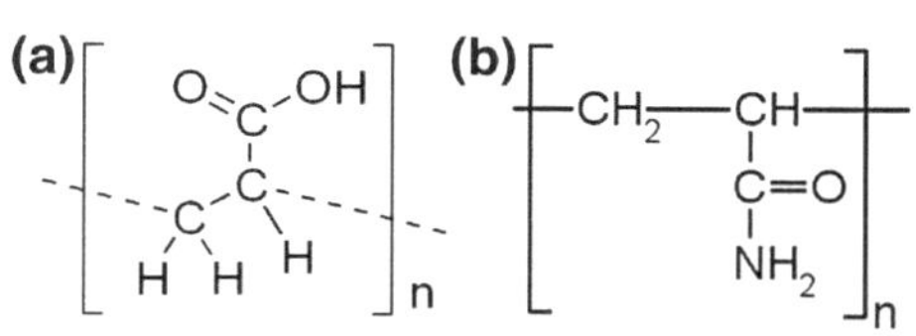

Fig. 9 Chemical structure of **a** polyacrylic acid and **b** polyacrylamide

1.2.3 Polyacrylamide

PAAm is synthesized from the monomer acrylamide (Fig. 9b), which can be synthesized as a simple linear-chain structure or in the cross-linked form using a suitable bifunctional crosslinking agent. In the linear-chain form, it can be used as a thickener and suspending agent, while in the gel form it is used as a highly water-absorbent polymer. Polyacrylamide gels having different pore sizes can be manufactured by adjusting the total acrylamide concentration for the size fractionation of a variety of proteins. PAAm is easy to synthesize, economical, and stable over a wide range of pH. It has been widely used for a range of applications ranging from microanalysis to macrofractionation for proteins, nucleic acid, and other biomolecules. High molecular weight of PAAm is used in various industries like minerals and water treatment. A major application of PAAm is its use as an efficient flocculating agent (Tripathy and Singh 2000). PAAm can also be used for encapsulation and as carriers for delivery of drugs and bioactive molecules (Sairam et al. 2006).

2 Need for Water Soluble Polymers with Better Properties

Water soluble polymers have created a lot of interest among industry and academia, as they are easily available, cheap, noninflammable, and more environmentally friendly than any other polymers. These polymers are presently used as stabilizers, thickeners, drug delivery systems, protective colloids, dispersants, flocculants, and different applications in oil recovery. However, one of the most important fields of application is the replacement of nondegradable and nonrenewable plastic packaging using biopolymers/water soluble polymers derived from renewable sources. The indiscriminate use of nondegradable plastic materials contributed lot of negative impact on our ecosystems. The environmental impact of ever accumulating plastic-based wastes has raised global concerns because the disposal methods for these materials are limited. The use of these environmentally friendly plastic materials can solve the waste disposal problem to certain extent. Several water soluble polymers like those derived from naturally occurring proteins, cellulose, starches, and other polysaccharides and those synthesized chemically from naturally derived monomers can be used for this application. Commercialization of some of these types of environmentally friendly packaging materials has already been started. However, most of these films cannot meet the requirements of cost-effective packaging films with mechanical and barrier properties similar to those of conventional plastics. Some of these polymers are extremely brittle and some of them are not having sufficient mechanical strength. Plasticizers such as glycerol and polyethylene glycol (PEG) were generally used to impart flexibility to these polymer films; however, their use considerably reduces the tensile strength and modulus.

Another area where water soluble polymers are having lots of potential is in the field of edible food packaging (Debeaufort et al. 1998). An edible film is a thin continuous layer of edible material formed or placed on food or food components, so that it forms an integral part of the food, which can be eaten along with the food. However, their widespread use as edible film is limited as they are having poor mechanical, gas/moisture barrier properties and strong hydrophilic nature. These drawbacks have to be rectified as these polymers are expected to exhibit physicochemical attributes analogous to the existing petrochemical polymers. Therefore, considerable research is needed to develop techniques for improving mechanical and barrier properties of water soluble polymer-based packaging materials. Developing new polymer nanocomposites using water soluble polymers is a promising option in improving the mechanical and barrier properties of these polymers.

3 Polymer Nanocomposites

A polymer nanocomposite is a multiphase material in which a polymer matrix is combined with nanometer-sized materials having one, two, or three dimensions smaller than 100 nm, or structures having nanoscale repeat distances between the different phases that make up the bulk polymer material (Thostenson et al. 2005; Lin et al. 2011a, b). In a polymer nanocomposite, the polymer matrix functions as a concentrated component that enables a phase continuum, while nanomaterials are deliberately added to induce an enhanced performance in the functional properties such as mechanical, barrier, electrical, triggered biodegradability, moisture resistance, and thermal properties of the polymer (Thakur et al. 2012, 2014c, d, e).

The interaction between nanometer-sized fillers with the polymer matrix enables them to act as molecular bridges in the polymer matrix, which can result in enhanced mechanical properties as compared to conventional microcomposites. These nanocomposites often exhibit remarkable improvement in material properties even though only small quantities of nanomaterials are added to the polymer matrix. Very large improvement in tensile strength, modulus, toughness, stiffness, barrier properties, thermal conductivity, etc., can be obtained by the addition of small amounts of nanomaterials in the polymer matrix (Maniar 2004). Since nanomaterials are used in small quantities, the resultant nanocomposites will be light in weight but exhibits excellent performance. The advantages of nanoscale particle incorporation can lead to a wide range of application possibilities where the analogous microcomposites would not yield the necessary property upgradation. The areas where novel polymer nanocomposites with improved properties find lots of potential include barrier properties for food packaging films, membrane separation applications, UV screens, inflammable materials, optics, electrical devices, photoconductors, biomedical applications, aerospace applications, biosensors, fuel cells, solar cells, and flexible energy storage devices (Hussain et al. 2006).

Polymer nanocomposites are manufactured by embedding nanomaterials in the matrix of common polymers to enhance certain properties and characteristics. Nanomaterials can result in significant mechanical and barrier properties advancements due to their ability to modify the properties of polymers by the formation of nanocomposites. At nanodimensions, quantum effects become more relevant in determining the properties of materials, leading to novel optical, electrical, barrier, and mechanical behaviors (Wautelet 2001). The nanomaterials are almost free from defects and their use as reinforcing fillers in polymer matrix results in overcoming several shortcomings of conventional micrometer-sized particles. Also by properly dispersing the nanoparticles in a polymer matrix, a very large matrix-filler interfacial area can be obtained which leads to changes in the molecular mobility, crystallinity, electrical, thermal, and mechanical properties (Schadler 2007). This type of property enhancements can also lead to the possibility of designing and creating novel materials and structures with unexpected physical and mechanical properties.

4 Mechanism of Polymer Reinforcement by Nanomaterials

Incorporation of nanomaterials having different size and shapes such as spheres, fibers, whiskers, or plates into polymer matrices enhances the mechanical properties like tensile strength, modulus, stiffness, and impact strength significantly. Also other physical properties like barrier, optical, thermal resistance, nonflammability, etc., can be improved by the introduction of nanomaterials. Nanomaterials exhibit some unique properties, which are completely different from their corresponding bulk materials. There are mainly three reasons for the improved performance of polymer nanocomposites. The first reason is the increased relative surface area (aspect ratios) and its associated quantum effects exhibited by nanoparticles. As the size of a particle decreases, the proportion of the number of atoms present on the surface will be more as compared to the atoms present in the bulk. The surface atoms exhibit very different properties compared to that of bulk atoms and hence, the properties of nanomaterials are determined by the properties of surface atoms, rather than that of bulk atoms. Thus the nanoparticles that possess a large surface area per unit mass exhibits, totally different quantum mechanical effects. As the size of the material reaches to nanometer size, most of the properties like mechanical, catalytic, electrical, optical properties, etc., can change.

The second reason for the reinforcing efficiency of the nanomaterials is their chemical characteristics like polymer-filler interactions, possibility of hydrogen bond formation and chain entanglement. The properties of polymer nanocomposites are greatly influenced by the size scale of its component phases and the degree of mixing between the two phases. Depending on the nature of the nanomaterials and the polymers as well as their method of preparation, the properties of polymer nanocomposites will vary (Park et al. 2001). In the case of polymer-layered silicate (PLS) nanocomposites, the entire properties depend on the physical state at which the silicate layers exist in the polymer. When the polymer chains do not penetrate between the silicate layers, a phase-separated nanocomposite will obtain and their properties will be in the same range as conventional microcomposites. If some amount of extended polymer chains were penetrated between the silicate layers, then a well-ordered multilayer morphology comprising alternating polymeric and inorganic layers will result, which gives good reinforcing effects. If the layered silicates exist in an exfoliated or delaminated and uniformly dispersed state within the polymer matrix, then maximum reinforcement can be expected (Alexandre et al. 2000). Similarly, in the case of fibrous or particulate-shaped nanomaterials, the dispersion of the nanoparticle and adhesion at the particle–matrix interface play a crucial role in determining the mechanical properties of the nanocomposites. In the absence of proper dispersion, then the nanocomposite will not exhibit properties better than that of conventional composites (Gorga et al. 2004). Also by optimizing the interfacial adhesion between the nanoparticle and the polymer matrix, one can tailor the properties of the overall nanocomposite. Greater adhesion between the polymer and nanomaterial will lead to an immobilization of polymer chains at the surface of nanomaterials, which result in better load transfer as it was observed in the case of carbon nanotubes (CNTs) filled polymer composites (Eitan et al. 2006). The segmental immobilization can happen at the interphase between polymer and nanoreinforcing material, leading to the formation of an interphase with properties different from that of polymer and nanomaterials. The immobilization of the polymer chains at the interphase causes mechanical stiffening of this interphase region which acts as the additional reinforcing component. The availability of very high surface area of nanomaterials results in the formation of a large amount of interphase regions in nanocomposites as compared to conventional polymer composites. The immobilized polymer exhibit different mechanical properties compared to the bulk polymer because of the restricted mobility introduced by the polymer-filler interaction. The segmental immobilization mechanism depends on two factors the size of the nanomaterials and their degree of interaction with the polymer chain.

The third reason for polymer reinforcement by nanomaterials is that the incorporation of rigid nanoscale materials in a polymer matrix can enhance the elastic modulus of nanocomposite considerably. This type of enhancement is mainly affected by three different factors like, (i) easy stress transfer from the polymer matrix to nanoparticles, due to the increased surface area available for the nanomaterials, (ii) the partial substitution of soft polymer matrix with comparatively rigid nanofillers (Kalfus et al. 2008) and (iii) the extent of spatial arrangement and hierarchial ordering of the nanoparticles (Yuan et al. 2006). Stress transfer between the polymer and the reinforcing nanomaterials depends on the aspect ratio of the filler, its orientation toward the applied stress and the level of adhesion between polymer and the nanomaterials. Hence, this stress transfer mechanism is a shape-dependant phenomenon rather than size dependant. For example, rod-like nanomaterials like carbon nanotubes (CNTs), cellulose nanocrystals (CNs), etc., having high aspect ratio are known to reinforce polymers better than flakes- or disk-shaped nanomaterials. The second mechanism that defines the reason for increased modulus, involves the partial substitution of softer polymer material with rigid nanofillers, which are uniformly distributed throughout the polymer matrix. This phenomenon is not dependant on the size of the nanomaterials. The third mechanism deals with the approaches to provide spatial and orientational control of the hierarchical morphology of nanomaterials within the polymer matrix, which can induce certain desired properties. The extreme high aspect ratio of certain nanomaterials like exfoliated clays, CNTs, CNs, etc., makes the hierarchy of morphology more important. The hierarchical structures can be induced in the nanocomposites by tailor-made nanoparticle orientation and deformation.

5 Water Soluble Polymer-Based Nanocomposites

Water soluble polymer-based nanocomposites belong to a fascinating interdisciplinary area that brings together environment friendly polymers with nanomaterials. They are formed by the combination of natural/tailor-made synthetic water soluble polymers and nanomaterials. Similar to conventional nanocomposites, these nanocomposite materials also exhibit improved structural and functional properties which make them suitable for different applications. The inherent properties of these polymers like biocompatibility and biodegradability open new application windows for these hybrid materials in the field of environment friendly/green nanocomposite materials. Research on these new classes of nanocomposites can be regarded as a new interdisciplinary field closely related to bio-inspired materials and systems that mimick naturally available materials. The development of novel nanocomposites represents a promising research area of taking advantages of the synergistic assembly of biopolymers with nanometer-sized materials. The extraordinary versatility in the development of these new materials originates from the large selection of biopolymers and fillers available to researchers. Several water soluble polymers such as PVA, derivatives of cellulose like HPMC, CMC, etc., certain other polysaccharides, water soluble proteins like gelatin, etc., are generally used for the development of polymer nanocomposites. Typical nanomaterials used for reinforcing water soluble polymers include, layered silicates, CNs, metallic nanoparticles, CNTs, cellulose nanofibers, and inorganic nanowires. Even though large amount of research is focussed on the use of nonrenewable nanomaterials, increased environmental concerns had created lot of interest on the use of bio-based and renewable materials for such applications. Among other nanomaterials, CNs have gained large amount of attention in the material technologists, due to their unique physical and chemical properties, their inherent renewability and sustainability in addition to their abundance.

6 Cellulose Nanocrystals

Cellulose, the most abundant biomass material in nature is known to occur in a wide variety of living species from the worlds of plants, animals, and bacteria as well as some amoebas. Cellulose is a fibrous, tough, and water insoluble polymer that plays an essential role in maintaining the structure of plant cell walls. It is a high molecular weight linear homopolymer of 1,4-anhydro-*D*-glucose units in which every unit is corkscrewed at 180° with respect to its neighbors. The repeating unit of cellulose is a dimer of glucose, known as cellobiose. The monomers are linked together by condensation reaction so that the sugar rings are joined by glycosidic oxygen bridges. In nature, cellulose chains have a degree of polymerization (DP) of approximately 10,000 glucose units in wood cellulose and 15,000 in native cellulose cotton. This structural material is naturally organized as microfibrils linked together to form cellulose fibers. Each monomer bears three hydroxyl groups. It is therefore obvious that these hydroxyl groups and their ability to form hydrogen bonds play a major role in directing the crystalline packing and also governing the physical properties of cellulose (O'Sullivan 1997). Cellulose is occurring naturally as assemblies of individual chain-forming fibers, as it is synthesized as individual molecules, which undergo spinning in a hierarchical order at the site of biosynthesis. Normally around 36 individual cellulose molecules assemble together into elementary fibrils, which are packed into larger units called microfibrils, and these are in turn assembled into the normal cellulose fibers. During biosynthesis, cellulose chains are aggregated in microfibrils with a diameter range (2–20 nm), depending on the source, as they found in different packing arrangements as dictated by the biosynthesis conditions. Biosynthesis of cellulose microfibrils involves a multistep process, which are highly specific to the organism producing the cellulose. Variations in this process can alter the morphology, aspect ratio, crystallinity, and crystal structure of the resulting microfibrils.

6.1 Sources of Cellulose

Different types of plants are the main sources of cellulose; however, certain bacteria and sea animals are also capable of producing cellulose fibers. A brief description about the main sources is as follows:

6.1.1 Plants

Plants are very attractive and potential sources of cellulose primarily because they are abundant and relatively cheap to harvest. Cellulose can be extracted from lignocellulosic fibers, which are available all over the world. A wide variety of plant materials like cotton, ramie, sisal, flax, wheat straw, tubers, sugar beet, soya bean, etc., are known for cellulose microfibril production. Wood is another main important source of extracting cellulose fibrils, where cellulose microfibrils were reinforced by intracellular amorphous materials made of hemicelluloses, lignin, resin, etc. Extraction of pure cellulose microfibrils from these lignocelluloseic materials involves chemical treatments such as alkali extraction and bleaching.

6.1.2 Tunicates

Tunicates are a family of sea animals which is known to produce cellulose microfibrils in large amount. These animals are having a thick, leathery mantle, which is a good source of cellulose. Large numbers of tunicate species are available in nature and the properties of cellulose obtained can vary from species to species. The cellulose microfibril structure and properties obtained from different species are often comparable, but small differences in the cellulose microfibril formation process may affect the final properties of microfibrils.

6.1.3 Algae

Several types of algae like red, green, and yellow are known to produce cellulose in their cell wall. Some of the most commonly used algae species are *Micrasterias denticulata*, valonia, cladophora, boergesenia, etc. The difference in the biosynthesis process happens in different species; cellulose microfibrils obtained may also differ from each other. Out of these different species, green algae are the most preferred species for cellulose extraction (Moon et al. 2011).

6.1.4 Bacteria

Certain microorganisms like *Glucanoacetobacter xylinum* mainly occurring as a contaminant in vinegar fermentation is well known for their ability to produce cellulose by utilizing a large variety of nitrogen and carbon sources (George et al. 2005a). Under favorable culturing conditions, this bacterium secretes cellulose microfibrils in the form of thick pellicles. The advantages of bacterial cellulose are that they are extremely pure and have better properties than plant-derived cellulose. Bacterial cellulose is unique when compared to cellulose obtained from other sources because microfibril biosynthesis process can be altered by changing the culturing conditions. This type of engineered biosynthesis can occur in different stages like cellulose culturing and pellicle management. The biosynthesis of bacterial cellulose occurs in a bioreactor, where the bacteria secrete cellulose microfibrils. Variations in the culture conditions like bacterium strain, nutrients, temperature, pH, rate of agitation, etc., can alter the microfibril morphology and the polymer network formation (Jonas et al. 1998). The shape of bacterial cellulose pellicle can also be altered by changing the shape of the bioreactor. Pellicle management refers to the mechanical and chemical process imparted on the pellicle to remove impurities from the pellicles. Dilute alkaline solutions of NaOH, KOH, etc., are capable of hydrolyzing and removing impurities present in the cellulose pellicle. After alkali treatment and washing in distilled water, the cellulose mass can be dried and processed into pure cellulose membranes (George et al. 2005b). Mechanical defibrillation treatments can also be used to break up to the entangled cellulose networks to obtain pure microfibrils.

6.2 Isolation of Cellulose Nanocrystals

CNs can be extracted from raw cellulose by following two main steps such as the purification of the source material to obtain cellulose microfibrils and the conversion of these microfibrils into nanocrystalline components. The purification and homogenization steps can vary depending on the cellulose source. For wood and plants, the most important step is the removal of the matrix materials like hemicellulose and lignin. Several mechanical processes like high-pressure homogenizers, cryocrushing, high-intensity ultrasonic treatments, and microfluidization techniques were utilized for the extraction of cellulose microfibrils. Generally, these mechanical processes produce high shear that leads to transverse cleavage of cellulose fibers along the longitudinal axis, resulting in the extraction of cellulose microfibrils. Each cellulose microfibril is devoid of chain folding and can be considered as a string of cellulose crystals, linked along the microfibril by disordered or paracrystalline regions (de Souza Lima et al. 2004).

The second step involves the conversion of cellulose microfibrils into nanocrystalline components, which can be performed by either enzymatic or acid hydrolysis. Strong acid hydrolysis is generally employed for the removal of the paracrystalline domains that are regularly distributed along the microfibrils to synthesize rod-like cellulose nanomaterials. Colloidal suspensions of cellulose were first synthesized by controlled sulfuric acid-catalyzed degradation of cellulose fibers in 1951 by Ranbi (1951). Later it was observed that degradation induced by boiling cellulose fibers in acidic solution reached a limit after a certain time of treatment. Transmission electron microscopy (TEM) images of dried suspensions revealed the presence of needle-shaped aggregated cellulose particles, while electron diffraction studies of these aggregates confirmed that they possess the same crystalline structure as the original cellulose fibers. Hydrochloric acid-assisted synthesis of cellulose particles followed by ultrasound sonication later led to the commercialization of microcrystalline cellulose (MCC), which is currently used for various applications (Battista et al. 1962). The cleavage of cellulose chains due to acid hydrolysis is mainly attributed to the differences in the kinetics of hydrolysis between amorphous and crystalline domains. Acid hydrolysis results in a rapid decrease in the degree of polymerization until it reaches a cut-off level called level-off degree of polymerization (LODP). The value of LODP depends on the cellulose origin as different LODP values were reported for different cellulose. Hydrolyzed cotton has a LODP of 250–300 for ramie fibers, 140–200 for bleached wood pulp, and up to 6000 for highly crystalline Valonia cellulose (Battista et al. 1956). A wide distribution of molecular weight and degree of polymerization exists in CNs obtained from animal sources. CNs obtained by the acid hydrolysis of cellulose from bacterial, tunicate, or valonia exhibits a high polydispersity in the molecular weight, without any evidence of the LODP, probably due to the absence of the regular distribution of the amorphous domains.

Prolonged hydrolysis will contribute to a further reduction in LODP or molecular weight. Hence, by stopping the acid hydrolysis after attaining LODP can result in the formation of CNs (Hakansson et al. 2005). This hypothesis was based on the assumption that disordered or paracrystalline cellulose regions are regularly distributed along the microfibrils and are susceptible to easy degradation compared to highly crystalline domains. The highly crystalline domains can withstand acid hydrolysis conditions for a comparatively longer time than the amorphous fractions and hence, they can be easily separated from the acid medium. The processing conditions such as time, temperature, etc., employed during the hydrolysis process are very critical in controlling the final properties of CNs. If the hydrolysis time employed is inadequate, then incomplete removal of amorphous fractions can occur, which results in reduction in crystallinity and change in particle morphology. Similarly, increasing the reaction time beyond a point can lead to depolymerization of crystalline cellulose, which decreases the aspect ratio of the nanocrystals and sometimes even lead to the formation of much smaller spherical particle.

6.3 Dimensions of Cellulose Nanocrystals

CNs are stiff rod-like particles consisting of cellulose chain segments in a perfect crystalline structure. Geometric dimensions like length, width, etc., may vary depending on the origin of cellulose microfibrils and acid hydrolysis conditions like time, temperature, purity, etc. The length of the rod-shaped particles can vary from tens of nanometers to several micrometers, while width ranges from 3 to 50 nm. CNs obtained from wood generally have a diameter of 3–5 nm and a length of 100–300 nm, likewise cotton gives CNs having a diameter of 5–10 nm and a length of 70–300 nm. Sea plants like Valonia produces CNs of 20 nm diameter and 1000–2000 nm length, while sea animals like tunicates produce nanocrystals with a dimension of 10–20 nm diameter and 500–1000 nm length (Lagerwall et al. 2014). Cellulose crystals obtained from bacterial cellulose by sulfuric acid hydrolysis is reported to have a diameter of 10–50 nm and a length of 100–1000 nm (Araki et al. 2001). The geometrical aspect ratio of CNs, defined as the length to diameter (L/D) ratio is very critical in determining phase separation and reinforcing capability of CNs. CNs having a high aspect ratio generally will not show phase separation and their reinforcing ability is very high. The aspect ratio is also a key factor in determining the percolation threshold value that controls the mechanical performances of polymer nanocomposites.

6.4 Properties of Cellulose Nanocrystals

The properties of CNs can be broadly described under three categories namely mechanical properties, liquid crystalline nature, and rheological properties. They are briefly discussed as follows;

6.4.1 Mechanical Properties

The dimensions in nanometer scale and extremely high mechanical properties make CNs ideal candidates to improve the mechanical properties of polymeric materials. The theoretical tensile strength of CNs was found to be more than that of steel and Kevlar. The tensile strength of CN is 7.5–7.7 GPa, while that of steel wire and Kevlar-49 is 4.1 and 3.5 GPa, respectively (Moon et al. 2011). Determining the mechanical properties of CNs are extremely challenging due to the small particle size and the nonavailability of metrology techniques to characterize such small particles. Additionally, several other factors can also influence the measured mechanical properties like crystal structure, percentage crystallinity, defects, etc. Indirect techniques like atomic force microscopy (AFM) and Raman spectroscopy were employed to estimate the modulus of CNs. AFM is capable of accurately applying force in minute levels and to measure the resulting deformation, which makes it a very useful tool for the mechanical testing of nanosize materials such as CNs. The tunicate CN obtained by sulfuric acid hydrolysis was deposited on a specially designed silicon wafer with grooves of 227 nm in width, and a three-point bending test was applied using an AFM cantilever (Iwamoto et al. 2009). Using this technique, the modulus of individual CN was measured in both axial and transverse directions. The axial elastic modulus of single nanocrystal prepared by acid hydrolysis was found to be around 150 GPa. Another technique was also developed using AFM to determine the transverse elastic modulus of CNs by comparing the experimental force–distance curves with three-dimensional finite elemental calculations (Lahiji et al. 2010). This experiment proved that the transverse modulus of an individual CN is in the range of 18–50 GPa. The deformation micromechanics of tunicate cellulose whiskers using Raman spectroscopy were useful in measuring the elastic modulus of nanocrystals. In this method, a dispersed sample of CN was deformed using a four-point bending test, and a shift in the characteristic band at 1095 cm^{-1} was used to get an indication about the stress in the material. Very negligible intensity change of the Raman band located at 1095 cm^{-1} was observed for samples oriented in parallel and perpendicular to the polarization direction of the laser. This suggests that the tunicate sample is a two-dimensional in-plane random network of fibers. The calculated modulus of tunicate nanocrystal using Raman shift and calibrations with strain from other materials was found around 143 GPa (Strucova et al. 2005). The lack of Raman band broadening in the spectra also revealed the fact that pure crystalline deformation was occurring without affecting crystalline/amorphous ratios. A molecular mechanics approach, using computer simulation and an empirical force field, was used to predict the modulus of a highly oriented chain of the material, which is in agreement with the experimental data.

6.4.2 Liquid Crystalline Nature of Cellulose Nanocrystals

It is well known that rod-like particles, at a critical concentration, spontaneously form ordered structures leading to the formation of a nematic phase. Rod-like CNs are also known to exhibit liquid crystalline properties. Their stiffness, aspect ratio, and the ability to align under certain conditions make CNs to exhibit nematic liquid crystalline behavior. CNs obtained by the hydrolysis of sulfuric acid often possess a negatively charged surface, which promotes a perfectly uniform dispersion of nanocrystals in water due to electrostatic repulsions. Even though the interactions between nanocrystals are strong, sulfuric acid hydrolysis makes them readily dispersible and this leads to the development of lyotropic behavior. The cellulose crystallites were having a helical twist similar to a screw, which induces crystal suspensions to attain a helical twist normal to the long axis of the rod. Thus, they can form a chiral, nematic, or cholesteric phase of stacked planes dependent on the concentration. Various factors such as size, charge, shape, dispersity, electrolyte, and external stimuli can affect the liquid crystallinity of CNs (Pan et al. 2010). Liquid crystallinity of nanocrystals coupled with the birefringent nature, leads to interesting optical phenomena.

6.4.3 Rheological Properties

CNs in dilute solutions were exhibiting shear thinning behavior and at higher concentrations they exhibit anomalous behavior. The main reason for such behavior is that the nanocrystals tend to align due to their rod-like shape at a critical shear rate. As the shear rate reaches a point, the chirality of nanocrystal suspension breaks down in favor of a simple nematic structure. Also the time constant of relaxation depends on the aspect ratio and nanocrystals with higher aspect ratios stays aligned for longer times even after shear. The type of acid used for hydrolysis can also control the rheological properties of nanocrystals. Sulfuric acid treated crystals showed some shear thinning which is independent of time, while HCl derived crystals showed much higher shear thinning behavior, thixotropy at higher concentrations and anti-thixotropy at lower concentrations (Araki et al. 1999).

7 Cellulose Nanocrystal Reinforced Water Soluble Polymer Nanocomposites

Polymer nanocomposites are a radical alternate to conventional polymer composites, where large amount of fillers are added to improve the properties. For polymer composite applications, the use of natural fibers are preferred to efficiently reduce the dependence on petrochemical-based plastics (Thakur et al. 2014d). Natural fibers obtained from various plant sources as such or in the form of extracted cellulose have been frequently used for this application (Singa et al. 2009). Cellulose in nanodimensions generated from cellulose fibers has much higher mechanical properties than those of natural fibers. Hence CNs have attracted a great deal of interest in the polymer nanocomposite field. Due to their nanodimensions, high surface area, low density, ability to functionalize, and sufficient strength, they proved to be a better reinforcing material than conventional fibers. Like any other multiphase material, the properties of nanocomposite depend on the morphological aspects and their interfacial interactions. There are four different factors that can affect the performance of CNs-based water soluble polymer nanocomposites (Miao et al. 2013). The first, and most critical, is the compatibility of CNs with polymer matrix. This is essential to allow uniform dispersion of reinforcing element into the matrix. The main challenge in attaining excellent performance lies in attaining homogenous dispersion of nanocrystals within the polymer matrix by avoiding the aggregation of nanocrystals. In the case of water soluble polymers, it is comparatively easy as both the phases are hydrophilic in nature. Also in some cases the formation of a percolated network of nanocrystals within the polymer matrix enhances the rigidity of polymers. The second factor is the molecular structure of the matrix, which influences the interaction between matrix and CN and their interfacial properties. This is also important in obtaining a good matrix-filler interaction. Here also since both the polymer matrix and reinforcing components are hydrophilic in nature, their interactions can be reasonably good. The third is the aspect ratio of CN particles, which is determined by the origin of the cellulose source and the manufacturing conditions. Since the reinforcing filler used is possessing nanodimensions, the reinforcing effect is better. This is the reason why CN-reinforced nanocomposites exhibit better properties than cellulose fibril/microfibril-reinforced counter parts. Typically, a higher aspect ratio gives more reinforcement potential, assuming aggregation does not occur. Fourth factor is the method of polymer nanocomposite fabrication. Solvent intercalation is the most widely used preparation method for these types of polymers. It is having both advantages and disadvantages. Ease of preparation, control over the nanocrystal aggregation, better dispersion, and less damage to the nanocrystals, cost effectiveness, etc., are the advantages while their inefficiency for large-scale production is a limitation. Several water soluble polymers-based nanocomposites have been prepared by solvent intercalation method and their details are briefly highlighted below.

7.1 Carragenan-Based Nanocomposites

Carrageenan is a polysaccharide-based water soluble polymer with a linear chain of partially sulfated galactans extracted from red seaweed and extensively used in foods, cosmetics, and pharmaceutical applications. Carrageenan is preferred for several applications because of its advantages like transparency, strength, gelling ability, and film-forming properties, but has some limitation like high moisture sensitivity, poor barrier properties, and lower elongation. Carrageenan is widely used as edible films and coatings in various fields of the food industry such as fresh and frozen meat, poultry, and fish to prevent dehydration, sausage casings, granulation-coated powders, dry solid foods, and oily foods (Sanchez-Garcia et al. 2010a). However, these films are characterized by their high moisture and gas permeability, which limits their widespread applications. To improve some of these limitations, CNs having cross-sectional dimension of 5 nm and length in the range of 25–50 nm was used (Sanchez-Garcia et al. 2010b). Nanocomposite film samples of carrageenan containing 1, 3, and 5 wt% of CN were prepared by solution casting technique, using water as a solvent. CN solution was mixed in a homogenizer for 2 min and then added with the carrageenan at ambient temperature for 30 min and subsequently, cast onto petri dishes to generate films of around 50 μm thickness at room temperature conditions.

Morphological analysis using TEM and polarized optical microscopy confirmed good dispersion of the nanocrystals in the polymer matrix, especially at low filler contents. Increasing the filler content above 3 wt% resulted in increasing the number of agglomerates due to the normal trend of CNs to self aggregate via hydrogen bonding. The main conclusion of the study was that the addition of just 3 wt% of CNs in Carrageenan resulted in around 71 % reduction in water vapor permeability. This type of large reduction in permeability is ascribed to a filler-induced water solubility reduction rather than a diffusion-driven tortuosity effect and it opens the possibility of enhancing the water barrier and resistance of carrageenan and subsequently to use these materials for edible food packaging and coating applications. As compared to other inorganic reinforcing fillers, CNs have additional advantages like environmental friendliness, low density, ease of recycling, and comparatively easy processability.

7.2 Alginate-Based Nanocomposites

Sodium salt of alginate is a water soluble polymer extracted from brown seaweed and is the only polysaccharide that naturally contains carboxyl groups in each constituent residue and possesses various functionalities. Its excellent properties such as biocompatibility, nontoxicity, and biodegradability, have led to its use in many areas such as biodegradable or edible food packaging films, pharmaceutical additives, enzyme carrier, and tissue engineering materials, etc. However, poor mechanical strength, gas barrier properties, and weak water resistance limit its application particularly in higher relative humidities. CNs were helpful in overcoming these limitations. Abdollahi et al. reported the use of CNs obtained from sulfuric acid hydrolysis into alginate biopolymer using solution casting method (Abdollahi et al. 2013a). Nanocomposite films were prepared by adding different amounts of the CNs to an alginate solution at various concentrations such as 1, 3, 5, and 10 % (w/w) and sonicated for 5 min. The mixture was stirred for 1 h and homogenized at 10,000 rpm for 5 min at room temperature. The nanocomposite solution was degassed under vacuum to remove the bubbles and the films were prepared by casting onto petri dishes and oven dried at 40 °C.

The effect of nanocrystal loading on the microstructural, physicomechanical, and optical properties of alginate nanocomposites was analyzed (Abdollahi et al. 2013a). The results proved that water solubility and water vapor permeability of these nanocomposites decreased by 40 and 17 %, respectively, upon increasing the CN content to 10 %. High degree of crystallinity coupled with the strong hydrogen bonding tendency with the alginate matrix and the tortuous path introduced by the nanocrystals reduced the water sensitivity of the nanocomposites. The tensile strength of these films increased from 18.03 to 22.4 MPa with 5 wt% of nanocrystal. However, further increase in the content of nanocrystals yielded negative results due to the partial agglomeration of the nanocrystals. These results demonstrated that CNs have good potential as reinforcing filler for overcoming the limitations of alginate film in food packaging applications. In another study, these authors also proved that CNs are more suitable to make a fully renewable and eco-friendly alginate nanocomposites compared to inorganic nanoclays (Abdollahi et al. 2013b).

In another report, nanocrystalline cellulose (NCC) obtained from a commercial bleached softwood kraft pulp, reinforced alginate nanocomposite films were reported (Huq et al. 2012). The NCC content in the matrix was varied from 1–8 % and it was observed that 5 wt% NCC containing films exhibit the highest tensile strength which was increased by 37 %. Addition of NCC (5 wt%) reduced water vapor permeability of the nanocomposite by 31 %. An increase in hydrogen bonding interactions between alginate and NCC were observed in Fourier Transform Infrared Spectroscopy (FTIR) analysis. The X-ray diffraction studies confirmed the formation of additional crystalline peaks in alginate films in the presence of the NCC, which may be an indication of the formation of a percolated nanocomposite structure. Thermal stability of alginate-based nanocomposite films was improved after incorporation of the NCC, which also indicated the strong interactions between NCC and alginate matrix. In this study it was also observed that a small amount (3–5 %) of CN can have a significant impact on the improvement of physicochemical and thermal properties of alginate-based films.

7.3 Nanocomposites Based on Cellulose Derivatives

CN containing nanocomposites based on water soluble cellulose derivatives falls in a new class of nanocomposites commonly known as all-cellulose nanocomposites. Here both the matrix and reinforcing phases are made up of cellulose even though they exhibit different morphological and structural characteristics. Water soluble cellulose derivatives were extensively used for the fabrication of CN containing nanocomposites. Water soluble MC-based biodegradable nanocomposite films containing cellulose nanofibers were reported (Khan et al. 2010). In this study, the MC solution (1 % w/w) was prepared separately using water with continuous stirring and aqueous CN (0.1–1 %) solution was mixed into it followed by sonication for 30 min at room temperature. Then 0.5 % vegetable oil, 0.25 % glycerol, and 0.025 % Tween 80 were also added into the MC solution. The mixture was then homogenized and films were then cast onto petri dishes and allowed to dry at room temperature. It was found that CN contributed to improve the mechanical and barrier properties of MC-based films. Water vapor permeability of the films containing NC was reduced by almost 25 % as compared to control, which indicated the enhanced moisture barrier properties. FTIR analysis elucidated the molecular interactions between MC and CN and helped to explain the improvements obtained in properties of these films.

In another interesting report, bacterial cellulose nanofibrils coated with HEC were developed by growing cultures of *Acetobacter aceti* in the presence of HEC (Zhou et al. 2009). These fibrils are made up of an assembled network of compartmentalized bacterial cellulose (BC) fibrils coated with a nanoscale coating of HEC. These HEC coated fibrils can self assemble to form loose bundles of fibril aggregates, which can be easily disintegrated by homogenization. Nanocomposite films prepared by press drying of the water suspension of the cellulose nanofibers, exhibited considerable improvement in tensile strength. The tensile strength of these nanocomposite films was 20 % higher than that of pure bacterial cellulose and wood cellulose nanopapers and 60 % higher than that of the conventional BC/HEC blends. These films are unique in performance due to their cellulose nanostructure, which is coated with hydrated HEC, an amorphous hydrophilic polysaccharide, which confers additional mechanical strength to the nanocomposite films owing to the use of isolated individual fibrils. Unmodified cellulose nanofibrils undergo fracture easily, while for the HEC-coated nanofibrils, the catastrophic fracture is delayed by the crack-deflecting function of nanocoating. This synthesis approach represents a low-energy process for building high-strength cellulose-based nanocomposite materials with hierarchical structures possessing enhanced mechanical properties and optical transparency.

HPMC is one of the most widely used water soluble cellulose ether as an emulsifier, protective colloid, stabilizer, suspending agent, thickener, or film former in various industrial applications. Bilbao-Saiz et al. (2011) systematically analyzed the effect of different cellulose nanofibers such as nanofibrillated cellulose, oxidized cellulose fibers and cellulose nanowhiskers on the properties of HPMC nanocomposite films. The film-forming solutions of HPMC nanocomposites were prepared by suspending different cellulose nanoparticles at various concentrations and it was casted into films on glass plates with a polyester film. The mechanical and moisture barrier properties were evaluated to probe the addition of cellulose nanoparticles affect on the properties of HPMC. The addition of nanofibrillated cellulose and oxidized cellulose fibers negatively affected on the performance of HPMC, while the addition of cellulose nanowhiskers resulted in an increase in tensile strength (22 %) and Young's modulus (55 %). Addition of cellulose whiskers also improved the water barrier properties of the composite films. Furthermore, the cellulose nanowhiskers only reduced 3–6 % of transparency of the HPMC films, allowing its application as edible, barrier and transparent films. The findings of this study indicated the potential of HPMC/cellulose whiskers nanocomposite films for sustainable packaging applications.

Research carried out by George et al. (2014), proved that addition of a combination of two different nanomaterials in a single polymer matrix can significantly

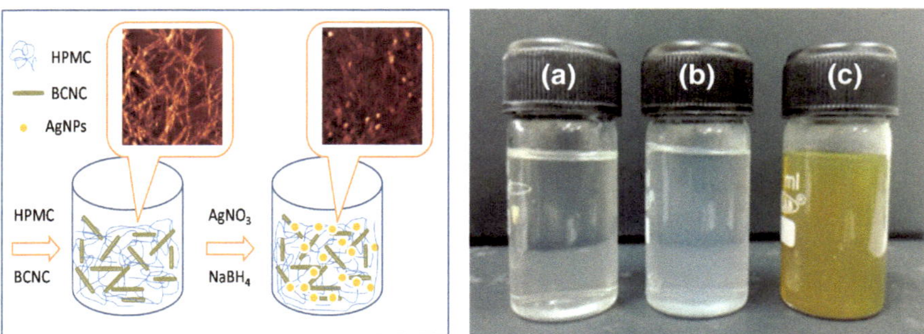

Fig. 10 Schematic representation of the synthesis of hybrid nanocomposites and the corresponding optical images of *a* HPMC, *b* HPMC–BCNC, and *c* HPMC–BCNC-AgNPs nanocomposite solutions. Reproduced from George et al. (2014) with permission. Copyright 2014. Elsevier Science Ltd

improves the properties of HPMC nanocomposite films. HPMC-based hybrid nanocomposites were developed by adding bacterial CNs (BCNC) in HPMC solution and silver nanoparticles (AgNPs) were synthesized in situ by the addition of $AgNO_3$ and $NaBH_4$ to this solution (Fig. 10). HPMC forms a clear solution with water, while the addition of BCNC had resulted in a cloudy appearance. The addition of $AgNO_3$ and $NaBH_4$ changed the color of the suspension to bright yellow, which confirmed the formation of silver nanoparticles.

The addition of BCNC into HPMC improved the tensile strength and modulus of the nanocomposites, but it reduced the elongation properties. However, the presence of AgNPs together with BCNC, helped to regain some of the lost elongation properties of these films. The addition of these nanomaterials also reduced the hydrophilicity of the nanocomposite considerably. A unique combination of two nanomaterials together was highly effective in overcoming certain limitations of nanocomposites, which uses only one type of nanomaterial. This type of hybrid nanocomposites is expected to be useful in developing eco-friendly polymer nanocomposites with superior properties.

CMC is a water soluble cellulose derivative with carboxymethyl groups attached to the cellulose backbone, which is widely used as thickening and stabilizing agent. Choi et al. (2006) reported the use of CNs as filler in CMC and compared the effects to MCC. The nanocomposite materials were prepared by mixing either MCC or CN in the range of 5–30 % with CMC by using 10 % glycerine as a plasticizer. The use of CN improved the strength and stiffness of the polymer nanocomposite as compared to MCC. Additionally, a simple heat treatment was found to be effective in rendering water resistance to the nanocomposite. The effect of nanosized cellulose on the properties of CMC-based composites proved that CMC-based materials can be used for the development of novel applications.

7.4 Starch-Based Nanocomposites

Starch-based polymers have attained great commercial importance as they are biodegradable, but some of the inherent properties like brittleness, high moisture and gas permeability, low heat distortion temperature (HDT), low melt viscosity for further processing, etc., restricts their use in a various applications. Modification of various properties of starch through innovative nanotechnology has already proven to be an effective way to overcome the limitations. Kaushik et al. (2010) prepared thermoplastic starch nanocomposites reinforced with cellulose nanofibers extracted from wheat straw by solution film casting. The mechanical properties of the nanocomposite films increased linearly with an increase in cellulose content. Starch nanocomposite with 15 % cellulose nanofiber content exhibited a strength improvement of 195 % over normal starch film. The addition of cellulose nanofiber also resulted in reducing the onset of degradation temperatures as well as water sorption as compared to the pure starch. Barrier properties also improved with the addition of cellulose nanofibers up to 10 % but further addition deteriorated properties due to the possible fiber agglomeration.

7.5 Gelatin-Based Nanocomposites

Gelatin is a water soluble protein obtained by partial hydrolysis of collagen, which is widely investigated for several applications due to their film-forming ability and environmental appeal. However, their applications are limited because of their poor mechanical properties and water sensitivity. Similar to other water soluble polymers, gelatin also can be reinforced by using CNs. George et al. (2012b) reported the use of bacterial CNs (BCNC) for preparing edible nanocomposite films with superior properties using gelatin as the polymer matrix. Bacterial cellulose, with a distinctly soft texture and high fiber content was selected for this study as they are edible and extremely pure. Gelatin-BCNC nanocomposites were prepared by the stochiometric addition of BCNC suspension to a calculated quantity of gelatin solution followed by stirring and ultrasonication for 30 min. The nanocomposite solutions were casted in polypropylene petri dishes and dried at 37 °C for 48 h. The formation of percolated networks of CNs within gelatin matrix contributed to improve the mechanical properties of these films. The addition of CNs also reduced the moisture affinity of gelatin, which is very favorable when these materials are used for edible packaging applications. Results of this study demonstrated that the use of CNs in the fabrication of edible, biodegradable, and high-performance gelatine-based nanocomposite films for various packaging applications at relatively low cost.

7.6 Polyvinyl Alcohol-Based Nanocomposites

Poly (vinyl alcohol) is the most widely studied synthetic water soluble polymer because of its application as controlled drug release materials, chemical separators, barrier membranes for food packaging applications, pharmaceutical components, manufacturing material for artificial human organs and as biomaterials (Hassan et al. 2000a, b). PVA is also widely preferred for the fabrication of nanocomposites due to its high efficiency to form strong hydrogen bonds and its superior capability to transfer load between the polymer and the reinforcing agents. Zimmermann et al. (2004) reported the use of cellulose nanofibrils having a diameter below 100 nm to reinforce PVA. The addition of 10 wt% of these fibrils resulted in a 2.5-fold increase in modulus and fivefold increase in tensile strength. In another study, nanocomposites were prepared from copolymers of PVA and polyvinyl acetate by using a colloidal aqueous suspension of cellulose whiskers prepared from cotton linter (Roohani et al. 2008). The degree of hydrolysis of the matrix was varied in order to vary the hydrophilic character of the polymer matrix and then the degree of interaction between the filler and the matrix. Results showed that stronger filler–matrix interactions occur in fully hydrolyzed PVA compared to partially hydrolyzed samples and the reinforcing effect was higher as the degree of hydrolysis of the matrix was high.

George et al. (2010) reported the use of BCNC synthesized by using a hydrochloric acid-assisted top down approach which is found to retain some of the natural properties of native cellulose even in nanodimensions. Using these nanocrystals, PVA nanocomposite films were prepared and characterized. The use of nanocrystals has significantly improved the mechanical properties and thermal stability of PVA nanocomposites. Results of this study demonstrated that CNs-reinforced PVA films can be used in the fabrication of high-performance polymer nanocomposite films for food packaging applications. A similar study from the same research group, recently reported the synthesis of more thermally stable BCNC by using an enzyme "cellulase" (George et al. 2011). Using these enzyme-processed nanocrystals, PVA nanocomposite films were prepared. The addition of these nanocrystals resulted in a remarkable improvement in the thermal stability as well as mechanical properties of nanocomposite films. These nanocomposites exhibited higher melting temperature (T_m) and enthalpy of melting (ΔH_m) than those of pure PVA, suggesting that the addition of nanocrystals modified the thermal properties of PVA. The effective load transfer from polymer chains to the reinforcing nanocrystals resulted in an improved tensile strength from 62.5 to 128 MPa, by the addition of just 4 wt% of NC. Similarly hybrid nanocomposites containing a combination of BCNC together with AgNP were also reported very recently (George et al. 2012a).

A new series of biomimetic, stimuli-responsive nanocomposites, which change their mechanical properties upon exposure to physiological conditions, was reported by Shanmuganathan et al. (2010). The materials were produced by introducing percolating networks of cellulose whiskers isolated from cotton into

PVAc. Below the glass-transition temperature (T_g), the tensile moduli of the dry nanocomposites increased two fold, while above T_g, it increased nearly 40 fold. Upon exposure to emulated physiological conditions, these nanocomposites exhibited a decrease in modulus, which clearly suggest the application of these nanocomposites as adaptive substrates for intracortical electrodes and other biomedical applications. Jalal Uddin et al. (2011) reported the fabrication of PVA nanocomposite fibers containing highly oriented cellulose nanowhiskers through gel spinning followed by hot drawing. These macroscopically homogeneous nanocomposites containing a small amount of cellulose whiskers (5 wt%) showed higher drawability compared to the neat PVA fiber. Investigation of the stress transfer mechanism indicated effective stress transfer between polymer and filler due to the strong interfacial interaction, which was supported by the outstanding mechanical performance of the fibers.

7.7 Polyacrylic Acid-Based Nanocomposites

PAA is widely used as a pH and electrically sensitive material due to the variation in the content of carboxylate anion. CN-loaded PAA nanocomposite fibrous membranes have been fabricated by electrospinning (Lu et al. 2009). The addition of CN into PAA matrix significantly reduced elecrtrospun fiber diameters. The Young's modulus and tensile stress of these nanocomposites were significantly improved up to 35-fold and 16-fold, respectively. Heat-induced esterification between the surface hydroxyl groups of cellulose and the carboxyl groups of acrylic acid produced covalent crosslinks at the interfaces. These types of crosslinks made the nanocomposite membranes insoluble in water, more thermally stable and superior in tensile strength. On addition of 20 % CNC, the crosslinked nanocomposite membrane exhibited a very impressive 77-fold increase in modulus and 58-fold increase in tensile stress.

7.8 Polyacrylamide-Based Nanocomposites

Polyacrylamide is one of the most commonly used water soluble polymer for the fabrication of hydrogels due to their ability to preserve their shape and mechanical strength, and adjustable mechanical, chemical, and biophysical properties. These hydrogels are widely used in various applications such as biomaterials, agriculture, drilling fluids, tissue engineering, etc. Zhou et al. (2011) studied the application of CNs in making PAAm nanocomposite hydrogels. They reported the synthesis of nanocomposite hydrogels containing different compositions of CN using an in situ free-radical polymerization. The addition of CN favored the formation of hydrogels and increased the crosslink density of hydrogels besides providing reinforcement to the hydrogel. The mechanical properties of the nanocomposite hydrogels were significantly improved due to the proper dispersion of CN in the polymer matrix as well as enhanced interfacial interaction between these two components. CN content of 6.7 wt% in the polymer found to impart the maximum mechanical properties to the hydrogels. In a similar study, Spagnol et al. (2012) reported the synthesis of superabsorbent hydrogels, based on poly(acrylamide-co-acrylate) reinforced with cellulose nanowhiskers. A series of hydrogels containing 1, 2, 5, 10, and 20 wt% of CN were proved to be more efficient in the water absorption than the poly(acrylamide-co-acrylate) hydrogels without CN. These hydrogels also exhibited a pH-responsiveness and cation-sensitivity character.

8 Applications of Cellulose Nanocrystals-Based Water Soluble Polymer Nanocomposites

Cellulosic nanomaterials have several advantages compared to other nanomaterials as they are biodegradable, abundant, cheap, easy to synthesize, and also they possess high specific strength and modulus. CNs were used as the load-bearing constituent in developing new and inexpensive biomaterials due to their high aspect ratio, good mechanical properties, and renewable nature. Compared to other inorganic reinforcing fillers, CNs have some additional advantages like their wide availability, nontoxic, low-energy consumption, ease of recycling including combustion, high sound attenuation, easy processability due to their nonabrasive nature, allowing high filling levels, and significant cost savings. In the area of polymer nanocomposites, cellulose nanocystals were used as model fillers with a defined morphology to impart sufficient strength and modulus. These types of nanocomposites can be used for making biomimetic foams of high mechanical performance, paper with high toughness, flexible panels for flat panel displays, water repellent papers, etc. High security paper containing CN can also be made by incorporating solidified liquid crystals of cellulose. These nanocomposite materials are also suitable for drug delivery systems, such as liposomes, micelles, microgels, etc. Considering the safety and efficacy of CNs, they can be used in various biomedical fields like fluorescence bioassay and bioimaging applications. These nanocrystals are also capable of making highly functional nanocomposites for applications such as smart coatings, pharmaceuticals, electronic materials, etc. CNs can also be used for stabilizing the nanoparticles of specific functionality which can be used for filtration and catalytic conversion applications. Potential applications of CN containing polymer nanocomposites also include biomedical implants, fibers and textiles, separation membranes, and electroactive polymers.

One potential area of application where CNs can be used extensively is in the field of high-performance nanocomposite packaging materials. The rod-shaped nanocrystals are known for reinforcing polymeric materials; however, not much information is available about their efficacy as barrier materials. Research in this area is evolving rapidly to enhance both mechanical and barrier properties of existing packaging materials and to overcome certain limitations like brittleness, lack of flexibility, durability, etc. Superior barrier membranes made up of bio-based nanocomposites can also find their applications in food and biomedical packaging areas, where lower permeability to oxygen, carbon dioxide, flavor, oil, and water vapor are very much needed. However, there are lot of scientific and technical challenges ahead before realizing truly eco-friendly packaging materials made up of water soluble polymers reinforced with CNs that satisfy all industry and consumer requirements.

References

Abdollahi M, Alboofetileh M, Behrooz R, Rezaei M, Miraki R (2013a) Reducing water sensitivity of alginate bio-nanocomposite film using cellulose nanoparticles. Intl J Biol Macromol 54:166–173

Abdollahi M, Alboofetileh M, Rezaei M, Behrooz R (2013b) Comparing physico-mechanical and thermal properties of alginate nanocomposite films reinforced with organic and/or inorganic nanofillers. Food Hydrocolloid 32(2):416–424

Alexandre M, Dubois P (2000) Polymer-layered silicate nanocomposites: preparation, properties and uses of a new class of materials. Mater Sci Eng Rep 28:1–63

Araki J, Kuga S (2001) Effect of trace electrolyte on liquid crystal type of cellulose microcrystals. Langmuir 17:4493–4496

Araki J, Wada M, Kuga S, Okana T (1999) Influence of surface charge on viscosity behavior of cellulose microcrystal suspension. J Wood Sci 45:258–261

Ashogbon AO, Akintayo ET (2014) Recent trend in the physical and chemical modification of starches from different botanical sources: a review. Starch 66:41–57

Baker MI, Walsh SP, Schwartz Z, Boyan BD (2012) A review of polyvinyl alcohol and its uses in cartilage and orthopedic applications. J Biomed Mater Res, Part B 100B:1451–1457

Barbucci R, Magnani A, Consumi M (2000) Swelling behavior of carboxymethyl cellulose hydrogels in relation to cross-linking, pH and charge density. Macromolecules 33:7457–7480

Battista OA, Coppick S, Howsmon JA, Morehead FF, Sisson WA (1956) Level off degree of polymerization. Ind Eng Chem 48:333–335

Battista OA, Smith PA (1962) Microcrystalline cellulose. J Ind Eng Chem 54:20–29

Bilbao-Sainz C, Bras J, Williams T, SÃ©nechal T, Orts W (2011) HPMC reinforced with different cellulose nano-particles. Carbohydr Polym 86(4):1549–1557

Choi Y, Simonsen J (2006) Cellulose nanocrystal-filled carboxymethyl cellulose nanocomposites. J Nanosci Nanotechnol 6(3):633–639

Cohen SG, Haas HC, Farney L, Valle C Jr (1953) Preparation and properties of some ether and ester derivatives of hydroxyethylcellulose. Ind Eng Chem 45:200–203

Debeaufort F, Voilley A (1995) Methyl cellulose-based edible films and coatings I. Effect of plasticizer content on water and 1-octen-3-ol sorption and transport. Cellulose 2:205–213

Debeaufort F, Quezada-Gallo JA, Voilley A (1998) Edible films and coatings: tomorrow's packagings: a review. Crit Rev Food Sci Nutr 38(4):299–313

de Souza Lima MM, Borsali R (2004) Rod like cellulose microcrystals: Structure, properties and applications. Macromol Rapid Comm 25:771–787

Eitan A, Fisher FT, Andrews R, Brinson LC, Schadler LS (2006) Reinforcement mechanisms in MWCNT-filled polycarbonate. Compos Sci Technol 66:1159–1170

George J, Bawa AS, Siddaramaiah (2010) Synthesis and characterization of bacterial cellulose nanocrystals and their PVA nanocomposites. Adv Mater Res 123:383–386

George J, Ramana KV, Bawa AS, Siddaramaiah (2011) Bacterial cellulose nanocrystals exhibiting high thermal stability and their polymer nanocomposites. Intl J Biol Macromol 48(1):50–57

George J, Ramana KV, Sabapathy SN, Bawa AS (2005a) Physico-mechanical properties of chemically treated bacterial (Acetobacter xylinum) cellulose membrane. World J Microbiol Biotechnol 21(8–9):1323–1327

George J, Ramana KV, Sabapathy SN, Jagannath JH, Bawa AS (2005b) Characterization of chemically treated bacterial (Acetobacter xylinum) biopolymer: some thermo-mechanical properties. Int J Biol Macromol 37(4):189–194

George J, Sajeevkumar VA, Ramana KV, Sabapathy SN, Siddaramaiah (2012a) Augmented properties of PVA hybrid nanocomposites containing cellulose nanocrystals and silver nanoparticles. J Mater Chem 22(42):22433–22439

George J, Siddaramaiah (2012b) High performance edible nanocomposite films containing bacterial cellulose nanocrystals. Carbohydr Polym 87(3):2031–2037

George J, Kumar R, Sajeevkumar VA, Ramana KV, Rajamanickam R, Abhishek V, Sabapathy SN, Siddaramaiah (2014) Hybrid HPMC nanocomposites containing bacterial cellulose nanocrystals and silver nanoparticles. Carbohydr Polym 105:285–292

Gomez-Guillen MC, Gimenez B, López-Caballero ME, Montero MP (2011) Functional and bioactive properties of collagen and gelatin from alternative sources: a review. Food Hydrocolloids 25:1813–1827

Gorga RE, Cohen RE (2004) Toughness enhancements in poly (methyl methacrylate) by addition of oriented multiwall carbon nanotube. J Polym Sci B Polym Phys 42(14):2690–2702

Gorgieva S, Kokol V (2011) Synthesis and application of new temperature responsive hydrogels based on carboxymethyl and hydroxyethyl cellulose derivatives for the functional finishing of cotton knitwear. Carbohydr Polym 85:664–673

Hakansson H, Ahlgren P (2005) Acid hydrolysis of some industrial pulps: effect of hydrolysis conditions and raw material. Cellulose 12:177–183

Hassan CM, Peppas NA (2000a) Structure and applications of Poly (vinyl alcohol) hydrogels produced by conventional cross linking or by freezing/thawing methods. Adv Polym Sci 153:37–65

Hassan CM, Peppas NA (2000b) Structure and applications of poly (vinyl alcohol) hydrogels produced by conventional crosslinking or by freezing/thawing methods. Biopolymers. PVA hydrogels, anionic polymerisation nanocomposites. Springer, Heidelberg, pp 37–65

Huq T, Salmieri S, Khan A, Khan RA, Le Tien C, Riedl B, Lacroix M (2012) Nanocrystalline cellulose (NCC) reinforced alginate based biodegradable nanocomposite film. Carbohydr Polym 90(4):1757–1763

Hussain F, Hojjati M, Okamoto M, Gorga RE (2006) Review article: polymer-matrix nanocomposites, processing, manufacturing and application: an overview. J Compos Mater 40:1511–1575

Iwamoto S, Kai W, Isogai A, Iwata T (2009) Elastic modulus of single cellulose microfibrils from tunicate measured by atomic force microscopy. Biomacromolecules 10:2571–2576

Jalal Uddin A, Araki J, Gotoh Y (2011) Toward "strong" green nanocomposites: polyvinyl alcohol reinforced with extremely oriented cellulose whiskers. Biomacromolecules 12(3):617–624

Jonas R, Farah LF (1998) Production and application of microbial cellulose. Polym Degrad Stab 59:101–106

Kalfus J, Jancar J (2008) Reinforcing mechanisms in amorphous polymer nano-composites. Compos Sci Technol 68(15):3444–3447

Kaushik A, Singh M, Verma G (2010) Green nanocomposites based on thermoplastic starch and steam exploded cellulose nanofibrils from wheat straw. Carbohydr Polym 82(2):337–345

Khan RA, Salmieri S, Dussault D, Uribe-Calderon J, Kamal MR, Safrany A, Lacroix M (2010) Production and properties of nanocellulose-reinforced methylcellulose based biodegradable films. J Agri Food Chem 58(13):7878–7885

Kolodziejska I, Kaczorowski K, Piotrowsia B, Sadowska M (2004) Modification of properties of gelatin from skins of baltic cod (Gadus morhua) with transglutaminase. Food Chem 86:203–209

Lagerwall JP, Schütz C, Salajkova M, Noh J, Park JH, Scalia G, Bergstrom L (2014) Cellulose nanocrystal-based materials: from liquid crystal self-assembly and glass formation to multifunctional thin films. NPG Asia Mater 6(1):e80

Lahiji RR, Xu X, Reifenberger R, Raman A, Rudie A, Moon RJ (2010) Atomic force microscopy characterization of cellulose nanocrystals. Langmuir 26:4480–4488

Lu P, Hsieh YL (2009) Cellulose nanocrystal-filled poly (acrylic acid) nanocomposite fibrous membranes. Nanotechnology 20(41):415604

Lin M-F, Thakur VK, Tan EJ, Lee PS (2011a) Surface functionalization of BaTiO3 nanoparticles and improved electrical properties of BaTiO3/polyvinylidene fluoride composite. RSC Adv 1:576–578

Lin M-F, Thakur VK, Tan EJ, Lee PS (2011b) Dopant induced hollow BaTiO3 nanostructures for application in high performance capacitors. J Mater Chem 21:16500–16504

Maniar KK (2004) Polymeric nanocomposites: a review. Polym Plast Technol Eng 43:427–443

Matsumoto T, Kawai M, Masuda T (1992) Influence of concentration and mannuronate/guluronate ratio on steady flow properties of alginate aqueous systems. Biorheology 29:411–417

Miao C, Hamad WY (2013) Cellulose reinforced polymer composites and nanocomposites: a critical review. Cellulose 20(5):2221–2262

Mischnick P, Momcilovic D (2010) Chemical structure analysis of starch and cellulose derivatives. Adv Carbohydr Chem Biochem 64:117–210

Moon RJ, Martini A, Nairn J, Simonsen J, Youngblood J (2011a) Cellulose nanomaterials review: structure, properties and nanocomposites. Chem Soc Rev 40(7):3941–3994

Moon RJ, Martini A, Nairn J, Simonsen J, Youngblood J (2011b) Cellulose nanomaterials review: structure, properties and nanocomposites. Chem Soc Rev 40:3941–3994

Necas J, Bartosikova L (2013) Carrageenan: a review. Vet Med-Czech 58(4):187–205

O'Sullivan AC (1997) Cellulose: the structure slowly unravels. Cellulose 4(3):173–207

Pan J, Hamad W, Straus SK (2010) Parameters affecting the chiral nematic phase of nanocrystalline cellulose films. Macromolecules 43(8):3851–3858

Park C, Park O, Lim J, Kim H (2001) The fabrication of syndiotactic polystyrene/organophilic clay nanocomposites and their properties. Polymer 42:7465–7475

Perez S, Bertoft E (2010) The molecular structures of starch components and their contribution to the architecture of starch granules: a comprehensive review. Starch 62:389–420

Ranby BG (1951) The colloidal properties of cellulose micelles. Discuss Faraday Soc 11:158–164

Roohani M, Habibi Y, Belgacem NM, Ebrahim G, Karimi AN, Dufresne A (2008) Cellulose whiskers reinforced polyvinyl alcohol copolymers nanocomposites. Eur Polym J 44(8):2489–2498

Sairam M, Babu VR, Vijaya B, Naidu K, Aminabhavi TM (2006) Encapsulation efficiency and controlled release characteristics of crosslinked polyacrylamide particles. Int J Pharm 320:131–136

Sanchez-Garcia MD, Hilliou L, Lagaron JM (2010a) Nanobiocomposites of carrageenan, zein, and mica of interest in food packaging and coating applications. J Agri Food Chem 58(11):6884–6894

Sanchez-Garcia MD, Hilliou L, Lagaron JM (2010b) Morphology and water barrier properties of nanobiocomposites of κ/ι-hybrid carrageenan and cellulose nanowhiskers. J Agri Food Chem 58(24):12847–12857

Sarker N, Walker LC (1995) Hydration-dehydration properties of methylcellulose and hydroxypropylmethylcellulose. Carbohydr Polym 27:177–185

Schadler LS, Brinson LC, Sawyer WG (2007) Polymer nanocomposites: a small part of the story. JOM 59(3):53–60

Schagerlf H, Richardson S, Momcilovic D, Brinkmalm G, Wittgren B, Tjerneld F (2006) Characterization of chemical substitution of hydroxypropyl cellulose using enzymatic degradation. Biomacromolecules 7:80–85

Shanmuganathan K, Capadona JR, Rowan SJ, Weder C (2010) Bio-inspired mechanically-adaptive nanocomposites derived from cotton cellulose whiskers. J Mater Chem 20(1):180–186

Singha AS, Thakur VK (2009a) Chemical resistance, mechanical and physical properties of biofibers-based polymer composites. Polym Plast Technol Eng 48:736–744
Singha AS, Thakur VK (2009b) Grewia optiva Fiber Reinforced Novel, low cost polymer composites. J Chem 6:71–76
Singha AS, Thakur VK (2009c) Synthesis, characterisation and analysis of hibiscus sabdariffa fibre reinforced polymer matrix based composites. Polym Polym Compos 17:189–194
Singha AS, Thakur VK (2009d) Fabrication and characterization of S. cilliare fibre reinforced polymer composites. Bull Mater Sci 32:49–58
Singha AS, Thakur VK (2009e) Physical, chemical and mechanical properties of hibiscus sabdariffa fiber/polymer composite. Int J Polym Mater 58:217–228
Singha AS, Thakur VK (2009f) Fabrication and characterization of h. sabdariffa fiber-reinforced green polymer composites. Polym-Plast Technol Eng 48:482–487
Singha AS, Thakur VK (2010a) Renewable resource-based green polymer composites: analysis and characterization. Int J Polym Anal Charact 15(3):127–146
Singha AS, Thakur VK (2010b) Mechanical, morphological, and thermal characterization of compression-molded polymer biocomposites. Int J Polym Anal Charact 15:87–97
Singha AS, Thakur VK (2010c) Synthesis, characterization and study of pine needles reinforced polymer matrix based composites. J Reinf Plast Compos 29:700–709
Spagnol C, Rodrigues FH, Neto AG, Pereira AG, Fajardo AR, Radovanovic E, Rubira AF, Muniz EC (2012) Nanocomposites based on poly(acrylamide-co-acrylate) and cellulose nanowhiskers. Eur Polym J 48(3):454–463
Strucova A, Davies GR, Eichhorn SJ (2005) Elastic modulus and stress-transfer properties of tunicate cellulose whiskers. Biomacromolecules 6:1055–1061
Tester RF, Karkalas J, Qi X (2004) Starch—composition, fine structure and architecture. J Cereal Sci 39(2):151–165
Thakur VK, Singha AS, Kaur I et al (2010a) Silane functionalization of saccaharum cilliare fibers: thermal, morphological, and physicochemical study. Int J Polym Anal Charact 15:397–414
Thakur VK, Singha AS, Mehta IK (2010b) Renewable resource-based green polymer composites: analysis and characterization. Int J Polym Anal Charact 15(3):137–146
Thakur VK, Yan J, Lin M-F et al (2012) Novel polymer nanocomposites from bioinspired green aqueous functionalization of BNNTs. Polym Chem 3:962–969
Thakur VK, Thakur MK (2014a) Processing and characterization of natural cellulose fibers/thermoset polymer composites. Carbohydr Polym 109:102–117
Thakur VK, Thakur MK (2014b) Recent trends in hydrogels based on *psyllium* polysaccharide: a review. J Clean Prod 82:1–15
Thakur VK, Thakur MK (2014c) Recent advances in graft copolymerization and applications of chitosan: a review. ACS Sustain Chem Eng 2:2637–2652
Thakur VK, Thakur MK, Gupta RK (2014a) Review: raw natural fiber-based polymer composites. Int J Polym Anal Charact 19(3):256–271
Thakur VK, Thakur MK, Raghavan P, Kessler MR (2014b) Progress in green polymer composites from lignin for multifunctional applications: a review. ACS Sustain Chem Eng 2(5):1072–1092
Thakur VK, Vennerberg D, Kessler MR (2014c) Green aqueous surface modification of polypropylene for novel polymer nanocomposites. ACS Appl Mater Interfaces 6:9349–9356
Thakur VK, Vennerberg D, Madbouly SA, Kessler MR (2014d) Bio-inspired green surface functionalization of PMMA for multifunctional capacitors. RSC Adv 4:6677–6684
Thakur VK, Thunga M, Madbouly SA, Kessler MR (2014e) PMMA-g-SOY as a sustainable novel dielectric material. RSC Adv 4:18240–18249
Thostenson ET, Li C, Chou TW (2005) Nanocomposites in context. Compos Sci Technol 65 (3):491–516
Tripathy T, Singh RP (2000) High performance flocculating agent based on partially hydrolysed sodium alginate-g-polyacrylamide. Eur Polym J 36(7):1471–1476
Wautelet M (2001) Scaling laws in the macro-, micro- and nanoworlds. Euro J Phys 22 (6):601–611

Yang JS, Xie YJ, He W (2011) Research progress on chemical modification of alginate: a review. Carbohydr Polym 84:33–39

Yang TH (2008) Recent applications of polyacrylamide as biomaterials. Recent Pat Mater Sci 1:29–40

Yuan Q, Misra RDK (2006) Polymer nanocomposites: current understanding and issues. Mater Sci Technol 22(7):742–755

Zhou C, Wu Q, Yue Y, Zhang Q (2011) Application of rod-shaped cellulose nanocrystals in polyacrylamide hydrogels. J Colloid Interf Sci 353(1):116–123

Zhou Q, Malm E, Nilsson H, Larsson PT, Iversen T, Berglund LA, Bulone V (2009) Nanostructured biocomposites based on bacterial cellulosic nanofibers compartmentalized by a soft hydroxyethylcellulose matrix coating. Soft Matter 5(21):4124–4130

Zimmermann T, Pohler E, Geiger T (2004) Cellulose fibrils for polymer reinforcement. Adv Eng Mater 6:754–761

Bionanocomposites of Regenerated Cellulose Reinforced with Halloysite Nanoclay and Graphene Nanoplatelets: Characterizations and Properties

Mohammad Soheilmoghaddam, Raheleh Heidar Pour, Mat Uzir Wahit and Harintharavimal Balakrishnan

Abstract In recent years, the development of environmentally friendly materials obtained from renewable resources has attracted immense interest due to the new sustainable development policies. Cellulose is a readily available, naturally occurring biodegradable, and biocompatible linear polysaccharide. Recently, room temperature ionic liquids have been used as solvents to produce regenerated cellulose (RC) due to their attractive properties such as good chemical and thermal stability, low flammability, low melting point, and ease of recycling. Polymer/nanofiller nanocomposites are believed to have strong potential to widen polymer applications due to enhanced performance. It is also widely accepted that the incorporation of small amount of nanofiller (less than 5 wt%) into bio-based matrixes to produce nano-biocomposite materials with enhanced mechanical, permeability, and thermal properties. The tubular silica-based naturally occurring nanofiller, halloysite nanotubes (HNT), has been investigated due to its high surface area, unique geometry, and its potential to make the hydrogen bonding with polymers to disperse well in the matrix. Graphene nanoplatelets (GNP) have also attracted enormous attention among polymer engineers over the last few years due to its unique electrical, thermal, and mechanical properties. Single layer two-dimensional GNP sheet is considered as the strongest material along with the high surface area and aspect ratio. The chapter aims to highlight the effect of the addition of two different types of nanofillers such as HNT and GNP to produce RC nanocomposites on selected properties.

Keywords Regenerated cellulose · Halloysite nanotube · Graphene nanoplates · Ionic liquids

M. Soheilmoghaddam · R.H. Pour · M.U. Wahit (✉)
Enhanced Polymer Research Group (EnPRO), Department of Polymer Engineering, Faculty of Chemical Engineering, Universiti Teknologi Malaysia (UTM), Johor, Malaysia
e-mail: mat.uzir@cheme.utm.my

M.U. Wahit · H. Balakrishnan
Centre for Composites (CfC), Universiti Teknologi Malaysia (UTM), 81300 Skudai, Johor Bahru, Johor, Malaysia

V.K. Thakur and M.K. Thakur (eds.), *Eco-friendly Polymer Nanocomposites*, Advanced Structured Materials 75, DOI 10.1007/978-81-322-2470-9_10

1 Introduction

1.1 Regenerated Cellulose

In recent years, bio-based materials have attracted much attention due to the expansion in environmental pollution caused by nonrenewable resources (Gross and Kalra 2002; Han et al. 2013; Nishio 2006; Singha et al. 2009a, b). An ongoing effort on aiming at the utilization of plant biomass for production of biofuels and platform chemicals is due to the fact that plants provide the largest reserves of renewable, high-energy content carbon materials on Earth (Rinaldi and Schüth 2009; Schüth et al. 2014; Thakur and Thakur 2014a, b, c; Thakur et al. 2014a, b). Cellulose is well known as one of the world's most abundant and renewable natural resources which is made up of layers upon layers of microfibrils with approximately 10–20 nm in diameter and may consist of up to 40 cellulose chains (Soheilmoghaddam et al. 2014a, b; Singha and Thakur 2009a, b, c; Thakur and Thakur 2014a). The lateral size of cellulose chains is about 0.3 nm. The degree of polymerization of native cellulose from various origins can fall in the range of 1000–30,000, which corresponds to chain lengths from 500 to 15,000 nm.

Cellulose is located within the fiber walls of plants (Thakur et al. 2013a, b). One fiber is an elongated vegetable cell. Fibers of various plants have different shapes and dimensions. Fiber of cotton and bast plants are enough long, with length in the range of centimeters, while wood fibers are short, typically 1–3 mm in length. Cotton fibers are twisted, while fibers of wood are generally untwisted. Fibers of the bast plants (flax, ramie, etc.) are strength and round (Ioelovich 2008; Thakur et al. 2014a).Cellulose fiber has high tensile strength and is associated with other natural polymers within the plant cell wall (Thakur et al. 2013c, d). Hemicellulose wraps the microfibril's cellulose core, a branched polymer composed of a mixture of primarily sugars (xylose, arabinose), and some hexoses (mannose, galactose, glucose). Cellulose is naturally occurring linear polymer in nature, which consists of β 1–4 glucopyranose units (Fig. 1), exhibiting many unique properties and applicability to a variety of uses such as paper, textiles, membranes, artificial fibers,

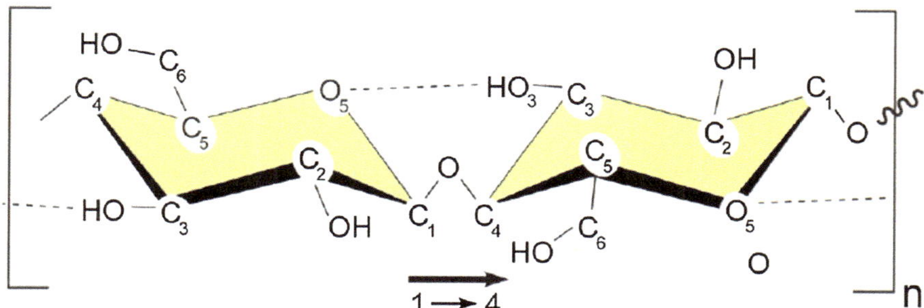

Fig. 1 Schematics of **a** single cellulose chain repeat unit, showing the directionality of the 1–4 linkage and intrachain hydrogen bonding (*dotted line*) (Moon et al. 2011)

biomedical, and the food industry (Thakur et al. 2013e). The repeat unit is comprised of two anhydroglucose rings (($C_6H_{10}O_5$)n; n = 10,000–15,000, where n is dependent on the cellulose source material) linked together through an oxygen covalently bonded to C1 of one glucose ring and C4 of the adjoining ring (1–4 linkage) and so-called the b 1–4 glucosidic bond (Azizi Samir et al. 2005).

These cellulose chains (molecules) are built up into "sheets" which are held together by side by side hydrogen bonding between the chains (Thakur et al. 2012a, b, c) There are two possibilities for hydroxyl groups in the cellulose molecules to form hydrogen bonds, i.e., by interaction between among suitably positioned hydroxyls in the same molecules (intermolecular) and by interaction between hydroxyl groups in neighboring cellulose molecules (intramolecular). Intermolecular hydrogen bonding is the main cause of the relative stiffness and rigidity of cellulose molecules, which is reflected in its high viscosity in solution, its high tendency to crystallize, and its ability to form fibrillar strands (Krassig 1996). Because of rigidity of molecules and close chain package via numerous inter- and intramolecular hydrogen bonds, cellulose is neither dissolved in common solvents nor melted to process. Then commercial regeneration of cotton or wood cellulose in the form of fiber and films has been based on solvent systems (Wu et al. 2009).

Nishikawa and Ono recorded the crystalline nature of cellulose using the X-ray diffraction patterns from fiber bundles from various plants. Cellulose is known to exist in at least four polymorphic crystalline forms, of which the structure and properties of cellulose I (native cellulose) and cellulose II (regenerated cellulose and mercerized cellulose) have been most extensively studied. As a first approximation, the crystal structure of cellulose I determined by X-ray diffraction can be described by monoclinic unit cell which contains two cellulose chains in a parallel orientation with a twofold screw axis (Klemm et al. 2005). Cellulose I has two polymorphs, a triclinic structure (Iα) and a monoclinic structure (Iβ), which coexist in various proportions depending on the cellulose source (Azizi Samir et al. 2005) (Nishiyama 2009). The Iα structure is the dominate polymorph for most algae (Yamamoto and Horii 1993) and bacteria (Yamamoto and Horn 1994), whereas Iβ is the dominant polymorph for higher plant cell wall cellulose and in tunicates.

Cellulose II as the most important from a technical and commercial point of view is formed from cellulose I by precipitating cellulose from solution into an aqueous medium at room or slightly elevated temperature, i.e., in technical spinning processes for man-made cellulose fibers. It is also obtained in the large-scale mercerization process of cotton, which proceeds via the formation of sodium cellulose by interaction of the polymer with aqueous sodium hydroxide and subsequent decomposition of this intermediate by neutralization or washing out of the sodium hydroxide. It is not yet understood how the parallel chain arrangement of cellulose I undergoes transition into the antiparallel orientation of cellulose II without an intermediate dispersion of cellulose molecules. The crystalline structure of cellulose I and cellulose II are shown in Fig. 2.

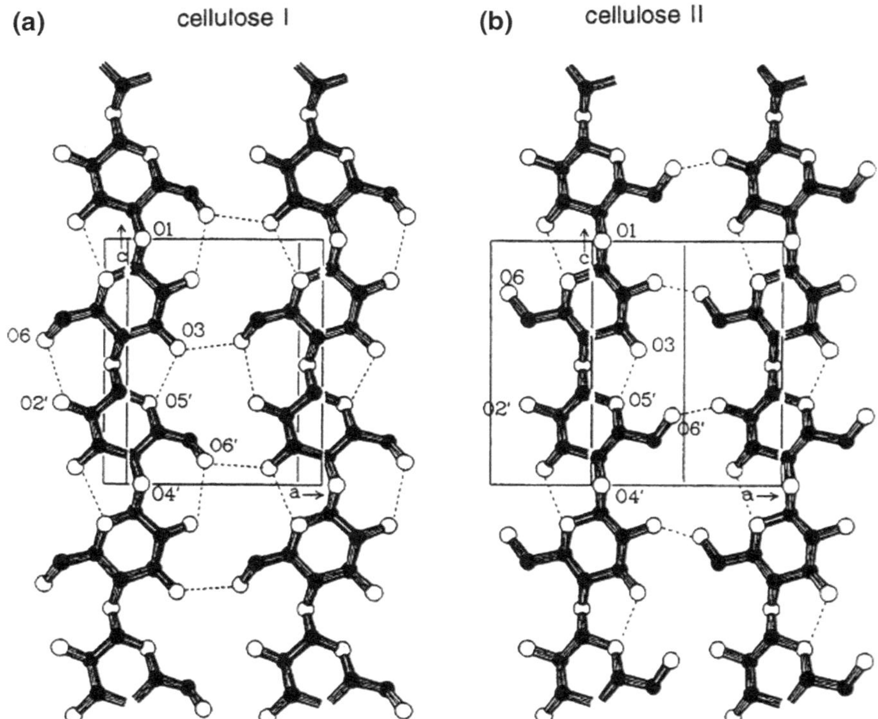

Fig. 2 Crystal structure of cellulose I and cellulose II: **a** projection of the unit cell (UC) along the **a**–**b** plane; **b** projection of the UC parallel to the (100) lattic plane (cellulose I) and the (010) latic plane (cellulose II) (Klemm et al. 2004)

1.2 Cellulose Dissolution

Most of the word's cellulosic fibers and films are produced via viscose process. The obtained fibers are called 'rayon' and the films are called 'cellophane.' In viscose process, at first cellulose is treated with sodium hydroxide solution (the steeping stage) in order to convert cellulose I structure to alkali cellulose and to enhance reactivity and to enable the penetration of carbon disulphide. For standard wood pulp, a sodium hydroxide concentration of about 18 % w/w is required for complete conversion of cellulose I. After steeping, cellulose is pressed under high pressure. Excess sodium hydroxide must be removed as best as possible to avoid any side reaction with carbon disulfide. The alkali cellulose press cake is then reduced to crumbs in the shredding stage. The crumbed alkali cellulose is aged at elevated temperature to adjust the degree of polymerization (DP) of cellulose. For regular stable fibers the process takes about 18 h. The DP-adjusted alkali cellulose reacts with carbon disulfide during the xanthation stage to form yellow-colored cellulose xanthate, which is soluble in dilute sodium hydroxide. Xanthation is critical to by-product formation. By-product formation is intensified by temperature and poor alkali cellulose quality (Bredereck and Hermanutz 2005; Hyden 1929). The obtained dope, known as viscose, is prepared by dissolving the xanthate crumbs in dilute sodium hydroxide solution under high shearing at temperature around 10 °C. The viscose dope must be filtered and aged before spinning or casting. Viscose contains many small air bubbles and particles that need to be removed by filtration. The solution also contains some gel-like materials that is retarded rather than removed by the filters. The viscose is unstable and tends to form more gel as it ages. Its flow characteristics make the material close to walls of any vessel or pipe move more slowly, get older, and gel more than mainstream viscose. So while filtration can hold back gels arising from incomplete mixing, new gels can form in the pipes work after the filtration. Therefore, the removal of gels occurs in several stages. Usually, three-stage filtration is used in viscose process.

The viscose process requires huge amount of water (i.e., 450–850 l for each kilogram of products). The effluent contains large amount of sodium sulfate, which has to be recovered. In addition, the hydrogen sulfide and carbon disulfide gases produced in viscose plants must be quantitatively removed because of strict legislation. Federal regulations on effluent concentrations of sodium salts, zinc salts, hydrogen sulfide, and carbon disulphide place strict restrictions on the viscose process. Investment in pollution control systems have became cost-determining and therefore processes using organic solvents with closed recycle–recovery loops to produce man-made cellulosic became increasingly important (Bredereck and Hermanutz 2005).

N-Methylmorpholine-*N*-oxide monohydrate (NMMO) is used as a solvent for direct dissolution of cellulose in industrial Lyocell process to produce cellulosic fibers and films. The first patent on dissolution of cellulose in different tertiary amine oxides was filed by Granacher and Sallmann in 1936 and 1939 (Graenacher and Sallmann 1939). About 30 years later, Johnson and his group (Johnson 1970; Wang and Wang 2014) issued patents on dissolution of different natural and synthetic polymers in different cyclic tertiary amine oxides. At the beginning of 1980s, intensive research on preparation of cellulosic fibers from amine oxide/cellulose solution had been carried out in United States (McCorsley 1979, 1981). Among several amine oxide tried, most promising results were obtained for *N*-Methylmorpholine-*N*-oxide as the cellulose solvent (Rosenau et al. 2001).

NMMO is produced by oxidation of the ternary amine *N*-Methylmorpholine with hydrogen peroxide. The melting point of the pure NMMO is at 170 °C. Hydration with one water molecule per NMMO molecule leads to the NMMO monohydrate with a melting point of about 74 °C and improved dissolution strength for cellulose. The importance of the NMMO process rests upon the fact that with its strong N–O dipoles, *N*-Methylmorpholine-*N*-oxide in aqueous solution is capable of physically dissolving cellulose without any derivatization (Fink et al. 2001; Rosenau et al. 2003). Nowadays, it is well known that NMMO power to dissolve cellulose originates in its capability to dissociate the hydrogen bond network of cellulose and to form solvent complexes by establishing new hydrogen bonds between the macromolecule and the solvent.

Ideally, Lyocell process should be entirely physical process that does not cause chemical changes in cellulose or solvent. However, there are several side reactions and considerable by-product formation in the cellulose/NMMO/water which cause detrimental effects, such as degradation of cellulose, decreased product performance, temporary or permanent discoloration of the resulting fibers, pronounced decomposition of NMMO, increased consumption of stabilizer, or even thermal runaway reaction. The first harmful effect of side reactions in the Lyocell system is a gradual decomposition of the solvent NMMO. Losses in NMMO are not only important from economical point of view, but also the primary decomposition products have been recognized as inducer of further decomposition. The second aspect and the most prominent result of side reactions is progressing formation of chromophore, starting from slight yellow tint of initially colorless solution up to a dark brown discoloration at elevated reaction times. Third, chromophores formation is accompanied by degradation of pulp, i.e., a progressing DP loss of the cellulose employed. This element of side reaction is most critical as it directly influence the fiber quality. An increased consumption of stabilizer is another facet of side reactions. The reaction products of stabilizer are potentially harmful compounds (Rosenau et al. 2003).

1.3 Ionic Liquids

Ionic liquids are a class of fascination, new liquid materials with unique combinations of properties. Ionic liquids have gained overwhelming interest over the past years, because they offer unique set of properties not achievable with any other material. This opens up opportunities in many different applications. They might be just a replacement for the material currently used as, for example, reaction media in chemical process or they are an enabling technology that allows totally new solution—as for example in manufacturing cellulose derivatives. Today, they are discussed as high-potential solution in a broad range of application segments: in chemical reaction and separation process, in processing metals and polymers, especially biopolymers like cellulose, as electrolytes in electronic devices and as functional or engineering fluids in many different applications (Kubisa 2005; Visser et al. 2000).

An ionic liquid is a salt with a melting temperature below the boiling point of water. Most ionic liquids are liquid at room temperature and often to substantially lower temperatures (Wilkes 2002). The story of ionic liquids is generally regarded as beginning with the first report of preparation of ethylammonium nitrate by Paul Walden in 1914. This compound has a melting point of 12 °C but owing to its high reactivity has not really found a use (Koel 2005). Ionic liquids attract remarkable interest in the early 1960s at the U.S. Air Force Academy as salt electrolytes for thermal batteries. The compounds used in the beginning had alkylpyridinium cations (Wilkes et al. 2008). Problem of these salts arose from their tendency to be reduced easily. Thus salts of more stable 1-alkyl-3-methylimidazolium type had been developed. Most of these substances melt below 100 °C and some of them are liquid at room temperature. These water-free systems consist completely of ions, making ionic liquids the solvents of choice for a variety of syntheses. Because of their low vapor pressure and the possible recycling, they are considered as green solvents.

Ionic liquids are defined as organic salts that melt below about 100 °C. They are composed entirely of ions, typically large organic cations and small organic anions (Han et al. 2009). These ionic liquids are either organic salts or mixture consisting of at least one organic component. The most popular five different well-known classes of ionic liquids are as follows: imidazolium, pyridinium, pyrrolidinium, quaternary ammonium, and tetra alkylphosphonium (Urszula 2008; Welton 1999). The physical and chemical properties of ionic liquids depend on the nature and size of the both cation and anion constituents. Their application in science and industries is merited because ionic liquids have some unique properties, such as negligible vapor pressure, good thermal stability, tunable viscosity and miscibility with water, inorganic and organic substances, high conductivity, high heat capacity, and solvents available to control reaction. Despite their high range of polarity and hydrogen bonding ability, these new solvents are liquid from 180 (glass transition) to 600 K. Possible choice of cation and anion that will result in formation of ionic liquids are numerous (Urszula 2008). Dissolution of cellulose in an ionic liquid reported for the first time in a US patent filed in 1934. The ionic liquid used was the *N*-ethylprydinum chloride in a presence of nitrogen base such as pyridine but this system displayed a relatively high melting point (mp: 118–20 °C) (Olivier-Bourbigou et al. 2010). Rogers and Seddon (2003) studied the dissolution of cellulose in ionic liquids and its regeneration. They used several ionic liquids, cellulose and different processing conditions and reported that the best case, with cellulose concentration in the range of 8–12 wt% and up to 25 % by microwave activation, where found with BMIMCl as the solvent. But it can be noted that degradation under microwave oven irradiation seems to be higher than under conventional heating condition (Gutowski et al. 2003; Swatloski et al. 2002). Rogers became a winner of the US Presidential Green Chemistry Challenge Awards in 2005, because of this great contribution (Zhu et al. 2006). These results opened up a new way for the development of a class of cellulose solvent systems and initiating a comprehensive research in this area. Generally, the ionic liquids ability to dissolve cellulose depends on the nature of the native cellulose (its degree of polymerization and its crystallinity), presence of impurities, mostly water that can significantly change the result and on the processing condition (temperature, reaction time, initial concentration of cellulose in ionic liquids, microwave heating) (Vitz et al. 2009). Today more than 20 ionic liquids are known which dissolve cellulose and were compared in terms of their kinetics of dissolution in a study. The screening results indicate that among the ionic liquids tested EMIMAc is the most efficient for dissolving cellulose. Moreover, it was observed that 1-allyl-3-methylimidazolium chloride (AMIMCl) is the most effective ionic liquid for dissolving wood chips (Zavrel et al. 2009). A screening of different anions and cations was also carried out by different researchers and recently some reviews on the dissolution of cellulose in ionic liquids have appeared (Feng and Chen 2008; Olivier-Bourbigou et al. 2010).

1.4 Mechanism of Cellulose Dissolution

The mechanism of cellulose dissolution in ionic liquids involves the oxygen and hydrogen atoms of cellulose-OH in the formation of electron donor–electron acceptor (EDA) complexes which interact with ionic liquids. For their interaction, the cellulose atoms serve as electron pair donor and hydrogen atoms act as an electron acceptor. In corresponding fashion, the cations in ionic solvents act as the electron donor center. Interaction between cellulose and ionic liquids and formation of EDA complexes are possible if the anion and cation are located close enough. The hydrogen bonds network between the glucosidic monomers in the cellulose is disrupted, resulting in the solubilisation (Feng and Chen 2008). The effect of temperature is also a nonnegligible parameter, above the critical temperature, the ion pairs in the [AMI][Cl] dissociated to Cl^- and AMI^+ ions. Then free Cl^- ions associated with cellulose hydroxyl protons and free cations complexed with cellulose hydroxyl oxygen, which led to an easier dissolution of cellulose.

1.5 Precipitation–Regeneration

Cellulose could be precipitated from the ionic liquid solution by the addition of water, or other precipitating solutions such as ethanol, methanol, or acetone. The regenerated cellulose can be separated by filtering or centrifugation and due to its nonvolatility; the ionic liquid can be recovered by removal of the antisolvent through distillation. Ionic liquids recovery is important for future cost-effective processing of cellulose material (Olivier-Bourbigou et al. 2010). Studies on biphasic system for recovery have been investigated using ionic liquid/water (Gutowski et al. 2003), ionic liquid/alcohol (Crosthwaite et al. 2004), ionic liquid/supercritical CO_2, or two immiscible ionic liquids (Arce et al. 2007).

Another study has been investigated using sugar or sugar derivatives as water–IL solution additives for preparing two-phase media. Solution of ionic liquid in water has gained more score for separation of ionic liquid in reasonable purity, from the aqueous phase. Regenerated cellulose can be prepared in different forms such as fibers, films, and monolits. Regenerated cellulose can have the same degree of polymerization and polydispersity compared to native cellulose. Although, this much depend on the processing condition of the treatment. Generally, by changing the regeneration condition, macro and microstructure especially the degree of crystallinity can be significantly changed and modulated. For example, cellulose regenerated after being dissolved in BMIMCl and AMIMCl have lower degree of crystallinity than native cellulose. The time of storage of the cellulose–ionic liquid solution also affect the regenerated cellulose microstructure. It has been shown that the resultant regenerated cellulose is mostly amorphous with greater accessibility of polysaccharide chains to cellulose and exhibits improved anzymatic hydrolysis kinetics (Lee et al. 2009; Olivier-Bourbigou et al. 2010; Swatloski et al. 2002; Zhu et al. 2006).

1.6 Main Properties Involved in the Dissolution Process

The basicity of the ionic liquids anion, their polar character and their ability to generate H-bonds are the main properties of ionic liquids relevant to the dissolution and functional modification of cellulose and carbohydrates. For example, the hydrogen bond basicity for 1-butyl-3-methylimidazolium acetate (BMIMAc) is around 1.09 when this value is 0.85 for the corresponding BMIMCl. But the most important contribution can be attributed to hydrogen bonding ability of the ionic liquid anion, such as chloride which gives H-bonding with the hydroxyl groups of the biomolecule. The ions of ionic liquids are large and asymmetric, probably free to interact with OH groups of the cellulose than that of classical chloride ions of LiCl in *N*,*N*-dimethylacetamide. Higher concentration of cellulose can be obtained in acetate-based ionic liquids compared to chloride-based ionic liquids due to the lower viscosity of the solutions which provides more efficient cellulose dissolution processes and shaping for fiber manufacture. In a typical process of cellulose dissolution, the media has to be heated, so that in these conditions the thermal stability is also an important aspect (Olivier-Bourbigou et al. 2010).

1.7 Nanofillers

Recently, a new class of organic–inorganic hybrid materials based on the ultra incorporation of nano-sized fillers (nanofillers) into a polymer matrix has been investigated. Nanotechnology is the aptitude to work on a scale of about 1–100 nm in order to understand, create, characterize and use material structure, devices, and system with unique properties derived from their base on the nanostructures. Nanocomposites could exhibit exclusive modifications in their properties, compared with conventional composites in terms of physical properties, including gas barrier, flammability resistance, thermal and environmental stability, solvent uptake, and rate of biodegradability of biodegradable (Chivrac et al. 2009).

Polymer nanocomposites are two-phase systems consisting of polymer matrix incorporated with high surface area reinforcing fillers. Such systems have attracted much interest from the materials community because they theoretically promise substantial development of mechanical properties at very low filler loadings.

Nanocomposites are a relatively new class of hybrid materials characterized by an ultra fine dispersion of nanofillers into a polymeric matrix. As the result of this dispersion, these materials possess unique properties, behaving much differently than conventional composites or microcomposites, and offering new technological and economical opportunities. The first studies on nanocomposites were carried out in 1961, when Blumstein performed the polymerization of vinyl monomer intercalated into montmorillonite structure. Since then, clay-based polymer nanocomposites have emerged as a new class of materials and attracted considerable interest and investment in research and development worldwide (Schaefer and Justice 2007).

The nanofillers incorporation into the polymer matrixes has been the particular interest because of their demonstrated significant enhancement, relative to an unmodified polymer resin, of a large number of physical properties, including barrier, flammability resistance, thermal and environmental stability, solvent uptake, and rate of biodegradability of biodegradable polymers. These improvements are generally attained at lower nanofiller content compared to that of conventional filler filled polymer. For these reasons nanocomposites are far lighter in weight than conventional composites, and make them competitive with other materials for specific applications. The main reason for these improved properties in nanocomposites is the strong interfacial interactions between matrix and nanofiller as opposed to conventional composites. Nanofillers generally have very high aspect ratios (e.g., 10–1000). A few weight percent of nanofiller that are properly dispersed throughout the matrix can create a much higher surface area for polymer filler interactions compared to conventional fillers. Thus, the incorporation of inorganic particular fillers has been shown to be an effective way for the improvement of the properties of polymeric materials (Sinha Ray and Bousmina 2005).

1.7.1 Halloysite Nanotube (HNT)

In recent years, there has been progress in the development and characterization of new materials based on clay mineral nanotubes (Abdullayev et al. 2012; De Silva et al. 2013; Ismail et al. 2008). Recently, halloysite nanotubes (HNT) have attracted interest as nanoparticles for polymers. HNT is a significant member of the kaolin group of clay minerals, with a chemical composition of $Al_2Si_2O_5(OH)_4 \cdot nH_2O$, where $n = 0$ and 2, for dehydrated form halloysite (7 Å d_{001} spacing) and hydrated mineral halloysite (10 Å d_{001} spacing) respectively (Guo et al. 2009; Joussein et al. 2005). HNTs have a highly unusual meso/macroscopic structure, which results in the formation of an appropriate hollow nanotubular geometry (Yuan et al. 2008). In addition, its typical dimensions are on the nanoscale (Joussein et al. 2005). The surface of HNT is made of siloxane and few hydroxyl groups which presents the potential for hydrogen bond formation between HNT and polymer matrix (Du et al. 2008; Ismail et al. 2008; Soheilmoghaddam and Wahit 2013).The combination of siloxane surface, tubular geometry, and also the higher stiffness of the HNT nanotubes can promote excellent nano reinforcement for polymer nanocomposites (Hedicke-Höchstötter et al. 2009). HNTs result from the wrapping of the clay layers

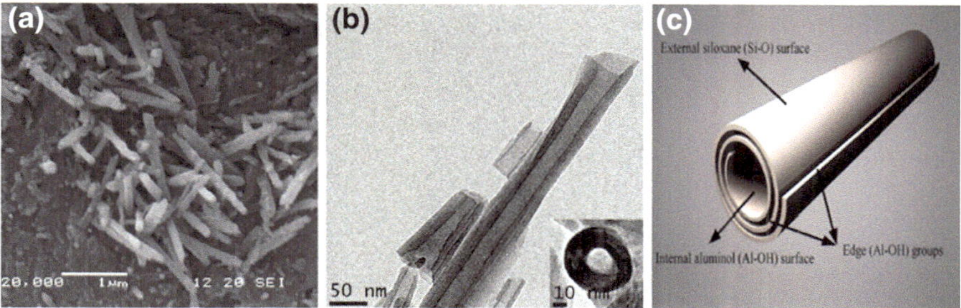

Fig. 3 **a** SEM, **b** TEM (Zhang et al. 2014), and **c** schematic (Soheilmoghaddam and Wahit 2013) images of HNT

around themselves to form hollow cylinders under appropriate geological conditions. The wrapping process is due to mismatch in the cyclicity among the oxygen sharing tetrahedral SiO_4 sheets and the neighboring octahedral $A1O_6$ sheets in the 1:1 layer (Hashemifard et al. 2011; Singh 1996; Yuan et al. 2008). Figure 3 represents multiwalled structure of HNTs.

1.7.2 Graphene Nanoplatelets

Recently discovered planar 2D form of carbon known as graphene has become one of the most exciting materials today because of its unique properties (Novoselov et al. 2004). Individual graphene sheets show high values of thermal conductivity (Balandin et al. 2008), Young's modulus (Lee et al. 2008), large surface area (Stoller et al. 2008), ballistic transport on submicron scales and massless Dirac fermion charge carrier abilities (Novoselov et al. 2005; Zhang et al. 2005). GNS is graphite derivative which are usually obtained by the rapid heating of an intercalated graphite compound (Debelak and Lafdi 2007). GNS is an atomically thick, two-dimensional (2-D) sheet composed of sp^2 carbon atoms arranged in a honeycomb structure (Fig. 4) with high mechanical properties (1 TPa in Young's modulus

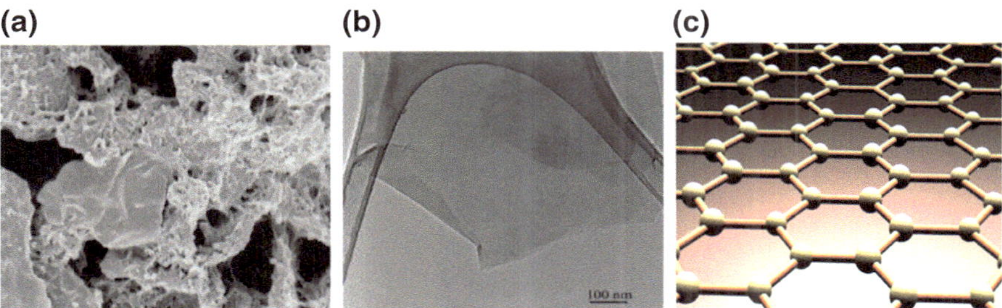

Fig. 4 **a** SEM (Oyefusi et al. 2014), **b** TEM (Tian et al. 2014), and **c** schematic (Soheilmoghaddam et al. 2014a, b) images of graphene nanosheet

and ultimate strength of 130 GPa). It is reported to be one of the strongest materials in the world (Lee et al. 2008; Wang et al. 2012). In addition to these unique mechanical properties, its high thermal resistance, chemical stability, gas impermeability, high surface area, and low cost make it excellent nanofiller for polymeric composites.

1.8 Nanocomposite Preparation Methods

Intercalation of polymers in layered hosts, such as layered silicate and graphene nanoplatelets, has proven to be a successful method to prepare PLS nanocomposites; the preparation methods are divided into three main groups according to the processing techniques and starting materials, which include solvent interaction process, in situ polymerization process, and melt interaction method (Sinha Ray and Okamoto 2003).

1.8.1 Solvent Intercalation Process

This elaboration process is based on a solvent system in which the polymer is soluble and the nanofiller is swellable. The polymer is first dissolved in an appropriate solvent. In parallel, the nanofiller is swollen and dispersed into the same solvent or another one to obtain a miscible solution. When the polymer and nanofiller solutions are mixed, the polymer chains intercalate and displace the solvent within the interlayer of the nanofiller. Upon solvent removal, the intercalated structure remains, resulting in a nanocomposite. In this method the nature of the solvents is critical in facilitating the insertion of polymer molecules between the nanofiller layers, polarity of medium being the determining factor for intercalation. Since some polysaccharides, such as cellulose or pectin, cannot be melt processed due to high thermal or thermomechanical degradation, the solvent process has been extensively used to produce polysaccharide/nanofiller hybrid materials.

1.8.2 In Situ Polymerization Process

In this method, the nanofillers are swollen within the monomer solution or a liquid monomer, so the polymer formation can occur between the intercalated sheets. Polymerization can be initiated either by radiation or heat, by the diffusion of a suitable initiator, or by an organic initiator or catalyst fixed through cation exchange inside the interlayer before the swelling step.

1.8.3 Melt Interaction Method

Both polymer and the nanofiller are introduced simultaneously into a melt mixing device (extruder, internal mixer, etc.). This process involves annealing a mixture of polymer and nanofiller above the softening point the polymer, statically or under the shear. During annealing, the polymer chains diffuse from the bulk polymer melt into the galleries between the nanofiller layers (Liu et al. 2012; Sinha Ray and Okamoto 2003).

1.9 Nanocomposites Structures

Depending on the components used (nanofiller and polymer matrix) and method of the preparation, three main types of composites may be obtained (Thakur et al. 2014c, d). When the polymer is unable to intercalate between the nanofiller layers, a phase separated composite is obtained, whose properties stay in the same range as traditional microcomposites. Beyond this classical family of composites, two types of nanocomposite can be recovered. Intercalated structure in which a single (and sometimes more than one) extended polymer chain is intercalated between the nanofiller layers resulting in a wall ordered multilayer morphology built up with alternating polymeric and inorganic layers. When the nanofiller layers are completely and uniformly dispersed in a continuous polymeric matrix, an exfoliated or delaminated structure is obtained.

2 Regenerated Cellulose Nanocomposites

2.1 Regenerated Cellulose/Halloysite Nanoclay

Incorporation of nanofillers into polymer matrix has been proved to be a powerful tool in order to increase the polymer properties (Lin et al. 2011a, b). It is widely accepted that addition of nanofiller into bio-based matrixes in order to fabricate nano-biocomposite materials could be a powerful solution to improve these properties (Alexandre and Dubois 2000; Bordes et al. 2009; Sinha Ray and Okamoto 2003). Studies on tubular silica-based naturally occurring nanoparticles as reinforcing material is still new (Ismail et al. 2008; Prashantha et al. 2011). Halloysite particles are readily obtainable and are much cheaper than other nanoparticles such as CNTs. More importantly, the unique crystal structure of HNTs resembles that of CNTs, and therefore halloysite particles may have the potential to provide cheap alternatives to expensive CNTs because of their tubular structure in nanoscale. Moreover, due to its similarity to other layered clay minerals such as MMT, halloysite has the potential to be further intercalated or exfoliated chemically or physically (Tang et al. 2011).

Table 1 The mechanical properties of RC and RC/HNT nanocomposite films (Soheilmoghaddam and Wahit 2013)

Samples	Young's modulus (GPa)	Tensile strength (MPa)	Elongation at break (%)
RC	1.8 ± 1.20	35.30 ± 1.12	5.26 ± 0.72
RC/HNT-2	2.1 ± 0.75	39.90 ± 0.80	5.95 ± 0.56
RC/HNT-4	2.9 ± 0.68	41.30 ± 0.93	6.63 ± 1.28
RC/HNT-6	3.4 ± 0.45	59.60 ± 0.75	8.34 ± 1.00
RC/HNT-8	4.1 ± 0.35	60.50 ± 1.33	8.10 ± 1.26

Soheilmoghaddam et al. (2013) have been reported a novel nanocomposite regenerated cellulose/halloysite nanotube (RC/HNT) biofilms using an environmentally 1-butyl-3-methylimidazolium chloride (BMIMCl) friendly ionic liquid by a simple green method. The Young's modulus and tensile strength of the nanocomposites significantly increased with increases in the HNT amount in the bionanocomposite films. The Young's modulus of nanocomposite increased by 120 % with 8 wt% of HNT (RC/HNT-8) incorporation as compared to that of the pure RC film. The tensile strength of RC/HNT-8 also increased by 71 % compared to the pure RC film. This significant enhancement in the mechanical properties of the RC/HNT bionanocomposite films was attributed to high aspect ratio of HNT and strong interaction between the hydroxyl groups of HNT and RC matrix. The elongation at break of the nanocomposite films was listed in Table 1. The elongation at break of RC/HNT nanocomposites increased when HNT was incorporated. The incorporation of 6 wt% HNT was resulted in 58 % increase in elongation at break; thereafter, the elongation at break was nearly constant. It was assumed that the improvement in stiffness and ductility of the nanocomposites is due to the good dispersion of HNTs inside the RC as well as the strong interaction between them.

Water barriers are also an important application of regenerated cellulose films composites. In the same study, the water absorption (%) of RC and RC/HNT nanocomposites for 2 and 24 h was investigated by Soheilmoghaddam et al. (2013) (Fig. 5). It was reported that the incorporation of HNT improved the water resistance of RC. Water absorption of the nanocomposite films decreased from 87.5 to 77 % as HNT content increased from 0 to 8 wt% after 24 h. This was attributed to the potential of hydrogen bond formation between HNT and RC matrix which can result in the formation of a strong structure which in turn reduces the diffusion of water molecules in the material.

In recent study, Hanid et al. (2014) developed the regenerated cellulose/halloysites (RC/HNT) nanocomposites with different nanofillers loading were fabricated by dissolving the cellulose in 1-ethyl-3-methylimidazolium chloride (EMIMCl) ionic liquid. The X-ray diffractometer (XRD) patterns of the films revealed that characteristic diffraction angle was displayed at $2\theta = 12°$ corresponds to $(1\bar{1}0)$ plane, while peak at $2\theta = 20.0°$ and $2\theta = 22.0°$ corresponded to (110) and (020) planes, respectively. These peaks are also known as the crystalline structure of cellulose II (Rahatekar et al. 2009). The transformation of crystalline structure from cellulose I to

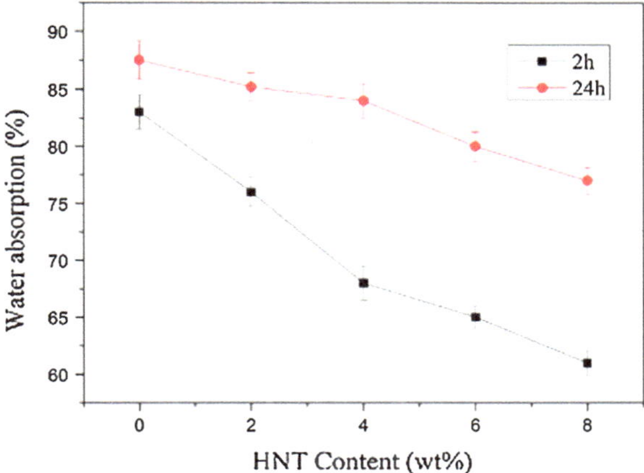

Fig. 5 Water absorption of RC and RC/HNT nanocomposites at HNT loadings (Soheilmoghaddam and Wahit 2013)

cellulose II was exhibited significant similarity to the diffraction angles of RC Peaks were located at lower angle, $2\theta = 12°$ and $2\theta = 22.0°$, indicated the presence of structures with limited intercalation and was attributed to the formation of nanocomposites. The appearance of diffraction peaks for HNT were expected at $2\theta = 20.2°$ and $2\theta = 25.0°$. The sharp peak at $2\theta = 20.2°$ for HNT still were observed due to the fact that the peak was overlapped with the peak for RC since both of the peak were appeared at the same diffraction angle. Even though without the incorporation of HNT nanotubes, peak $2\theta = 20.2°$ was clearly observed in RC, which was the pure RC. Nevertheless, it was observed clearly that the peak at $2\theta = 25.0°$ was totally disappeared with the addition of HNT in the polymer matrix as illustrated of regenerated cellulose with 1 wt% HNT content (RC/HNT-1), RC/HMT-3 and RC/HNT-5 in Fig. 6, indicated that all the HNT nanotubes were almost fully intercalated. The dispersion and intercalation of the HNT nanotubes was observed in the TEM images.

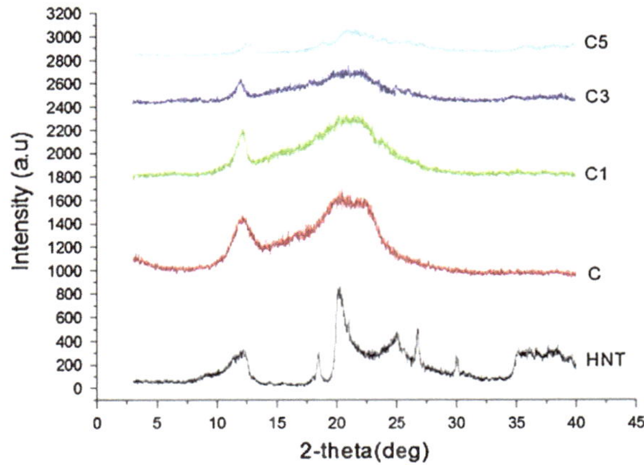

Fig. 6 XRD patterns of C, C1, C3, C5, and HNT (Hanid et al. 2014)

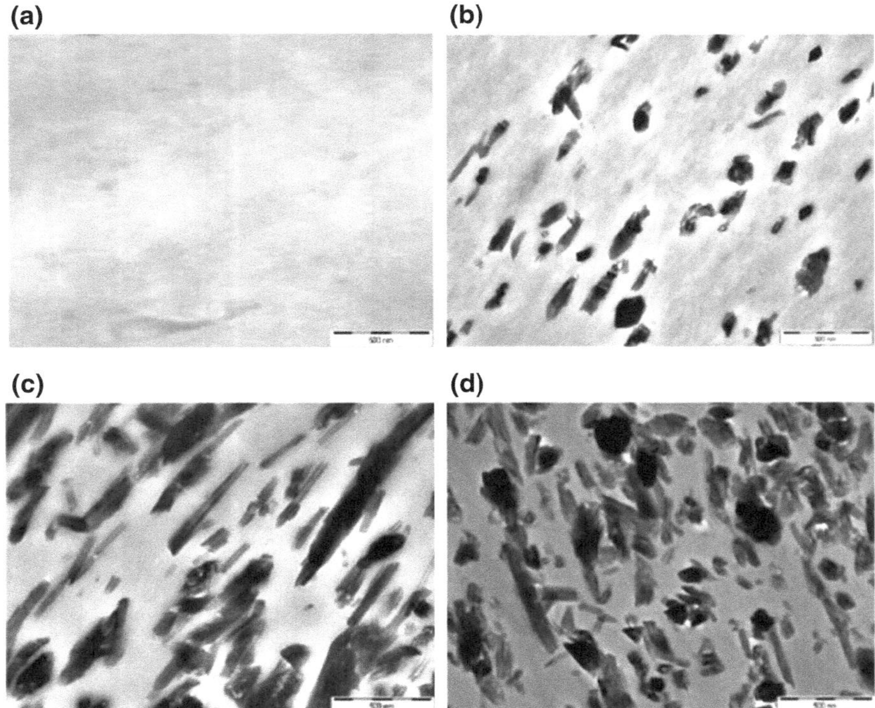

Fig. 7 Transmission electron micrographs of halloysite nanocomposites **a** C1, **b** C1, **c** C3, and **d** C5 (magnification: 25 K×) (Hanid et al. 2014)

The TEM image of pure RC and its nanocomposites was studied by Baiti et al. (Hanid et al. 2014) as shown in Fig. 7. It was seen that the HNT tubes were tend to align by themselves compared to the RC/HNT-3 with higher HNT loading (3 wt%) in Fig. 7c where it was clearly seen that the zigzag structures were formed between the HNTs. Some of them were formed by edge-to-edge interactions between the HNTs, while others were formed face to edge. As the HNT loading was increased up to 5 wt% as illustrated in Fig. 7d of C5, they did not only increase the formation of the zigzag structures, but also have the tendency to agglomerate.

The thermal properties of RC/HNT nanocomposite files have characterized by Soheilmoghaddam et al. (2013) using the thermogravimetric analysis (TGA) results (Fig. 8). The data revealed that HNT loading has a significant effect on the thermal stability of RC nanocomposites. Thermal stability was enhanced to a considerable extent with the incorporation of 8 wt% HNT into the RC matrix. The temperature corresponding to 20 % weight loss for the pure sample was 310 °C, and this temperature became 346 °C in the presence of 8 wt% HNT. Similarly, at 50 % weight loss, the T_{50} value of RC films (366 °C) was improved by about 27–393 °C with 8 wt% of HNT loading. Besides, activation energy for the thermal degradation

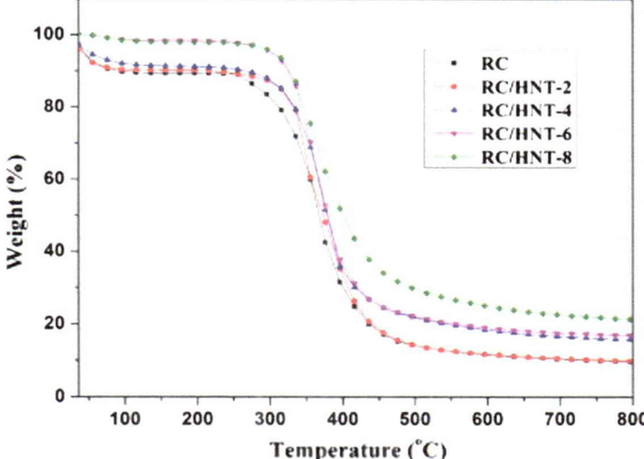

Fig. 8 TGA curves of RC/HNT nanocomposites with various HNT loadings under nitrogen environment (Soheilmoghaddam et al. 2013)

of the RC and RC/HNT nanocomposite films was determined. The value of calculated activation energy of the pure RC was 37.91 kJ mol^{-1}. This amount increased to 45.95, 49.78, 61.65 kJ mol^{-1} for RC/HNT-2, RC/HNT-4, RC/HNT-6, respectively, and finally reached 63.18 kJ mol^{-1} for RC/HNT-8. Increases in activation energy values with increases in HNT loading indicated higher thermal stability for RC/HNT nanocomposite films which demonstrated a certain level of interaction between RC and HNTs (Fig. 8).

The effect of HNT content mechanical properties and moisture absorption of RC and RC/HNT nanocomposites was investigated by Soheilmoghaddam et al. (2013). It was reported that the addition of HNTs increased the tensile strength of the nanocomposite films up to a loading of 6 wt%. The RC/HNT composite containing 6 wt% HNTs exhibited a remarkable 55.3 % increase in tensile strength at room temperature compared to the pure RC film. The increase in tensile strength, especially in 6 wt%, was explained by various factors such as good HNT dispersion inside the RC, inter-tubular interactions between HNTs and RC, the edge-to-edge and face-to-edge interactions between HNTs which create zigzag structures, and the three-dimensional orientations of HNTs inside the nanocomposites. The enhancement in elongation at break values of RC/HNT nanocomposites as compared to RC film was observed. The elongation at break values of the films increased from 7.4 to 10.8 % as the HNT content increased from 0 to 8 wt%. This increment, especially in 8 wt% HNT content (45.9 %) in comparison with controlled RC film, was a unique behavior of system which was attributed to the homogenous dispersion of HNTs inside the RC and some interaction with the polymer matrix.

In addition, the moisture content as a function of time was determined in order to measure the equilibrium moisture content of RC and nanocomposite films. The moisture absorption of composites containing different mass fractions of HNT in the environment maintained at RH = 75 % and 25 °C temperature. It was observed that the equilibrium water content of nanocomoposites was less than that of RC film. Approximately 7.1 % reduction in total moisture absorbency was achieved by a 2 wt% HNT in the film, which increased to more than 19.5 and 26.4 % when the HNT content increased to 4 and 6 wt% respectively. The minimum moisture absorption value (35.6 %) was achieved at the 8 wt% level of HNT in the film. This was attributed to the formation of hydrogen bonds between the hydroxyls of the HNT layers and RC matrix. This strong structure reduced the destructive effect of free water molecules on the hydrogen bonds between the HNT and RC molecules.

2.2 Regenerated Cellulose/Graphene Nanocomposites

Over the last few years, polymer/exfoliated graphite nanosheets (GNS) nanocomposites have attracted considerable attention from both the scientific and engineering communities as a result of the substantial property enhancements obtained from relatively low nanofiller loadings (Chandrasekaran et al. 2013; Hatui et al. 2014; Kalaitzidou et al. 2007; Rath and Li 2011). Nevertheless, the reinforcing efficiency is dependent on the degree of dispersion of the GNS in the polymer matrix (Hatui et al. 2014; Wang et al. 2012). Recently, regenerated cellulose (RC) reinforced with low graphene content using BMIMCL ionic liquid as the medium for the dispersion of GNS has been successfully prepared by our group (Soheilmoghaddam et al. 2014a, b). The effect of graphene nanoplatelets (GNP) on the mechanical properties of RC films was studied (Fig. 9). The tensile strength and elongation at break of the nanocomposite films were improved as compared to the pure RC film. The tensile strength for the pure RC was 38.1 MPa while its elongation at break was 5.2 %. The incorporation of 0.75 wt% GNS increased the tensile strength and elongation at break to 97.5 MPa and 7.8 % respectively. This was 155 % increase in tensile strength and 50 % increase in elongation at break (Fig. 9a, b). The increase in GNS content also led to a significant increase in Young's modulus, from 1.8 GPa for pure RC to 4.9 GPa for RC/GNS containing 1 wt% GNS. This was around 2.7 times higher than the pure RC films (Fig. 9c). The enhancement in stiffness and ductility of the nanocomposites was explained due to the homogeneous dispersion of GNS all through the RC matrix and the strong interfacial adhesion between them.

Soheilmoghaddam et al. (2014a, b) also reported the equilibrium content of RC and RC/GNS as a function the time conditioned at 75 % RH. It was observed that the water resistance of the nanocomposites improved with GNS incorporation. Water absorption of pure RC was 147 % but decreased by 40 % for the 0.75 wt% GNS nanocomposite. The presence of GNS in the RC nanocomposite films enhanced their water resistance. This was attributed to the less hydrophilicity of GNS in the RC matrix and also the potential hydrogen bond formation between GNS and RC. This partially blocks the hydrophilic site of RC owing to the presence of less OH group interaction with the water molecules.

In the same study by Soheilmoghaddam et al. (2014a, b) the oxygen (O_2) permeability values of RC film and RC/GNS nanocomposite films was reported. The oxygen permeability values for the nanocomposite films decreased with the

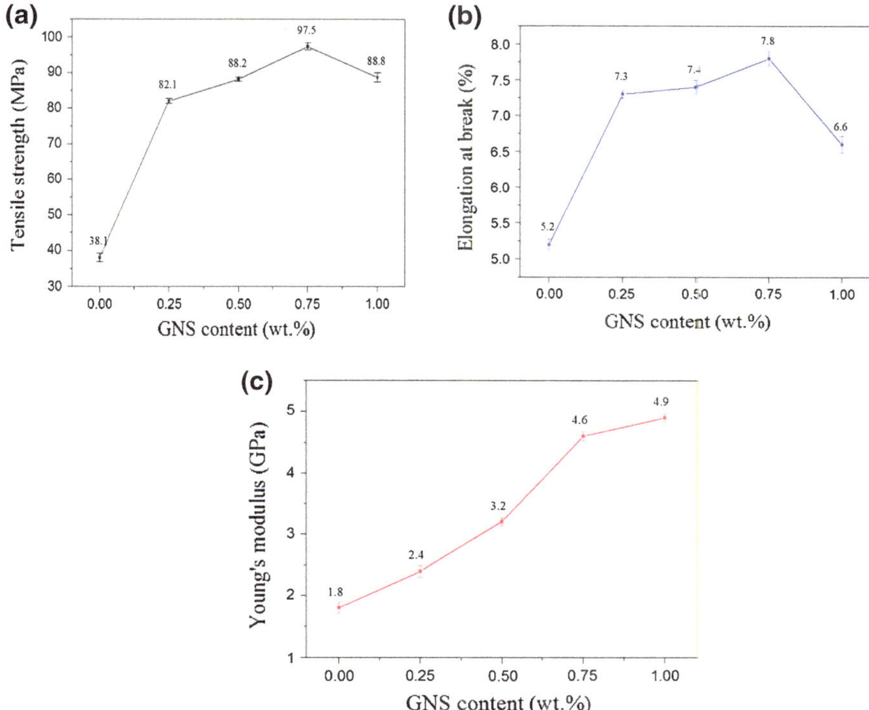

Fig. 9 **a** Tensile strength, **b** elongation at break, **c** Young's modulus of RC/GNS nanocomposite films (Soheilmoghaddam et al. 2014a, b)

addition of GNS into the RC matrix. Oxygen permeability of the nanocomposites reduced from 1.1 × 10^{-18} (m^3 m/m^2 s Pa) for pure RC to 0.64 × 10^{-18} (m^3 m/m^2 s Pa) for the nanocomposite with 0.75 wt% GNS. The well-dispersed GNS with high aspect ratio forced the oxygen molecules to travel longer distances across the films as it follows a tortuous path around them. This improved in the oxygen barrier property of the composites was explained by the more compact structure of the nanocomposites compared to the pure RC films.

The work of Zhang et al. (2012) had investigated the properties of regenerated cellulose/graphene composite by dispersing graphene into cellulose solution in DMAC/LiCl. To reveal the dispersion of graphene in cellulose matrix, TEM was used to watch the microstructure of the nanocomposite films. The TEM image of cellulose nanocomposite with 1.6 wt% graphene revealed that the graphene nanosheets were homogeneously dispersed in cellulose matrix. The exfoliation of graphene was achieved in cellulose matrix (Fig. 10).

In addition, the effect of graphene content on mechanical properties of RC/graphene nanocomposites has been investigated in the same research (Zhang et al. 2012). Abundant thin stacks of a few sheets of monolayer graphene in RC had significant

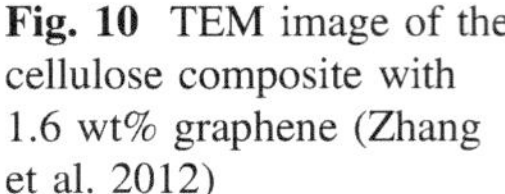

Fig. 10 TEM image of the cellulose composite with 1.6 wt% graphene (Zhang et al. 2012)

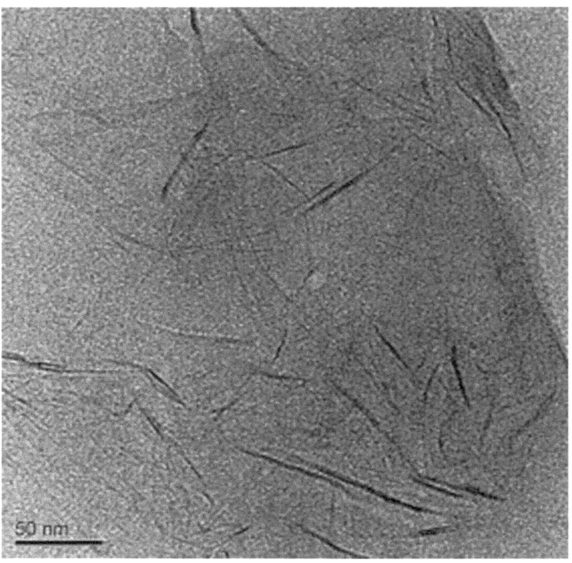

effect on the improvement of mechanical properties of RC. The addition of graphene significantly increased the Young's modulus and tensile strength. For example, the Young's modulus of the nanocomposite film with 1.6 wt% of graphene was 7.2 GPa, corresponding to an increase of 110 % compared to 3.4 GPa of pure cellulose film. The tensile strength was increased to 148 MPa, corresponding to an increase of 66 % compared to 89 MPa of neat RC film. The elongation at break did not change much with addition of graphene. It was obvious that the addition of graphene has significant improvement on the mechanical behavior of the neat RC film. This was attributed to the strong interaction between graphene and the RC matrix, which restricts the movement of polymer chains.

Enhancement in mechanical and thermal properties of regenerated cellulose/graphene composite fibers has been reported by Tian et al. (2014). In this research the wet spinning method was applied to fabricate regenerated cellulose fibers filled with low graphene loading. The addition of graphene reinforced the ultimate tensile strength. The tensile strength of the RC/G 0.2 fiber was 360 MPa, which is about 50 % larger than the neat RC fibers (240 MPa). Remarkably, this improvement was achieved at a very low graphene loading (0.1–0.2 wt%). Young's module of cellulose fibers with incorporation of 0.1 and 0.2 wt% graphene increased the Young's Modulus from 1.25 to 1.34 GPa for RC/G 0.1, while 25 % improvement for RC/G 0.2 in comparison to that of neat fibers. The increment of tensile strength was attributed to the good coherence between the regenerated cellulose matrix and graphene fillers, and the presence of strong interactions. Thermal properties of the composite fibers were elucidated here by the thermal stability (TGA) and dynamic heat transfer. It was observed that the neat fibers and the composite fibers showed thermal degradation and significant weight loss with temperature. Furthermore, the onset and the endset of thermal degradation temperature were determined from the intersection of two tangents, and the corresponding TGA analytical values of samples (onset, endset, decomposition temperatures at 10 and 50 % weight loss, char content). The onset temperature of neat RC fiber was 265 °C, while for the composite fibers it increased to 298 °C for RC/G 0.1 fiber and 309 °C for RC/G 0.2 fiber, and the incorporation of 0.1 and 0.2 wt% graphene in neat regenerated cellulose increased the 50 % decomposition temperature by 6 and 25 °C, respectively. All the results indicated that the thermal stability of the neat fibers was enhanced by the low graphene filler.

Mahmoudian et al. (2012) investigated the preparation of regenerated cellulose (RC)/GNP nanocomposites via room temperature ionic liquid, 1-ethyl-3-methylimidazolium acetate (EMIMAc) using solution casting method. It was observed that the addition of GNPs increased the tensile strength of RC films. The RCG nanocomposite films with 3 wt% GNPs exhibited 34 % increase in tensile strength compared to RC. The tensile strength remained unchanged with further increase in GNPs loading which may be attributed to the restacking of graphene sheets after a certain amount due to increased filler–filler attraction forces within GNPs resulting in agglomeration that reduces the surface area available for stress transfer (Hermanutz et al. 2008; Vadukumpully et al. 2011). The Young's modulus of RC films improved significantly by increasing GNP loadings. The RCG nanocomposite film containing 3 wt% GNP exhibited 52 % increase in modulus compared to the pure regenerated cellulose films. The presence of graphene sheets in cellulose matrix offers resistance to the segmental movement of the polymer chains which led to enhancement in modulus.

The elongation at break values of RCG nanocomposites, however, decreased with the presence of graphene nanoplatelets. The elongation at break values decreased from 8.1 to 4.4 % with 3 wt% GNP loading in RCG nanocomposite film. This was attributed to the interaction between graphene nanoplatelets and polymer matrix, which restricts the movement of cellulose chains. The water absorption (%) of RC and RCG nanocomposites for 2 and 24 h was calculated. It was seen that the presence of graphene nanoplatelets in cellulose improve water resistance of RCG nanocomposite films.

The water absorption of the nanocomposite films after 24 h decreased from 95 to 71 % as the GNPs content increased from 0.5 to 5 wt%. The lower water absorption of RCG nanocomposites versus that of pure RC was due to the lower water affinity of GNPs than that of cellulose and also strong interactions between cellulose and graphene nanoplatelets. It was deduced that the addition of GNPs decreases the water absorption of the RC films due to the presence of impermeable graphene nanoplatelets in the nanocomposites which can lower the rate of water transport in the polymer matrix.

The effect of GNPs content on the permeability values of RC and RCG nanocomposites with oxygen (O_2) and carbon dioxide (CO_2) gases was investigated by Mahmoudian et al. (2012). The permeability values decreased gradually with increasing GNPs contents due to the impermeable graphene nanoplatelets which reduce the cross-sectional area available for permeation and provide torturous paths.

It was reported that the permeability of CO_2 through RCG nanocomposites is higher compared to O_2. It was believed that the observed variation of permeability between gases is very much dependent on the size of the gas molecules and its solubility toward cellulose as the CO_2 is smaller kinetic diameter combined with its greater solubility toward cellulose.

3 Conclusions and Future Perspective

Regenerated cellulose is one of the most promising natural polymeric resources as alternative for petroleum-based polymers. It can be used in packaging, membranes and medical products as well as in electronic devices. However, only about 2 % of cellulose renewed annually by nature has been used as raw material in manufacturing of cellulose material and goods. The main problem to the more extensive use of cellulose is the lack of suitable solvents for cellulose regeneration. To date, most of the solvents used to prepare regenerated cellulose materials have many drawbacks such as toxicity, volatility, side reactions, high cost, and difficult solvent recovery. In recent years, ionic liquids (IL) have begun to be used in cellulose materials processing as environmentally friendly solvents. However, high price of IL at the current market, enable regenerated cellulose to compete with petroleum-based plastics. Another challenge is the limited ductility (elongation at break) of regenerated cellulose products which has urged researchers to develop regenerated cellulose blends or nanocomposites to improve elongation of RC. RC films are now commercially used as biodegradable material. However, the major reason for the low utilization of regenerated cellulose films is their relatively higher water absorption and gas permeability than conventional thermoplastic packaging materials. Therefore, it appears necessary to improve the properties of this polymer to make it fully competitive with common thermoplastics. Despite the setback, the properties of these nanocomposites up to now confirm that the promise of a dream material for future can be fulfilled. Further research is therefore necessary to develop a proper understanding of the formulation/structure/property relationship/toughening mechanisms and interactions involved in RC nanocomposites system.

References

Abdullayev E, Joshi A, Wei W, Zhao Y, Lvov Y (2012) Enlargement of halloysite clay nanotube lumen by selective etching of aluminum oxide. ACS Nano 6(8):7216–7226

Alexandre M, Dubois P (2000) Polymer-layered silicate nanocomposites: preparation, properties and uses of a new class of materials. Mater Sci Eng R Rep 28(1–2):1–63

Arce A, Earle MJ, Katdare SP, Rodríguez H, Seddon KR (2007) Phase equilibria of mixtures of mutually immiscible ionic liquids. Fluid Phase Equilib 261(1–2):427–433

Azizi Samir MAS, Alloin F, Dufresne A (2005) Review of recent research into cellulosic whiskers, their properties and their application in nanocomposite field. Biomacromolecules 6(2):612–626

Balandin AA, Ghosh S, Bao W, Calizo I, Teweldebrhan D, Miao F, Lau CN (2008) Superior thermal conductivity of single-layer graphene. Nano Lett 8(3):902–907
Bordes P, Pollet E, Avérous L (2009) Nano-biocomposites: biodegradable polyester/nanoclay systems. Prog Polym Sci 34(2):125–155
Bredereck K, Hermanutz F (2005) Man–made cellulosics. Rev Prog Color Relat Top 35(1):59–75
Chandrasekaran S, Seidel C, Schulte K (2013) Preparation and characterization of graphite nano-platelet (GNP)/epoxy nano-composite: mechanical, electrical and thermal properties. Eur Polymer J 49(12):3878–3888
Chivrac F, Pollet E, Avérous L (2009) Progress in nano-biocomposites based on polysaccharides and nanoclays. Mater Sci Eng R Rep 67(1):1–17
Crosthwaite JM, Aki SNVK, Maginn EJ, Brennecke JF (2004) Liquid phase behavior of imidazolium-based ionic liquids with alcohols. J Phys Chem B 108(16):5113–5119
De Silva RT, Pasbakhsh P, Goh KL, Chai S-P, Ismail H (2013) Physico-chemical characterisation of chitosan/halloysite composite membranes. Polym Testing 32(2):265–271
Debelak B, Lafdi K (2007) Use of exfoliated graphite filler to enhance polymer physical properties. Carbon 45(9):1727–1734
Du M, Guo B, Lei Y, Liu M, Jia D (2008) Carboxylated butadiene–styrene rubber/halloysite nanotube nanocomposites: interfacial interaction and performance. Polymer 49(22):4871–4876
Feng L, Chen Z (2008) Research progress on dissolution and functional modification of cellulose in ionic liquids. J Mol Liq 142(1–3):1–5
Fink HP, Weigel P, Purz HJ, Ganster J (2001) Structure formation of regenerated cellulose materials from NMMO-solutions. Prog Polym Sci 26(9):1473–1524
Graenacher C, Sallmann R (1939) Cellulose solutions and process of making same. Google Patents
Gross RA, Kalra B (2002) Biodegradable polymers for the environment. Science 297 (5582):803–807
Guo B, Chen F, Lei Y, Liu X, Wan J, Jia D (2009) Styrene-butadiene rubber/halloysite nanotubes nanocomposites modified by sorbic acid. Appl Surf Sci 255(16):7329–7336
Gutowski KE, Broker GA, Willauer HD, Huddleston JG, Swatloski RP, Holbrey JD, Rogers RD (2003) Controlling the aqueous miscibility of ionic liquids: aqueous biphasic systems of water-miscible ionic liquids and water-structuring salts for recycle, metathesis, and separations. J Am Chem Soc 125(22):6632–6633
Han J, Zhou C, French AD, Han G, Wu Q (2013) Characterization of cellulose II nanoparticles regenerated from 1-butyl-3-methylimidazolium chloride. Carbohydr Polym 94(2):773–781
Han S, Li J, Zhu S, Chen R, Wu Y, Zhang X, Yu Z (2009) sotential applications of ionic liquids in wood related industries
Hanid NA, Wahit MU, Guo Q, Mahmoodian S, Soheilmoghaddam M (2014) Development of regenerated cellulose/halloysites nanocomposites via ionic liquids. Carbohydr Polym 99:91–97
Hashemifard SA, Ismail AF, Matsuura T (2011) Mixed matrix membrane incorporated with large pore size halloysite nanotubes (HNT) as filler for gas separation: experimental. J Colloid Interf Sci 359(2):359–370
Hatui G, Bhattacharya P, Sahoo S, Dhibar S, Das CK (2014) Combined effect of expanded graphite and multiwall carbon nanotubes on the thermo mechanical, morphological as well as electrical conductivity of in situ bulk polymerized polystyrene composites. Compos A Appl Sci Manuf 56:181–191
Hedicke-Höchstötter K, Lim GT, Altstädt V (2009) Novel polyamide nanocomposites based on silicate nanotubes of the mineral halloysite. Compos Sci Technol 69(3–4):330–334
Hermanutz F, Gähr F, Uerdingen E, Meister F, Kosan B (2008) New developments in dissolving and processing of cellulose in ionic liquids. Macromol Symp 262(1):23–27
Hyden WL (1929) Manufacture and properties of regenerated cellulose films. Ind Eng Chem 21 (5):405–410
Ioelovich M (2008) Cellulose as a nanostructured polymer: a short review
Ismail H, Pasbakhsh P, Fauzi MNA, Abu Bakar A (2008) Morphological, thermal and tensile properties of halloysite nanotubes filled ethylene propylene diene monomer (EPDM) nanocomposites. Polym Test 27(7):841–850

Johnson DL (1970) Method of preparing polymers from a mixture of cyclic amine oxides and polymers. Google Patents

Joussein E, Petit S, Churchman J, Theng B, Righi D, Delvaux B (2005) Halloysite clay minerals: a review. Clay Miner 40(4):383–426

Kalaitzidou K, Fukushima H, Drzal LT (2007) Multifunctional polypropylene composites produced by incorporation of exfoliated graphite nanoplatelets. Carbon 45(7):1446–1452

Klemm D, Heublein B, Fink H-P, Bohn A (2005) Cellulose: fascinating biopolymer and sustainable raw material. Angew Chem Int Ed 44(22):3358–3393

Klemm D, Philipp B, Heinze T, Heinze U, Wagenknecht W (2004) General considerations on structure and reactivity of cellulose: section 2.1–2.1.4. Comprehensive cellulose chemistry. Wiley-VCH Verlag GmbH & Co. KGaA, Weinheim, pp 9–29

Koel M (2005) Ionic liquids in chemical analysis. Crit Rev Anal Chem 35(3):177–192

Krassig HA (1996) Cellulose, structure, accessibility and reactivity: polymer monographs. Gordan and Breach Science Publishers, Netherlands, p 361

Kubisa P (2005) Ionic liquids in the synthesis and modification of polymers. J Polym Sci Part A Polym Chem 43(20):4675–4683

Lee C, Wei X, Kysar JW, Hone J (2008) Measurement of the elastic properties and intrinsic strength of monolayer graphene. Science 321(5887):385–388

Lee SH, Doherty TV, Linhardt RJ, Dordick JS (2009) Ionic liquid-mediated selective extraction of lignin from wood leading to enhanced enzymatic cellulose hydrolysis. Biotechnol Bioeng 102 (5):1368–1376

Liu D-T, Xia K-F, Cai W-H, Yang R-D, Wang L-Q, Wang B (2012) Investigations about dissolution of cellulose in the 1-allyl-3-alkylimidazolium chloride ionic liquids. Carbohydr Polym 87(2):1058–1064

Lin M-F, Thakur VK, Tan EJ, Lee PS (2011a) Dopant induced hollow BaTiO3 nanostructures for application in high performance capacitors. J Mater Chem 21:16500–16504

Lin M-F, Thakur VK, Tan EJ, Lee PS (2011b) Surface functionalization of BaTiO3 nanoparticles and improved electrical properties of BaTiO3/polyvinylidene fluoride composite. RSC Adv 1:576–578

Mahmoudian S, Wahit MU, Imran M, Ismail AF, Balakrishnan H (2012) A facile approach to prepare regenerated cellulose/graphene nanoplatelets nanocomposite using room-temperature ionic liquid. J Nanosci Nanotechnol 12(7):5233–5239

McCorsley CC (1979) Process for making amine oxide solution of cellulose. Google Patents

McCorsley CC (1981) Extrusion, molecular orientation. Google Patents

Moon RJ, Martini A, Nairn J, Simonsen J, Youngblood J (2011) Cellulose nanomaterials review: structure, properties and nanocomposites. Chem Soc Rev 40(7):3941–3994

Nishio Y (2006) Material functionalization of cellulose and related polysaccharides via diverse microcompositions. In: Klemm D (ed) Polysaccharides II, vol 205. Springer, Berlin, pp 97–151

Nishiyama Y (2009) Structure and properties of the cellulose microfibril. J Wood Sci 55 (4):241–249

Novoselov KS, Geim AK, Morozov SV, Jiang D, Katsnelson MI, Grigorieva IV, Dubonos SV, Firsov AA (2005) Two-dimensional gas of massless Dirac fermions in graphene. Nature 438 (7065):197–200

Novoselov KS, Geim AK, Morozov SV, Jiang D, Zhang Y, Dubonos SV, Grigorieva IV, Firsov AA (2004) Electric field effect in atomically thin carbon films. Science 306(5696):666–669

Olivier-Bourbigou H, Magna L, Morvan D (2010) Ionic liquids and catalysis: recent progress from knowledge to applications. Appl Catal A 373(1–2):1–56

Oyefusi A, Olanipekun O, Neelgund GM, Peterson D, Stone JM, Williams E, Carson L, Regisford G, Oki A (2014) Hydroxyapatite grafted carbon nanotubes and graphene nanosheets: promising bone implant materials. Spectrochim Acta Part A Mol Biomol Spectrosc 132:410–416

Prashantha K, Schmitt H, Lacrampe MF, Krawczak P (2011) Mechanical behaviour and essential work of fracture of halloysite nanotubes filled polyamide 6 nanocomposites. Compos Sci Technol 71(16):1859–1866

Rahatekar SS, Rasheed A, Jain R, Zammarano M, Koziol KK, Windle AH, Gilman JW, Kumar S (2009) Solution spinning of cellulose carbon nanotube composites using room temperature ionic liquids. Polymer 50(19):4577–4583

Rath T, Li Y (2011) Nanocomposites based on polystyrene-b-poly(ethylene-r-butylene)-b-polystyrene and exfoliated graphite nanoplates: effect of nanoplatelet loading on morphology and mechanical properties. Compos A Appl Sci Manuf 42(12):1995–2002

Rinaldi R, Schüth F (2009) Acid hydrolysis of cellulose as the entry point into biorefinery schemes. ChemSusChem 2(12):1096–1107

Rogers RD, Seddon KR (2003) Ionic liquids: solvents of the future? Science 302(5646):792–793

Rosenau T, Hofinger A, Potthast A, Kosma P (2003) On the conformation of the cellulose solvent N-methylmorpholine-N-oxide (NMMO) in solution. Polymer 44(20):6153–6158

Rosenau T, Potthast A, Sixta H, Kosma P (2001) The chemistry of side reactions and byproduct formation in the system NMMO/cellulose (Lyocell process). Prog Polym Sci 26(9):1763–1837

Schaefer DW, Justice RS (2007) How nano are nanocomposites? Macromolecules 40(24):8501–8517

Schüth F, Rinaldi R, Meine N, Käldström M, Hilgert J, Rechulski MDK (2014) Mechanocatalytic depolymerization of cellulose and raw biomass and downstream processing of the products. Catal Today 234:24–30

Singh B (1996) Why does halloysite roll? A new model. Clays Clay Miner 44(2):191–196

Singha AS, Thakur VK, Mehta IK, Shama A, Khanna AJ, Rana RK, Rana AK (2009a) Surface-modified Hibiscus sabdariffa fibers: physicochemical, thermal, and morphological properties evaluation. Int J Polym Anal Charact 14(8):695–711

Singha AS, Thakur VK, Mishra BN (2009b) Study of Grewia Optiva fiber reinforced urea-formaldehyde composites. J Polym Mater 26:81–90

Singha AS, Thakur VK (2009a) Grewia optiva fiber reinforced novel, low cost polymer composites. J Chem 6:71–76

Singha AS, Thakur VK (2009b) Synthesis, characterisation and analysis of Hibiscus Sabdariffa fibre reinforced polymer matrix based composites. Polym Polym Compos 17:189–194

Singha AS, Thakur VK (2009c) Fabrication and characterization of H. sabdariffa fiber-reinforced green polymer composites. Polym Plast Technol Eng 48:482–487

Sinha Ray S, Bousmina M (2005) Biodegradable polymers and their layered silicate nanocomposites: in greening the 21st century materials world. Prog Mater Sci 50(8):962–1079

Sinha Ray S, Okamoto M (2003) Polymer/layered silicate nanocomposites: a review from preparation to processing. Prog Polym Sci 28(11):1539–1641

Soheilmoghaddam M, Pasbakhsh P, Wahit MU, Bidsorkhi HC, Pour RH, Whye WT, De Silva RT (2014a) Regenerated cellulose nanocomposites reinforced with exfoliated graphite nanosheets using BMIMCL ionic liquid. Polymer 55(14):3130–3138

Soheilmoghaddam M, Sharifzadeh G, Pour RH, Wahit MU, Whye WT, Lee XY (2014b) Regenerated cellulose/β-cyclodextrin scaffold prepared using ionic liquid. Mater Lett 135:210–213

Soheilmoghaddam M, Wahit MU (2013) Development of regenerated cellulose/halloysite nanotube bionanocomposite films with ionic liquid. Int J Biol Macromol 58:133–139

Soheilmoghaddam M, Wahit MU, Mahmoudian S, Hanid NA (2013) Regenerated cellulose/halloysite nanotube nanocomposite films prepared with an ionic liquid. Mater Chem Phys 141(2–3):936–943

Stoller MD, Park S, Zhu Y, An J, Ruoff RS (2008) Graphene-based ultracapacitors. Nano Lett 8(10):3498–3502

Swatloski RP, Spear SK, Holbrey JD, Rogers RD (2002) Dissolution of cellose with ionic liquids. J Am Chem Soc 124(18):4974–4975

Tang Y, Deng S, Ye L, Yang C, Yuan Q, Zhang J, Zhao C (2011) Effects of unfolded and intercalated halloysites on mechanical properties of halloysite–epoxy nanocomposites. Compos A Appl Sci Manuf 42(4):345–354

Thakur VK, Singha AS, Thakur MK (2012a) Rapid synthesis of MMA grafted pine needles using microwave radiation. Polym-Plast Technol Eng 51:1598–1604

Thakur VK, Singha AS, Thakur MK (2012b) Biopolymers based green composites: mechanical, thermal and physico-chemical characterization. J Polym Environ 20:412–421
Thakur VK, Singha AS, Thakur MK (2012c) Graft copolymerization of methyl acrylate onto cellulosic biofibers: synthesis, characterization and applications. J Polym Environ 20:164–174
Thakur VK, Thakur MK, Gupta RK (2013a) Development of functionalized cellulosic biopolymers by graft copolymerization. Int J Biol Macromol 62:44–51
Thakur VK, Thakur MK, Gupta RK (2013b) Rapid synthesis of graft copolymers from natural cellulose fibers. Carbohydr Polym 98:820–828
Thakur VK, Thakur MK, Gupta RK (2013c) Synthesis of lignocellulosic polymer with improved chemical resistance through free radical polymerization. Int J Biol Macromol 61:121–126
Thakur VK, Thakur MK, Gupta RK (2013d) Graft copolymers from natural polymers using free radical polymerization. Int J Polym Anal Charact 18:495–503
Thakur VK, Thakur MK, Gupta RK (2013e) Graft copolymers from cellulose: synthesis, characterization and evaluation. Carbohydr Polym 97:18–25
Thakur VK, Thakur MK (2014a) Processing and characterization of natural cellulose fibers/ thermoset polymer composites. Carbohydr Polym 109:102–117
Thakur VK, Thakur MK (2014b) Recent trends in hydrogels based on psyllium polysaccharide: a review. J Clean Prod 82:1–15
Thakur VK, Thakur MK (2014c) Recent advances in graft copolymerization and applications of chitosan: a review. ACS Sustain Chem Eng 2(12):2637–2652
Thakur VK, Thakur MK, Gupta RK (2014a) Review: raw natural fiber-based polymer composites. Int J Polym Anal Charact 19(3):256–271
Thakur VK, Thakur MK, Raghavan P, Kessler MR (2014b) Progress in green polymer composites from lignin for multifunctional applications: a review. ACS Sustain Chem Eng 2(5):1072–1092
Thakur VK, Vennerberg D, Kessler MR (2014c) Green aqueous surface modification of polypropylene for novel polymer nanocomposites. ACS Appl Mater Interf 6:9349–9356
Thakur VK, Vennerberg D, Madbouly SA, Kessler MR (2014d) Bio-inspired green surface functionalization of PMMA for multifunctional capacitors. RSC Adv 4:6677–6684
Tian M, Qu L, Zhang X, Zhang K, Zhu S, Guo X, Han G, Tang X, Sun Y (2014) Enhanced mechanical and thermal properties of regenerated cellulose/graphene composite fibers. Carbohydr Polym 111:456–462
Urszula D (2008) General review of ionic liquids and their properties. Ionic liquids in chemical analysis. CRC Press, Boca Raton, pp 1–71
Vadukumpully S, Paul J, Mahanta N, Valiyaveettil S (2011) Flexible conductive graphene/poly (vinyl chloride) composite thin films with high mechanical strength and thermal stability. Carbon 49(1):198–205
Visser AE, Swatloski RP, Rogers RD (2000) pH-dependent partitioning in room temperature ionic liquids provides a link to traditional solvent extraction behavior. Green Chem 2(1):1–4
Vitz J, Erdmenger T, Haensch C, Schubert US (2009) Extended dissolution studies of cellulose in imidazolium based ionic liquids. Green Chem 11(3):417–424
Wang R, Wang L (2014) Substituted cyclic compounds and methods of use. Google Patents
Wang X, Jin J, Song M (2012) Cyanate ester resin/graphene nanocomposite: curing dynamics and network formation. Eur Polymer J 48(6):1034–1041
Welton T (1999) Room-temperature ionic liquids. Solvents for synthesis and catalysis. Chem Rev 99(8):2071–2084
Wilkes J, Wasserscheid S, Welton T (2008) Ionic liquids in synthesis, vol 1. Willey-VCH Verlags Gmbh & Co KgaA, New Delhi
Wilkes JS (2002) A short history of ionic liquids-from molten salts to neoteric solvents. Green Chem 4(2):73–80
Wu RL, Wang XL, Wang YZ, Bian XC, Li F (2009) Cellulose/soy protein isolate blend films prepared via room-temperature ionic liquid. Ind Eng Chem Res 48(15):7132–7136
Yamamoto H, Horii F (1993) CPMAS carbon-13 NMR analysis of the crystal transformation induced for Valonia cellulose by annealing at high temperatures. Macromolecules 26 (6):1313–1317

Yamamoto H, Horn F (1994) In situ crystallization of bacterial cellulose I. Influences of polymeric additives, stirring and temperature on the formation celluloses I α and I β as revealed by cross polarization/magic angle spinning (CP/MAS)13C NMR spectroscopy. Cellulose 1(1):57–66

Yuan P, Southon PD, Liu Z, Green MER, Hook JM, Antill SJ, Kepert CJ (2008) Functionalization of halloysite clay nanotubes by grafting with γ-aminopropyltriethoxysilane. J Phys Chem C 112(40):15742–15751

Zavrel M, Bross D, Funke M, Büchs J, Spiess AC (2009) High-throughput screening for ionic liquids dissolving (ligno-) cellulose. Bioresour Technol 100(9):2580–2587

Zhang X, Liu X, Zheng W, Zhu J (2012) Regenerated cellulose/graphene nanocomposite films prepared in DMAC/LiCl solution. Carbohydr Polym 88(1):26–30

Zhang Y, Ouyang J, Yang H (2014) Metal oxide nanoparticles deposited onto carbon-coated halloysite nanotubes. Appl Clay Sci 95:252–259

Zhang YH, Hu DJ, Song W (2005) Study on CCD online system for monitoring a grinding wheel in process of point grinding on curve surface. Binggong Xuebao/Acta Armamentarii 26 (2):201–204

Zhu S, Wu Y, Chen Q, Yu Z, Wang C, Jin S, Ding Y, Wu G (2006) Dissolution of cellulose with ionic liquids and its application: a mini-review. Green Chem 8(4):325–327

Cellulose Nanofiber for Eco-friendly Polymer Nanocomposites

Ida Idayu Muhamad, Mohd Harfiz Salehudin and Eraricar Salleh

Abstract Nanocomposite is the reinforced composite material consists of nanoscale reinforcing fillers and matrix polymer. Fillers are dispersed within nanoscale and require just less amount than conventional reinforcing fillers, but the properties of composites are greatly improved. There would be only insignificant deterioration of properties in case of recycling; therefore, it is able to be an eco-friendly composite material. Recent studies show that interests in cellulose nanocomposites consists of nanocellulose fiber and matrix polymer are enhanced more and more in recent years. Especially, cellulose nanocomposites are best representative eco-friendly material as compared with nanocomposites reinforced with inorganic nanoscale fillers such as nanoclay, montmorillonite, mica, and silica. Natural filler such as cellulose nanofiber from palm empty fruit bunch (OPEFB) has drawn bigger attention as it promotes eco-friendly character. In current study, cellulose nanofiber (CNF) was prepared through pretreatment to remove noncellulosic content and then undergoes acid hydrolysis process. Starch-based nanocomposite film was formed by incorporation of 2–10 % CNF per weight of starch into the film matrix. The nanocomposite film that formed appears translucent and easy to handle. However, the film becomes more opaque as percentage of CNF incorporation increased. It was observed that films with the addition of up to 2 % CNF showed higher tensile strength and thermal stability, better barrier properties to water vapor than control films. Further study on the effect of CNF was carried out on Starch/Chitosan composite packaging film to determine the influence of CNF toward antimicrobial properties of the composite film as applied packaging for perishable food. The effects of CNF contents on the tensile, dynamic mechanical and thermal properties as well as the barrier properties of the Starch/Chitosan nanocomposite

I.I. Muhamad (✉) · E. Salleh
Bioprocess Engineering Department, Faculty of Chemical Engineering,
Universiti Teknologi Malaysia, 81310 Johor Bahru, Malaysia
e-mail: idayu@cheme.utm.my

M.H. Salehudin
P.G Programme in Bioprocess Engineering, Bioprocess Engineering
Department, Faculty of Chemical Engineering, Universiti Teknologi Malaysia,
81310 Johor Bahru, Malaysia

V.K. Thakur and M.K. Thakur (eds.), *Eco-friendly Polymer Nanocomposites*,
Advanced Structured Materials 75, DOI 10.1007/978-81-322-2470-9_11

were also investigated. It also embarks a potential of cellulose nanofiber as filler for antimicrobial packaging as it enhances the results on antimicrobial efficacy toward food shelf life.

Keywords Nanocomposite · Eco-friendly · Starch · Cellulose nanofiber · Antimicrobial

1 Nanotechnology in Polymers

Nanotechnology is found to be a potential technology that emerged in wider areas such as medicine, electronics, and food technology by manipulating the nanoparticles for various purposes. A nanomaterial is defined as a discrete entity that has at least one of its dimensions is 100 nm or less (Som et al. 2010). It may include any of the following nano forms: nanoparticles, nanotubes, fullerenes, nanoclays, nanocor, nanofibres, nanowhiskers, and nanosheets (Lin et al. 2011a, b). The applications of nanomaterial are broad. Some of them were used as nanosensor in smart food packaging technology (Duncan 2011). These also could provide an antimicrobial mechanism by introducing nano-bullet in active packaging. The most popular purposes of nanomaterials are very wide. These have been used as nanoreinforcement in polymer composite in fact, many studies on nanoreinforcement were reported (Thakur et al. 2012a, b, c, d, e). Nanoreinforcements that been studied are such as clay and silicates (Tunç and Duman 2011), cellulose-based (Samir et al. 2004; Liu et al. 2013; Jonoobi et al. 2011; Nakagaito and Yano 2008; Azeredo et al. 2012) carbon nanotubes (Salajkova et al. 2013), starch nanocrystal (Kristo and Biliaderis 2007), and chitin/chitosan nanoparticles (Lu et al. 2004; Wongpanit et al. 2007). However, cellulose nanofiber that also known as nanocellulose becoming such an emerging source of reinforcement material nowadays. Generally, this particular nanofiber is used as filler or fabricator for material reinforcement and toughness. In addition, futuristic nanotechnology background has found that highly crystalline nanofiber invention is interesting as it has unique properties and sizes different from synthetic nanofibers (Abdul Khalil et al. 2012a). Cellulose nanofibers have a high potential to be used as reinforcing material in many different areas. Differs from other nanofiller such as nanoclays, the usage of cellulose nanofiber becomes such a serious deal because it is natural and the source is readily available yet abundant. Furthermore, this green cellulose nanofiber is not only renewable but also known as a multifunctional raw material and expected to replace many nonrenewable materials in the future. Natural cellulose nanofibers come from different sources and categories. Natural fibers which also known as lignocellulosic fibers are subdivided based on their origins, which are plants, animal, or minerals (John and Thomas 2008). In general, natural fiber which origin of the plant is usually referred as cellulosic fiber since the fibers usually often contain a natural polyphenolic polymer and lignin in their structure (Phiriyawirut 2012).

The utilization and manipulation of natural fiber source is not outlandish. In history, different class of natural fiber were used as a main source in making paper, silk, cupboard, and others.

Natural biobased materials such as cellulose, psyllium, soy, etc., possesses biodegradability, easy availability, environmental friendliness, flexibility, easy processing, and impressive physicomechanical properties which favored by research community around the world (Thakur and Thakur 2014a, b, c; Thakur et al. 2014a, b, c, d, e). Different kinds of natural polymers including celulose has served humankind for over centuries before low-cost petroleum-based synthetic takes place. A number of synthetic materials have well established and their applications are varied because of their high specific strength and modulus with longer durability for a number of applications ranging from automotive to energy storage (Satyanarayana et al. 2009; Thakur et al. 2011a, b, c, d, e). As time passed, massive production of synthetic polymer was established to supply wide sectors. However, the use of great amount of this synthetic polymer led to ugly consequences as it gives problem on material disposal and depletion of nonrenewable source itself. At early 1990s, natural polymer was resurfaced as it gives a better prospect. There are reasons that leads to this situation. Today, humankind are aware of eco-friendliness that subsequenty drags people's attention more on reducing environmental impact caused by polymers and composites. Another point is, limited petroleum source forces people to find another alternative to decrease the pressure on petroleum-based products. Thus, the potential of maximizing the use of renewable material suddenly becomes interesting and equally important. Other than that, sudden increase in research study on this particular field especially in cellulose-based material makes all the data on the properties and morphologies of natural fiber generously available thus gives better understanding of the structure and interaction between biopolymer and cellulose. Such achievements are possible to reach through modern instruments that assist better understanding of natural biocomposite, hence accelerates the development of new materials. Natural fibers are naturally occurring composite consist mainly cellulose fibrils embedded in lignin matrix (John and Thomas 2008). Cellulose on the other hand is the abundant organic compound that is a structural component of the cell walls of many plants. Mainly in the forest, it is present in the wood as the most important source. In addition, cellulose fiber has a unique pecking order. It comprised of nanofibers assemblies with diameter ranges from 2 to 20 mm and a length of more than a few micrometers (Abdul Khalil et al. 2012b). Other cellulose-containing materials include agriculture residues (fruit bunch, leaf, trunk), grasses, and vegetables (jute). Other than cellulose, plant cell wall contains hemicelluloses, lignin, and small amount of extractives.

From recent studies, cellulose is found to be the most common organic polymer and considered as almost infinite source of raw material to fulfill the increasing demand for eco-friendly and biocompatible products. It represents about 1.5 billion tons of the total annual biomass production (Klemm et al. 2005). As lignocellulosic fiber becomes vital, many possible plants were tested. Natural fiber not only can be harvested from cotton, wool, and other ancient source, but it covers different varieties and different sources all over the world. Available natural fiber that has been conducted in previous research are wood (Abe and Yano 2009a; Chen et al. 2011; Jonoobi et al. 2011), cotton (de Morais Teixeira et al. 2010), jute fiber (Soykeabkaew et al. 2004), potato tuber cells (Abe and Yano 2009a), wheat straw and soy hulls (Alemdar and Sain 2008), pineapple fiber (Cherian et al. 2010), sisal (Morán et al. 2007), oil palm empty fruit bunch (Fahma et al. 2010; Jonoobi et al. 2011; Abdul Khalil et al. 2011), coconut husk (Rosa et al. 2010), banana biomass (Zuluaga et al. 2007), hemp and jute (Soykeabkaew et al. 2004), sugar beet (Dufresne et al. 1997), peas hull (Chen et al. 2009), prickly pear fruits (Habibi et al. 2008), *Hibiscus sabdariffa* (Singha and Thakur 2008a, b), pine needles (Thakur and Singha 2011; Singha and Thakur 2009a, b, 2010a, b), and rice husk (Arayapranee et al. 2005). In fact, scientists now are looking at the various possibilities by combining biofibers such as sisal, flax, hemp, jute, banana, wood, and various grasses with polymer matrices from nonrenewable and renewable resources to form composite materials to increase its value. The combination of those fibers with natural or synthetic polymer generally tends to produce material with great properties. Previously, different types of fiber which in micro size were incorporated into material (e.g. polymer) to make cupboard, disposable plate, etc., but with the help of nanotechnology the fiber could be grinded chemically, biologically, and mechanically into nanosize. Because of its small size, nanosized fiber is more suitable to be incorporated into various products including thin film. It is even crucial to develop such filler for reinforcement purpose as the natural-based polymer, for an examples starch-based polymer, has weak mechanical strength. Previous reports show that the nanosized fiber exhibits unique characteristics that could reinforce the film. There are also many research were conducted in nanosize fiber making.

1.1 Cellulose Nanofiber

Cellulose is a structural component of most plants that is readily abundant. Hierarchy structure of cellulose is very unique. It comprised of nanofibers with diameter size about 2–20 nm and a length of more than few microns that assemble to form biofiber (Abdul Khalil et al. 2012b; Dean et al. 2007; Wang and Sain 2007; de Moura et al. 2009). In the microfiber, the cellulose fibrils are aligned providing maximum tensile, flexural strength, and rigidity to the fiber. Hierarchy structure of cellulose is illustrated by schematic model of microfibril in Fig. 1. It explains structural integration of cellulose, hemicelluloses, and lignin in the fiber which shows how single cellulose chain is arranged in a bundle to form elementary fibril.

The elementary fibril is then arranged in parallel to form parallel elementary fibrils. Finally, four microfibrils held together by hemicellulose and lignin. It is crucial to understand the structural hierarchy of cellulose, individualized nanofibres so that cellulose with high crystallinity could be obtained. The structure of the plan could be break down into high crystalline compound of the fiber at the same time reduce the amount of amorphous material present (Azeredo 2009). Studies showed the crystalline parts such as in whiskers, also known as nanocrystal, nanorods or rodlike cellulose microcrystal, or cellulose crystal can be isolated by several treatment.

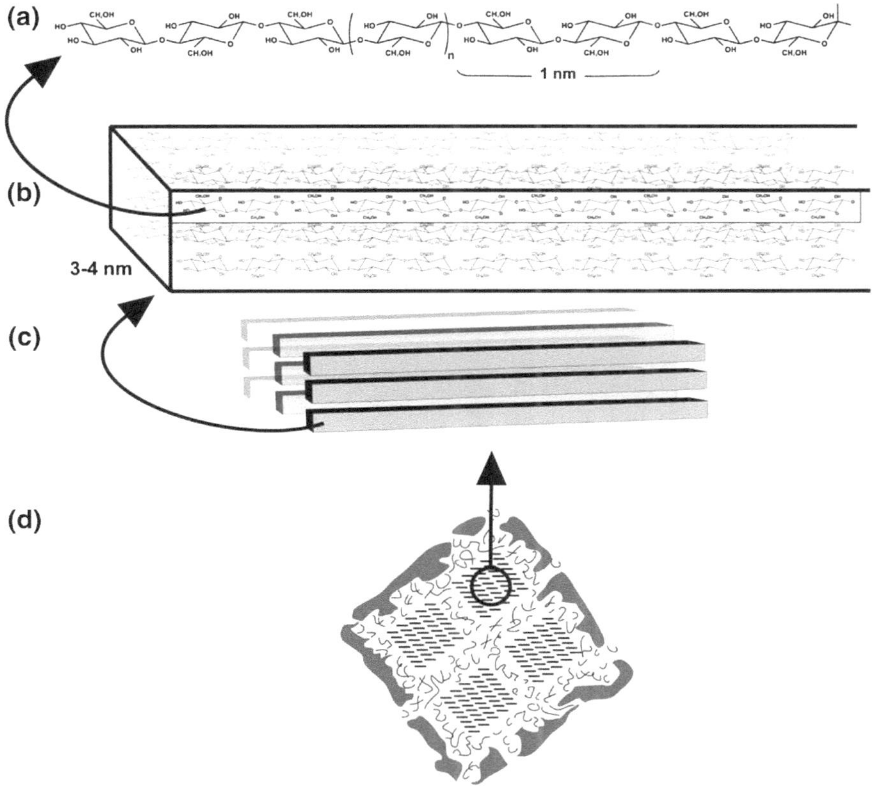

Fig. 1 Structural disintegration of cellulose. **a** a cellulose chain; **b** an elementary fibril containing bundles of cellulose chains; **c** parallel elementary fibrils; **d** four microfibrils held together by hemicellulose and lignin, internal structure of a cellulose microfibril (Adapted from Ramos 2003)

Based on Fahma et al. (2010) nanofiber from palm oil EFB, which also known as cellulose nanofiber, are produce by hydrolyzing oil palm empty fruit bunches fiber with sulfuric acid and through other chemical and physical process. Method describes EFB cellulose nanofiber isolation comprises two process steps, pretreatment and bleaching, and acid hydrolysis, will be described further.

2 Eco-friendly Cellulose from Natural Fiber as Reinforcement Material

A number of natural fibers have been found as emerging materials that can be used as fillers for composites improvement and reinforcement. Extracted cellulose, for an example, have been frequently used for the preparation of high-performance

Table 1 Variety of natural fibers (Abdul Khalil et al. 2012a, b)

Plant biofiber	Category/part	Example
Wood	Soft and hard wood	Pine, teak, rubber wood, acacia
	Recycled wood fibers	Newspaper, magazine fiber
Non-wood	Grass	Bamboo fiber, elephant grass
	Seed/fruit	Cotton, coir, EFB, Palm press fiber, jathropa, cocoa
	Leaf	Henequen, sisal, pineapple leaf fiber, pandaus helicopus
	Bast	Kenaf, flax, jute, hemp
	Straw	Rice, wheat, corn straw
	Trunk	Oil palm, coconut tree, date palm, banana stem

composites (Thakur et al. 2014a). Those abundant sources not only renewable but also have marketing appeals that already penetrated into Asian marketplace long time ago. In India, for instance, jute fiber are widely used as a common reinforcement material.

However, there are thousands of tons of agricultural wastes produced without proper utilization which found to be useful to prepare polymer composite for commercial purposes for example EFB, sisal fiber, wheat straw, and others (Abdul Khalil et al. 2012b). Numbers of natural cellulose fibers have been used to reinforce polymer composites. In fact, various types of natural fibers were investigated for incorporation in plastics. Basically, natural fibers are diverted into two categories which are wood and non-wood. Table 1 shows varieties of natural fibers that available on current research.

Among the constituents present in any cellulosic fiber, cellulose is the most essential component (John and Thomas 2008; Dufresne et al. 2000; Phiriyawirut 2012; Wu et al. 2009). The utilization of EFB are expected to have a great prospect, as cellulose materials are the most abundant biobased raw materials and can self assembled into well-defined architectures in multiple scales, from micro to nanosize as a reinforcing agent in composite materials. Theoretically, the nanocellulose can be a reinforcement material because Cellulose I or native cellulose typically has a structure in which the molecular chains pack in layers that held together by weak van der Waals forces. The cellulose layers consist of parallel chains of anhydroglucopyranose units are held together by intermolecular hydrogen bonds (Rowell et al. 2005). These hydrogen bonds occurred because cellulose structure consists of high-affinity ligands of hydroxyl group and the affinity becomes higher as the size of cellulose structure becomes smaller. These unique characteristics of cellulose nanofibers make it excellently compatible to be incorporated into biopolymer matrix and provide more strength to the composites. The common cellulosic material in current research that been utilized as reinforcement material is given below.

2.1 Kenaf Fiber

Kenaf (Hibiscus cannabinus, L. family Malvaceae) is seen as an herbaceous annual plant that can be grown under a wide range of weather condition. It is also a dicotyledonous plant meaning that the stalk has three layers; an outer cortical also referred to as ("bast") tissue layer called phloem, an inner woody ("core") tissue layer xylem, and a thin central pith layer which consist of sponge-like tissue with mostly nonferrous cells (Hossain et al. 2011). In history, it has been used traditionally as a rope, canvas, and sacking. It is clearly known as a cellulosic source with economical and ecological advantages. Kenaf exhibits low density, non-abrasiveness during processing, high specific mechanical properties, and biodegradability (Nishino et al. 2003). Kenaf bast fiber has been reported to have superior flexural strength and excellent tensile strength that makes it the material of choice for wide range of extruded, molded, and nonwoven products such as polymer composite.

Kenaf fiber could also disintegrate into nanosize. Incorporation of cellulose nanofiber from kenaf into polymer matrix can improve the mechanical properties. It was reported that the incorporation of cellulose nanofiber from kenaf into (polyvinyl alcohol) PVA polymer as low as 10 % could provide remarkable reinforcing potential (Bhatnagar 2005). Basically, it has great potential in high-end applications such as using as high-strength components in aerospace and automotive sector due to its lightweight and high strength.

2.2 Hibiscus sabdariffa

Hibiscus sabdariffa or in common name roselle, is an annual fibre plant, has been found to be an important source of fibres for a number of application since ancient times. The bast fibre has high potential as a reinforcing fibre in polymer composites. *Hibiscus sabdariffa* plant fibre is abundantly found in the Himalayan region, especially in Himachal Pradesh. It is traditionally used by the locals as socks, boots, mats, ropes, bags, etc.

Numerous research has been conducted which reveal the application of this fiber source as reinforcing material in polymer composites. For instance, it has been used for reinforcement in urea formaldehyde resin-based composite which also suggested that *Hibiscus sabdariffa* fibre has immense scope in the fabrication of natural fibre reinforced polymer composites for vast number of industrial applications (Singha and Thakur 2008a). The graft copolymerization of *Hibiscus sabdariffa* also have been found to be more moisture resistant and also showed better chemical and thermal resistance (Thakur et al. 2011a).

2.3 Rice Husk

An agricultural waste from rice milling of paddy generates a rice husk which comes from the fields. It can be used as a combustion source for electrical power plants for the boiler in paddy processing from which finally rice husk ash (RHA) is obtained. RHA is about 25 % by weight of rice husk when burnt in boilers. The amount of silica and carbon black in the ash varies depending on the combustion conditions.

The usage of this fiber can be found in construction and automobile industries. For instance, rubbers are rarely used in their unmodified form with respect to their applications. They are often mixed with fillers to improve their process ability, mechanical strength, and to reduce cost. Carbon black and silica are well-known filler that highly is commercialized. Rice husk ash in rubber compounding has drags intense interest because of its low cost, environmental preservation benefit, and an increased emphasis on the use of renewable resources (Arayapranee et al. 2005). Numerous trials have been done by the researchers which use Rice Husk as a filler for polymeric materials. Sae-oui et al. (2002) investigated the effects of filler loading on the properties of RHA-filled natural rubber (NR) materials compared with those of commercial fillers.

They found that both grades of RHA, low- and high-carbon contents, provided inferior mechanical properties (tensile strength, modulus, hardness, abrasion resistance, and tear strength) compared with those of reinforcing filler such as silica and carbon black. Other finding reported that incorporation of RHA into polypropylene led to an increased flexural modulus of the composites (Fuad et al. 1998).

2.4 Oil Palm Empty Fruit Bunch (OPEFB) Fiber

Malaysia is the second biggest producer of palm oil in the world after Indonesia. By-product such as empty fruit bunch, a non-wood waste product from palm oil industry was generated abundantly. Among the variety of sources of fiber available in the world, empty fruit bunch is selected as the best fibrous material because it contains high percentage of cellulose and comparatively low lignin than other part of palm oil (Sreekala et al. 1997). *Elaeis guineensis* is the species of palm oil that widely grows for commercial purpose. Mature trees are single-stemmed and could reach 20 m tall. It has leaves pinnate with 3–5 m long. In a year, a young tree produces about 30 leaves a year; whereas, trees that are older than 10 years produced about 20 leaves a year. The flowers are produced in dense bunch; each individual flower is small, with three petals and three sepals. The oil palm fruit takes about 5–6 months to mature from pollination to maturity. It grows in large bunches, reddish in color, and about the size of a large plum. Every single fruit comprises an oily, fleshy pericarp (outer layer) and a palm kernel (seed) that also rich in oil. When it ripes, each bunch of fruit weighs about 40–50 kg. In palm oil processing industry, the fruit is separated from the bunch leaving an empty fruit bunch as shown in Fig. 3. In general, EFB fiber could be derived into two components, which are mesocarp and fruit bunch. Fruit bunches however; are more fibrous than mesocarp. After the oil palm fruit was separated from its bunch, the empty fruit bunch undergoes pressing process to eliminate excess moisture and other constituents such as oil. It was then shredded to produce a microfiber strand.

Oil palm processing generates 53.4 % mesocarp fiber (MF), 6.4 % palm kernel shell (PKS), 21 % empty fruit bunch (EFB), and 58.3 % palm oil mills effluent (POME) from every wet fresh fruit bunch (FFB) basis (Hambali et al. 2010). The overall palm oil market is dominated by two countries; Indonesia and Malaysia and these two accounts more than 85 % of the worldwide production (Sulaiman et al. 2011). The source of palm oil fruit bunch is very much related to palm oil industries. Palm oil producing countries such as Malaysia have more reserve oil palm plantation which lead to abundant source of empty fruit bunches. Previously, EFB was considered as a under-utilized source. Once it was viewed as an embarrassing liability of palm oil waste but later due to dynamic research and innovation, the ability of oil palm fiber as valuable source embarks. In Malaysia itself, the production of empty fruit bunch reaches 21.34 million tons in year the 2011, which make Malaysia a major fiber producer (Mun 2011).

EFB fiber contains three major compositions namely cellulose, hemicelluloses, and lignin with 44.4, 30.9, and 14.2 % respectively (Sulaiman et al. 2011). Cellulose fiber is cylindrical rodlike microfibril that in is crystalline and amorphous form. It is a natural polymer that consist of *D*-anhydroglucose ($C_6H_{11}O_5$) repeating units joined by 1,4-*β*-*D*-glycosidic linkages. Each repeating unit contains three hydroxyl groups that are able to form hydrogen bond. There are two types of cellulose, higher order and lower order. Higher order is when the solid cellulose forms a microcrystalline structure such as crystalline regions. The lower order, however, forms amorphous region. Degree of polymerization (DP) of cellulose is highest, i.e., 10,000. Cellulose is resistance to strong alkali (17 wt%) and oxidizing agent. However, it is easy to hydrolyze using acid to form water-soluble sugar. Another element in natural fiber is hemicellulose that acts as supportive matrix for cellulose. Hemicelluloses are bunch of polysaccharides containing sugar 5 and 6 carbon ring sugar. The polymerization of cellulose is 50–300 (DP). Hemicellulose exhibits strong hydrophobicity. Unlike cellulose, hemicellulose is soluble in alkali and even easier to hydrolyze in acid. Lignin, on the other hand, is aliphatic and aromatic complex hydrocarbon that contains methoxyl and hydroxyl group. In one building unit it contains 5 methoxyl and hydroxyl group that makes its molecular weight relatively high.

Lignin is insoluble in nearly all solvents and cannot be broken down to smaller units. Lignin is amorphous and shows thermoplastic behavior. It is naturally hydrophobic and cannot be hydrolyzed with acid. However, it is soluble in hot alkali and readily oxidized and condensable with phenol. Lignin has melting temperature of 170 °C and glass transition temperature of 90 °C. Natural fibers such as palm oil fiber is natural composite consist of hollow cellulose fibrils held together by lignin and hemicellulose matrix (John and Thomas 2008; Jayaraman 2003). It is rich with lignocellulosic contents that include cellulose, hemicelluloses, lignin, and other insignificant materials such as waxes, ashes, and others. Each amount of element is, however, varied according to the fiber sources (Mohanty et al. 2002).

Structural constitution of natural fiber signifying that each microfibril is formed by the arrangement of elementary fibrils, which are made up of crystalline and amorphous parts as illustrated in Fig. 1. Cellulose mostly located in the secondary wall of biofiber. Biofiber cell wall membrane is not homogenous. Each microfibril has complex-layered structure consisting of thin primary wall. Primary cell wall that roughly contains 6000 glucose unit is generally the first layer deposited during cell growth that encircling three layers of secondary wall. Mechanical properties of the fiber, however, are determined by thick middle layer of secondary wall that mostly consists of cellulose pecking.

3 EFB Cellulose Nanofiber Isolation

Cellulose EFB could disintegrate in order to produce cellulose nanofiber (CNF). In history, there are various techniques studied in order to obtain pure, new, and improved EFB fiber. The purity and quality of the final product is distinguished from the very beginning of each process integration. There are several processes in nanofiber isolation that include pretreatment, bleaching, and extraction of cellulose nanofiber suspension which can be done through hydrolysis using chemical or mechanical process.

3.1 Pretreatment

Pretreatment is a series of processes designed before the isolation process. The main purpose of introduction of pretreatment is to remove unnecessary components. Pretreatment includes bleaching and washing to remove ashes, waxes, and non-cellulosic compound, which is crucial to produce pure and high quality end product, mainly cellulose. Based on Suradi et al. (2009) pretreatment of the cellulose fiber could improve the strength of the fiber thus increases mechanical properties of fiber-based composite as well. It is also proven that pretreatment process is able to remove a bit of lignin contents. Lignin is not welcome to be a part of polymer matrix because it negatively influences the properties of polypropylene (Abdul Khalil et al. 2012b). During pretreatment process, lignin seal will be broken and crystalline structure of cellulose is disrupted that made lignin removal possible (Mosier et al. 2005). There are several pretreatment processes available that have been developed from years of study. However, alkali/base treatment is frequently used method because it is more simple and effective. Other pretreatment processes and the purpose of each treatment process are shown in Table 2.

Table 2 Pretreatment and its purpose

Method	Purpose	References
Dip the fiber in 2.5 M sodium hydroxide (NaOH) solution, at 80 °C for 48 h	To remove lignin and hemicelluloses	Wyman et al. (2005)
Filter and wash the fiber with distilled water and few drop of acetic acid	To neutralize alkali EFB fibers	
Wash again the fiber repetitively with distilled water until ph 7	To remove excess base	
Dry the fiber at 50 °C	To dry the fiber	
Dry, cut, and sieve the fiber	To get uniform size	Carvalheiro et al. (2008) and Sun et al. (2004)
Soak the fiber in water with temperature 65–80 °C, for 12 h	To eliminate impurities and large particles	
Soak the fibers in 10–30 % (w/v) of Sodium Hydroxide (NaOH) and maintained in water bath at temperature 75 °C, speed 40 rpm about 3 h	To remove lignin content	
Wash the fiber with distilled water for a few times	To remove NaOH	
Fiber treated with distilled water contain 2 % Sodium Peroxide (H_2O_2) in water bath at temperature 45 °C, speed 40 rpm for 8 h	To remove more lignin content and active the OH group of the cellulose	
Wash the fiber with distilled water	To remove excess H_2O_2	
Treat the fiber with acetic acid 10 % (v/v) for 30 min at room temperature	To neutralize the excess NaOH	
Repeatedly wash the fiber with distilled water	To remove acid residue from the fiber	
Fiber was dried in an oven at 70 °C for overnight	To dry the fiber	
Soak the fiber five times in acidified sodium chlorite solution (pH 4–5) at 75 °C for an hour	To remove lignin	Abe and Yano (2009, 2010), Fahma et al. (2010) and Hamzah et al. (2011)
Treat the fiber in 3 wt% potassium hydroxide (KOH) at 80 °C for 2 h	To leach hemicellulose, residual starch, and pectin	
Filter and rinse the fiber with distilled water	Neutralize the sample	
Soak the EFB in hot distilled water at EFB 80 °C for 1 h	To remove ashes and large particles	
Add 2.5 mol/L of sodium hydroxide (NaOH) into the fiber and autoclave for 15 min at temperature of 121 °C	To explode the fiber component so lignin and hemicelluloses could be removed effectively	
Separate the fiber and bleach it using sodium hypochlorite solution (6–5 % chlorine)	To further remove lignin and produce whiter product	
Using tap water, wash the fiber several time until the pH reached 7	To eliminate hyprochlorite solution from fiber	
Dry the fiber in the oven at 37 °C for 1 h	To dry the fiber	

Fig. 2 **a** Empty fruit bunch fiber before pretreatment. **b** Empty fruit bunch fiber after bleaching processes. **c** Surface Electron microscopy of empty fruit bunch fiber before pretreatment (×500 magnification). **d** Surface electron microscopy of treated fibre (×500 magnification)

As discussed before, pretreatment and bleaching processes are very important to produce high cellulose strands. As a result, chemical and physical treatments have altered the morphology of the fiber. The visual evolutions of empty fruit bunch fiber before pretreatment (a) and bleaching processes (b) are shown in Fig. 2. Initial color of the EFB fiber is brown and it turned a little less intense after pretreatment and bleaching. As the bleaching process continues for second and third time, the fiber became whiter which indicates the removal of lignin.

The morphology of the EFB fiber was further analyzed using scanning electron microscopy (SEM). Figure 2c illustrates SEM image of the untreated (raw) EFB while Fig. 2d illustrates SEM image of treated EFB. In terms of size, the diameter of original fiber size ranged between 150 and 250 µm with 209 µm in average. From the analysis, the diameter of treated and bleached fiber ranged between 15.55 and 27.48 µm with an average of 17.6 µm. The images also reveal that physical changes have occurred in the pretreated EFB fiber due to the removal of a large fraction of hemicellulose and lignin during the pretreatment process (Hamzah et al. 2011). Figure 2d indicates that after the pretreatment and bleaching, the fibers were distorted and break down into smaller strand. The pretreatment process removes most of the hemicelluloses and lignin component of EFB fiber thus leaving the cellulose unbinds and is detached. In contrast, the untreated fiber which shown in Fig. 2c is still in its pecking structure (rigid form) and highly ordered fibrils because the microfibrils are still held together by hemicellulose and lignin.

3.2 Bleaching

Bleaching of fiber could result in whiter and purer product. It could also improve aging resistance that prevents the fiber from yellowing and brittle that might link to lignin contents in the fiber. In several stages, different chemicals are used for bleaching, e.g., hydrogen peroxide (H_2O_2), chlorine dioxide (ClO_2), ozone (O_3) or peracetic acid (Ek et al. 2009). Sulfite pulps are more readily bleached and are obtained in higher yields (Young 1994). The main purpose of bleaching process isto create whiter cellulose by removing the lignin.

3.3 Preparation of Cellulose Nanofiber Suspension

Method of cellulose nanofibers isolation from lignocellulosic source are varied that include mechanical, chemical, chemomechanical, and enzymatic isolation processes (Jonoobi et al. 2011). Previous study reported that cellulose nanofiber can be obtained using novel isolation process such as high-pressure homogenizers which were practiced by Herrick et al. (1983) and Turbak et al. (1983). Technology advances and developments leading researchers to innovate processing technique in order to obtain new and improved nanocellulose (Table 3). In fact, numerous studies investigate the isolation of nanocellulose from various raw materials, mainly of plant origin using different techniques (Dufresne et al. 2000; Abe and Yano 2009; Abe et al. 2007; Alemdar and Sain 2008; Jonoobi et al. 2011; Zhao et al. 2007).

Table 3 Various techniques of extracting nanofibers (Abdul Khalil et al. 2011)

Extraction processes and methods	References
Mechanical treatments, e.g., cryocrushing grinding	Chakraborty et al. (2006), Abe et al. (2007, 2009), Abe and Yano (2009, 2010), Nogi et al. (2009)
High-pressure homogenizing	Herrick et al. (1983), Nakagaito and Yano (2004, 2005, 2008), Turbak et al. (1983)
Chemical treatments, e.g., acid hydrolysis	Araki et al. (2000), Hafraoui et al. (2007), Jonoobi et al. (2010), Liu et al. (2010)
Biological treatments, e.g., enzyme-assisted hydrolysis	Hayashi et al. (2005), Henriksson et al. (2007), Paakko et al. (2007)
TEMPO-mediated oxidation on the surface of microfibrils and a subsequent mild mechanical treatment	Saito et al. (2006, 2007, 2009), Iwamoto et al. (2010)
Synthetic and electrospinning methods	Frenot et al. (2007), Kim et al. (2007), Ma et al. (2005)
Ultrasonic technique	Cheng et al. (2007, 2009, 2010), Wang and Cheng (2009), Zhao et al. (2007)

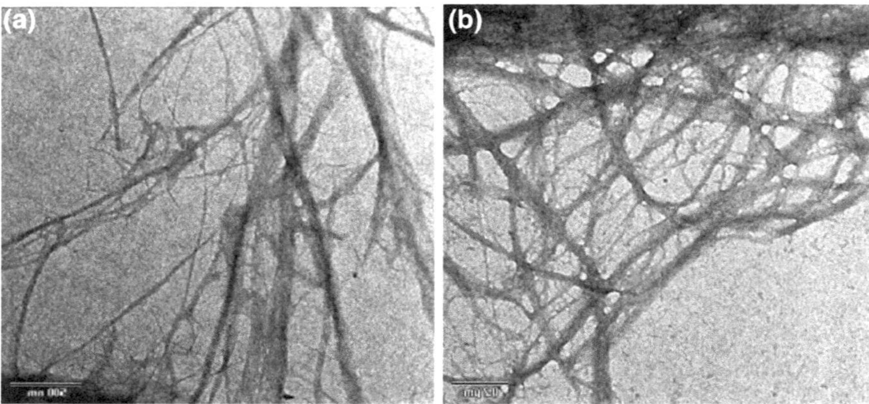

Fig. 3 Transmission Electron Microscopy (TEM) images of **a** CNF, **b** CNF after homogenization process

In this chemomechanical process, treated fiber that contains high purity of cellulose was acid-hydrolyzed in order to break down the pecking structure of the cellulose fiber into nanosized individual cellulose strand. The nanostructure of the fiber can be verified using Transmission Electron Micrograph (TEM). From the image shown in Fig. 3a, cellulose fiber that undergoes acid hydrolysis produced thin fiber strands with diameter range between 28 and 87 nm. Based on Samir et al. (2004), the amorphous region of microfibril cleaved by sulfuric acid hydrolysis makes the diameter of the fiber to reduce from microns to nanometers. The process was continued by homogenizing the nanofiber. Figure 2b shows that after homogenizing the CNF using ultra-turrax for 30 s, the cellulose was fibrillated into web-like form. In the end, the CNF was kept in suspension form.

4 Eco-friendly Biopolymer

Interest in eco-friendly or "green" polymeric materials is growing due to concerns with increasing carbon emission and the limited nature of petroleum and natural gas resources. These "green" materials contain polymers and possibly fillers which are both biodegradable and from renewable resources. The lifecycle of renewable polymeric materials is a carbon-neutral process, and their use may reduce carbon dioxide accumulation in the atmosphere and dependence on petroleum-derived materials. Sources of biopolymer are enormous; it could be derived either from renewable source, genetic manipulation, or microbial synthesis. In general, eco-friendly polymer composites can be divided into two different types, namely totally-renewable composites and partially renewable composites, depending upon the nature and origin of the filler (reinforcement material) and the base (polymer matrix) (Singha and Thakur 2010a, b; Ramanaiah et al. 2011) that can be portrayed in Fig. 4.

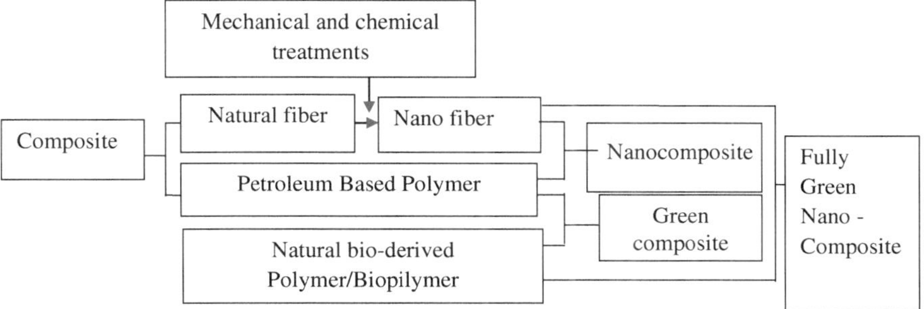

Fig. 4 The connection between composite, green composite

The most common polymer derived from renewable sources is polysaccharides and protein/lipid. Polysaccharides come from plant that is exists as starch, and as chitosan/chitin in animal. Starch and cellulose are most abundant source of biopolymer. It is known as highly biodegradable polymer, low cost, and very much relevant to various applications (Vilpoux and Averous 2004). In order to prepare a strictly "biodegradable polymer" such as thermoplastic starch, starch granules is mixed with plasticizers, for an example water or glycol, and the heat is applied to gelatinize it. The hydrophilic and moisture sensitivity characteristics of starch limit its application. Because of this reason, some trial has been done to overcome this issues by blending starch with other biopolymer such as polyhyroxyacetone (PHA), polylactides (PLA), PVA, and others (Avérous 2004). At the very beginning of the composite development era, the petroleum-based polymer was combined with natural bio-derived polymer to create biodegradable polymer composite which has specific purposes. For an example, incorporating biopolymer into synthetic polymer may not exhibit better mechanical properties, but it could assist biodegradability of the composite. Other technique is assortment of natural fiber with petroleum-based polymer to create other types of composite. The addition of natural fiber has different purposes either to give mechanical and physical enhancement of the composite or biodegradability or both. Petroleum source becomes depleted and biopolymers are expected to replace petroleum-based polymer. Polymer called as green composite such as starch-based film is fully constructed by green materials. However, there are certain drawbacks using biopolymer majorly in mechanical properties. Researchers have run several attempts on increasing the weak properties of the polymer by combining it with natural biofiber. Introduction of natural biofiber into the petroleum-based polymer not only could enhance biodegradability but also mechanically strengthen the composite.

Natural filler comes with various properties and sizes. Interestingly, some of the filler could disintegrate into smaller structures that finally could give novel effects on the properties of the composite. As for an example, the natural fiber could be hydrolyzed or disintegrated into nanosized fiber that finally incorporated into polymer matrix to create nanocomposite. Figure 4 shows the possible pathways for green and natural nanocomposite development. It also reveals the difference between nanocomposite, green composite, and natural and fully green nanocomposite.

5 Eco-friendly Starch-Based Nanocomposite

Starch-based nanobiocomposites are a new class of composite composed of nanosized filler (nanofiller) incorporated into a biobased matrix (Chang et al. 2010b). Such an association between natural biopolymers and nano-objects, with the aim to obtain synergic effects, is one of the most innovating routes to enhance the properties of these bio-matrices (Alexandre and Dubois 2000). Starch is originated from a variety of crops such as potato, wheat, rice, and corn. The source is abundant, and readily available at low cost (Chang et al. 2010a). Previously, starch is used to produce biodegradable films in aiming to partially or entirely replace plastic polymers due to its ideal feature that is low cost, abundant, and renewable with satisfactory mechanical properties (Xu et al. 2005). Starch is a biopolymer that contains major components of amylose and amylopectin that function as binder and barriers in packaging materials. At the beginning, water and glycols are commonly used as plasticizer in bioplastic making that make the starch behaves like thermoplastic instead of thermoset (Huang et al. 2004; Ma et al. 2008).

Starch-based active packaging with nanoreinforcement is an outcome of advancements in nanotechnology, food safety, and biodegradable polymer. Through current nanotechnology developments, EFB can be treated in chemical and physical pathways to nanosized cellulose/nano fiber before incorporating it into biodegradable film polymeric system in order to enhance it (Jonoobi et al. 2011). In terms of biodegradability, starch is used as polymer matrix as it comes from renewable source that can be degraded biologically. Recent progress in packaging technology takes account of food safety. It makes the film to release antimicrobial agent that could extent food shelf life, hence improve food safety.

Back then, starch-based active packaging becomes a remarkable invention as it provides many advantages. According to Avérous (2004), starch is sustainable source, widely available, renewable, and cheap. Conventional packaging especially from petroleum source on the other hand is not biodegradable with limited non-renewable source. In biodegradable plastic making, starch originated from variety of crops such as corn, rice, wheat, and potato, a sustainable source of biopolymer (Ma et al. 2008). For these reasons, starch generates a great interest and it is considered as a promising alternative to synthetic polymers for packaging applications (Savadekar and Mhaske 2012). General procedure to process starchy materials involves the granular disruption by the combination of temperature, shear, and a plasticizer, that usually needs water and/or glycerol (Avérous 2004).

However, starch is naturally brittle and water sensitive (Liu et al. 2010). Various approaches have been done to overcome these problems. Plasticizer (usually glycerol) is commonly added to the starch matrix to improve its flexibility. Plasticizer incorporation also causes the plasticized starch to melt at a lower temperature (below the decomposition temperature of starch), leading to a significant improvement in its process ability. Based on Lu et al. (2006), nanofiller geometry and nature have an ability to create new and improved nanocomposites. Some improvement includes mechanical strength, barrier properties, thermal stability, and transparency. In general, nanocomposite biopackaging are two-phase material that one of the phases has at least one dimension in nanometer range 1–100 nm (Abdul Khalil et al. 2012a). Native starch-based polymer gave such drawback in term of mechanical properties and water uptake. Addition of reinforcing agents to the starch matrix has been proven as an effective approach to obtain high-performance starch-based composite materials. It was found that after the incorporation of nanofiller into the starch-based matrix, it tend to exhibit some improvement in term of physical, mechanical, and barrier properties such as moisture and water uptake barrier (Duncan 2011).

Other finding claims that the incorporation of nanofiller into starch matrix could provide strong interfacial adhesion between nanofiller and starch matrix that finally improve the mechanical strength of the composite (Savadekar and Mhaske 2012). It also shows improvement on barrier properties as the existence of nanofiller caused the water and oxygen molecule pathway become obstructed. It also enables tortuous paths, which elongate the passageway of water and oxygen molecule hence slowing the rate of moisture transfer in or out of the food packaging system. Illustration of the "tortuous pathway" created by incorporation of nanofiller is shown in Fig. 5.

In a film composed only of polymer (a) the gas molecule migrates perpendicular to the film orientation. This characteristic is important in food packaging application because these barrier properties could increase the mean gas diffusion length, and thus extend food shelf life. However in nanocomposite film, the molecule pathway navigates around impenetrable nanofiller particles. The molecule should pass through interfacial zones that have different permeability characteristics than those of the pure biobased polymer. As the reinforcement of nanosize is scientifically proven, various types of nanofiller were intensively studied. The sources of nanofiller may come from organic or inorganic material such as clays (Lee et al. 2007; Cyras et al. 2008; Ludueña et al. 2007), natural fibers (Lee et al. 2009; Morán et al. 2007; Masoodi et al. 2012; Jonoobi et al. 2011; Bhatnagar 2005), Cellulose whisker (Cao et al. 2008) and microcrystalline cellulose (Ma et al. 2008;

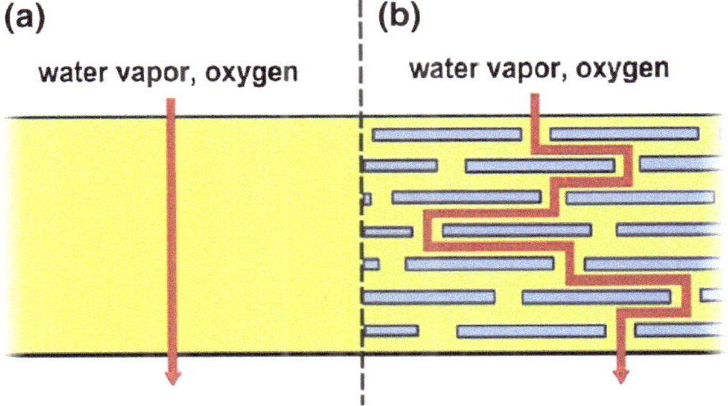

Fig. 5 Illustration of molecule migration path between, **a** Starch-based film, **b** starch-based nanocomposite film 5

Kumar et al. 2010). However, as green technology becomes a priority in order to promote sustainable development, cellulosic nanomaterials from natural fibers might be the key for the polymer/biopolymer reinforcement.

5.1 Effect of Cellulose Nanofiber Incorporation Toward Film Opacity

The opacity plays an important role in food packaging application. High translucent packaging is favorable because it can provide product visibility for product presentation purposes. Optical visibility of the film could be measured either as light transmittance (Tr) or opacity. In general, high transparency indicates that the film is less opaque. Optical visibility was expressed as opacity determined by calculating the area using the graph showing absorbance versus wavelength. Savadekar and Mhaske (2012) claimed that the incorporation of CNF increases the film's opacity. Figure 6 shows a graph of the film opacity versus CNF incorporation. Initially, the films are more transparent at low CNF content.

With the addition of 2 % CNF, the film shows lower opacity than control film. It may indicate that the addition of cellulose nanofiber into the thermoplastic starch causes more homogeneity of the molecules contained in the matrix. Melo et al. (2011) suggested that xanthan was dispersed homogenously in the matrix with nanoclays addition forming more translucent film. Furthermore, based on Yano (2005), reinforcing elements with diameters less than one tenth of visible light wavelengths did not cause light scattering. Cellulose nanofiber used in this study is typically in the size range, and therefore unless significant nanofiber agglomeration occurs, highly transparent nanocomposite films should be expected. There are also some opinion stated that the decrement of opacity especially at lower filler incorporation is caused by discontinuous phase in the composite samples after drying process (Ban et al. 2006).

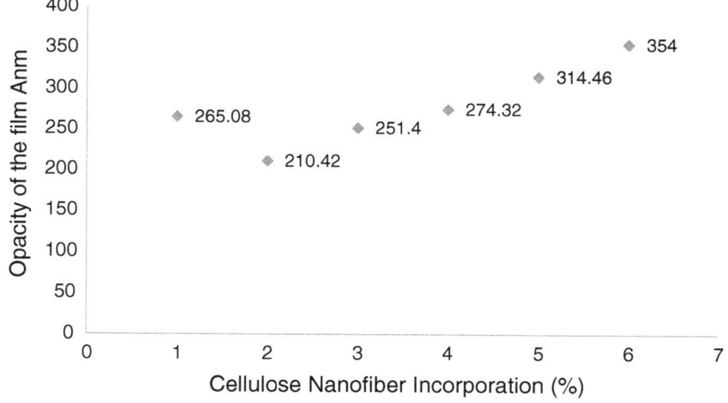

Fig. 6 Graph opacity of the film versus cellulose nanofiber (CNF) incorporation

Based on statistical analysis, ANOVA, the opacity of nanocomposite film was significantly increased when CNF was added by 6 % and more. It is because the light scattering might occur due to the inhomogenous dispersion of CNF. It is supported by the film morphology that significant nanofiber agglomeration has occurred when more CNF was added into the film. Those agglomerations could cause light scattering and finally increase the film opacity.

5.2 *Effect of Cellulose Nanofiber Toward Mechanical Properties*

Mechanical properties are one of the crucial criteria in packaging film that include tensile strength, percentage elongation, and Young's modulus. It also reflects the ability of the packaging materials in order to maintain good integrity of the product (Pitak and Rakshit 2011). In basic concept, the addition of natural nanocellulose into packaging matrix could enhance the mechanical properties of the film. Soykeabkaew et al. (2012) claimed that the sufficient amount of nanocellulose in the film matrix could increase the Young's modulus, and expends its elongation at break. Table 4 shows tensile strength, elongation at break, and Young's modulus of starch-based film and starch-based film incorporated with different percentage of cellulose nanofiber.

5.2.1 Tensile Strength (TS)

CNF incorporation could improve the tensile strength of the film (Siqueira et al. 2010). Strength of starch film incorporated with varied CNF ranges between 2 and 10 % was observed. Effect of CNF loading in starch-based film toward the film's tensile strength was tabulated in Table 4. From the result, the tensile strength of control film (0 % CNF incorporation) was 3.66 MPa and the tensile strength increased to 4.68 MPa when 2 % of CNF was added into the film.

Table 4 Tensile strength, elongation percentage, and young modulus of formed films

Film	Tensile strength (Mpa)	Elongation percentage (%) E	Young's modulus (Mpa)
Starch	3.66 ± 0.0105	1.79 ± 0.1743	207.56 ± 20.1717
Starch + 2 %CNF	4.68 ± 0.0545	1.15 ± 0.0125	381.94 ± 7.9990
Starch + 4 %CNF	3.73 ± 0.0102	0.98 ± 0.0102	381.63 ± 4.3156
Starch + 6 %CNF	3.45 ± 0.1258	0.94 ± 0.0039	367.32 ± 13.5589
Starch + 8 %CNF	3.33 ± 0.0616	0.92 ± 0.0008	362.47 ± 7.0133
Starch + 10 % CNF	2.90 ± 0.0847	0.89 ± 0.0046	326.11 ± 10.5417

However, the addition of CNF beyond 2 % made the tensile strength reduced. Addition of 4 % CNF dropped to 3.74 MPa and the tensile strength keeps reducing as the CNF incorporation increased. It showed that 2 % of EFB cellulose nanofiber incorporation in the film matrix gives the highest tensile strength. High tensile strength values at 2 % addition of CNF can be caused by two factors. The first factor is favorable nanocrystal–polymer interaction and the second is reinforcement effects that occurred through effective stress transfer at the nanocrystal–polymer interface (Khan et al. 2012). In order to reach stable and strong interfacial adhesion, increase effective stress transfer, the CNF should homogeneously dispersed in the starch matrix.

This hypothesis was proven by the FESEM analysis in Fig. 8b at next subchapter 5.4 on film morphology that shows the surface of the film is more even and smooth for 2 % of CNF incorporation compared with other percentage of CNF incorporation. It also indicates that the CNF was dispersed homogeneously in the film matrix hence tensile strength could be optimized. As reported by Khan et al. (2012), aggregates tend to form after certain concentration of nanocellulose was reached. Referring to FESEM morphology in Fig. 8b, the incorporation of CNF beyond 2 %, makes the surface of the film become rough and inhomogeneous which also indicates that the nanofiber poorly dispersed in the film matrix. The structure of the film surface and the tensile strength somehow connected where the tensile strength was reduced as more aggregates were formed on the film.

The presence of aggregates by any means, causes the distribution of stress around the film to be distorted hence reduces the tensile strength of the film. The formation of aggregates that caused by the addition of an improper amount of CNF was shown in Fig. 8f. It shows that lumpy structure was created on the surface of the film. The overall images also show that when large amounts of CNF was incorporated, the surfaces of the film become rough and uneven which indicates the nanofiber was poorly dispersed in the film. In basic idea, excessive addition of CNF could decrease the strength of the film due to the unequal bonding between the molecules in the film. In details, the improper amount of CNF addition in starch matrix will make existed molecules poorly interfered with each other hence reduce the film's strength.

In general, the successfulness of CNF reinforcement depends on the dispersion and the arrangement of CNF in the matrix. Nanofiber itself is nonhomogeneous entity that makes it almost impossible to reach equal dispersion in the matrix. The unequal dispersion of CNF unfortunately, could cause uneven stack of CNFs in the starch matrix thus, weakens the whole structure of the film. It is possible to infer that the dispersion of CNF depends on the proportion of the CNF incorporation.

5.2.2 Elongation at Break (EB)

Elasticity of the film can be described as the EB. As shown Table 4, EB for control starch (0 % CNF) is at 1.79 %. Addition of CNF into starch-based film decreases the value of EB. This infers that CNF incorporation may reduce the elasticity of the film. The value of EB decreases from 1.79 % for controlled sample to 1.15, 0.98, 0.94, 0.92, and 0.89 % for a sample with 2, 4, 6, 8, and 10 % CNF loading, respectively.

Previous findings proved that the addition of CNF into the polymer matrix reduced the EB values. As found by Li et al. (2010), the EB value decreased from 20 to 6 % due to the incorporation of cellulose nanocrystals using chitosan as a polymer matrix. Declining EB value indicates that the incorporation of CNF into the starch matrix formed strong interactions between filler and the matrix. It subsequently restricted the motion of the matrix, hence decreased the EB value (Azeredo et al. 2010; Samir et al. 2004).

5.2.3 Young's Modulus

Young's modulus, also known as the tensile modulus or elastic modulus, is a measurement of the stiffness of elastic material. Within the limits of elasticity, the ratio of the linear stress to the linear strain finally can be described as Young's modulus (Y) = (Stress/Strain). It is a quantity used to characterize materials, which measure the resistance of a material to elastic (recoverable) deformation under load. As shown in Table 4, the values for Young's modulus significantly increased from 207.56 MPa for starch-based film without CNF incorporation and 381.94 MPa for 2 % CNF incorporated. Young's modulus shows the same pattern as the tensile strength where 2 % of CNF incorporation shows the optimum condition then the value starting to drop as more of CNF incorporated into the film. The Young's modulus keep decreasing to 381.63, 367.32, 362.47, and 326.11 MPa for 4, 6, 8, and 10 % of CNF incorporated film, respectively. However, from the statistical analysis obtained, the decreasing Young's modulus value after 4 % of CNF addition is not significant.

The addition of 2 % of CNF into the film matrix results in high Young's modulus because the tensile strength (stress) is high while the elongation (strain) of the film is low. Thus, the Young's modulus (Y), which is described as unit of stress per unit of strain becomes high. As the conclusion, the high Young's modulus of the film with 2 % of CNF incorporation describes the ideal mechanical characteristic of the film that possessed high tensile strength, at the same time, able to provide satisfactory elasticity to the film.

5.3 Water Uptake Rate

Starch, a natural polymer source has many advantages such as biodegradable, abundance, and readily available at low cost. However, on the negative side, it has low resistance to water which reduces its potential as polymer material. In nature, when the starch-based film in contact with water, it will absorb the water and swell. As mentioned by Duanmu et al. (2007), the addition of filler is an effective way in

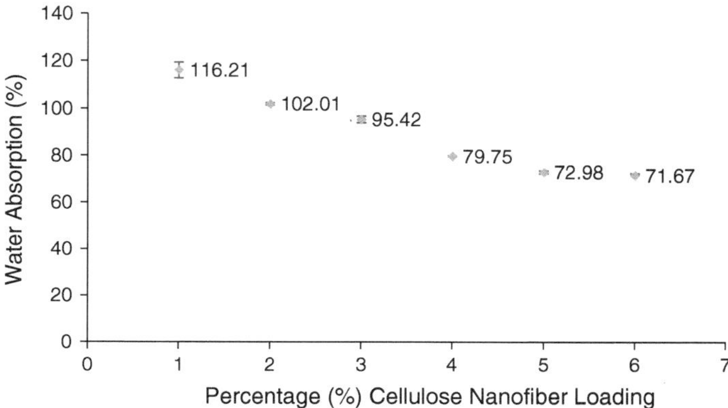

Fig. 7 Percentage of water absorption with filler incorporation

order to reduce the sensitivity toward moisture and reinforce mechanical properties of starch-based polymer. Appropriate CNF addition could repel the water from being absorbed into the film matrix that will be discussed next. In experiment conducted by Salehudin et al. (2014a), water absorption of the formed film with varied composition of CNF was observed for 5 h.

In general, water absorption percentage increases with time and become constant at 3–4 h for all tested film. Figure 7 illustrates the effect of CNF incorporation into starch-based film toward the water uptake. Without the addition of CNF, the film can absorb water that is 116 % of its own weight. The addition of 2 % CNF shows decrement in the amount of water uptake to 102 % and the values keep on decreasing when more CNF incorporated into the film. The water absorption difference between 8 and 10 % of CNF is small. From ANOVA post hoc test for homogenous subset, it shows that the water absorption between 8 and 10 % is not significantly different.

It also predicted that the water absorption might remain constant over 10 % of CNF incorporation. From data obtained, it could be inferred that the resistance of the nanocomposite film towards water absorption was increased with the addition of CNF. The presence of CNF within the film's matrix significantly reduced the amount of water absorbed by the film.

This results on the enhancement of material integrity. Same result also documented by Wan et al. (2009) showed that the water resistance of starch-based film incorporated with bacterial cellulose is higher than the neat starch-based film. In addition, the water uptake of the CNF reinforced films depends on the nature of the matrix and filler (Khan et al. 2012). Decreasing water uptake trend at equilibrium is because of that highly crystalline nanocelulose is less hydrophilic than starch and thus the preference toward the water is less. Other than that, the formation of strong filler-matrix interactions also contributes to the decreasing trend of the water uptake rate (Li et al. 2010). Furthermore, the nanocellulose added into the film acted as an interpenetrated network within the matrix and prevented the swelling of the starch films when exposed to water.

5.4 *Field Emission Scanning Emission Microscopy (FESEM) Analysis*

Savadekar and Mhaske (2012) claimed that the incorporation of nanocellulose fiber effectively enhances the structure of the film matrix (starch), since there are no observable starch granular structures observed after composite processing. Proper amount of CNF provides better dispersion in the matrix with no obvious CNF agglomeration. Based on Savadekar and Mhaske (2012), an optimum ratio of CNF to starch matrix (0.2 wt% of CNF) provide strong interfacial adhesion between CNF and starch matrix and also due to its good distribution of nanofiber within the biopolymer matrix. As mentioned earlier, the properties of the film, for an example mechanical strength, are related to its surface morphology. It showed that the agglomeration that formed due to improper amount of CNF could decrease the mechanical strength of the film. Thus, it is important to study the surface morphology of the film to understand CNF dispersion behavior in the matrix. Figure 8a–f shown the FESEM images at high magnification for 0, 2, 4, 6, 8 and 10 % of CNF incorporated film matrix, respectively. The film without incorporation of Cellulose Nanofiber (0 %), exhibited smooth surface. Addition of 2 % of CNF displayed homogenous and dense structure that indicates good dispersion level of the CNF in the starch matrix is achieved. The individual dispersion of CNF is difficult to be observed. However, as the CNF addition increased (as shown in Fig. 8c–f) the film surface become relatively rough. In 10 % of CNF, the lumpy structure was formed which indicates that agglomeration of CNF has occurred in that region. It can be said that excessive incorporation of nanocellulose fiber exhibits unsatisfied interaction between the nanocellulose fiber and starch matrix, and subsequently promotes the CNF agglomeration in the starch matrix.

6 Active or Antimicrobial Nanocomposite Packaging Film

Chitosan is one of the natural antimicrobial agents that widely used in active packaging. It either incorporated directly or indirectly into the film matrix. Effect of CNF incorporation into starch/chitosan composite film towards antimicrobial properties was previously studied by Salehudin et al. (2014b). Starch/chitosan composite (S:C 9:1) film was prepared with starch and chitosan only while starch/chitosan nanocomposite film (S:C 9:1 + 2 %CNF) was prepared by adding 2 % of CNF, chitosan and starch. Significant enhancement on the antimicrobial properties and food shelf life for both films was observed.

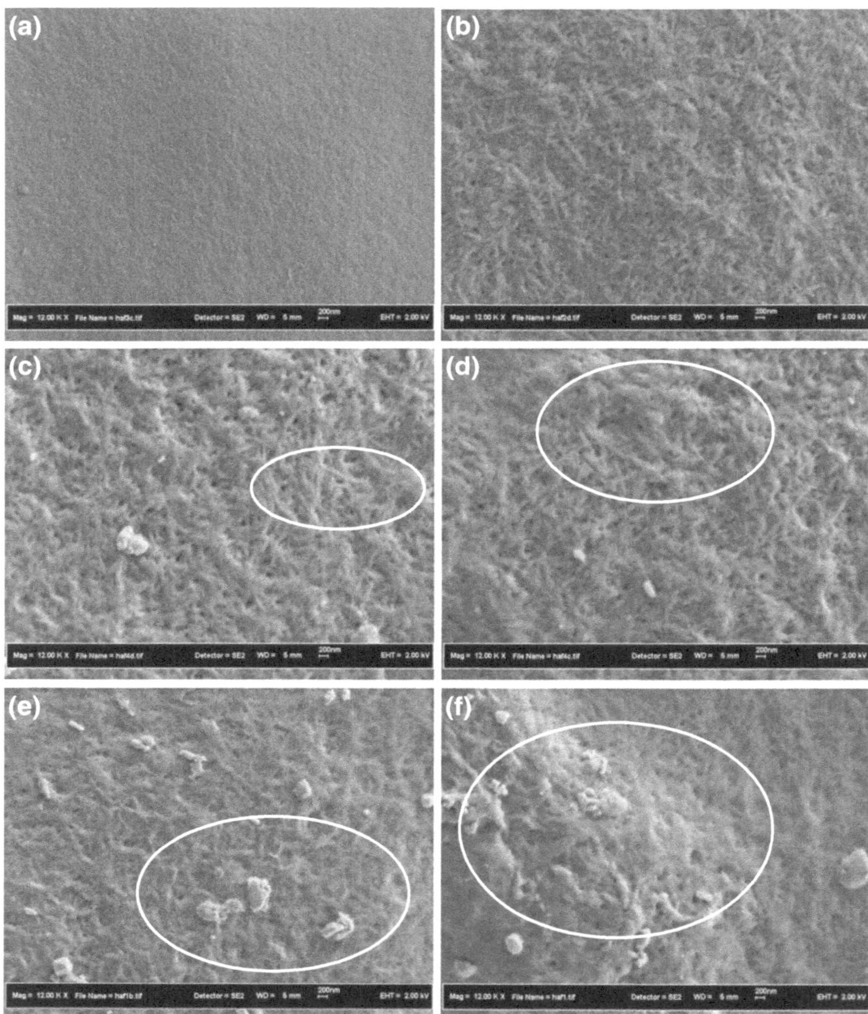

Fig. 8 FESEM images for **a** 0 %, **b** 2 %, **c** 4 %, **d** 6 %, **e** 8 % and **f** 10 % of CNF incorporation into starch film matrix (High Magnification; ×12,000). *Circle* indicates agglomeration occurred on the surface of the film

6.1 Film Appearance

Based on (Salleh et al. 2007), the ratio 9:1 (weight of starch to weight of chitosan) present optimum bacterial inhibition hence the same ratio was applied in this experiment. As shown in the previous results, 2 % of CNF incorporation describes good packaging criteria as it provides high tensile strength, excellent dispersion of

CNF, and good optical transparency. As typical biopackaging, a thin transparent film was formed. Starch/chitosan nanocomposite film (S:C 9:1 + 2 %CNF) that incorporated with CNF showed less translucent than starch/chitosan composite film (S:C 9:1) but still relatively adequate to be classify as good packaging film. This may due to the good dispersion of CNF in the composite film. The composite film that incorporated with CNF also represents more tough structure and easy to handle where composite film without incorporation of CNF was easy to break when peeled off from the casting space.

6.2 Antimicrobial Properties of Formed Films

Antimicrobial characterization of formed film was obtained by the agar diffusion method and liquid culture test. The agar diffusion test was intended to impersonate the antibacterial activities of the film on the solid food surface and liquid culture test on the other hand infers the antibacterial activities of the film on liquid-form food.

6.2.1 Agar Diffusion Test

Figures 9 shows the inhibition of control starch film, starch/chitosan composite film (S:C 9:1), and Starch/Chitosan nanocomposite (S:C 9:1 + 2 %CNF) film toward bacteria *Bacillus subtilis* (left) and *Escherichia coli* (right). Control starch film showed no inhibition toward both bacteria *Bacillus subtilis* and *Escherichia coli*; in fact, there were colonies of bacteria formed under the film. It shows that the control starch film is not able to inhibit bacteria. In general, Starch/Chitosan composite (S:C 9:1) film demonstrates inhibition toward both Gram-positive and Gram-negative bacteria, *Bacillus subtilis* and *Escherichia coli*.

There are no bacterial growth under both composite film and inhibition zone appeared on agar solid medium. However, the zone of inhibition forms is not completely clear. This is due to lack of inhibitory character that could be explained by the limitation of the diffusion of chitosan in agar medium (Coma et al. 2001). The antimicrobial activity occurred because chitosan is positively charged amino group that interacts with negatively charged microbial cell wall. Cell wall then disrupted and leaked out the proteinaceous and other intracellular constituents of the microorganisms (Hadwiger et al. 1986).

In general, starch/chitosan composite (S:C 9:1) film demonstrated better inhibition toward Gram-negative bacteria. This is because the hydrophilicity in gram-negative bacteria is significantly higher than in gram-positive bacteria, making them most sensitive to chitosan (Chung et al. 2004). The results demonstrate that the inhibition diameter for *Bacillus subtilis* bacteria appears larger in starch/chitosan nanocomposite (S:C 9:1 + 2 %CNF) film compared to starch/chitosan film (S:C 9:1).

In Gram-positive bacteria; i.e., *Bacillus subtilis,* the addition of 2 % CNF has increased the inhibition diameter to 33.21 % compared to starch–chitosan film. This

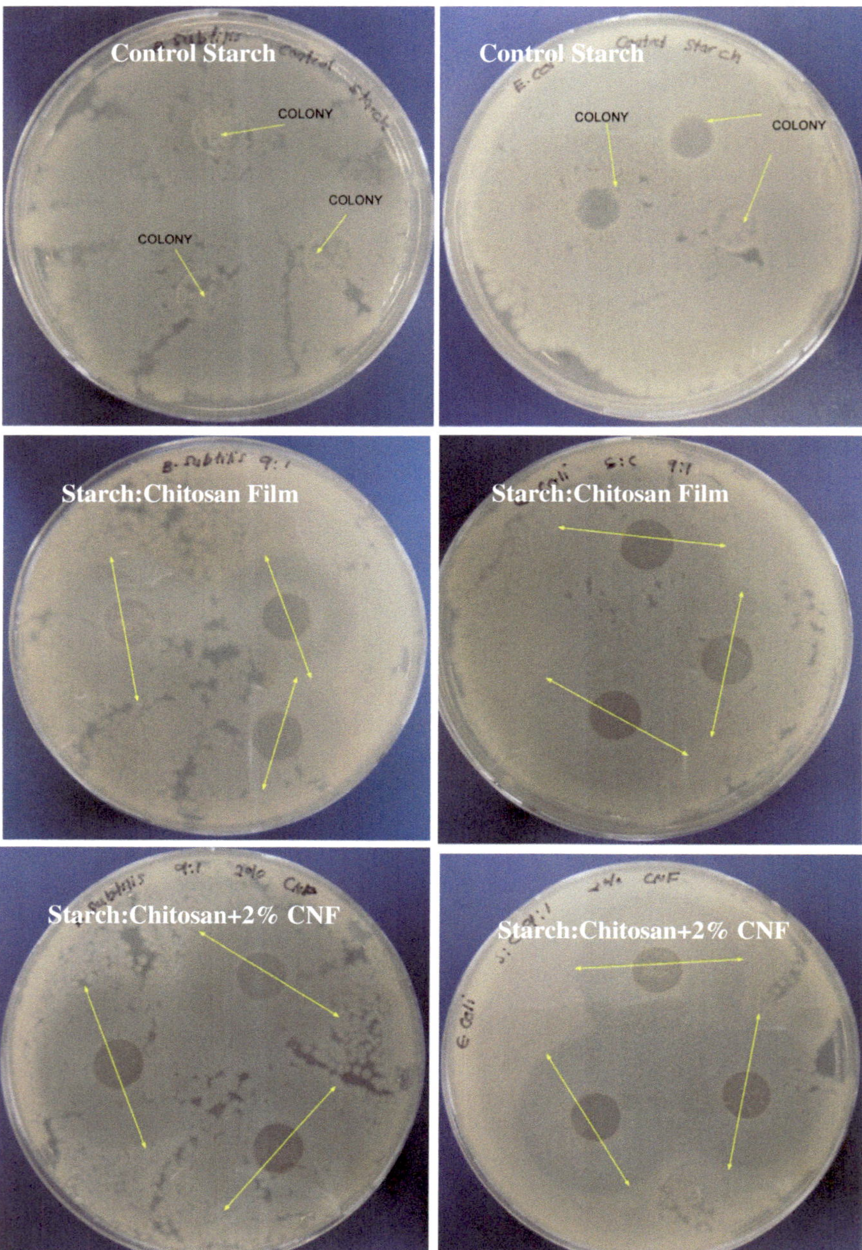

Fig. 9 Bacterial inhibition of forming film toward *Bacillus subtilis* (*left*) and *Escherichia coli* (*right*) through the agar diffusion test

is because the addition of CNF caused uniform dispersion of chitosan particles in the fiber matrix hence increased the antimicrobial activity (Shi et al. 2013). Theoretically, good dispersion of antimicrobial agent could provide more active site for chitosan–bacteria contact hence bacteria killing is more effective.

In Gram-negative bacteria, *Escherichia coli*, inhibition diameter for starch/chitosan nanocomposite (S:C 9:1 + 2 %CNF) film shows relatively small improvement on inhibition diameter about 3.53 % compared to starch/chitosan composite (S:C 9:1) film which indicates that the film is still able to show antimicrobial activity toward *Escherichia coli.*

6.2.2 Liquid Culture Test

As the previous results suggest that the agar diffusion method has a limitation on diffusivity in characterizing the antimicrobial properties of chitosan, thus diffusion in liquid culture medium was also used. In addition, the antimicrobial activity might differ in liquid compared to solid and semi-solid environment (Cao et al. 2008). Thus, in liquid culture, the behavior of bacteria inhibition might also be different as in agar diffusion method. It takes into account the bacteria mobility, antimicrobial release characteristic, and so on (Sanchez-Garcia et al. 2010). In this method, higher optical density value infers that more bacteria live in the culture media. The high-growth pattern of the bacteria in the liquid culture shows that the bacteria grow productively because of the film fails to give inhibition. Figure 10a shows the growth pattern for Gram-positive bacteria, *Bacillus subtilis* of three formed films namely the control starch, starch/chitosan composite(S:C 9:1), and starch/chitosan nanocomposite (S:C 9:1 + 2 %CNF) film.

As shown in the graph, addition of 2 % of CNF in active composite film shows the growth of bacteria was inhibited as early as 2 h of incubation. It shows that the

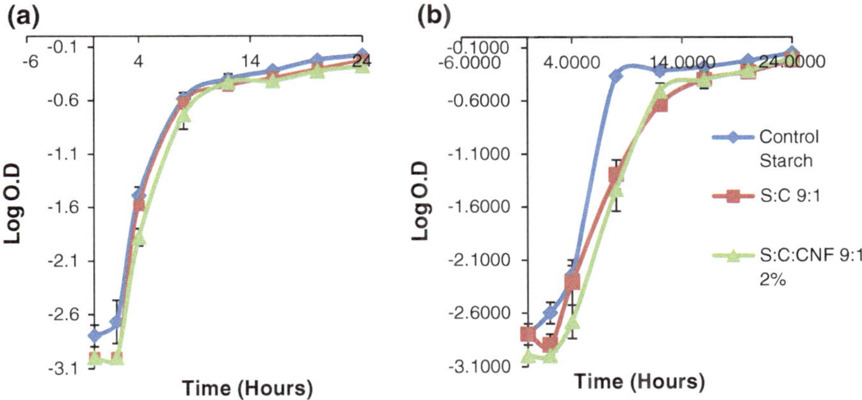

Fig. 10 Growth curve (Log OD) of *Bacillus subtilis* (**a**) and *Escherichia coli* (**b**) versus time (hour)

growth rate of *Bacillus subtilis* bacteria in active film incorporated with 2 % of CNF is low. It indicates the antimicrobial inhibition of the composite film that contains CNF is more effective. As mentioned before, the effectiveness of the film to give antimicrobial inhibition is caused by the good dispersion of the chitosan in the film. Good dispersion of antimicrobial agent could provide more active site for chitosan–bacteria contact hence bacteria killing is more effective (Shi et al. 2013). Based on Wang et al. (2007), chitosan nanocomposites showed stronger antimicrobial activity than pure chitosan, particularly against Gram-positive bacteria. They also found that the antibacterial activity of the nanocomposites increased with the increase of the amount of clay in the nanocomposite.

Wang et al. (2007) suggested two stages of antibacterial mechanism of the nanocomposites: (i) adsorption of the bacteria and immobilization on the clay surface, (ii) accumulation of chitosan on the clay surface and inhibiting bacterial growth. Applying those findings, it could be suggested that the nanofiber may influence the structure of starch/chitosan composite. The chitosan may accumulate on the surface of cellulose nanofiber and hence inhibit bacterial growth more efficiently compared to starch/chitosan composite packaging.

The addition of cellulose nanofiber into the starch/chitosan composite film did not increase or decrease Gram-negative bacteria, *Escherichia coli,* inhibition. The reason is that the antimicrobial mechanism of chitosan may differ between Gram-positive and Gram-negative bacteria. Gram-negative bacteria have thicker cell membrane compared to Gram-positive bacteria hence it is not easy to kill (Y et al. 1997). It also has an outer membrane, which constitutes the outer surface of the wall. Jiang et al. (1997), Y et al. (1997) has observed cell structure of the bacteria with the presence of chitosan and found that the cell wall of Gram-positive bacteria was weakened or even broken, while the cytoplasm of Gram-negative bacteria was concentrated and the interstices of the cell were clearly enlarged. This study indicates that the mechanisms of the antimicrobial activity of chitosan were different between Gram-positive and Gram-negative bacteria.

Environment condition may also affect the inhibition mechanism that link to the release of antimicrobial agent from the film matrix. Hence, there is a potential that different bacteria induce the release of certain antimicrobial as in previous finding showed that *Bacillus subtilis* provides an environment where it may induced greater release rate of the chitosan out of the film hence improve bactericidal activity (Wang et al. 2007). In conclusion, the addition of 2 % of CNF not significantly reduce nor increase the Gram-negative bacteria inhibition but give better bacterial inhibition on Gram-positive bacteria.

6.3 Storage Study on Food

The food that is not consumed immediately after production such as vegetables need to be contained in package to retain its shelf life and quality. The role of food packaging is to protect the food from dust, light, oxygen, pathogenic microorganism, moisture, and other harmful substance from contaminating it. Food storage and shelf life were studied for three different films which are starch-based film (Control starch), starch/chitosan composite film (S:C 9:1), and starch/chitosan nanocomposite film (S:C 9:1 + 2 %CNF). The food that was selected in this study is tomato wrapped with the film casted earlier. Quality parameters of the foods such as weight loss, colour, and visual observation were investigated in 28 days of storage period.

6.3.1 Weight Loss of Stored Food

During the storage period, moisture and quality of the food will reduce. It is necessary to keep the moisture of the food at certain level in order to maintain its quality before reaching the consumers. In order to do so, the food must be packed. In common method, synthetic plastic is used to pack food where the moisture transfer out of the system is very low. Biopolymer-based film especially from starch, are easy to transfer moisture where it could be a slight problem. In this study, the starch-based film is enhanced by the addition of chitosan and CNF that may give different results on food moisture loss. Hence, it is crucial to study the effect of CNF incorporation in the film toward food moisture loss. Figure 11 shows the effect of packaging types on tomato weight loss in room condition. In general, it shows that tomato wrapped with starch/chitosan nanocomposite (S:C 9:1 + 2 % CNF) film gives the lowest weight loss.

On the last day of storage, the nanocomposite film (S:C 9:1 2 %CNF) significantly showed the smallest weight loss at 10.73 % followed by starch/chitosan composite (S:C 9:1) film 11.76 %, and control starch film 11.83 %. From the graph,

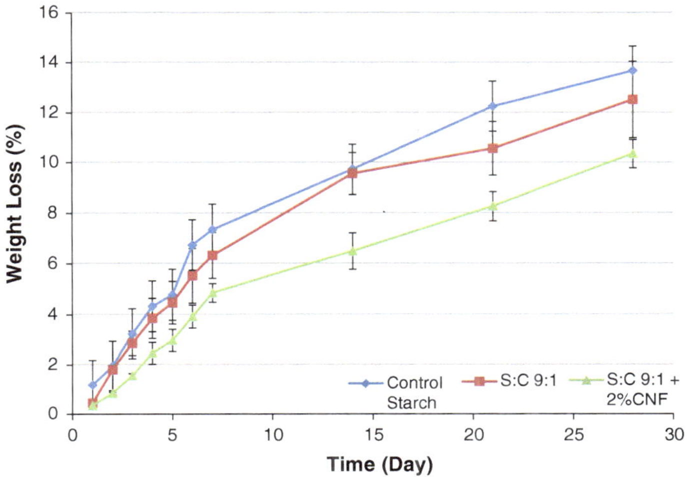

Fig. 11 Percentage weight loss of tomato

it also demonstrated that the weight loss of tomato wrapped with nanocomposite film was reduced at a slow pace. It means that films contain CNF can retain the food's moisture better than control starch film and starch/chitosan composite (S:C 9:1) film. The tomato's weight loss for starch/chitosan nanocomposite film was reduced to 9.52 % compared to starch/chitosan composite (S:C 9:1) film and 10.23 % compared to control starch film.

The results demonstrate that the different film composition affects moisture loss of the package food. The addition of chitosan could enhance the ability of the film to retain food's moisture. However, finding shows moisture uphold was furthered enhanced by CNF addition. Based on Lagaron and Garcia (2008), direct addition of filler in polymer is one of the efficient approaches to achieve desired barrier and mechanical properties. As reference, there are several studies that conducted on nanocomposite polymer. It was found that the addition of nanomaterial in polymer could exhibit excellent barrier properties against gases (e.g., O_2 and CO_2) and water vapor (Rhim et al. 2013). Another finding also mentioned that the addition of cellulose nanofiber found as an effective way to improve water vapor barrier of the films, in which the water vapor permeability (WVP) decreased significantly when 10 % of CNF was incorporated in thermoplastic starch matrix (Koh et al. 2008). From the results obtained, the addition of CNF significantly reduced tomato weight loss. It signifies that the addition of CNF in the starch matrix does improve the barrier properties of the film, preventing moisture or water molecules to be transported pass across the film. Nanosized filler such as CNF possibly affects the barrier properties of the bionanocomposite film in two particular ways. First is by creating a tortuous path for molecule diffusion (Choudalakis and Gotsis 2009). Second is by interfacial region changes that occur between nanoparticulate fillers and matrix (Duncan 2011). The tortuous path of molecule diffusion can be explained in Fig. 5 previously. In Fig. 5a, it shows that original starch-based packaging which does not contain filler allows any molecules to migrate via a pathway that is perpendicular to the orientation of the film. The addition of CNF, however, changes the molecular pathway from perpendicular to tortuous path as shows in Fig. 5b. This is because, CNF is essentially less permeable than a starch matrix, and hence gas molecules ought to diffuse around impenetrable filler instead of taking a straight line path that lies perpendicular to the film surface. As a result, the mean path for gas diffusion becomes longer. Even though tortuosity is the primary mechanism by which nanofillers influence the barrier properties of nanocomposite film, sometimes it is not always the case. There is another factor influencing it. The other way that nanoparticulate fillers could influence the barrier properties is by causing changes to the polymer matrix at the interracial regions. When the polymer–nanoparticle interaction occurs, polymer matrix (strand) in this case starch strand located near nanoparticles becomes partially immobile (Duncan 2011). As a result, the rate of the gas or molecules traveling through these interfacial zones becomes attenuated due to altered density and or size of the holes which previously has been observed directly via the use of positron annihilation lifetime spectroscopy (Wang et al. 2005). Those findings subsequently demonstrate why nanomaterials were chosen as fillers for polymer composites compared to microscale fillers. Most reason relies that nanoparticles actually have much higher aspect ratios due to their high-surface area per volume. Hence, the interfacial volume element in a nanocomposite film is significantly greater than that of a microcomposite polymer (Mogri and Paul 2001; Hiltner et al. 2005). As a result, the interaction between nanomaterial itself and nanomaterial–matrix become stronger.

6.3.2 Color Changes of Stored Food

The quality of the food that being stored could be identified by observing its colour change. In this experiment, the influences of different packaging systems on quality loss of tomato were observed using colorimetric technique. In this experiment, different packaging systems; control starch film, starch/chitosan composite (S:C 9:1) film, and starch/chitosan nanocomposite (S:C 9:1 + 2 %CNF) were applied on tomato and colour change of the tomato was observed for 4 weeks. Table 5 shows changes of the tomato colour (a*) with storage time for three different packages, control starch, starch/chitosan composite (S:C 9:1), and starch/chitosan nanocomposite (S:C 9:1 + 2 %CNF) film.

Color Retention of a*

As mentioned earlier a* denotes green to red colour. As shown in Table 5, the value for a* coordinate was lying between 4 and 10 which is in red region. The increasing value of a* coordinates mean that the tomato colour has become more intense. Basically, it shows the control starch film has the highest red colour intensity followed by and starch/chitosan composite (S:C 9:1) film and starch/chitosan

Table 5 Colour transition of L (Darkness to Lightness), a*(greenness to redness) and b* (blueness to yellowness) of wrapped tomato at room condition

Day												
Film		0	1	2	3	4	5	6	7	14	21	28
Colour change L	Starch	42.0	42.6	37.5	37.3	37.0	38.1	38.1	38.6	37.9	37.9	37.6
	S:C 9:1	41.1	40.6	41.8	41.6	41.9	42.4	42.6	42.0	41.0	41.2	40.8
	S:C 9:1 + 2 % CNF	42.1	39.4	39.3	40.8	41.7	41.3	42.6	41.3	41.7	41.3	41.5
Colour change a	Starch	5.8	6.5	6.7	7.3	8.5	9.5	9.4	9.6	9.8	9.7	9.8
	S:C 9:1	5.2	5.1	5.9	6.7	6.6	7.2	6.5	7.1	7.2	7.4	7.6
	S:C 9:1 + 2 % CNF	5.4	4.7	5.0	6.2	6.1	6.9	6.2	6.7	6.0	6.3	7.0
Colour change b	Starch	11.7	11.8	11.8	12.3	12.2	12.7	12.0	12.8	11.7	11.5	11.9
	S:C 9:1	11.7	11.6	12.1	13.1	12.3	12.7	12.3	12.5	12.5	13.0	12.7
	S:C 9:1 + 2 % CNF	12.5	12.3	13.4	13.2	13.1	13.9	13.7	13.4	14.5	14.5	12.4

nanocomposite (S:C 9:1 + 2 %CNF) film. Colour transition of starch film is the highest as the tomato decays more rapidly. The values of a* rapidly increased from 5.8 on day 0 to 9.5 at day 6 then the value increased gradually until the end of the experiment (28th days). It was because there are no antimicrobial agents present in the control starch film to inhibit microbes from spoiling the tomato. Hence, the colour changes more rapidly. However, in starch/chitosan composite (S:C 9:1 film), the a* colour coordinate changed from 5.2 on day 0 to 7.2 in day 6 then the value increased slowly until the end of the experiment. It showed that the colour intensity of the tomato wrapped with and starch/chitosan composite (S:C 9:1) film was less than control starch film. It is caused by the presence of chitosan that inhibit the bacterial activities thus slow down the rotting process of the tomato. When the tomato wrapped with starch/chitosan nanocomposite (S:C 9:1 + 2 %CNF) film, the colour coordinates a* further reduced which signify the red colour intensity was also reduced. Certaintly, and starch/chitosan nanocomposite (S:C 9:1 + 2 %CNF) film packaging shows its ability to delay quality loss/preserve food better. This result may due to a lower water loss and a more controlled rate of respiration that in turn means less obvious ripening process (Conte et al. 2009). The shelf life of freshly produced goods is greatly related to transpiration rate (TR), a physiological process that subsequently influences the relative humidity (RH) and also condensation inside the food package (Sousa-Gallagher et al. 2013). Fascinatingly, the incorporation of CNF was found to be able to counterbalance the effects between dehydration and respiration activity and hence give better preservation on the tomato.

Color Retention of b*

General idea stated that b* denotes for blueness to yellowness; however in hue principle, it also in some way represents greenness and redness to express intensity of a specific colour. Take the tomato for example, the a* colour coordinate is located at $a* = 12$ (in between −60 and 60 at x axis) which in red region. Coordinate $b*$ otherwise is located at $b* = 10$ (in between −60 and 60 at y axis). The hue interprets both point $a*$ and $b*$ as degree (h^o).

Calculation of the hue degree $h°$ is expressed as

$$\text{Tan } h^\circ = \frac{b*}{a*} \tag{1}$$

From the calculation, $h° = 1.\ 19$ is located in the red region but lies more toward $+b*$ axis. In theory, if the $b*$ value increases, the degree of h also increased. The colour region on a hue colour wheel is actually lying more toward $+b*$ axis where the red colour is less intense. Hence, it could be affirmed that the higher $b*$ value denotes lower red colour intensity. As stated in Table 5, the colour drastically changed in tomato that was wrapped with control starch film. The colour transition of tomato from light red (yellowish red) to dark red (blue red) was a little bit severe in control starch film which indicates that the process of ripening and rotting is faster compared to other films. It then followed by starch/chitosan composite (S:C 9:1) film with moderate colour change. The colour transition changed the least in the tomato that wrapped with and starch/chitosan nanocomposite (S:C 9:1 + 2 % CNF) film. It shows that the colour retention has increased significantly when 2 % of CNF was added into starch/chitosan composite. The colour change patterns in b^* coordinate is same as the colour change pattern in a^* coordinate. Colour change in b^* coordinates (blueness and yellowness) showed the same pattern as a^* colour coordinate transition which also support that the starch/chitosan nanocomposite (S: C 9:1 + 2 %CNF) film gave the longest food shelf life. It then followed by starch/ chitosan composite (S:C 9:1) film and lastly the control starch film.

Retention of Darkness to Lightness L*

Another sort of colour identification and classification is L* denotes as lightness. The higher value of L* means the sample tested is whiter or brighter. In general, it shows that the tomato that was wrapped with control starch film darkened more rapidly compared to other films. However, the change in lightness L* of S starch/ chitosan (S:C 9:1) and starch/chitosan nanocomposite (S:C 9:1 + 2 %CNF) film is similar to a certain extent. Lightness, L* alone cannot differentiate the colour difference in this case, thus a^* and b^* coordinates are more proper to be used.

6.3.3 Visual Observation of Microflora Appearance

Storage study was conducted for 4 weeks at room temperature 25–27 °C to measure the effectiveness of the film against the food spoilage. Tomato spoilage, for an example, is due to decaying process that caused by microbes, i.e., mold. In first 2 weeks, there are no obvious changes in all three groups of wrapped tomatoes. In the tomato packed with control starch film, the colour has changed from orange-red to red-orange colour where the other packages show no obvious change on the tomato colour. Tomato surface was deformed at the 3rd weeks for control film, starch/chitosan (S:C 9:1), and and starch/chitosan nanocomposite (S:C 9:1 + 2 % CNF) but more severely in control film and starch/chitosan (S:C 9:1). At fourth weeks (28th days), the tomatoes started to mold especially those wrapped with control film and S:C 9:1 film. The average diameter of the mold appears on the tomato wrapped with control starch and starch/chitosan composite are 0.93 cm and 0.57 cm, respectively. On the other hand, tomato that wrapped with S:C 9:1 + 2 % CNF film shows no mold infection. In general, the tomato that wrapped with control starch film has shown more severe result as it burst and completely out of shape. In starch/chitosan nanocomposite (S:C 9:1 + 2 %CNF) film shows no mold was present on the surface of the tomatoes. From the observation of the packaging film itself, it shows that the control starch film and starch/chitosan composite film were swollen and burst where starch/chitosan nanocomposite (S:C 9:1 + 2 %CNF) maintains its shape. These result indicates that the addition of CNF into the film matrix could maintains the integrity of food packaging material hence enhances its barrier and mechanical properties.

In general, CNF incorporation in the antimicrobial film has effect on mold retardation as observed in day 21 and 28. Tomato packed with starch/chitosan nanocomposite (S:C 9:1 + 2 %CNF) showed the least mold spoilage. The first reason is that the film already contains antimicrobial agent. However, it seems that the starch/chitosan nanocomposite (S:C 9:1 + 2 %CNF) relatively gives longer storage time compared to starch/chitosan composite (S:C 9:1) film. The reason why starch/chitosan nanocomposite gives better preservation is the antimicrobial agent contact or migration between packaging and food surface. In active nanocomposite film, the release/migration of antimicrobial agent might be better compared to active packaging without nanocellulose incorporation. As reported by Lavoine et al. (2014), addition of microfibrillated cellulose (MFC) exhibits excellent behavior of additive release. In this case, the addition of cellulose nanofiber might create good release mechanism of chitosan from the starch-based matrix. However, study on release kinetic of chitosan in packaging film that with or without CNF incorporation needs to be done as verification to the statement that been given above. In general, starch/chitosan nanocomposite (S:C 9:1 + 2 %CNF) demonstrates characteristic for good active packaging that possessed good mechanical and antimicrobial properties for food preservation.

7 Conclusion

The future of eco-friendly materials as polymer building block becomes brighter not only because the source is infinite, but it could compete and may replace synthetic material that is available today. Since there are numerous research on eco-friendly materials such as cellulose nanofiber, characteristics, morphology, and possible interaction could be possibly predicted. In nanofiber production, various methods are available that gave different results in term of quality and yields. In general, source of nanofibers is varied either from animal or plant. In this scope, the cellulose nanofiber from oil palm empty fruit bunch (OPEFB) was isolated by pretreatment and then hydrolysis using strong acid. An original EFB fiber initially have diameter around 150–250 μm and disrupted into smaller fiber that size from 16 to 28 μm. The size of the fiber could be further reduced to nanosized through hydrolysis process. Through TEM analysis, the fiber size ranged between 30–100 nm.

Eco-friendly starch-based nanocomposite or by other name, CNF reinforced biopolymer was formulated by adding cellulose nanofiber into starch-based matrix. The effect on cellulose nanofiber incorporation into starch-based film was studied. Basically, starch-based film incorporated with cellulose nanofiber shows an improvement on mechanical strength that includes tensile strength, percentage of elongation, and Young's modulus. Two percentage of cellulose nanofiber addition is an optimum ratio for the nanobiocomposites film to give the maximum mechanical strength. Mechanical strength is strongly related to nanofiber and starch interaction in the polymer system. High mechanical strength could be attained by excellent nanofiber–polymer (starch) interaction and effective stress transfer in the polymer matrix. Additionally, the cellulose nanofiber should disperse homogeneously in the starch matrix in order to reach stable and strong interfacial adhesion. As shown in FESEM analysis, the surface morphology of the film is more even for 2 % of cellulose nanofiber incorporation. It shows that the cellulose nanofiber dispersed homogenously in the film matrix hence tensile strength could be optimized. At 10 % of filler incorporation, the cellulose nanofiber agglomerated causing a lumpy structure on the film surface and reduces the strength of the film. Incorporation of cellulose nanofiber into the film increases its opacity. As the cellulose nanofiber addition increases, the opacity of the film also increased. Nanofiber agglomeration occurred when more cellulose nanofiber was added. Consequently, it causes light scattering and increases the film's opacity. Water absorption test showed that the addition of cellulose nanofiber increases film's moisture resistance as crystalline nanocellulose is less hydrophilic than starch. Antimicrobial nanobiocomposite or active nanopackaging was developed by incorporating chitosan into the selected film formulation that provides the excellent characteristic. In comparison, an active packaging film incorporated with cellulose nanofiber gave better inhibition toward Gram-positive bacteria *Bacillus subtillis*, than Gram-negative bacteria, *Escherichia coli.* Incorporation of cellulose nanofiber into active packaging also could improve the value of active packaging for food preservation. Tomato wrapped using active packaging incorporated with cellulose nanofber, (S:C 9:1 + 2 % CNF) showed the least moisture reduction percentage. It also shows that the packaging S:C 9:1 + 2 % CNF could prolonged the tomato storage time as the colour transition of the tomato was slower compared to the active packaging without incorporation of cellulose nanofiber. The decaying and molding was observed in both active packaging (S:C 9:1) and active nanopackaging (9:1 + 2 % CNF) but it seems more severe on the tomato that wrapped with packaging (S:C 9:1). As a conclusion, incorporation of cellulose nanofiber could enhance physical and mechanical properties of starch-based film. Active packaging reinforced with 2 % of cellulose nanofiber on the other hand, showed significant improvement on antimicrobial properties and on food storage life. It reveals the potential of cellulose nanofiber as filler for antimicrobial packaging as it enhances antimicrobial efficiency toward food's shelf life.

8 Future Perspective

The source of nanofiber is varied either from animal or plant that contains different amount of chemical composition. Thus, in nanofiber production, different natural fibers need to undergo several chemical and physical treatments. In addition, every extraction method gave different results in term of quality and yields, hence other techniques of cellulose nanofiber production need to be explored. Furthermore, as found in previous research, combination of chemical, enzymatic, and physical method, also by adding pretreatment into process integration could significantly reduce energy consumption.

In this experiment, chitosan was incorporated into the film formulation that contain constant amount of cellulose nanofiber, 2 %. As stated in this study, active packaging with incorporation of cellulose nanofiber showed significant enhancement on antimicrobial properties, and food shelf life compared to the active packaging without the incorporation of cellulose nanofiber. However, further addition of cellulose nanofiber into active packaging toward antimicrobial properties and food shelf life remains uncertain. It may shows different results as the interaction between cellulose nanofiber and biopolymer matrix is a bit complex. Those interactions may link to antimicrobial agent release and subsequently affecting antimicrobial properties and food shelf life. It is possible to state that the addition of cellulose nanofiber might be able to control the antimicrobial agent release. It also infers that the amount of cellulose nanofiber might as well affect the antimicrobial release. Thus, in further study, the amount of cellulose nanofiber incorporation into the active packaging can be increased and varied. On top of that, the release studies need to be done to understand more about the effect of cellulose nanofiber addition toward antimicrobial migration behavior out of the packaging film.

References

Abdul Khalil HPS, Marliana MM, Issam AM, Bakare IO (2011) Exploring isolated lignin material from oil palm biomass waste in green composites. Mater Des 32:2604–2610

Abdul Khalil HPS, Bhat AH, Ireana Yusra AF (2012a) Green composites from sustainable cellulose nanofibrils: a review. Carbohydr Polym 87:963–979

Abdul Khalil HPS, Bhat IUH, Jawaid M, Zaidon A, Hermawan D, Hadi YS (2012b) Bamboo fibre reinforced biocomposites: a review. Mater Des 42:353–368

Abe K, Yano H (2009) Comparison of the characteristics of cellulose microfibril aggregates isolated from fiber and parenchyma cells of Moso bamboo (Phyllostachys pubescens). Cellulose 17:271–277

Abe K, Yano H (2010) Comparison of the characteristics of cellulose microfibril aggregates isolated from fiber and parenchyma cells of moso bamboo (Phyllostachys pubescens). Cellulose 17(2):271–277

Abe K, Iwamoto S, Yano H (2007) Obtaining cellulose nanofibers with a uniform width of 15 nm from wood. Biomacromolecules 8:3276–3278

Alemdar A, Sain M (2008) Biocomposites from wheat straw nanofibers: morphology, thermal and mechanical properties. Compos Sci Technol 68:557–565

Alexandre M, Dubois P (2000) Polymer-layered silicate nanocomposites: preparation, properties and uses of a new class of materials. Mater Sci Eng: R: Rep 28:1–63

Araki J, Wada M, Kuga S, Okano T (2000) Birefringent glassy phase of a cellulose microcrystal suspension. Langmuir 16(6):2413–2415

Arayapranee W, Na-Ranong N, Rempel GL (2005) Application of rice husk ash as fillers in the natural rubber industry. J Appl Polym Sci 98:34–41

Avérous L (2004) Biodegradable multiphase systems based on plasticized starch: a review. J Macromol Sci, Part C: Polym Rev 44:231–274

Azeredo HMCd (2009) Nanocomposites for food packaging applications. Food Res Int 42: 1240–1253

Azeredo HM, Mattoso LH, Avena-Bustillos RJ, Filho GC, Munford ML, Wood D, McHugh TH (2010) Nanocellulose reinforced chitosan composite films as affected by nanofiller loading and plasticizer content. J Food Sci 75:N1–N7

Azeredo HMC, Miranda KWE, Rosa MF, Nascimento DM, de Moura MR (2012) Edible films from alginate-acerola puree reinforced with cellulose whiskers. LWT–Food Sci Technol 46:294–297

Ban W, Song J, Argyropoulos DS, Lucia LA (2006) Improving the physical and chemical functionality of starch-derived films with biopolymers. J Appl Polym Sci 100:2542–2548

Bhatnagar A (2005) Processing of Cellulose Nanofiber-reinforced Composites. J Reinf Plast Compos 24:1259–1268

Cao X, Chen Y, Chang PR, Muir AD, Falk G (2008) Starch-based nanocomposites reinforced with flax cellulose nanocrystals. Express Polym Lett 2:502–510

Carvalheiro F, Duarte LC, Gírio FM (2008) Hemicellulose biorefineries: a review on biomass pretreatments. J Sci Ind Res 67:849–864

Chakraborty A, Sain M, Kortschot M (2006) Reinforcing potential of wood pulp derived microfibres in a PVA matrix. Holzforschung 60(1):53–58

Chang PR, Jian R, Yu J, Ma X (2010a) Fabrication and characterisation of chitosan nanoparticles/plasticised-starch composites. Food Chem 120:736–740

Chang PR, Jian RJ, Yu JG, Ma XF (2010b) Starch-based composites reinforced with novel chitin nanoparticles. Carbohydr Polym 80:420–425

Chen Y, Liu C, Chang PR, Cao X, Anderson DP (2009) Bionanocomposites based on pea starch and cellulose nanowhiskers hydrolyzed from pea hull fibre: affect of hydrolysis time. Carbohydr Polym 76:607–615

Chen W, Yu H, Liu Y, Chen P, Zhang M, Hai Y (2011) Individualization of cellulose nanofibers from wood using high-intensity ultrasonication combined with chemical pretreatments. Carbohydr Polym 83:1804–1811

Cheng Q, Wang S, Han Q (2010) Novel process for isolating fibrils from cellulose fibers by high-intensity ultrasonication. II. Fibril characterization. J Appl Polym Sci 115(5):2756–2762

Cheng Q, Wang S, Rials T, Lee S (2007) Physical and mechanical properties of polyvinyl alcohol and polypropylene composite materials reinforced with fibril aggregates isolated from regenerated cellulose fibers. Cellulose 14(6):593–602

Cherian BM, Leão AL, de Souza SF, Thomas S, Pothan LA, Kottaisamy M (2010) Isolation of nanocellulose from pineapple leaf fibres by steam explosion. Carbohydr Polym 81:720–725

Choudalakis G, Gotsis AD (2009) Permeability of polymer/clay nanocomposites: a review. Eur Polymer J 45:967–984

Chung YC, Su YP, Chen CC, Jia G, Wang HL, Wu JC, Lin JG (2004) Relationship between antibacterial activity of chitosan and surface characteristics of cell wall. Acta Pharmacol Sin 25:932–936

Coma V, Sebti I, Pardon P, Deschamps A, Pichavant FH (2001) Antimicrobial edible packaging based on cellulosic ethers, fatty acids, and nisin incorporation to inhibit ilisteria innocua and staphylococcus aureus. J Food Prot 64:470–475

Cyras VP, Manfredi LB, Ton-That M-T, Vázquez A (2008) Physical and mechanical properties of thermoplastic starch/montmorillonite nanocomposite films. Carbohydr Polym 73:55–63

de Morais Teixeira E, Corrêa A, Manzoli A, de Lima Leite F, de Oliveira C, Mattoso L (2010) Cellulose nanofibers from white and naturally colored cotton fibers. Cellulose 17:595–606

de Moura MR, Aouada FA, Avena-Bustillos RJ, McHugh TH, Krochta JM, Mattoso LHC (2009) Improved barrier and mechanical properties of novel hydroxypropyl methylcellulose edible films with chitosan/tripolyphosphate nanoparticles. J Food Eng 92:448–453

Dean K, Yu L, Wu DY (2007) Preparation and characterization of melt-extruded thermoplastic starch/clay nanocomposites. Compos Sci Technol 67:413–421

Duanmu J, Gamstedt EK, Rosling A (2007) Hygromechanical properties of composites of crosslinked allylglycidyl-ether modified starch reinforced by wood fibres. Compos Sci Technol 67:3090–3097

Dufresne A, Cavaillé J-Y, Vignon MR (1997) Mechanical behavior of sheets prepared from sugar beet cellulose microfibrils. J Appl Polym Sci 64:1185–1194

Dufresne A, Dupeyre D, Vignon MR (2000) Cellulose microfibrils from potato tuber cells: processing and characterization of starch–cellulose microfibril composites. J Appl Polym Sci 76:2080–2092

Duncan TV (2011) Applications of nanotechnology in food packaging and food safety: barrier materials, antimicrobials and sensors. J Colloid Interface Sci 363:1–24

Ek M, Gellerstedt G, Henriksson G (2009) Pulping chemistry and technology, vol 2

Elazzouzi-Hafraoui S, Nishiyama Y, Putaux J-L, Heux L, Dubreuil F, Rochas C (2007) The shape and size distribution of crystalline nanoparticles prepared by acid hydrolysis of native cellulose. Biomacromolecules 9(1):57–65

Fahma F, Iwamoto S, Hori N, Iwata T, Takemura A (2010) Isolation, preparation, and characterization of nanofibers from oil palm empty-fruit-bunch (OPEFB). Cellulose 17:977–985

Frenot A, Henriksson MW, Walkenstrom P (2007) Electrospinning of cellulosebased nanofibers. J Appl Polym Sci 103(3):1473–1482

Fuad MYA, Ismail Z, Ishak ZAM, Omar AKM (1998) Rice husk ash. In: Pritchard G (ed) Plastics additives. Springer, Berlin, pp 561–566

Habibi Y, Heux L, Mahrouz M, Vignon MR (2008) Morphological and structural study of seed pericarp of Opuntia ficus-indica prickly pear fruits. Carbohydr Polym 72:102–112

Hadwiger LA, Kendra DF, Fristensky BW, Wagoner W (1986) Chitosan both activates genes in plants and inhibits RNA synthesis in fungi. In: Muzzarelli R, Jeuniaux C, Gooday G (eds) Chitin in nature and technology. Springer, US, pp 209–214

Hambali E, Thahar A, Komarudin A (2010) The potential of oil palm and rice biomass as bioenergy feedstock. In: 7th biomass Asia workshop. Jakarta, Indonesia

Hamzah F, Idris A, Shuan TK (2011) Preliminary study on enzymatic hydrolysis of treated oil palm (Elaeis) empty fruit bunches fibre (EFB) by using combination of cellulase and β 1-4 glucosidase. Biomass Bioenergy 35:1055–1059

Hayashi N, Kondo T, Ishihara M (2005) Enzymatically produced nano-ordered short elements containing cellulose I crystalline domains. Carbohydr Polym 61(2):191–197

Henriksson M, Henriksson G, Berglund LA, Lindstrom T (2007) An environmentally friendly method for enzyme-assisted preparation of microfibrillated cellulose (MFC) nanofibers. Eur Polym J 43(48):3434–3441

Herrick FW, Casebier RL, Hamilton JK, Sandberg KR (1983) Microfibrillated cellulose: morphology and accessibility. J Appl Polym Sci: Appl Polym Symp 37:797–813

Hiltner A, Liu RYF, Hu YS, Baer E (2005) Oxygen transport as a solid-state structure probe for polymeric materials: a review. J Polym Sci, Part B: Polym Phys 43:1047–1063

Hossain MD, Hanafi MM, Jol H, Hazandy AH (2011) Growth, yield and fiber morphology of kenaf (Hibiscus cannabinus L.) grown on sandy bris soil as influenced by different levels of carbon. Afr J Biotechnol 10:10087–10094

Huang M-F, Yu J-G, Ma X-F (2004) Studies on the properties of Montmorillonite-reinforced thermoplastic starch composites. Polymer 45:7017–7023

Iwamoto S, Kai W, Isogai T, Saito T, Isogai A, Iwata T (2010) Comparison study of TEMPO-analogous compounds on oxidation efficiency of woodcellulose for preparation of cellulose nanofibrils. Polym Degrad Stab 95(8):1394–1398

Jayaraman K (2003) Manufacturing sisal–polypropylene composites with minimum fibre degradation. Compos Sci Technol 63:367–374

John M, Thomas S (2008) Biofibres and biocomposites. Carbohydr Polym 71:343–364

Jonoobi M, Harun J, Mathew AP, Hussein MZ, Oksman K (2010) Preparation of cellulose nanofibers with hydrophobic surface characteristics. Cellulose 17(2):299–307

Jonoobi M, Khazaeian A, Tahir PM, Azry SS, Oksman K (2011) Characteristics of cellulose nanofibers isolated from rubberwood and empty fruit bunches of oil palm using chemo-mechanical process. Cellulose 18:1085–1095

Khan ARAK, Salmieri Stephane, Le Tien Canh, Riedl Bernard, Bouchard Jean, Chauve Gregory, Tan Victor, Kamal Musa R, Lacroix Monique (2012) Mechanical and barrier properties of nanocrystalline cellulose reinforced chitosan based nanocomposite films. Carbohydr Polym 90:1601–1608

Kim EG, Kim BS, Kim DS (2007) Physical properties and morphology of polycaprolactone/starch/pine-leaf composites. J Appl Polym Sci 103(2):928–934

Klemm D, Heublein B, Fink HP, Bohn A (2005) Cellulose: fascinating biopolymer and sustainable raw material. Angew Chem Int Ed Engl 44:3358–3393

Koh HC, Park JS, Jeong MA, Hwang HY, Hong YT, Ha SY, Nam SY (2008) Preparation and gas permeation properties of biodegradable polymer/layered silicate nanocomposite membranes. Desalination 233:201–209

Kumar S, Negi YS, Upadhyaya JS (2010) Studies on characterization of corn cob based nanoparticles. Adv Mater Lett 1:246–253

Kristo E, Biliaderis CG (2007) Physical properties of starch nanocrystal-reinforced pullulan films. Carbohydr Polym 68:146–158

Lagaron JM, Garcia S (2008) Thermoplastic nanobiocomposites for rigid and flexible food packaging applications. Woodhead Publishers, Boca Raton

Lavoine N, Desloges I, Khelifi B, Bras J (2014) Impact of different coating processes of microfibrillated cellulose on the mechanical and barrier properties of paper. J Mater Sci 49:2879–2893

Lee SY, Xu YX, Hanna MA (2007) Tapioca starch-poly (lactic acid)-based nanocomposite foams as affected by type of nanoclay. Int Polym Process 22(5):429–435

Lee B-H, Kim H-S, Lee S, Kim H-J, Dorgan JR (2009) Bio-composites of kenaf fibers in polylactide: role of improved interfacial adhesion in the carding process. Compos Sci Technol 69:2573–2579

Li L-H, Deng J-C, Deng H-R, Liu Z-L, Li X-L (2010) Preparation, characterization and antimicrobial activities of chitosan/Ag/ZnO blend films. Chem Eng J 160:378–382

Lin M-F, Thakur VK, Tan EJ, Lee PS (2011a) Dopant induced hollow $BaTiO_3$ nanostructures for application in high performance capacitors. J Mater Chem 21:16500–16504

Lin M-F, Thakur VK, Tan EJ, Lee PS (2011b) Surface functionalization of $BaTiO_3$ nanoparticles and improved electrical properties of $BaTiO_3$/polyvinylidene fluoride composite. RSC Adv 1:576–578

Liu H, Liu D, Yao F, Wu Q (2010) Fabrication and properties of transparent polymethylmethacrylate/cellulose nanocrystals composites. Bioresour Technol 101:5685–5692

Liu K, Lin X, Chen L, Huang L, Cao S, Wang H (2013) Preparation of microfibrillated cellulose/chitosan-benzalkonium chloride biocomposite for enhancing antibacterium and strength of sodium alginate films. J Agric Food Chem 61:6562–6567

Lu Y, Weng L, Zhang L (2004) Morphology and properties of soy protein isolate thermoplastics reinforced with chitin whiskers. Biomacromolecules 5:1046–1051

Lu Y, Weng L, Cao X (2006) Morphological, thermal and mechanical properties of ramie crystallites—reinforced plasticized starch biocomposites. Carbohydr Polym 63:198–204

Ludueña LN, Alvarez VA, Vazquez A (2007) Processing and microstructure of PCL/clay nanocomposites. Mater Sci Eng, A 460–461:121–129

Ma X, Chang PR, Yu J (2008) Properties of biodegradable thermoplastic pea starch/carboxymethyl cellulose and pea starch/microcrystalline cellulose composites. Carbohydr Polym 72:369–375

Ma Z, Kotaki M, Ramakrishna S (2005) Electrospun cellulose nanofiber as affinity membrane. J Memb Sci 265(1–2):115–123

Masoodi R, El-Hajjar RF, Pillai KM, Sabo R (2012) Mechanical characterization of cellulose nanofiber and bio-based epoxy composite. Mater Des 36:570–576

Melo Cd, Garcia PS, Grossmann MVE, Yamashita F, Dall'Antônia LH, Mali S (2011) Properties of extruded xanthan-starch-clay nanocomposite films. Braz Arch Biol Technol 54:1223–1333

Mogri Z, Paul DR (2001) Water-vapor permeation in semicrystalline and molten poly(octadecyl acrylate). J Polym Sci, Part B: Polym Phys 39:979–984

Mohanty AK, Misra M, Drzal LT (2002) Sustainable bio-composites from renewable resources: opportunities and challenges in the green materials world. J Polym Environ 10:19–26

Morán JI, Alvarez VA, Cyras VP, Vázquez A (2007) Extraction of cellulose and preparation of nanocellulose from sisal fibers. Cellulose 15:149–159

Mosier N, Wyman C, Dale B, Elander R, Lee YY, Holtzapple M, Ladisch M (2005) Features of promising technologies for pretreatment of lignocellulosic biomass. Bioresour Technol 96:673–686

Mun TK (2011) Development of Malaysia Biomass Industry Technical Coach EU-Malaysia Biomass Sustainable Production Initiative. In: (Biomass-SP) Briefing session to financial institution on green technology financing 6. Cyberjaya

Nakagaito AN, Yano H (2004) The effect of morphological changes from pulp fiber towards nano-scale fibrillated cellulose on the mechanical properties of high-strength plant fiber based composites. Appl Phys A 78:547–552

Nakagaito AN, Yano H (2005) Novel high-strength biocomposites based on microfibrillated cellulose having nano-order-unit web-like network structure. Appl Phys A 80:155–159

Nakagaito AN, Yano H (2008) The effect of fiber content on the mechanical and thermal expansion properties of biocomposites based on microfibrillated cellulose. Cellulose 15:555–559

Nishino T, Hirao K, Kotera M, Nakamae K, Inagaki H (2003) Kenaf reinforced biodegradable composite. Compos Sci Technol 63:1281–1286

Paakko M, Ankerfors M, Kosonen H, Nykanen A, Ahola S, Osterberg M (2007) Enzymatic hydrolysis combined with mechanical shearing and high-pressure homogenization for nanoscale cellulose fibrils and strong gels. Biomacromolecules 8(6):1934–1941

Phiriyawirut M (2012) Cellulose microfibril from banana peels as a nanoreinforcing fillers for zein films. Open J Polym Chem 02:56–62

Pitak N, Rakshit SK (2011) Physical and antimicrobial properties of banana flour/chitosan biodegradable and self sealing films used for preserving fresh-cut vegetables. LWT—Food Sci Technol 44:2310–2315

Ramanaiah K, Ratna Prasad AV, Chandra Reddy KH (2011) Mechanical properties and thermal conductivity of typha angustifolia natural fiber-reinforced polyester composites. Int J Polym Anal Charact 16:496–503

Ramos LP (2003) The chemistry involved in the steam treatment of lignocellulosic materials. Quim Nova 26:863–871

Rhim J-W, Park H-M, Ha C-S (2013) Bio-nanocomposites for food packaging applications. Prog Polym Sci

Rosa MF, Medeiros ES, Malmonge JA, Gregorski KS, Wood DF, Mattoso LHC, Glenn G, Orts WJ, Imam SH (2010) Cellulose nanowhiskers from coconut husk fibers: effect of preparation conditions on their thermal and morphological behavior. Carbohydr Polym 81:83–92

Rowell RM, Pettersen R, Han JS, Rowell JS, Tshabalala MA (2005) Cell wall chemistry. In: Rowell RM (ed) Handbook of wood chemistry and wood composites, 37, CRS Press, Florida

Sae-Oui P, Rakdee C, Thanmathorn P (2002) Use of rice husk ash as filler in natural rubber vulcanizates: in comparison with other commercial fillers. J Appl Polym Sci 83:2485–2493

Saito T, Hirota M, Tamura N, Kimura S, Fukuzumi H, Heux L (2009) Individualization of nano-sized plant cellulose fibrils by direct surface carboxylation using TEMPO catalyst under neutral conditions. Biomacromolecules 10(7):1992–1996

Saito T, Kimura S, Nishiyama Y, Isogai A (2007) Cellulose nanofibers prepared by TEMPO-mediated oxidation of native cellulose. Biomacromolecules 8(8):2485–2491

Saito T, Nishiyama Y, Putaux JL, Vignon M, Isogai A (2006) Homogeneous suspensions of individualized microfibrils from TEMPO-catalyzed oxidation of native cellulose. Biomacromolecules 7(6):1687–1691

Salajkova M, Valentini L, Zhou Q, Berglund LA (2013) Tough nanopaper structures based on cellulose nanofibers and carbon nanotubes. Compos Sci Technol 87:103–110

Salehudin MH, Salleh E, Muhamad II, Mamat SNH (2014a) Starch based biofilm reinforced with empty fruit bunch (EFB). Mater Res Innov 18(S6), S6-322–S6-325

Salehudin MH, Salleh E, Muhamad II, Mamat SNH (2014b) Starch based active packaging film reinforced with empty fruit bunch (EFB) cellulose nanofiber. Procedia Chem 9:23–33

Salleh E, Muhamad II, Khairuddin N (2007) Inhibition of Bacillus subtilis and Escherichia coli by antimicrobial starch-based film incorporated with lauric acid and chitosan. In: Proceedings of the 3rd CIGR section VI international symposium on food and agricultural products: processing and innovation. Naples, Italy

Samir MASA, Alloin F, Paillet M, Dufresne A (2004) Tangling effect in fibrillated cellulose reinforced nanocomposites. Macromolecules 37:4313–4316

Sanchez-Garcia MD, Lopez-Rubio A, Lagaron JM (2010) Natural micro and nanobiocomposites with enhanced barrier properties and novel functionalities for food biopackaging applications. Trends Food Sci Technol 21:528–536

Satyanarayana KG, Arizaga GGC, Wypych F (2009) Biodegradable composites based on lignocellulosic fibers—an overview. Prog Polym Sci 34:982–1021

Savadekar NR, Mhaske ST (2012) Synthesis of nano cellulose fibers and effect on thermoplastics starch based films. Carbohydr Polym 89:146–151

Shi Q, Vitchuli N, Nowak J, Jiang S, Caldwell JM, Breidt F, Bourham M, Zhang X, McCord M (2013) Multifunctional and durable nanofiber-fabric-layered composite for protective application. J Appl Polym Sci 128:1219–1226

Singha AS, Thakur VK (2008a) Mechanical properties of natural fibre reinforced polymer composites. Bull Mater Sci 31:791–799

Singha AS, Thakur VK (2008b) Effect of fibre loading on urea-formaldehyde matrix based green composites. Iran Polym J 17:861–873

Singha AS, Thakur VK (2009a) Grewia optiva fiber reinforced novel, low cost polymer composites. E-J Chem 6:71–76

Singha AS, Thakur VK (2009b) Synthesis, characterisation and analysis of hibiscus sabdariffa fibre reinforced polymer matrix based composites. Polym Polym Compos 17:189–194

Singha AS, Thakur VK (2010a) Synthesis, characterization and study of pine needles reinforced polymer matrix based composites. J Reinf Plast Compos 29:700–709

Singha AS, Thakur VK (2010b) Mechanical, morphological, and thermal characterization of compression-molded polymer biocomposites. Int J Polym Anal Charact 15:87–97

Siqueira G, Bras J, Dufresne A (2010) Cellulosic bionanocomposites: a review of preparation, properties and applications. Polymers 2:728–765

Som C, Berges M, Chaudhry Q, Dusinska M, Fernandes TF, Olsen SI, Nowack B (2010) The importance of life cycle concepts for the development of safe nanoproducts. Toxicology 269:160–169

Sousa-Gallagher MJ, Mahajan PV, Mezdad T (2013) Engineering packaging design accounting for transpiration rate: model development and validation with strawberries. J Food Eng 119:370–376

Soykeabkaew N, Supaphol P, Rujiravanit R (2004) Preparation and characterization of jute- and flax-reinforced starch-based composite foams. Carbohydr Polym 58:53–63

Soykeabkaew N, Laosat Nittaya, Ngaokla Atitaya, Yodsuwan Natthawut, Tunkasiri Tawee (2012) Reinforcing potential of micro- and nano-sized fibers in the starch-based biocomposites. Compos Sci Technol 72:845–852

Sreekala MS, Kumaran MG, Thomas S (1997) Oil palm fibers: morphology, chemical composition, surface modification, and mechanical properties. J Appl Polym Sci 66:821–835

Sulaiman F, Abdullah N, Gerhauser H, Shariff A (2011) An outlook of Malaysian energy, oil palm industry and its utilization of wastes as useful resources. Biomass Bioenergy 35:3775–3786

Sun J-X, Sun R, Sun X-F, Su Y (2004) Fractional and physico-chemical characterization of hemicelluloses from ultrasonic irradiated sugarcane bagasse. Carbohydr Res 339:291–300

Suradi SS, Yunus RM, Beg MDH,Yusof ZAM (2009) Influence pre-treatment on the properties of lignocellulose based composite. In: National conference on postgraduate research (ncon-pgr). UMP Conference Hall, Malaysia: Centre for Graduate Studies, Universiti Malaysia Pahang

Thakur VK, Singha AS (2011) Physicochemical and mechanical behavior of cellulosic pine needle-based biocomposites. Int J Polym Anal Charact 16:390–398

Thakur VK, Thakur MK (2014a) Processing and characterization of natural cellulose fibers/ thermoset polymer composites. Carbohydr Polym 109:102–117

Thakur VK, Thakur MK (2014b) Recent advances in graft copolymerization and applications of chitosan: a review. ACS Sustain Chem Eng 2:2637–2652

Thakur VK, Thakur MK (2014c) Recent trends in hydrogels based on psyllium polysaccharide: a review. J Clean Prod 82:1–15

Thakur VK, Singha AS, Misra BN (2011a) Graft copolymerization of methyl methacrylate onto cellulosic biofibers. J Appl Polym Sci 122:532–544

Thakur VK, Tan EJ, Lin M-F, Lee PS (2011b) Polystyrene grafted polyvinylidenefluoride copolymers with high capacitive performance. Polym Chem 2:2000–2009

Thakur VK, Tan EJ, Lin M-F, Lee PS (2011c) Poly(vinylidene fluoride)-graft-poly(2-hydroxyethyl methacrylate): a novel material for high energy density capacitors. J Mater Chem 21:3751–3759

Thakur VK, Singha AS, Kaur I et al (2011d) Studies on analysis and characterization of phenolic composites fabricated from lignocellulosic fibres. Polym Polym Compos 19:505–511

Thakur VK, Yan J, Lin M-F et al (2012a) Novel polymer nanocomposites from bioinspired green aqueous functionalization of BNNTs. Polym Chem 3:962–969

Thakur VK, Singha AS, Thakur MK (2012b) Biopolymers based green composites: mechanical, thermal and physico-chemical characterization. J Polym Environ 20:412–421

Thakur VK, Lin M-F, Tan EJ, Lee PS (2012c) Green aqueous modification of fluoropolymers for energy storage applications. J Mater Chem 22:5951–5959

Thakur VK, Ding G, Ma J et al (2012d) Hybrid materials and polymer electrolytes for electrochromic device applications. Adv Mater 24:4071–4096

Thakur VK, Singha AS, Thakur MK (2012e) Surface modification of natural polymers to impart low water absorbency. Int J Polym Anal Charact 17:133–143

Thakur VK, Thakur MK, Raghavan P, Kessler MR (2014a) Progress in green polymer composites from lignin for multifunctional applications: a review. ACS Sustain Chem Eng 2:1072–1092

Thakur VK, Thunga M, Madbouly SA, Kessler MR (2014b) PMMA-g-SOY as a sustainable novel dielectric material. Rsc Adv 4:18240–18249

Thakur VK, Grewell D, Thunga M, Kessler MR (2014c) Novel composites from eco-friendly soy flour/SBS triblock copolymer. Macromol Mater Eng 299:953–958

Thakur VK, Thakur MK, Gupta RK (2014d) Review: raw natural fiber-based polymer composites. Int J Polym Anal Charact 19:256–271

Tunç S, Duman O (2011) Preparation of active antimicrobial methyl cellulose/carvacrol/montmorillonite nanocomposite films and investigation of carvacrol release. LWT—Food Sci Technol 44:465–472

Turbak AF, Snyder FW,Sandberg KR (1983) Microfibrillated cellulose, a new cellulose product: properties, uses, and commercial potential

Vilpoux O, Averous L (2004) Starch-based plastics in technology, use and potentialities of Latin American starchy tubers. NGO Raízes and Cargill Foundation-Sao Paolo-Brazil. Book N0. 3., Chap. 18. pp 521–553

Wang S, Cheng Q (2009) A novel process to isolate fibrils from cellulose fibers by high-intensity ultrasonication. Part 1. Process optimization. J Appl Polym Sci 113(2):1270–1275

Wan YZ, Luo H, He F, Liang H, Huang Y, Li XL (2009) Mechanical, moisture absorption, and biodegradation behaviours of bacterial cellulose fibre-reinforced starch biocomposites. Compos Sci Technol 69:1212–1217

Wang B, Sain M (2007) Isolation of nanofibers from soybean source and their reinforcing capability on synthetic polymers. Compos Sci Technol 67:2521–2527

Wang X, Du Y, Luo J, Lin B, Kennedy JF (2007) Chitosan/organic rectorite nanocomposite films: structure, characteristic and drug delivery behaviour. Carbohydr Polym 69:41–49

Wang ZF, Wang B, Qi N, Zhang HF, Zhang LQ (2005) Influence of fillers on free volume and gas barrier properties in styrene-butadiene rubber studied by positrons. Polymer 46:719–724

Wongpanit P, Sanchavanakit N, Pavasant P, Bunaprasert T, Tabata Y, Rujiravanit R (2007) Preparation and characterization of chitin whisker-reinforced silk fibroin nanocomposite sponges. Eur Polym J 43:4123–4135

Wu RL, Wang XL, Li F, Li HZ, Wang YZ (2009) Green composite films prepared from cellulose, starch and lignin in room-temperature ionic liquid. Bioresour Technol 100:2569–2574

Wyman CE, Dale BE, Elander RT, Holtzapple M, Ladisch MR, Lee YY (2005) Comparative sugar recovery data from laboratory scale application of leading pretreatment technologies to corn stover. Bioresour Technol 96:2026–2032

Xu YX, Kim KM, Hanna MA, Nag D (2005) Chitosan–starch composite film: preparation and characterization. Ind Crops Prod 21:185–192

Y JY, Q BY, W WZ, Q XL,G JJ (1997) Application and explore. 22

Yano HSJ, Nakagaito AN, Nogi M, Matsuura T, Hikita M, Handa K (2005) Optically transparent composites reinforced with networks of bacterial nanofibers. Adv Mater 17:153

Young R (1994) Comparison of the properties of chemical cellulose pulps. Cellulose 1:107–130

Zhao H-P, Feng X-Q,Gao H (2007) Ultrasonic technique for extracting nanofibers from nature materials. Appl Phys Lett 90(7):073112

Zuluaga R, Putaux J-L, Restrepo A, Mondragon I, Gañán P (2007) Cellulose microfibrils from banana farming residues: isolation and characterization. Cellulose 14:585–592

Cellulose Acetate Nanocomposites with Antimicrobial Properties

Adina Maria Dobos, Mihaela-Dorina Onofrei and Silvia Ioan

Abstract The chapter overviews recent progress made in the area of cellulose acetate nanocomposites, considering their high-volume applications and easy processing ability. Based on their structural details, the review provides data concerning the manufacturing, characterization, and new developments in this area, with particular emphasis on biomedical applications. Stress is laid on the importance of antimicrobial activity, correlated with different bacteria characteristics, on also considering that their interaction mechanisms create inhibitory effects against microbial growth in a solid medium, and decide their areas of applicability. In this context, the presented aspects show that cellulosic materials can be designed and fine-tuned to acquire certain properties required in different biomedical areas.

Keywords Nanocomposite · Cellulose derivative · Antimicrobial activity

1 Introduction

The interest in nanocomposites has increased markedly in recent years, due to the great potential associated with this relatively new group of materials. Nanocomposites, defined as compounds with reinforcement in the nanometer range (<100 nm), in at least one dimension, can be divided into three groups, depending on the geometrical shape of their nanoreinforcement:

A.M. Dobos · M.-D. Onofrei · S. Ioan (✉)
"Petru Poni" Institute of Macromolecular Chemistry, 700487 Iasi, Romania
e-mail: ioan_silvia@yahoo.com; sioan@icmpp.ro

A.M. Dobos
e-mail: necula_adina@yahoo.com

M.-D. Onofrei
e-mail: myha1976@yahoo.com

V.K. Thakur and M.K. Thakur (eds.), *Eco-friendly Polymer Nanocomposites*,
Advanced Structured Materials 75, DOI 10.1007/978-81-322-2470-9_12

- *One-dimensional nanocomposites* are polymers reinforced with sheets, having nanometric thickness. Examples of nanoreinforcement from this group include different kinds of clays (Maiti et al. 2002) and graphites (Fukushima et al. 2006).
- *Two-dimensional nanocomposites* are reinforced with tubes or whiskers with diameter in the nanometer range; examples are carbon nanotubes (Xie et al. 2000), cellulose (Ranby 1952), and chitin whiskers (Gopalan and Dufresne 2003).
- *Three-dimensional nanocomposites* are reinforced by spherical particles in the nanometer range. Examples of nanoreinforcement belonging to this group are nanosized metals (Watkins and McCarthy 1995), metal oxides (Thompson et al. 2003), silica (Tian et al. 2006), and carbon (Murugaraj et al. 2006).

Depending on the nanoreinforcement used and composition of nanocomposite, materials with improved abrasion and wear resistance, dimensional and thermal stability, electrical conductivity, permittivity and breakdown strength, mechanical barrier, and flammability properties may be designed (Schadler 2003; Lin et al. 2011a, b; Thakur et al. 2012a). Concretely, some literature data show that renewable resources such as natural fibers in the field of fiber-reinforced materials with their new range of applications represent an important basis in obtaining eco-friendly materials (Thakur et al. 2012b, c, d, e). Tensile strength, compressive strength, and wear resistance of urea-formaldehyde resin (UF) (Singha and Thakur 2008a, b) and resorcinol–formaldehyde (Singha and Thakur 2010a) increase to a significant extent when reinforced with Saccaharum cilliare fiber. In the perspectives of using natural fibers in polymer composites, the synthesis of pine needles-reinforced polymer composites using phenol-formaldehyde as a novel polymer matrix through compression molding techniques was realized (Singha and Thakur 2010b; Thakur et al. 2010b).

Also, such properties may be improved, often at a nanoreinforcement content of 5–10 wt%, due to properties and nanometric size of the reinforcement (Thakur et al. 2014a). The likelihood of finding defects, such as grain boundaries, voids, dislocations, and imperfections, in nanosized particles is greatly reduced compared to their micro- or macrosized counterpart, which leads to significantly enhanced properties of the composite.

Additionally, by shifting from micro- to nanoscale, the interfacial area of the reinforcement increases tremendously, and the number of possible interaction sites between the reinforcement and the matrix increases dramatically. For example, at a 10 vol.% loading, 15-nm particles have an interparticle spacing of only 10 nm (Schadler 2003). Even if the interfacial region between the nanoparticle and the polymer measures only a few nanometers, the entire polymer matrix is affected by the nanoparticle interface. As a consequence, the entire polymer matrix will have a different behavior than the bulk. The chain mobility of the polymer at the interface can be either facilitated—if the interaction is weak or restricted—if the interaction is strong. Therefore, by controlling the degree of interaction between the polymer and the nanoreinforcement, the size of the interfacial region and, hence, the properties of the entire matrix can be controlled (Schadler 2003). Another unique property of nanocomposites is their ability to preserve optical clarity, or transparency, in the material, after addition of the nanoreinforcement. This property is generated by the reinforced matrix with relatively low nanoreinforcement content and small sizes of the reinforcement particles. In general, particles much smaller than the wavelength of light do not scatter light, and thus preserving optical clarity (Schadler 2003). Another major difference from traditional composites is that nanosized reinforcements do not create large stress concentrations and thus do not compromise the ductility of the polymer (Schadler 2003; Bondeson 2007).

Nanoscale dispersion of the filler into the polymer matrix generates a tremendous amount of interfacial contacts between the organic and inorganic components. The polymer matrices generally used for the synthesis of polymer nanocomposites are nonbiodegradable, and environmentally hazard. To generate more environmental friendly materials, as well as to reduce the dependence on fossil-based resources, several biopolymers have been tested in recent years. In this context, the most important properties of polysaccharide hydrogels relevant to their biomedical/environmental applications are identified, especially for the use of polysaccharide hydrogels as drug delivery/flocculant and superabsorbent systems (Thakur and Thakur 2014b). As the properties of such polymers are sometimes inferior to those of the commercial nonbiodegradable ones, nanocomposites of such biopolymers have been prepared to improve performance. The polymers which find utility in composite technology for replacing nondegradable polymers include starch, cellulose, poly(lactic acid), poly(hydroxy alkanoates), pectin, chitosan, lignin (Thakur et al. 2014a, b, c), etc. The other polymers which, though having fossil-based sources, are still biodegradable, include poly (caprolactone), poly(butylene succinate), etc. Significant improvements of the mechanical and barrier thermal properties have been reported in such bionanocomposites, as compared to pure polymers. An important observation to be made refers to the fact that, following clay incorporation polymer biodegradation increased in most of the cases, being always produced, in spite of some occasional decreases. Thus, such bionanocomposites represent potential high-value materials of the future.

Starting from these considerations, the present chapter summarizes several types of cellulose nanocomposite (including the cellulose acetate-based ones), classified according to nanoreinforcement nature and surface modification processes. These recent scientific information can add to the basic knowledge in the field of polymeric materials, and to the diversification of their applications, especially in the biomedical domain, as due to their antimicrobial properties.

2 Processing and Characterization of Cellulose Nanocomposites

2.1 One-Dimensional Cellulose Nanocomposites

Cellulose-based nanocomposites have been reported world-wide due to their cost-effectiveness, high-volume application, easy processability, renewable nature, and simple recycling. In this context, natural cellulose in the form of wood fibers and microfibrillated cellulose has been effectively used as a binder for the elaboration of single cell components (i.e., anode, cathode, and separator)—through either casting (Jabbour et al. 2010) or filtration (Jabbour et al. 2012), and also of complete flexible cells—through filtration (Jabbour et al. 2013; Leijonmarck et al. 2013a, b). Literature shows that self-standing cell components and complete cells can be obtained using filtration methods—namely, the conventional well-established papermaking technologies with high mass production capacity adapted to battery manufacturing (Biermann 1996; Beneventi et al. 2014). Other quick and flexible manufacturing techniques, such as spray coating (Singh et al. 2012), printing and textile impregnation (Hu et al. 2010a, b, 2011), have been successfully used with conventional solvent-based formulations for the elaboration of electrodes and complete cells. These processes are still developed at laboratory scale and none of them can couple the use of water-based electrodes and biosourced binders with flexible and high production capacity technologies. Thereafter, conventional coating of metal foils with organic slurries remains the reference procedure for electrodes manufacturing. In this respect, the aim of the present work was to obtain new spray coating, proposed for a rapid and reliable large-scale production of self-standing Li-ion battery electrodes, using truly natural microfibrillated cellulose as a binder. The graphite/carbon black microfibrillated cellulose slurry, spray-coated on a wet paper substrate, subsequently pressed and dried on a conventional pilot paper machine, formed a bilayer electrode with excellent mechanical properties and cycling performances versus the Li metal, comparable to those of anodes with standard composition, (i.e., Young Modulus of 2.5 GPa and specific capacity of 350 mAh g^{-1} at 0.1 C), yet a Coulombic efficiency (cca. 98 % in the first 50 cycles) which needs to be improved to maintain good cycling performances in Li-ion systems. This study demonstrates that the manufacturing techniques for industrial papers and materials permit the application in Li-ion battery production technology.

2.2 Two-Dimensional Cellulose Nanocomposites

Badawi and Ashraf (2013) obtained carbon nanotubes–cellulose acetate nanocomposites membranes for water desalination. To this end, cellulose acetate (CA) (Mw = 52,000 Da) membranes were prepared by phase inversion (PI) using acetone as a solvent. The reasearches have analyzed the effect of thickness, CA content, coagulation bath temperature (PI temperature), solvent evaporation, addition of a nonsolvent (deionized water), and of multiwalled carbon nanotubes (MWCNTs) on the morphology and performance (permeation rates and salt rejection rates) of membranes. Membranes morphologies were analyzed by scanning electron microscopy (SEM), while their permeations and salt rejection rates were investigated using a 1000 ppm NaCl solution. Optimum conditions for developing CA-based nanocomposites were attained, entailing a 15 wt% CA content, 20 wt% H_2O nonsolvent additive, low functionalized carbon nanotubes (CNTs) contents (0.0005, 0.005, and 0.01 wt%), PI at room temperature, and a sonication time for CNTs proper dispersion in less than 1 min. MWCNTs/CA nanocomposites membranes were prepared. MWCNTs were first functionalized by oxidation-purification in a strong acidic medium, to enhance their dispersion within the polymer matrix, and the success of MWCNTs functionalization was recorded by Fourier transform infrared (FTIR) spectroscopy. Nanocomposites morphologies were also characterized by several methods, such as SEM and nitrogen adsorption. SEM images showed large MWCNTs networks, randomly oriented and properly dispersed, and a significant decrease in the number of macrovoids with the increase of the CNT content for the same final thickness of nanocomposites. This was verified by analysis of pore sizes (differential volumes and surface areas), which were found to decrease with the increase in the CNT content. Nanocomposites permeations and salt rejection rates were investigated using a 1000 ppm NaCl solution, and it was found out that permeation improved significantly with the addition of CNTs, the improvement being considerably higher with the addition of a small amount of CNT. Salt rejection was found to decrease with the presence of CNTs. The novelty of the present study is obtaining of CA-CNT nanocomposites with improved permeation values—19.57 L/m^2 h, along with a minimal decrease of salt retention performance—69.4 % at an operating pressure of 24 bars (Badawi and Ashraf 2013).

Still in the context of two-dimensional cellulose nanocomposites, Huang and co-workers (2013) synthesized cellulose films reinforced by chitin whiskers, considered as basic elements for obtaining cytocompatible nanocomposite films by blending α-chitin whiskers and a cellulose solution in NaOH/urea. The structure and properties of chitin/cellulose composite films were analyzed by FTIR, XRD, ^{13}C NMR, SEM, UV-Vis, TGA methods, and tensile tests. The results obtained reveal that chitin whiskers are homogeneously dispersed, leading to good miscibility and properties of the chitin/cellulose composite films. By varying the chitin whisker content, the tensile strength and elastic modulus of films can be controlled. Seeding of HeLa and T293 type cells on the surfaces of nanocomposite films evidenced their nontoxicity to both cell types and also that addition of chitin whiskers promotes cell adhesion and proliferation. In the same context, other researchers synthesized cellulose nanofiber/chitin whisker/silk sericin bionanocomposite sponges. Chitin whisker, (CTW) has been frequently preferred as another component of the bionanocomposite sponges, due to its ability to enhance tissue repairing of wounds (Siqueira et al. 2010). Chitin is the second most abundant biopolymer found in the exoskeleton of crustaceans, in the inner structures of invertebrates, cuticles of insects, and cell walls of fungi (Morganti et al. 2006), its chemical structure being quite similar to that of cellulose. It includes *N*-acetyl glucosamine as a repeating unit linked through β-(1 $\rightarrow$ 4) linkage. Chitin and cellulose differ in their side groups at C_2 position. While chitin contains an acetyl amide group (ANHCOCH$_3$), cellulose contains a hydroxyl group (AOH). Apart from its biocompatibility, chitin has received more and more attention in biomedical fields, due to its ability to accelerate the wound healing process (Khoushab and Yamabhai 2010). *N*-acetyl glucosamine is able to regulate collagen synthesis at the level of fibroblasts and to facilitate the granulation and repair of damaged skin tissues (Chan et al. 2000). Considering chitin in the form of whiskers, CTWs can interact effectively with the substances present in the living tissues, over to their enormous surface area, which is actually the main advantage of CTWs over the usual chitin powder (Siqueira et al. 2010). Furthermore, its ability to be hydrolyzed by lysozyme—an enzyme present in the wound-makes chitin releasable from the sponge, permitting to manifest its competence (Fischer et al. 2005; Wongpanit et al. 2007; Ang-atikarnkul et al. 2014).

2.3 Three-Dimensional Cellulose Nanocomposites

Polymer nanocomposites containing metal nanoparticles have attracted a great deal of attention, due to their unique optical, electrical, and catalytic properties (Shiraishi and Toshima 2000; Ottaviani et al. 2002; Sarma et al. 2002). They can be prepared through mechanical mixing of a polymer with metal nanoparticles, in situ polymerization of a monomer in the presence of metal nanoparticles, or in situ reduction of metal salts or complexes in a polymer. The properties of these polymer nanocomposites are strongly dependent on the size, content, dispersivity, and structure of the metal nanoparticles incorporated within them. Recently, considerable attention has been paid to polymer nanofibers, as they exhibit outstanding properties, such as high specific surface area and high porosity, which recommends them for a wide range of applications, as separation filters, wound-dressing materials, tissue scaffolds, sensors, etc. (Huang et al. 2001, 2003; Li et al. 2002; Matthews et al. 2002; Wang et al. 2002; Park and Park 2005). Functional nanofibers with optical, electrical, or catalytic properties may by produced by incorporating metal nanoparticles into them for extended applications. Li and cooorkers (2004), developed gold (Au) nanostructures on electrospun titania nanofibers by photocatalytically reducing $HAuCl_4$. As known, attachment of Au or of other metal nanoparticles onto titania nanofibers significantly improves their photocatalytic activity and photoelectrochemical response (Subramanian et al. 2004). These titania nanofibers can serve as catalysts and/or catalyst supports. Rutledge and his co-workers (Wang et al. 2004) produced superparamagnetic polymer nanofibers by electrospinning magnetic nanoparticle suspensions in poly(ethylene oxide) and poly (vinyl alcohol) solutions. At room temperature, these magnetite nanofibers exhibited superparamagnetic behavior. Yang and colaborators (2003) electrospun poly(acrylonitrile) (PAN) nanofibers containing Ag nanoparticles, the average diameters of PAN nanofibers and Ag nanoparticles being of 400 and 100 nm, respectively.

3 Modification of Surface Cellulose Fibers for Specific Applications

In order to develop composites with superior mechanical properties, environmental performance is necessary to increase the hydrophobicity of the cellulose fibers and to improve the matrix–fibers interface (Thakur et al. 2014b, c). Lack of good interfacial adhesion, low melting point, and poor resistance to moisture make the use of plant cellulose fiber-reinforced composites less attractive. Pretreatments of the cellulose fiber can clean and chemically modify the surface, stop the moisture absorption process, and increase surface roughness (Kalia et al. 2008. 2009; Singha et al. 2009; Thakur and Thakur 2014a). Among the various pretreatment techniques, silylation, mercerization, peroxide benzoylation, graft copolymerization, and bacterial cellulose treatment represent optimum methods for surface modification of natural fibers (Thakur et al. 2010a, 2011, 2012a, b, c, d, e).

3.1 Silylation, Mercerization, and Surface Chemical Modifications

According to literature data, silane-coupling agents usually improve the degree of cross-linking in the interface region and offer a perfect bonding. Among the various agents, silane-coupling agents were found to be effective in modifying the natural fiber–matrix interface. Efficiency of silane treatment was higher for the alkaline-treated fiber than for the untreated fiber, because more reactive sites can be generated for silane reaction. Silanes were effective in improving the interface properties (Culler et al. 1986; Ghatge and Khisti 1989; Coutinho et al. 1997; Gonzalez et al. 1997; Thakur et al. 2010a).

Mercerization is the most common method to produce high-quality fibers (Ray et al. 2001). Scheme 1 shows the probable mechanism of mercerization of cellulose fibers. Mercerization leads to fibrillation, which causes breaking down of the composite fiber bundle into smaller fibers. Mercerization reduces fiber diameter, which leads to the development of a rough surface topography, resulting in better fiber–matrix interface adhesion and improved mechanical properties (Joseph et al. 2000). Moreover, mercerization increases the number of possible reactive sites and allows better fiber wetting.

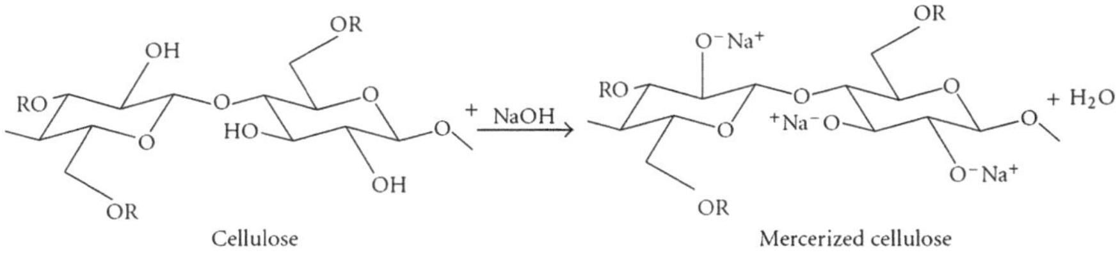

Cellulose + NaOH → Mercerized cellulose + H_2O

Scheme 1 Coutinho et al. (1997)

3.2 Surface Modification by Polymer Grafting

Desirable and targeted properties for cellulose fibers can be obtained through graft copolymerization, in order to meet the requirements of specialized applications. Graft copolymerization is one of the best methods recommended for modifying the properties of cellulose fibers. Different binary vinyl monomers and their mixtures have been graft copolymerized onto cellulosic materials for modifying the properties of numerous polymer backbones (Kaith et al. 2005; Kalia et al. 2009).

3.3 Bacterial Modification

Coating of bacterial cellulose onto cellulose fibers provides new means for controlling the interaction between fibers and polymer matrices. Fibers coating with bacterial cellulose does not only facilitate a good distribution of bacterial cellulose within the matrix, but also results in an improved interfacial adhesion between fibers and the matrix. This enhances the interaction between fibers and the polymer matrix through mechanical interlocking (Pommet et al. 2008; Eichhorn et al. 2010). Surface modification of cellulose fibers using bacterial cellulose is one of the best methods applicable for greener surface treatment of fibers. Bacterial Cellulose, BC, has gained attention in the research area for the encouraging characteristics, such as its significant mechanical properties in both dry and wet states, porosity, water absorbancy, moldability, biodegradability, and excellent biological affinity (Shoda and Sugano 2005). Due to all these properties, BC has a wide range of potential applications.

4 Applicability Areas of Cellulose Nanocomposites

The potential applicability of nanocellulose has been widely extended. Applications of nanocellulose are mainly considered to be in paper and packaging products, although construction, automotive, furniture, electronics, pharmacy, and cosmetics are also mentioned. For companies producing electroacoustic devices, nanocellulose is used as a membrane for high-quality sound. Additionally, nanocellulose is applied in membranes for combustible cells (hydrogen); additives for high-quality electronic paper (e-paper); ultrafiltrating membranes (water purification); membranes used to retrieve minerals and oils (Brown 1988). The high strength and stiffness, as well as the small dimensions of nanocellulose, confer specific properties to the composite materials reinforced with fibers containing it, required by various applications in electronic, pharmaceutical, and biomedical domains.

4.1 Audio Speaker Diaphragm

Literature shows that nanocellulose diaphragms of about 20 microns thickness were obtained by dehydration and compression in a suitable matrix. Advantages of these types of cellulosic diaphragms were that they allows the sound going through at the same speed as the diaphragms made of aluminum or titanium (Jonas and Farah 1998; Iguchi et al. 2000). In these types of headphones, sounds can be heard very clearly and bass notes are remarkably deep (Fig. 1).

4.2 Digital Displays

In recent years, researchers have tested a method for obtaining electronic paper. Nanocellulose is dimensionally stable and, as it resembles paper, it is most suitable for the basic structure of electronic paper (Shah and Brown 2005). Shah and Brown (2005) presented the concept in a device that holds many advantages, such as high paper-like reflectivity, flexibility, contrast, and biodegradability. Figure 2 illustrates the fabrication process of the display device using nanocellulose. Briefly, the procedure involves inserting a color in microbial cellulose nanostructure. After

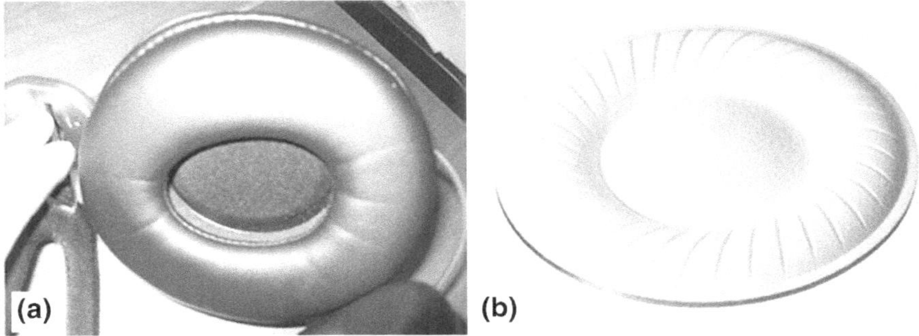

Fig. 1 SONY headphones (**a**) from nanocellulose diaphragm (**b**)

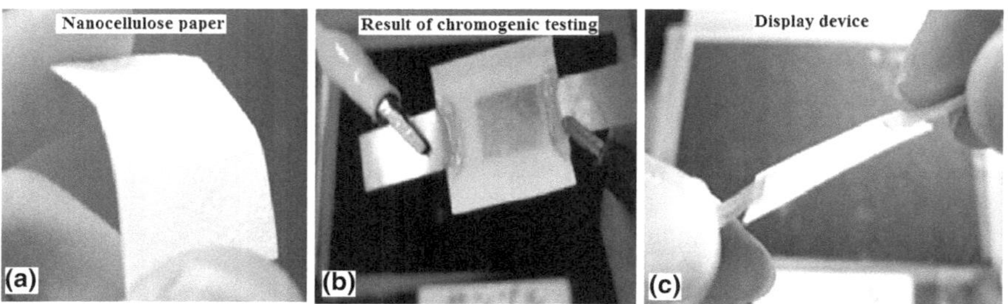

Fig. 2 The procedure for obtaining an electronic display: **a** cellulose nanostructure; **b** inserting a color in microbial cellulose nanostructure; **c** electronic display

integration, a simple pixel can reversibly switch from the ON to the OFF state. The pixel size is controlled by the minimum addressing resolution of the backplane drive circuits.

Other authors, such as Yano and co-workers (2005) have indicated the extraordinary potential of nanocellulose as a reinforcement material in optically transparent plastics, for instance, or as a substrate for bendable displays. In their opinion, the composite remained optically transparent even at high fiber contents.

Legnani et al. (2009) has developed a biodegradable and biocompatible flexible organic light emitting diode (FOLED) based on a nanocellulose (NC) membrane as a substrate. For OLED applications, nanocomposite substrates based on nanocellulose (NC) and Boehmite-siloxane systems with improved optical transmittance in the visible region were used as flexible substrates. Optical transmittance in visible range is enhanced by nanocomposites formations. At wavelength of 550 nm, a 66 % transmittance was obtained for the NCnanocomposite/ITO (Indium Tin Oxide) substrate. This transmittance value was compared with that of 40 %, obtained at the same wavelength for the NC/ITO substrate.

4.3 Electronic Usages

Evans and collaborators (2003) found that, for catalyzed nanocellulose, introduction of metals in its structure forms a finely divided homogeneous catalyst layer. The obtained results suggested that nanocellulose possesses reducing groups able to initiate the precipitation of palladium, gold, and silver from aqueous solution, and that the newly obtained structure was suitable for the construction of membrane electrode assemblies. Also, Olson and co-authors (2010) showed that freeze-dried cellulose nanofibril aerogels can be used as templates for obtaining lightweight porous magnetic aerogels, which can be compacted into a stiff magnetic nanopaper.

4.4 Pharmaceutical Applications

Cellulose and its derivatives are known for a long time as having applications in the pharmaceutical domain. They are more effective when blended with other pharmaceutical excipients, such as drug-loaded tablets form dense matrices suitable for the oral administration of drugs. The hydrophilic matrices based on polysaccharides and natural polymers represent the most famous biomaterials for controlled-release dosage forms. Simultaneously, the hydrophilic polymer matrix has become one of the most popular approaches in formulating extended release dosage forms (Alderman 1984; Heller 1987; Longer and Robinson 1980), as due to their relative flexiblily, so that a well-designed system usually gives reproducible release profiles. Drug release represents the process by which a drug product is released from the polymer matrix and is subjected to absorption, distribution, metabolism, and excretion, becoming available for pharmacological action. Literature shows that crystalline nanocellulose offers several potential advantages as a drug delivery excipient. For it and also for other types of cellulose in powerful granulated form, the rate of tablet disintegration and drug release may be controlled by microparticle inclusion, excipient layering or tablet coating (Watanabe et al. 2002; Baumann et al. 2009). A large amount of drug can be fixed on the crystalline cellulose surface, due to its large surface area and negative charging. This means a potential for high payloads and optimal control of drug dosing. Literature on pharmaceutical formulations also refers to other nanocrystalline materials, such as nanocrystalline clays, which have been shown to bind and subsequently release drugs in a controlled manner via ion exchange mechanisms (Shaikh et al. 2007). The hydroxyl group on the surface of crystalline cellulose represents the site for surface modification of the material with a series of chemical groups, by various chemical methods. The modified surface may be used to modulate the loading and release of drugs, such as nonionized and hydrophobic drugs, that would not normally bind to nanocellulose. In this context, Lonnberg and collaborators (2008) suggested that poly(caprolactone) chains might be conjugated onto nanocrystalline cellulose. Additionally, mention should be made of advantages of celluloses application, comparatively with other nanomaterials, such as a relatively low cost, natural abundance, and renewable and sustainable characteristics.

4.5 Veterinary Field Utilizations

In the case of chronic wounds, healing is affected by the difficulty of forming a new tissue in the same place. Apparently, this impairment reflects both a reduced capacity to synthesize new tissues and the antagonistic activities of the high levels of proteinases within the chronic wound environment. Historically, wound dressings have largely been passive devices that offer to the wound an interim barrier function and establish a moist healing environment. A new generation of devices, designed to interact with the wound and to promote new tissue formation, is currently developed and tested. This study shows that the oxidized regenerated cellulose (ORC)/collagen system (used for wound healing) is capable to promote fibroblast migration, in vitro proliferation, and acceleration of wound repair in the diabetic mouse. ORC/collagen was found to promote both human dermal fibroblasts proliferation and cell migration. In vivo studies have shown that the (ORC)/collagen system heals and closes the wounds of diabetics more easily than standard treatments. In addition, this treatment leads to a measurable improvement in the histological appearance of wound tissues. It can be expected that this system, tested on diabetic mouse, can be extended to human chronic wound healing (Hart et al. 2002).

On the other hand, the biocompatibility of a scaffold for tissue-engineered constructs is essential for healing. The bacterial cellulose (BC) used in such cases consists of completely pure cellulose nanofibrils synthesized by *Acetobacter xylinum*. BC has high mechanical strength and can be shaped into three-dimensional structures. Cellulose-based materials induce negligible foreign body and inflammatory responses and are considered as biocompatible. The in vivo biocompatibility of BC has been never evaluated in a systematic manner. Thus, in the development of tissue-engineered constructs with a BC scaffold, in vivo biocompatibility should be (first) evaluated. Helenius and collaborators (2006) were the first who systematically studied in vivo biocompatibility of nanocellulose systematically implanted subcutaneously. Implants were evaluated as to their chronic inflammation, foreign body response, cell ingrowth, and angiogenesis, by histology, immunohistochemistry, and electron microscopy. No microscopic signs of inflammation (i.e., a high number of small cells around the implants or blood vessels) were observed, and no fibrotic capsule or giant cells were present. Fibroblasts-infiltrated BC was well integrated into the host tissue, without causing adverse chronic inflammatory reactions. Thus, the biocompatibility and potential of nanocellulose to be used as a scaffold in tissue engineering was proved.

In parallel, another work (Silva et al. 2009) evaluated the biological behavior of synthetic hydroxyapatite (HAP-91) implanted in dental cavities and covered by nanocellulose. Also, nanocellulose associated to HAP promoted faster bone regeneration. The same system was studied for healing of pig skin (Costa and de Souza 2005). By comparing the daily healing bandages with those made of cellulose films, no differences were observed, which means that the healing process was similar. In dogs, to which the peritoneum had been replaced, after 45, 90, and 180 days, fibroblasts and blood vessels appeared in large numbers, while collagen and fibroblasts penetrated nanocellulose, and the implanted nanocellulose formed a net along the conjunctive tissue, little neovascularization being observed (Nemetz 2001).

4.6 Dental Field Applications

Nanocellulose was also tested in dental tissue regeneration. The literature of the field (Novaes and Novaes 1997) demonstrated that the microbial cellulose produced by the *Glucanacetobacter xylinus* strain can be used to regenerate human dental tissues. Nanocellulose products, such as Gengiflex and Gore-Tex, were applied in dental industry to support periodontal tissue recovery. A description was given of a complete restoration of an osseous defect, in association with a Gengiflex therapy. The benefits include reestablishment of esthetics and function of the mouth, reducing the number of surgeries. The bandage, called Gengiflex, consists of two layers: the inner layer, composed of microbial cellulose, provides rigidity to the membrane, and the chemically modified outer alkali cellulose layer (Novaes and Novaes 1992). Salata et al. (1995) compared the biological performances of Gengiflex and Gore-Tex membranes using the in vivo nonhealing bone-defect model proposed by Dahlin et al. (1988). The study showed that membrane made of Gore-Tex (a composite containing polytetrafluoroethylene, urethane, and nylon) was associated with significantly reduced inflammation, both membranes generating the same effect upon bone reconstructions. Compared with the control sites, better reconstruction was present in the bone defects protected by Gore-Tex or microbial cellulose membrane. Gore-Tex is better tolerated by the tissues than Gengiflex.

Recently, in a similar vein, Macedo et al. (2004) compared bacterial cellulose and polytetrafluoroethylene (PTFE) used to treat bone defects. In their study, two osseous defects (8 mm in diameter) were performed in each hind-foot of four adult rabbits, using surgical burs with constant sterile saline solution irrigation. The effects obtained on the right hind-feet were due to protection with PTFE barriers, while Gengiflex membranes were used over the wounds created in the left hind-feet. After 3 months, histological evaluation of the treatments revealed that the defects covered with PTFE were completely repaired with bone tissue, whereas incomplete lamellar bone formation was detected in defects treated with Gengiflex membranes, resulting in voids and lack of continuity of bone deposition.

The conclusion to be drawn is that nanocellulose, with its characteristics—nanofibers size and distribution, mechanical properties, compatibility, and ability to regenerate—, has been considered an indispensable biomaterial in health area. Also, the nanocellulose composite scaffolds are biocompatible, appearing as a promising biomaterial, suitable for cell adhesion/attachment, which recommends them for wound-dressing or tissue engineering scaffolds (Kalia et al. 2011).

5 Cellulose Nanocomposite with Antimicrobial Activity

Recent investigations have evidenced the unique, superior, and important properties of nanomaterials, which make them a significant group of products used in the development of new devices for various chemical, biological, physical, biomedical, and pharmaceutical applications. Literature has recorded promising results regarding the antimicrobial activity of various drugs—in nanoparticles form (Dobos et al. 2014). In the context of cellulose biomedical applications, are presented ways of obtaining metal nanoparticles in the cellulose matrix. Researchers have shown that some metals, namely zinc, silver, gold, platinum, have the lowest side effects when coming into contact with the human and animal body.

Electrospun bionanocomposites represent a recent research topic, with a huge impact in different fields, such as tissue engineering, replacement or repair of bone, dental applications in wound healing or controlled release of drugs (Zeng et al. 2003; Rujitanaroj et al. 2008; Schneider et al. 2009; Cui et al. 2010; Shafei and Abou-Okeil 2011). Cellulose and its derivatives are widely used as versatile materials, because of their low cost and easy processability in different forms. Unlike starch, these materials have low water solubility, allowing better control over scaffold design, textile, filtering, etc. However, their applications are still limited, due to the absence of antimicrobial activity. Recently, it has been found out that these limitations can be overcome by the incorporation of antimicrobial nanoparticles into them (Son 2004; Son et al. 2006). In this context, ZnO appears as direct wide band gap semiconductor ($E_g = 3.4$ eV). The high exciton binding energy (60 meV) of this material makes it a promising candidate, e.g., for stable room temperature luminescent. Moreover, ZnO is a bactericide and inhibits both Gram-positive and Gram-negative bacteria (Dai et al. 2003; Tam et al. 2008).

Microbial contamination is a life-threatening issue in food industry, synthetic textiles, packaging, and healthcare products (Li et al. 2008a, b). Among the various kinds of pathogenic microorganisms, *Staphylococcus aureus* and *Escherichia coli* are closely related species, that commonly cause a wide variety of infections and diseases (Panacek et al. 2009; Bae et al. 2010). Therefore, the development of antimicrobial surface coating of cellulosic materials has attracted increased attention for preventing microbial contamination (Necula et al. 2010, 2011; Ioan and Dobos-Necula 2012). The use of inorganic antimicrobial agents has gained importance because of their ability to withstand adverse processing conditions, as compared to organic antimicrobial agents (Wu et al. 2009a, b; Yuvaraj et al. 2010).

Some authors have presented the ability to prepare nanobiocomposites of cellulose acetate and ZnO as a way to develop new biocidal agents (Anitha et al. 2013). The self-cleaning effect caused by superhydrophobicity is needed for many applications, such as self-cleaning and antifouling coatings. Till date, various methods have been adopted by researchers to produce superhydrophobic surfaces. Therefore, developing a simple approach to fabricate a superhydrophobic surface without further coating of a low surface energy material is important and scientifically challenging. Characterization of cellulose acetate membranes containing ZnO nanoparticle was designed to optimize the results for obtaining the optical, bactericidal, and water repellent properties required in specific applications. Studies lead to several conclusions: formation of a hydrogen bond between ZnO and CA, which means a higher compatibility between components, was observed by FTIR spectra analysis; the ZnO-embedded fibrous membrane showed improved water repellency property than the pure CA; antimicrobial properties of the fibrous CA and ZnO impregnated CA membranes (Figs. 3 and 4) against both Gram-positive and Gram-negative bacteria under normal light conditions.

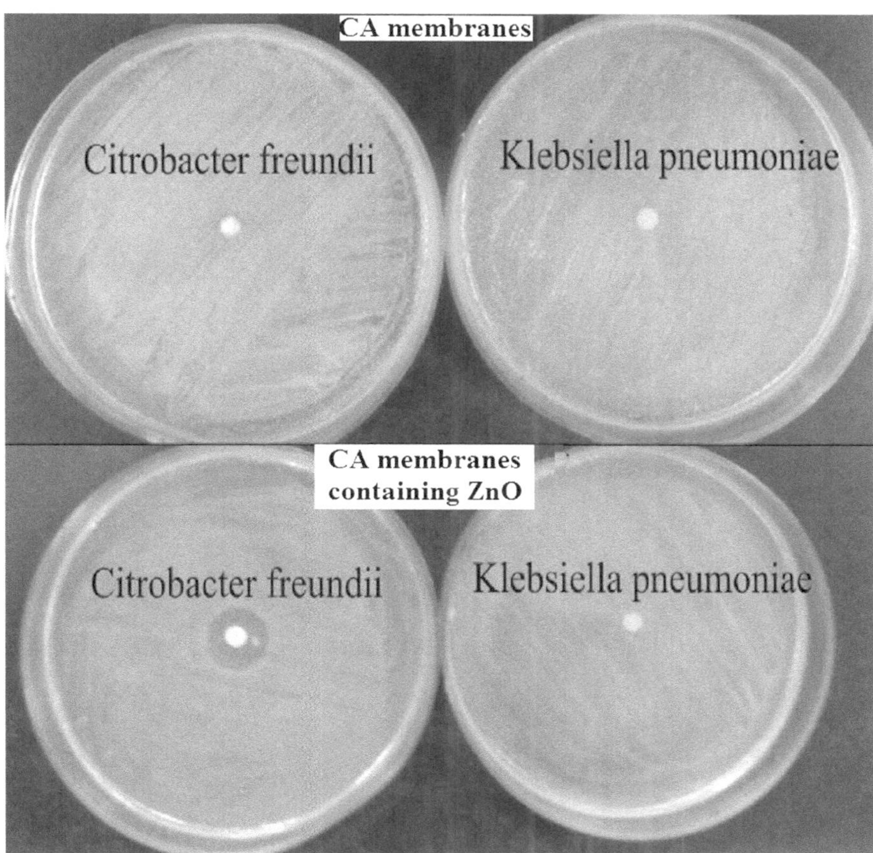

Fig. 3 Inhibition areas of CA and CA–ZnO against *Citrobacter freundii* and *Klebsiella pneumoniae* bacteria

Evaluation of antimicrobial activity adopted the Kirby Bauer technique, the simple CA fibrous membrane being used as a control. According to the data obtained for simple CA, no detectable inhibition zones were seen. On the other hand, for the ZnO containing samples significant inhibition zones were visible. The diameters of the inhibition zone around the membranes measured after one day were of 27 mm, 22 mm, 14 mm for *Methicillin-resistant S. aureus* (MRSA), *E. coli,* and *C. freundii*, respectively. However, no antibacterial activity was shown against *K. pneumoniae*. It is seen that the composite membrane had a stronger influence on *MRSA* than *E. coli,* a result agreeing with other studies. The observed differences can be attributed to the nature of the cell wall structure and of the forces developed within the polymeric matrix. In this context, one should mention that surface wettability can be changed from hydrophilic into hydrophobic. A new approach to fabricate a hydrophobic surface without further surface treatment targets the electrospinning technique,

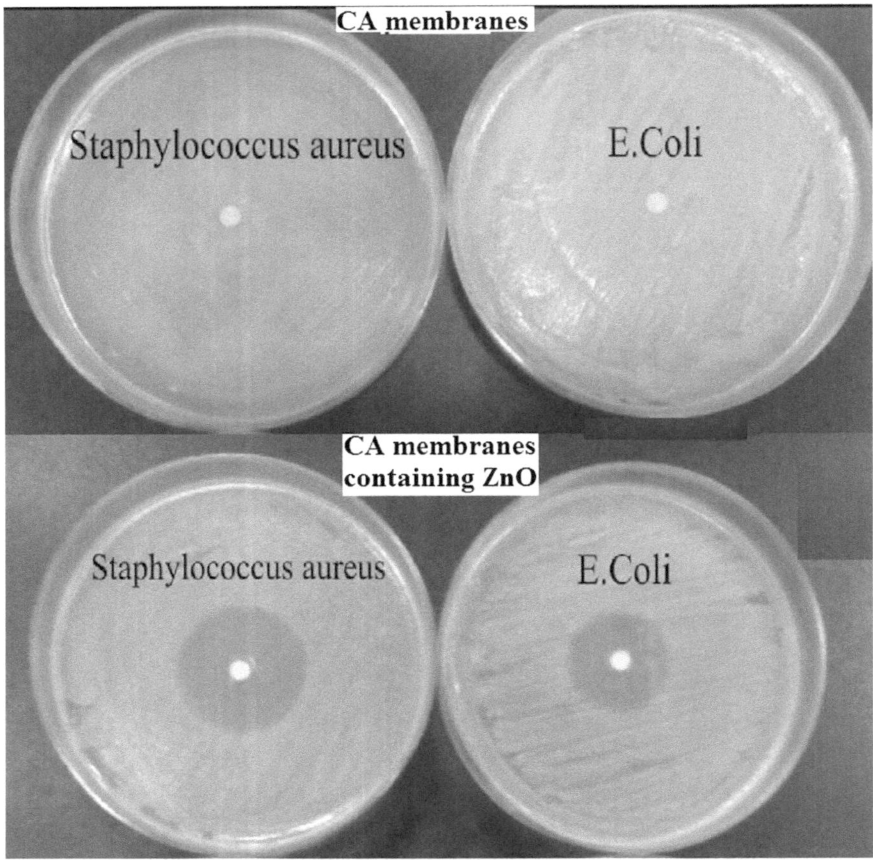

Fig. 4 Inhibition areas of CA and CA–ZnO against *Staphylococcus aureus* and *Escherichia coli* bacteria

considered as most suitable for the synthesis of composite fibrous membranes, since agglomeration of nanoparticles is prevented and the contact area between the surface and the microorganisms is increased (Anitha et al. 2013).

On the other hand, among other metal, silver have attracted intensive research interest for centuries, because of their important biological applications, especially as to their bactericidal effect, namely is the capability of killing about 650 types of diseases caused by microorganisms (Martin 1996; Southward et al. 2001; Fong and Wood 2006). Silver has a significant potential for preventing infections, in the treatment of skin diseases, healing wounds (Martin 1995) and also anti-inflammatory diseases and in other medical areas, due to its antimicrobial activity as absence of adverse effects upon the human body.

The literature describes the methods applied for obtaining cellulose fibers with antimicrobial activity by linking silver ions to the cellulose matrix or to other polymers, widely used in the field of medical services, such as: latex, polyethylene, polypropylene (Johnson et al. 2006; Wu et al. 2009a, b; Zhu et al. 2009). The main properties of silver that make it scientifically interesting are:

- first, numerous Gram-positive and Gram-negative bacteria, viruses, fungi are sensitive to silver ions (Wright 1999; Klasen 2000; Dunn and Edwards-Jones 2004; Percival et al. 2005). Silver is considered to be "oligodynamic," which means that its bactericidal effect is manifested even when occurring in small quantities (Clement and Jarrett 1994; Furr et al. 1994; Percival et al. 2005). In its active form, the mechanism through which it acts on bacteria, is dictated by manner in which it interferes with DNA and causes conformational changes of cell membranes (Lansdown 2002a, b);
- second, when used in appropriate concentrations, it is nontoxic for mammalian cells (Lansdown 2002a, b; Ewald et al. 2006; Burd et al. 2007);
- most bacteria show very low tolerance in the presence of silver ions (Percival et al. 2005). Some of the most important compounds that contain silver are: silver sulphadiazine, metallic silver, silver acetate, and silver proteins. They are used to prevent infections in burned areas and eye diseases. In recent years, bandages were treated with silver to get grafted silver materials; however, problems with cytotoxicity against the human body are still present (Van Den Plas et al. 2008; Madhumathi et al. 2010; Wiegand et al. 2009).

Kim and co-authors (2009) have proposed a method for preparing silver-impregnated cellulose, based on associating ionic silver with water-soluble polymers, and thus introducing the Ag^{+}–polymer complex into the cellulose matrix. *Candida albicans*, *Micrococcus luteus*, *Pseudomonas putida*, and *E. coli* were used for antimicrobial testing. The effect of silver on the cytotoxicity to human cells was analyzed using silver-impregnated cellulose on human fibroblasts. In the cellulose capsule with certain amount of silver, *M. luteus* and *P. putida* were not detected, while for a content of 0.00035 Ag w/v %, a reduced amount of the *C. albicans* and *E. coli* bacteria were detected. The amount of 0.01 mg/g silver in the cellulose capsules is sufficient for assuring antimicrobial activity, and this antimicrobial activity increases with the increasing of silver content. When silver is present in the polymer matrix in excessive amounts, (more than 0.35 w/v %), it becomes toxic for the human body. The way in which silver binds to the cellulose substrate is shown in environmental scanning electron microscopy images (Fig. 5).

As one can observe, for a solution with concentrations ranging between 0.035 and 0.35 w/v % range, silver crystals were formed on the cellulose surface; whereas, for a solution with a content of 0.0035 w/v % or less, no physical binding was observed between silver and the cellulosic substrate. The observed silver crystals had a hexagonal shape, their size varying between 1 and 10 μm in diameter. Formation of these crystals was analyzed using energy dispersive X-Ray analysis method. Finally, analysis has demonstrated that the chemically bonded ionic silver

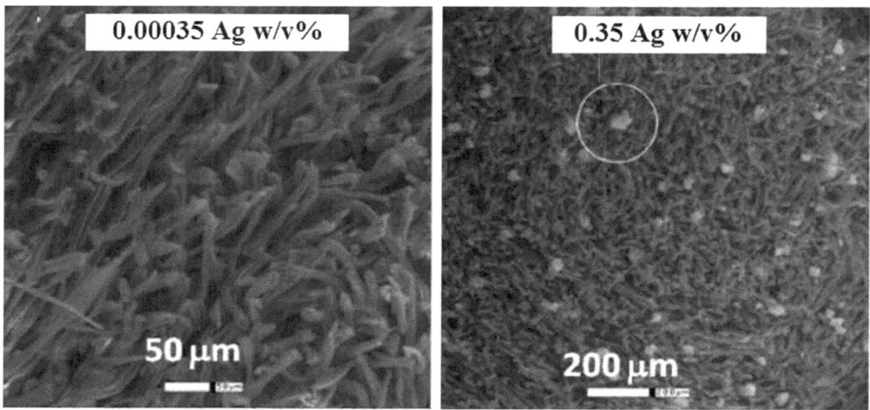

Fig. 5 ESEM images of silver-containing cellulose, obtained for two silver contents (0.00035 w/v % and 0.35 w/v %), at different scan areas. Circle exemplifies a silver nanoparticle

was uniformly distributed over cellulose matrix surface, and that silver chemically binds to cellulose by means of the mechanism presented in Scheme 2.

There are also and other mechanisms involving the possibility to binding silver ions, e.g., through interaction with thiol groups (Fuhrmann and Rothstein 1968; Bragg and Rannie 1974; Richards 1981; Belly and Kydd 1982; Thurmann and Gerba 1988, 1989; Furr et al. 1994). This mechanism includes the interactions of silver ions with the thiol groups of enzymes and proteins, known as playing a role in antimicrobial activity, although other cell components might be also involved. Hydrogen bonding, the agents inducing hydrogen bonds breaking, as well as the affinity of the silver ions for thiol groups were in attention of researcher (Russell and Hugo 1994). Some authors stated that silver salts and other heavy metals, such as copper, act through their binding to fungal enzymes functional groups (Lukens 1983). Silver ions determine the release of potassium ions from microorganisms, so

Scheme 2 Mechanism of silver binding to the cellulosic substrate (Kim et al. 2009)

that the microbial cytoplasmic membrane becomes an important site for Ag^+ activity (Miller and McCallan 1957; Fuhrmann and Rothstein 1968; Schreurs and Rosenburgh 1982). In addition to its effect on the enzyme, the silver ion also produces and other changes in microorganisms. For example, silver nitrate inhibits the growth of *Cryptococcus neoformans*, being localized in the form of granules inside the cell wall. Silver ions also inhibit cell division and damage the cell envelope and contents of *Pseudomonas aeruginosa* (Brown and Smith 1976). In addition, bacterial cells increase in size, and the cytoplasmic membrane, cytoplasmic contents, and outer cell layers exhibit structural abnormalities (Richards 1984).

In another train of thoughts, silver nanoparticles (AgNPs) had been incorporated in textile fabrics, polymers, dental material, medical device ,and burn dressing to eliminate microorganisms. During AgNPs synthesis, stabilizers play a main role in controlling the size of particles, as well as their dispersion stability. Polymers, often used as stabilizers, due to their effectiveness in preventing agglomeration and precipitation of particles, provide an excellent distribution of particles. The silver nanoparticles using cellulose acetate (CA) as a dispersion medium have been investigated for elucidating the interactions between silver complexes and cellulose acetate molecules, the formation of Ag nanoparticles with their size control, and their reaction chemistry in the cellulose acetate polymer (Kwon et al. 2005). Formation of silver nanoparticles and the antimicrobial activity in the cellulose acetate phthalate (CAP) matrix was also observed and plotted in Figs. 6 and 7, respectively, where diameters of the inhibition zones are clearly higher than those of cellulose acetate even for the pure samples (without silver nanoparticles) (Necula et al. 2010). Moreover, the silver nitrate incorporated in cellulose acetate with 1.73, 1.88, or 1.90 substitution degrees, used as a dispersion medium, created composite structures (Necula et al. 2011; Ioan and Dobos-Necula 2012). The different distribution of the obtained silver nanoparticles in the polymer matrix, as well as the influence of the hydrophobic character of cellulose acetate, induced by the substitution degree, were established by atomic force microscopy and environmental

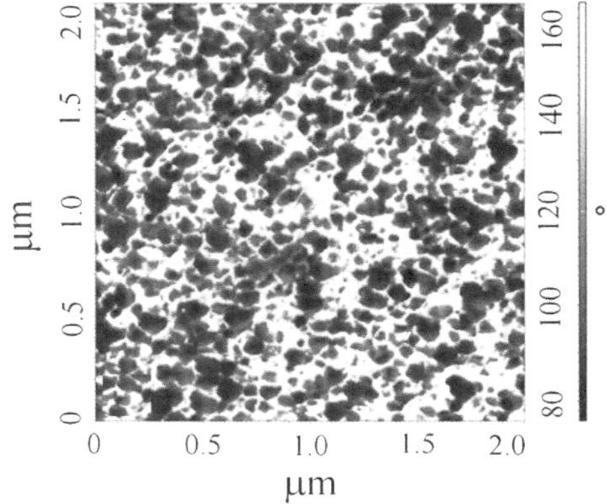

Fig. 6 2-D AFM phase images at 2 × 2 μm^2 scan area for silver-containing cellulose acetate phthalate films prepared from solutions in 2-methoxyethanol

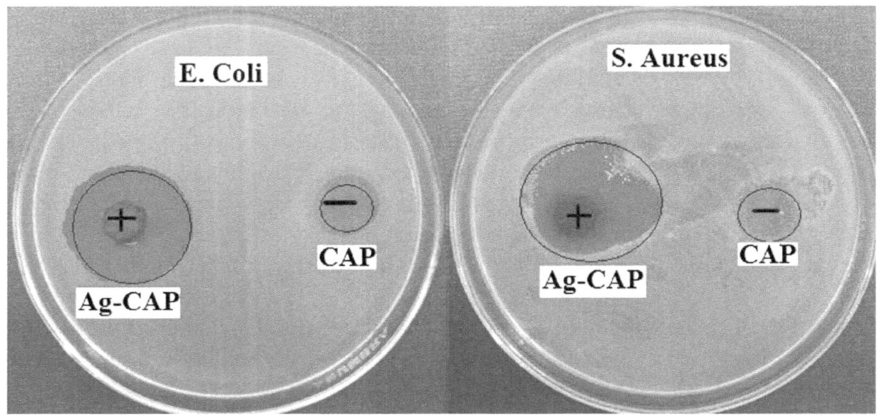

Fig. 7 Antimicrobial activity of Ag-CAP films against *Escherichia coli* and *Staphylococcus aureus*. In each figures, the inhibition areas on the right side were recorded for CAP as a control sample

scanning electron microscopy. A more uniform distribution and higher sizes of the silver nanoparticles were observed in the more hydrophobic cellulose acetate samples (Fig. 8). The antimicrobial activity of these composites was investigated against *Escherichia coli* (*E. coli*) and *Staphylococcus aureus* (*S. aureus*) (Necula et al. 2010). As shown in Fig. 9, *E. coli* is much more sensitive to the Ag-$CA_{1.88}$ film, and least sensitive to Ag-$CA_{1.90}$, comparatively with *S. aureus*.

Other studies performed on silver nanostructures with distinct morphologies have demonstrated that AgNPs undergo a shape-dependent interaction with bacteria (Pal et al. 2007). Transmission electron microscopy evaluations of bacteria surface showed that spherical NPs exhibit enhanced antibacterial activity than, for example, Ag nanorods. This effect was explained by the higher reactivity of these

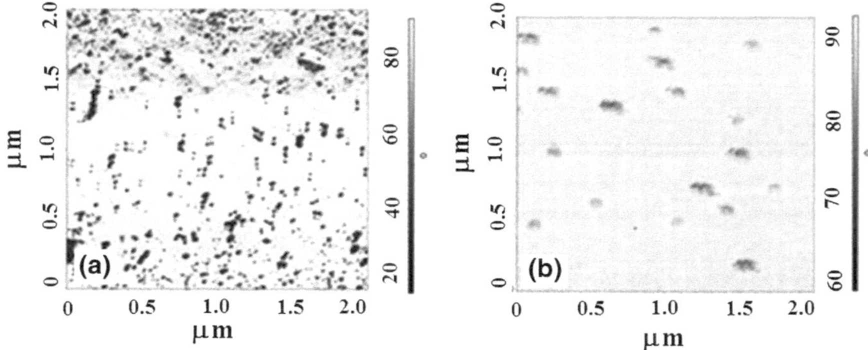

Fig. 8 2D-AFM phase images at 2 × 2 μm^2 for **a** Ag-$CA_{1.88}$ and **b** Ag-$CA_{1.90}$ films prepared from solutions in 2-methoxyethanol

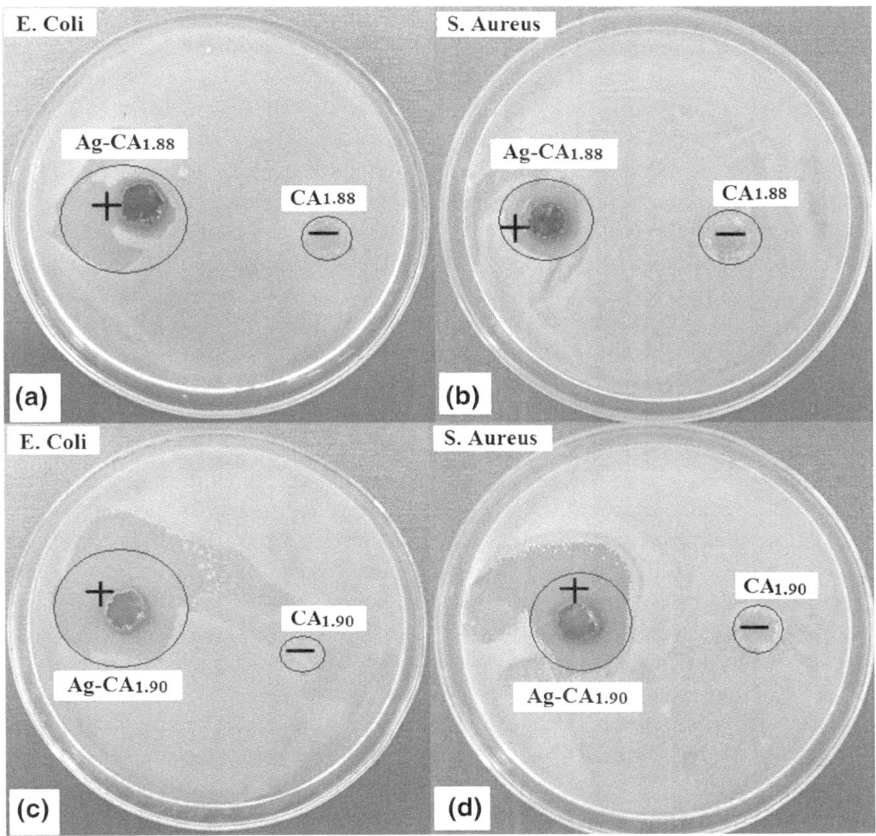

Fig. 9 Antimicrobial activity of Ag-CA$_{1.88}$ and Ag-CA$_{1.90}$ films against *Escherichia coli* and *Staphylococcus aureus*. In each figures, the inhibition areas on the right side were recorded for CAP as a control sample

nanostructures, due to their higher atomic density. The higher reactivity of NPs leads to a more rapid release of the metallic ions and to an enhancement of antibacterial activity for this type of nanostructures.

Considering the nanocomposites obtained from 100 mg vegetable cellulose/cooper nanoparticles (VC/CuNPs) and 50 mg bacterial cellulose/NPs (BC/NPs), the total amount of inorganic filler is almost the same. The VC/CuNPs and BC/CuNPs nanocomposites present a similar antibacterial activity for *S. aureus*. On the other hand, the bactericidal effect of nanocomposites prepared with VC fibers is superior for *K. pneumoniae*, which is quite unexpected, once the CuNPs on the VC fibers are not individualized, forming a film generated by the fast oxidation under normal ambient conditions (Pinto et al. 2012). Explanation consists in preferential deposition of copper on the VC fibers' surfaces, in contrast with the BC matrix which, due to its three-dimensional internal organization, acts as a protective cage for

CuNPs. In this case, the release of Cu ions is limited, compared to the more open structure of the VC nanocomposite. Similar observations have been reported for materials based on cellulose and AgNPs, in which, for VC/Ag nanocomposites, the release of Ag^+ was superior compared to BC analogs (Pinto et al. 2009, 2013).

The use of copper in combination with biopolymers, such as cellulose, permits to obtain nanostructures less examined in the literature, compared to other composite materials (Pinto et al. 2009; Vargas-Reus 2012). The copper nanocomposites from the cellulose matrix were obtained by in situ and ex situ methods. Studies evidence the antimicrobial activity of copper nanocomposites against *Staphylococcus aureus* and *Klebsiella pneumonia,* the most visible effect being recorded for *Klebsiella pneumonia.* Although there is some debate in the literature about the relative effect of nanoparticles on the type of bacteria, this study is in line with reports that suggest that Gram-negative bacteria are more affected by copper-based nanoparticles (Ruparelia et al. 2008). In Gram-negative bacteria, the peptidoglycan layer is thinner (between 2 and 3 nm) than in Gram-positive ones (around 30 nm), and externally to this layer there is an outer asymmetrical membrane composed of proteins, phospholipids, and lipopolysaccharides (Morones et al. 2005). Some authors have explained the higher antibacterial effect of Gram-negative bacteria as a consequence of the interactions occurring between the bacterial outer membrane and the solid surfaces, for copper either (Ruparelia et al. 2008; Jadhav et al. 2011) or silver nanoparticles (Sondi and Salopek-Sondi 2004; Ruparelia et al. 2008). Such particle–microorganism interactions promote the formation of irregular pores in the outer membrane of Gram-negative bacteria, due to the direct interaction of the released nanoparticles or metallic ions, modifying their permeability and causing the release of cell components. These structural changes cause membrane degradation and eventually death of bacteria (Amro et al. 2000; Sondi and Salopek-Sondi 2004). Several studies have reported that the extent of bacterial growth inhibition in this type of materials depends on the inorganic content of the medium, for both metal NPs and metal oxides used as fillers (Pinto et al. 2009; Vargas-Reus 2012). For the investigated nanocomposites, two spherical samples of BC/CuNPs have been evaluated by varying the Cu content. Increasing of the Cu content from 0.93 to 4.95 % (w/w) resulted in a significant bactericidal effect against the *S. aureus*. A similar trend was also observed for the *K. pneumoniae* strain, verifying a direct dependence of the antibacterial action on the Cu content. Note that a complete killing effect was observed for the BC/CuNPs nanocomposites with 4.95 % (w/w) Cu. Similarly to silver materials, the antibacterial activity of Cu nanostructures has been associated with the release of ionic species and formation of reactive oxygen species (Morones et al. 2005; Hajipour et al. 2012). Increase of the copper amount in nanocomposites results in a higher release of cations, thus increasing the antibacterial activity of the corresponding cellulose-based nanocomposites. It is interesting to note that, despite the influence of the copper content, the samples with the highest copper nanowires content (BC/CuNWs sample) did not present the highest antibacterial effect. For both studied bacteria, BC/CuNWs present a significantly lower antibacterial activity versus the nanocomposites with CuNPs, with a similar copper amount. Assuming that the antibacterial effect is mainly due to cationic release, the lower efficiency of the CuNWs-containing composites, compared to those incorporating CuNPs, is probably due to the lower surface reactivity of nanowires, which determines a lower amount of soluble and oxidized copper species. In fact, the former materials have already oxidized the copper phases, so that the surface chemistry is thus markedly distinct from that of CuNWs. In the case of individualized nanoparticles and composite materials, no knowledge is available on the effect of the morphology of copper nanostructures on their antibacterial properties.

The antimicrobial effect of cellulose acetate nanoparticles has been studied in the literature by Hassanien et al. (2013), who obtained a polyvinyl alcohol/cellulose acetate/polyethylene glycol (PVA/CA/PEG) antibacterial membrane with potential water purification applications. Advantages of this membrane consist, on one hand, in some properties, namely: chemically inert, mechanically strong, creep resistant, and lower energy consumption and, on the other, in desalting and removal of bacteria and viruses from feed water. The membrane must be obtained in a configuration area larger than the volume (Lonsdale and Padall 1972). To render the surface of ultra-filtration membranes biocidal, cellulose membranes were modified with PVA, a naturally occurring polycationic biocide. Moreover, to analyze the antimicrobial activity against *Escherichia coli* bacteria, the membranes with pores of different sizes were obtained. Increased antimicrobial activity occurs when membranes with smaller pore sizes are used, which suggests that the surface location of the grafted membrane chains was more suitable for a higher antimicrobial activity of the surface. Membranes modified with polyethylene glycol showed higher antimicrobial efficiency against Gram-positive (*Staphylococcus sp.*) and Gram-negative (*Pseudomonas sp.*) bacteria (Hassanien et al. 2013). Bactericidal activity was determined in terms of growth inhibition, by equality: *bacterial* activity $= \frac{1}{4}N_1 - \frac{N_2}{N_1}$, where N_1 and N_2 are the numbers of viable colonies developed on the control and modified membranes, respectively. The unmodified membrane was used as a control sample during measurements. The tests conducted to positive result, which means that the bacteria used for experimental purposes interact with the membrane and create an inhibition zone—indicating that the composite membrane kills these bacteria.

Materials with antimicrobial activity, including surgical drapes, instrument wraps, and surgical packs that reduce the risk of postoperative infection (Hilal et al. 2003) and tooth fillings (Danese 2002) are well-accepted, especially in the medical field. Microbiological evaluations have been carried out on a variety of textile (Stashak et al. 2004) and food packaging materials (Imazato et al. 2003) treated with antimicrobial agents. They have been treated against a broad spectrum of microorganisms, including odor-causing bacteria, as well as bacteria and fungi which cause rot and mildew. Most such materials are based on compositions that release biocidal molecules or ions. However, the application of polymer biocides has opened new frontiers in the development of nonleaching antibacterial surfaces (Appendini and Hotchkiss 2002; Bajaj 2002). Recently, attempts have been made to render membrane surfaces antimicrobial by graft copolymerization.

Polycondensation with amine-containing polymers, potentially antimicrobial agents, was also realized (Tiller et al. 2002; Lea et al. 2004; Hassanien et al. 2013).

All mentioned studies demonstrate that cellulose derivatives, implicitly nanocomposites based on cellulose derivatives, play an important role in our lives, finding application in all areas, including biomedical field.

6 Conclusions and Future Perspectives

The present chapter summarizes the existing literature data concerning the nanocomposites based on cellulose derivatives, including their obtaining method, structural analysis, and properties. The survey demonstrates that cellulose derivatives, including cellulose acetate, appears as basic elements in the production of different materials with excellent biodegradability, biocompatibility, or antimicrobial activity properties, applicable in various fields, such as food packaging, biomedical areas, etc. As seen, a large number of bionanocomposites with antimicrobial properties can be synthesized, which demonstrates their safe use in the human body.

References

Alderman DAA (1984) A review of cellulose ethers in hydrophilic matrices for oral controlled-release dosage forms. Int J Pharm Technol Prod Man 5:1–9

Amro NA, Kotra LP, Wadu-Mesthrige K, Bulychev A, Mobashery S, Liu G (2000) High-resolution atomic force microscopy studies of the Escherichia coli outer membrane: structural basis for permeability. Langmuir 16:2789–2796

Ang-atikarnkul P, Watthanaphanit A, Rujiravanit R (2014) Fabrication of cellulose nanofiber/chitin whisker/silk sericin bionanocomposite sponges and characterizations of their physical and biological properties. Compos Sci Technol 96:88–96

Anitha S, Brabu B, Thiruvadigal John D, Gopalakrishnan C, Natarajanc TS (2013) Optical, bactericidal and water repellent properties of electrospun nano-composite membranes of cellulose acetate and ZnO. Carbohyd Polym 97:856–863

Appendini P, Hotchkiss JH (2002) Review of antimicrobial food packaging. Innov Food Sci Emerg Technol 3:113–126

Badawi E, Ashraf N (2013) Carbon nanotubes-cellulose acetate nanocomposites: membranes for water desalination—thesis

Bae E, Park HJ, Lee J, Kim Y, Yoon J, Park K, Choi K, Yi J (2010) Bacterial cytotoxicity of the silver nanoparticle related to physicochemical metrics and agglomeration properties. Environ Toxicol Chem 29:2154–2160

Bajaj PJ (2002) Finishing of textile materials. Appl Polym Sci 83:631–659

Baumann MD, Kang CE, Stanwick JC, Wang Y, Kim H, Lapitsky Y, Shoichet MS (2009) An injectable drug delivery platform for sustained combination therapy. J Cont Rel 138:205–213

Belly RT, Kydd GC (1982) Silver resistance in microorganisms. Dev Ind Microbiol 23:567–577

Beneventi D, Chaussy D, Curtil D, Zolin L, Bruno E, Bongiovanni R, Destro M, Gerbaldi C, Penazzi N, Tapin-Lingua S (2014) Pilot-scale elaboration of graphite/microfibrillated cellulose anodes for Li-ion batteries by spray deposition on a forming paper sheet. Chem Eng J 243:372–379

Biermann C (1996) Handbook of pulping and papermaking, 2nd edn. Academic Press, London
Bondeson D (2007) Biopolymer-based Nanocomposites: processing and properties. Thesis for the degree of philosophiae doctor Trondheim. ISBN 978-82-471-1254-0 (printed ver); ISBN 978-82-471-1268-7 (electronic ver); ISSN 1503-818
Bragg PD, Rannie DJ (1974) The effect of silver ions on the respiratory chain of Escherichia coli. Can J Microbiol 20:883–889
Brown RM Jr (1998) Microbial cellulose: a new resource for wood, paper, textiles, food and specialty products. Position Paper
Brown TA, Smith DG (1976) The effects of silver nitrate on the growth and ultrastructure of the yeast Cryptococcus albidus. Microbios Lett 3:155–162
Burd A, Kwok CH, Hung SC, Chan HS, Gu H, Lam WK, Huang L (2007) A comparative study of the cytotoxicity of silver-based dressings in monolayer cell, tissue explant, and animal models. Wound Repair Regen 15:94–104
Chan MW, Schwaitzberg SD, Demcheva M, Vournakis J, Finkielsztein S, Connolly RJ (2000) Comparison of poly-N-acetyl glucosamine (P-GlcNAc) with absorbable collagen (Actifoam), and fibrin sealant (Bolheal) for achieving hemostasis in a swine model of splenic hemorrhage. J Trauma 48:454–457
Clement JL, Jarrett PS (1994) Antibacterial silver. Met Based Drugs 1:467–482
Costa HO, de Souza FC (2005) Evaluation of the tissue regeneration of the burned pig′s skin followed by BiotissueTM grafting. Acta ORL/T'ecnicas emOtorrinolaringologia 23:192–196
Coutinho FMB, Costa THS, Carvalho DL (1997) Polypropylene-wood fiber composites: effect of treatment and mixing conditions on mechanical properties. J Appl Polym Sci 65:1227–1235
Culler SR, Ishida H, Koenig JL (1986) Silane interphase of composites: effect of process conditions on gamma—aminopropyltriethoxysilane. Polym Compos 7:231–238
Cui W, Li X, Xie C, Chen J, Zou J, Zhou S, Weng J (2010) Controllable growth of hydroxyapatite on electrospun poly(dl-lactide) fibers grafted with chitosan as potential tissue engineering scaffolds. Polymer 51:2320–2328
Dahlin C, Linde A, Gottlow J, Nyman S (1988) Healing of bone defects by guided tissue regeneration. Plastic Reconstruct Surgery 81:672–676
Dai L, Chen XL, Wang WJ, Zhou T, Hu BH (2003) Growth and luminescence characterization of large-scale zinc oxide nanowires. J Phys Condens Matter 15:2221–2226
Danese PN (2002) Antibiofilm approaches: prevention of catheter colonization. Chem Biol 9:873–880
Dobos AM, Onofrei MD, Ioan S (2014) Liquid crystals and cellulose derivatives composites. In: Thakur VK, Kessle MR (eds) Green biorenewable biocomposites from knowledge to industrial applications, CRC Press, ISBN: 9781771880329
Dunn K, Edwards-Jones V (2004) The role of Acticoat with nanocrystalline silver in the management of burns. Burns 30(1):S1–S9
Eichhorn SJ, Dufresne A, Aranguren M, Marcovich NE, Capadona JR, Rowan SJ, Weder C, Thielemans W, Roman M, Renneckar S, Gindl W, Veigel S, Keckes J, Yano H, Abe K, Nogi M, Nakagaito AN, Mangalam A, Simonsen J, Benight AS, Bismarck A, Berglund LA, Peijs T (2010) Review: current international research into cellulose nanofibres and nanocomposites. J Mater Sci 45:1–33
Evans BR, O'Neill HM, Malyvanh VP, Lee I, Woodward J (2003) Palladium-bacterial cellulose membranes for fuel cells. Biosens Bioelectron 18:917–923
Ewald A, Gluckermann SK, Thull R, Gbureck U (2006) Antimicrobial titanium/silver PVD coatings on titanium. Biomed Eng Online 5:22–31
Fischer TH, Thatte HS, Nichols TC, Bender-Neal DE, Bellinger DA, Vournakis JN (2005) Synergistic platelet integrin signaling and factor XII activation in poly-N-acetyl glucosamine fiber-mediated hemostasis. Biomaterials 26:5433–5443
Fong J, Wood F (2006) Nanocrystalline silver dressings in wound management: a review. Int J Nanomedicine 1:441–449
Fuhrmann GF, Rothstein A (1968) The mechanism of the partial inhibition of fermentation in yeast by nickel ions. Biochem Biophys Acta 163:331–338

Fukushima H, Drzal LT, Rook BP, Rich MJ (2006) Thermal conductivity of exfoliated graphite nanocomposites. J Therm Anal Calorim 85:235–238
Furr JR, Russell AD, Turner TD, Andrews A (1994) Antibacterial activity of actisorb plus, actisorb and silver nitrate. J Hosp Infect 27:201–208
Ghatge ND, Khisti RS (1989) Performance of new silane coupling agents along with phenolic nobake binder for sand core. J Polym Mater 6:145–149
Gopalan Nair K, Dufresne A (2003) Crab shell chitin whisker reinforced natural rubber nanocomposites. 1. Processing and swelling behavior. Biomacromolecules 4:657–665
Gonz′alez L, Rodr′ıguez A, de Benito JL, Marcos- Fern′andez A (1997) Applications of an azide sulfonyl silane as elastomer crosslinking and coupling agent. J Appl Polym Sci 63: 1353–1359
Hajipour MJ, Fromm KM, Ashkarran AA, de Aberasturi DJ, Larramendi IR, Rojo T, Serpooshan V, Parak Wolfgang J, Mahmoudi M (2012) Antibacterial properties of nanoparticles. Trends Biotechnol 30:499–511
Hassanien AM, El-Hashash MA, Mekewi MA, Guirguis DB, Ramadan AM (2013) Fabrication of polyvinyl alcohol/cellulose acetate (PVA/CA/PEG), antibacterial membrane for potential water purification application. Hydrol Current Res 4:1–6
Hart J, Silcock D, Gunnigle S, Cullen B, Light ND, Watt PW (2002) The role of oxidised regenerated cellulose/collagen in wound repair: effects in vitro on fibroblast biology and in vivo in a model of compromised healing. Int J Biochem Cell Biol 34:1557–1570
Helenius G, Bäckdahl H, Bodin A, Nannmark U, Gatenholm P, Risberg B (2006) In vivo biocompatibility of bacterial cellulose. J Biomed Mater Res A 76:431–438
Heller J (1987) Use of polymers in controlled release of active agents, In: Robinson JR, Lee VHL (eds) Controlled drug delivery. Fundamentals and applications, 180–210
Hilal N, Al-Khatib L, Atkin BP, Kochkodan V, Potapchenko N (2003) Photochemical modification of membrane surfaces for (bio) fouling reduction: a nano-scale study using AFM. Desalination 158:65–72
Huang Y, Zhang L, Yang J, Zhang X, Xu M (2013) Structure and properties of cellulose films reinforced by chitin whiskers. Macromol Mater Eng 298:303–310
Huang L, Nagapudi K, Chaikof EL (2001) Engeenered collagen-PEO nanofibers and fabrics. J Biomat Sci Polym Ed 12:979–993
Huang ZM, Zhang YZ, Kotaki M, Ramakrishna S (2003) A review on polymer nanofibers by electrospinning and their applications in nanocomposites. Compos Sci Technol 63:2223–2253
Hu L, Wu H, Cui Y (2010a) Printed energy storage devices by integration of electrodes and separators into single sheets of paper. Appl Phys Lett 96:183502–183504
Hu L, Pasta M, La Mantia F, Cui L, Jeong S, Dawn Deshazer H, Choi JW, Han SM, Cui Y (2010b) Stretchable, porous, and conductive energy textiles. Nano Lett 10:708–714
Hu L, La Mantia F, Wu H, Xie X, McDonough J, Pasta M, Cui Y (2011) Lithium-ion textile batteries with large areal mass loading. Adv Energy Mater 1:1012–1017
Iguchi M, Yamanaka S, Budhiono A (2000) Bacterial cellulose—amasterpiece of nature's arts. J Mater Sci 35:261–270
Imazato S, Ebi N, Takahashi Y, Kaneko T, Ebisu S, Russell RR (2003) Antibacterial activity of bactericide-immobilized filler for resin-based restoratives. Biomaterials 24:3605–3609
Ioan S, Dobos-Necula AM (2012) Silver nanoparticles in cellulose derivative matrix. In: Thakur VK, Singha AS (eds) Nanotechnology in polymers. Studium Presss, USA, chapter 11, pp 191–248
Jabbour L, Gerbaldi C, Chaussy D, Zeno E, Bodoardo S, Beneventi D (2010) Microfibrillated cellulose–graphite nanocomposites for highly flexible paperlike Li-ion battery electrodes. J Mater Chem 20:7344–7347
Jabbour L, Destro M, Gerbaldi C, Chaussy D, Penazzi N, Beneventi D (2012) Aqueous processing of cellulose based paper-anodes for flexible Li-ion batteries. J Mater Chem 22:3227–3233
Jabbour L, Destro M, Chaussy D, Gerbaldi C, Penazzi N, Bodoardo S, Beneventi D (2013) Flexible cellulose/LiFePO4 paper-cathodes: toward eco-friendly wallpaper Li ion batteries. Cellulose 20:571–582

Jadhav S, Gaikwad S, Nimse M, Rajbhoj A (2011) Copper oxide nanoparticles: synthesis, characterization and their antibacterial activity. J Cluster Sci 22:121–129

Johnson JR, Kuskowski MA, Wilt TJ (2006) Systematic review: antimicrobial urinary catheters to prevent catheter-associated urinary tract infection in hospitalized patients. Ann Int Med 144:116–126

Jonas R, Farah LF (1998) Production and application of microbial cellulose. Polym Degrad Stab 59:101–106

Joseph K, Mattoso CLH, Toledo RD (2000) Natural fiber reinforced thermoplastic composites.In: Frollini E, Leao AL, Mattoso CLH (eds) Nature Polymeric Agrofibers Compos, chapter 4, Embrapa. S˜an Carlos, Brazil, pp 159–201

Kaith BS, Singha AS, Kumar S, Misra BN (2005) $FASH_2O_2$ initiated graft copolymerization of methylmethacrylate onto flax and evaluation of some physical and chemical properties. J Polym Mat 22:425–432

Kalia S, Kaith BS, Sharma S, Bhardwaj B (2008) Mechanical properties of flax-g-poly(methyl acrylate) reinforced phenolic composites. Fibers Polym 9:416–422

Kalia S, Kaith BS, Kaur I (2009) Pretreatments of natural fibers and their application as reinforcing material in polymer composites-a review. Polym Eng Sci 49:1253–1272

Kalia S, Dufresne A, Cherian BM, Kaith BS, Av′erous L, Njuguna J, Nassiopoulos E (2011) Cellulose-based bio- and nanocomposites: a review. Int J Polym Sci 2011:35, Article ID 837875

Khoushab F, Yamabhai M (2010) Review: chitin research revisited. Mar Drugs 8:1988–2012

Kim J, Kwon S, Ostler E (2009) Antimicrobial effect of silver-impregnated cellulose: potential for antimicrobial therapy. J Biol Eng 3:1–9

Klasen HJ (2000) A historical review of the use of silver in the treatment of burns II. Renewed Interest silver. Burns 26:131–138

Kwon J-W, Yoon SH, Lee SS, Seo KW, Shim IS (2005) Preparation of silver nanoparticles in cellulose acetate polymer and the reaction chemistry of silver complexes in the polymer. Bull Korean Chem Soc 26:837–840

Lansdown AB (2002a) Silver. I: Its antibacterial properties and mechanism of action. J Wound Care 11:125–130

Lansdown AB (2002b) Toxicity in mammals and how its products aid wound repair. J Wound Care 11:173–177

Lea SB, Koepsel RR, Morley SW, Matyjaszewski K, Sun Y, Russell Alan J (2004) Permanent, nonleaching antibacterial surfaces. 1. Synthesis by atom transfer radical polymerization. Biomacromolecules 5:877–882

Legnani C, Barud HS, Quirino WG, Caiut JMA, Ribeiro SJL, Achete CA, Cremona M (2009) Transparent nanocomposite bacterial cellulose used as flexible substrate for OLED. In: Proceedings of the 11th international conference on advanced materials. Rio de Janeiro, Brazil

Leijonmarck S, Cornell A, Lindbergh G, Wågberg L (2013a) Flexible nano-paperbased positive electrodes for Li-ion batteries—preparation process and properties. Rapid Commun Nano Energy 2:794–800

Leijonmarck S, Cornell A, Lindbergh G, Wagberg L (2013b) Single-paper flexible Liion battery cells through a papermaking process based on nano-fibrillated cellulose. J Mater Chem A 1:4671–4677

Li WJ, Laurencin CT, Caterson EJ, Tuan RS, Ko FK (2002) Electrospun nanofibrous structure: a novel scafold for tissue engineering. J Biomed Mater Res Part A 60:613–621

Li D, McCann JT, Gratt M, Xia Y (2004) Photocatalytic deposition of gold nanoparticles on electrospun nanofibers of titania. Chemi Phys Lett 394:387–391

Li Q, Mahendra S, Lyon DY, Brunet L, Liga MV, Li D, Alvarez Pedro JJ (2008a) Antimicrobial nanomaterials for water disinfection and microbial control: potential applications and implications. Water Res 42:4591–4602

Li XH, Shao CL, Liu YC, Chu XY, Wang CH, Zhang BX (2008b) Photoluminescence properties of highly dispersed ZnO quantum dots in polyvinylpyrrolidone nanotubes prepared by a single capillary electrospinning. J Chem Phys 129:114708–114715

Lin M-F, Thakur VK, Tan EJ, Lee PS (2011a) Surface functionalization of BaTiO3 nanoparticles and improved electrical properties of $BaTiO_3$/polyvinylidene fluoride composite. RSC Adv 1:576–578

Lin M-F, Thakur VK, Tan EJ, Lee PS (2011b) Dopant induced hollow $BaTiO_3$ nanostructures for application in high performance capacitors. J Mater Chem 21:16500–16504

Longer MA, Robinson JR (1980) Sustained-release drug delivery systems. In: Remington JP (ed) Remington's pharmaceutical sciences, 18th edn. Mack publishing, easton, pp 1676–1693

Lonsdale HK, Padall HK (1972) Reverse osmosis membrane research. Plenum press, New York 155

Lonnberg H, Fogelstr¨om L, Samir MASA, Berglund L, Malmstr¨om E, Hult A (2008) Surface grafting of microfibrillated cellulose with poly(ε-caprolactone)—synthesis and Characterization. Europ Polym J 44:2991–2997

Lukens RJ (1983) Chemistry of fungicidal action. Mol Biol Biochem Biophys 10

Macedo NL, Matuda FS, Macedo LGS, Monteiro ASF, Valera MC, Carvalho YR (2004) Evaluation of two membranes in guided bone tissue regeneration: histological study in rabbits. Braz J Oral Sci 3:395–400

Madhumathi K, Sudheesh Kumar PT, Abhilash S, Sreeja V, Tamura H, Manzoor K, Nair SV, Jayakumar R (2010) Development of novel chitin/nanosilver composite scaffolds for wound dressing applications. J Mater Sci Mater Med 21:807–813

Maiti P, Yamada K, Okamoto M, Ueda K, Okamoto K (2002) New polylactide/layered silicate nanocomposites: role of organoclays. Chem Mater 14:4654–4661

Matthews JA, Wnek GE, Simpson DG, Bowlin GL (2002) Electrospinning of collagen nanofibers. Biomacromolecules 3:232–238

Martin CR (1995) Template synthesis of electronically conductive polymer nanostructures. Acc Chem Res 28:61–68

Martin CR (1996) Membrane-based synthesis of nanomaterials. Chem Mater 8:1739–1746

Miller LP, McCallan EA (1957) Toxic action of metal ions to fugus spores. Agric Food Chem 5:116–122

Morganti P, Muzzarelli RAA, Muzzarelli C (2006) Multifunctional use of innovative chitin nanofibrils for skin care. J Appl Cosmetology 24:105–114

Morones JR, Elechiguerra JL, Camacho A (2005) The bactericidal effect of silver nanoparticles. Nanotechnology 16:2346–2353

Murugaraj P, Mainwaring DE, Jakubov T, Mora-Huertas NE, Khelil NA, Siegele R (2006) Electron transport in semiconducting nanoparticle and nanocluster carbon– polymer composites. Solid State Commun 137:422–426

Necula M, Dunca S, Stoica I, Olaru N, Olaru L, Ioan S (2010) Morphological properties and antibacterial activity of nano-silver-containing cellulose acetate phthalate films. Int J Polym Anal Charact 15:341–350

Necula AM, Stoica I, Olaru N, Doroftei F, Ioan S (2011) Silver nanoparticles in cellulose acetate polymers. Rheological and morphological properties. J Macromol Sci Part B: Phys 50:639–651

Nemetz AP, Loures DRR, Coelho JCU (2001) Efeito estrutural da utilização de celulose biossintética e politetrafluoretileno expandido como substitutos do peritônio em cães Arquivos Brasileiros De Cirugia Digestiva 14:139–142

Novaes AB Jr, Novaes AB (1992) IMZ implants placed into extraction sockets in association with membrane therapy (Gengiflex) and porous hydroxyapatite: a case report. Int J Oral Maxillofac Implants 7:536–540

Novaes AB Jr, Novaes AB (1997) Soft tissue management for primary closure in guided bone regeneration: surgical technique and case report. Int J Oral Maxillofac Implants 12:84–87

Ottaviani MF, Valluzzi R, Balogh L (2002) Internal structure of silver-poly(amidoamine) dendrimer complexes and nanocomposites. Macromolecules 35:5105–5115

Olson DG, Tripathi SA, Giannone RJ (2010) Deletion of the Cel48S cellulase from clostridium thermocellum. Int J Polym Sci Proc Nat Acad Sci USA 107:17727–17732

Pal S, Tak YK, Song JM (2007) Does the antibacterial activity of silver nanoparticles depend on the shape of the nanoparticle? a study of the gram-negative bacterium Escherichia coli. Appl Environ Microb 73:1712–1720

Panacek A, Kolar M, Vecerova R, Pruceka R, Soukupova J, Krystof V, Hamal P, Zboril R, Kvítek L (2009) Antifungal activity of silver nanoparticles against Candida spp. Biomaterials 30:6333–6340

Park HS, Park YO (2005) Filtration properties of electrospun utrafine fiber webs. Korean J Chem Eng 22:165–172

Percival SL, Bowler PG, Russell D (2005) Bacterial resistance to silver in wound care. J Hosp Infect 60:1–7

Pinto RJB, Marques PAAP, Neto CP, Trindade T, Daina S, Sadocco P (2009) Antibacterial activity of nanocomposites of silver and bacterial or vegetable cellulosic fibers. Acta Biomater 5:2279–2289

Pinto RJB, Neves MC, Pascoal Neto C, Trindade T (2012) Growth and chemical stability of copper nanostructures on cellulosic fibers. Eur J Inorg Chem 2012:5043–5049

Pinto RJB, Daina S, Sadocco P, Neto CP, Trindade T (2013) Antibacterial activity of nanocomposites of copper and cellulose. BioMed Res Int 280512:6

Van Den Plas D, De Smet K, Lens D, Sollie P (2008) Differential cell death programmes induced by silver dressings in vitro. Eur J Dermatol 18:416–421

Pommet M, Juntaro J, Heng JYY, Mantalaris A, Lee AF, Wilson K, Kalinka G, Shaffer Milo SP, Bismarck A (2008) Surface modification of natural fibers using bacteria: depositing bacterial cellulose onto natural fibers to create hierarchical fiber reinforced nanocomposites. Biomacromolecules 9:1643–1651

Ray D, Sarkar BK, Rana AK, Bose NR (2001) Effect of alkali treated jute fibres on composite properties. Bull Mat Sci 24:129–135

Ranby BG (1952) The cellulose micelles. Tappi 35:53–58

Richards RME (1981) Antimicrobial action of silver nitrate. Microbios 31:83–91

Richards RME, Odelola HA, Anderson B (1984) Effect of silver on whole cells and spheroplasts of a silver resistant Pseudomonas aeruginosa. Microbios 39:151–258

Rujitanaroj P, Pimpha N, Supaphol P (2008) Wound-dressing materials with antibacterial activity from electrospun gelatin fiber mats containing silver nanoparticles. Polymer 49:4723–4732

Ruparelia JP, Chatterjee AK, Duttagupta SP, Mukherji S (2008) Strain specificity in antimicrobial activity of silver and copper nanoparticles. Acta Biomat 4:707–716

Russell AD, Hugo WB (1994) Antimicrobial activity and action of silver. Prog Med Chem 31:351–371

Salata LA, Craig GT, Brook IM (1995) In vivo evaluation of a new membrane (Gengiflex) for guided bone regeneratien (GBR). J Dental Res 74:825–831

Sarma TK, Chowdhury D, Paul A, Chattopadhyay A (2002) Synthesis of Au-nanoparticle conductive polyaniline composite using H_2O_2 as oxidizing as well as reducing agent. Chem Comun 14:1048–1049

Schadler LS (2003) Polymer-based and polymer-filled nanocomposites. In: Ajayan PM, Schadler LS, Braun PV (eds) Nanocomposite science and technology. Weinheim Wiley-VCH Verlag, New York

Schneider A, Wang XY, Kaplan DL, Garlick JA, Egles C (2009) Biofunctionalized electrospun silk mats as a topical bioactive dressing for accelerated wound healing. Acta Biomat 5:2570–2578

Schreurs WJA, Rosenburgh H (1982) Effect of silver ions on transport and retention of phosphate by Escherichia coli. J Bacteriol 152:7–13

Shafei AE, Abou-Okeil A (2011) ZnO/carboxymethyl chitosan bionanocomposite to impart antibacterial and UV protection for cotton fabric. Carbohydr Polym 83:920–925

Shaikh S, Birdi A, Qutubuddin S, Lakatosh E, Baskaran H (2007) Controlled release in transdermal pressure sensitive adhesives using organosilicate nanocomposites. Annals Biomed Eng 35:2130–2137

Shah J, Brown RM Jr (2005) Towards electronic paper displays made from microbial cellulose. Appl Microbiol Biotechnol 66:352–355
Shiraishi Y, Toshima N (2000) Oxidation of ethylene catalyzed by colloidal dispersion of poly (sodium acrylate) protected silver nanoclusters. Coll Surf A 169:59–66
Shoda M, Sugano Y (2005) Recent advances in bacterial cellulose production. Biotechnol Bioproc Eng 10:1–8
Silva EC (2009) Hidroxiapatita Sintetica em alveolo dentario apos exodontia em Felis catus: estudo clınico, radiologico e histomorfometrico. M.S. Dissertation, Universidade Federal
Singha AS, Thakur VK (2008a) Saccaharum cilliare fiber reinforced polymer composites. E-J Chem 5:782–791
Singha AS, Thakur VK (2008b) Effect of fibre loading on urea-formaldehyde matrix based green composites. Iran Polym J 17:861–873
Singha AS, Thakur VK, Mehta IK, Shama A, Khanna AJ, Rana RK, Rana AK (2009) Surface-modified hibiscus sabdariffa fibers: physicochemical, thermal, and morphological properties evaluation. Int J Polym Anal Charact 14:695–711
Singha AS, Thakur VK (2010 a) Mechanical, morphological, and thermal characterization of compression-molded polymer biocomposites. Int J Polym Anal Charact 15:87–97
Singha AS, Thakur VK (2010b) Synthesis, characterization and study of pine needles reinforced polymer matrix based composites. J Reinf Plast Compos 29:700–709
Singh N, Galande C, Miranda A, Mathkar A, Gao W, Mohana Reddy AL, Vlad A, Ajayan PM (2012) Paintable battery. Sci Rep 2:481
Siqueira G, Bras J, Dufresne A (2010) Cellulosic bionanocomposites: a review of preparation, properties and applications. Polymers 2:728–765
Son WK, Youk JH, Lee TS, Park WH (2004) Preparation of antimicrobial ultrafine cellulose acetate fibers with silver nanoparticles. Macromol Rapid Commun 25:1632–1637
Son WK, Youk JH, Park WH (2006) Antimicrobial cellulose acetate nanofibers containing silver nanoparticles. Carbohydr Polym 65:430–434
Sondi I, Salopek-Sondi B (2004) Silver nanoparticles as antimicrobial agent: a case study on E. coli as a model for Gramnegative bacteria. J Colloid Int Sci 275:177–182
Southward RE, Thompson DW (2001) Reflective and conductive silvered polyimide films for space applications prepared via a novel single-stage self-metallization technique. Mat Des 22:565–576
Stashak TS, Farstvedt E, Othic A (2004) Update on wound dressings: indications and best use. Clin Technol Equine Pract 3:148–163
Subramanian V, Wolf EE, Kamat PV (2004) Catalysis with TiO2/Gold nanocomposites. Efect of metal particle size on the Fermi level equilibration. J Am Chem Soc 126:4943–4950
Tam KH, Djurisic AB, Chan CMN, Xi YY, Tse CW, Leung YH, Chanb WK, Leungc FCC, Au DWT (2008) Antibacterial activity of ZnO nanorods prepared by a hydrothermal method. Thin Solid Films 516:6167–6174
Thakur VK, Singha AS, Kaur I, Nagarajarao RP, Liping Y (2010a) Silane functionalization of saccaharum cilliare fibers: thermal, morphological, and physicochemical study. Int J Polym Anal Charact 15:397–414
Thakur VK, Singha AS, Mehta IK (2010b) Renewable resource-based green polymer composites: analysis and characterization. Int J Polym Anal Charact 15:137–146
Thakur VK, Singha AS, Misra BN (2011) Graft copolymerization of methyl methacrylate onto cellulosic biofibers. J Appl Polym Sci 122:532–544
Thakur VK, Yan J, Lin M-F et al (2012a) Novel polymer nanocomposites from bioinspired green aqueous functionalization of BNNTs. Polym Chem 3:962–969
Thakur VK, Singha AS, Thakur MK (2012b) In-air graft copolymerization of ethyl acrylate onto natural cellulosic polymers. Int J Polym Anal Charact 17:48–60
Thakur VK, Singha AS, Thakur MK (2012c) Biopolymers based green composites: mechanical, thermal and physico-chemical characterization. J Polym Environ 20:412–421
Thakur VK, Singha AS, Thakur MK (2012d) Graft Copolymerization of Methyl Acrylate onto Cellulosic Biofibers: synthesis, characterization and applications. J Polym Environ 20:164–174

Thakur VK, Singha AS, Thakur MK (2012d) Modification of natural biomass by graft copolymerization. Int J Polym Anal Charact 17:547–555

Thakur VK, Thakur MK (2014a) Processing and characterization of natural cellulose fibers/thermoset polymer composites. Carbohydr Polym 109:102–117

Thakur VK, Thakur MK (2014b) Recent trends in hydrogels based on psyllium polysaccharide: a review. J Cleaner Product 82:1–15

Thakur VK, Vennerberg D, Kessler MR (2014a) Green aqueous surface modification of polypropylene for novel polymer nanocomposites. ACS Appl Mater Interfaces 6:9349–9356

Thakur VK, Thakur MK, Raghavan P, Kessler MR (2014b) Progress in green polymer composites from lignin for multifunctional applications: a review. ACS Sust Chem Eng 2:1072–1092

Thakur VK, Thakur MK, Gupta RK (2014c) Review: raw natural fiber–based polymer composites. Int J Polym Anal Charact 19:256–271

Thompson CM, Herring HM, Gates TS, Connell JW (2003) Preparation and characterization of metal oxide/polyimide nanocomposites. Compos Sci Technol 63:1591–1598

Thurmann RB, Gerba CP (1988) Molecules mechanisms of viral inactivation by water disinfectants. Adv Appl Microbiol 33:75–105

Thurmann RB, Gerba CP (1989) The molecules mechanisms of copper and silver ion disinfection of bacteria and viruses. Crit Rev Environ Control 18:295–315

Tian X, Zhang X, Liu W, Zheng J, Ruan C, Cui P (2006) Preparation and properties of poly (ethylene terephthalate)/silica nanocomposites. J Macromol Sci Part B Phys 45:507–513

Tiller JC, Lee SB, Lewis K, Klibanov AM (2002) Polymer surfaces derivatized with poly(vinyl-N-hexylpyridinium) kill airborne and waterborne bacteria. Biotechnol Bioeng 79:465–471

Vargas-Reus MA, Memarzadeh K, Huang J, Ren GG, Allaker RP (2012) Antimicrobial activity of nanoparticulate metal oxides against peri-implantitis pathogens. Int J Antimicrob Agents 40:135–139

Wang X, Drew C, Lee SH, Senecal KJ, Kumar J, Samuelson LA (2002) Electrospun nanofibrous membranes for highly sensitive optical sensors. Nano Lett 2:1273–1275

Wang M, Singh H, Hatton TA, Rutledge GK (2004) Field responsive superparamagnetic composite nanofibers by electrospinning. Polymer 45:5505–5514

Watanabe Y, Mukai B, Kawamura KI, Ishikawa T, Namiki M, Utoguchi N, Fujii M (2002) Preparation and evaluation of press-coated aminophylline tablet using crystalline cellulose and polyethylene glycol in the outer shell for timed-release dosage forms. In: Zasshi Y (ed) The Pharmaceutical Society of Japan, 122:157–162

Wongpanit P, Sanchavanakit N, Pavasant P, Bunapresert T, Tabata Y, Rujiravanit R (2007) Preparation and characterization of chitin whisker-reinforced silk fibroin nanocomposite sponges. Eur Polym J 43:4123–4135

Watkins JJ, McCarthy TJ (1995) Polymer/metal nanocomposite synthesis in supercritical CO2. Chem Mater 7:1991–1994

Wiegand C, Heinze T, Hipler UC (2009) Comparative in vitro study on cytotoxicity, antimicrobial activity, and binding capacity for pathophysiological factors in chronic wounds of alginate and silver-containing alginate. Wound Repair Regen 17:511–521

Wright JB, Lam K, Hansen D, Burrell RE (1999) Efficacy of topical silver against fungal burn wound pathogens. Am J Infect Control 27:344–350

Wu Y, Jia W, An Q, Liu Y, Chen J, Li G (2009a) Multiaction antibacterial nanofibrous membranes fabricated by electrospinning: an excellent system for antibacterial applications. Nanotechnology 20:245101

Wu J, Hou S, Ren D, Mather PT (2009b) Antimicrobial properties of nanostructured hydrogel webs containing silver. Biomacromolecules 10:2686–2693

Xie S, Li W, Pan Z, Chang B, Sun L (2000) Mechanical and physical properties on carbon nanotubes. J Phys Chem Solids 61(7):1153–1158

Yang QB, Li DM, Hong YL, Li ZY, Wang C, Qiu SL, Wei Y (2003) Preparation and characterization of a PAN nanofibre containing Ag nanoparticles via electrospinning. Synthetic Met 137:973–974

Yano H, Sugiyama J, Nogi M, Matsuura T, Hikita M, Handa K (2005) Optically transparent composites reinforced with networks of bacterial nanofibers. Adv Mat 17:153–155

Yuvaraj D, Kaushik R, Narasimha RK (2010) Optical, field-emission and antimicrobial properties of ZnO nanostructured films deposited at room temperature by activated evaporation. ACS Appl Matter Interfaces 2:1019–1024

Zhu C, Xue J, He J (2009) Controlled in-situ synthesis of silver nanoparticles in natural cellulose fibers toward highly efficient antimicrobial materials. J Nanosci Nanotechnol 9:3067–3074

Zeng J, Xu X, Chen X, Liang Q, Bian X, Yang L, Jing X (2003) Biodegradable electrospun fibers for drug delivery. J Controlled Release 92:227–231

Eco-friendly Electrospun Polymeric Nanofibers-Based Nanocomposites for Wound Healing and Tissue Engineering

Ibrahim M. El-Sherbiny and Isra H. Ali

Abstract Recently, nanofibers have been investigated with remarkable increased applicability in different fields due to their numerous advantages such as large surface area and controlled morphology. Nanofibers can be prepared using three main techniques, namely electrospinning, phase separation, and self-assembly. Of these, electrospinning is the most commonly used technique and also seems to exhibit the most desirable results. This chapter provides an overview of the electrospinning of eco-friendly polymers and polymeric nanocomposites for biomedical applications with an emphasis on their applications in wound healing and tissue regenration. Controlling the characteristics of the developed electrospun nanocomposites via tailoring the collectors used during electrospinning as well as carefully changing their surface chemistry for the proper design of wound healing nanofibrous dressings and tissue engineering nanofibrous scaffolds will be explored. Also, the challenges associated with the use of electrospun polymeric nanofibers-based nanocomposites for wound healing and tissue engineering will be described.

Keywords Electrospinning · Nanofibers · Nanocomposites · Wound healing · Tissue engineering

I.M. El-Sherbiny (✉) · I.H. Ali
Center for Materials Science, University of Science and Technology (UST), Zewail City of Science and Technology, Giza, Egypt
e-mail: ielsherbiny@zewailcity.edu.eg

I.H. Ali
e-mail: igalal@zewailcity.edu.eg

V.K. Thakur and M.K. Thakur (eds.), *Eco-friendly Polymer Nanocomposites*, Advanced Structured Materials 75, DOI 10.1007/978-81-322-2470-9_13

1 Introduction

1.1 Wound Healing and Dressings

Since the ancient era, many materials such as animal fats, honey, and plant fibers have been used as cover materials to prevent incidence of wound external infections. Recently, with the discovery of new materials and evolution of new fabrication techniques, it has been a necessity to design and develop new wound dressings possessing exceptional features. They should be capable of healing wounds at high rate, protecting wounds from external infection, avoiding wound contamination, preventing scars formation, and guaranteeing the preferable wounds remedy (Zahedi et al. 2009).

Wound dressings are classified into three main categories which are passive, interactive, and active (bioactive) dressings. Passive wound dressings such as tulle and gauze are those used only to cover the wounds underneath. In other words, they just act like a barrier between wounds and the external environment to prohibit any external infection. On the other hand, interactive wound dressings are made of polymeric foams, hydrogel films, or hyaluronic acid (HA). Although they are permeable to atmospheric oxygen and water vapor, they are highly capable of preventing of bacterial permeation. The active wound dressings are those containing bioactive materials possessing wound healing activity and antibacterial activity besides their protecting effects. Bioactive materials include, but not limited to, alginate, chitosan, collagen, and hydrocolloids (Muzzarelli 2009; Zahedi et al. 2009).

It was reported that fabricating wound dressings in a nanofibrous form is more advantageous than fabricating it as a solid film. Nanofibrous wound dressing would be distinguished by possessing larger aspect ratio, i.e., larger surface area to volume ratio, in addition to high porosity. Increasing the surface area of the dressing would consequently increases its potency as healing enhancer and antibacterial material. Moreover, it was found that as the diameter of the nanofibers decreases, both the proliferation and the spreading tendency of the dermal fibroblasts increase leading to an increase in the wound healing rate (Kumbar et al. 2008; Ladd et al. 2008).

Porosity is also important in order to offer sufficient air and liquid diffusion which are necessary for cellular regeneration and wound healing (Ladd et al. 2008; Yang et al. 2009).

Therefore, electrospinning is considered to be an ideal fabrication technique for wound dressings. Nanofibrous wound dressings with well-controlled nanofeatures mandatory for wound healing can be obtained through electrospinning (Ladd et al. 2008; Yang et al. 2009; Zahedi et al. 2009).

1.2 Tissue Engineering

Tissue engineering is a tremendously processing interdisciplinary research field that targets compensating the damaged or lost tissues or organs (Murugan and Ramakrishna 2006; Pham et al. 2006).

The issue of damaged or lost organ due to accidents or certain diseases has been usually settled through organ transplant. However, this solution showed several drawbacks that include for instance; (i) immunological problems where the patient's body may reject the transplanted organ, (ii) infectious diseases, and/or (iii) shortage of donors (Niklason 2001; Pham et al. 2006).

Another approach toward organ loss fixation is using artificial or mechanical organs like kidney dialyzers. Although those systems have performed their functions successfully, they possessed some limitations such as; (i) being expensive, and/or (ii) lacking the ability to perform the whole function of the entire organ. For instance, kidney dialyzers can perform a certain kidney function, however, cannot act completely as the entire natural kidney (Niklason 2001).

Tissue engineering field is a more promising approach to compensate the damaged organs since it aims at designing and fabricating of three-dimensional scaffolds using biocompatible and biodegradable biomaterials capable of mimicking the extracellular matrices (ECM) nanoscaled lineaments. These scaffolds would be implanted at the injured site along with the corresponding cells and appropriate growth factors in order to provoke cellular differentiation, growth, and proliferation till natural ECM restoration and total recovery of the damaged tissue or lost organ occurs (Silva et al. 2012; Dhandayuthapani et al. 2011; Pham et al. 2006). Tissue engineering can be verified through following three main strategies which are: (i) incorporation of corresponding cells within the damaged site, (ii) delivery of the appropriate growth factors, and/or (iii) implantation of both the corresponding cells and scaffold within the damaged tissues. The third strategy is the most commonly used strategy for tissue regeneration (Nerem 1992; Vacanti and Langer 1999).

Diverse traditional techniques have been used to fabricate tissue engineered scaffolds which include: drawing, melt molding, template synthesis, freeze drying, gas foaming, particulate leaching, solid-free forming, self-assembly and phase separation (Feng et al. 2002; Glicklis et al. 2000; Liu et al. 1999; Madihally and Matthew 1999; Martin 1996; Mikos et al. 1994; Mooney et al. 1996; Murugan and Ramakrishna 2006; Ondarçuhu and Joachim 1998; Schoof et al. 2001; Tang et al. 2014; Thomson et al. 1996). Although each of these techniques possesses some advantages, all of them fail to present an ideal scaffold with well-controlled mandatory scaffold nanofeatures like fiber spatial arrangement, dimensions, size distribution, geometry and pore size. This led to inadequate nutrient supply and oxygen transport to the cells and consequently inadequate cell growth and proliferation ending to failure of tissue compensation (Ma 2004; Murugan and Ramakrishna 2006; Sachlos and Czernuszka 2003; Yang et al. 2001, 2004). Moreover, those aforementioned traditional techniques show some limitations in selecting the polymers used in scaffold fabrication. For instance, although some techniques like melt molding may allow incorporation of naturally derived polymers such as collagen, the use of high temperature may lead to denaturation of the protein structure of such polymers. Other techniques like particulate leaching and solvent casting can fabricate highly porous scaffolds but with poor interconnectivity and shaping capability (Ladd et al. 2008; Yang et al. 2001).

Since, the cells inside the human body normally live, proliferate, and grow within sophisticated fibrous ECM enriched with pores and ridges at the nanoscale (Desai 2000; Murugan and Ramakrishna 2006), it is mandatory to fabricate the scaffold with well-controlled features that mimic the naturally occurring ECM inside the human body (Murugan and Ramakrishna 2006). Electrospinning technique seems to be the most promising abundant technique capable of forming quiet well-controlled nanofibers with highly controlled features (nanofibers dimensions, porosity, etc.) to be used as scaffolds (Huang et al. 2003; Li et al. 2002; Pham et al. 2006; Sill and von Recum 2008).

1.3 Eco-friendly Composites and Biomaterials

Upon selecting the biomaterials used for either wound healing or tissue regeneration applications, some general specifications should be taken into consideration. The selected biomaterial should possess high biocompatibility while low cytotoxicity and inflammatory reactions (Sill and von Recum 2008).

Another criterion that has attracted scientists attention during selecting biomaterials is choosing those having no harmful effects on the environments. These materials are called green materials or eco-friendly biomaterials (Thakur et al. 2013a, b, c, d). These distinctive biomaterials are considered to be advantageous since many environmental problems would be avoided unlike the other petroleum derived nonbiodegradable materials (Thakur et al. 2012a, b, 2014a, b, c, d, e).

Biocompatible biomaterials can be divided into two main categories; natural biomaterials and synthetic biocompatible biomaterials. They can be used to form either the composite matrices or reinforcements during fabrication of wound dressings or tissue engineered scaffolds.

1.3.1 Natural Biomaterials

Diverse raw natural fibers have been utilized during composite fabrication due to being characterized with numerous unique properties (Singha and Thakur 2010; Thakur et al. 2013a, b). The formed composites would be characterized by possessing high biocompatibility and toughness, low density and considerable strength as well as good thermal stability (Thakur et al. 2014a, b, c; Thakur and Singha 2010).

Cellulose, a natural polysaccharide fibrous structure is considered to be the highest abundant polymer on earth. Cellulose is characterized by possessing certain hierarchal with multilevel arrangement leading to the possibility of facile formation of diverse microfibrillated cellulose (MFC) and cellulosic nanocrystals (CNC). MFC and CNC have been recently been incorporated as fillers in composites fabrication in order to enhance their mechanical properties (Ni et al. 2012; Siqueira et al. 2010). Cellulose fibers can be obtained from different biorenewable resources such as higher and lower plants mainly, some microorganisms like bacteria and fungi, some amoeba, and sea animals (Siqueira et al. 2010). Cellulose fibers have been one of the most promising biomaterials to be used in biocomposites since it is characterized by possessing fundamental features such as availability, biodegradability, easy processing, being eco-friendly, flexibility, and good physico-mechanical properties. Therefore, cellulose can be used in diverse applications besides biomedical applications, e.g. automotive and electronic applications (Thakur et al. 2011; Thakur and Thakur 2014a).

Chitin is the second abundant natural polymer after cellulose. It can be obtained from crabs and shrimps mainly. However, chitin can also be found as a main structural component in arthropods exoskeletons or in the cellular wall of some microorganisms such as fungi and yeast. The extracted chitin cannot be used in its native state in biomedical applications due to the presence of proteins and pigments. Chitin undergoes grading processes in which it is deacetylated in presence of alkaline medium in order to obtain chitosan, the most important chitin derivative (Rinaudo 2006). Chitosan has been found to be quiet promising biomaterial for different biomedical applications including drug delivery, tissue engineering, and wound healing applications. This reiterates that chitosan is biocompatible, biodegradable, and immunostimulant biopolymer with low cytotoxicity. Chitosan has the capability of binding to macromolecules and delivering them. Moreover, chitosan has been characterized by its wide antimicrobial activity against diverse types of bacterial fungi and virus (Jayakumar et al. 2010; Mi et al. 2002; Rabea et al. 2003a).

In addition, several natural polysaccharides have been reported to be utilized as fibers in composite formations. For instance, psyllium was found to be one of the good biomaterials to be used in hydrogels owing to its biodegradability, biocompatibility, and swellability. Psyllium hydrogels were found to be bioactive and stimuli responsive, thus they are capable to be used for several biomedical as well as environmental applications, e.g., water purification (Thakur and Thakur 2014b). Another example is the Hibiscus Sabdariffa fibers which showed good swellability, physicochemical, and thermal properties. It was reported that these fibers have the ability to be tailored superficially through chemical modifications to match the targeted application (Singha et al. 2009). Other reported naturally derived biomaterials are collagen, gelatin, starch, etc. (Murugan and Ramakrishna 2006).

1.3.2 Synthetic Biomaterials

It was reported that some synthetic polymers are biocompatible and biodegradable, thus can be used as safe biomaterials in biomedical applications. Synthetic polymers are degraded hydrolytically unlike enzymatically degraded natural polymers.

Although, they are biologically inert, they are characterized by possessing several unique features not abundant in most of the natural polymers. Most of the synthetic polymers are characterized by high mechanical properties and tendency to be easily tailored thus easily surface modified when compared to natural polymers (Thakur and Kessler 2014a, b). Moreover, they can be sterilized, processed easily, and tuned to be free of immunogenicity. Biodegradable synthetic polymers with different biodegradability degrees were reported to be utilized in different biomedical applications such as drug delivery vehicles, tissue engineered scaffolds, and transient implants (Murugan and Ramakrishna 2006; Nair and Laurencin 2007). Although there are several synthetic polymers used in biomedical research, only few among them are approved by FDA to be used in tissue engineering applications such as poly(glycolic acid), poly(lactic acid), and poly(lactic-co-glycolic acid) (Murugan and Ramakrishna 2006).

2 Electrospinning

Electrospinning is cost effective, eco-friendly, facile, and simple technique that depends mainly on the applied electrostatic forces and possess high capability of producing well-controlled features of nanofibers whose dimensions range from tens of nanometers up to few micrometers (Jafari et al. 2011; Steyaert et al. 2012). The produced electrospun nanofibers can possess diverse morphology features (dimensions, porosity, etc.) according to the selected electrospinning parameters. In other words, any alteration of the ambient variables, controlled variable as well as electrospinning solutions parameters (concentration, conductivity, flow rate, viscosity, etc.) would impact the morphology and the characteristics of the produced electrospun nanofibers. Thus, the characteristics of the electrospun nanofibers and in particular their morphology can be manipulated through modifying some of those parameters as will be discussed in detail in Sect. 2.3 (Ladd et al. 2008; Murugan and Ramakrishna 2006; Pham et al. 2006).

2.1 Electrospinner Components

Electrospinner is composed of four main components; the pump, syringe and spinneret, power supply, and the collector as illustrated in Fig. 1 (Murugan and Ramakrishna 2006).

2.1.1 Pump

The role of the pump is to exert pressure and offer controlled flow rate in order to propel the polymer out of the syringe though the spinneret. The electrospinner user

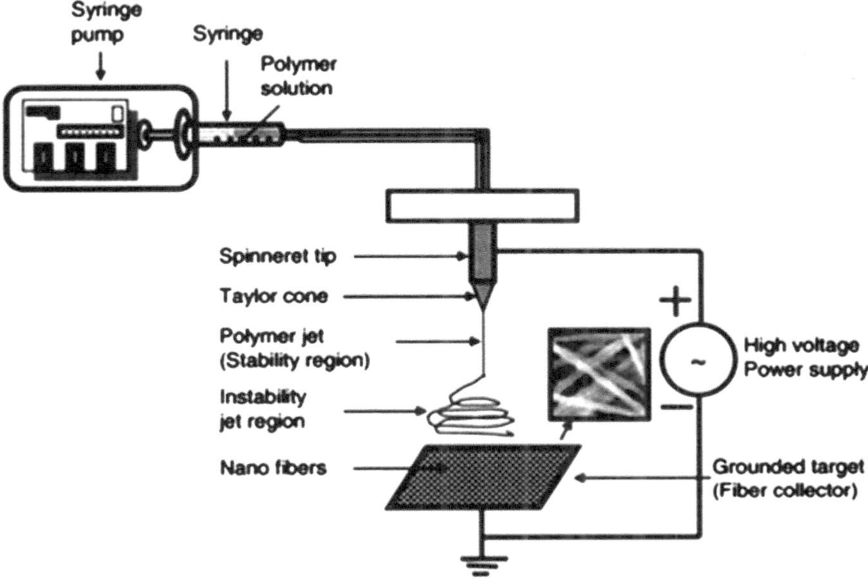

Fig. 1 Schematic drawing of electropinner framework. Adopted from Murugan and Ramakrishna (2006)

can change and control the flow rate of the polymer solution as well as the diameter spinneret through pump programming (Murugan and Ramakrishna 2006).

2.1.2 Syringe and Its Spinneret

The syringe acts as a polymer solution reservoir and it is directly coupled to the spinneret through which ejection of the polymer solution occurs under the influence of the adjusted constant flow rate programmed by the pump (Murugan and Ramakrishna 2006).

2.1.3 Power Supply

The function of the power supply is to electrify the polymer solution inside the syringe with the desired amount of electric current. The supplied electric current can be changed and adjusted till reaching the critical electric current at which conversion of the polymer solution into an electrified polymer jet occurs. In general, this can be verified using low electric current (high applied voltage). The applied voltage can be manipulated up to 30 keV according to the desired morphology and features of the produced electrospun nanofibers (Murugan and Ramakrishna 2006).

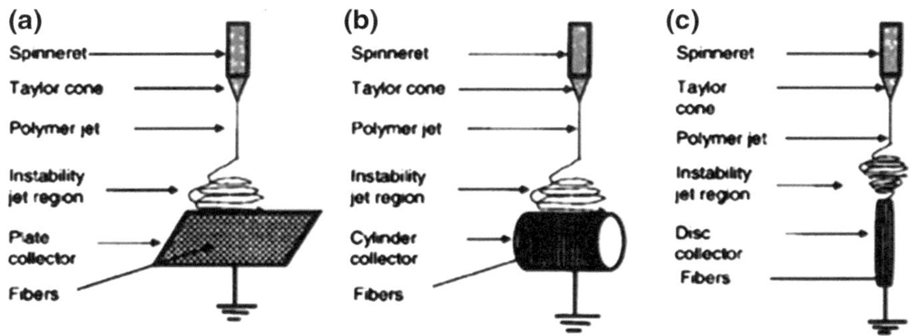

Fig. 2 Different types of electrospinner collectors: **a** stationary, **b** drum, and **c** disc. Adopted from Murugan and Ramakrishna (2006)

2.1.4 Collectors

Collectors with different shapes and types (Fig. 2) are considered to be the receivers of electrospun nanofibers after being ejected out of the spinneret. The orientation of the electrospun nanofibers can be controlled via using specific collectors with certain shape and type. The three main types of collectors are the stationary (plate), cylinder, and the disc (Murugan and Ramakrishna 2006; Theron et al. 2001).

2.2 Electrospinner Operation

Electrospinner operation and setup appear to be very straightforward, however, the spinning is a highly sophisticated mechanism (Murugan and Ramakrishna 2006). The concept of electrospinning process resembles that of electrospraying technique. Both techniques depend on applying high voltage to polymer solution to form either polymer particles or nanofibers in case of using electrospray dryer or electrospinner, respectively. However, several electrospinning parameters have to be adjusted in order to obtain reliable nanofibers as will be discussed shortly (Sill and von Recum 2008).

Upon connecting the high applied voltage (low electric current) between the tip of the spinneret and the collector situated precisely in front of the tip of the spinneret, an intense electrification of the first polymer droplet forced out of the syringe under the influence of the adjusted controlled flow rate occurs (Qian et al. 2011).

Under the influence of the continuous applied voltage, the prompting charges start to accumulate on the ejected polymer droplet surface leading to its elongation till formation of a conical shape commonly known as Taylor cone (Murugan and Ramakrishna 2006). The cause of this distortion is the two electrostatic forces—electrostatic repulsion occurring between the Columbic forces that supplied the high external electric field, and the accumulated charges on the polymer droplet surface (Kolambkar et al. 2011). The electrostatic forces would exceed both the surface tension and the viscoelasticity of the polymer droplet upon exceeding the applied voltage to a critical threshold, leading to ejection of an ultrafine charged jet out of what is called the Taylor cone. This charged jet is considered to be an instability region since it would be exposed to several successive whipping and stretching processes in addition to the prompt evaporation of the solvent(s) used till deposition of the ultrafine fibers on the collector occurs (Murugan and Ramakrishna 2006; Pham et al. 2006).

The migration voyage of the polymer droplet after being forced through the spinneret till its charge loss and deposition on the surface of the collector can be divided into four distinct stages (Ladd et al. 2008). The first stage is the stage at which the Taylor cone is formed as discussed before pointing at the surface of the targeted collector (Greiner and Wendorff 2007).

The second stage is the stage of successive process of stretching and whipping, evaporation of most of the solvent(s) used, migration of most of the distance between the spinneret and the collector, and finally a decrease in the fiber diameter. The third stage is known as the splay stage since it involves the protrusion of several strands out of the main polymer jet.

However, the multiple stranded splay would contain a major and single strand that whip and spiral speedily toward the collector. The fourth stage is the final stage at which total evaporation of the solvent(s) occurs leading to the deposition of ultrafine nanofibers on the surface of the targeted collector (Ladd et al. 2008).

2.3 Electrospinning Parameters

Electrospinning parameters are the variables which can be controlled and adjusted in order to manipulate the morphology and features of the electrospun nanofibers. Electrospinning parameters are classified into three main categories which are: (i) processing (controlled) variables, (ii) ambient (uncontrolled) variables, and (iii) the solution variables (Huang et al. 2003; Pham et al. 2006; Zander 2013). Electrospinning parameters are summarized in Table 1.

2.3.1 Processing (Controlled) Variables

They are the electrospinning parameters that could be changed and controlled during electrospinning process in order to manipulate the morphology and the features of the produced electrospun nanofibers as well as their orientation (Sill and von Recum 2008). These include the collector design, movement and structure, the electric field strength (applied voltage), geometry and style of the syringe needle tip, the gap between the spinneret tip and the collector, i.e., its location away from the collector,

Table 1 A summary of electrospinning parameters

Parameters category	Parameters	Change in parameter	Effect on nanofibers morphology
Processing	Collectors	Stationary	Random unaligned nanofibers
		Parallel rods	Aligned nanofibers
		Rotatory disc or drum	Oriented nanofibers
		Cross-linked mesh	Patterned nanofibers
	Applied voltage	↑	↑ Diameter, beads
	Spinneret design	Coaxial spinneret	Hollow or core–shell nanofibers
		Multiple jet spinneret	↑ nanofibers yield or spinning different nanofibers with certain ratio
	Distance between spinneret and collector	Should be sufficient to allow solvent evaporation if increased or decreased	Beads formation
	Solution flow rate	↓	↓ Nanofibers diameter, ↓ beads
		↑	↑ Nanofibers diameter, ↑ beads
Ambient	Air velocity	Presence of vacuum	↑ Electric field
			↑ Nanofibers diameter
	Temperature	↑	↓ Solution viscosity
			↓ Nanofibers diameter
	Humidity	↑	↑ Porous Nanofibers
Solution	Polymer concentration (Solution viscosity)	↑	↓ Beads
			↑ Uniformity of nanofibers
	Polymer molecular weight	↓	Brittle beaded nanofibers
		↑	↑ Solution viscosity
			↓ Beads
			↑ Uniformity of nanofibers
	Solution conductivity	↓	↑ Heavily beaded nanofibers
		↑	↓ Beads
			↑ Uniformity of nanofibers
	Dielectric constant and dipole moment	↑	↑ Uniformity of nanofibers
			↑ Nanofibers yield
	Surface tension	↓ in presence of good solvents	↓ Beads
		↑ in presence of nonsolvents	↑ Beads
	Volatility	↓	Smooth nanofibers
			↓ Porosity
		↑	Porous nanofibers
			↑ Porosity

in addition to the solution flow rate (Huang et al. 2003; Ladd et al. 2008; Pham et al. 2006; Reneker and Chun 1996).

Collector Conductivity, Design, Structure, and Movement

The orientation and the properties of the received electrospun nanofibers depend mainly on the design, movement pattern, and composition of the used collector. Smooth nanofibers are obtained upon using different shapes of the collector. However, nanofibers orientation can be controlled according to the geometry and way of movement of the collector (Pham et al. 2006). This was reported upon developing nanofibers using different patterns of collectors with different geometries and movements as shown in Fig. 3 (Zander 2013).

It was also reported that using stationary plate collector would produce unaligned nanofibers with totally random orientation, while parallel conductive rods are able to produce highly aligned electrospun nanofibers due to the gap present between the two rods that is able to suspend the electrospun nanofibers between the two rods (Chew et al. 2005; Pham et al. 2006; Zander 2013). Highly oriented nanofibers can also be produced through using either revolving discs or revolving drums (Teo and Ramakrishna 2006) (Fig. 4). Using a grid of conductive cross-linked network would lead to the production of nanofibers networks with different patterns either at micro- or macroscale according to the used grid pattern (Ner et al. 2009; Wang et al. 2009; Zhang et al. 2009). In addition, it was found that using those porous grids would produce less packed nanofibrous membranes when

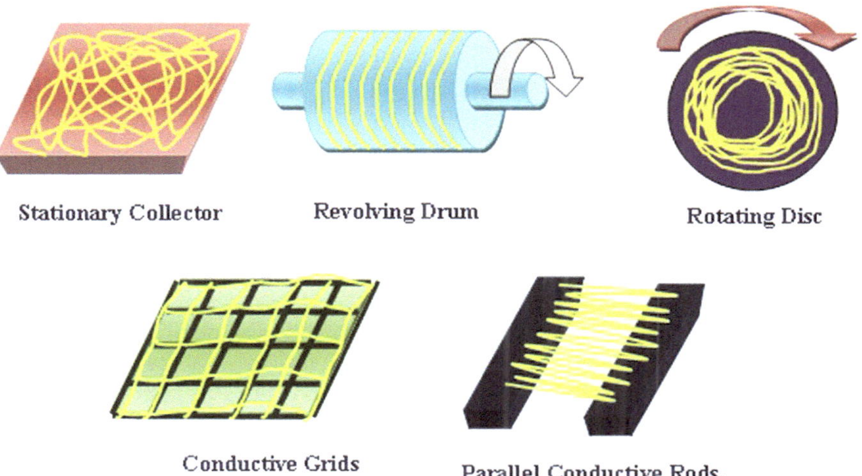

Fig. 3 Different patterned electrospinning collectors. Adopted with modification from Zander (2013)

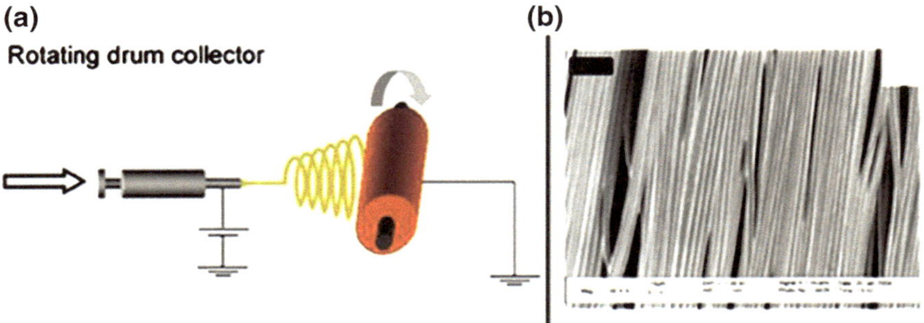

Fig. 4 Modified electrospinner. **a** Rotating drum collector, and **b** the produced aligned nanofibers. Adopted from Pham et al. (2006)

compared to those obtained upon using stationary solid collector (Liu and Hsieh 2002).

It was also found that using other additional materials with the collector during electrospinning process has an impact on the properties of the resulting nanofibers. For instance, it was reported that using water reservoir would lead to the shrinkage of hydrophobic nanofibers made of polymers such as PLLA and PGA, while collecting them in methanol led to their swelling (Kim et al. 2005).

Conductivity of the used collector was found to have an influence on the packing density of the obtained nanofibers. For instance, the highly conductive collectors would be able to withdraw and disseminate the charges out of the produced nanofibers leading to increasing nanofibers packing density. On the other hand, less conductive collectors would lead to charge withdrawal failure out of the nanofibers, nanofibers repulsion, and consequently formation of more porous nanofibrous mats (Liu and Hsieh 2002).

Electric Field Strength (Applied Voltage)

The effect of the applied voltage as a processing parameter was extensively studied using various polymers, mainly poly(ethylene oxide) (PEO), and it was found that the effect of the applied voltage can differ upon using different polymers. Generally, it was reported that in case of most polymers, a beaded-free polymer jet originates out of the spinneret tip at low electric field as long as it exceeds the critical electric field value which overcomes the surface tension of the polymer solution (Deitzel et al. 2001; Sill and von Recum 2008). On the other hand, upon increasing the applied voltage, the polymer solution jet starts to decrease in volume and move back toward the spinneret tip leading to formation of more beads. Excessive increase in the applied electric field would lead to the total disappearance of the Taylor cone generating beads intensively (Deitzel et al. 2001; Sill and von Recum 2008).

It was reported in several studies that increasing the applied voltage would increase the diameter of the produced nanofibers and generate multiple polymer jets at the same time during spinning of many polymers (Demir et al. 2002). These polymers include some natural polymers such as chitosan (Geng et al. 2005) and gelatin (Ki et al. 2005) in addition to synthetic polymers such as PDLA (Zong et al. 2002) and PEO (Deitzel et al. 2001; Fong et al. 1999). On the other hand, some other polymers tend to decrease their nanofibers' diameters upon their exposure to an increased applied voltage during spinning such as silk-like polymers containing some specific functionalities like bisophenol-polysulphone (Yuan et al. 2004) or fibronectin (Buchko et al. 1999).

Geometry, Style, and Movement of the Syringe Needle Tip

Spinneret needle has been modified in some setups to produce nanofibers with new properties. For instance, coaxial spinneret which consists of two capillary spinnerets one inside another has been developed for some certain purposes and applications (Fig. 5a). This modified spinneret is capable of producing hollow nanofibers (Fig. 5b) (Li and Xia 2004). Core–shell nanofibers can also be obtained using this modified coaxial spinneret through spinning two different and immiscible polymer solutions (Zander 2013) (Fig. 5c) or emulsion polymer solution (Fig. 5d).

Another modified spinneret was designed with multiple single jets in order to enhance nanofibers' spinning rate of PEO (Theron et al. 2005) or to prepare nanofibrous mats composed of two different and immiscible polymers such as PVA and cellulose acetate. A multiple jet composed of four tips was used in order to blend PVA and cellulose acetate nanofibers in different ratios (Pham et al. 2006). In another study, a collector moving transversely was used in front of a multiple jet composed of two tips in order to produce a mixture of electrospun PEO and polyurethane (PU) nanofibers and a uniform distribution of this mixture was achieved due to the motion of the collector (Kidoaki et al. 2005).

Distance Between Spinneret Tip and the Collector

The distance at which the collector is located relative to the spinneret tip is considered one of the critical parameters during electrospinning since it has an influence on nanofibers morphology and formation. Optimum sufficient distance should be maintained to guarantee formation of dry free-beaded nanofibers by reaching the target (collector) (Geng et al. 2005). Failure in reaching the optimum distance adjustment by either increasing or decreasing the distance would lead to formation of beaded nanofibers (Ki et al. 2005; Lee et al. 2004).

Solution Flow Rate

Few studies were performed to investigate the effect of the change in the solution flow rate on the morphology of the electrospun nanofibers. The results showed

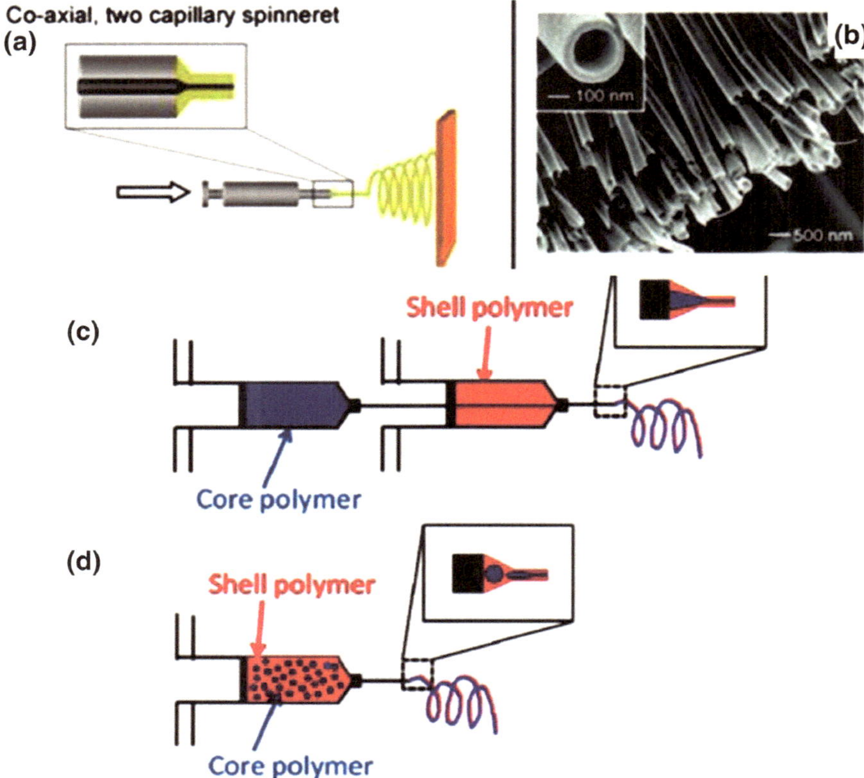

Fig. 5 **a** Coaxial spinneret of a modified electrospinner, **b** the obtained hollow nanofibers, **c** spinning of two immiscible polymer solutions in form of core–shell nanofibers using a coaxial spinneret, **d** spinning of polymer emulsion solution through single nozzle spinneret. Adopted with modification from Pham et al. (2006) and Zander (2013)

that decreasing the flow rate at which the polymer solution is pumped out of the syringe through the spinneret led to formation of nanofibers with narrower diameter and devoid of beaded regions when compared to those produced at higher flow rate (Zong et al. 2002). Appearance of many beads at higher flow rates can be attributed to the insufficient time required for the nanofibers to dry before their deposition on the collector surface (Jarusuwannapoom et al. 2005a; Yuan et al. 2004; Zhang et al. 2005; Zuo et al. 2005).

2.3.2 Ambient (Uncontrolled) Variables

Ambient parameters are variables with ability to affect the morphology and the formation of the electrospun nanofibers like the processing ones. However, unlike the processing ones, ambient parameters are almost impossible to be well controlled during the electrospinning process. These ambient parameters include the air velocity, the atmosphere humidity and the temperature measured at the time of electrospinning (Pham et al. 2006).

Air Velocity

Electrospinning was done in presence of vacuum which led to an increase in the abundant electric field and consequently an increase in the diameter of the produced nanofibers (Reneker and Chun 1996).

Atmosphere Temperature

A study was done to examine the effect of changing the temperature during conducting electrospinning of polyamine-6 solution. The temperature was raised from 25 up to 60 °C to detect the change in the morphology of the produced nanofibers. It was found that as the temperature increased, the viscosity of the polymer solution decreased and consequently the nanofibers' diameter decreased (Mit-uppatham et al. 2004).

Atmosphere Humidity

The effect of increasing humidity was studied during the electrospinning process of polystyrene (PS) nanofibers. It was reported that increasing humidity ended up with formation of nanofibers with superficial circular pores (Fig. 6). Moreover, extra humidity led to those pores coherence (Casper et al. 2004).

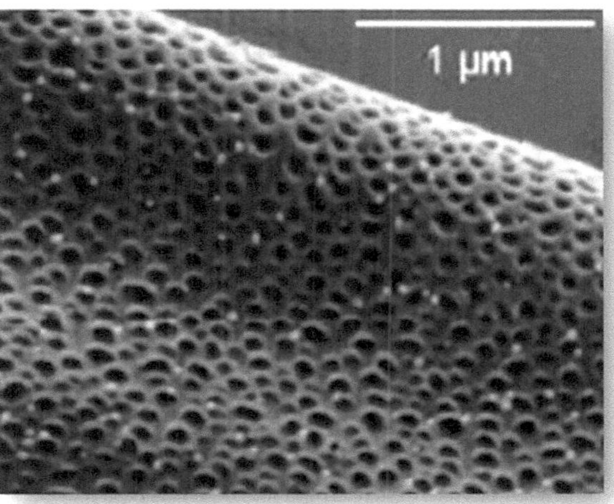

Fig. 6 Effect of high humidity on the surface of electrospun nanofibers. Adopted from Zander (2013)

2.3.3 Polymer Solution Variables

Polymer solution parameters are certain variables related to the physical properties of the polymer solution used for electrospinning nanofibers such as polymer concentration, polymer solution viscosity, polymer molecular weight, solution charge density, conductivity, volatility, surface tension, dielectric constant, and dipole moment. These variables are hard to be altered since changing one of those variables would consequently change some of the others. An example of that is the alteration in the polymer solution viscosity upon changing its conductivity (Pham et al. 2006).

Polymer Concentration (Solution Viscosity)

Polymer concentration determines also the solution viscosity. Polymer solution viscosity is considered to be a remarkable parameter since it highly affects both morphology and diameter of the produced nanofibers. It was found that the viscosity of the polymer solution used during electrospinning and the amount of beaded regions found in the corresponding electrospun nanofibers are inversely proportional. In other words, it was reported that as the polymer solution viscosity (polymer concentration) decreases, the number of beads increases and vice versa as illustrated in Fig. 7a (Huang et al. 2003). This was reported with a wide range of polymers which include natural polymers such as collagen, dextran and gelatin as well as synthetic polymers such as poly(DL-lactic acid) (PDLA), PEO, PLGA, PLLA, poly(methyl methacrylate) (PMMA) and PS (Deitzel et al. 2001; Fong et al. 1999; Gupta et al. 2005; Huang et al. 2001; Jarusuwannapoom et al. 2005b; Jiang et al. 2004; Ki et al. 2005; Kim et al. 2005a, b; Koski et al. 2004; Lee et al. 2004; Son et al. 2004; Zhang et al. 2005; Zong et al. 2002). For instance, the beads disappeared completely in polycaprolactone (PCL) nanofibers upon increasing the PCL solution concentration from 5 to 9 % with keeping all of the other variables constant (Fig. 7b) (Deitzel et al. 2001; Fong et al. 1999).

It was also observed that increasing the solution viscosity leads to elongation of the rounded beaded regions, then forming uniform free-beaded nanofibers (Fig. 8) (Zander 2013).

However, it was reported that exaggerated increase in solution viscosity would lead to complete drying of the first polymer droplet at the tip of the spinneret leading to prevention of evolving of the polymer jet and finally blocking the electrospinning process (Demir et al. 2002; Duan et al. 2004; Zong et al. 2002).

Moreover, it was reported that each polymer solution should be prepared with certain optimum concentration and viscosity in order to have good electrospinnability. For instance, in the case of polyethylene oxide (PEO), the value of $(\eta)c$ has to exceed 10, where (η) is the solution intrinsic viscosity and (c) is the polymer concentration. In the case of PVA, it will show good spinnability as long as its $(\eta)c$ exceeds only five (Koski et al. 2004; Son et al. 2004).

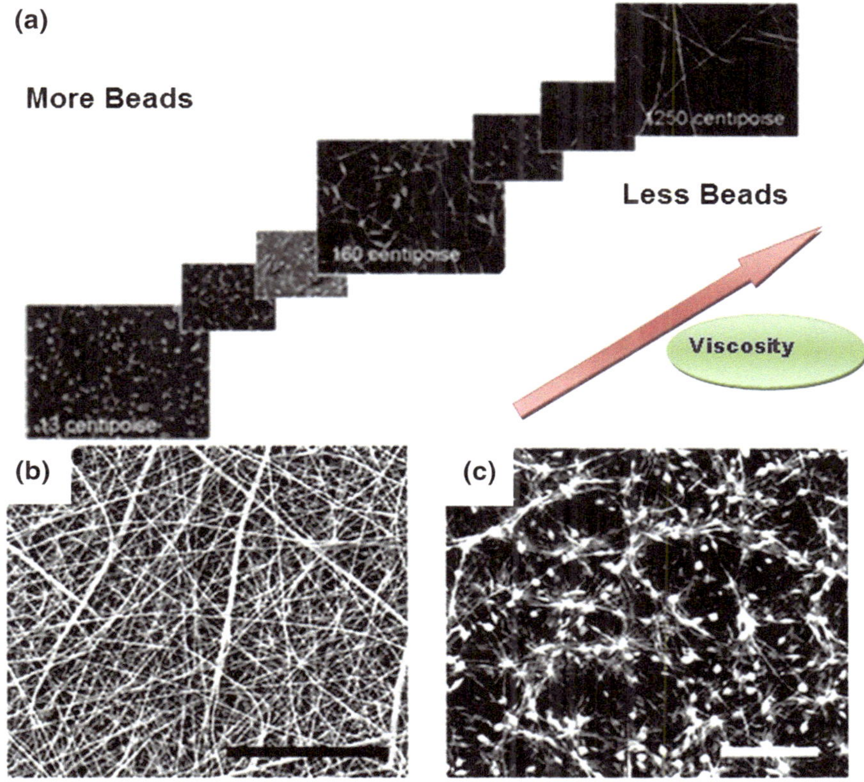

Fig. 7 **a** SEM micrographs showing the inverse relationship between the polymer solution viscosity used during electrospinning and the amount of beaded regions detected in the corresponding electrospun nanofibers, **b**, **c** PCL electrospun nanofibers made from 9 to 5 % PCL solutions, respectively. Adopted from Huang et al. (2003) and Pham et al. (2006)

Polymer Molecular Weight

The effect of polymer molecular weight on the morphology of the produced nanofibers was studied using different polymer systems (Pham et al. 2006). Acetic acid solutions of chitosan with two different molecular weights (30 and 398 KDa) were electrospun separately. All of the other variables were kept constant in order to investigate the effect of polymer molecular weight change solely. It was found that 30 KDa chitosan solution produced highly beaded brittle nanofibers while 398 KDa chitosan solution yielded rugged electrospun nanofibers (Geng et al. 2005). Spinning of polyamide-6 with different molecular weights led to the formation of nanofibers with different diameters (Mit-uppatham et al. 2004). PMMA was synthesized with wide range of molecular weights from 12.47 up to 365.7 KDa. Each molecular weight was dissolved in dimethylformamide separately in order to be electrospun.

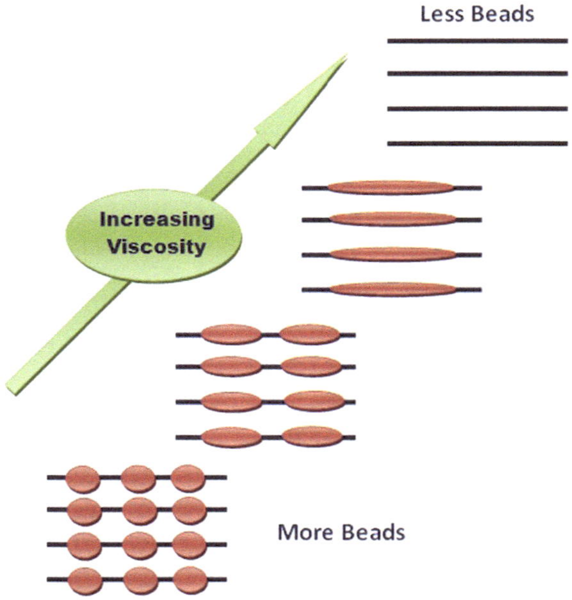

Fig. 8 Schematic diagram illustrating the disappearance of the beaded regions with increasing viscosity of polymer solution. Adopted from Zander 2013

The spinning parameters were fixed for all of the molecular weights to be: (i) 3 ml/hr flow rate, (ii) 15 cm distance between the spinneret and the collector, and (iii) 10 keV applied voltage. It was observed that as the molecular weight increases, the number of beads decreases and the yielded nanofibers tend to be more uniform (Gupta et al. 2005). This may be attributed to the increase of the viscosity of the polymer solution upon increasing the molecular weight of the polymer dissolved in the solution that would consequently lead to decreasing the amount of beads in nanofibers as discussed previously (Mit-uppatham et al. 2004; Pham et al. 2006).

Polymer Solution Charge Density (Solution Conductivity)

It was reported that less beaded and more uniform nanofibers could be obtained through increasing the conductivity of the used polymer solution during electrospinning (Fong et al. 1999; Huang et al. 2001; Jiang et al. 2004; Kim et al. 2005; Mit-uppatham et al. 2004; Zong et al. 2002; Zuo et al. 2005). Three approaches were applied to increase the polymer solution conductivity through addition of: (a) alcohols, (b) salts, or (c) surfactants (Pham et al. 2006).

(a) *Addition of alcohols*
It was reported that addition of alcohol to poly(hydroxybutrate-co-valerate) (PHBV) solution has increased its conductivity and consequently the uniformity of the produced nanofibers. On the other hand, reduction of polymer

solution conductivity by addition of certain substances like tetrachloromethane led to formation of highly beaded electrospun nanofibers (Zuo et al. 2005).

(b) *Addition of salts*

It was observed that addition of salts to various polymer solutions like collagen type I-PEO, PEO, PDLA, polyamide-6, and poly(acrylic acid) (PAA) led to increasing the solution conductivity and consequently formation of more uniform nanofibers (Fong et al. 1999; Huang et al. 2001; Kim et al. 2005a, b; Mit-uppatham et al. 2004; Zong et al. 2002). In another study, PLLA solution conductivity was increased by addition of pyridium formiate which showed a considerable reduction in the amount of the beads within the produced nanofibers. The volatility of this added salt is considered to be a great advantage since it will be evaporated without affecting the properties of the produced nanofibers (Zeng et al. 2003).

(c) *Addition of surfactants*

In a study done by Lin et al. (2004), several surfactants were added separately to PS solutions and examined for their effect on the produced nanofibers. It was noted that only small amount, not exceeding 10^{-6} M of cationic ones such as dodecyltrimethylammonium bromide and tetrabutylammonium chloride caused inhibition of beads formation. However, adding nonionic surfactants such as TritonX-405 to PS solution did not inhibit beads formation within the electrospun nanofibers. Hence, it was concluded that increasing the polymer solution conductivity would help in decreasing the beads formed while increasing the nanofibers uniformity.

Dielectric Constant and Dipole Moment

It is hard to isolate and study the effect of dielectric and dipole moment only on fiber formation. Therefore, only few experiments have been done to investigate the effect of this parameter.

PS solutions were prepared using 18 different solvents then electrospun to be investigated. It was observed that solutions prepared using solvents with higher dipole moments have produced more uniform nanofibers (Jarusuwannapoom et al. 2005a). In addition, it was reported that polymer solutions prepared using solvents with higher dipole moments led to a higher yield of nanofibers compared to that prepared using solvents with lower dipole moments (Wannatong et al. 2004).

2.3.4 Surface Tension

Different amounts of triethylbenzyl ammonium chloride were added to PHBV solutions in order to prepare polymer solutions with different surface tension and same conductivity. It was noted that change in the surface tension has an impact on bead formation (Zuo et al. 2005). It was also observed that there is no an obvious correlation between the surface tension of the polymer solution and the uniformity of the produced nanofibers. For instance, alcohol was added to each of PEO and PVA solutions to decrease the solution surface tension.

Although the surface tension was lowered in both solutions, it was found that PEO nanofibers showed decreased amount of beadings while PVA showed increased amount of beadings within the electrospun nanofibers (Fong et al. 1999; Zhang et al. 2005). This may be attributed to the fact that alcohol is a solvent for PEO but a nonsolvent for PVA (Pham et al. 2006).

Another example was studied where polydimethyl siloxane was added into PU urea solutions. However, it was found that changing surface tension of the solution in this case has no effect on the produced nanofibers (Demir et al. 2002).

Solution Volatility

Selection of appropriate solvent for the polymer to be electrospun is a critical step. The selected solvent should be volatile enough to evaporate rapidly during the spinning process allowing complete drying of the nanofibers upon reaching the collector. During the journey of the polymer jet in the atmosphere from the tip of the spinneret till the surface of the collector, a phase separation takes place. This phase separation is highly controlled by the degree of volatility of the used solvent (Sill and von Recum, 2008).

It was found that the degree of the solvent volatility has a great influence on the nanofibers morphology. For instance, PS was dissolved in several solvent systems composed of dimethylformamide (DMF) and tetrahydrofuran (THF) in different ratios. Upon using 100 % THF (highly volatility) as a solvent, highly porous nanofibers were formed with very high surface area due to the high porosity. On the contrary, using 100 % DMF (low volatility) as a solvent caused the formation of very smooth nanofibers completely devoid of microtexture properties. It was also observed that between the two extremities, as the ratio of DMF/THF increased, i.e., as the volatility of the solvent system decreased, the pore size increased while its depth decreased leading to the decrease in the pore density of the obtained nanofibers (Megelski et al. 2002).

3 Applications of Eco-friendly Electrospun Polymeric Nanofibers

Nanofibers exhibit diverse mechanical, physical, and chemical properties, and therefore they have been used as potential candidates in various applications. For instance, several US patents reported the use of nanofibers in a wide range of biomedical applications. In addition, nanofibers were incorporated in some other applications like composite formation, electromagnetic shielding and liquid crystal devices (Fig. 9) (Huang et al. 2003).

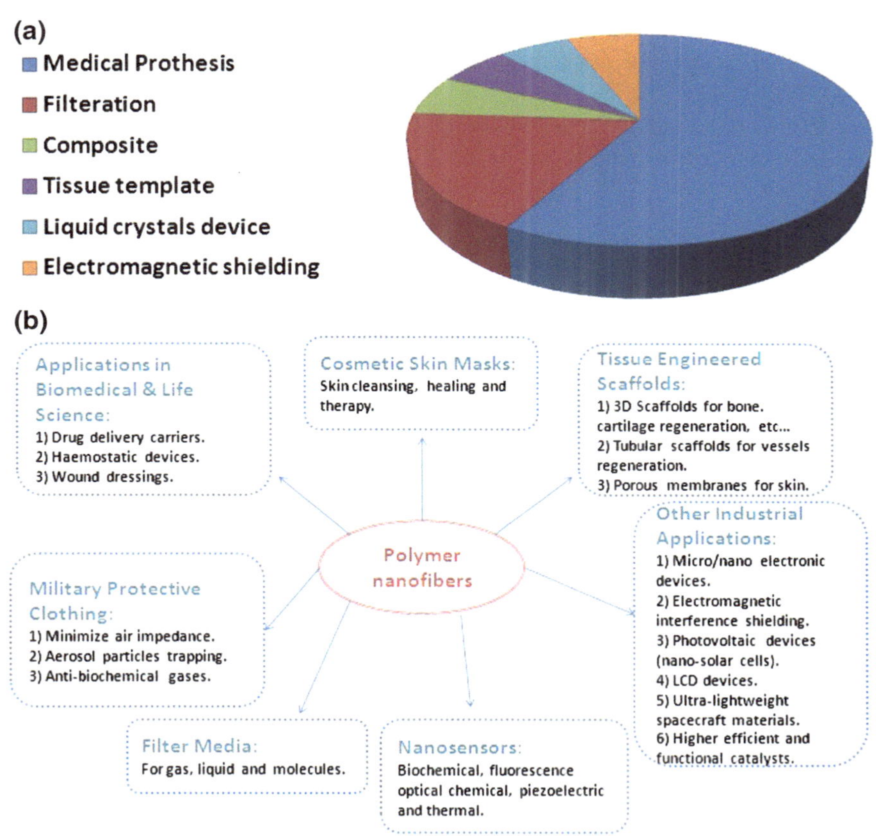

Fig. 9 **a** Pie chart representing electrospun nanofibers applications in US patents, and **b** different applications of electrospun polymer nanofibers. Adopted with modifications from Huang et al. (2003)

3.1 Dermal Reconstruction and Wound Healing

Skin is one of the most body tissues susceptible to numerous burns and injuries that lead to large surface areas of wounds needed to be covered by skin substitutes. The primary traditional approach is autologous skin grafting. However, this procedure demands several grafting steps to cover all of the injured or burnt areas in addition to probability of morbidity of donor sites. In addition, in some complicated patients' cases, donor sites may not be available (Ladd et al. 2008).

Another traditional approach is using skin substitutes. However, this approach showed several drawbacks that include: (i) poor compatibility with the patient tissues, (ii) scar formation, and (iii) wound contractions (Zhong et al. n.d.).

One of the most promising approaches to be used instead of skin substitutes is the fabrication of tissue engineered scaffolds to be used as wound dressings. These wound dressings can be designed to possess the capability to enhance skin regeneration and cutaneous wound healing. Tissue engineered scaffolds would be more advantageous over the traditional skin substitutes due to lack of previously mentioned drawbacks of the traditional approaches (Zhong et al. n.d.). In other words, scientists considered fabrication of 3D tissue engineered scaffolds as a promising approach since they have several advantages that include: (i) possessing three-dimensional structural design resembling the natural ECM within the human body (ii) prompting cell adhesion, differentiation and growth, (iii) acting as a wound dressing forming a physical barrier between the wounds and the external environment in order to avoid any external infection and (iv) acting as a carrier for both skin fibroblasts and keratinocytes enhancing dermal regeneration rate (Jayakumar et al. 2011; Ma 2004; Nair and Laurencin 2007; Zhong et al. n.d.).

3.1.1 Natural Polymers-Based Nanofibers for Dermal Reconstruction and Wound Healing

Electrospinning of biodegradable eco-friendly polymer nanofibers has been used as a promising approach to fabricate scaffolds for dermal construction and wound healing (Priya et al. 2008). Several studies have been done in vitro in order to test the capability of the fabricated electrospun scaffolds to promote skin tissues regeneration (Kumbar et al. 2008; Powell and Boyce 2009; Shalumon et al. 2009; Sun et al. 2005; Yang et al. 2009).

Chitosan-based biomaterials were reported to be one of the most promising materials to be used in fabricating electrospun scaffolds for skin tissue regeneration (Jayakumar et al. 2011). Chitosan-based nanofibers proved their ability to enhance mouse fibroblasts attachment, growth, and proliferation (Shalumon et al. 2009). In addition, chitosan containing quaternary ammonium groups exhibited a potent effect against both bacteria and fungi. This was attributed to the attacking ability of these charged cationic moieties toward the bacterial cytoplasmic membrane. Electrospun nanofibrous membranes containing chitosan also showed antibacterial activity against both gram positive and gram negative bacteria (Muzzarelli 2009; Rabea et al. 2003b; Sun 2011). Chitosan was tested for its biocompatibility by assessing its cytotoxicity and proved to be nontoxic biomaterial (Mi et al. 2002; Shalumon et al. 2009).

Since chitosan is soluble in aqueous media which has low volatility, it is difficult to electrospin chitosan solely. Therefore, researchers tended to mix chitosan with other highly spinnable biocompatible polymers in order to obtain electrospun nanofibers containing chitosan. For instance, chitosan was blended with polyvinyl alcohol (PVA) and electrospun successfully to be tested for their ability to enhance fibroblasts growth. The produced nanofibers were cross-linked to decrease their solubility before being tested for their biocompatibility and tissue regeneration effect. They showed promising results to be used as scaffolds and wound dressings (Powell and Boyce 2009; Liu and Kim 2012).

Some electrospun chitosan-based nanofibers were reinforced with silver nanoparticles in order to increase their activity as antibacterial wound dressings against bacteria. For example, chitosan/gelatin nanofibers were electrospun successfully after silver nanoparticles have been added to the polymer solution before spinning (Zhuang et al. 2010). Another example was reinforcing chitosan/PVA nanofibers with silver nanoparticles. The antibacterial activity was tested for the obtained nanofibers and it was observed that this antibacterial activity of the nanofibers was highly elevated after reinforcement with silver nanoparticles (Hang et al. 2010).

Collagen is another natural spinnable polymer which has shown good biocompatibility, biodegradability and tissue regeneration activity. However, several studies added other biocompatible polymers to collagen in order to enhance collagen properties within the produced nanofibers. For instance, a study was done by preparing collagen/PCL nanofibers in order to enhance the mechanical properties of the produced nanofibers in addition to limiting the nanofibers size reduction after being wet. It was found that the ideal scaffolds were those that contained 10 % PCL since they showed good results for both fibroblasts and keratinocytes proliferation. In addition, the mechanical properties have been enhanced as well as the dermal and epidermal differentiation (Powell and Boyce 2009).

Moreover, PLGA and collagen I were electrospun successfully together in order to fabricate skin regeneration scaffolds compromising both the good mechanical properties of PLGA and the dermal and epidermal cells proliferation activity of collagen. This scaffold was characterized by being highly porous which was mandatory for migration and proliferation of fibroblasts within the scaffold (Yang et al. 2009).

3.1.2 Synthetic Polymers-Based Nanofibers for Dermal Reconstruction and Wound Healing

Several synthetic biodegradable and biocompatible polymers have been successfully electrospun and tested as scaffolds for skin regeneration and wound healing. PS was electrospun and tested for its skin regeneration activity using both fibroblasts and keratinocytes. The air–liquid culture technique; a more commonly utilized technique for skin regeneration testing, was used instead of submerged culture technique. It was observed that upon culturing the tested cells with electrospun PS nanofibrous mats, the cells tended immediately to organize themselves within the scaffold without any additional interference (Sun et al. 2005).

Electrospun PLGA nanofibers were fabricated and tested for their efficiency to be used as wound dressings. The study was concerned with observing the effect of change in fiber diameter of the cocultured fibroblasts. The range of the tested nanofibers' diameter was from 150 nm up to 6000 nm. The gene expression, morphology as well as proliferation activity of the fibroblasts were tested with all of the different diameters. It was observed that fibroblasts had the highest proliferation rate as well as the best spreading morphology when they were cocultured with PLGA nanofibers ranging from 350 to 1100 nm. In addition, these range of nanofibers' diameter showed the highest collagen III expression (Kumbar et al. 2008).

3.2 Tissue Engineering

3.2.1 Natural Polymers-Based Nanofibers for Tissue Engineering

It is preferable to use natural polymers or at least to incorporate them within the fabricated nanofibrous scaffolds in order to promote the scaffolds biocompatibility and biofunctionality (Almany and Seliktar 2005).

Collagen has been extensively used and tested with various types of cells. Collagen showed promising results and high biocompatibility in several studies. This is attributed to already presence of collagen as the chief constituent of the ECM inside the human body (Matthews et al. 2002; Torres-Giner et al. 2009).

Chitosan which is the second abundant polysaccharide, is considered to be one of the most attractive polymers to be electrospun for tissue regeneration field (Jayakumar et al. 2011; Sun 2011). Chitosan has proven to be efficient in stimulating the growth of diverse types of cells such as endothelial cell, fibroblasts, hepatocytes, neural cells and others (Jafari et al. 2011; Jiankang et al. 2009; Nisbet et al. 2008; Qian et al. 2011).

Gelatin is naturally derived polymer that can be obtained through acidic or alkaline hydrolysis of collagen. Due to its high hydrophilicity, gelatin has the ability to enhance both cell proliferation and fluid diffusion within the fabricated scaffold (Lee et al. 2012; Sundar et al. 2010). Gelatin has been used as a cell proliferation enhancing material in numerous nanofibrous systems (Jafari et al. 2011; Li et al. 2006; Zhuang et al. 2010). Gelatin has been also blended successfully with several polymers to fabricate tissue engineered nanofibrous scaffolds. An example, is its incorporation with chitosan to obtain chitosan/gelatin hybrid nanofibers which were tested successfully with variety of different cells like fibroblasts, dermal cells, hepatocytes, and porcine endothelial cells (Dhandayuthapani et al. 2010; Jafari et al. 2011; Jiankang et al. 2009; Qian et al. 2011).

Gelatin was also added to several synthetic biocompatible polymers in order to enhance their proliferation features. For instance, gelatin was added to PCL to obtain gelatin/PCL-blended nanofibers. The nanofibers were tested successfully to proliferate Schwann's cells (Gupta et al. 2009). Another example is blending gelatin with polyaniline (PAN). Gelatin/PAN nanofibers were reported to enhance myocardial proliferation (Li et al. 2006).

3.2.2 Synthetic Biodegradable Polymers-Based Nanofibers for Tissue Engineering

Numerous synthetic polymers were reported to be biodegradable to some considerable extent. In other words, they can be broken down by time either enzymatically or hydrolytically. Almost all of the synthetic polymers are biologically inert (Nair and Laurencin 2007). However, they have high capability to be tailored and chemically modified to possess some biological functionalities (Nair and Laurencin

2007; Teo and Ramakrishna 2006). Besides, most of the synthetic polymers are degraded hydrolytically not enzymatically unlike the naturally derived polymers. Therefore, synthetic polymers are preferred to be used as tissue engineered implants (Katti et al. 2002).

Although several synthetic polymers have been reported to be electrospun successfully and tested with different animal cells, not all of them were approved by FDA as safe biomaterials for biomedical applications. The most commonly known ones are polylactic acid (PLA), polyglycolic acid (PGA), and their copolymer polylactic-co-glycolic acid (PLGA) (Teo and Ramakrishna 2006).

Electrospun PLGA nanofibrous mats were reported to be one of the most perfect scaffolds for tissue regeneration owing to their high porosity that reached 90 % (Li et al. 2002). Electrospun PLGA nanofibers reinforced with $CaCl_2$ and H_3PO_4 were successfully fabricated to obtain a mineralized biomaterial suitable for bone tissue regeneration (Liao et al. 2008). PLGA nanofibers showed better results for chondrocytes maintenance, proliferation and viability when compared to PLGA membranes (Shin et al. 2006).

PLA and its copolymers have been used in enhancing bone tissue regeneration (Jang et al. 2009). For instance, PLA has been blended with either collagen or elastin to enhance mechanical properties of nanofibrous scaffolds used for blood vessels tissue regeneration (Lee et al. 2008). In addition, the osteogenic differentiation effect of PLA has been reported in previous studies (Badami et al. 2006).

PCL is used mainly to support the tissue engineered scaffolds by increasing their mechanical properties. For instance, PCL is added to either collagen or gelatin to fabricate nanofibers enhancing neural cells growth (Gupta et al. 2009; Prabhakaran et al. 2009). PCL/collagen nanofibrous systems were also reported to be useful in enhancing skeletal cellular differentiation (Choi et al. 2008). PCL was reported to be one of the suitable polymers to be used in cardiac tissue regeneration due to its high mechanical properties (Sohier et al. 2014).

PU nanofibrous scaffolds have been reported to be highly resembling the natural connective tissues inside the human body. Therefore, they were tested for ligaments regeneration.

4 Conclusion

Electrospinning has been considered recently to be one of the most promising eco-friendly techniques to fabricate nanomaterials. It is cheap and facile technique. Besides, it is counted as an environmentally friendly technique since if all of the solvents used during electrospinning are eco-friendly, no harm can occur to the external environment.

Although electrospinning appear to be a highly sophisticated technique dealing with many parameters and variables, most of these variables can be manipulated in order to produce nanofibrous materials with certain nanofeatures according to the targeted application.

Modifications in spinneret nozzles and collectors have been done in order to produce nanofibers with more utilized features. For instance, coaxial spinnerets were developed to form core–shell nanofibers. Rotating discs and drums were developed in order to obtain oriented nanofibers, while meshes were used for obtaining nanofibrous patterns at the macroscale or microscale.

Electrospinning technique has proved to be efficient in nanofibers production for diverse applications such as biomedical applications, liquid crystal devices, nanofiltration, etc. However, it has been very promising technique for tissue engineered scaffolds fabrication. This reiterates the fact that through this technique, it is possible to fabricate scaffolds by resembling the natural ECM inside human body. The features of electrospun nanofibers can be manipulated so that: (i) the dimensions of nanofibers could resemble the natural tissue fibers dimensions in the body and (ii) the scaffolds porosity could be manipulated to mimic that of the natural tissues inside the body. High porosity is mandatory in scaffolds for better fluid perfusion and cell proliferation.

Moreover, electrospinning was found to be ideal for fabrication of wound dressings. In addition to possessing wound healing effect, the materials used in electrospun nanofibers could be manipulated to show antibacterial activity against several bacteria and fungi to prevent wound external infection. This can be done either through using polymers possessing antibacterial such as chitosan or by reinforcing the nanofibers with an antibacterial agent such as silver nanoparticles. Wound dressings fabricated by electrospinning did not possess the drawbacks of skin substitutes like scars formation and wound contractions.

References

Almany L, Seliktar D (2005) Biosynthetic hydrogel scaffolds made from fibrinogen and polyethylene glycol for 3D cell cultures. Biomaterials 26(15):2467–2477. doi:10.1016/j.biomaterials.2004.06.047

Silva A, Silva-Freitas É, Carvalho J, Pontes T, Araújo-Neto R, Silva K, Carriço A, Egito E (2012) Advances in applied biotechnology (M Petre, Ed). InTech. doi:10.5772/1096

Badami AS, Kreke MR, Thompson MS, Riffle JS, Goldstein AS (2006) Effect of fiber diameter on spreading, proliferation, and differentiation of osteoblastic cells on electrospun poly(lactic acid) substrates. Biomaterials 27(4):596–606. doi:10.1016/j.biomaterials.2005.05.084

Buchko CJ, Chen LC, Shen Y, Martin DC (1999) Processing and microstructural characterization of porous biocompatible protein polymer thin films. Polymer 40(26):7397–7407. doi:10.1016/S0032-3861(98)00866-0

Casper CL, Stephens JS, Tassi NG, Chase DB, Rabolt JF (2004) Controlling surface morphology of electrospun polystyrene fibers: effect of humidity and molecular weight in the electrospinning process. Macromolecules 37(2):573–578. doi:10.1021/ma0351975

Chew SY, Wen J, Yim EKF, Leong KW (2005) Sustained release of proteins from electrospun biodegradable fibers. Biomacromolecules 6(4):2017–2024. doi:10.1021/bm0501149

Choi JS, Lee SJ, Christ GJ, Atala A, Yoo JJ (2008) The influence of electrospun aligned poly (epsilon-caprolactone)/collagen nanofiber meshes on the formation of self-aligned skeletal muscle myotubes. Biomaterials 29(19):2899–2906. doi:10.1016/j.biomaterials.2008.03.031

Deitzel J, Kleinmeyer J, Harris D, Beck Tan N (2001) The effect of processing variables on the morphology of electrospun nanofibers and textiles. Polymer 42(1):261–272. doi:10.1016/S0032-3861(00)00250-0

Demir M, Yilgor I, Yilgor E, Erman B (2002) Electrospinning of polyurethane fibers. Polymer 43 (11):3303–3309. doi:10.1016/S0032-3861(02)00136-2

Desai TA (2000) Micro- and nanoscale structures for tissue engineering constructs. Med Eng Phys *22*(9):595–606. http://yadda.icm.edu.pl/yadda/element/bwmeta1.element.elsevier-8b9c2a18-0fa6-3146-8a5e-f0b5a004a799

Dhandayuthapani B, Krishnan UM, Sethuraman S (2010) Fabrication and characterization of chitosan-gelatin blend nanofibers for skin tissue engineering. J Biomed Mater Res B Appl Biomater 94(1):264–272. doi:10.1002/jbm.b.31651

Dhandayuthapani B, Yoshida Y, Maekawa T, Kumar DS (2011) Polymeric scaffolds in tissue engineering application: a review, 2011(ii). doi:10.1155/2011/290602

Duan B, Dong C, Yuan X, Yao K (2004) Electrospinning of chitosan solutions in acetic acid with poly(ethylene oxide). J Biomater Sci Polym Ed 15(6):797–811. doi:10.1163/156856204774196171

Feng L, Li S, Li H, Zhai J, Song Y, Jiang L, Zhu D (2002) Super-Hydrophobic surface of aligned polyacrylonitrile nanofibers. Angew Chem 114(7):1269–1271. doi:10.1002/1521-3757(20020402)114:7<1269:AID-ANGE1269>3.0.CO;2-E

Fong H, Chun I, Reneker D (1999) Beaded nanofibers formed during electrospinning. Polymer 40 (16):4585–4592. doi:10.1016/S0032-3861(99)00068-3

Geng X, Kwon O-H, Jang J (2005) Electrospinning of chitosan dissolved in concentrated acetic acid solution. Biomaterials 26(27):5427–5432. doi:10.1016/j.biomaterials.2005.01.066

Glicklis R, Shapiro L, Agbaria R, Merchuk JC, Cohen S (2000) Hepatocyte behavior within three-dimensional porous alginate scaffolds. Biotechnol Bioeng 67(3):344–53. http://www.ncbi.nlm.nih.gov/pubmed/10620265

Greiner A, Wendorff JH (2007) Electrospinning: a fascinating method for the preparation of ultrathin fibers. Angew Chem Int Ed Engl 46(30):5670–5703. doi:10.1002/anie.200604646

Gupta D, Venugopal J, Prabhakaran MP, Dev VRG, Low S, Choon AT, Ramakrishna S (2009) Aligned and random nanofibrous substrate for the in vitro culture of Schwann cells for neural tissue engineering. Acta Biomater 5(7):2560–2569. doi:10.1016/j.actbio.2009.01.039

Gupta P, Elkins C, Long TE, Wilkes GL (2005) Electrospinning of linear homopolymers of poly (methyl methacrylate): exploring relationships between fiber formation, viscosity, molecular weight and concentration in a good solvent. Polymer 46(13):4799–4810. doi:10.1016/j.polymer.2005.04.021

Hang AT, Tae B, Park JS (2010) Non-woven mats of poly(vinyl alcohol)/chitosan blends containing silver nanoparticles: fabrication and characterization. Carbohydr Polym 82(2):472–479. doi:10.1016/j.carbpol.2010.05.016

Huang L, Nagapudi K, P.Apkarian R, Chaikof EL (2001) Engineered collagen–PEO nanofibers and fabrics. J Biomater Sci, Polym Ed 12(9):979–993. doi:10.1163/156856201753252516

Huang Z-M, Zhang Y-Z, Kotaki M, Ramakrishna S (2003) A review on polymer nanofibers by electrospinning and their applications in nanocomposites. Compos Sci Technol 63(15):2223–2253. doi:10.1016/S0266-3538(03)00178-7

Jafari J, Emami SH, Samadikuchaksaraei A, Bahar MA, Gorjipour F (2011) Electrospun chitosan-gelatin nanofiberous scaffold: fabrication and in vitro evaluation. Bio-Med Mater Eng 21 (2):99–112. doi:10.3233/BME-2011-0660

Jang J-H, Castano O, Kim H-W (2009) Electrospun materials as potential platforms for bone tissue engineering. Adv Drug Deliv Rev 61(12):1065–1083. doi:10.1016/j.addr.2009.07.008

Jarusuwannapoom T, Hongrojjanawiwat W, Jitjaicham S, Wannatong L, Nithitanakul M, Pattamaprom C, Supaphol P (2005a) Effect of solvents on electro-spinnability of polystyrene solutions and morphological appearance of resulting electrospun polystyrene fibers. Eur Polym J 41(3):409–421. doi:10.1016/j.eurpolymj.2004.10.010

Jarusuwannapoom T, Hongrojjanawiwat W, Jitjaicham S, Wannatong L, Nithitanakul M, Pattamaprom C, Supaphol P (2005b) Effect of solvents on electro-spinnability of polystyrene solutions and morphological appearance of resulting electrospun polystyrene fibers. Eur Polym J 41(3):409–421. doi:10.1016/j.eurpolymj.2004.10.010

Jayakumar R, Menon D, Manzoor K, Nair SV, Tamura H (2010) Biomedical applications of chitin and chitosan based nanomaterials—a short review. Carbohydr Polym 82(2):227–232. doi:10.1016/j.carbpol.2010.04.074

Jayakumar R, Prabaharan M, Sudheesh Kumar PT, Nair SV, Tamura H (2011) Biomaterials based on chitin and chitosan in wound dressing applications. Biotechnol Adv 29(3):322–337. doi:10.1016/j.biotechadv.2011.01.005

Jiang H, Fang D, Hsiao BS, Chu B, Chen W (2004) Optimization and characterization of dextran membranes prepared by electrospinning. Biomacromolecules 5(2):326–333. doi:10.1021/bm034345w

Jiankang H, Dichen L, Yaxiong L, Bo Y, Hanxiang Z, Qin L, Yi L (2009) Preparation of chitosan-gelatin hybrid scaffolds with well-organized microstructures for hepatic tissue engineering. Acta Biomater 5(1):453–461. doi:10.1016/j.actbio.2008.07.002

Katti D, Lakshmi S, Langer R, Laurencin C (2002) Toxicity, biodegradation and elimination of polyanhydrides. Adv Drug Deliv Rev 54(7):933–961. doi:10.1016/S0169-409X(02)00052-2

Ki CS, Baek DH, Gang KD, Lee KH, Um IC, Park YH (2005) Characterization of gelatin nanofiber prepared from gelatin–formic acid solution. Polymer 46(14):5094–5102. doi:10.1016/j.polymer.2005.04.040

Kidoaki S, Kwon IK, Matsuda T (2005) Mesoscopic spatial designs of nano- and microfiber meshes for tissue-engineering matrix and scaffold based on newly devised multilayering and mixing electrospinning techniques. Biomaterials 26(1):37–46. doi:10.1016/j.biomaterials.2004.01.063

Kim B, Park H, Lee S-H, Sigmund WM (2005a) Poly(acrylic acid) nanofibers by electrospinning. Mater Lett 59(7):829–832. doi:10.1016/j.matlet.2004.11.032

Kim HS, Kim K, Jin HJ, Chin I-J (2005b) Morphological characterization of electrospun nano-fibrous membranes of biodegradable poly(l-lactide) and poly(lactide–co–glycolide). Macromol Symp 224(1):145–154. doi:10.1002/masy.200550613

Kolambkar YM, Dupont KM, Boerckel JD, Huebsch N, Mooney DJ, Hutmacher DW, Guldberg RE (2011) An alginate-based hybrid system for growth factor delivery in the functional repair of large bone defects. Biomaterials 32(1):65–74. doi:10.1016/j.biomaterials.2010.08.074

Koski A, Yim K, Shivkumar S (2004) Effect of molecular weight on fibrous PVA produced by electrospinning. Mater Lett 58(3–4):493–497. doi:10.1016/S0167-577X(03)00532-9

Kumbar SG, Nukavarapu SP, James R, Nair LS, Laurencin CT (2008) Electrospun poly(lactic acid-co-glycolic acid) scaffolds for skin tissue engineering. Biomaterials 29(30):4100–4107. doi:10.1016/j.biomaterials.2008.06.028

Ladd MR, Hill TK, Yoo JJ, Lee SJ (2008) Electrospun nanofibers in tissue engineering

Lee JS, Choi KH, Ghim H Do, Kim SS, Chun DH, Kim HY, Lyoo WS (2004) Role of molecular weight of atactic poly(vinyl alcohol) (PVA) in the structure and properties of PVA nanofabric prepared by electrospinning. J Appl Polym Sci 93(4):1638–1646. doi:10.1002/app.20602

Lee SJ, Liu J, Oh SH, Soker S, Atala A, Yoo JJ (2008) Development of a composite vascular scaffolding system that withstands physiological vascular conditions. Biomaterials 29 (19):2891–2898. doi:10.1016/j.biomaterials.2008.03.032

Lee EJ, Khan SA, Park JK, Lim K-H (2012) Studies on the characteristics of drug-loaded gelatin nanoparticles prepared by nanoprecipitation. Bioprocess Biosyst Eng 35(1–2):297–307. doi:10.1007/s00449-011-0591-2

Li D, Xia Y (2004) Direct fabrication of composite and ceramic hollow nanofibers by electrospinning. Nano Lett 4(5):933–938. doi:10.1021/nl049590f

Li W-J, Laurencin CT, Caterson EJ, Tuan RS, Ko FK (2002) Electrospun nanofibrous structure: a novel scaffold for tissue engineering. J Biomed Mater Res, 60(4):613–21. http://www.ncbi.nlm.nih.gov/pubmed/11948520

Li M, Guo Y, Wei Y, MacDiarmid AG, Lelkes PI (2006) Electrospinning polyaniline-contained gelatin nanofibers for tissue engineering applications. Biomaterials 27(13):2705–2715. doi:10.1016/j.biomaterials.2005.11.037

Liao S, Murugan R, Chan CK, Ramakrishna S (2008) Processing nanoengineered scaffolds through electrospinning and mineralization suitable for biomimetic bone tissue engineering. J Mech Behav Biomed Mater 1(3):252–260. doi:10.1016/j.jmbbm.2008.01.007

Lin T, Wang H, Wang H, Wang X (2004) The charge effect of cationic surfactants on the elimination of fibre beads in the electrospinning of polystyrene. Nanotechnology 15(9):1375–1381. doi:10.1088/0957-4484/15/9/044

Liu G, Ding J, Qiao L, Guo A, Dymov BP, Gleeson JT, Saijo K (1999) Polystyrene-block-poly(2-cinnamoylethyl methacrylate) nanofibers—preparation, characterization, and liquid crystalline properties. Chem Eur J 5(9):2740–2749. doi:10.1002/(SICI)1521-3765(19990903)5:9<2740:AID-CHEM2740>3.0.CO;2-V

Liu H, Hsieh Y-L (2002) Ultrafine fibrous cellulose membranes from electrospinning of cellulose acetate. J Polym Sci, Part B: Polym Phys 40(18):2119–2129. doi:10.1002/polb.10261

Liu Y, Kim H-I (2012) Characterization and antibacterial properties of genipin-crosslinked chitosan/poly(ethylene glycol)/ZnO/Ag nanocomposites. Carbohydr Polym 89(1):111–116. doi:10.1016/j.carbpol.2012.02.058

Ma PX (2004) Scaffolds for tissue fabrication. Mater Today 7(5):30–40. doi:10.1016/S1369-7021(04)00233-0

Madihally SV, Matthew HWT (1999) Porous chitosan scaffolds for tissue engineering. Biomaterials 20(12):1133–1142. doi:10.1016/S0142-9612(99)00011-3

Martin CR (1996) Membrane-based synthesis of nanomaterials. Chem Mater 8(8):1739–1746. doi:10.1021/cm960166s

Matthews JA, Wnek GE, Simpson DG, Bowlin GL (2002) Electrospinning of collagen nanofibers. Biomacromolecules 3(2):232–238. doi:10.1021/bm015533u

Megelski S, Stephens JS, Chase DB, Rabolt JF (2002) Micro- and nanostructured surface morphology on electrospun polymer fibers. Macromolecules 35(22):8456–8466. doi:10.1021/ma020444a

Mi F-L, Tan Y-C, Liang H-F, Sung H-W (2002) In vivo biocompatibility and degradability of a novel injectable-chitosan-based implant. Biomaterials 23(1):181–191. doi:10.1016/S0142-9612(01)00094-1

Mikos AG, Thorsen AJ, Czerwonka LA, Bao Y, Langer R, Winslow DN, Vacanti JP (1994) Preparation and characterization of poly(l-lactic acid) foams. Polymer 35(5):1068–1077. doi:10.1016/0032-3861(94)90953-9

Mit-uppatham C, Nithitanakul M, Supaphol P (2004) Ultrafine electrospun polyamide-6 fibers: effect of solution conditions on morphology and average fiber diameter. Macromol Chem Phys 205(17):2327–2338. doi:10.1002/macp.200400225

Mooney DJ, Baldwin DF, Suh NP, Vacanti JP, Langer R (1996) Novel approach to fabricate porous sponges of poly(d, l-lactic-co-glycolic acid) without the use of organic solvents. Biomaterials 17(14):1417–1422. doi:10.1016/0142-9612(96)87284-X

Murugan R, Ramakrishna S (2006) Nano-featured scaffolds for tissue engineering: a review of spinning methodologies. Tissue Eng 12(3):435–447. doi:10.1089/ten.2006.12.435

Muzzarelli RAA (2009) Chitins and chitosans for the repair of wounded skin, nerve, cartilage and bone. Carbohydr Polym 76(2):167–182. doi:10.1016/j.carbpol.2008.11.002

Nair LS, Laurencin CT (2007) Biodegradable polymers as biomaterials. Prog Polym Sci 32(8–9):762–798. doi:10.1016/j.progpolymsci.2007.05.017

Ner Y, Asemota C, Olson JR, Sotzing GA (2009) Nanofiber alignment on a flexible substrate: hierarchical order from macro to nano. ACS Appl Mater Interf 1(10):2093–2097. doi:10.1021/am900382f

Nerem RM (1992) Tissue engineering in the USA. Med Biol Eng Comput 30(4):CE8–CE12. doi:10.1007/BF02446171

Ni H, Zeng S, Wu J, Cheng X, Luo T, Wang W (2012) Cellulose nanowhiskers: preparation, characterization and cytotoxicity evaluation. 22:121–127. doi:10.3233/BME-2012-0697

Niklason LE (2001) Prospects for organ and tissue replacement. JAMA 285(5):573. doi:10.1001/jama.285.5.573

Nisbet DR, Crompton KE, Horne MK, Finkelstein DI, Forsythe JS (2008) Neural tissue engineering of the CNS using hydrogels. J Biomed Mater Res B Appl Biomater 87(1):251–263. doi:10.1002/jbm.b.31000

Ondarçuhu T, Joachim C (1998) Drawing a single nanofibre over hundreds of microns. Europhys Lett (EPL) 42(2):215–220. doi:10.1209/epl/i1998-00233-9

Pham QP, Sharma U, Mikos AG (2006) Electrospinning of polymeric nanofibers for tissue engineering applications: a review. Tissue Eng 12(5):1197–1211. doi:10.1089/ten.2006.12.1197

Powell HM, Boyce ST (2009) Engineered human skin fabricated using electrospun collagen-PCL blends: morphogenesis and mechanical properties. Tissue Eng Part A 15(8):2177–2187. doi:10.1089/ten.tea.2008.0473

Prabhakaran MP, Venugopal JR, Ramakrishna S (2009) Mesenchymal stem cell differentiation to neuronal cells on electrospun nanofibrous substrates for nerve tissue engineering. Biomaterials 30(28):4996–5003. doi:10.1016/j.biomaterials.2009.05.057

Priya SG, Jungvid H, Kumar A (2008) Skin tissue engineering for tissue repair and regeneration. Tissue Eng Part B, Rev 14(1):105–18. doi:10.1089/teb.2007.0318

Qian Y-F, Zhang K-H, Chen F, Ke Q-F, Mo X-M (2011) Cross-linking of gelatin and chitosan complex nanofibers for tissue-engineering scaffolds. J Biomater Sci Polym Ed 22(8):1099–1113. doi:10.1163/092050610X499447

Rabea EI, Badawy ME-T, Stevens CV, Smagghe G, Steurbaut W (2003a) Chitosan as antimicrobial agent: applications and mode of action. Biomacromolecules 4(6):1457–1465. doi:10.1021/bm034130m

Rabea EI, Badawy ME-T, Stevens CV, Smagghe G, Steurbaut W (2003b) Chitosan as antimicrobial agent: applications and mode of action. Biomacromolecules 4(6):1457–1465. doi:10.1021/bm034130m

Reneker DH, Chun I (1996) Nanometre diameter fibres of polymer, produced by electrospinning. Nanotechnology 7(3):216–223. doi:10.1088/0957-4484/7/3/009

Rinaudo M (2006) Chitin and chitosan: properties and applications. Prog Polym Sci 31(7):603–632. doi:10.1016/j.progpolymsci.2006.06.001

Sachlos E, Czernuszka JT (2003) Making tissue engineering scaffolds work. Review: the application of solid freeform fabrication technology to the production of tissue engineering scaffolds. Eur Cells Mater 5:29–39, discussion 39–40. http://www.ncbi.nlm.nih.gov/pubmed/14562270

Schoof H, Apel J, Heschel I, Rau G (2001) Control of pore structure and size in freeze-dried collagen sponges. J Biomed Mater Res 58(4):352–357. doi:10.1002/jbm.1028

Shalumon KT, Binulal NS, Selvamurugan N, Nair SV, Menon D, Furuike T, Jayakumar R (2009) Electrospinning of carboxymethyl chitin/poly(vinyl alcohol) nanofibrous scaffolds for tissue engineering applications. Carbohydr Polym 77(4):863–869. doi:10.1016/j.carbpol.2009.03.009

Shin HJ, Lee CH, Cho IH, Kim Y-J, Lee Y-J, Kim IA, Shin J-W (2006) Electrospun PLGA nanofiber scaffolds for articular cartilage reconstruction: mechanical stability, degradation and cellular responses under mechanical stimulation in vitro. J Biomater Sci Polym Ed 17(1–2):103–119. doi:10.1163/156856206774879126

Sill TJ, von Recum HA (2008) Electrospinning: applications in drug delivery and tissue engineering. Biomaterials 29(13):1989–2006. doi:10.1016/j.biomaterials.2008.01.011

Singha AS, Thakur VK, Mehta IK, Shama A, Khanna AJ, Rana RK, Rana AK (2009) Surface-modified Hibiscus sabdariffa fibers: physicochemical, thermal, and morphological properties evaluation. Int J Polym Anal Charact 14(8):695–711. doi:10.1080/10236660903325518

Singha AS, Thakur VK (2010) Mechanical, morphological, and thermal characterization of compression-molded polymer biocomposites. Int J Polym Anal Charact 15:87–97. doi:10.1080/10236660903474506

Siqueira G, Bras J, Dufresne A (2010) Cellulosic bionanocomposites: a review of preparation, properties and applications, (i):728–765. doi:10.3390/polym2040728

Sohier J, Carubelli I, Sarathchandra P, Latif N, Chester AH, Yacoub MH (2014) The potential of anisotropic matrices as substrate for heart valve engineering. Biomaterials 35(6):1833–1844. doi:10.1016/j.biomaterials.2013.10.061

Son WK, Youk JH, Lee TS, Park WH (2004) The effects of solution properties and polyelectrolyte on electrospinning of ultrafine poly(ethylene oxide) fibers. Polymer 45(9):2959–2966. doi:10.1016/j.polymer.2004.03.006

Steyaert I, Van der Schueren L, Rahier H, de Clerck K (2012) An alternative solvent system for blend electrospinning of polycaprolactone/chitosan nanofibres. Macromol Symp 321–322 (1):71–75. doi:10.1002/masy.201251111

Sun K (2011) Preparations, properties and applications of chitosan based nanofibers fabricated by electrospinning. Expr Polym Lett 5(4):342–361. doi:10.3144/expresspolymlett.2011.34

Sun T, Mai S, Norton D, Haycock JW, Ryan AJ, MacNeil S (2005) Self-organization of skin cells in three-dimensional electrospun polystyrene scaffolds. Tissue Eng 11(7–8):1023–1033. doi:10.1089/ten.2005.11.1023

Sundar S, Kundu J, Kundu SC (2010) Biopolymeric nanoparticles. Sci Technol Adv Mater 11 (1):014104. doi:10.1088/1468-6996/11/1/014104

Thakur VK, Kessler MR (2014a) Free radical induced graft copolymerization of ethyl acrylate onto SOY for multifunctional materials. Mater Today Commun. doi:10.1016/j.mtcomm.2014.09.003

Thakur VK, Kessler MR (2014b) Synthesis and characterization of AN–g–SOY for sustainable polymer composites. ACS Sustain Chem Eng 2:2454–2460. doi:10.1021/sc500473a

Tang Y, Su Y, Yang N, Zhang L, Lv Y (2014) Carbon nitride quantum dots: a novel chemiluminescence system for selective detection of free chlorine in water. Anal Chem. doi:10.1021/ac5005162

Teo WE, Ramakrishna S (2006) A review on electrospinning design and nanofibre assemblies. Nanotechnology 17(14):R89–R106. doi:10.1088/0957-4484/17/14/R01

Thakur VK, Singha AS (2010) Natural fibres-based polymers: part i—mechanical analysis of Pine needles reinforced biocomposites. Bull Mater Sci 33:257–264. doi:10.1007/s12034-010-0040-x

Thakur VK, Singha AS, Kaur I et al (2010a) Silane functionalization of Saccaharum cilliare fibers: thermal, morphological, and physicochemical study. Int J Polym Anal Charact 15:397–414. doi:10.1080/1023666X.2010.510106

Thakur VK, Singha AS, Mehta IK (2010b) Renewable resource-based green polymer composites: analysis and characterization. Int J Polym Anal Charact 15(3):137–146. doi:10.1080/10236660903582233

Thakur VK, Singha AS, Misra BN (2011) Graft copolymerization of methyl methacrylate onto cellulosic biofibers. J Appl Polym Sci 122(1):532–544. doi:10.1002/app.34094

Thakur VK, Singha AS, Thakur MK (2012a) Green composites from natural fibers: mechanical and chemical aging properties. Int J Polym Anal Charact 17(6):401–407. doi:10.1080/1023666X.2012.668665

Thakur VK, Singha AS, Thakur MK (2012b) Rapid synthesis of MMA grafted pine needles using microwave radiation. Polym-Plast Technol Eng 51:1598–1604. doi:10.1080/03602559.2012.721443

Thakur VK, Singha AS, Thakur MK (2013a) Natural cellulosic polymers as potential reinforcement in composites: physicochemical and mechanical studies. Adv Polym Technol 32:E427–E435. doi:10.1002/adv.21290

Thakur VK, Singha AS, Thakur MK (2013b) Fabrication and physico-chemical properties of high-performance pine needles/green polymer composites. Int J Polym Mater Polym Biomater 62:226–230. doi:10.1080/00914037.2011.641694

Thakur VK, Singha AS, Thakur MK (2013c) Ecofriendly biocomposites from natural fibers: mechanical and weathering study. Int J Polym Anal Charact 18:64–72. doi:10.1080/1023666X.2013.747246

Thakur VK, Singha AS, Thakur MK (2013d) Synthesis of natural cellulose-based graft copolymers using methyl methacrylate as an efficient monomer. Adv Polym Technol 32: E741–E748. doi:10.1002/adv.21317

Thakur VK, Thakur MK (2014a) Recent advances in graft copolymerization and applications of chitosan: a review. ACS Sustain Chem Eng 2:2637–2652. doi:10.1021/sc500634p

Thakur VK, Thakur MK (2014b) Processing and characterization of natural cellulose fibers/ thermoset polymer composites. Carbohydr Polym 109:102–117. doi:10.1016/j.carbpol.2014.03.039

Thakur and Thakur (2014c) Recent trends in hydrogels based on psyllium polysaccharide: a review. J Clean Prod 82:1–15. doi:10.1016/j.jclepro.2014.06.066

Thakur VK, Thakur MK, Gupta RK (2014a) Review: raw natural fiber-based polymer composites. Int J Polym Anal Charact 19(3):256–271. doi:10.1080/1023666X.2014.880016

Thakur VK, Thakur MK, Raghavan P, Kessler MR (2014b) Progress in green polymer composites from lignin for multifunctional applications: a review. ACS Sustain Chem Eng 2(5):1072–1092. doi:10.1021/sc500087z

Thakur VK, Vennerberg D, Madbouly SA, Kessler MR (2014c) Bio-inspired green surface functionalization of PMMA for multifunctional capacitors. Rsc Adv 4:6677–6684. doi:10.1039/c3ra46592f

Thakur VK, Thunga M, Madbouly SA, Kessler MR (2014d) PMMA–g–SOY as a sustainable novel dielectric material. RSC Adv 4:18240–18249. doi:10.1039/c4ra01894j

Thakur VK, Grewell D, Thunga M, Kessler MR (2014e) Novel composites from eco-friendly soy flour/SBS triblock copolymer. Macromol Mater Eng 299:953–958. doi:10.1002/mame.201300368

Theron A, Zussman E, Yarin AL (2001) Electrostatic field-assisted alignment of electrospun nanofibres. Nanotechnology 12(3):384–390. doi:10.1088/0957-4484/12/3/329

Theron SA, Yarin AL, Zussman E, Kroll E (2005) Multiple jets in electrospinning: experiment and modeling. Polymer 46(9):2889–2899. doi:10.1016/j.polymer.2005.01.054

Thomson RC, Yaszemski MJ, Powers JM, Mikos AG (1996) Fabrication of biodegradable polymer scaffolds to engineer trabecular bone. J Biomater Sci Polym Ed 7(1):23–38. doi:10.1163/156856295X00805

Torres-Giner S, Gimeno-Alcañiz JV, Ocio MJ, Lagaron JM (2009) Comparative performance of electrospun collagen nanofibers cross-linked by means of different methods. ACS Appl Mater Interfaces 1(1):218–223. doi:10.1021/am800063x

Vacanti JP, Langer R (1999) Tissue engineering: the design and fabrication of living replacement devices for surgical reconstruction and transplantation. Lancet 354:S32–S34. doi:10.1016/S0140-6736(99)90247-7

Wang Y, Wang G, Chen L, Li H, Yin T, Wang B, Yu Q (2009) Electrospun nanofiber meshes with tailored architectures and patterns as potential tissue-engineering scaffolds. Biofabrication 1 (1):015001. doi:10.1088/1758-5082/1/1/015001

Wannatong L, Sirivat A, Supaphol P (2004) Effects of solvents on electrospun polymeric fibers: preliminary study on polystyrene. Polym Int 53(11):1851–1859. doi:10.1002/pi.1599

Yang F, Murugan R, Ramakrishna S, Wang X, Ma Y-X, Wang S (2004) Fabrication of nano-structured porous PLLA scaffold intended for nerve tissue engineering. Biomaterials 25 (10):1891–1900. doi:10.1016/j.biomaterials.2003.08.062

Yang S, Leong KF, Du Z, Chua CK (2001) The design of scaffolds for use in tissue engineering. Part I. Traditional factors. Tissue Eng 7(6):679–689. doi:10.1089/107632701753337645

Yang Y, Zhu X, Cui W, Li X, Jin Y (2009) Electrospun composite mats of poly[(D, L-lactide) – co– glycolide] and collagen with high porosity as potential scaffolds for skin tissue engineering. Macromol Mater Eng 294(9):611–619. doi:10.1002/mame.200900052

Yuan X, Zhang Y, Dong C, Sheng J (2004) Morphology of ultrafine polysulfone fibers prepared by electrospinning. Polym Int 53(11):1704–1710. doi:10.1002/pi.1538

Zahedi P, Rezaeian I, Ranaei-Siadat S-O, Jafari S-H, Supaphol P (2009) A review on wound dressings with an emphasis on electrospun nanofibrous polymeric bandages. Polym Adv Technol n/a–n/a. doi:10.1002/pat.1625

Zander N (2013) Hierarchically structured electrospun fibers. Polymers 5(1):19–44. doi:10.3390/polym5010019

Zeng J, Haoqing H, Schaper A, Wendorff JH, Greiner A (2003) Poly-L-lactide nanofibers by electrospinning—influence of solution viscosity and electrical conductivity on fiber diameter and fiber morphology. E-Polymers 3(1):102–110. http://www.degruyter.com/view/j/epoly.2003.3.issue-1/epoly.2003.3.1.102/epoly.2003.3.1.102.xml

Zhang C, Yuan X, Wu L, Han Y, Sheng J (2005) Study on morphology of electrospun poly(vinyl alcohol) mats. Eur Polymer J 41(3):423–432. doi:10.1016/j.eurpolymj.2004.10.027

Zhang K, Wang X, Jing D, Yang Y, Zhu M (2009) Bionic electrospun ultrafine fibrous poly(L-lactic acid) scaffolds with a multi-scale structure. Biomed Mater (Bristol, England) 4(3):035004. doi:10.1088/1748-6041/4/3/035004

Zhong SP, Zhang YZ, Lim CT (n.d.). Tissue scaffolds for skin wound healing and dermal reconstruction. Wiley Interdisc Rev Nanomed Nanobiotechnol 2(5):510–25. doi:10.1002/wnan.100

Zhuang X, Cheng B, Kang W, Xu X (2010) Electrospun chitosan/gelatin nanofibers containing silver nanoparticles. Carbohydr Polym 82(2):524–527. doi:10.1016/j.carbpol.2010.04.085

Zong X, Kim K, Fang D, Ran S, Hsiao BS, Chu B (2002) Structure and process relationship of electrospun bioabsorbable nanofiber membranes. Polymer 43(16):4403–4412. doi:10.1016/S0032-3861(02)00275-6

Zuo W, Zhu M, Yang W, Yu H, Chen Y, Zhang Y (2005) Experimental study on relationship between jet instability and formation of beaded fibers during electrospinning. Polym Eng Sci 45(5):704–709. doi:10.1002/pen.20304

Soy Protein- and Starch-Based Green Composites/Nanocomposites: Preparation, Properties, and Applications

Rekha Rose Koshy, Siji K. Mary, Laly A. Pothan and Sabu Thomas

Abstract With the environmental appeal around the planet for a sustainable development, there is the need to develop new materials from renewable resources, which can be degraded in a short time in the environment, thereby maintaining the proper balance of the carbon cycle. Biopolymers from various natural botanical resources can act as a substitute for petroleum-based synthetic polymers because of their low cost, ease of availability, and biodegradability along with other organic wastes to soil humic materials. Materials which are biodegradable and fully sustainable are termed as "Green Composites". This development not only solves the white pollution problem but also stops the overdependence on petroleum products. Development of Green Composites made from soy protein and starch has been a great challenge for the scientific community, since these materials do not possess all the desirable characteristics of the synthetic polymers, being mostly often, highly hydrophilic and also presenting poor mechanical properties to be used as engineering's materials. Cellulose macro- and nano-fibers can be used as reinforcement in composite materials to enhance mechanical, thermal, and biodegradation properties of the composites. In this chapter we will be dealing mainly with the preparation, properties, and applications of cellulose fiber-reinforced green composites based on soy protein and starch.

Keywords Green composites · Biopolymers · Soy protein · Starch

R.R. Koshy · S.K. Mary · L.A. Pothan (✉)
Department of Chemistry, Bishop Moore College, Mavelikara, Kerala 690101, India
e-mail: lapothan@gmail.com

S. Thomas (✉)
School of Chemical Sciences, Mahatma Gandhi University, Kottayam, Kerala 686560, India
e-mail: sabuchathukulam@yahoo.co.uk

S. Thomas
International and Inter University Centre for Nanoscience and Nanotechnology, Mahatma Gandhi University, Kottayam, Kerala 686560, India

V.K. Thakur and M.K. Thakur (eds.), *Eco-friendly Polymer Nanocomposites*, Advanced Structured Materials 75, DOI 10.1007/978-81-322-2470-9_14

1 Introduction

Development of advanced polymer composite materials having superior mechanical properties have opened up new horizons in the engineering, medical, cosmetics, pharmaceutical, and chemical fields (Singha and Thakur 2009a, b, c, d, e). Advantages such as corrosion resistance, electrical insulation, easy processability at relatively less energy, higher stiffness and strength, fatigue resistance and lower weight than metals have made polymer composites widely acceptable in all walks of life (Lin et al. 2011a, b; Thakur et al. 2010a, b). Polymers and fiber-reinforced polymer composites have been extensively used in applications ranging from aerospace to autos (Li et al. 2013, 2014) and from circuit boards to sporting goods (Xu et al. 2014). However, most of the fibers and resins are derived from petroleum feed stocks and do not degrade for several decades under normal environmental conditions. In addition, composites made from thermosetting resins cannot be reprocessed or recycled (Mitra 2014; Singha and Thakur 2008a, b, c, d). Disposal of these nonbiodegradable plastic wastes is a huge eco-technological problem addressing worldwide environmental pollution and climate change, which calls for a market stimulus and mandatory policy for biomass-sourced polymer production (Song and Zheng 2014).

Extensive efforts are being made to develop biodegradable composites using renewable resources in an attempt to replace the nonbiodegradable synthetic polymers used for composites (Thakur et al. 2013a, b, c, d, e). Both matrix and reinforcing materials derived from renewable resources have been used to develop biodegradable composites (Reddy and Yang 2011). Biodegradable materials are naturally comprised of polymers that should be capable of being ultimately degraded by microorganisms (bacteria, fungi, and algae) through composting processes to produce natural breakdown compounds such as carbon dioxide, water, methane, and biomass (Nur Hanani et al. 2014). The natural polymer offers a number of advantages over synthetic polymer such as complete degradation, increased soil fertility, low accumulation of bulky plastic materials in the environment, and reduction in the cost of waste management (Thakur et al. 2012a, b, c, d, e). Natural polymers can be obtained from three kinds of renewable resources: (a) from plants (starch) (Chinma et al. 2013), (soy protein) (Echeverría et al. 2014), and (cellulose) (Deepa et al. 2011; Sirviö et al. 2014); (b) from animals (chitosan) (Venkatesan et al. 2014), (keratin) (Flores-Hernández et al. 2014), (silk) (Mitra 2014); and (c) by microbial fermentation (polyhydroxyalkanoates) (PHA) and (polyhydroxybutyrate) (PHB) (Gupta and Nayak 2014).

Proteins and polysaccharides produced on an annual kiloton scale in the world have attracted extensive research attention as potentially the most significant eco-materials as well as edible and biodegradable films and coatings (Thakur and Thakur 2014a, b, c). Edible films made from polysaccharides and proteins act as excellent barriers to nonpolar substances such as O_2, CO_2, and lipid, particularly at low relative humidities. In contrast to synthetic polymers having a simpler and more random structure, biopolymers are complex molecular assemblies with precise and defined three-dimensional structures. Proteins are naturally occurring linear, random heteropolymers built from up to 20 amino acid residues and polysaccharides are formed from mono or disaccharide repeating units connected by glycosidic bonds (Song and Zheng 2014).

Biopolymers are commonly used in the medical field, where they are used mainly for surgical sutures and drug delivery systems. They also have a great potential for use as food packaging. A notable example of a biopolymer packaging material is cellulose, which has been widely used in the form of paper and cardboard as exterior food packaging. Cellulose fiber-reinforced polymer composites have received much attention because of their low density, nonabrasive, combustible, nontoxic, low cost, and biodegradable properties (Thakur et al. 2010a, b). The reinforcing ability of cellulose is mainly due to its semicrystalline nature and the extended chain conformation of cellulose molecules in the crystalline regions (Mariano et al. 2014). A lot of research works (Thakur et al. 2014a, b; Deepa et al. 2011; Liang et al. 2013; Singha et al. 2009a, b; Sareena et al. 2014) have been performed all over the world on the use of cellulose fibers as a reinforcing material for the preparation of various types of composites.

1.1 Green Composites

Green composites combine plant fibers with natural resins to create natural composite materials (Thakur et al. 2014a). Natural fibers emerge as low cost, light weight, and apparently environmentally superior alternative to synthetic fibers. Natural fibers can be defined as bio-based fibers or fibers from vegetable and animal origin. This includes all plant fibers (cotton, jute, hemp, kenaf, flax, coir, abaca, ramie, etc.) and protein-based fibers (wool, silk). Of these fibers, jute, flax, hemp, and sisal are the most commonly used fibers for polymer composites. When they are dumped, they get decomposed by the action of microorganisms and are converted to water and carbon dioxide. These are absorbed into the plant systems. Biodegradable polymers produced from renewable resources such as plants, animals, and microbes through biochemical reactions offer a convenient and environment-friendly solution to the problem of plastic wastes (Mitra 2014).

The potential of nanocomposites in various sectors of research and application is promising and is attracting increasing investments (Boufi et al. 2014; Siqueira et al. 2010). In the nanocomposite industry, a reinforcing particle is usually considered as one where at least one of its linear dimensions is smaller than 100 nm (Kumar et al. 2011). The addition (loading) of nanomaterials to polymers leads to significant changes in their physical, mechanical as well as thermal properties. Due to the hierarchical structure and semicrystalline nature of polysaccharides (cellulose, starch, and chitin), nanoparticles can be extracted from these naturally occurring polymers (Dufresne 2010).

Many diverse topics, including composite reinforcement, barrier properties, flame resistance, electro-optical properties, cosmetic applications, and bactericidal properties, exist in the field of polymer nanocomposites. This is a result of the uniform dispersion of the nanomaterials in the polymer matrices. The fabricated polymer nanocomposites have been shown to possess excellent mechanical properties. However, most of them are nonbiodegradable. Recently, many polymer nanocomposites using bio-based and biodegradable polymers and nanomaterials with excellent mechanical and thermal properties as well as other functionalities have been reported (Fernandes et al. 2013; Siro and Plackett 2010). These bio-based and biodegradable polymer nanocomposites have the potential to replace traditional nonbiodegradable plastic materials in many applications, including tennis/badminton/squash racket frames, ski pole, circuit board, automobile inside panels, etc. The food industry also seems to receive large benefits from nanotechnology, with potential uses already identified in virtually every segment of the sector (i.e., agriculture, food processing, food packaging, and nutrient supplements) (Unalan et al. 2014).

Extensive research has been undertaken in blending different polymers to obtain new products having some of the desired properties of each component. Among protein- and polysaccharide-based green materials, those made from soy protein (Maruthi et al. 2014; Ghidelli et al. 2014; Behera et al. 2012) and starch (Katerinopoulou et al. 2014; Flores-Hernández et al. 2014) have been extensively studied for and their physiochemical properties been analyzed. The literature review clearly shows that development of biodegradable biopolymer-based materials based on these materials can not only solve the "white pollution" problem but also ease the overdependence on petroleum resources. This chapter provides a brief overview of the preparation, properties, and application of cellulose fiber-reinforced soy protein-based and starch-based biocomposites.

2 Soy Protein-Based Green Composites

Soy protein, the major component of the soybean, is biodegradable, environmental friendly, and readily available from an abundant renewable resource (Thakur et al. 2014c, d). Soy protein has been consumed as food material for thousands of years, called as "Gold that Grows," the "Cinderella Crop," the "Balance of Payments Champions," the "Queen of the Commodity Exchanges," etc. It had been considered an interesting starting material for the development of new materials as devices for biotechnological and biomedical utilization (Gupta and Nayak 2014). The major soybean proteins have molecular weights ranging from 200 to 600 kDa. Most soy proteins (90 %) are globulins, which can be fractionated into 2S, 7S, 11S, and 15S according to their sedimentation coefficients. 7S and 11S are the main fractions being about 37 and 31 % of the total extractable protein and have the capacity of polymerization (Kokoszka et al. 2010). The behavior of protein is determined by its amino acid composition and its molecular size. Soy protein contains 18 amino acids including those that contain polar functional groups, such as carboxyl, amine, and hydroxyl groups that are capable of chemically reacting. These reactive groups can be utilized for chemical modification to improve the mechanical and physical properties of the protein and thus improve the properties of the composites made from them (Huang and Netravali 2009; Thakur and Kessler 2014a, b). Protein conformation also affects functionality; in globular proteins the more polar charged groups are oriented toward the surface. Moreover, noncovalent forces (hydrophobic interactions, hydrogen bonding, electrostatic attractions, etc.) are involved in protein–protein and protein–solvent interactions which influence the overall functional properties.

In recent years, soy products such as soy whole flour (SF), soy protein concentrate (SPC), and soy protein isolate (SPI) have been considered as alternatives to petroleum polymers because of their abundance, low cost, perfect adhesion, and good biodegradability (Maruthi et al. 2014). SF contains about 40–60 % protein, combined with fats and carbohydrates. Soy protein concentrate contains about 60–70 % protein. SPI contains more than 90 % of protein and is the most widely used soybean product for film processing (Ciannamea et al. 2014). Moreover, SPI-based films are clearer, smoother, and more flexible compared to other plant protein-based films, and they have impressive gas barrier properties compared to those prepared from lipids and polysaccharides. When SPI films are not moist, their O_2 permeability was 500, 260, 540, and 670 times lower than that of films based on low-density polyethylene, methylcellulose, starch, and pectin, respectively (Song et al. 2011). Thus, in addition to their large availability, soy protein-based materials have interesting barrier and release properties ideal for packaging applications.

A downside to the sustainable and extensive use of soy protein-based materials is their intrinsic reactivity and thus lower inertia when compared to most conventional petrochemical-based plastics. They are known to be sensitive to microbial spoilage and also to water due to hydrophilic nature of many amino acids constituting their primary structure and to the substantial amount of hydrophilic plasticizer required to impart thermo-processability and film flexibility. As a consequence, their mechanical properties and water vapor barrier properties in high moisture conditions are poor compared to synthetic films such as low-density polyethylene.

One route to significantly modulate the functional properties of soy protein-based material is developing biocomposites structures by incorporating cellulose fibers having high specific strength, elastic modulus, and low density. Natural cellulosic fibers such as flax, jute, ramie, hemp, sisal, and pineapple have attracted attention as reinforcements for composites due to many advantages such as annually renewable, sustainable, low cost, high specific modulus, lightweight, biodegradable, and biocompatible features. With increasing environmental and sustainable concerns about petroleum-based polymer materials, development of biodegradable bio-based materials based on soy protein and cellulose fiber will catch more and more attention (Song et al. 2011).

2.1 Preparation

To form SPI films, the protein structures of the native state would need to be denatured to reform new configurations via new linkages within the protein molecule. Denaturation of the protein can be induced by changes in pH, electrical force, mechanical force, or heat. Changing pH conditions away from the isoelectric point (4.2–4.6 for SPI) can cause the protein to unfold and to increase its solubility. Casting leverages this phenomenon by evaporating solubilized SPI solution on a flat surface forming film. Mechanical forces like pressure and shear are known to break bonds and induce flow in an extruder which is important for increasing intermolecular entanglement. Heat denaturation of protein typically occurs above a certain threshold temperature. For SPI, this is generally between 65 and 70 °C. As the proteins unfold, sulfhydryl and hydrophobic groups are exposed and disulfide bonds are reformed, thereby forming new structural arrangements (Chan 2012). There are two common technologies to prepare protein films: wet (or solvent) process and dry process.

2.1.1 Wet Process

Many biopolymers could not be melt processed because of their degradation on or before melting (softening). In this case, solution blending is the preferred technology. Wet process, also called solution casting, is based on the dispersion or solubilisation of proteins in a solvent medium (Guerrero and de la Caba 2010). For a continuous process for solvent casting of films, parameters that need to be determined for continuous film production are air temperature, surface properties of the substrate upon which the films are formed, flow rate, and drying time. Films can be dried under ambient conditions, with hot air, infrared energy, or microwave energy (Wang et al. 2014). Cross-linked porous structures can be developed by means of combining a sol–gel process with the freeze-drying technique. The properties of the final films depend, except for the film composition, on the protein–filler interaction and miscibility, depending on the composition, penalty of plasticizer to the protein and the filler, and the micromorphology (phase-segregated and co-continuous structures). To elaborate, the protein is typically solubilized using large amounts of aqueous solvent and dehydrated on a flat surface. The solubility of the protein is enhanced by applying heat around 80 °C under alkaline conditions. With elevated pH, soy protein is known to form β-sheet structures through intermolecular hydrogen bonds during drying. Alkaline casting also yields mechanically stronger performing films due to higher protein solubility. Casting around neutral pH yields heterogeneous films with insolubilized protein particles causing uneven film matrices leading to weak mechanical properties.

2.1.2 Dry Process

The combination of thermal and mechanical inputs can be used to disrupt intra–and intermolecular interactions of biopolymers by extrusion or mechanical mixing, common plastic processing techniques. Plasticized materials could be extruded (Zhang and Mittal 2010), injection or compression molded (Reddy and Yang 2011) into articles in which the protein component forms networks through disulfide bonds or amide bonds between free-carboxyl and amino side groups on the protein chains. Feed composition, temperature, and screw speed influence the extruder dimensions and the heat denaturing and cross-linking of the protein in the produced sheets. The molding temperature has significant effects on the cross-linking density of protein, the intermolecular cross-linking between protein and filler, and thereby on the mechanical properties of the processed materials.

2.2 Properties

2.2.1 Mechanical Properties

While many of the biodegradable resins such as poly (lactic acid), poly (hydroxy alkanoates), etc., are hydrophobic and do not bond well to plant-based fibers, others such as protein and soy-based resins which contain polar groups show good adhesion and form composites with good mechanical properties. It was observed that the incorporation of cellulose fibers into the SPI matrix results in strengthening of the materials, as a result of stiffness of the fibers and the strong interactions caused by hydrogen bonds between the cellulose and the SPI. Soy protein molecules are ductile and can undergo bending, torsional, and tensile deformations without visible damage (Mitra 2014).

Reddy and Yang (2011) showed that the thermoplastic soy protein can act as a binder and provide composites with much better flexural and tensile properties than similar composites developed using polypropylene as the matrix polymer. Fully biodegradable, environment-friendly SPI–jute fiber composites were fabricated using water without any chemicals as the plasticizer and the mechanical properties of the composites were evaluated. It was observed that water decreases the melting temperature and also provides a much higher melting enthalpy to soy proteins. Composites developed using 60 % soy proteins and 40 % jute fibers and compression molded at 170 °C and 15 min had optimum properties. At the optimized conditions, the soy protein composites have more than twice the flexural strength and more than 80 and 90 % higher tensile strength and tensile modulus, respectively, than polypropylene composites.

Effect of pH values on the tensile properties of hemp yarn/soy protein concentrate (SPC) was investigated by (Kim and Netravali 2011). Unidirectional hemp yarn-reinforced green composites were fabricated with soy protein concentrate (SPC) at different pH (7, 10, and 12) and the composites were denoted as

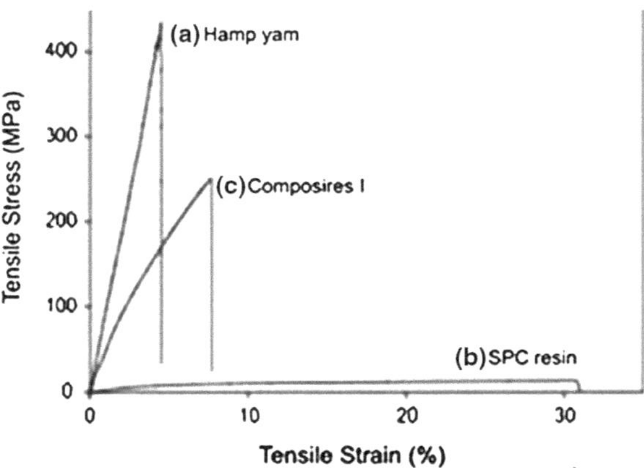

Fig. 1 Typical stress–strain plots of (**a**) hemp yarn, (**b**) SPC resin at pH 7, and (**c**) composite I; reproduced and reprinted with permission from Kim and Netravali (2011). Copyright 2011 Elsevier

Composites I–III. It was observed that tensile properties of the composites slightly differed depending on the pH values at which the SPC resins were prepared. The composite I showed the lowest fracture stress of 253 MPa, fracture strain of 7.7 %, and toughness of 11.8 MPa while the composite II showed the highest fractures stress, strain, and toughness of 277 MPa, 9.7 %, and 15.4 MPa, respectively. The tensile properties of the composite III were in between the composites I and II. This was attributed to the higher interfacial shear strength (IFSS) between yarn and SPC resin at pH 10. Figure 1 shows typical stress–strain plots of (a) hemp yarn, (b) SPC resin at processed pH7 and (c) composite I.

Soy proteins have also been modified to improve the mechanical properties significantly and thus to make the material more useful as resin. Kenaf fiber/soy protein flour-based biocomposites modified with poly (carboxylic acid) resin were prepared by Liang et al. (2013). It was found that the mechanical properties and density of the composites were affected by the substrate and adhesive nature, and the interactions between the substrate and adhesives. The hot pressing time had significant effects on the properties of the composites. A long pressing time promoted interaction between the protein and fiber surface and led to a higher mechanical strength.

Advanced green composites made from liquid crystalline (LC-linen yarn fibers) and SPC modified with agar and nanoclay were prepared by Huang and Netravali (2009) and its tensile testing was done to analyze the effect of liquid crystalline treatment of cellulose in the matrix. The composite showed much higher tensile strength (tensile failure stress of 616 MPa and Young's modulus of 13.7 GPa) than the linen yarn-reinforced composite because of the significantly higher tensile properties of LC-cellulose fiber. An increase of 8–10 % over the SPC resin composite was observed. The LC-cellulose fibers, because of the presence of three hydroxyl groups on each glucose monomeric unit in cellulose, are expected to have excellent hydrogen bonding with soy protein resin resulting in good interfacial properties in the composites. This composite was found to have a high failure strain compared to most advanced composites that use fibers like graphite and kevlar. As a result, these composites possess excellent fracture toughness making them useful in many applications where toughness is critical. Also, this composite was found to be five to six times stronger than steel on "per weight" basis.

Micro/nanosized bamboo fibril (MBF)-reinforced soy protein concentrate (SPC) resin composites were prepared by Huang and Netravali (2009). Mechanical studies of these composites showed that the addition of MBF significantly increased the fracture stress and Young's modulus of the reinforced specimens. With the incorporation of 30 parts of MBF, 15 parts of glycerol in SPC showed fracture stress of 59.3 MPa and Young's modulus of 1816 MPa compared to the fracture stress of 20.2 MPa, and Young's modulus of 596 MPa obtained for SPC containing no MBF. This was attributed to the highly ordered and crystalline nature of the cellulose microfibrils.

2.2.2 Thermal Properties

Differential scanning calorimetry (DSC) and thermogravimetric analysis (TGA) have been widely used to characterize the thermal properties of food proteins, including heat-induced denaturation. The denaturation process is an intramolecular change involving the destruction of internal order, and in some cases, the complete unfolding of peptide chains with the formation of so-called "random coils". Heating can change soybean protein from its native state to a denatured one, accompanied by unfolding and disruption of the intramolecular bonding, which is observable as an endothermic peak. The major components of soy proteins are globular proteins, 7S (about 35 %) and 11S (about 52 %). Temperature, pressure, and time are the main parameters in soy protein processing in order to denature the protein, unfold globular structure, and permit interaction and entanglement between protein chains to modify material properties. Similar unfolding could be obtained with high temperature and short time or low temperature and long time, although time–temperature relation must be optimized in order to avoid color change from light to dark yellow. Therefore, intra- and inter-molecular interactions will be significantly influenced by the processing temperature, pressure, and time employed.

DSC thermograms for soy proteins show two characteristic denaturation temperatures for 7 and 11S globulins (Routray et al. 2013). The first peak at around 75 °C corresponds to the lower molecular fraction (7S) denaturation, and the second one at around 225 °C is related to the high molecular fraction (11S). Figure 2 shows typical DSC thermograms for soy proteins, in which the two characteristic denaturation temperatures for 7 and 11S globulins are shown. Denaturation temperatures of 7 and 11S globulins are strongly dependant on moisture content, shifting to higher values at low moisture contents. Thermogravimetric analysis indicates that soy protein films exhibit substantial thermal degradation at temperatures above 180 °C, so that the processing temperature has to be chosen between 150 and 180 °C (Guerrero et al. 2010).

The effect of plasticizer content on the thermal degradation of soy protein films has been investigated by Guerrero and de la Caba (2010) and Guerrero et al. (2010).

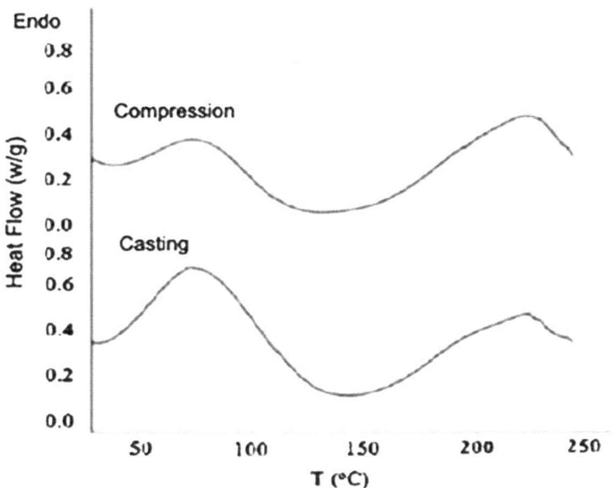

Fig. 2 DSC thermograms of SPI with 30 % glycerol. Reprinted with permission from Guerrero et al. (2010). Copyright 2010 Elsevier

Thermal studies by TGA revealed that for pure SPI there is a small weight loss at temperatures below 100 °C, which is due to the loss of moisture. Above 100 °C, the rate of weight loss is small until 200 °C, but starts to become significant above 225 °C. For the plasticized films, the weight loss starts to become significant above 200 °C. The higher weight loss of plasticized samples was attributed to the high vapor pressure of glycerol.

Only limited studies on the improvement of the thermal properties of proteins by altering their molecular structure or conformation with the addition of natural fibers has been documented in recent literature.

Thermal behavior of soy protein–jute fiber composites using water without using any chemicals as the plasticizer was studied by Reddy and Yang (2011). Figure 3 shows the DSC thermogram of soy protein with and without water as plasticizer at a heating rate of 20 °C/min. It was observed that water effectively plasticized soy protein and decreased the melting temperature from 150 °C to about 130 °C. However, the optimum temperature for composite fabrication was between 150 and 180 °C. The higher temperatures for composite fabrication are necessary to ensure the complete removal of water used as plasticizer from the soy proteins and jute fibers. The melting peak of soy protein with water as plasticizer had a melting enthalpy of 1.4 kJ/g compared to 76 J/g for the nonplasticized soy proteins. The low melting temperature and high melting enthalpy of the soy proteins with water as the plasticizer allows the use of soy protein as the matrix without damaging the soy proteins.

Effect of modification of fillers on the thermal properties of soy protein composites was studied by Wang et al. (2013). They prepared edible films containing different ratios of SPI with wheat-bran cellulose (WC), microcrystalline wheat-bran cellulose (MWC), and ultrasonic/microwave-modified MWC (MMWC) by casting and thermal properties of the films were analyzed. It was found that different proportions and particle size of the fillers (WC, MWC, and MMWC) affected the thermal properties of each of the films. The SPI/MMWC film demonstrated the best

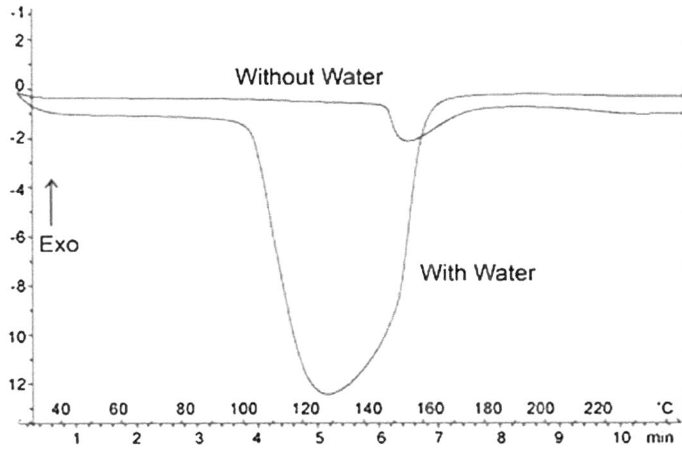

Fig. 3 DSC thermogram of soyprotein with and without water as plasticizer. Reproduced and reprinted with permission from Reddy and Yang (2011), Copyright 2011 Elsevier

properties. This is because of the fact that ultrasonic/microwave-assisted treatment led to the disaggregation of side-by-side hydrogen bonding leading to more free hydroxyls on the surface of MMWC, enabling the formation of stronger intramolecular hydrogen bonds with SPI that yielded a blend film with a homogeneous, compact structure. Glass transition temperature (Tg) of the SPI, WC, MWC, and MMWC was observed to be 160.6, 174.4, 189.7, and 205.5 °C, respectively, while the crystallization temperature (T_c) was 230.1, 251.6, 264.2, and 286.7 °C, respectively.

Modification of soy protein matrix and fabrication of composites with fibers also show good thermal properties. Green composites using cross-linked soy flour (CSF) and flax yarns were prepared by Chabba et al. (2005) and its thermal properties were analyzed. It was found that modification of SF with glutaraldehyde (GA) followed by fabrication of composites with flax yarns can give good thermal properties. It was found that the TGA thermogram for the CSF resin consistently showed less weight loss at all temperatures compared to SF resin, indicating a higher thermal stability than SF resin. The thermogram showed that CSF resin is very stable up to 120 °C, the processing temperature for yarn-reinforced composites. The improved thermal stability of CSF resin is attributed to the cross-linking between GA and SF.

2.2.3 Barrier Properties

In order to be used as a food packaging material, biopolymer packaging films must have mass transfer properties suitable for their use. Significant food quality losses can occur due to transfer of moisture, gases, aroma, flavor, or color to and from the surrounding environment. For example, moisture losses in fruits and vegetables upon long-term storage may result in subsequent weight loss and shrinkage (Ghidelli et al. 2014). Crispy potato chips become soggy when exposed to a humid environment (Rossi Marquez et al. 2013); in multiphase foods, quality may degrade when moisture moves from one component of the food to another, such as from the moist filling of a pie to its dry crust. Similarly, undesirable gas transfers can also cause quality problems. Nuts and other oil-containing products may experience oxidative rancidity when there is too much oxygen diffusion from the environment into the food, causing loss of nutritional content and deterioration of texture, flavor, color, and aroma, which can eventually lead to the reduction of the product value and shelf life. Therefore, control of mass transfer in food packaging systems is necessary in order to maintain food quality.

It is in this context that biopolymer packaging becomes a new alternative to modified atmosphere packaging (MAP) (Ghanbarzadeh et al. 2014). This can be a better option for situations when controlling or modifying the atmospheric conditions during storage are not feasible, or is limited by cost or equipment requirements. Additionally, biopolymer films can supplement the natural barriers of food products such as fruits and vegetables and fish products (Ghidelli et al. 2014). The film can function as a replacement barrier in cases where washing and handling have partially removed or altered the natural coating.

Soy protein-based edible films have received considerable attention due to their excellent film-forming abilities, low cost and barrier properties against oxygen, lipid and aroma permeation under low to intermediate humidity conditions, and sequence of amino acid residues. Compared to synthetic films, protein-based films exhibit poor water resistance and lower mechanical strength. Yet, proteins are still generally superior to polysaccharides in their ability to form films with greater mechanical and barrier properties. Protein-based films possess better oxygen and carbon dioxide barrier properties and mechanical properties than polysaccharide films (Nur Hanani et al. 2014).

However, due to its inherent hydrophilic nature, this material presents two major disadvantages: fragility in the wet state and poor properties of moisture barrier. These effects can be minimized using physical, chemical, or enzymatic treatments including: blending with hydrophobic additives such as neutral lipids, fatty acids, or waxes, changing drying conditions (Guerrero et al. 2011); enzymatic treatment with horseradish peroxidase; heat curing, UV irradiation, and cross-linking (Wihodo and Moraru 2013). The e-amino group of lysine was considered the primary reactive site between proteins and cross-linkers. The cross-linkers most extensively used for proteins are aldehydic compounds such as glutaraldehyde, formaldehyde and glyoxal, and epoxy and phenolic compounds (Wihodo and Moraru 2013). However, the cytotoxicity of these compounds restricts their use for food covering. Hence study on barrier properties of fiber-reinforced soy protein composites is the need of the hour.

The effects of the cellulose whisker content on the water absorption properties of the glycerol-plasticized SPI composites were investigated by Wang et al. (2006) and it was observed that the water uptake of the SPI/cellulose whisker composites decreases with an increase of the cellulose whisker content. While the SPI sheet without cellulose whiskers absorbs nearly 40 wt% water, for SPI with 30 wt% cellulose whisker, the water content is only about 25 wt%, indicating an improvement in the water resistivity. This was related to the rigid cellulose network resulting from three-dimensional hydrogen bonding between the cellulose whiskers formed during the film formation leading to an enhanced water resistance of the composite.

The effect of ultrasonic/microwave-assisted treatment on the water vapor permeability (WVP) and oxygen permeability (OP) of SPI/microcrystalline wheat-bran cellulose film was studied by Wang et al. (2014). It was observed that ultrasonic/microwave-assisted treatment increases the water holding capacity by producing more free hydroxyls on the surface and forming a compact structure through a stronger interaction between hydrogen and SPI. It has also been shown that (i) small fibers were more uniformly dispersed inside the matrix than larger ones, (ii) the WVP decreased if the filler was less permeable, and (iii) barrier properties were enhanced by good dispersion in the matrix and a high aspect ratio. Regarding oxygen permeability, it was observed that when the ratio (SPI: Fillers) was varied from 10:0 to 5:5, the OP decreased. Such decrease was hypothesized to be attributed to the hydrogen-bonded network structure. This closely packed arrangement prevented O_2 transport within the film. It has been observed that the OP strongly depends on the interaction between the polymer matrixes due to the formation of intermolecular hydrogen bonds between soy protein and the filler. This made the structure of blend films compressed and prolonged the transport path of O_2. This was also associated with the state of aggregation, the relative orientation, and the dispersion of the fibers in the matrix.

The effect on WVP caused by chemical cross-linking of soy protein with natural cross-linker, genipin (Gen) was analyzed by González et al. (2011). This novel cross-linker is obtained from the enzymatic hydrolysis of Genipa with β-glucosidase. Genipa is extracted from the fruit of a type of jasmine called Gardenia Jasminoides Ellis. The water vapor transmission rate (WVTR) (kg $s^{-1}m^{-2}$) and WVP (kgmPa^{-1} $s^{-1}m^{-2}$) were calculated from Eqs. (1) and (2), respectively:

$$WVTR = \frac{F}{A} \tag{1}$$

$$WVP = (WVTR \times \epsilon)/[Sp \times (RH1 - RH2)] \tag{2}$$

where F is the slope of the graph of variation of mass versus time (kg s^{-1}), A is the test area (cup mouth area), ε is the film thickness (m), Sp is the saturation pressure (Pa) at the test temperature, RH1 is the relative humidity in the humidity chamber, and RH2 is the relative humidity inside the cell test. It was observed that WVP values for films with 2.5 % (w/w of SPI) of Gen diminished approximately 29.5 % of WVP with respect to the film without Gen. This decrease can be attributed to an increase in the film density generated by the higher intermolecular cross-linking degree into the matrices. WVP values of films between 2.5 and 10 % (w/w of SPI) of Gen did not vary significantly ($P \geq 0.05$) due to the intramolecular cross-linking possible with the amount of Gen. This amount of Gen did not probably avoid the permeation of water since the water penetrates through the matrix by the spaces between the protein structures.

3 Soy Protein-Based Bionanocomposites

The incorporation of nanofillers into soy protein materials significantly enhances tensile strength and modulus, and in some cases, contributes to a simultaneous increase in strength and elongation (Echeverría et al. 2014). As novel functional renewable materials, natural cellulose and chitin nano- or microfibers have received increasing interest because of their abundance, biocompatibility, and specific properties. In addition to their high mechanical strength, they have several useful advantages such as high aspect ratio and facile chemical modification, showing a potential application in the material and food industry (Yuan et al. 2014). The single most important factor affecting the high performance of soy protein-based nanocomposites is strengthening the interfacial adhesion between the soy protein matrix and nanofillers. Chemical modification on the nanofiller surface (Ifuku et al. 2010) is expected to improve the miscibility between filler and soy protein matrix. The decrease in tensile strength that results from plasticization is therefore, to some extent recovered. The nanocomposite shows great potential for partly solving the two prevailing problems of low strength and water sensitivity, which greatly hamper the development and application of soy protein-based plastics. It also contributes to the development of high performance/novel functionality soy protein-based materials.

Chitin nanowhiskers obtained from crab shells were incorporated into SPI to improve the thermomechanical properties and to decrease water sensitivity of the SPI (Lu et al. 2004). A relatively uniform distribution of the chitin whisker in the SPI matrix can be observed when the chitin content is lower than 15 wt%. With an increase of chitin whiskers in the SPI matrix from 0 to 30 wt%, enhanced thermal, mechanical, and water resistance of the soy protein composite was observed. The water diffusion coefficients (*D*) of the nanocomposites decrease from 2.56×10^{-10} to 1.23×10^{-10} $cm^2\ s^{-1}$. The tensile strength and Young's modulus of the composites increase from 3.3 to 8.4 MPa and from 26.4 to 158 MPa with increasing chitin content from 0 to 20 wt%, whereas the elongation at break of the filled composites decreases from 205 to 29 %. The improvement in all of the properties of these novel SPI/chitin whisker nanocomposites may be ascribed to three-dimensional networks of intermolecular hydrogen bonding interactions between filler and filler and between filler and SPI matrix.

Cellulose whisker alone and in combination with nanoclay were used as reinforcing agents for the preparation of green nanocomposites based on jute fabric and glutaraldehyde (GA) cross-linked SF (Iman et al. 2013). Cellulose whiskers (CWs) were extracted from ordinary filter paper by acid hydrolysis. The addition of the clay nanoparticles did not cause any structural modification in the SF composite but increased the conjugation of the system as evidenced by the bathochromic shift due to $n \rightarrow \pi^*$ transition in the UV Visible spectrum. This showed that the green nanocomposites comprising SF, jute, glutaraldehyde, cellulose whisker, and nanoclay are cross-linked by –N–C = O and –CN linkages, among others. The nanoclay-incorporated composites showed significant improvements in the mechanical properties with respect to those without nanoclay. This enhancement in mechanical properties of clay-loaded composites is because of their intrinsic high modulus and high aspect ratio of the nanoclay, which generates a large surface area for polymer adsorption.

The environment-friendly, fully biodegradable green composites, based on bamboo micro/nanofibrils and silane, (3-isocyanatopropyl) triethoxysilane (ITES)-modified Soy Protein Concentrate resins have excellent properties fracture stress of 82 MPa, Young's modulus of around 3.2 GPa, and toughness of 4.3 MPa (Huang et al. 2009). The incorporation of the cellulose whiskers obtained from cotton linter pulp into the SPI matrix led to an improvement in the properties like increase in tensile strength and Young's modulus of the SPI/cellulose whisker composites from 5.8 to 8.1 MPa and from 44.7 to 133.2 MPa (Wang et al. 2006). Cellulose nanofibrous mats (CNM)-reinforced soybean protein isolate (SPI) composite with high visible light transmittance showed strong interfacial interactions at the cellulose nanofiber /SPI interfaces. The incorporation of 20 wt% cellulose nanofibers in the SPI matrix resulted in great improvement of mechanical strength and Young's modulus by, respectively, 13 and 6 times more than neat SPI film. More interestingly, this composite was translucent with light transmittance of over 75 % at 700 nm. Furthermore, the swelling ratio of this Inter Penetrating Network (IPN)-like CNM/SPI composite decreased from 106 to 22 % as CNM content increased from 0 to 20 wt% (Chen et al. 2008).

Contrary to most cases with soy protein as matrix, one recent study focused on applying soy protein nanoparticle aggregates to modify styrene-butadiene elastomer. Soy protein nanoparticle aggregates were prepared by alkaline hydrolysis of SPI. The results showed that compact soy protein nanoparticle aggregates interacted more strongly with the polymer matrix and dispersed more uniformly than crude soy protein, and hence produced better modulus retention for the nanocomposites (Jong et al. 2008).

4 Applications

Since late 1980s, edible films from natural polymers have become an ecologically important alternative to common synthetic polymer for food or drug (drug encapsulation for controlled release) packaging. Edible films made from biopolymers can be used for individual coating of small food products that are normally not individually packaged, such as cherries (Fagundes et al. 2014), nuts (Riveros et al. 2013), and mushrooms (Riveros et al. 2013). Edible films can also be placed within a dual texture food, for example in between the dry crust and the moist filling of a pizza, to prevent moisture migration and maintain the texture of each of the layers (Wihodo and Moraru 2013). Other uses of biopolymer films include pouches for dried soups or powdered beverages (Wihodo and Moraru 2013), or food overwrap or shrink-wrap (Slomkowski et al. 2014).

Soy protein-based green composite is an ideal candidate for the preparation of edible films (Pan et al. 2014; Bai et al. 2012; Thakur et al. 2014c, d). These soy-based plastics could be employed as short-term use or one-time use plastic products in place of the nonbiodegradable materials currently used. Soy protein films are also used as edible food coatings for the preservation and protection of minimally processed food and raw material, during processing, manufacturing, handling, and storage (González et al. 2011). Soy protein-based edible films have received considerable attention due to their excellent film-forming abilities, low cost, good barrier properties against oxygen, lipid, and aroma permeation under low to intermediate humidity conditions. This type of proteins produce smoother, clearer, and more flexible films compared to those from other sources. Once disposed off, soy-based plastics do not take long to biodegrade and is safer for the environment than more durable petroleum-based plastics (González et al. 2011). Furthermore, soy protein could be used for food packaging purposes since it meets food grade standards (Guerrero et al. 2010).

Biopolymer protein films have unique capability in improving food preservation properties based on their interaction with packaging. Such techniques are known as "active packaging systems" (Espitia et al. 2014). An active packaging can be defined as a type of material that changes its packaging conditions to extend shelf life, interacting directly with the food, enhancing security, and maintaining quality. In particular, the antimicrobial packaging is one of the most innovative and promising active packaging types developed over the last decade, which includes systems capable of inhibiting microorganism action and loss of food quality. The purpose is to create inside the package an environment that will delay or prevent the growth of microorganisms on the product surface, which can extend shelf life and improve the safety of packaged foods. Previous literature have shown that protein coating on films in composite structure would require relatively lower amounts of added antimicrobial agents to reach the desired effect as compared to synthetic polymers or other biopolymer films.

Biopolymer protein films can also be used as carrier agents for many types of additives (Espitia et al. 2014). Additives, such as antioxidants, anti-browning agents, nutraceuticals, texture enhancers, flavor, and color ingredients can also be added to enhance the functional and organoleptic properties of the films and/or the packaged foods (Wihodo and Moraru, 2013).

Soy protein-based green composites are not only applied as an environmental friendly material in the fields of adhesives (Kumar et al. 2002), plastics (Kumar et al. 2011), and textile fibers (Kobayashi et al. 2014), but also as biodegradable membranes (Maruthi et al. 2014). Furthermore, the nutritional and health benefits of soy protein draw attention to the application in the field of biomedical materials (Silva et al. 2014), such as tissue engineering scaffolds (Chien and Shah 2012), wound dressing material (Ong et al. 2008), and drug delivery system (Chen et al. 2006). It has also been considered as an interesting starting material for the development of new materials as devices for biotechnological and biomedical utilization (Maruthi et al. 2014). The possibility of building high speed, featherweight circuit boards in computers using waste chicken feathers and soy-based resin has also been explored (Lodha and Netravali 2005). With improved mechanical and physical properties, soy protein-based green composites have the potential for a wider application range such as computer casings, packaging, and panels for auto interiors (Mitra 2014).

5 Starch-Based Green Composites

Starch, the second most abundant biomass material in nature has been extensively studied as a structural or functional polymer because of its renewability, abundance, and biodegradability. It is the dominant carbohydrate reserve materials of higher plants and is relatively an inexpensive biopolymer (Chung et al. 2010). It is mainly found in plants roots, stalks, crop seeds, and staple crops such as rice, corn, wheat, tapioca, and potato. Most synthetic polymers are produced from petrochemicals and are harmful to nature and their synthesis produces hazardous waste and these materials are not easily degradable, causing environmental problems. For these reasons, biodegradable polymers, particularly those from agro polymers such as polysaccharides (e.g., starch, cellulose, etc.) have gained increasing attention. Agro polymers are mainly extracted from plants, are renewable, biocompatible, and biodegradable, which make them superior to synthetic polymers and particularly useful in disposable plastics, food, and biomedical applications. The following section includes a brief review of the processing, performance properties (thermomechanical and barrier), and applications of green composites based on starch.

5.1 Processing Methods

In green polymer composites, matrix is filled with natural organic fillers, i.e., fillers coming from renewable sources and biodegradable. The most widely known and used natural organic fillers are wood flour and fibers. In green composites, starch can be used as either the continuous polymeric phase (matrix) or the dispersed phase (filler) or both. Various plant fibers viz. cotton, flax, sisal, kenaf, jute, hemp, etc. (La Mantia 2011) are added as fillers to starch to prepare green composites. Native starch can be chemically or/and physically modified and used as destructured starch for food or nonfood applications. Starch is gelatinized with heat combined with high water content, which is the destructuring agent. Gelatinization is the basic endothermic process of converting the semicrystal structure of native starch to amorphous state before transforming to thermoplastic starch. The gelatinization process depends mainly on the water content and heating temperature of the starch. Most starch applications require water addition and partial or complete gelatinization. By decreasing the moisture content (to less than 20 wt%), the melting temperature tends to be close to the degradation temperature. To overcome this issue, a nonvolatile (at the process temperature) plasticizer, such as glycerol or other polyols (sorbitol, polyethylene glycol), is added to decrease the melting and process temperature. The addition of glycerol reduces both the melting and the glass transition temperature (Tg) (Ibrahim et al. 2014). The common methods that are used to prepare starch-based composite are mentioned below.

5.1.1 Baking Process

The technology involves the deposition of starch/fiber in heated mold (≥150 °C) to initiate the baking process. Inside the mold, the dough gets heated rapidly and flows to fill the void spaces. The product is baked until it is solidified and dried. The commercial potential for starch/fiber composites made by baking process appears very promising. The composite films confer moisture resistance to the products and also improve mechanical properties. The products made by baking process are functional for many packaging applications.

For instance, Soykeabkaew et al. (2004) prepared cellulose fiber-reinforced starch-based composite foams (SCFs) by baking process. SCFs were prepared successfully by baking starch-based batters incorporating either jute or flax fibers inside a hot mold. Starch is an alternative material for making foams. Batters of starch and water can readily be baked in a closed, heated mold where the starch granules gelatinize and the evaporation of water causes the starch to foam out and take up the shape of the mold. Foams made from pure starch have major drawbacks on their brittleness and sensitivity to moisture and water. Since both the fibers and the starch matrix were naturally polar and hydrophilic, strong interaction between them was expected.

5.1.2 Melt Processing

Thermoplastic starch reinforced composite with sisal and hemp fiber was prepared by melt processing (Gironès et al. 2012). Composites reinforced with different amounts of either sisal or hemp strands have been prepared. Cornstarch and glycerol were manually premixed in polyethylene bags. The polymeric matrix is mixed with 30 % w/w of glycerol. The resulting blend was further processed, together with the corresponding amount of hemp and sisal fiber reinforced, for 6 min at 120 °C and 60 rpm. In those formulations containing latex, this was added in a proportion of 2.5 % w/w during this processing stage. After processing the composites, these were granulated and thermo-pressed in order to obtain film plates.

5.1.3 Casting and Evaporation Method

In this method starch, glycerol, filler, and distilled water were mixed together to obtain composite solution with homogeneous dispersion. The mixture was then heated under reflux and stirred to ensure complete gelatinization of starch. The mixture was casted in a PVC dish and dried to obtain the films.

Glycerol-plasticized bionanocomposite films of cassava starch/bacterial cellulose (Acetobacter xylinum) were prepared by casting method (Woehl et al. 2010). Starch was dispersed in water in a concentration of 40 g L/1, to which glycerol was added at a ratio of 30 % (w/w) in relation to the dry starch mass. Bacterial cellulose nanofibers, treated or not with T. reesei endoglucanases for various incubation times, were then added at a ratio of 2.5 % (w/w) relative to the dry starch mass, under strong magnetic stirring. The mixture was then heated under reflux and stirred at 90–95 °C for 30 min to ensure complete gelatinization of starch. The mixture was casted in a PVC dish and dried in a vacuum oven at 60 °C for 24 h to obtain the films.

A series of thermoplastic starch/hemp cellulose nanocrystals (HCNs) films with different concentrations of HCNs over the range from 5 to 30 % wt, were prepared by casting and evaporation method (Chen et al. 2008). Starch, glycerol, CN suspension, and distilled water were mixed together to obtain composite solution with homogeneous dispersion. The glycerol content was fixed at 36 % w/w on the basis of the dry starch matrix. A homogeneous dispersion of the HCNs in the thermoplastic matrix was observed in all of the composite films. This reveals that there exists new strong interaction between fillers and matrix and interaction among HCNs themselves were destroyed.

Chitosan-reinforced thermoplastic starch films can also be prepared by casting and evaporation method. Xu et al. (2005) prepared chitosan/starch composite by combining chitosan (deacetylated degree, 90 %) solution and two thermally gelatinized cornstarches (waxy starch and regular starch with 25 % amylose). Chitosan/starch composite films were made from blends of either regular or waxy cornstarch and chitosan. The composite films had increasing tensile strengths and elongation at breaks, and decreasing water vapor transmission rates with increasing starch to chitosan ratios. Chitosan solutions (2 %, w/v) were prepared by dispersing 10 g of chitosan (deacetylated degree of 90 %) in 500 ml of lactic acid solution (1 %, v/v). After the chitosan was dissolved completely, the solutions were filtered with cheesecloth by vacuum aspiration. Starch solutions with concentrations of 1, 2, 3, and 4 % (w/v) were prepared by dispersing 25 % amylose cornstarch or waxy starch in distilled water and heating the mixtures on hotplates with stirring until it gelatinized, and then cooling to 25 °C. A series of chitosan/starch composite films were prepared by mixing 100 ml of 2 % chitosan solution with 100 ml of 1, 2, 3, 4 % starch solutions. Glycerin was added as 25 % (w/w) of the total solid weight in solution. The mixtures were cast onto flat, level Teflon-coated glass plates. After drying the films at room temperature for at least 72 h, they were peeled from the plates.

5.1.4 Solution Impregnation Process

In this method starch and plasticizer were mixed together prior to the addition of water. Solution was stirred and heated. Then dried fiber sheets were immersed in the solution. The impregnated sheets were taken out of the solution and dried.

Bacterial Cellulose (BC) fiber-reinforced thermoplastic starch was prepared via solution impregnation method. (Wan et al. 2009). Starch and glycerin, 30 % w/w of glycerine to starch, were mixed prior to the addition of water. Solutions containing starch and glycerol were stirred at 80 °C for 30 min. The pH of the resulting solutions was kept at 3–4 and the concentration varied from 10 to 20 (w/v) %. Then air-dried BC sheets were immersed in the solutions. The immersed mats were maintained in reduced pressure at 0.03 MPa for 12 h and kept under ambient pressure at 25 °C over 96 h. The impregnated sheets were taken out of the solutions, and air dried. The films made from BC and starch were fully biodegradable, which renders them advantageous in terms of environmental protection.

5.1.5 Melt Mixing and Thermocompression

In this technique the components of the samples were premixed at room temperature, blended at higher temperature, and then injection molded.

Thermoplastic pehuen starch (TPS) and TPS/poly (lactic acid) (PLA)/polyvinyl alcohol (PVA) composites, reinforced with 5 and 10 % of pehuen husk, were prepared by melt blending (Castano et al. 2012). Before melt blending, the respective components of the samples were premixed by hand at room temperature. TPS composites were blended at 120 °C with a rotor speed of 60 rpm for 15 min. TPS/PLA/PVA blend and TPS/PLA/PVA composite were mixed at 160 °C and 40 rpm for 15 min and finally, the composites were injection molded to obtain the films.

Starch films reinforced with cellulose nanocrystals obtained from sugarcane bagasse were prepared by Slavutsky et al. (2014). The cellulose is obtained from bagasse by a pulping process. Film-forming solution was prepared by mixing starch (4 %), glycerol (20 % dry weight), water, and an appropriate amount of CNC suspension in order to obtain a CNC concentration of 3 % dry weight. The resulting dispersion was kept 60 min in an ultrasonic bath. Dispersions were gelatinized in a shaking water bath at 78–80 °C during 10 min. This procedure ensured disintegration of starch granules and formation of a homogeneous dispersion. The resulting dispersion, while still hot, was poured on polystyrene plates, placed in an air-circulating oven at 35 °C and 53 % RH for 15 h. After that, plates were removed from the oven and films were peeled off from the plates.

Films based on thermoplastic cornstarch (TPS) and chitosan/chitin were prepared by melt mixing and thermocompression method (Lopez et al. 2014) has also been reported. Chitosan (5 and 10 g/100 g starch) or chitin (10 g/100 g starch) was premixed with starch and then, glycerol (30 g/100 g starch) and distilled water (45 g/100 g starch) were added. Samples were mixed and conditioned at 25 °C during 24 h. Mixtures were processed in a Brabender Plastograph at 140 °C and 50 rpm for 15 min. TPS films were obtained by thermocompression using a thermostatized hydraulic press. Processing conditions were 140 °C for 6 min, increasing the pressure every 2 min (80, 140 and 180 kg/cm^2). An aluminum frame as mold of 1 mm thickness and a relation of 1.9 g sample per cm^3 were used. Material was cooled under pressure up to approximately 50 °C, then the pressure was released and obtained films were removed from the frames.

5.2 *Properties of Starch-Based Green Composites*

The thermoplastic properties of starch are directly related to its water content; in its dry state, the glass transition temperature (Tg) of starch is above its degradation temperature. Starch is not truly thermoplastic but can be converted into a continuous polymeric material by mixing with enough water or nonaqueous plasticizers such as glycerol. The developed thermoplastic starch is hydrophilic in nature that makes them susceptible to moisture attack. Retrogradation, crystallization of the mobile starch chains, and changes in dimensional stability lead to an undesired change in thermomechanical properties of TPS which can be improved by adding different fillers, including nanosized particles. The fillers enhance mechanical, thermal, and barrier properties. The properties depend on the effectiveness of interactions at the interfacial region, that is, on both the surface area and the dispersion of particles.

5.2.1 Thermal Properties

The thermal behavior of starch-based composites includes glass transition and thermal decomposition, Glass transition (Tg) is characterized by dynamic mechanical analysis (DMA), dynamic mechanical thermal analysis (DMTA), and differential scanning calorimetry (DSC). Thermal decomposition is determined by using thermogravimetric analysis (TGA) and derivative thermogravimetric analysis (DTG).

Tg is the critical temperature at which a material changes its behavior from "glassy" (hard and brittle) to "rubbery" (elastic and flexible). The use of plasticizers would lead to change in properties of TPS, of course this change is a function of the amount of plasticizer used. For example, glass transition temperature (Tg) of cassava starch is reported as 131.9 °C and it decreases with increasing addition of glycerol. At 30 % glycerol content, value of Tg is 62.2 °C. DMA was used to evaluate glass transitions associated to the TPS-chitosan films. Figure 4 shows the multifrequency DMA spectrum (1, 3, 5, 10, 15, and 20 Hz) for TPS-chitosan10 films of TPS-chitosan films. Figure 5 shows DMA spectrum of films based on thermoplastic cornstarch, DMA spectrum presented two thermal transitions due to the presence of two separate phases. Two separate transitions correspond to starch-

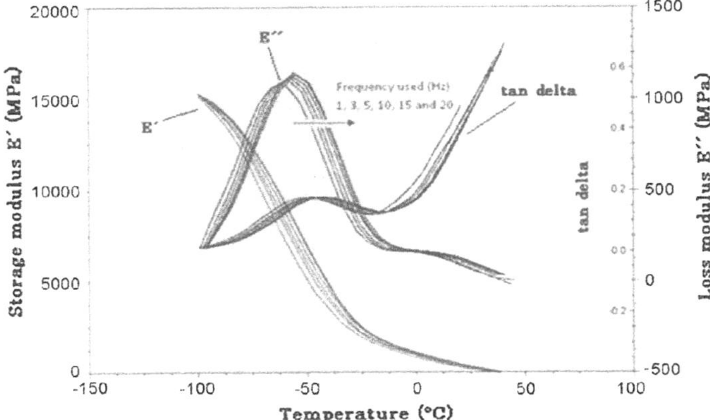

Fig. 4 DMA spectra of films based on thermoplastic cornstarch with 10 g chitosan/100 g starch. Dependence of E′ (storage modulus), E″ (loss modulus), and tan δ with temperature at a constant frequency of 1, 3, 5, 10, 15, and 20 Hz. Reproduced and reprinted with permission from Lopez et al. (2014), Copyright 2014 Elsevier

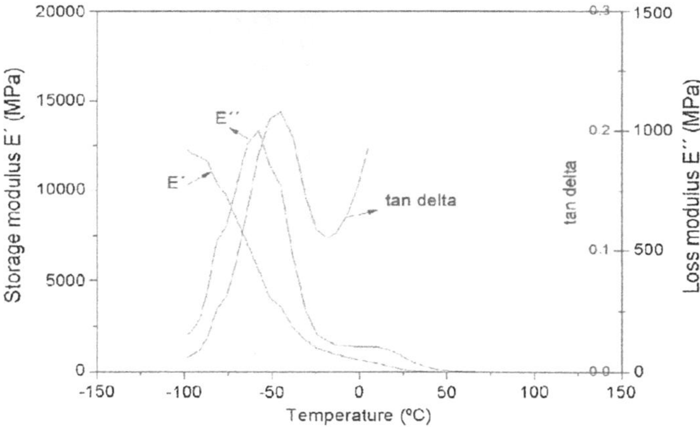

Fig. 5 DMA spectra of films based on thermoplastic cornstarch. Dependence of E′ (storage modulus), E″ (loss modulus), and tan δ with temperature at a constant frequency of 3 Hz. Reproduced and reprinted with permission from Lopez et al. (2014), Copyright 2014 Elsevier

rich phase and glycerol-rich phase associated to the presence of distinctive domains enriched in glycerol or in starch (Lopez et al. 2014).

Gironès et al. (2012) studied the effect of incorporation of sisal or hemp strands in Thermoplastic starch (TPS) from cornstarch. Figure 6 shows the DMTA analysis of thermoplastic cornstarch reinforced with various hemp fiber amounts. The addition of fibers caused an increase in the glass transition temperature (Tg) of the TPS as determined by DMTA.

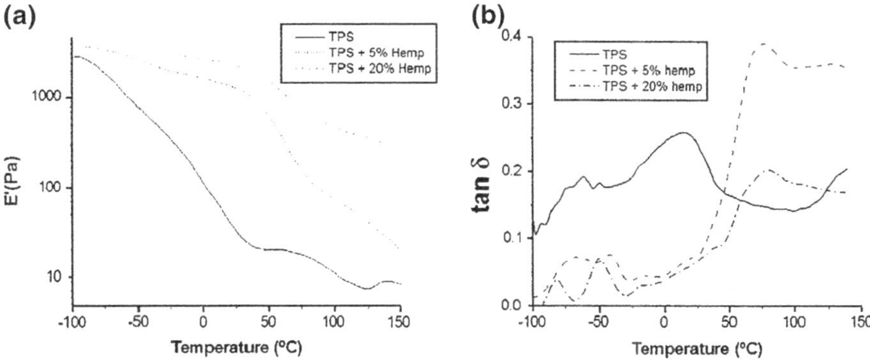

Fig. 6 DMTA analysis of thermoplastic cornstarch reinforced with various hemp fiber amounts. **a** E′ versus Temp; **b** tan δ curves. Reprinted with permission from Gironès et al. (2012), Copyright 2012 Elsevier

TPS showed a biphasic behavior, presenting two different transitions in the tan δ curves. The first one, appearing at 80 to 60 °C corresponds to the phase rich in glycerol. The second transition is assigned to starch occurred at higher temperatures. As it can be seen, an increase proportional to the amount of fibers added to the TPS matrix is observed in E′ throughout the temperatures analyzed, evidencing the stiffness provided by the reinforcing fibers. Decrease observed in the relative intensity of the tan δ signal with increasing fiber content, indicated that fibers promote the inhibition of the relaxation process, thus leading to more rigid systems and consequently increasing the temperature of glass transition (Gironès et al. 2012).

Thermal degradation of the thermoplastic starch (TPS)-based matrix is an important issue to identify the limits of processing, treatment, or operating temperatures. Ibrahim et al. (2014) studied TGA/DTA curves obtained for the hot pressed TPS-based matrix by emulsion technique, where the percentage loss of the sample weight and the derivative of the weight loss due to the volatilization of the degradation products are monitored as a function of temperature. It can be seen that there are two weight loss phases. The first phase is a weight loss of 9.4 % up to 174 ° C due to the presence of hydrated/adsorbed water and glycerin. The second weight loss phase corresponds to the thermal decomposition of starch (burning of starch organic matter). The total weight loss in the second phase was about 88.4 % between 175 and 530 °C, while most of starch weight loss (about 60 %) was between 280 and 340 °C that was associated with peak derivative of the weight loss at 316 °C. The remaining 2.3 % is due to the presence of inorganic materials in the hot pressed TPS-based matrix that included during matrix processing or sample preparation. In the temperature range between 280 and 340 °C, where most of weight loss takes place, the decomposition process was associated with two endothermic peaks: the first one (1) at about 295 °C due to the evaporation (boiling) of glycerin content, and the second one (2) at about 320 °C which is much greater due to the higher energy needed to burn and cause decomposition of starch content (Ibrahim et al. 2014).

Thermal properties of thermoplastic starch composites reinforced with pehuen husk showed the potential of this biofiber as an excellent reinforcement for composite materials. TPS composites showed a good interaction between the fibers and the plasticized starch matrix due to the natural affinity between husk and starch in the pehuen seed. TPS/PLA/PVA blend showed partial miscibility or co-continuous phase and TPS/PLA/PVA composites presented also discontinuities at the biofiber–polymeric matrix interface. The incorporation of biofiber improved the thermal stability of the composites, increasing the initial decomposition temperature. The biofiber hinders the "out-diffusion" of the volatile molecules (e.g., glycerol), retarding the decomposition process of starch composites. On the other hand, the degree of crystallinity of composites decreases when pehuen husk content increases (Castano et al. 2012).

5.2.2 Mechanical Properties

Under the action of high temperature and shear stress, starch can be processed into a thermoplastic starch by breaking its structure (semicrystalline form), causing destructuring of the starch chains, and leading to intermolecular rearrangement, changes the tensile properties also. TPS absorbs moisture when exposed to humidity and has inferior but useful mechanical properties compared to synthetic polymeric materials. One of the methods to overcome these negative attributes, as well as to improve the strength properties in general, has been to incorporate TPS by abundant and relatively inexpensive lignocellulosic materials, such as hard and soft wood fibers or other plant fibers (jute, coconut, banana, cane bagasse, etc.). When natural fibers are used as reinforcements in TPS, an obvious improvement in the mechanical properties and performance of the composite is expected. This is due to the chemical similarities between the fibers and starch, which provides good compatibility, particularly of the cellulose chains.

Soykeabkaew et al. (2004) studied the effect of moisture content on the mechanical properties of SCFs. Both the flexural strength and the flexural modulus of elasticity appeared to be markedly improved with the addition of 5–10 % by weight of jute and flax fibers. At a fixed fiber content of 10 % by weight, both the flexural strength and the flexural modulus of elasticity were found to increase with increasing aspect ratio of the fibers. The improvement in the mechanical properties of SCFs was attributable to the strong interaction between fibers and the starch matrix. Jute fibers had greater reinforcing effect than flax fibers. Orientation of fibers was shown to have a strong effect on both the flexural strength and the flexural modulus of elasticity of SCFs, with the highest values being observed on specimens having fibers oriented in the longitudinal direction (fibers oriented perpendicularly to the crack propagation direction).

Gironès et al. (2012) studied the mechanical properties of sisal and hemp fiber-reinforced thermoplastic cornstarch. The fiber increased the stiffness of the material. Results obtained in the tensile test displayed a continuous increase in both tensile modulus and ultimate strength that was proportional to the amount of reinforcement. Thus, Young's modulus for composites reinforced with a 20 % w/w of hemp strands presented values 18 times higher than that of nonreinforced composites. In this same case, the presence of reinforcing fibers trebled the ultimate tensile strength of the composite. Reinforcing fibers did not only introduce higher strain resistance but also caused a decrease in the flexibility of the material. In consequence, the maximum deformation of the material was greatly diminished. Hemp reinforced composites presented better mechanical properties than those containing sisal.

Chen et al. (2008) studied the mechanical properties of composite films from the suspension of hemp cellulose nanocrystals (HCNs) and thermoplastic starch. The films exhibited significant increase in the tensile strength and Young's modulus, with increasing HCN content from 0 to 30 wt% of HCNs. In addition to the improvement in mechanical properties, the incorporation of HCNs into the PS matrix also led to a decrease in the water sensitivity of the final composite materials. Therefore, the CNs played an important role in improving the mechanical properties and water resistance of the starch-based material.

5.2.3 Barrier Properties

Water transport in edible films based on hydrophilic materials such as starch, is a complex phenomenon due to the strong interaction of sorbed water molecules with the polymeric structure.

Slavutsky et al. (2014) prepared starch/cellulose nanocrystals (CNCs) films and their water barrier properties were studied. The measured film solubility, contact angle, and water sorption isotherm indicated that reinforced starch/CNC films have a lower affinity to water molecules than starch films. Permeability, diffusivity, and solubility coefficients indicated that the permeation process was controlled by the water diffusion and was dependent on the tortuous pathway formed by CNC incorporation. The decrease in surface hydrophilicity and the improvement in water vapor barrier properties with the addition of CNC showed that these nanocomposites present excellent potential as a new biomaterial for application in food packaging and conservation.

Soon-Do Yoon et al. (2012) studied the water barrier properties of starch/PVA composite films by adding nanosized poly (methyl methacrylate-co-acrylamide) particles. It can be found that the water vapor absorption of the films decreased with the increase of PMMA-co-AAm contents and MMA mole ratio of PMMA-co-AAm. The reason of the increase of water content for films added PMMA-co-AAm particles with the difference particle sizes was because of the formation of voids and the agglomeration between copolymer particles.

Bourtoom and Chinnan (2009) studied the effect of lipid types (oleic acid, palm oil, and margarine) and their concentrations (0, 10, 20, 30, 40, and 50 wt%) on the water vapor permeability (WVP) on rice starch–chitosan composite film. WVP of rice starch–chitosan composite film decreased with the addition of lipids. Oleic acid-incorporated films provided the films with smoother surface and lower WVP than margarine and palm oil, respectively. Wrapping cracker samples in the rice starch–chitosan composite film incorporated with oleic acid could maintain the hardness and provide longer shelf life and lower moisture content than synthetic polyvinyl chloride film (PVC).

5.3 Applications of Starch-Based Green Composites

5.3.1 Food Packaging Applications

Packaging plays a variety of important role in the food industry. The major role of packaging is to protect food from spoilage by microbial contamination, physical damage, or biochemical reactions. Packaging also provides ease in handling, storage efficiency, attractiveness, and product information for food. The ideal food packaging material serves all of these purposes and is also cost efficient. Different types of materials, including plastics, cardboard, and metal, are used for food packaging depending on specific needs. The use of plastics in food packaging, including films, is common and increasing because of low cost and functional advantages over other materials. Though plastics are one of the cheapest sources available for food packaging, their reliance on petroleum and their long-term impact on the environment have spurred research in recent years on alternative packaging based on renewable and biodegradable materials. Starch is one such inexpensive, abundantly available, and renewable material that can be used for making biodegradable packaging films. Starch-based bionanocomposites are hybrid nanostructured materials with improved mechanical, thermal, and gas barrier properties. The use of these materials for food packaging not only protects the food and increases its shelf life but can also be considered a more environmentally friendly solution because it reduces the requirement to use plastics as packaging materials (Fig. 7).

Polysaccharides-based coatings have low oxygen permeability, which can reduce the respiration rate of minimally processed products. Starch is the most important polysaccharide used in the formulation of biodegradable films and edible coatings. Cassava starch-based coatings are tasteless, odorless, and transparent, not changing the taste, aroma, and appearance of the product. Although the starch is a cheap and abundant material, able to form a continuous polymer matrix, it exhibits a strong hydrophilic character, constituting poor barriers to water vapor. The addition of lipids can reduce water vapor permeability, but it can affect coatings transparency and mechanical properties, despite the lipid offers after taste, which may impair the sensory characteristic of food. The results with cassava starch, glycerol, carnauba wax, and stearic acid showed that high wax content and low glycerol concentration form poor films, with rigid structure and poor gases and water vapor barrier. Films, composed of 3 % (w/w) of cassava starch, 1.5 % (w/w) glycerol, 0.2 % (w/w) of carnauba wax, and 0.8 % (w/w) of stearic acid presented coatings with good barrier properties, and films with good mechanical, thermal, physical, and structural properties, suitable for use as coatings. It can be potentially

Fig. 7 Toast packaged with biodegradable film based on cassava starch formulate with glycerol, sucrose, and inverted sugar as plasticizers (Souza et al. 2012)

used in minimally processed fruit for product quality maintenance and it will be evaluated in a further work (Chiumarelli and Hubinger 2014)

Cassava starch has also been extensively used to produce biodegradable films and the results indicated that these carbohydrates are promising materials in this regard. Films developed from starch are described as isotropic, odorless, tasteless, colorless, nontoxic, and biologically degradable. The association of cassava starch with plasticizers as glycerol, sucrose, and inverted sugar can promote alterations in the films, justifying the study of these additives to develop a potential and ecological alternative to the synthetic packaging of several food products. As a natural biopolymer, besides its biodegradable character, starch would be a promising alternative for the development of new food packaging materials because of its attractive combination of availability and price (Souza et al. 2012).

5.3.2 Biomedical Application

Studies with biodegradable starch-based polymers have recently demonstrated that these materials have a range of properties, which make them suitable for use in several biomedical applications, ranging from bone plates and screws to drug delivery carriers and tissue engineering scaffolds. Biodegradable starch-based polymers have recently been proposed as having great potential for several applications in the biomedical field such as bone replacement implants, bone cements, drug delivery systems, and tissue engineering scaffolds. The development of new processing technique and the reinforcement with various fillers result in materials with mechanical properties matching those of bone. However, other conditions should be met for a material to be considered suitable for any biomedical use. The evaluation of the in vitro cytotoxicity of a biomaterial is the initial step on a biocompatibility study, and is usually performed using immortalized cell lines being often a qualitative analysis, based on the morphological examination of cell damage and growth when in direct or indirect contact with the materials.

Marques et al. (2002) studied the biocompatibility of starch-based polymers. The materials used for this study were: (i) a 50/50 (wt%) blend of cornstarch and ethylene vinyl alcohol (SEVA-C), (ii) SEVA-C reinforced with 30 % (wt) of hydroxyapatite, (iii) a 50/50 (wt%) blend of cornstarch and cellulose acetate (SCA), and (iv) SCA reinforced with 30 % (wt) of hydroxyapatite. In the composites the average size of 90 % of the HA particles was found to be below 6.5 mm. Cytotoxicity tests with the extract of the materials were performed in order to evaluate the presence and or release of toxic leachables and degradation products. Cell material interactions on the surface of the polymers were observed by scanning electron microscopy (SEM) and related to the materials formulations. The short-term effect of leachables from starch-based polymers was quantified by exposing L929 cell to the degradation products released by those materials after immersion in culture medium.

A confluent monolayer was present for the two polymers after 7 days of contact, even though SEVA-C composite presented a slightly higher degree of cytotoxicity than the unreinforced SEVA-C. After 7 days of contact an almost confluent monolayer was observed, there was a delayed proliferation of the cells but they maintained good morphology. From SEM micrograph (Fig. 8), it was possible to

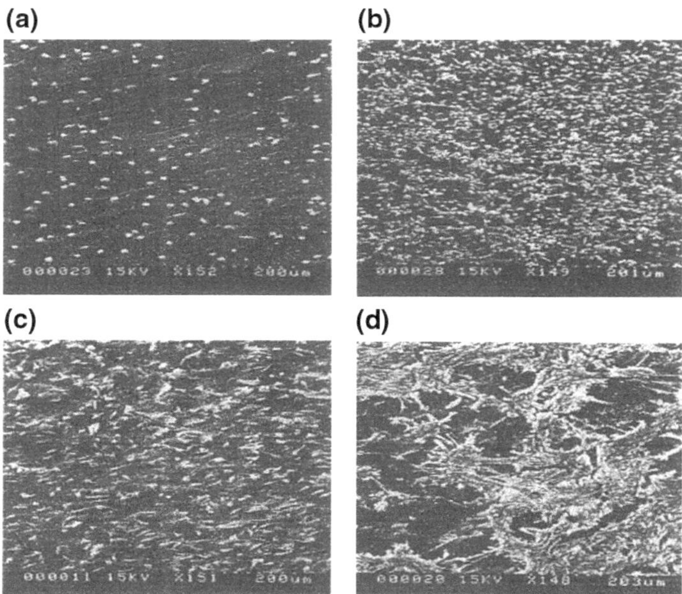

Fig. 8 SEM micrograph showing the L929 fibroblasts adhesion to the surface of SEVA-C (**a**, **b**) and SCA (**c**, **d**) polymers. **a**, **C** One day of growth; **b**, **d** Seven days of growth. Reprinted with permission Marques et al. (2002), Copyright 2002 Elsevier

see cells adhered on the surface of the polymers. On SEVA-C and SCA, cells cover most of the surface after 7 days of growth.

The differences in cytotoxicity, between SEVA-C polymers and composites, can be explained by some thermal degradation of the polymeric chains that occurred during the preparation of the composites. During the processing, the polymers with reinforcement are subjected to more severe thermal and shear cycles (extrusion compounding and injection molding) that always provoke some thermal degradation (due to viscous heat dissipation), generating low molecular weight fragments. These fragments are easily leached to the solution during the extract preparation and consequently the concentration of degradation products for the same extraction period is higher for the reinforced polymers. Also it has been reported that there is a preferential attack by the degradation fluids at the polymer/reinforcement interface. This higher degradation rate may explain the obtained cytotoxicity data. The same type of behavior would be predicted for SCA polymers and their composites. However, the opposite was observed and this kind of behavior could be due to the pre-purification stage (in water) performed for these materials. During this procedure the low molecular weight chains that were originated by some thermal degradation during processing, are leached out to the solution and the cellular response to the polymer and composite become more similar.

6 Conclusion

With increasing environmental and sustainable concerns about petroleum-based polymer materials, development of biodegradable bio-based materials will catch more and more attention. The good biodegradability, biocompatibility, and processability of green composites based on soy protein and starch have generated great interest among researchers to develop materials based on these bioplastics to replace at least partially petroleum-based composites. Performance of the green composites is dependent on the inherent properties of the matrix and reinforcement and their interface characteristics. We can tailor the properties of the green composites by optimizing the processing parameters and by employing suitable physical or chemical modifications to improve the interface. Identifying the thrust areas for the application of green composites and manufacture of prototypes and fabrication of useful products has an important role in green composite research. The promising results shown by these materials enable continuity of this research line and as a consequence additional exploration of biodegradation performance and different semi-industrial processes is actually underway, in order to take into account the excellent characteristics of both natural components: matrix and reinforcement. Thus the green composites will play a major role in replacing nonbiodegradable synthetic materials in the near future.

References

Bai H, Xu J, Liao P, Liu X (2012) Mechanical and water barrier properties of soy protein isolate film incorporated with gelatin. J Plast Film Sheeting 29(2):174–188

Behera AK, Avancha S, Basak RK, Sen R, Adhikari B (2012) Fabrication and characterizations of biodegradable jute reinforced soy based green composites. Carbohydr Polym 88(1):329–335

Boufi S, Kaddami H, Dufresne A (2014) Mechanical performance and transparency of nanocellulose reinforced polymer nanocomposites. Macromol Mater Eng 299(5):560–568

Bourtoom T, Chinnan MS (2009) Improvement of water barrier property of rice starch–chitosan composite film incorporated with lipids. Food Sci Technol Int 15(2):149–158

Castano J, Rodriguez-Llamazares S, Carrascoa C, Bouza R (2012) Physical, chemical and mechanical properties of pehuen cellulosic husk and its pehuen-starch based composites. Carbohydr Polym 90:1550–1556

Chabba S, Matthews GF, Netravali aN (2005) Green composites using cross-linked soy flour and flax yarns. Green Chem 7(8):576–581

Chan RT (2012). A study on the extrusion of soy protein film incorporated with soy-derived cellulose fibers. Thesis

Chen G, Liu H (2008) Electrospun cellulose nanofiber reinforced soybean protein isolate composite film. J Appl Polym Sci 110(2):641–646

Chen L, Remondetto GE, Subirade M (2006) Food protein-based materials as nutraceutical delivery systems. Trends Food Sci Technol 17(5):272–283

Chen CX, Chang PR, Stumborg M, Huneeault MA (2008) Green composites reinforced with hemp nanocrystals in plasticized starch. J Appl Polym Sci 109:3804–3810

Cherian BM, Leao AL, de Souza SF, Costa LMM, de Olyveira GM, Kottaisamy M, Nagarajan ER, Thomas S (2011) Cellulose nanocomposites with nanofibres isolated from pineapple leaf fibers for medical applications. Carbohydr Polym 86(4):1790–1798

Chien KB, Shah RN (2012) Novel soy protein scaffolds for tissue regeneration: material characterization and interaction with human mesenchymal stem cells. Acta Biomater 8 (2):694–703

Chinma CE, Ariahu CC, Alakali JS (2013). Effect of temperature and relative humidity on the water vapour permeability and mechanical properties of cassava starch and soy protein concentrate based edible films. J Food Sci Technol. doi: 10.1007/s13197-013-1227-0

Chiumarelli M, Hubinger MD (2014) Evaluation of edible films and coatings formulate with cassava starch, glycerol, carnauba wax and stearic acid. Food Hydrocolloids 38:20–27

Chung YL, Ansari S, Estevez L, Hayrapetyan S, Giannelis EP, Lai HM (2010) Preparation and properties of biodegradable starch–clay nanocomposites. Carbohydr Polym 79:391–396

Ciannamea EM, Stefani PM, Ruseckaite Ra (2014) Physical and mechanical properties of compression molded and solution casting soybean protein concentrate based films. Food Hydrocolloids 38:193–204

Deepa B, Abraham E, Cherian BM, Bismark A, Blaker JJ, Pothan LA, Leao AL, de Souza SF, Kottaisamy M (2011) Structure, morphology and thermal characteristics of banana nano fibers obtained by steam explosion. Bioresour Technol 102(2):1988–1997

Dufresne A (2010) Processing of Polymer Nanocomposites Reinforced with Polysaccharide Nanocrystals. Molecules 15:4111–4128

Echeverría I, Eisenberg P, Mauri AN (2014) Nanocomposites fi lms based on soy proteins and montmorillonite processed by casting. J Membr Sci 449:15–26

Espitia PJP, Du W-X, Avena-Bustillos R de J, Soares N de FF, McHugh TH (2014). Edible films from pectin: Physical-mechanical and antimicrobial properties—a review. Food Hydrocolloids 35:287–296

Fagundes C, Palou L, Monteiro AR, Pérez-Gago MB (2014) Effect of antifungal hydroxypropyl methylcellulose-beeswax edible coatings on gray mold development and quality attributes of cold-stored cherry tomato fruit. Postharvest Biol Technol 92:1–8

Fernandes EM, Pires RA, Mano JF, Reis RL (2013) Bionanocomposites from lignocellulosic resources: properties, applications and future trends for their use in the biomedical field. Prog Polym Sci 38(10–11):1415–1441

Flores S, Fama L, Rojas AM, Goyanes S, Gerschenson L (2007) Physical properties of tapioca-starch edible films: influence of film making and potassium sorbate. Food Res Int 40:257–265

Flores-Hernández C et al (2014) All green composites from fully renewable biopolymers: chitosan-starch reinforced with keratin from feathers. Polymers 6(3):686–705

Ghanbarzadeh B, Oleyaei SA, Almasi H (2014) Nano-structured materials utilized in biopolymer based plastics for food packaging applications. Crit Rev Food Sci Nutr 54(2):1–78

Ghidelli C, Mateos M, Rojas-Argudo C, Pérez-Gago MB (2014) Extending the shelf life of fresh-cut eggplant with a soy protein–cysteine based edible coating and modified atmosphere packaging. Postharvest Biol Technol 95:81–87

Gironès J, López JP, Mutj P, Carvalho AJF, Curvelo AAS, Vilaseca F (2012) Natural fiber-reinforced thermoplastic starch composites obtained by melt processing. Compos Sci Technol 72:858–863

González A, Strumia MC, Ines C, Igarzabal A (2011) Cross-linked soy protein as material for biodegradable films: synthesis, characterization and biodegradation. J Food Eng 106:331–338

Guerrero P, de la Caba K (2010) Thermal and mechanical properties of soy protein films processed at different pH by compression. J Food Eng 100(2):261–269

Guerrero P, Retegi a, Gabilondo N, de la Caba K (2010) Mechanical and thermal properties of soy protein films processed by casting and compression. J Food Eng 100(1):145–151

Guerrero P, Nur Hanani Za, Kerry JP, de la Caba K (2011) Characterization of soy protein-based films prepared with acids and oils by compression. J Food Eng 107(1):41–49

Gupta P, Nayak KK (2014) Characteristics of protein-based biopolymer and its application. Polym Eng Sci 4(2):3721–3735

Huang X, Netravali A (2009) Biodegradable green composites made using bamboo micro/nano-fibrils and chemically modified soy protein resin. Compos Sci Technol 69(7–8):1009–1015

Ibrahim H, Farag M, Megahed H, Mehanny S (2014) Characteristics of starch-based biodegradable composites reinforced with date palm and flax fibers. Carbohydr Polym 101:11–19

Ifuku S, Morooka S, Morimoto M, Saimoto H (2010) Acetylation of chitin nanofibers and their transparent nanocomposite films. Biomacromolecules 11:1326–1330

Imam M, Bania KK, Maji TK (2013) Green jute-based cross-linked soy flour nanocomposites reinforced with cellulose whiskers and nanoclay. Ind Eng Chem Res 52:6969–6983

John Wiley & Sons (2014) Cellulose based composites: new green nanomaterials (Google eBook)

Jong L, Peterson SC (2008) Effects of soy protein nanoparticle aggregate size on the viscoelastic properties of styrene–butadiene composites. Compos A Appl Sci Manuf 39(11):1768–1777

Kalia S, Dufresne A, Cherian BM, Kaith BS, Averous L, Njuguna J, Nassiopoulos E (2011). Cellulose based bio and nano composites-a review. Int J Polym Sci. http://dx.doi.org/10.1155/2011/837875

Katerinopoulou K, Giannakas A, Grigoriadi K, Barkoula NM, Ladavos A (2014) Preparation and characterization of acetylated corn starch-(PVOH)/clay nanocomposite films. Carbohydr Polym 102:216–222

Kim JT, Netravali AN (2011) Development of aligned-hemp yarn-reinforced green composites with soy protein resin: effect of pH on mechanical and interfacial properties. Compos Sci Technol 71(4):541–547

Kobayashi S, Takada K, Nakamura R (2014) Processing and characterization of hemp fiber textile composites with micro-braiding technique. Compos A Appl Sci Manuf 59:1–8

Kokoszka S, Debeaufort F, Hambleton A, Lenart A, Voilley A (2010) Protein and glycerol contents affect physico-chemical properties of soy protein isolate-based edible films. Innovative Food Sci Emerg Technol 11(3):503–510

Kumar R, Choudhary V, Mishra S, Varma IK, Mattiason B (2002) Adhesives and plastics based on soy protein products. Ind Crops Prod 16(3):155–172

Kumar P, Sandeep KP, Alavi S, Truong VD (2011) A review of experimental and modeling techniques to determine properties of biopolymer-based nanocomposites. J Food Sci 76 (1):2–14

La Mantia FP, Morreale (2011) Green composites: a brief review. Compos A 42:579–588

Li Y, Li N, Gao J (2013) Tooling design and microwave curing technologies for the manufacturing of fiber-reinforced polymer composites in aerospace applications. Int J Adv Manuf Technol 70 (1–4):591–606

Lin M-F, Thakur VK, Tan EJ, Lee PS (2011a) Dopant induced hollow $BaTiO_3$ nanostructures for application in high performance capacitors. J Mater Chem 21:16500–16504

Lin M-F, Thakur VK, Tan EJ, Lee PS (2011b) Surface functionalization of $BaTiO_3$ nanoparticles and improved electrical properties of BaTiO3/polyvinylidene fluoride composite. RSC Adv 1:576–578

Liang K, Gao Q, Shi SQ (2013) Kenaf fiber/soy protein based biocomposites modified with poly (carboxylic acid) resin. J Appl Polym Sci 128(2):1213–1218

Liu Y, Du H, Liu L and Leng J (2014). Shape memory polymers and their composites in aerospace applications: a review. Smart Mater Struct 23(2):22

Lodha P, Netravali AN (2005) Characterization of stearic acid modified soy protein isolate resin and ramie fiber reinforced green composites. Compos Sci Technol 65(7–8):1211–1225

Lopez O, Garcia MA, Villar MA, Gentili A, Rodriguez MS, Alberteng L (2014) Thermo-compression of biodegradable thermoplastic corn starch films containing chitin and chitosan. Food Sci Technol 57:106–115

Lu Y, Weng L, Zhang L (2004) Morphology and properties of soy protein isolatethermoplastics reinforced with chitin whiskers. Biomacromolecules 5:1046–1051

Mariano M, Kissi NE, Dufresne A (2014) Cellulose nanocrystals and related nanocomposites: review of some properties and challenges. J Polym Sci B Polym Phys 52:791–806

Marques AP, Reisa RL, Hunt JA (2002) The biocompatibility of novel starch-based polymers and composites: in vitro studies. Biomaterials 23:1471–1478

Maruthi Y, Sudhakar H, Rao US, Babu PK, Rao KC, Subha MCS (2014) Blend membranes of sodium alginate and soya protein for pervaporation dehydration of isopropanol. Adv Polym Sci Technol 4(2):12–21

Mitra B (2014) Environment friendly composite materials: biocomposites and green composites. Defence Sci J 64(3):244–261

Neethirajan S, Jayas DS (2011) Nanotechnology for the food and bioprocessing industries. Food Bioprocess Technol 4:39–47

Nur Hanani ZA, Roos YH, Kerry JP (2014) Use and application of gelatin as potential biodegradable packaging materials for food products. Int J Biol Macromol 71:1–9

Ong S-Y, Wu J, Moochhala SM, Tan M-H, Lu J (2008) Development of a chitosan-based wound dressing with improved hemostatic and antimicrobial properties. Biomaterials 29 (32):4323–4332

Pan H, Jiang B, Chen J, Jin Z (2014) Blend-modification of soy protein/lauric acid edible films using polysaccharides. Food Chem 151:1–6

Privas E, Leroux F, Navard P (2013) Preparation and properties of blends composed of lignosulfonated layered double hydroxide/plasticized starch and thermoplastics. Carbohydr Polym 96:91–100

Ramaraj B (2007) Crosslinked poly(vinyl alcohol) and starch composite films: study of their physico-mechanical, thermal and swelling properties. J Appl Polym Sci 103:1127–1132

Reddy N, Yang Y (2011) Completely biodegradable soyprotein–jute biocomposites developed using water without any chemicals as plasticizer. Ind Crops Prod 33(1):35–41

Riveros CG, Mestrallet MG, Quiroga PR, Nepote V, Grosso NR (2013) Preserving sensory attributes of roasted peanuts using edible coatings. Int J Food Sci Technol 48(4):850–859

Rossi Marquez G, Di Pierro P, Esposito M, Mariniello L, Porta R (2013) Application of transglutaminase-crosslinked whey protein/pectin films as water barrier coatings in fried and baked foods. Food Bioprocess Technol 7(2):447–455

Routray M, Rout SN, Mohanty GC, Nayak PL (2013) Preparation and characterization of soy protein isolate films processed by compression and casting. J Chem Pharm Res 5(11):752–761

Sareena C, Sreejith MP, Ramesan MT, Purushothaman E (2014) Biodegradation behaviour of natural rubber composites reinforced with natural resource fillers—monitoring by soil burial test. J Reinf Plast Compos 33(5):412–429

Silva NHCS, Vilela C, Marrucho IM, Freire CSR, Neto CP, Silvestre AJD (2014) Protein-based materials: from sources to innovative sustainable materials for biomedical applications. J Mater Chem B 2(24):3715–3740

Singha AS, Thakur VK (2008a) Effect of fibre loading on urea-formaldehyde matrix based green composites. Iran Polym J 17:861–873

Singha AS, Thakur VK (2008b) Synthesis and characterization of pine needles reinforced RF matrix based biocomposites. J Chem 5:1055–1062

Singha AS, Thakur VK (2008c) Saccaharum cilliare fiber reinforced polymer composites. E-J Chem 5:782–791

Singha AS, Thakur VK (2008d) Fabrication and study of lignocellulosic hibiscus sabdariffa fiber reinforced polymer composites. Bioresources 3:1173–1186

Singha AS, Thakur VK (2009a) Morphological, thermal, and physicochemical characterization of surface modified pinus fibers. Int J Polym Anal Charact 14(3):271–289

Singha AS, Thakur VK (2009b) Fabrication and characterization of H. sabdariffa fiber-reinforced green polymer composites. Polym-Plast Technol Eng 48:482–487

Singha AS, Thakur VK (2009c) Grewia optiva fiber reinforced novel, low cost polymer composites. J Chem 6:71–76

Singha AS, Thakur VK (2009d) Fabrication and characterization of S. cilliare fibre reinforced polymer composites. Bull Mater Sci 32:49–58

Singha AS, Thakur VK (2009e) Synthesis, characterization and analysis of Hibiscus Sabdariffa fibre reinforced polymer matrix based composites. Polym Polym Compos 17:189–194

Singha AS, Thakur VK, Mehta IK, Shama A, Khanna AJ, Rana RK, Rana AK (2009a) Surface-modified Hibiscus sabdariffa fibers: physicochemical, thermal, and morphological properties evaluation. Int J Polym Anal Charact 14(8):695–711

Singha AS, Thakur VK, Mishra BN (2009b) Study of Grewia optiva fiber reinforced urea-formaldehyde composites. J Polym Mater 26:81–90

Siqueira G, Bras J, Dufresne A (2010) Cellulosic bionanocomposites: a review of preparation. Prop Appl Polym 2(4):728–765

Siro I, Plackett D (2010) Microfibrillated cellulose and new nanocomposite materials: a Review. Cellulose 17:459–494

Sirviö JA, Kolehmainen A, Liimatainen H, Niinimäki J, Hormi OEO (2014) Biocomposite cellulose-alginate films: promising packaging materials. Food Chem 1513:43–51

Slavutskya AM, Bertuzz MA (2014) Water barrier properties of starch films reinforced with cellulose nanocrystals obtained from sugarcane bagasse. Carbohydr Polym 110:53–61

Slomkowski S, Penczek S, Duda A (2014) Polylactides-an overview. Polym Adv Technol 25 (5):436–447

Song Y, Zheng Q (2014) Ecomaterials based on food proteins and polysaccharides. Polym Rev 54 (3):514–571

Song F, Tang D, Wang X, Wang Y (2011) Biodegradable Soy Protein Isolate-Based Materials: a Review. Biomacromolecules 12(10):3369–3380

Souza AC, Benze R, Ferrão ES, Ditchfield C, Coelho ACV, Tadini CC (2012) Cassava starch biodegradable films: influence of glycerol and clay nanoparticles content on tensile and barrier properties and glass transition temperature. Food Sci Technol 46:110–117

Soykeabkaew N, Supaphol P, Rujiravanit R (2004) Preparation and characterization of jute- and flax-reinforced starch-based composite foams. Carbohydr Polym 58:53–63

Thakur VK, Kessler MR (2014a) Free radical induced graft copolymerization of ethyl acrylate onto SOY for multifunctional materials. Mater Today Commun. doi: 10.1016/j.mtcomm.2014.09.003

Thakur VK, Kessler MR (2014b) Synthesis and characterization of AN-g-SOY for sustainable polymer composites. ACS Sustain Chem Eng. 2:2454–2460
Thakur VK, Thakur MK (2014a) Recent trends in hydrogels based on psyllium polysaccharide: a review. J Clean Prod 82:1–15
Thakur VK, Thakur MK (2014b) Recent advances in graft copolymerization and applications of chitosan: a review. ACS Sustaina Chem Eng 2:2637–2652
Thakur VK, Thakur MK (2014c) Processing and characterization of natural cellulose fibers/ thermoset polymer composites. Carbohydr Polym 109:102–117
Thakur VK, Singha AS, Kaur I, Nagarajarao RP, Liping Y (2010a) Silane functionalization of Saccaharum cilliare fibers: thermal, morphological, and physicochemical study. Int J Polym Anal Charact 15(7):397–414
Thakur VK, Singha AS, Mehta IK (2010b) Renewable resource-based green polymer composites: analysis and characterization. Int J Polym Anal Charact 15(3):137–146
Thakur VK, Singha AS, Thakur MK (2012a) Rapid synthesis of MMA grafted pine needles using microwave radiation. Polym-Plast Technol Eng 51:1598–1604
Thakur VK, Singha AS, Thakur MK (2012b) Graft copolymerization of methyl acrylate onto cellulosic biofibers: synthesis, characterization and applications. J Polym Environ 20:164–174
Thakur VK, Singha AS, Thakur MK (2012c) Biopolymers based green composites: mechanical, thermal and physico-chemical characterization. J Polym Environ 20:412–421
Thakur VK, Singha AS, Thakur MK (2012d) Modification of natural biomass by graft copolymerization. Int J Polym Anal Charact 17:547–555
Thakur VK, Singha AS, Thakur MK (2012e) In-air graft copolymerization of ethyl acrylate onto natural cellulosic polymers. Int J Polym Anal Charact 17:48–60
Thakur VK, Thakur MK, Gupta RK (2013a) Development of functionalized cellulosic biopolymers by graft copolymerization. Int J Biol Macromol 62:44–51
Thakur VK, Thakur MK, Gupta RK (2013b) Synthesis of lignocellulosic polymer with improved chemical resistance through free radical polymerization. Int J Biol Macromol 61:121–126
Thakur VK, Thakur MK, Gupta RK (2013c) Rapid synthesis of graft copolymers from natural cellulose fibers. Carbohydr Polym 98:820–828
Thakur VK, Thakur MK, Gupta RK (2013d) Graft copolymers from cellulose: Synthesis, characterization and evaluation. Carbohydr Polym 97:18–25
Thakur VK, Thakur MK, Gupta RK (2013e) Graft copolymers from natural polymers using free radical polymerization. Int J Polym Anal Charact 18:495–503
Thakur VK, Thakur MK, Gupta RK (2014a) Review: raw natural fiber-based polymer composites. Int J Polym Anal Charact 19:256–271
Thakur VK, Thakur MK, Raghavan P, Kessler MR (2014b) Progress in green polymer composites from lignin for multifunctional applications: a review. ACS Sustain Chem Eng 2(5):1072–1092
Thakur VK, Thunga M, Madbouly SA, Kessler MR (2014c) PMMA-g-SOY as a sustainable novel dielectric material. RSC Adv 4:18240–18249
Thakur VK, Grewell D, Thunga M, Kessler MR (2014d) Novel composites from eco-friendly soy flour/SBS triblock copolymer. Macromol Mater Eng 299:953–958
Unalan IU, Cerri G, Marcuzzo E, Cozzolino CA, Farris S (2014) Nanocomposite films and coatings using inorganic nanobuilding blocks (NBB): current applications and future opportunities in the food packaging sector. R Soc Chem 4:29393–29428
Venkatesan J, Bhatnagar I, Kim SK (2014) Chitosan-alginate biocomposite containing fucoidan for bone tissue engineering. Marine drugs 12(1):300–316
Wan YZ, Luo H, He F, Liang H, Huang Y, Li XL (2009) Mechanical, moisture absorption, and biodegradation behaviours of bacterial cellulose fiber-reinforced starch biocomposites. Compos Sci Technol 69:1212–1217
Wang Y, Cao X, Zhang L (2006) Effects of cellulose whiskers on properties of soy protein thermoplastics. Macromol Biosci 6(7):524–531
Wang Z, Sun X, Lian Z, Wang X, Zhou J, Ma Z (2013) The effects of ultrasonic/microwave assisted treatment on the properties of soy protein isolate/microcrystalline wheat-bran cellulose film. J Food Eng 114(2):183–191

Wang Z, Zhou J, Wang X, Zhang N, Sun X, Ma Z (2014) The effects of ultrasonic/microwave assisted treatment on the water vapor barrier properties of soybean protein isolate-based oleic acid/stearic acid blend edible films. Food Hydrocolloids 35:51–58

Wihodo M, Moraru CI (2013) Physical and chemical methods used to enhance the structure and mechanical properties of protein films: a review. J Food Eng 114(3):292–302

Woehl MA, Canestraro CD, Mikowski A, Sierakowski MR (2010) Bionanocomposites of thermoplastic starch reinforced with bacterial cellulose nanofibers: effect of enzymatic treatment on mechanical properties. Carbohydr Polym 80:866–873

Xu YX, Kim KM, Hanna MA, Nag D (2005) Chitosan–starch composite film: preparation and characterization. Ind Crops Prod 21:185–192

Xu B, Lin Z, Huo Z, Cao L, Wang Y, Gaosun W, Mai K, Wang Y (2014). Preparation and characterization of polypropylene composites with nonmetallic materials recycled from printed circuit boards. J Thermoplast Compos Mater. doi:10.1177/0892705713518788

Yang S, Huang C (2008) Plasma treatment for enhancing mechanical and thermal properties of biodegradable PVA/starch blends. J Appl Polym Sci 109:2452–2459

Yoon SD, Park Mi-Hwa, Byun Hun-Soo (2012) Mechanical and water barrier properties of starch/ PVA composite films by adding nano-sized poly(methyl methacrylate-co- acrylamide) particles. Carbohydr Polym 87(1):676–686

Yuan Y, Sun Y, Wan Z, Yang X, Wu J, Yin S, Wang J, Guo J (2014) Chitin microfibers reinforce soy protein gels cross-linked by transglutaminase. J Agric Food Chem 62:4434–4442

Zhang H, Mittal G (2010) Biodegradable protein-based films from plant resources: a review. Environ prog Sustain Energy 29(2):203–220

Multicomponent Polymer Composite/Nanocomposite Systems Using Polymer Matrices from Sustainable Renewable Sources

Carmen-Alice Teacă and Ruxanda Bodîrlău

Abstract Green composites are increasingly promoted for sustainable development considering the growing awareness of environmental and waste management issues. Recent advances in natural fiber development, and nanocomposites research area generate significant opportunities for obtaining materials from renewable resources with improved properties and suitable for different applications. Green composites are made from both renewable resource-based polymers (biopolymers) and bio-fillers (including nano-type fillers), with a positive environmental impact. Green composites based on biopolymer matrix (plasticized starch) have been obtained by combination with various bio-fillers (beech wood sawdust, fir tree needles, beech wood lignin). Their structure and properties were further investigated through Fourier Transform Infrared (FTIR) spectroscopy, X-ray diffraction, scanning electron microscopy (SEM), and TG/DTG/DTA simultaneous thermal analysis methods, as well as by water uptake and opacity measurements. The results are presented in this chapter.

Keywords Composites · Thermoplastic starch · Beech wood sawdust · Fir tree needles · Beech wood lignin · Properties

1 Introduction

The development of advanced polymer materials for various sustainable applications requires obtaining polymer composites from low-cost, environmental friendly, renewable, and biodegradable resources (with natural biopolymers in their structure), with the main focus on enhanced multifunctional properties (e.g., thermal stability, impact resistance, water resistance, photo-stability) and reduced impact as pollution effects under environmental factors action (Thakur et al. 2014a, b; Faruk et al. 2014; Lee et al. 2014; Al-Oqla and Sapuan 2014; Isogai 2013; Lee et al. 2012a, b;

C.-A. Teacă (✉) · R. Bodîrlău
Advanced Research Center for Bionanoconjugates and Biopolymers, "Petru Poni" Institute of Macromolecular Chemistry, 41A Gr. Ghica-Voda Alley, 700487 Iaşi, Romania
e-mail: cateaca14@yahoo.com

V.K. Thakur and M.K. Thakur (eds.), *Eco-friendly Polymer Nanocomposites*,
Advanced Structured Materials 75, DOI 10.1007/978-81-322-2470-9_15

Huber et al. 2012; Klemm et al. 2011; Blaker et al. 2011; Eichhorn 2011; Habibi et al. 2010; Bledzki and Jaszkiewicz 2010; Satyanarayana et al. 2009). Polymer composites, based on polymer matrices (thermoplastics, thermosets, and elastomers) and nano-reinforced fillers (layered silicates/clay/carbon nanotubes; natural biopolymers with nano-type size dimension, e.g., nanocellulose, nanostarch, nanochitin), exhibit significantly enhanced properties including mechanical, thermal, gas barrier, and good flame retardancy, due to good dispersion of fillers within polymer matrices (Reddy et al. 2013; Koga et al. 2013; Majeed et al. 2013; Savadekar and Mhaske 2012; Lee et al. 2012c; Ifuku et al. 2011; Kord 2011; Megiatto et al. 2010; Ramires et al. 2010; Bledzki et al. 2010; Chang et al. 2010; Rojas et al. 2009; Cai et al. 2008). In addition to the improved physical and chemical properties, these can be prepared through simple processes with reduced costs (Thakur et al. 2014a, b, c, d, e, f). Some important issues should be investigated in regard to use of environmentally friendly polymers, improvement of adhesion, and compatibility between the polymer matrices and different fillers, and impact of chemical modification of composites structural components on the desired properties, envisaging mainly the mechanical reinforcing ones (Thakur et al. 2012a, b, c, d).

The development of nanocomposites derived from renewable sources with nanocellulose as reinforcing filler represents a significant research area (Isogai 2013; Klemm et al. 2011; Blaker et al. 2011; Hubbe et al. 2008). Promising results were obtained in nanocellulose applications including hybrid composite materials, films, dispersions, and foams. Incorporation of various nano-type reinforcements (such as cellulose nanofibers) derived from biodegradable biomass sources within different polymer matrices constitutes a sustainable and significant strategy for obtainment of nanocomposites with improved properties, mainly mechanical ones (Abdul Khalil et al. 2012). These biodegradable nano-reinforcing fillers can generate significant opportunities for the development of novel green nanocomposite materials to be suitable for various applications (e.g., for electronic devices). There are some comprehensive articles which have reviewed various issues related to different nanocelluloses and their obtainment methodologies and potential applications (Isogai 2013; Klemm et al. 2011; Habibi et al. 2010; Klemm et al. 2009). New efficient methods for production of various nanocellulose filler types include enzymatic, chemical, and physical methodologies for their isolation from wood sources and forest/agricultural waste residues (Nair et al. 2014; Qamhia et al. 2014; Abraham et al. 2013; Chen et al. 2011a, b; Abraham et al. 2011; Syverud et al. 2011; Kaushik and Singh 2011; Cherian et al. 2010; Klemm et al. 2009; Fukuzumi et al. 2009; Stelte and Sanadi 2009; Jonoobi et al. 2009; Moran et al. 2008; Abe et al. 2007; Janardhnan and Sain 2006; Chakraborty et al. 2005).

All-green composites are made from both renewable resource-based polymer (biopolymer) and bio-filler (Faruk et al. 2012; Soykeabkaew et al. 2012; Abdul Khalil et al. 2012; Singha et al. 2010; Liu et al. 2010; Dobircau et al. 2009; Pandey et al. 2009). Composites with natural biopolymers in composition (as polymer matrix and/or reinforcement) are recyclable, renewable, biodegradable, and may reduce energy consumption to a great extent when used in interior applications (Azwa et al. 2013; Singha and Thakur 2008a, b, c, d). All-green composites may be successfully applied in automotive, interior building, and packaging areas (Singha and Thakur 2009a, b, c).

Bio-based natural polymers including cellulose, starch, chitin, and various polysaccharides and proteins, and their derivatives are suitable for a wide range of properties and applications. For example, biopolymers, such as starch and cellulose, have been mainly used for food, furniture, and clothing applications. Use of non-woody materials such as agricultural residues and forest wastes for making panel products (Singha et al. 2009a, b; Singha and Thakur 2010a, b; Thakur and Singha 2010; Yang et al. 2003) have attracted more interest due to the lignocellulose-based sources similar to wood particles and limited wood resources.

The use of natural fibers, derived from a number of renewable resources, as reinforcing fibers in both thermoplastic and thermoset matrix composites provides significantly environmental advantages and generates improved properties over conventional materials such as lightness, resistance to corrosion, good mechanical behavior, and ease of processing (Thakur and Thakur 2014a, b; Thakur et al. 2011, 2012a; Faruk et al. 2012; Kalia et al. 2011).

Starch processing techniques include methods similar to those widely used for conventional synthetic thermoplastics such as solution casting, internal mixing, extrusion, injection molding, and compression molding (Liu et al. 2009). Usually, these methods are green, environmental friendly involving water which is indispensable for the thermal processing of starch. Different plasticizers are commonly used for improving starch properties, mainly those related to mechanical behavior and moisture resistance (Qiao et al. 2011; Dai et al. 2010a, b; Zhang et al. 2008).

Fabrication of composites based on organic or inorganic reinforcement is an efficient way to improve the performance of starch based films. Application of fillers such as clays (Chang et al. 2012; Gao et al. 2012; Majdzadeh-Ardakani et al. 2010; Zeppa et al. 2009; Zhang et al. 2007; Pandey and Singh 2005), natural fibers (Müller et al. 2014; Teacă et al. 2014; Bodîrlău et al. 2014; Mueangta and Hanchana 2013; Kaewtatip and Thongmee 2012; Gironès et al. 2012; Soykeabkaew et al. 2012; Prachayawarakorn et al. 2011; Sreekumar et al. 2010), cellulose with nano-sized dimension fibers (Hietala et al. 2013; Moran et al. 2013; Savadekar and Mhaske 2012; Liu et al. 2010; Kaushik et al. 2010), or microcrystalline cellulose (Bodîrlău et al. 2014; Ma et al. 2008a, b; Kumar and Singh 2008) are reported. Microfibrillated cellulose (MFC) and bacterial cellulose (BC) have also been reported as good reinforcements for starch films (Martins et al. 2009).

2 Polymer Matrices from Sustainable Renewable Sources

2.1 Starch

Starch is a unique bio-based polymer which occurs in nature as energy storage granules. Starch is a natural polysaccharide abundantly available in nature from various botanical sources including wheat, rice, corn, and potato. Its chemical structure is very complex, starch comprising two different polysaccharides, namely one linear, generally called amylose, and highly branched one, which is called amylopectin. Starch presents significant advantages including reduced costs, abundance, and environmental friendly through completely biodegradable behavior and no toxic effects after degradation. Starch can be processed in the form of nanocrystals and nanoparticles by using different approach methods (acid hydrolysis, enzymatic pretreatment of starch to reduce the acid hydrolysis duration, and precipitation by using ethanol, respectively) due to its versatile semicrystalline structure which is favorable for their production (LeCorre et al. 2011, 2012; Ma et al. 2008). These are suitable to be incorporated as fillers in order to improve mechanical and barrier properties of biocomposites (Averous and Pollet 2014; Reddy et al. 2013; Xie et al. 2013). Starch is usually converted into a molten state, namely thermoplastic starch, in the presence of different plasticizers (e.g., water, glycerol) by using conventional processing techniques (Ojijo and Ray 2013). Starch-based materials have many drawbacks due to their reduced processing ability and properties (low water resistance, reduced mechanical properties). Polymer composites comprising starch with nanosized dimension have suitable applications as edible films or food packaging (Habibi and Dufresne 2008).

2.2 Wood

Nowadays, ecological concern has resulted in a renewed interest in natural composite materials and issues such as recyclability and environmental protection are becoming increasingly important for the introduction of new materials and products. At this moment, eco-design concept is applied to more and more materials and products. These environmental issues in combination with the low cost of plant fibers such as flax, hemp, kenaf, palm, coconut, sisal, and wheat straw have generated considerable interest to be used as reinforcements in engineering composites (Al-Oqla and Sapuan 2014; Fernandes et al. 2013a, b; Wu et al. 2013; Faruk et al. 2012; Barone 2009). Plant fibers are currently being evaluated as environmental friendly and low-cost alternatives for glass fibers, whereas wood particles and derivatives are being used as an alternative to mineral fillers in commodity plastics like PE (polyethylene) and PP (polypropylene), PU (polyurethane) resin or for the upgrading of post-consumer recycled PVC (Bledzki et al. 2010; Ramires et al. 2010; Lei and Wu 2010; Sailaja and Deepthi 2010; Bledzki and Jaszkiewicz 2010; Bodîrlău et al. 2009).

Natural fibers have a number of advantages including: they are renewable, abundant, cheap, lightweight, biodegradable, non-abrasive to processing equipment, CO_2 neutral (when burned), flexible and give a 'safer' crash behavior, can be incinerated with energy recovery, show less concern with safety and health (no skin irritations), exhibit good mechanical properties, and have unique acoustic and thermal insulating properties. The combination of interesting mechanical and physical properties together with their environmentally friendly feature has triggered various research activities in the area of "green composites" (Faruk et al. 2012; Thakur et al. 2012a; Pandey et al. 2010).

Composites containing wood derivatives (WPC) have attracted a significant interest in the past decades, thanks to the specific advantages they can exhibit in comparison with the classic mineral filler/plastic composites (Sailaja and Deepthi 2010; Bodîrlău et al. 2009). These include mainly the improved environmental performance, due to the use of biodegradable materials and the reduction in the use of non-renewable (oil based) resources throughout the whole life cycle of the composite materials. Significant interest has derived from the outdoor performance of these composites, in particular their resistance to photo-oxidation processes. The wood particles and cellulose fibers which have high strength and modulus—with good adhesion and uniform dispersion—can impart better mechanical properties to the polymer matrix in order to obtain a composite with better properties than those of the unfilled polymer matrix (Csizmadia et al. 2013; Nair et al. 2013).

The enthusiasm in this field is explained by the number of potential applications. For example wood-polymer composites (WPC), mostly manufactured through extrusion and injection moulding processes are used in automotive (dashboards or screen-doors of the vehicles) and construction applications (interior floor coverings, profiles for doors and windows, ornamental panels, external shutters, pavements, garage, or entrance doors). Replacement of wood products for building applications such as particleboard and fiberboard materials or 'injection mouldable' wood is another area where natural fibers are generating increased interest.

The main disadvantage of natural fibers is related to their poor moisture resistance (rotting) and dimensional stability (swelling), which can lead to de-bonding and micro-cracking in the composite materials. However, the moisture resistance of natural fibers in composite structures can be improved through fiber treatments like chemical modifications (Csizmadia et al. 2013; Rowell 2012a; Gregorova et al. 2011; Bettini et al. 2010; Singha et al. 2009a, b; Stark and Gardner 2008). These modifications influence the thermal and mechanical properties of resulted composite materials. It is expected that addition of chemically modified wood as filler within the polymer matrix in composite materials will have a positive influence, which will be reflected in their behavior to accelerated weathering and attack of biological agents (Rowell 2012b).

2.3 *Softwood Needles*

Pine needles represent a significant waste biomass generated by *Pinus* softwood species. These needles contain significant amounts of carbohydrates (cellulose, hemicelluloses) and reduced quantity of lignin, besides predominant extractives.

The tensile strength of the pine needles from *Pinus pinaster* species (maritime pine) was recently investigated and this property makes them suitable for inclusion as reinforcement in polymer matrix composites subjected to low stress or non-load bearing applications including fiberboard and thermal or acoustic insulation (Dong et al. 2014).

Composites containing both pine needles and eucalyptus wood particles exhibited superior mechanical properties to the composites made only with the pine needles when these were used with the composite adhesive system: monomer isocyanate for pretreatment and isocyanate prepolymer for particle–particle bonding (Chauhan et al. 2013).

Isocyanate bonded pine needle composites were investigated for their dimensional stability, flammability characteristics, biological resistance, and thermoacoustic properties (Chauhan et al. 2012). Chauhan and co-workers (2012) concluded that such composites can be suitable for use as panel products in buildings applications.

Alkali treated (mercerized) pine needle furnishes bonded with isocyanate prepolymer can be effectively used as panel products for wood substitute in buildings applications, mainly under wet conditions when a good bonding with resin adhesive is required (Gupta et al. 2010).

Pine needles reinforced polymer composites using phenol-formaldehyde and urea-formaldehyde resins as polymer matrix were obtained through compression molding technique and investigated for mechanical properties and thermal behavior (Thakur and Singha 2011; Singha and Thakur 2010a, b, c).

3 Composites Comprising Starch as Polymer Matrix and Natural Fillers

3.1 Obtainment

3.1.1 Materials

A commercially corn starch (S) was used as continuous polymer matrix of the composite films. Glycerol was purchased from Fluka (98 % purity, Fluka Chemical, Germany) and used as plasticizer. The natural fillers, including fir tree needles, and beech wood flour, were obtained from local sources. Lignin was separated from beech wood using Tappi method. The fir tree needles of 300–350 mm length were collected from the Christmas tree, being previously stored at room temperature (moisture content around 10 %). Fillers were ground in a Retsch PM 200 planetary ball mill. Particles passed through 0.40 mm sieve were used for preparation of composites. The particles distribution size was determined for starch matrix and fillers as presented below.

Beech wood chemical composition was previously investigated and presented (Bodîrlău et al. 2007), this wood biomass resource comprising cellulose, lignin and ethanol extractives.

3.1.2 Method Used for Obtainment Natural Filler/Plasticized Starch Composites

Five gram corn starch was added in a solution of distilled water (100 ml) and glycerol (1 ml). The natural filler loading level was based on the amount of starch, being 15, 30, and 45 % respectively. A reference composite film was also prepared and comprised only glycerol thermoplasticized corn starch matrix without filler (coded S). The mixture was heated at 90 °C for 30 min for the complete gelatinization of corn starch with constant stirring. Natural filler/plasticized starch composite films were prepared through the fast coating technique namely the doctor blade technique. Films were obtained by dropping and spreading the mixture on glass plates using a blade with a slit width of 0.8 mm. After degassing in a vacuum oven at 50 °C for 24 h up to constant weight, the films were air cooled and detached from the glass surfaces to be investigated. Films with a thickness of ~0.2 mm (determined by means of a digital micrometer) were obtained. The composite film samples, coded as presented in Table 1, were pre-conditioned in a climate chamber at 25 °C and 50 % RH for at least 48 h prior to the testing. Water content of the composite films was around 9 wt%.

3.2 Investigation Methods

3.2.1 FTIR Spectroscopy Investigation

FTIR spectra of the components as well as of the composite film samples were recorded on a Bruker Vertex 70 spectrophotometer. The spectral resolution was 4 cm^{-1} and the scanning range varied from 400 to 4000 cm^{-1}.

Table 1 Code names used for composite films under study

Sample code	Composition
S	Glycerol plasticized starch film without filler
S/FTN15	Glycerol plasticized starch film with 15 % fir tree needles
S/FTN30	Glycerol plasticized starch film with 30 % fir tree needles
S/FTN45	Glycerol plasticized starch film with 45 % fir tree needles
S/BS15	Glycerol plasticized starch film with 15 % beech wood sawdust
S/BS30	Glycerol plasticized starch film with 30 % beech wood sawdust
S/BS45	Glycerol plasticized starch film with 45 % beech wood sawdust
S/BL15	Glycerol plasticized starch film with 15 % beech wood lignin
S/BL30	Glycerol plasticized starch film with 30 % beech wood lignin
S/BL45	Glycerol plasticized starch film with 45 % beech wood lignin

3.2.2 X-ray Diffraction Analysis

The crystalline structure of the starch and glycerol plasticized starch was studied by X-ray diffraction using a Bruker AD8 ADVANCE X-ray diffractometer with Cu Kα radiation at 60 kV and 50 mA, at room temperature. Scattered radiation was detected in the diffraction angle 2θ ranging from 10 to 30° at a rate of 2° min^{-1}.

3.2.3 TG/DTG/DTA Simultaneous Thermal Analysis

TG/DTG/DTA analysis was performed using a Netzsch STA 449 F1 Jupiter system under nitrogen atmosphere. The measurements were performed while heating the samples (≈5 mg) placed in Al_2O_3 crucibles at a rate of 10 °C/min from room temperature up to 600 °C and using nitrogen as a purging gas at a flow rate of 50 ml/min. TG curves recorded with a ± 0.5 °C precision were analyzed using a Netzsch Proteus analysis software.

3.2.4 Water Uptake Measurements

Thin rectangular film strips with dimensions of 10 mm × 10 mm × 0.2 mm were used as specimens in order to determine the water absorption. The films were supposed to be thin enough so that the molecular diffusion could be considered as one-dimensional and were vacuum-dried at 90 °C overnight. After weighing, film specimens were conditioned at 25 °C in a desiccator containing sodium sulfate in order to ensure a relative humidity (RH) of 95 %. Further, samples were removed at specific time intervals and gently blotted with tissue paper in order to remove the excess of water on the surface. Water uptake values were calculated with Eq. 1, as follows:

$$\text{water uptake }(\%) = [(W_t - W_0)/W_0] \times 100 \tag{1}$$

where W_t and W_0 represent the specimen weight at time t and before exposure to 95 % RH, respectively. The determinations were performed in triplicate.

3.2.5 Opacity Measurements

The optical properties of the films (opacity) were measured by a JENWAY 6405 UV–VIS spectrophotometer. This property is usually defined as the area under the absorbance spectrum between 400 and 800 nm according to the ASTM D 1003-00 method (ASTM D 1003-00 Standard Test Method for Haze and Luminous Transmittance of Transparent Plastics). The film samples were cut into rectangular pieces (1 × 2.5 cm, 0.2 mm thickness), fixed on the inner side of a spectrophotometer cell (with dimension of 1 cm) and the absorbance spectra recorded. The measurement of the films opacity was repeated thrice .

3.2.6 Particle Size Distribution Analysis

Particle size distribution analysis was performed using a laser diffractometer (Mastersizer 2000, Malvern Instruments). The film samples (corn starch, fir tree needles, and beech wood lignin, respectively) were diluted in water (concentration of 0.05 %) at 2000 rpm until an obscuration rate of 12.06 % was obtained. Three samples were measured in quintuplicate. The Mie theory was applied by considering the following optical properties: RI for starch of 1.527, and RI for lignocellulose of 1.36, respectively, and absorption of 0.001.

3.2.7 Scanning Electron Microscopy (SEM) Investigation

Samples were investigated using a scanning electron microscope (FEI QUANTA 200 ESEM). Air dried samples were fixed onto aluminum stubs through carbon adhesive disks and their fractured surface was observed with a low-vacuum secondary electron detector using the accelerating voltage of 25.0 kV. The samples were analyzed at room temperature and at an internal pressure of 0.50 torr.

4 Natural Filler/Plasticized Starch Composites—Structure and Properties

4.1 FTIR Spectroscopy Investigation

FTIR spectroscopy was used to investigate the interactions between polymer components, starch, and natural fillers respectively, in composite films. The infrared spectra of the components and different composite formulations are presented in Fig. 1a, b.

The neat starch and starch-based composite films present similar features in the two FTIR spectral regions, except the peaks related to the absorption of the hydroxyl groups, noticed at 1650 and 1017 cm^{-1}, respectively (Fang et al. 2002).

The peak at 3330 cm^{-1} (Fig. 1a) corresponds to the mixed hydroxyl groups from cellulose and hemicelluloses present in wood filler. The peak around 1076 cm^{-1} recorded for beech wood sawdust filler is a characteristic peak which can be attributed to the C–O stretching in cellulose, and hemicelluloses components, respectively. The peak noticed around 1643 cm^{-1} is characteristic to the absorption of conjugated carbonyl group C=O of lignin in beech wood sawdust filler.

The peaks around 1608 and 1506 cm^{-1} represent the stretching of aromatic ring in lignin present in beech wood sawdust filler. The peaks at 2930, 1456 cm^{-1} are

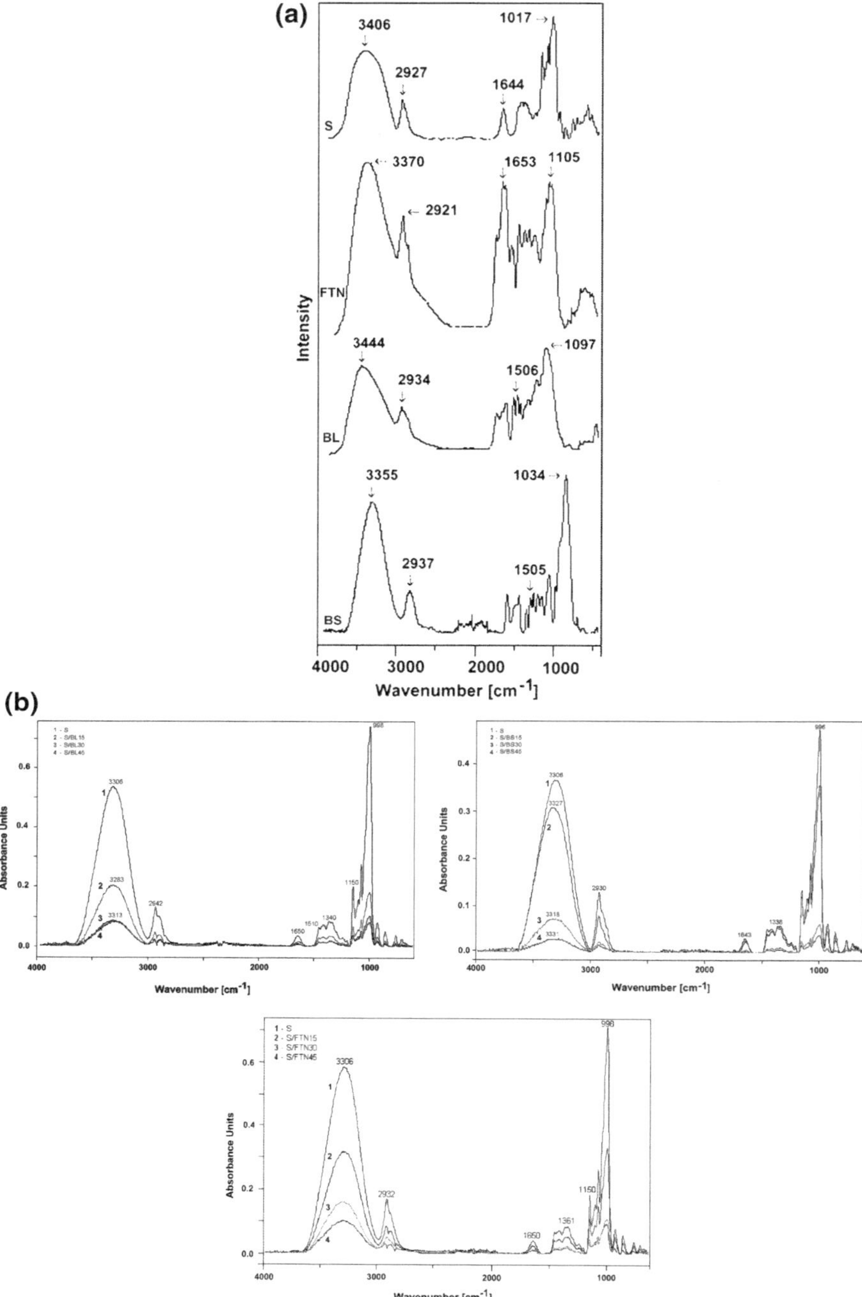

Fig. 1 FTIR spectra recorded for polymer components and composite films

the characteristic peaks of methylene groups, being attributed to the symmetric stretching, asymmetric stretching, in-plane deformation of C–H bond in cellulose and hemicelluloses. The peaks at 1735 and 1058 cm^{-1} correspond to carbonyl (C=O) stretching and C–O stretching mainly in cellulose and hemicelluloses of wood filler (Pandey 1999).

A less pronounced band between 1642 and 1740 cm^{-1} with a maximum at 1650 cm^{-1} corresponds to conjugated carbonyl stretching. Between 1200 and 800 cm^{-1}, the spectra of composites films comprising different amounts of lignin differ in the resolution between the two bands at 1150 and 1120 cm^{-1}, corresponding to aromatic C–H in-plane deformation and to C–O and C–C stretching of the carbohydrates associated to lignin.

Hardwood lignins present in their structure both guaiacyl- and syringyl-propane units. Lignin contains mainly glycerol-aryl ether linkages, but there are also present various types of C–C bonds which may act as cross-linkages between relative short, linear chains of phenyl-propane units.

The characteristic bands at 1605, 1513, 1460, and 1425 cm^{-1} correspond to aromatic ring vibration of the phenyl-propane skeleton from lignin chemical structure. The peaks noticed at 1150 and 1105 cm^{-1} can be assigned to the syringyl C–H in beech wood lignin (Pandey 1999).

Figure 1b presents the FTIR spectra of plasticized starch/natural filler composite films. Through addition of glycerol, new hydrogen bonds are formed. The bands at 1650 and 1456 cm^{-1} are specific to the O–H bending of water and CH_2, respectively. The bands at 1150 and 1076 cm^{-1} , respectively, can be assigned to the stretching vibration of C–O in C–O–H groups, while the band at 998 cm^{-1} is attributed to the stretching vibration of C–O in C–O–C groups. All absorption peaks shifted to a higher wave-number value than those noticed for starch. This indicated that new hydrogen bonds occurred between polymer components in the films. The absorption band located at 1700 cm^{-1} is related to C–O stretching. The band at 1217 cm^{-1} corresponds to C–O–C stretching of aryl ether. The band noticed at 1080 cm^{-1} for S, assigned to the C–O stretching is shifted to 1077 cm^{-1} in the composite films comprising starch and lignin-based fillers (BS, and FTN, respectively). All these changes in wavenumber values indicate important interactions between the hydroxyl, carbonyl and ether groups from the starch matrix and lignin based polymer components. These modifications in spectra can be attributed to the hydrogen bonding occurrence during production of the composite films.

4.2 X-ray Diffraction Analysis

The corn starch, a semi-crystalline polysaccharide, exhibits sharp diffraction peaks of A-type starch at 15.3°, 17.5°, and 23.1° (2θ), respectively—see Fig. 2.

During gelatinization processing, glycerol and water molecules altered the structure of native starch granules, by replacing the starch intermolecular and intramolecular hydrogen bonds, and modifying the crystalline part of starch. Thus,

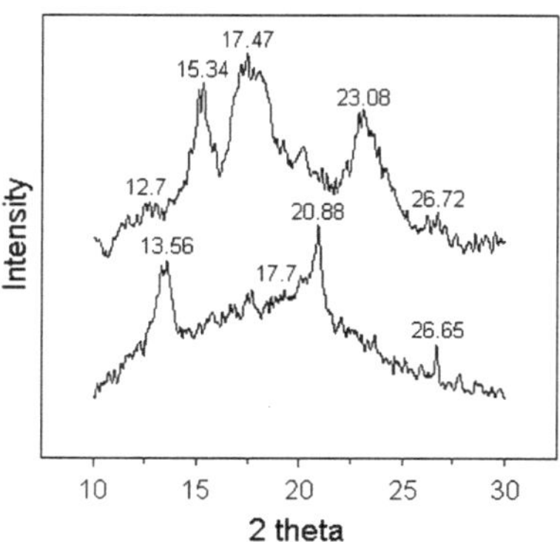

Fig. 2 X-ray diffraction patterns recorded for initial corn starch polymer matrix and glycerol plasticized corn starch film

there is no obvious starch crystallinity in glycerol-plasticized starch matrix (see Fig. 2).

The plasticized starch exhibits a V-type starch pattern (related to single helical complexes) with diffraction peaks at 13.56°, 17.7°, 20.88°, and 26.65° (2θ), respectively (Vermeylen et al. 2006).

4.3 TG/DTG/DTA Simultaneous Thermal Analysis

TG/DTG/DTA investigation of the composite samples was performed in order to evaluate their thermal decomposition behavior. TG curves obtained for starch-based composite materials (S, S/BS45, S/FTN45, respectively S/BL45) are presented in Fig. 3a–d. The thermal decomposition process of the initial plasticized starch composite films without natural filler presents three main stages of reaction, in agreement with previous reports (Chung and Mei 2007). The first stage corresponds to the removal of water from composite films. The second stage is due to the decomposition of the glycerol-rich phase, which also contains starch molecules, the third stage corresponding to the partially decomposed starch which undergoes oxidation processes.

The TG thermal derivate (DTG) curves evidence a wide peak that occurs around 100 °C, which can be associated with the maximum value for the water loss rate. DTG signals evidenced two different stages: one around 200 °C, attributed to the decomposition of the glycerol-rich phase occurs and another around 320 °C, which can be associated with the degradation of the starch-rich phase. These results are in agreement with the FTIR spectroscopy evidences related to the presence of hydrogen bonding (glycerol-starch components) in the composite materials. The degradation process which occurs around 320 °C, associated with starch matrix degradation, is similar in all the composite samples.

As it can be observed, the thermal decomposition for the starch-lignin based fillers composite materials can be divided into three individual stages: moisture

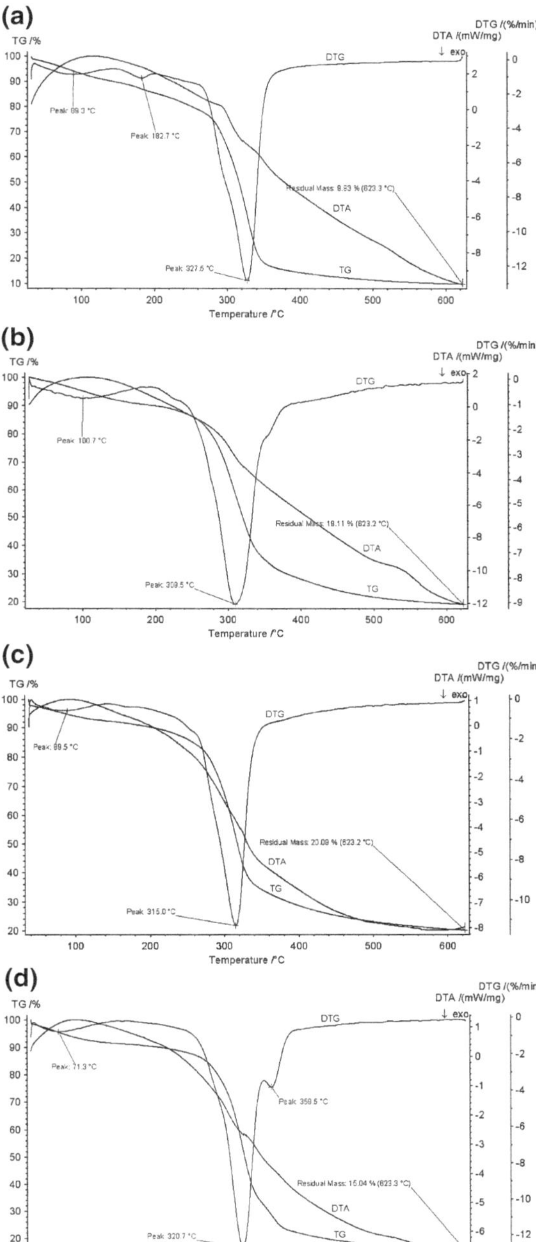

Fig. 3 TG/DTG/DTA curves recorded for: **a** S composite film (plasticized starch without natural filler in composition); **b** S/FTN45 composite film; **c** S/BL45 composite film; **d** S/BS45 composite film

Table 2 Some parameters evaluated from TG/DTG/DTA simultaneous thermal analysis of natural filler/glycerol plasticized starch matrix composites

Composite sample	DTA			DTG				
	T_{on} (°C)	T_{peak} (°C)	T_{end} (°C)	T_{on} (°C)	T_{peak} (°C)	T_{end} (°C)	T_{end}-T_{on} (°C)	W_T (%)
S	286	290	314	174	327.5	370	206	9.93 (at 623.3 °C)
S/FTN15	318	321	330	230	318.2	380	150	14.53 (at 623.1 °C)
S/FTN45	311	322	328	238	309.5	379	141	19.11 (at 623.2 °C)
S/BS15	300	315	324	245	322.7	390	145	13.35 (at 623.2 °C)
S/BS45	310	322	338	243	320.7	387	144	15.04 (at 623.3 °C)
S/BL15	318	323	327	258	322.8	377	119	12.99 (at 623.2 °C)
S/BL45	315	325	326	239	315.0	355	116	20.09 (at 623.2 °C)

DTA data: T_{on} onset temperature; T_{peak} maximum peak at melting point; T_{end} final melting temperature

DTG data: T_{on} initial decomposition temperature; T_{peak} maximum decomposition temperature; T_{end} final decomposition temperature; T_{end}-T_{on} decomposition temperature interval; W_T weight loss in the 0–600 °C temperature range

evolution, starch matrix decomposition, and decomposition of the lignin component from BS and FTN fillers, respectively. The various interactions between the polymer molecules, as well as the cross-linking processes during composites production, influenced the thermal degradation of starch/BS and starch/FTN composite materials. The thermal decomposition process is evidenced in a large temperature range, the major volatilization stage occurring between 200 °C and 600 °C. Thus, the thermal peak observed at 100 °C is attributed to water removal, while the peak at around 300 °C can be assigned to the decomposition of starch polymer matrix. The third thermal degradation peak recorded at around 500 °C is related to the decomposition of lignin component from BS and FTN fillers respectively. However, lignin is more difficult to decompose, as its weight loss occurred within a relatively large temperature range (from 160 to 900 °C) (Yang et al. 2007). Some thermal parameters are presented in Table 2 presented below.

4.4 Water Uptake Measurements

In a multiple biopolymer system, each biopolymer contributes both to the film properties and to the biopolymer-biopolymer interactions thus affecting the overall multi-component polymer system properties. Usually, these interactions have great

Table 3 Water absorption values recorded for natural filler/glycerol plasticized starch matrix composites

Composite sample	Time, h/Water uptake (%)			
	5	30	50	125
S	52.70	58.09	56.36	54.43
S/FTN15	49.51	56.52	58.16	59.91
S/FTN30	45.01	48.47	56.87	56.89
S/FTN45	44.24	44.80	48.59	53.81
S/BL15	41.61	51.12	51.57	52.25
S/BL30	38.88	48.34	47.65	47.79
S/BL45	45.98	45.96	45.98	46.15
S/BS15	32.52	40.28	49.80	53.03
S/BS30	24.55	27.20	33.55	41.42
S/BS45	27.33	32.97	39.49	52.03

impact on final properties than individual actions. Natural fillers comprising cellulose (namely beech wood flour and fir tree needles) were not expected to confer significantly water resistance to the starch composite materials. As one can observe, the addition of these fillers in composite formulations improved to some extent this property.

Because films comprising only starch are sensitive to changes in moisture that generates significant changes in their mechanical properties, controlling the films' moisture content is of high significance. The water resistance conferred by cellulose containing fillers (BS, FTN) is significant for new applications that are moisture sensitive such as gas selective membranes, ion channels, etc. Due to the enhanced hydrophilic features of starch polymer matrix, a neat starch film will always display reduced water resistance. Thus, pure starch films are not suitable for most non-absorbency applications. However, incorporation of various amounts of cellulose-based fillers (BS, FTN) within the starch polymer matrix was able to reduce to a less extent the water absorbency of these films as shown in Table 3. Although cellulose is typically a strong hydrophilic polymer, its high crystallinity and tight microfibrils structure within the cellulose fibers component tend to reduce water absorbency when compared with that of initial amorphous starch. Nevertheless, there are hemicelluloses (mainly xylan) in cellulose-based fillers (BS, FTN) and these are more hydrophobic than cellulose polymer component, fact which could explain the results of water sorption measurements.

The water uptake-time evolution (data not represented here) exhibit two well-distinct zones. For lower period of time values, the kinetics of absorption is very fast due to the hydrophilic polymer components, whereas at longer period of time, after 30 h, the kinetics of absorption is reduced, leading to a relative uniform plateau, corresponding to the equilibrium of water absorption. Therefore, the swelling of the films is reduced by addition of cellulose-based fillers into the plasticized starch matrix. The slight improvement of the surface through bulk-hydrophobation phenomena of the starch-based films, resulted from addition of cellulose-based fillers, should be considered.

The presence of natural fillers causes a slightly increased water resistance and could improve the application property of these composite films. This behavior could be due to the presence of hydrogen bonding interactions between starch polymer matrix and crystalline part of wood and fir tree needles fillers (which comprise cellulose in their chemical composition). The hydrogen bonds network occurred within the composite films tends to stabilize the starch polymer matrix when it is subjected to high moisture content environment.

4.5 Opacity Measurements and SEM Investigation

Optical properties are important issues to be investigated for starch composite films to be suitable for various applications including food packaging or coating. This provides information on the size of dispersed filler particles within starch polymer matrix, when the particle sizes larger than the visible wavelength could generate opaque composite materials.

Figure 4a–c evidences the influence of different natural fillers on the transparency properties of starch-based composite films. The transparency can be decreased with increase in the natural filler content of the films, as one can observe from Fig. 4. Light interacts with the film surface, being reflected, absorbed, or transmitted. Thus, color, gloss, and transparency properties are exhibited by the films. Usually, light diffusion occurs mainly at the interface of the different film polymer components, impairing its further transmission.

Natural fillers used in this study are not water soluble (two of them contain significant amounts of cellulose in their chemical composition) and randomly disperse within the starch plasticized matrix, and appears as a discontinuous phase after composite films drying (as evidenced by the scanning electron microscopy SEM investigation presented below). Therefore, these natural fillers provide more interfacial area within the composite films structure. As a result, light diffusion is enhanced, thus generating a reduced transparency property for the composite films. Overall, plasticized starch-based composite films exhibit some relative transparency mainly at low amount values for natural fillers in composite formulations.

This evolution of optical properties can be associated with the particle size of the natural fillers. The larger the particle size the lesser the light transmission, therefore the results of light transmittance, as seen in Fig. 5, which presents the particle size analysis.

As it can be observed from Fig. 6, natural fillers are relative homogeneously dispersed within the plasticized starch matrix. The fillers are well dispersed in the

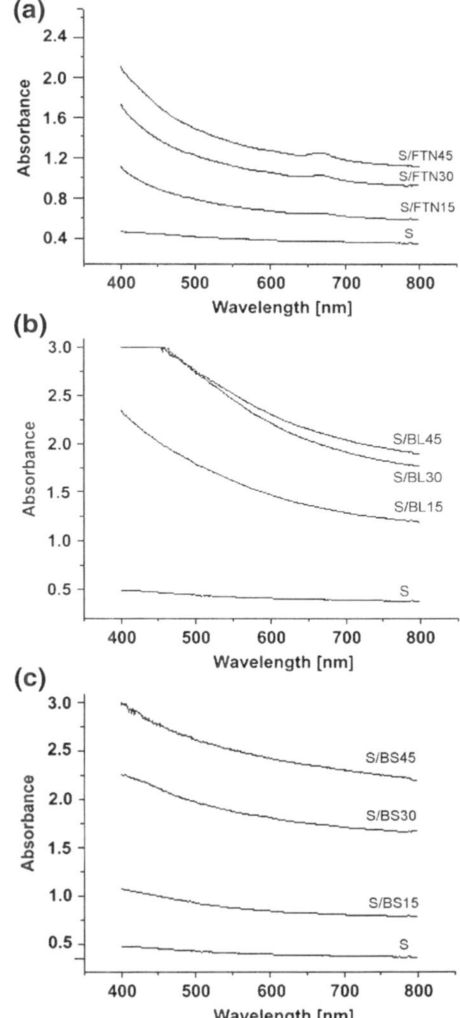

Fig. 4 Optical properties (opacity) determined for S composite film (plasticized starch without natural filler in composition) in comparison with: **a** S/FTN composite films; **b** S/BL composite films; **c** S/BS composite films

mixture comprising water, glycerol, and starch during preparation of composite formulations (plasticization of the starch polymer matrix). Glycerol used as plasticizer, allows the starch polymer to gain increased mobility, thus facilitating further packing of the starch macromolecules into crystal lattices. Both water and glycerol contribute to a large extent to the uniform dispersion of the natural fillers (FTN, BL, BS) within composite films as shown in Fig. 6.

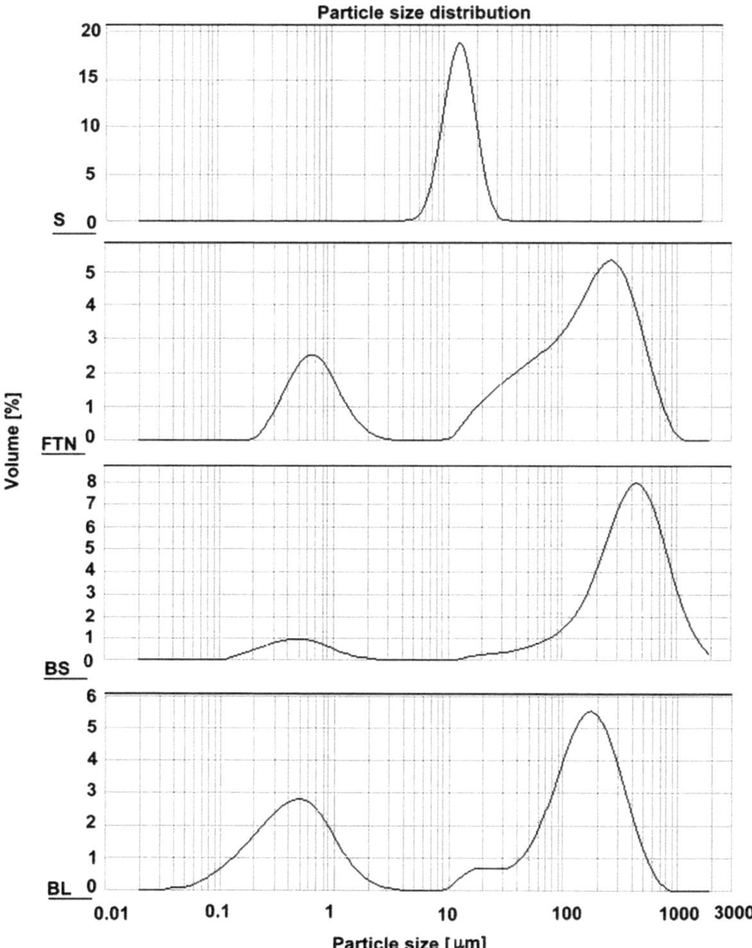

Fig. 5 Particle size analysis results for starch polymer matrix and natural fillers used in composite formulations

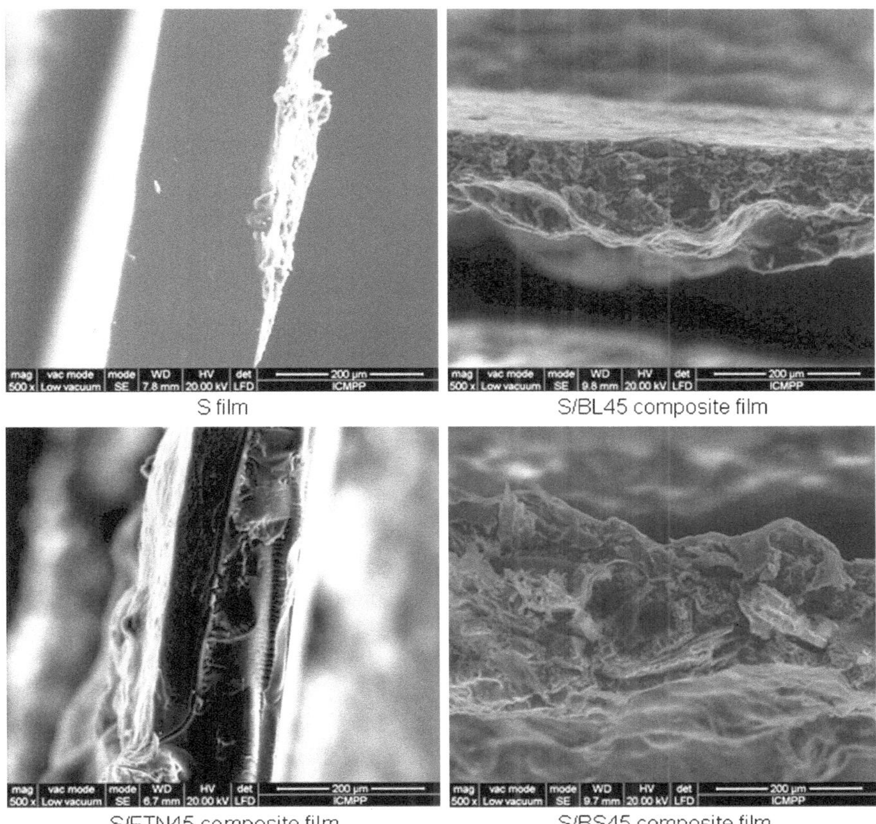

Fig. 6 SEM micrographs recorded for: **a** S composite film (plasticized starch without natural filler in composition); **b** S/BL45 composite film; **c** S/FTN45 S/BL45 composite film; **d** S/BS45 composite film

5 Conclusions and Future Perspectives

Multi-component polymer composite systems comprising natural fillers in composition may exhibit improved and sustainable durability when exposed to a high moisture content environment. The new effective methods applied to modify the natural fillers or the entire composite structure can significantly contribute to this enhancement in overall properties. Modifications of the resulted composite formulations may involve surface coatings, including development of efficient protective paints or coating layers. Incorporation of various lubricants, photo-stabilizers, and/or biocides can be successfully applied to generate such improvements, including also mechanical properties. Nevertheless, addition of nanoparticles within various natural filler composite formulations may greatly impart enhanced resistance to humid environments, photo-degradation and biodegradation (biological attack) phenomena in outdoor applications. Characterization of composite samples at nanoscale range can explain significant changes occurred during exposure and may provide efficient methods in order to prevent such changes.

Acknowledgments This work was supported by a grant of the Romanian National Authority for Scientific Research, CNCS-UEFISCDI—Project number PN-II-ID-PCE-2011-3-0187 (We gratefully acknowledge the project director, Dr. Dan Roşu, for overall guidance and fruitful discussion on the experimental data presented in this chapter).

References

Abdul Khalil HPS, Bhat AH, Irena Yusra AF (2012) Green composites from sustainable cellulose nanofibrils: a review. Carbohyd Polym 87:963–979

Abe K, Iwamoto S, Yano H (2007) Obtaining cellulose nanofibers with a uniform width of 15 nm from wood. Biomacromolecules 8:3276–3278

AL-Oqla FM, Sapuan SM (2014) Natural fiber reinforced polymer composites in industrial applications: feasibility of date palm fibers for sustainable automotive industry. J Clean Prod 66:347–354

Abraham E, Deepa B, Pothan LA, Cintil J, Thomas S, John MJ, Anandjiwala R, Narine SS (2013) Environmental friendly method for the extraction of coir fibre and isolation of nanofiber. Carbohyd Polym 92:1477–1483

Abraham E, Deepa B, Pothan LA, Jacob M, Thomas S, Cvelbar U, Anandjiwala R (2011) Extraction of nanocellulose fibrils from lignocellulosic fibres: a novel approach. Carbohyd Polym 86:1468–1475

Azwa ZN, Yousif BF, Manalo AC, Karunasena W (2013) A review on the degradability of polymeric composites based on natural fibres. Mater Design 47:424–442

Averous L, Pollet E (2014) Nanobiocomposites based on plasticized starch, Chapter 8, *Starch polymers from genetic engineering to green applications*, Edited by P. Halley and L. Averous, Elsevier, pp 211–239

Barone JR (2009) Lignocellulosic fiber-reinforced keratin polymer composites. J Polym Environ 17:143–151

Bettini SHP, Bonse BC, Melo EA, Munoz PAR (2010) Effect of sawdust surface treatment and compatibilizer addition on mechanical behavior, morphology, and moisture uptake of polypropylene/sawdust composites. Polym Eng Sci 50:978–985

Blaker JJ, Lee KY, Bismarck A (2011) Hierarchical composites made entirely from renewable resources. J Biobased Mater Bioenergy 5:1–16

Bledzki A, Jaszkiewicz A (2010) Mechanical performance of biocomposites based on PLA and PHBV reinforced with natural fibers-a comparative study to PP. Compos Sci Technol 70:1687–1696

Bledzki AK, Mamun AA, Volk J (2010) Barley husk and coconut shell reinforced polypropylene composites: the effect of fibre physical, chemical and surface properties. Compos Sci Technol 70:840–846

Bodîrlău R, Spiridon I, Teacă CA (2007) Chemical investigation of wood tree species in temperate forest in east-northern Romania. BioResources 2:41–57

Bodîrlău R, Teacă CA, Spiridon I (2009) Preparation and characterization of composites comprising modified hardwood and wood polymers/poly(vinyl chloride). BioResources 4:1285–1304

Bodîrlău R, Teacă CA, Spiridon I (2014) Green composites comprising thermoplastic corn starch and various cellulose-based fillers. BioResources 9:39–53

Cai X, Riedl B, Zhang S, Wan H (2008) The impact of the nature of nanofillers on the performance of wood polymer nanocomposites. Compos A 39:727–737

Chakraborty A, Sain M, Kortschot M (2005) Cellulose microfibrils: a novel method of preparation using high shear refining and cryocrushing. Holzforschung 59:102–107

Chang PR, Jian RJ, Yu JG, Ma XF (2010) Starch-based composites reinforced with novel chitin nanoparticles. Carbohyd Polym 80:420–425

Chang PR, Wu D, Anderson DP, Ma X (2012) Nanocomposites based on plasticized starch and rectorite clay: structure and properties. Carbohyd Polym 89:687–693

Chauhan M, Gupta M, Singh B, Bhattacharyya SK, Singh AK, Gupta VK (2013) Pretreatment of pine needles/wood particles and their composites with isocyanate prepolymer adhesive. Polym Eng Sci 53:1740–1750

Chauhan M, Gupta M, Singh B, Singh AK, Gupta VK (2012) Pine needle/isocyanate composites: dimensional stability, biological resistance, flammability, and thermo-acoustic characteristics. Polym Compos 33:324–335

Chen W, Yu H, Liu Y, Hai Y, Zhang M, Chen P (2011a) Isolation and characterization of cellulose nanofibers from four plant cellulose fibers using a chemical-ultrasonic process. Cellulose 18:433–442

Chen W, Yu H, Liu Y, Chen P, Zhang M, Hai Y (2011b) Individualization of cellulose nanofibers from wood using high-intensity ultrasonication combined with chemical pretreatments. Carbohyd Polym 83:1804–1811

Cherian BM, Leao AL, de Souza SF, Thomas S, Pothan LA, Kottaisamy M (2010) Isolation of nanocellulose from pineapple leaf fibres by steam explosion. Carbohyd Polym 81:720–725

Chung YL, Hsi-Mei L (2007) Properties of cast films made of HCl—methanol modified corn starch. Starch/Stärke 59:583–592

Csizmadia R, Faludi G, Renner K, Móczó J, Pukánszky B (2013) PLA/wood biocomposites: improving composite strength by chemical treatment of the fibers. Compos A 53:46–53

Dai H, Chang PR, Geng F, Yu J, Ma X (2010a) Preparation and properties of starch-based film using N, N-bis-(2-hydroxyethyl)formamide as a new plasticizer. Carbohyd Polym 79:306–311

Dai H, Chang PR, Yu J, Geng F, Ma X (2010b) N-(2-hydroxypropyl)formamide and N-(2-hydroxyethyl)-N-methylformamide as two new plasticizers for thermoplastic starch. Carbohyd Polym 80:139–144

Dobircau L, Sreekumar PA, Saiah R, Leblanc N, Terrié C, Gattin R, Saiter JM (2009) Wheat flour thermoplastic matrix reinforced by waste cotton fibre: agro-green-composites. Compos A 40:329–334

Dong C, Parsons D, Davies IJ (2014) Tensile strength of pine needles and their feasibility as reinforcement in composite materials. J Mater Sci 49:8057–8062

Eichhorn SJ (2011) Cellulose nanowhiskers: promising materials for advanced applications. Soft Matter 7:303–315

Fang JM, Fowler PA, Tomkinson J, Hill CAS (2002) The preparation and characterization of a series of chemically modified potato starches. Carbohyd Polym 47:245–252

Faruk O, Bledzki AK, Fink HP, Sain M (2014) Progress report on natural fiber reinforced composites. Macromol Mater Eng 299:9–26

Faruk O, Bledzki AK, Fink HP, Sain M (2012) Biocomposites reinforced with natural fibers: 2000-2010. Prog Polym Sci 37:1552–1596

Fernandes EM, Mano JF, Reis RL (2013a) Hybrid cork–polymer composites containing sisal fibre: morphology, effect of the fibre treatment on the mechanical properties and tensile failure prediction. Compos Struct 105:153–162

Fernandes EM, Correlo VM, Mano JF, Reis RL (2013b) Novel cork–polymer composites reinforced with short natural coconut fibres: effect of fibre loading and coupling agent addition. Compos Sci Technol 78:56–62

Fujisawa S, Ikeuchi T, Takeuchi M, Saito T, Isogai A (2012) Superior reinforcement effect of TEMPO-oxidized cellulose nanofibrils in polystyrene matrix: optical, thermal, and mechanical studies. Biomacromolecules 13:2188–2194

Fukuzumi H, Saito T, Iwata T, Kumamoto Y, Isogai A (2009) Transparent and high gas barrier films of cellulose nanofibers prepared by TEMPO-mediated oxidation. Biomacromolecules 10:162–165

Gao W, Dong H, Hou H, Zhang H (2012) Effects of clays with various hydrophilicities on properties of starch–clay nanocomposites by film blowing. Carbohyd Polym 88:321–328

Gironès J, López JP, Mutjé P, Carvalho AJF, Curvelo AAS, Vilaseca F (2012) Natural fiber-reinforced thermoplastic starch composites obtained by melt processing. Compos Sci Technol 72:858–863

Gregorova A, Hrabalova M, Kovalcik R, Wimmer R (2011) Surface modification of spruce wood flour and effects on the dynamic fragility of PLA/wood composites. Polym Eng Sci 51:143–150

Gupta M, Chauhan M, Khatoon N, Singh B (2010) Composite boards from isocyanate bonded pine needles. J Appl Polym Sci 118:3477–3489

Habibi Y, Dufresne A (2008) Highly filled bionanocomposites from functionalized polysaccharide nanocrystals. Biomacromolecules 9:1974–1980

Habibi Y, Lucia LA, Rojas OJ (2010) Cellulose nanocrystals: chemistry, self-assembly, and applications. Chem Rev 110:3479–3500

Hietala M, Mathew AP, Oksman K (2013) Bionanocomposites of thermoplastic starch and cellulose nanofibers manufactured using twin-screw extrusion. Eur Polym J 49:950–956

Hubbe MA, Rojas OJ, Lucia LA, Sain M (2008) Cellulosic nanocomposites: a review. BioResources 3:929–980

Huber T, Müssig J, Curnow O, Pang S, Bickerton S, Staiger MP (2012) A critical review of all-cellulose composites. J Mater Sci 47:1171–1186

Ifuku S, Morooka S, Nakagaito AN, Morimoto M, Saimoto H (2011) Preparation and characterization of optically transparent chitin nanofiber/(meth)acrylic resin composites. Green Chem 13:1708–1711

Isogai A (2013) Wood nanocelluloses: fundamentals and applications as new bio-based nanomaterials. J Wood Sci 59:449–459

Janardhnan S, Sain MM (2006) Isolation of cellulose microfibrils–an enzymatic approach. BioResources 1:176–188

Jonoobi M, Harun J, Shakeri A, Misra M, Oksman K (2009) Chemical composition, crystallinity, and thermal degradation of bleached and unbleached kenaf bast (*Hibiscus cannabinus*) pulp and nanofibers. BioResources 4:626–639

Kalia S, Avérous L, Njuguna J, Dufresne A, Cherian BM (2011) Natural fibers, bio- and nanocomposites. Int J Polymer Sci http://dx.doi.org/10.1155/2011/735932; Article ID 735932

Kaewtatip K, Thongmee J (2012) Studies on the structure and properties of thermoplastic starch/ luffa fiber composites. Mater Design 40:314–318

Kaushik A, Singh M, Verma G (2010) Green nanocomposites based on thermoplastic starch and steam exploded cellulose nanofibrils from wheat straw. Carbohyd Polym 82:337–345

Kaushik A, Singh M (2011) Isolation and characterization of cellulose nanofibrils from wheat straw using steam explosion coupled with high shear homogenization. Carbohyd Res 346:76–85

Klemm D, Kramer F, Moritz S, Tom Lindström, Ankerfors M, Gray D, Dorris A (2011) Nanocelluloses: a new family of nature-based materials. Angew Chem Int Ed 50:5438–5466

Klemm D, Schumann D, Kramer F, Heßler N, Koth D, Sultanova B (2009) Nanocellulose materials—different cellulose, different functionality. Macromol Symp 280:60–71

Koga H, Saito T, Kitaoka T, Nogi M, Suganuma K, Isogai A (2013) Transparent, conductive, and printable composites consisting of TEMPO-oxidized nanocellulose and carbon nanotube. Biomacromolecules 14:1160–1165

Kord B (2011) Nanofiller reinforcement effects on the thermal, dynamic mechanical and morphological behavior of HDPE/rice husk flour composites. BioResources 6:1351–1358

Kumar AP, Singh RP (2008) Biocomposites of cellulose reinforced starch: Improvement of properties by photo-induced crosslinking. Bioresour Technol 99:8803–8809

LeCorre D, Vahanian E, Dufresne A, Bras J (2012) Enzymatic pretreatment for preparing starch nanocrystals. Biomacromolecules 13:132–137

LeCorre D, Bras J, Dufresne A (2011) Evidence of micro- and nanoscaled particles during starch nanocrystals preparation and their isolation. Biomacromolecules 12:3039–3046

Lee KY, Tang M, Williams CK, Bismarck A (2012a) Carbohydrate derived copoly(lactide) as the compatibilizer for bacterial cellulose reinforced polylactide nanocomposites. Compos Sci Technol 72:1646–1650

Lee KY, Buldum G, Mantalaris A, Bismarck A (2014) More than meets the eye in bacterial cellulose: biosynthesis, bioprocessing, and applications in advanced fiber composites. Macromol Biosci 14:10–32

Lee KY, Ho KKC, Schlufter K, Bismarck A (2012b) Hierarchical composites reinforced with robust short sisal fibre preforms utilising bacterial cellulose as binder. Compos Sci Technol 72:1479–1486

Lee KY, Bharadia P, Blaker JJ, Bismarck A (2012c) Short sisal fibre reinforced bacterial cellulose polylactide nanocomposites using hairy sisal fibres as reinforcement. Compos A 43:2065–2074

Lei Y, Wu Q (2010) Wood plastic composites based on microfibrillar blends of high density polyethylene/poly(ethylene terephthalate). Bioresour Technol 101:3665–3671

Liu D, Zhong T, Chang PR, Li K, Wu Q (2010) Starch composites reinforced by bamboo cellulosic crystals. Bioresour Technol 101:2529–2536

Liu H, Xie F, Yu L, Chen L, Li L (2009) Thermal processing of starch-based polymers. Prog Polym Sci 34:1348–1368

Ma X, Chang PR, Yu J (2008a) Properties of biodegradable thermoplastic pea starch/ carboxymethyl cellulose and pea starch/microcrystalline cellulose composites. Carbohyd Polym 72:369–375

Ma XF, Jian RJ, Chang PR, Yu JG (2008b) Fabrication and characterization of citric acid-modified starch nanoparticles/plasticized-starch composites. Biomacromolecules 9:3314–3320

Majdzadeh-Ardakani K, Navarchian AH, Sadeghi F (2010) Optimization of mechanical properties of thermoplastic starch/clay nanocomposites. Carbohyd Polym 79:547–554

Majeed K, Jawaid M, Hassan A, Abu Bakar A, Abdul Khalil HPS, Salema AA, Inuwa I (2013) Potential materials for food packaging from nanoclay/natural fibres filled hybrid composites. Mater Design 46:391–410

Martins IMG, Magina SP, Oliveira L, Freire CSR, Silvestre AJD, Neto CP, Gandini A (2009) New biocomposites based on thermoplastic starch and bacterial cellulose. Compos Sci Technol 69:2163–2168

JD Megiatto Jr, Ramires EC, Frollini E (2010) Phenolic matrices and sisal fibers modified with hydroxy terminated polybutadiene rubber: impact strength, water absorption, and morphological aspects of thermosets and composites. Ind Crop Prod 31:178–184

Moran JI, Vazquez A, Cyras VP (2013) Bio-nanocomposites based on derivatized potato starch and cellulose, preparation and characterization. J Mater Sci 48:7196–7203

Moran JI, Alvarez VA, Cyras VP, Vazquez A (2008) Extraction of cellulose and preparation of nanocellulose from sisal fibers. Cellulose 15:149–159

Mueangta S, Hanchana A (2013) Effect of jute and kapok fibers on properties of thermoplastic cassava starch composites. Mater Design 47:309–315

Müller P, Renner K, Móczó J, Fekete E, Pukánszky B (2014) Thermoplastic starch/wood composites: interfacial interactions and functional properties. Carbohyd Polym 102:821–829

Nair SS, Hurley DC, Wang S, Young TM (2013) Nanoscale characterization of interphase properties in maleated polypropylene-treated natural fiber-reinforced polymer composites. Polym Eng Sci 53:888–896

Nair SS, Zhu JY, Deng Y, Ragauskas AJ (2014) Characterization of cellulose nanofibrillation by microgrinding. J Nanopart Res 16:2349. doi:10.1007/s11051-014-2349-7

Ojijo V, Ray SS (2013) Processing strategies in bionanocomposites. Prog Polym Sci 38:1543–1589

Pandey KK (1999) A study of chemical structure of soft and hardwood and wood polymers by FTIR spectroscopy. J Appl Polym Sci 71:1969–1975

Pandey JK, Ahn SH, Lee CS, Mohanty AK, Misra M (2010) Recent advances in the application of natural fiber based composites. Macromol Mater Eng 295:975–989

Pandey JK, Chu WS, Kim CS, Lee CS, Ahn SH (2009) Bio-nano reinforcement of environmentally degradable polymer matrix by cellulose whiskers from grass. Compos B 40:676–680

Pandey JK, Singh RP (2005) Green nanocomposites from renewable resources: effect of plasticizer on the structure and material properties of clay-filled starch. Starch/Stärke 57:8–15

Prachayawarakorn J, Ruttanabus P, Boonsom P (2011) Effect of cotton fiber contents and lengths on properties of thermoplastic starch composites prepared from rice and waxy rice starches. J Polym Environ 19:274–282

Qamhia II, Sabo RC, Elhajjar RF (2014) Static and dynamic characterization of cellulose nanofibril scaffold-based composites. BioResources 9:381–392

Qiao X, Tang Z, Sun K (2011) Plasticization of corn starch by polyol mixtures. Carbohyd Polym 83:659–664

Ramires EC, Megiatto JD Jr, Gardrat C, Castellan A, Frollini E (2010) Biobased composites from glyoxal-phenolic resins and sisal fibers. Bioresour Technol 101:1998–2006

Reddy MM, Vivekanandhan S, Misra M, Bhatia SK, Mohanty AK (2013) Biobased plastics and bionanocomposites: current status and future opportunities. Prog Polym Sci 38:1653–1689

Rojas OJ, Montero GA, Habibi Y (2009) Electrospun nanocomposites from polystyrene loaded with cellulose nanowhiskers. J Appl Polym Sci 113:927–935

Rowell RM (2012a) Chemical modification of wood to produce stable and durable composites. Cell Chem Technol 46:443–448

Rowell RM (ed) (2012b) Handbook of wood chemistry and wood composites, Second Edition, CRC Press, Boca Raton, USA, p 703

Sailaja RRN, Deepthi MV (2010) Mechanical and thermal properties of compatibilized composites of polyethylene and esterified lignin. Mater Design 31:4369–4379

Satyanarayana KG, Arizaga GGC, Wypych F (2009) Biodegradable composites based on lignocellulosic fibers-an overview. Prog Polym Sci 34:982–1021

Savadekar NR, Mhaske ST (2012) Synthesis of nano cellulose fibers and effect on thermoplastics starch based films. Carbohyd Polym 89:146–151

Singha AS, Thakur VK (2008a) Saccaharum cilliare fiber reinforced polymer composites. E-J Chem 5:782–791

Singha AS, Thakur VK (2008b) Effect of fibre loading on urea-formaldehyde matrix based green composites. Iran Polym J 17:861–873

Singha AS, Thakur VK (2008c) Synthesis and characterization of pine needles reinforced RF matrix based biocomposites. J Chem 5:1055–1062

Singha AS, Thakur VK (2008d) Fabrication of Hibiscus Sabdariffa fibre reinforced polymer composites. Iran Polym J 17:541–553

Singha AS, Thakur VK (2009a) Study of mechanical properties of urea-formaldehyde thermosets reinforced by *Pine needle* powder. BioResources 4:292–308

Singha AS, Thakur VK (2009b) Grewia optiva fiber reinforced novel, low cost polymer composites. J Chem 6:71–76

Singha AS, Thakur VK (2009c) Synthesis, characterization and analysis of Hibiscus Sabdariffa fibre reinforced polymer matrix based composites. Polym Polym Compos 17:189–194

Singha AS, Thakur VK (2010a) Synthesis and characterization of short *Grewia optiva* fiber based polymer composites. Polym Compos 31:459–470

Singha AS, Thakur VK (2010b) Mechanical, morphological, and thermal characterization of compression-molded polymer biocomposites. Int J Polym Anal Charact 15:87–97

Singha AS, Thakur VK (2010c) Synthesis, characterization and study of pine needles reinforced polymer matrix based composites. J Reinf Plast Compos 29:700–709

Singha AS, Thakur VK, Mehta IK (2010) Renewable resource based green polymer composites: analysis and characterization. Int J Polym Anal Charact 15:137–146

Singha AS, Thakur VK, Mehta IK, Shama A, Khanna AJ, Rana RK, Rana AK (2009a) Surface-modified *Hibiscus sabdariffa* fibers: physicochemical, thermal, and morphological properties evaluation. Int J Polym Anal Charact 14:695–711

Singha AS, Thakur VK, Mishra BN (2009b) Study of grewia optiva fiber reinforced urea-formaldehyde composites. J Polym Mater 26:81–90

Soykeabkaew N, Laosat N, Ngaokla A, Yodsuwan N, Tunkasiri T (2012) Reinforcing potential of micro- and nano-sized fibers in the starch-based biocomposites. Compos Sci Technol 72:845–852

Sreekumar PA, Gopalakrishnan P, Leblanc N, Saiter JM (2010) Effect of glycerol and short sisal fibers on the viscoelastic behavior of wheat flour based thermoplastic. Compos A Appl Sci Manuf 41:991–996

Stark NM, Gardner DJ (2008) Outdoor durability of wood-polymer composites (Chapter 7). In: Wood-polymer composites Edited by Oksman Niska K, Sain M, Woodhead Publishing Limited Cambridge, England and CRC Press LLC, Boca Raton USA, pp. 142–165

Stelte W, Sanadi AR (2009) Preparation and characterization of cellulose nanofibers from two commercial hardwood and softwood pulps. Ind Eng Chem Res 48:11211–11219

Syverud K, Chinga-Carrasco G, Toledo J, Toledo PG (2011) A comparative study of *Eucalyptus* and *Pinus radiata* pulp fibres as raw materials for production of cellulose nanofibrils. Carbohyd Polym 84:1033–1038

Teacă CA, Bodîrlău R, Roşu D, Roşu L, Varganici CD (2014) Multicomponent bio-based polymer systems comprising starch and wood polymers—structure and thermal behavior. J Biobased Mater Bioenergy 8:253–260

Thakur VK, Singha AS (2010) Mechanical and water absorption properties of natural fibers/ polymer biocomposites. Polym Plast Technol Eng 49:670–694

Thakur VK, Singha AS (2011) Physico-chemical and mechanical behavior of cellulosic pine needles based biocomposites. Int J Polym Anal Charact 16:390–398

Thakur VK, Singha AS, Misra BN (2011) Graft copolymerization of methyl methacrylate onto cellulosic biofibers. J Appl Polym Sci 122:532–544

Thakur VK, Singha AS, Thakur MK (2012a) Biopolymers based green composites: mechanical, thermal and physico-chemical characterization. J Polym Environ 20:412–421

Thakur VK, Singha AS, Thakur MK (2012b) In-air graft copolymerization of ethyl acrylate onto natural cellulosic polymers. Int J Polym Anal Charact 17:48–60

Thakur VK, Singha AS, Thakur MK (2012c) Graft copolymerization of methyl acrylate onto cellulosic biofibers: synthesis, characterization and applications. J Polym Environ 20:164–174

Thakur VK, Singha AS, Thakur MK (2012d) Surface modification of natural polymers to impart low water absorbency. Int J Polym Anal Charact 17:133–143

Thakur VK, Thakur MK (2014a) Processing and characterization of natural cellulose fibers/ thermoset polymer composites. Carbohyd Polym 109:102–117

Thakur VK, Thakur MK (2014b) Recent advances in graft copolymerization and applications of chitosan: a review. ACS Sustain Chem Eng 2:2637–2652

Thakur VK, Thakur MK, Raghavan P, Kessler MR (2014a) Progress in green polymer composites from lignin for multifunctional applications: a review. ACS Sustain Chem Eng 2:1072–1092

Thakur VK, Thakur MK, Gupta RK (2014b) Review: raw natural fiber–based polymer composites. Int J Polym Anal Charact 19:256–271

Thakur VK, Vennerberg D, Kessler MR (2014c) Green aqueous surface modification of polypropylene for novel polymer nanocomposites. ACS Appl Mater Interfaces 6:9349–9356

Thakur VK, Vennerberg D, Madbouly SA, Kessler MR (2014d) Bio-inspired green surface functionalization of PMMA for multifunctional capacitors. RSC Adv 4:6677–6684

Thakur VK, Thunga M, Madbouly SA, Kessler MR (2014e) PMMA-g-SOY as a sustainable novel dielectric material. RSC Adv 4:18240–18249

Thakur VK, Grewell D, Thunga M, Kessler MR (2014f) Novel composites from eco-friendly soy flour/SBS triblock copolymer. Macromol Mater Eng 299:953–958

Vermeylen R, Derycke V, Delcour JA, Goderis B, Reynaers H, Koch MHJ (2006) Structural transformations during gelatinization of starches in limited water: combined wide- and small-angle x-ray scattering study. Biomacromolecules 7:1231–1238

Wu Y, Wang S, Zhou D, Zhang Y, Wang X, Yang R (2013) Biodegradable poly(vinyl alcohol) nanocomposites made from rice straw fibrils: mechanical and thermal properties. J Compos Mater 47:1449–1459

Xie F, Pollet E, Halley PJ, Averous L (2013) Starch-based nano-biocomposites. Prog Polym Sci 38:1590–1628

Yang HS, Kim DJ, Kim HJ (2003) Rice straw-wood particle composite for sound absorbing wooden construction materials. Bioresour Technol 86:117–121

Yang H, Yan R, Chen H, Lee DH, Zheng C (2007) Characteristics of hemicellulose, cellulose and lignin pyrolysis. Fuel 86:1781–1788

Zeppa C, Gouanvé F, Espuche E (2009) Effect of a plasticizer on the structure of biodegradable starch/clay nanocomposites: thermal, water-sorption, and oxygen-barrier properties. J Appl Polym Sci 112:2044–2056

Zhang JS, Chang PR, Wu Y, Yu JG, Ma XF (2008) Aliphatic amidediol and glycerol as a mixed plasticizer for the preparation of thermoplastic starch. Starch/Stärke 60:617–623

Zhang QX, Yu ZZ, Xie XL, Naito K, Kagawa Y (2007) Preparation and crystalline morphology of biodegradable starch/clay nanocomposites. Polymer 48:7193–7200

Green Synthesis of Polymer Composites/ Nanocomposites Using Vegetable Oil

Selvaraj Mohana Roopan and Gunabalan Madhumitha

Abstract Vegetable triglycerides are among the first renewable resources exploited by man primarily in coating applications because their unsaturated varieties polymerize as thin films in the presence of atmospheric oxygen. Nowadays, use of the vegetable oils is spotlight of the chemical industries and as they are using these as a renewable platform for further ability. In order to overcome disadvantages such as poor mechanical properties of polymers from renewable resources, or to offset the high price of synthetic biodegradable polymers, various blends and composites have been developed over the last decade. The progress of blends from three kinds of polymers from renewable resources (1) natural polymers, such as starch, protein, and cellulose; (2) synthetic polymers from natural monomers, such as polylactic acid; and (3) polymers from microbial fermentation. In this chapter we have discussed about the different types of polymer composites obtained from the vegetable oil and applications of the polymer composites.

Keywords Vegetable oil · Polymer composites · Industrial and biomedical application

1 Introduction

1.1 Polymer Composites

Biocomposites can be obtained from most of the sources like natural and also in laboratories using various chemicals process including ceramics, polymers, and various natural composites (Shida et al. 2014; Singha and Thakur 2008a, b, c; Thakur et al. 2014a, b, c, d, e, f). During recent decade's polymers from renewable

S.M. Roopan (✉) · G. Madhumitha
Chemistry of Heterocycles & Natural Product Research Laboratory,
Organic Chemistry Division, School of Advanced Sciences, VIT University,
Vellore 632014, Tamilnadu, India
e-mail: mohanaroopan.s@gmail.com; mohanaroopan.s@vit.ac.in

V.K. Thakur and M.K. Thakur (eds.), *Eco-friendly Polymer Nanocomposites*,
Advanced Structured Materials 75, DOI 10.1007/978-81-322-2470-9_16

sources have attracted most of the researchers to carry out their research progress due to its cost-effective and huge availability (Cinita et al. 2014; Thakur et al. 2012a, b, c, d). Polymers are widely used for technical purposes because they are expected to exhibit some properties like gas permeability, electrical conductivity, thermal stability, flexibility, resistance to chemicals, biocompatibility, biodegradability, adhesion to metallic substances (Lin et al. 2011a, b; Singha and Thakur, 2009a, b, c, d, e). Nowadays, researchers are focused to synthesize biopolymers which are nontoxic to environment and also have several advantages compared to organic polymers like easily degradable and mainly cheaper than petroleum polymers (Seniha et al. 2014; Thakur et al. 2010a, b).

In many industries polymer composites with vegetable fibers which are easily degradable was one of the major attractive fields to the use of compatibility between polymer matrix and fiber (Arrakhiz et al. 2013; Thakur and Thakur 2014a, b, c). Most of the biocomposites consist matrix material as biodegradable polymer and also it was expected to degrade the internal part of the composite by embedding with natural reinforcing fibers (Reis et al. 2014; Singha and Thakur 2010a, b). Polymers that are derived from various living organisms like trees, algae, plants are stated to be biopolymer which is eco-friendly. Due to its environmental concerns, much attention was paid for synthesis of biopolymer and its application studies (Shida et al. 2014).

Polymer composites contain several matrices such as elastomers, thermosets, thermoplastics, which contains several materials like aliphatic and aromatic polyamides, PTFE, polyolefins, polyester, aminoplast, phenoplast, rubber materials including butyl rubber, and other rubbers. Mostly, these bio-composite polymeric materials were used in industries like construction materials, fibrous fillers, dental filling, car tires, and various coating industries. These properties of polymer can able to change by intramolecular interaction of polymer (Mikitaev et al. 2009).

Rather than that of traditional polymer composites like POM and PTFE, nowadays high-performance polymers are found which contain mainly tribological applications to withstand high service temperature which was a major issue (Friedrich et al. 1995). These polymer composites will join with nanoparticles and play a major role in electrical conductivity improvement. To prepare these composites with a complex shape and patterns by simple processing techniques are used which reduce the synthesizing cost (Kymakis et al. 2002).

In olden days artificial composites are prepared by mechanical blending of polymers and glasses but in these materials polymers are not homogenously dispersed, to overcome this issue we can prepare polymer by using vegetable oils (Okada and Usuki 1995). These vegetable oil synthesized polymers have thermo-physical and mechanical properties from soft plastic and rubbers for hardening of the material (Xia and Larock 2010).

Due to its high usage and low toxicity biopolymer composite was mostly used in several industrial applications. We have chosen vegetable oil which was one of the cheapest sources in renewable sources that is present and readily available worldwide. So this chapter deals about the polymer composite which was derived from vegetable oil as a natural source.

The first generation of bio-based polymer composites focused on deriving polymers from agricultural feedstocks such as corn, potatoes, and other carbohydrate feedstocks. However, the focus has shifted in recent years due to a desire to move away from food-based resources and significant breakthroughs in biotechnology. Bio-based polymers similar to conventional polymers are produced by bacterial fermentation processes by synthesizing the building blocks (monomers) from renewable resources, including lignocellulosic biomass (starch and cellulose), fatty acids, and organic waste. Natural bio-based polymers are the other class of bio-based polymers which from plant and animal origins are presented. And starch-based biodegradable polymers have some advantages to be medical polymer materials (Dave et al. 1999)

(a) Good biocompatibility
(b) Biodegradable and its degradation products are nontoxic
(c) Proper mechanical properties
(d) Degradation as requirement

Starch-based biodegradable polymers have been widely investigated in bone tissue engineering.

Starch-based biodegradable bone cements can provide immediate structural support and degrade from the site of application. Moreover, they can be combined with bioactive particles, which allow new bone growth to be induced in both the interface of cement bone and the volume left by polymer degradation (Sinha et al. 2001).

1.2 Vegetable Oil

One of the most important renewable resources was vegetable oil mainly in chemical industry to synthesize monomers and polymers which can play a role in day-to-day life (Biermann et al. 2000; Gunner et al. 2006). Vegetable oils are already used in many industries like coatings, lubricants, and agrochemicals as already reported in recent articles and reviews. This vegetable oil plays a major role in synthesis of polymer composites mainly in four continents, namely Asia, North America, South America, and Europe (Cunningham and Yapp 1974; Bussell 1974; Hodakowski et al. 1975; Trecker et al. 1976; Salunkhe et al. 1992; Force and Starr 1988).

1.3 Various Sources for Vegetable Oils

To isolate vegetable oils several sources are available like soybean, palm, rapeseed/canola, sunflower, tallow, lard, butterfat, groundnut, cottonseed, oconut, palm kernel, olive, corn, fish, linseed, sesame, castor, etc. (Biermann et al. 2000). To extract vegetable oils from the above sources two methods of extraction techniques are available like solvent extraction and mechanical extraction. This solvent extraction contains majorly organic solvents mainly hexane and this method is more efficient while comparing to mechanical method but it contains drawbacks like use of volatile organic solvents. Mechanical extraction is the process of shearing the cell and oil bodies to extract oil from the shell, while doing this process heat will be generated so there is high possibility for denaturation of proteins due to heat liberated in this process (Norris 1996). The most common vegetable-based polymers are mostly isolated from soybean, fish, corn, tung, linseed and castor oils, and also it contains various constituents of fatty acids which is shown in Fig. 1 (Johnson and Fritz 1989).

Structure of vegetable oil contains a mixture of triglycerides with long chain of fatty acids as shown in Fig. 2. Generally, length of fatty acids varies from 14 to 22 carbons with 0–3 double bonds chain present in it (Guner et al. 2006; Belgacem and Gandini 2008; Sharma and Kundu 2006). Until now most researchers found more than 1000 of fatty acids but around 20 only present in vegetable oils. List of fatty acids present in vegetable oil was listed below in Table 1 (Belgacem and Gandini 2008; Barnwal and Sharma 2005).

1.4 Properties of Vegetable Oils

The properties of vegetable oils were classified as chemical and physical properties which mostly depend on the distribution of fatty acid and also position of aliphatic chains that present within double bonds. The specific gravity of vegetable oils is lower than expected which was less than density of water. This chemical and physical property mainly consists of viscosity, specific gravity, refractive index, and melting point which were tabulated in Table 2 (Guner et al. 2006; Barnwal and Sharma 2005).

1.5 Applications of Vegetable Oils

Vegetable oil consists of several applications in food industries and also nowadays vegetable oils are used for cooking purposes (Dyer et al. 2008). Vegetable oil is also used for production of biofuels, paintings, lubricants, coatings, inks, pharmaceuticals, agrochemicals, and materials used for construction (Williams and Hillmyer 2008). Instead of using petroleum-based paraffin waxes nowadays peoples are like to use the renewable resources, for example, vegetable oils are also used in

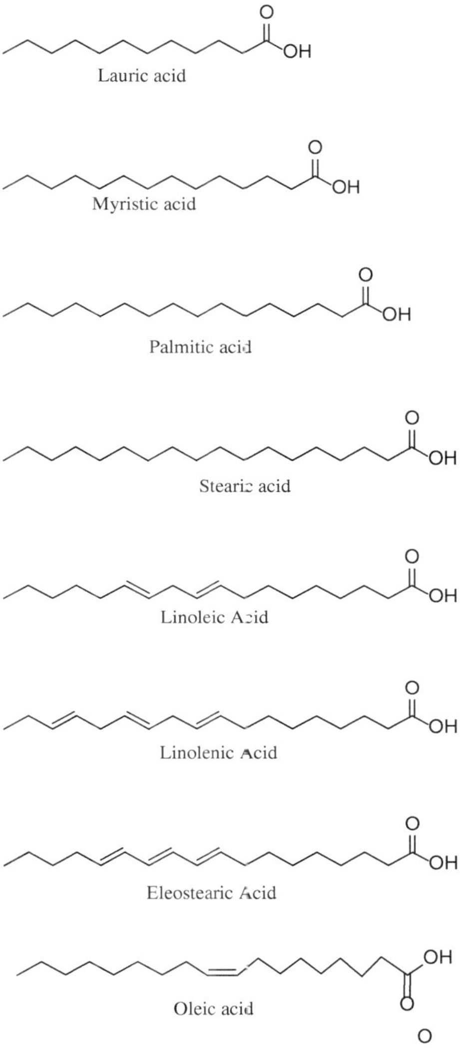

Fig. 1 Constituents of fatty acids present in vegetable oils

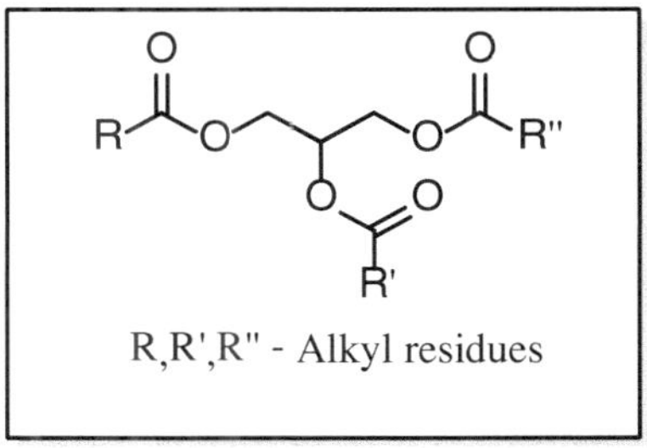

Fig. 2 General structure of triglyceridesas major component of vegetable oils

Table 1 List of fatty acids present in vegetable oil (Belgacem and Gandini 2008; Barnwal and Sharma 2005)

Common name	Systematic name	Formula	Structure (C:DB)*
Lauric acid	Dodecanoic acid	$C_{12}H_{24}O_2$	12:0
Myristic acid	Tetradecanoic acid	$C_{14}H_{28}O_2$	14:0
Palmitic acid	Hexadecanoic acid	$C_{16}H_{32}O_2$	16:0
Stearic acid	Octadecanoic acid	$C_{18}H_{36}O_2$	18:0
Arachidic acid	Eicosanoic acid	$C_{20}H_{40}O_2$	20:0
Behenic acid	Docosanoic acid	$C_{22}H_{44}O_2$	22:0
Lignoceric acid	Tetracosanoic acid	$C_{24}H_{48}O_2$	24:0
Palmitoleic acid	*cis*-9-Hexadecenoic acid	$C_{16}H_{30}O_2$	16:1
Oleic acid	*cis*-9-Octadecenoic acid	$C_{18}H_{34}O_2$	18:1
Linoleic acid	*cis,cis*-9,12-Octadecadienoic acid	$C_{18}H_{32}O_2$	18:2
Linolenic acid	*cis,cis,cis*-9,12,15-Octadecatrienoic acid	$C_{18}H_{30}O_2$	18:3
α-Eleostearic acid	*cis,trans,trans*-9,11,13-Octadecatrienoic acid	$C_{18}H_{30}O_2$	18:3
Erucic acid	*cis*-13-Docosenoic acid	$C_{22}H_{42}O_2$	22:1
Ricinoleic acid	12-Hydroxy-*cis*-9-octadecenoic acid	$C_{18}H_{34}O_3$	18:1
Vernolic acid	12,13-Epoxy-*cis*-9-octadecenoic acid	$C_{18}H_{32}O_3$	18:1
Licanic acid	4-Oxo-*cis,trans,trans*-,11, 13-octadecatrienoic acid	$C_{18}H_{28}O_3$	18:3

Table 2 The chemical and physical properties of vegetable oil (Guner et al. 2006; Barnwal and Sharma 2005)

Sources	Viscosity at RT	Specific gravity	Refractive index	Melting point (°C)
Castor oil	293.4	0.951–0.966	1.473–1.480	−20 to −10
Linseed oil	29.60	0.925–0.932	1.480–1.483	−20
Palm oil	30.92	0.890–0.893	1.453–1.456	33–40
Soybean oil	28.49	0.917–0.924	1.473–1.477	−23 to −20
Sunflower oil	33.31	0.916–0.923	1.473–1.477	−18 to −16

candle industry to avoid a by-product of the fractionation and refining of the fossil fuel crude oil. Although candles made from paraffin are low priced for consumers, the rising petroleum prices and due to some environmental concerns have increased people's interest in using products that are from renewable sources (Wang and Wang 2007).

2 Vegetable Oil-Based Polymer Composites

Renewable resources can distribute an interesting sustainable platform to substitute partially, and to some extent totally, petroleum-based polymers through the design of bio related polymers Fig. 3, that can compete or even surpass the existing petroleum-based composites on a cost performance basis with a positive environmental impact (Belgacem and Gandini 2008; Williams and Hillmyer 2008; Gallezot 2012). Vegetable oils are recognized as one of the most important classes of renewable resources for the chemical industry in the synthesis of fine chemicals, monomers, and polymers (Frederick et al. 2004). Vegetable oils have been used for the production of polymer composites incorporating organic or inorganic particles or fibers, both synthetic and natural, and sized from the macro- to the micro- and to the nanoscale (Mosiewicki and Aranguren 2013; Vaidya 2012; Baillie 2004).

2.1 Chain Growth Polymer Composites

2.1.1 Polymers by Free Radical Polymerization

The multiple C=C bonds of vegetable oils are susceptible to be polymerized by a free radical mechanism. Little attention has been given to free radical polymerization of triglyceride double bonds because of the presence of chain-transfer processes (Lu and Larock 2009).

The different steps of the formation of a cross-linked network of a “dried” paint having different characteristics. The first step is the hydroperoxide formation with simultaneous formation of conjugated double bonds if the fatty acids contain the pentadienyl moiety. The second step, the hydroperoxide decomposition, results in the formation of peroxy and alkoxy radicals whose recombination originates ether and peroxide dimmers (Thakur et al. 2010a, b). Concomitantly, alkoxy radicals give rise to β-scission reactions which lead to the degradation of the macromolecule and the formation of various oxidation products. Subsequently, the species with conjugated double bonds are prone to radical addition and further oxidation reactions which lead to a network formation linked via ester bonds and ether cross-links Figs. 4 and 5 (Gorkum and Bouwman 2005; Thakur et al. 2014c).

2.1.2 Polymers by Cationic Polymerization

The carbon-carbon double bonds of vegetable oils are also susceptible to cationic polymerization. Since vegetable oils are multifunctional monomers which is caused by multiple C=C double bonds, cationic polymerization of vegetable oils resulted eventually in cross-linked polymers. Most of Lewis acids, such as $AlCl_3$, $TiCl_4$, $ZnCl_2$, $SnCl_4$ and $BF_3{\cdot}OEt_2$, are used as initiator systems for cationic polymerization

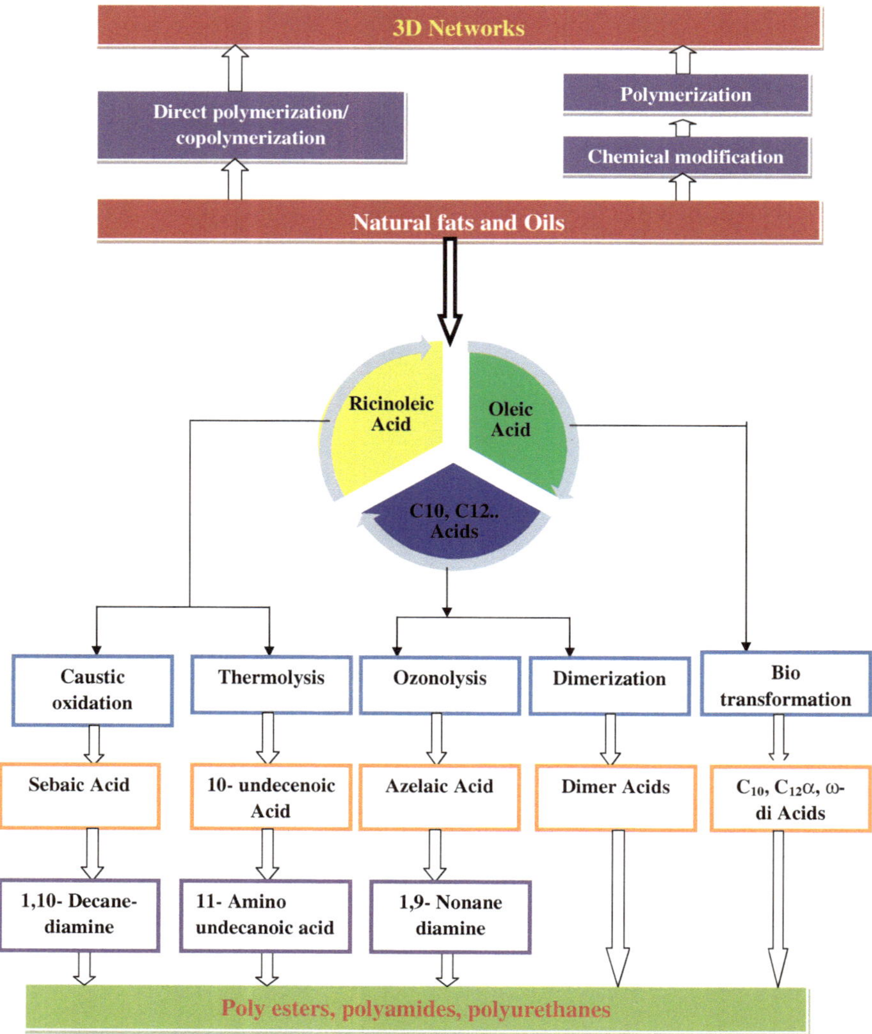

Fig. 3 Polymers isolated from vegetable oils

under mild conditions and $BF_3{\cdot}OEt_2$ has proved to be the most efficient among them (Lu and Larock 2009).

2.1.3 Polymers by Anionic Polymerization

In anionic polymerizations, the initiator system might be of different classes such as radical anions (e.g., sodium naphthalenide), alkali metals, and alkyllithium

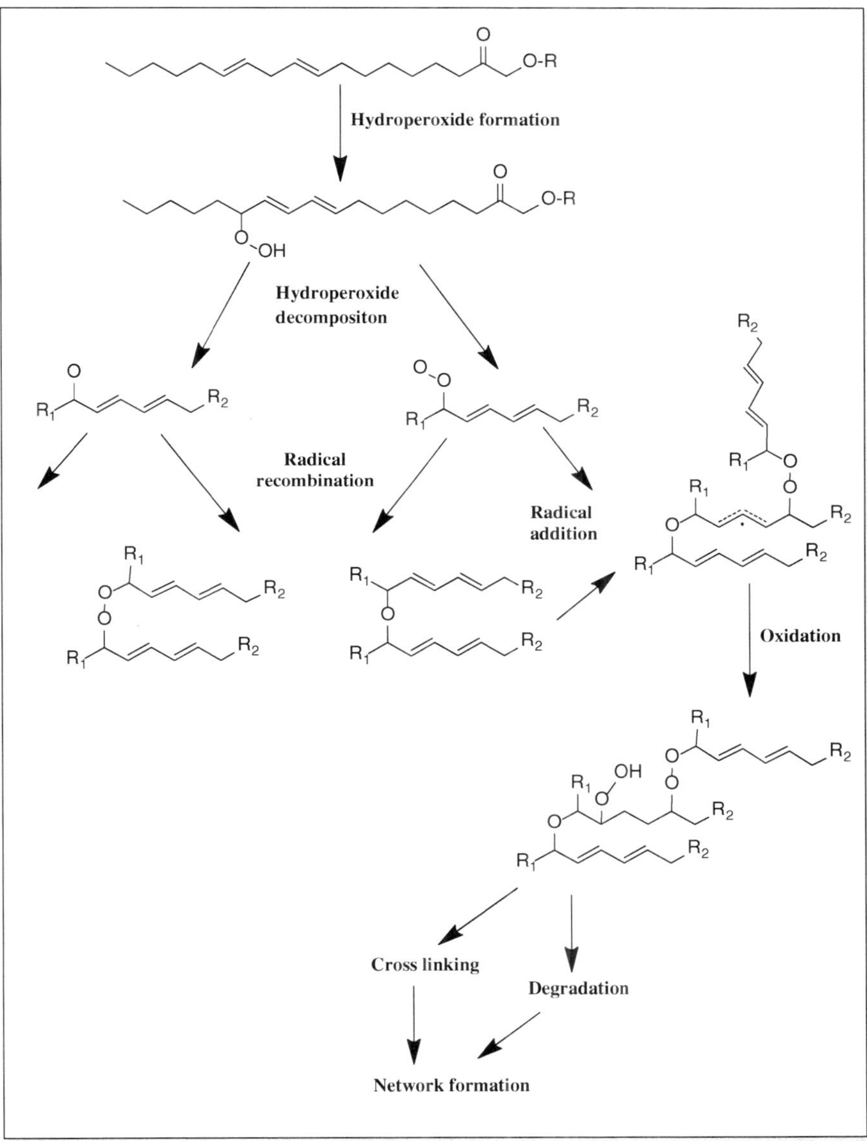

Fig. 4 Cross-linked network formation from oxidative drying

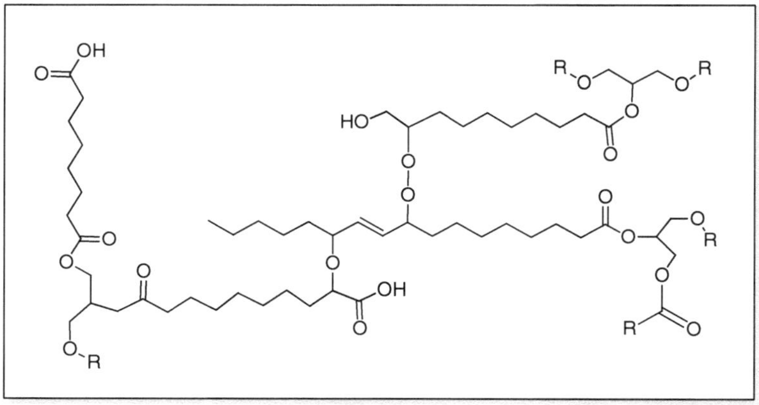

Fig. 5 Network structure of crossed-linkage

compounds (e.g., *sec*-butyllithium). For the anionic polymerization of lauryl methacrylate at −40 °C in THF in the presence of additives (e.g., dilithium salt of triethylene glycol, LiCl, or $LiClO_4$) and obtained polymers with a narrow molecular weight distribution (PDI < 1.10). Furthermore, diblock copolymers of methyl methacrylate with lauryl methacrylate resulted in materials with improved properties for application as a new class of thermoplastic elastomers (Meier et al. 2007).

2.2 Step Growth Polymer Composites

Step growth polymer composites can be generated from vegetable oils by three main routes: (i) Polycondensation of hydroxycarboxylic acids, (ii) Polycondensation of a dicarboxylic acid and a diol, or (iii) Ring opening polymerization of lactones. Emphasis in the present chapter is placed on aliphatic polyesters derived from the simple procedure based on the direct polycondensation of aliphatic α, ω-alkanedioic acids, and α, ω-alkanediols. These can be generated by olefin metathesis of unsaturated fatty acids, followed by hydrogenation to the saturated product (Güner et al. 2006).

An unusually long carbon chain fatty acid of interest as a building block in polyester synthesis is erucic acid [*cis*-13-docosenoic acid], a monounsaturated ω-9 fatty acid readily available from rapeseed or crambe oils. The major contributions involving this precursor comprise the work of mecking and coworkers on the preparation of linear saturated semicrystalline polyesters with long-chain hydrocarbon segments based on the complete incorporation of unsaturated fatty acids. The isomerizing carbonylation of ethyl erucate yielded the saturated diethyl-1,23-tricosanedioate, which was then polymerized with tricosane-1,23-diol (formed by the reduction of the 1,23-diacid), dodecane-1,12-diol, and hexane-1,6-diol to obtain polyester 23,23, polyester 12,23, and polyester 6,23 with high melting points respectively (Bantchev et al. 2009; Lluch et al. 2010).

2.3 Vegetable-Oil-Based Polyurethane

2.3.1 Synthesis of Multi-isocyanates

Polyurethanes are the most abundant polymeric materials which have versatile properties suitable for medical applications due to their extreme biocompatibility and mechanical properties (Lamba et al. 1998; Zdrahala and Zdrahala 1999). Polyurethanes are synthesized from the reaction of polyols with derivatives of petroleum multi-isocyanate, both of which could be derived from triglycerides and their derivatives (More et al. 2013). Most of the vegetable and industrially used method for multi-isocyanate production is the reaction of highly toxic gaseous phosgene with amines or their corresponding salts (Singha et al. 2009). Triglycerides or fatty acid as starting materials and different chemical approaches have been explored to synthesize novel multi-isocyanate. By the reaction of double bonds in soybean oil triglycerides with iodine isocyanate, soybean oil iodoisocyanate was synthesized (Thakur et al. 2012a, b, c, d). The number of isocyanate groups per triglyceride was approximately three. Soybean triglycerides are brominated at the allylic position with NBS from reacting with AgNCO to synthesize soybean triglycerides multi-isocyanates (Thakur et al. 2014a). Approximately 60–70 % of the bromine was replaced by NCO groups and it increased the yield up to increasing the amount of AgNCO (Çayl and Kusefoglu 2008). Fatty acids derivated from triglycerides were also important starting materials for multi-isocyanates. Oleic acid was used in the preparation of 1,7-heptamethylene di-isocyanate and 1,16-diisocyanatohexadec-8-ene. First oleic acid was converted to diacid that were then further converted to di-isocyanates (Hojabri et al. 2009).

Polyurethanes are based on 1,7-heptamethylene di-isocyanate showing similar physical properties as those made from petroleum-derived 1,6-hexamethylene di-isocyanate. Polyurethanes based on 1,16-diisocyanatohexadec-8-ene showed an even higher tensile strength than those commonly derived from petroleum-based 1,7-heptamethylene di-isocyanate (Hojabri et al. 2010a, b; Singha and Thakur 2010a). This development was possibly due to the higher degree of hydrogen bonding corresponding with the longer alkane chains. The long-chain di-isocyanate polyurethanes had a lower Young's modulus and higher elongation. This improvement was presumably due to the improved flexibility of the long di-isocyanate chain. Fatty-acid-based di-isocyanate was also synthesized starting from diesters (Hojabri et al. 2010a, b; Thakur et al. 2010a).

Dimethyl sebacate was hydrazinolysed in absolute ethanol to form diacyl hydrazide, which was then converted into diacylazide. Di-isocyanate was synthesized by the Curtius rearrangement of diacylazide. The determining step is the formation of diacylazide from diacylhydrazide because the side reaction formation of secondary amide was observed in this process (Turunc and Meier 2010).

2.3.2 Synthesis of Polyols

So many protocols have been followed for the synthesis of vegetable-oil-based polyols, and the most widely studied reactions are summarized in thiolene coupling reactions that can directly form hydroxyl groups through double bonds by reacting with 2-mercaptoethanol normally, thioene coupling reactions were very fast under mild reaction conditions (Chen et al. 2010). Recently, reviewed the chemistry approach to obtain monomers and polymers from vegetable oil, thiolene addition reactions of fatty acids and derivatives have been successfully embraced to synthesize diols suitable for polyurethane technology. This group is also applicable to the thiolene additions to 10-undecenoate derivatives to obtain monomers for production of thermoplastic polyurethanes (Gonzalez et al. 2011). Rapeseed oil was photochemically reacted with 2-mercaptoethanol via a thiolene coupling reaction. The obtained polyols have an average hydroxyl functionality of 3.6 (the number of hydroxyl groups per molecule). Neither solvents nor photoinitiator was needed and the product was easily purified. The number of double bonds per chain in the vegetable oil had a marked effect on the reaction, and several side reactions were observed such as disulfide formation and intermolecular combinations (Turunc and Meier 2010).

3 Applications of Vegetable-Oil-Based Polymer

Vegetable-oil-based polymers have several applications which play a major role in industries like painting, adhesive, and also in biomedical fields which are elaborated below.

3.1 Industrial Applications

From the olden days, these vegetable oils are used for paintings (cave paintings), the ancient human used vegetable oils as a drying source for paintings (Johannes et al. 1996). These vegetable oils have several derivatives like polymer, monomer which play a busy role in environmental-friendly applications like waterborne coating materials which reduces use of organic solvents which are toxic to globe (Yongshang and Richard 2007). The major drawback of isolating waterborne coating materials from vegetable oils was one of the toughest processes but these are excellent co-monomer for the synthesis of latex, adhesives, etc., by these processes we can synthesize pressure-sensitive adhesives conventionally compared to that of normally synthesized adhesives which can withstand most of the parameters and reusability (Mohammad and Sharif 2012).

3.2 *Biomedical Applications*

Recently, polymer composites were not used only for industrial applications but also in biomedical applications like glues, pharmacological patches (Stephen et al. 2009), surgical sealants (Cathryn et al. 2005), wound healing devices, and also as a drug carriers for tissue engineering (Delora et al. 2006). Vegetable oil moiety can enhance the biodegradation of material when it was incorporated and also they attracted by some of the biomedical applications to softer parts of the human beings, for example, like retinal, nerve, and vascular regions (Shida et al. 2012; Thakur et al. 2014d). There are some hard materials like polymeric soybean oil–polystyrene membranes, polyurethane networks based on soybean and sunflower oils (Beauty et al. 2013) were tested against scaffolds for tissue engineering which shows cell adherence and proliferation (Gerard et al. 2013).

4 Conclusions

This study states that the uses of renewable sources which are vegetable oils, polymers synthesis is stepping forward. In this study we were reviewed about the different types of syntheses and the importance of polymer composites such as crossed-linked polymers and step-growth polymers from the vegetable oils. The polymer composites have achieved importance in industrial and biomedical applications. To our knowledge from various sources we have reviewed the preparation and applications of polymers from vegetable oil.

References

Arrakhiz FZ, Achaby ME, Malha M, Bensalah MO, Fassi F, Bouhfid R, Benmoussa K, Qaiss A (2013) Mechanical and thermal properties of natural fibers reinforced polymer composites: doum/low density polyethylene. Mater Des 43:200–205

Baillie C (2004) Green composites: polymer composites and the environment. Woodhead Publishing

Bantchev GB, Kenar JA, Biresaw G, Han MG (2009) Free radical addition of butanethiol to vegetable oil double bonds. J Agri Food Chem 57:1282–1290

Barnwal BK, Sharma MP (2005) Prospects of biodiesel production from vegetable oils in India. Renew Sustain Energy Rev 9:363–378

Beauty D, Pronobesh C, Manabendra M, Brigitte V, Niranjan K (2013) Bio-based biodegradable and biocompatible hyperbranched polyurethane: a scaffold for tissue engineering. MacromolBiosci 13:126–139

Belgacem MN, Gandini A (2008) Monomers, polymers and composites from renewable resources. Elsevier, Amsterdam. ISBN 978-0-08-045316-3

Biermann U, Friedt W, Lang S, Luhs W, Machmuller G, Metzger JO, Klass MR, Schafer HJ, Schneider MP (2000) New syntheses with oils and fats as renewable raw materials for the chemical industry. Angew Chem Int Ed 39:2206–2224

Bussell GW (1974) US Patent, 3,855,163

Cathryn AS, Jeffery YS, Yadong W, William CF, Robert SL, Joseph PV, Tessa AH (2005) Biocompatibility analysis of poly (glycerol sebacate) as a nerve guide material. Biomater 26:5454–5464

Çayl G, Kusefoglu S (2008) Biobasedpolyisocyanates from plant oil triglycerides: synthesis, polymerization, and characterization. J Appl Polym Sci 109:2948–2955

Chen Z, Chisholm BJ, Patani R, Wu JF, Fernando S, Jogodzinski K (2010) Soybased UV-curable thiol-ene coatings. J Coat Technol Res 7:603–613

Cinita M, Diego M, Kleber RP, Mirta IA, Mirna AM (2014) Nanocomposites with superpara-magneticbehavior based on a vegetable oil and magnetite nanoparticles. Euro Poly J 53:90–99

Cunningham A, Yapp A (1974) US Patent, 3,827,993

Dave AM, Mehta MH, Aminabhavi TM, Kulkarni AR, Soppimath KS (1999) A review on controlled release of nitrogen fertilizers through polymeric membrane devices. Polym-plast technol 38:675–711

Delara M, Jian Y, Karen YL, Antonio RW, Guillermo AA (2006) Hemocompatibility evaluation of poly(glycerol-sebacate) in vitro for vascular tissue engineering. Biomater 27:4315–4324

Dyer JM, Stymme S, Green AG, Carlsson AS (2008) High-value oils from plants. Plant J 54:640–655

Force CG, Starr FS (1988) US Patent, 4,740,367

Frederick T, Wallenberger T, Norman E (2004) Natural fibers, plastics and composites. Springer

Friedrich K, Lu Z, Hager AM (1995) Recent advances in polymer composites' tribology. Wear 190:139–144

Gallezot P (2012) Conversion of biomass to selected chemical products. Chem Soc Rev 41:1538–1558

Gerard L, Juan CR, Marina G, Virginia C (2013) Renewable polymeric materials from vegetable oils: a perspective. doi:10.1016/j.mattod.2013.08.016

Gonzalez-Paz R, Lluch C, Lligadas G, Ronda J, Galia M, Cadiz V (2011) A green approach toward oleic- and undecylenic acid-derived polyurethanes. J Polym Sci Pol Chem 49:2407–2416

Gorkum R, Bouwman E (2005) The oxidative drying of alkyd paint catalyzed by metal complexes. Coord Chem Rev 249:1709–1728

Guner FS, Yagci Y, Erciyes AT (2006) Polymers from triglycerides oils. Prog Poly Sci 31:633–670

Hodakowski LE, Osborn CL, Harris EB (1975) US Patent, 4,119,640

Hojabri L, Kong X, Narine SS (2009) Fatty acid-derived diisocyanate and biobased polyurethane produced from vegetable oil: synthesis, polymerization, and characterization. Biomacromolecules 10:884–891

Hojabri L, Kong X, Narine SS (2010a) Novel long chain unsaturated diisocyanate from fatty acid: synthesis, characterization, and application in bio-based polyurethane. J Polym Sci Pol Chem 48:3302–3310

Hojabri L, Kong X, Narine SS (2010b) Functional thermoplastics from linear diols and diisocyanates produced entirely from renewable lipid sources. Biomacromolecules 11:911–918

Johannes TP, Derksen F, Petrus C, Peter K (1996) Renewable resources in coatings technology: a review. Progress Org Coat 27:45–53

Johnson RW, Fritz EE (1989) Fatty acids in industry. New York

Kymakis E, Alexandou I, Amaratunga GAG (2002) Single- walled carbon nanotube- polymer composites: electrical optical and structural investigations. Synth Met 127:50–62
Lamba NM, Woodhouse KA, Cooper SL (1998) Polyurethanes in biomedical applications. CRC Press, Boca Raton, FL
Lin M-F, Thakur VK, Tan EJ, Lee PS (2011a) Dopant induced hollow $BaTiO_3$ nanostructures for application in high performance capacitors. J Mater Chem 21:16500–16504
Lin M-F, Thakur VK, Tan EJ, Lee PS (2011b) Surface functionalization of $BaTiO_3$ nanoparticles and improved electrical properties of $BaTiO_3$/polyvinylidene fluoride composite. RSC Adv 1:576–578
Lluch C, Ronda JC, Galia M, Lligadas G, Cadiz V (2010) Rapid approach to biobased telechelics through two one-pot thiol—ene click reactions. Bio mac mol 11:1646–1653
Lu Y, Larock RC (2009) Novel polymeric materials from vegetable oils and vinyl monomers: preparation, properties, and applications. Chem Sus Chem 2:136–147
Meier MAR, Metzger JO, Schubert US (2007) Plant oil renewable resources as green alternatives in polymer science. Chem Sus Rev 36:1788–1802
Mikitaev A, Alexey KR, Elena GN (2009) Polymer nanocomposites: variety of structural forms and applications. Nova Science Publishers, 319 pp
Mohammad YS, Sharif A (2012) Waterborne vegetable oil epoxy coatings: Preparation and characterization. Prog Org Coat 75:248–252
More AS, Lebarbé T, Maisonneuve L, Gadenne B, Alfos C, Cramail H (2013) Novel fatty acid based di-isocyanates towards the synthesis of thermoplastic polyurethanes. Eur Polym J 49:823–833
Mosiewicki MA, Aranguren MI (2013) A short review on novel biocomposites based on plant oil Precursors. Eur Polym J 44:1243–1256
Norris FA (1996) Extraction of fats and oils. Bailey's industrial oil and fat products. vol 2. New York
Okada A, Usuki A (1995) The chemistry of polymer-clay hybrids. Mat Sci Engg C3:109–115
Reis JML, Mota EP (2014) Mechanical behaviour of piassava fiber reinforced castor oil polymer mortars. Comp Strut 111:468–472
Salunkhe DK, Chavan JK, Adsule RN, Kadam SS (1992) World oilseeds: chemistry, technology and utilization. Van Nostrand Reinhold, New York
Seniha GN, Yusuf Y, Tuncer E (2014) Polymers from triglyceride oils. Sep Purify Technol 133:260–275
Sharma V, Kundu PP (2006) Addition polymers from natural oils—a review. Prog Poly Sci 31:983–1008
Shida M, Lijing S, Ping W, Ruina L, Zhiguo S, Songping Z (2012) Soybean oil-based polyurethane networks as candidate biomaterials: Synthesis and biocompatibility. Euro J Lipid Sci Tech 114:1165–1174
Shida M, Ping W, Zhiguo S, Songping Z (2014) Vegetable-oil-based polymers as future polymeric biomaterials. Acta Biomater 10:1692–1704
Singha AS, Thakur VK (2008a) Saccaharum cilliare fiber reinforced polymer composites. E-J Chem 5(4):782–791
Singha AS, Thakur VK (2008b) Fabrication and study of lignocellulosic hibiscus sabdariffa fiber reinforced polymer composites. Bioresources 3:1173–1186
Singha AS, Thakur VK (2008c) Synthesis and characterization of pine needles reinforced rf matrix based biocomposites. E-J Chem 5:1055–1062
Singha AS, Thakur VK (2009a) Fabrication and characterization of H. Sabdariffa fiber-reinforced green polymer composites. Polym-Plast Technol Eng 48:482–487
Singha AS, Thakur VK (2009b) Fabrication and characterization of S. Cilliare fibre reinforced polymer composites. Bull Mater Sci 32:49–58
Singha AS, Thakur VK (2009c) Synthesis, characterisation and analysis of hibiscus sabdariffa fibre reinforced polymer matrix based composites. Polym Polym Compos 17:189–194

Singha AS, Thakur VK (2009d) Grewia optiva fiber reinforced novel, low cost polymer composites. J Chem 6:71–76
Singha AS, Thakur VK (2009e) Physical, chemical and mechanical properties of hibiscus sabdariffa fiber/polymer composite. Int J Polym Mater 58:217–228
Singha A S, Thakur VK (2010a) Mechanical, morphological, and thermal characterization of compression-molded polymer biocomposites. Int J Polym Anal Charact 15(2):87–97
Singha AS, Thakur VK (2010b) Synthesis, characterization and study of pine needles reinforced polymer matrix based composites. J Reinf Plast Compos 29:700–709
Singha AS, Thakur VK, Mehta IK, Shama A, Khanna AJ, Rana RK, Rana AK (2009) Surface-modified hibiscus sabdariffa fibers: physicochemical, thermal, and morphological properties evaluation. Int J Polym Anal Charact 14(8):695–711
Sinha VR, Kumria R (2001) Polysaccharides in colon-specific drug delivery. Int J Pharm 224:19–38
Stephen R, William LN, Santiago R, Sunita S, Jing Y, Henry K, Robert L, Michael JY (2009) Engineering retinal progenitor cell and scrollable poly (glycerol-sebacate) composites for expansion and subretinal transplantation. Biomater 30:3405–3414
Thakur VK, Thakur MK (2014a) Recent advances in graft copolymerization and applications of chitosan: a review. ACS Sustain Chem Eng 2:2637–2652
Thakur VK, Thakur MK (2014b) Recent trends in hydrogels based on psyllium polysaccharide: a review. J Cleaner Prod 82:1–15
Thakur VK, Thakur MK (2014c) Processing and characterization of natural cellulose fibers/ thermoset polymer composites. Carbohydr Polym 109:102–117
Thakur VK, Singha AS, Kaur I, Nagarajarao RP, Liping Y (2010a) Silane functionalization of Saccaharum cilliare fibers: thermal, morphological, and physicochemical study. Int J Polym Anal Charact 15(7):397–414
Thakur VK, Singha AS, Mehta I K (2010b) Renewable resource-based green polymer composites: analysis and characterization. Int J Polym Anal Charact 15(3):137–146
Thakur VK, Singha AS, Thakur MK (2012a) In air graft copolymerization of ethyl acrylate onto natural cellulosic polymers. Int J Polym Anal Charact 17(1):48–60
Thakur VK, Singha AS, Thakur MK (2012b) Surface modification of natural polymers to impart low water absorbency. Int J Polym Anal Charact 17:133–143
Thakur VK, Singha AS, Thakur MK (2012c) Biopolymers based green composites: mechanical, thermal and physico-chemical characterization. J Polym Environ 20:412–421
Thakur VK, Singha AS, Thakur MK (2012d) Graft copolymerization of methyl acrylate onto cellulosic biofibers: synthesis, characterization and applications. J Polym Environ 20:164–174
Thakur VK, Thunga M, Madbouly SA, Kessler MR (2014a) PMMA-g-SOY as a sustainable novel dielectric material. RSC Adv 4:18240–18249
Thakur VK, Grewell D, Thunga M, Kessler MR (2014b) Novel composites from eco-friendly soy flour/SBS triblock copolymer. Macromol Mater Eng 299:953–958
Thakur VK, Vennerberg D, Kessler MR (2014c) Green aqueous surface modification of polypropylene for novel polymer nanocomposites. ACS Appl Mater Interfaces 6:9349–9356
Thakur VK, Vennerberg D, Madbouly SA, Kessler MR (2014d) Bio-inspired green surface functionalization of PMMA for multifunctional capacitors. RSC Adv 4:6677–6684
Thakur VK, Thakur MK, Gupta RK (2014e) Review: raw natural fiber–based polymer composites. Int J Polym Anal Charact 19(3):256–271
Thakur VK, Thakur MK, Raghavan P, Kessler M R (2014f) Progress in green polymer composites from lignin for multifunctional applications: a review. ACS Sustain Chem Eng 2(5):1072–1092
Trecker DJ, Borden GW, Smith OW (1976) US Patent, 3,931,075
Turunc O, Meier MA (2010) Fatty acid derived monomers and related polymers via thiol-ene (Click) additions. Macromol Rapid Commun 31:1822–1826
Vaidya R (2012) International conference on environmental. Biomed Biotech 41:55
Wang L, Wang T (2007) Chemical modification of partially hydrogenated vegetable oil to improve its functional properties for candles. J American Oil Chem Soc 84:1149–1159

Williams CK, Hillmyer MA (2008) Polymers from renewable resources: a perspective for a special issue of polymer reviews. Polym Rev 48:1–10

Xia Y, Larock RC (2010) Vegetable oil-based polymeric materials: synthesis, properties, and applications. Green Chem 1893–1909

Yongshang L, Richard CL (2007) New hybrid latexes from a soybean oil-based waterborne polyurethane and acrylics via emulsion polymerization. Biomacro 8:3108–3144

Zdrahala RJ, Zdrahala IJ (1999) Biomedical applications of polyurethanes: a review of past promises, present realities, and a vibrant future. J Biomater Appl 14:67–90

Hierarchically Fabrication of Amylosic Supramolecular Nanocomposites by Means of Inclusion Complexation in Phosphorylase-Catalyzed Enzymatic Polymerization Field

J Kadokawa

Abstract This chapter reviews hierarchically fabrication of eco-friendly supramolecular nanocomposites by means of inclusion complexation by amylose in phosphorylase-catalyzed enzymatic polymerization field. Amylose is a well-known polysaccharide and forms inclusion nanocomplex with various hydrophobic small molecules. A pure amylose is produced by the enzymatic polymerization using α-D-glucose 1-phosphate as a monomer and maltooligosaccharide as a primer catalyzed by phosphorylase. The author has found that a propagating chain of amylose in the enzymatic polymerization twines around hydrophobic polymers present in the reaction system to form inclusion nanocomplexes. The author named this polymerization system as 'vine-twining polymerization' because it is similar as the way that vine of plant grows twining around a rod. Amylosic supramolecular nanocomposite materials such as hydrogel and film were hierarchically fabricated by means of the vine-twining polymerization approach in the presence of copolymers covalently grafting with hydrophobic guest polymers. The enzymatically produced amyloses induced complexation with guest polymers in intermolecular graft copolymers, which acted as cross-linking points to form a supramolecular nanocomposite hydrogel. By using a film-formable main-chain in the graft copolymer, a supramolecular nanocomposite film was also obtained through hydrogelation. A supramolecular polymeric nanocomposite was successfully fabricated by the vine-twining polymerization using a primer-guest conjugate. The product in the vine-twining polymerization system formed a polymeric continuum of an inclusion nanocomplex, where the enzymatically produced amylose chain elongated from the conjugate included the guest segment in the other conjugate.

J. Kadokawa (✉)
Graduate School of Science and Engineering, Kagoshima University,
Korimoto 1-21-40, Kagoshima 890-0065, Japan
e-mail: kadokawa@eng.kagoshima-u.ac.jp

V.K. Thakur and M.K. Thakur (eds.), *Eco-friendly Polymer Nanocomposites*,
Advanced Structured Materials 75, DOI 10.1007/978-81-322-2470-9_17

Keywords Amylose · Supramolecule · Nanocomplex · Enzymatic polymerization · Inclusion

Polysaccharides are widely distributed in nature and have been regarded as structural materials and energy providers (Schuerch 1986; Singha and Thakur 2009a, b, c, d, e). Besides the use of natural polysaccharides in traditional purposes, they have also been expected as the component in new functional eco-friendly materials because they are the most abundant organic substances and posses the large diversity of the structures (Rouilly and Rigal 2002; Thakur et al. 2014a, b, c, d, e). Amylose is one of the representative natural polysaccharide and acts as an energy resource as one component of starch (Lenz 1993). This is composed of glucose residues linked through α-(1 → 4)-glucosidic bonds. Amylose has recently been recognized as a good candidate for a high-performance eco-friendly material because it acts as a host molecule and forms supramolecules by inclusion of various guest molecules with relatively low molecular weight (inclusion nanocomplexes) owing to the helical conformation (Sarko and Zugenmaier 1980). The driving force for complexation of guest molecules in the cavity is mainly hydrophobic interaction as the inside of the amylose helix has a hydrophobic nature due to the presence of hydrophilic hydroxy groups in the glucose residues on outer part of the helix. Accordingly, hydrophobicity is generally in demand as the property of guest molecules for complexation by amylose. Although inclusion nanocomplexes of amylose with polymeric guest compounds are also interesting as functional eco-friendly polymeric materials, only limited studies have been reported regarding the direct complexation of amylose and polymeric molecules (Frampton et al. 2008; Ikeda et al. 2006; Kaneko et al. 2011a; Kida et al. 2007, 2008; Shogren 1993; Shogren et al. 1991). Because the driving force for complexation of guest molecules by amylose is the weak hydrophobic interaction as aforementioned, the amylose cavity does not have a sufficient ability to directly include the long chains of polymeric guests. The author has considered for the direct complexation of amylose with polymeric guests in the polymerization field, that produces amylose as the product, to efficiently construct inclusion nanocomplexes (Kadokawa and Kaneko 2013).

A pure amylose is presently not available from starch because the complete separation of natural amylose from amylopectin as the other component of starch is difficult. One of the efficient methods to provide such a pure amylose is the enzymatic polymerization approach catalyzed by phosphorylase (EC 2.4.1.1) (Kobayashi and Makino 2009; Ziegast and Pfannemüller 1987). Phosphorylase is an enzyme that catalyzes the reversible phosphorolysis of α-(1 → 4)-glucans such as amylose at a non-reducing end in the presence of inorganic phosphate to produce α-D-glucose 1-phosphate (G-1-P) and those with one smaller degree of polymerization (DP) (Kitaoka and Hayashi 2002; Nakai et al. 2013; Seibel et al. 2006). Because the phosphorylase-catalyzed reaction exhibits reversibility, the enzyme also catalyzes the reversible reaction of the phosphorolysis, that is, a glucosylation using G-1-P as a glycosyl donor, to form α-(1 → 4)-glucosidic linkage with liberating inorganic phosphate. As a glycosyl acceptor, maltooligosaccharides with DPs higher than the smallest one recognized by phosphorylase, are used. In the

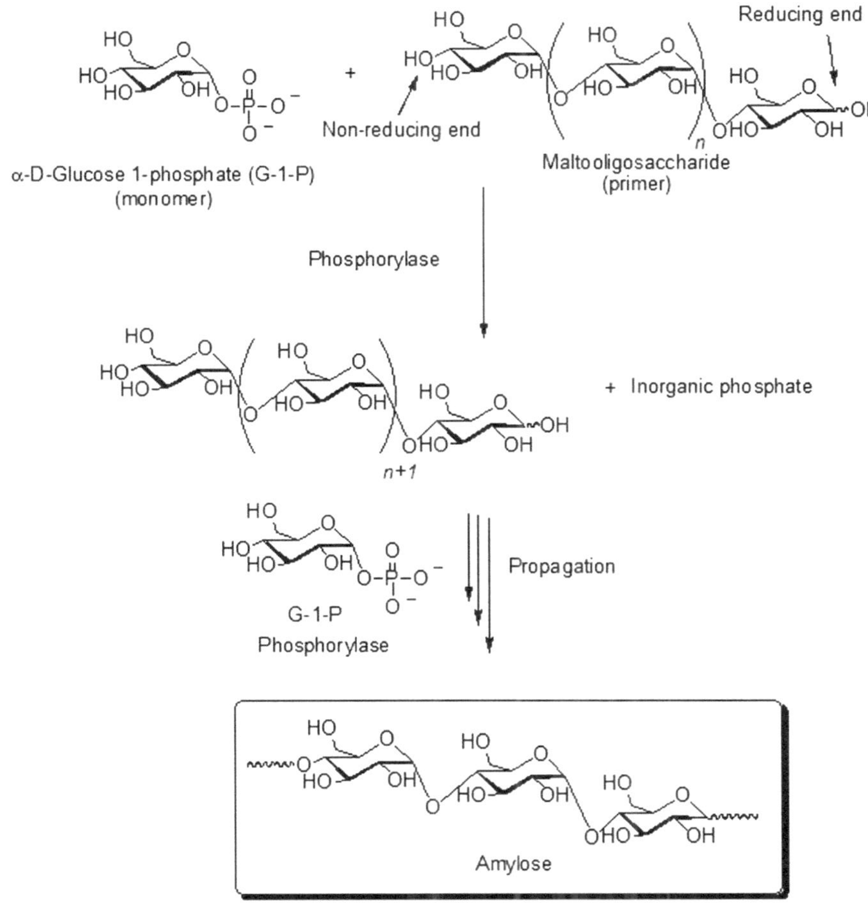

Fig. 1 Phosphorylase-catalyzed enzymatic polymerization to produce amylose

glucosylation, a glucose unit is transferred from G-1-P to the non-reducing end of the acceptor to form α-(1 → 4)-glucosidic linkage. When the excess molar ratio of G-1-P to the acceptor is present in the reaction system, as shown in Fig. 1, the successive glucosylations occur at the elongating non-reducing ends as a propagation of polymerization to produce α-(1 → 4)-glucan, i.e., amylose (Fujii et al. 2003; Ohdan et al. 2006; Yanase et al. 2006). Because the first glucosylation as an initiation occurs at the non-reducing end of the acceptor, it calls a primer of the polymerization. The DPs of the produced amyloses can be controlled by the G-1-P/primer feed molar ratios, and their distributions are typically narrow ($M_w/M_n < 1.2$) (Ohdan et al. 2006; Takata et al. 1998).

By means of the phosphorylase-catalyzed enzymatic polymerization of G-1-P, the author has developed an efficient method for the progress of inclusion complexation of amylose with polymeric guest molecules (Kadokawa 2011, 2012, 2013, 2014; Kadokawa and Kobayashi 2010; Kaneko and Kadokawa 2005). The representation of the inclusion process in the polymerization system is similar as the way that vine of plant grows twining around a rod. Accordingly, the author has proposed that this polymerization method for the formation of amylose-polymer inclusion nanocomplexes is named 'vine-twining polymerization.' Furthermore, the vine-twining polymerization approach has been applied to the architecture of eco-friendly supramolecular nanocomposites. In this chapter, the author reviews hierarchically fabrication of such amylosic nanocomposites by means of supramolecular complexation of amylose with guest polymers in the phosphorylase-catalyzed enzymatic polymerization field.

1 Vine-Twining Polymerization to Form Inclusion Supramolecules

As so-called the vine-twining polymerization, the author found that inclusion in the cavity of amylose toward appropriate guest polymers took place to form inclusion nanocomplexes when the phosphorylase-catalyzed enzymatic polymerization was performed in the presence of such guests dispersed in aqueous polymerization solvent (Fig. 2). Hydrophobicity is required in nature of guest polymers for complexation by amylose because the driving force for binding by amylose cavity is hydrophobic interaction with guest molecules as aforementioned. A first example of the vine-twining polymerization was achieved using poly(tetrahydrofuran) (PTHF)

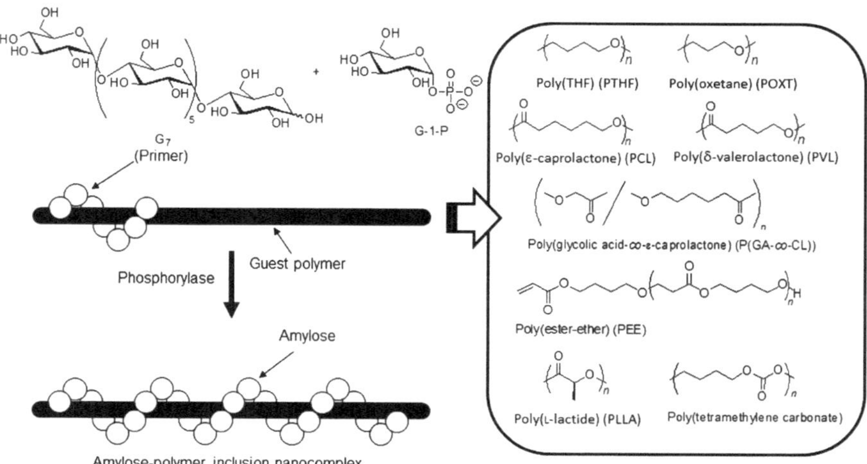

Fig. 2 Image of 'vine-twining polymerization' using various guest polymers to form inclusion nanocomplexes

as a hydrophobic guest polymer (Kadokawa et al. 2001b). When the phosphorylase-catalyzed enzymatic polymerization of G-1-P from maltoheptaose (G_7) as a primer was conducted in the presence of PTHF dispersed in sodium citrate buffer as a polymerization solvent, the product was gradually precipitated with the progress of the polymerization. The structure of the precipitated product was determined by the ^{1}H NMR and powder X-ray diffraction (XRD) measurements to be an inclusion nanocomplex between amylose and PTHF. The formation of the inclusion nanocomplex was not observed by mixing amylose and PTHF in a buffer solvent, strongly suggesting its formation during the progress of the enzymatic polymerization in the above system. The other hydrophobic polyether, that is, poly(oxetane) (POXT) with three methylene lengths has also been found to act a guest polymer for complexation by amylose in the vine-twining polymerization system (Kadokawa et al. 2002). When poly(ethylene glycol) as a hydrophilic polyether with two methylene lengths was used as a guest polymer, on the other hand, amylose produced by the enzymatic polymerization did not induce complexation. The above results suggested that hydrophobicity of guest polymers strongly affected whether complexation by amylose takes place in the vine-twining polymerization system.

On the basis of significance in hydrophobicity of guest polymers on the formation of inclusion nanocomplexes, hydrophobic polyesters, that is, poly(ε-caprolactone) (PCL) and poly(δ-valerolactone) (PVL) were also employed as guest polymers in the vine-twining polymerization (Kadokawa et al. 2001a, 2003). The phosphorylase-catalyzed enzymatic polymerization of G-1-P from G_7 was conducted in the presence of PCL or PVL in sodium citrate buffer and the precipitated products were characterized by ^{1}H NMR and XRD measurements to be inclusion nanocomplexes. It was also found that a copolyester, poly(glycolic acid-*co*-ε-caprolactone) (P(GA-co-CL)) was suitable for complexation by amylose in the vine-twining polymerization, whereas a homopolyester, poly(glycolic acid) (PGA), was not formed an inclusion nanocomplex in the same system (Nomura et al. 2011). The homopolyester, PGA, showed high crystallinity and low dispersibility in aqueous buffer, resulting in the difficulty in complexation by amylose.

As another guest polyester, a hydrophobic poly(ester-ether) (PEE, $–CH_2CH_2C(=O)OCH_2CH_2CH_2CH_2O–$), which was composed of ester and ether linkages alternatingly, was used in complexation by amylose in the vine-twining polymerization (Kadokawa et al. 2003). The structure of the precipitated product was confirmed by ^{1}H NMR and XRD measurements to be an inclusion nanocomplex. When a hydrophilic analogue of the above PEE, which had a shorter methylene length ($–CH_2CH_2C(=O)OCH_2CH_2O–$) was employed, no complexation occurred. This result supported that hydrophobicity of guest polymers strongly affected complexation by amylose in the vine-twining polymerization.

When the vine-twining polymerization was conducted using optically active polyesters, poly(lactide)s (PLAs), which had three kinds of the stereoisomers, i.e., poly(L-lactide) (PLLA), poly(D-lactide) (PDLA), and poly(DL-lactide) (PLDLA), as guest polymers, the author found that amylose, produced by the enzymatic polymerization, perfectly recognized the chirality in PLAs on complexation and consequently it only formed an inclusion nanocomplex with PLLA (Kaneko et al. 2011b).

When the vine-twining polymerization was conducted using poly(tetramethylene carbonate) as a hydrophobic guest polycarbonate in an acetone/aqueous buffer mixed solvent, the ^{1}H NMR and XRD measurements of the precipitated product supported to be the structure of an inclusion nanocomplex (Kaneko et al. 2008a). Unlike this polycarbonate, however, polycarbonates having longer methylene lengths were less suitable for the formation of inclusion nanocomplexes in the same system. Such strongly hydrophobic polycarbonates were not sufficiently dispersed in the acetone/aqueous buffer mixed solvent, resulting in difficulty in complexation by amylose. These results suggested that by taking moderate hydrophobicity into consideration, the structure of guest polymers should be designed. On the basis of the significant change of inclusion ability in guest polymers depending on their hydrophobicity, therefore, the author has also found that amylose precisely exhibits selectivity of complexation toward the guest polymers with subtle different chemical structures (Kaneko et al. 2007, 2008b, 2009a, b).

2 Hierarchically Fabrication of Amylosic Supramolecular Nanocomposite Hydrogel and Film by Vine-Twining Polymerization Approach

The aforementioned vine-twining polymerization approach has been applied to the fabrication of amylosic supramolecular nanocomposite materials such as hydrogel and film. For example, the author has assumed that if a graft copolymer having hydrophobic graft chains is used as a guest in the vine-twining polymerization, supramolecular cross-linking points are obtained owing to complexation of amylose with the graft chains between graft copolymers (Kaneko et al. 2010). Accordingly, such a graft copolymer with an appropriate structure has been designed and chemically synthesized. The graft copolymer has to exhibit water-solubility as a component of hydrogel, whereas hydrophobicity is necessary for complexation by amylose. Taking these antagonistic properties for the graft copolymer structure into consideration, poly(acrylic acid sodium salt-*graft*-δ-valerolactone) (PAA-Na-*g*-PVL) was designed, in which PAA-Na is a strong hydrophilic polymer to contribute to enhancing water-solubility of the graft copolymer and PVL has been found to allow complexation by amylose in the vine-twining polymerization (Kadokawa et al. 2003). When the vine-twining polymerization was carried out by the phosphorylase-catalyzed enzymatic polymerization of G-1-P from a maltooligosaccharide primer in the presence of PAA-Na-*g*-PVL in sodium acetate buffer, the reaction mixture turned into the hydrogel (Fig. 3). The occurrence of complexation by amylose, which was produced with the progress of the polymerization, toward PVL graft chains was confirmed by the XRD measurement of the powdered sample obtained by lyophilization of the produced hydrogel. The XRD pattern of the sample in Fig. 4a showed the typical diffraction peaks at ca. 13° and 20° corresponding to the structure of an amylosic

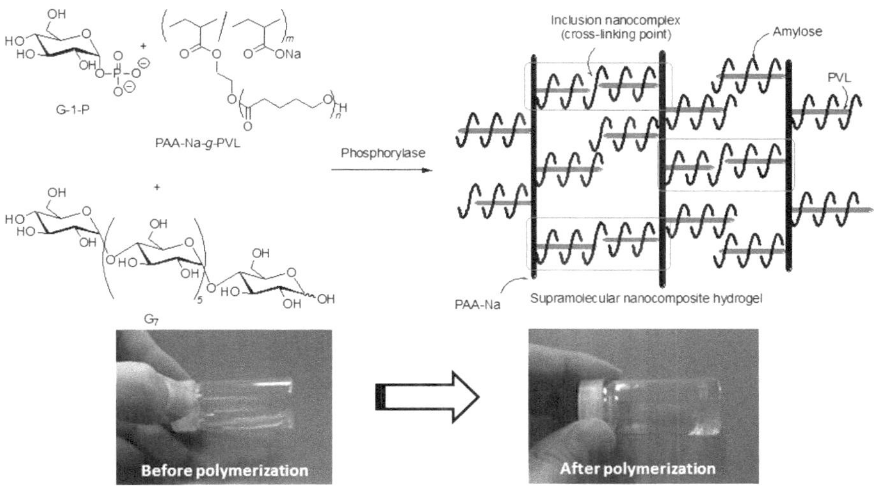

Fig. 3 Preparation of supramolecular nanocomposite hydrogel through inclusion complexation by vine-twining polymerization using PAA-Na-*g*-PVL

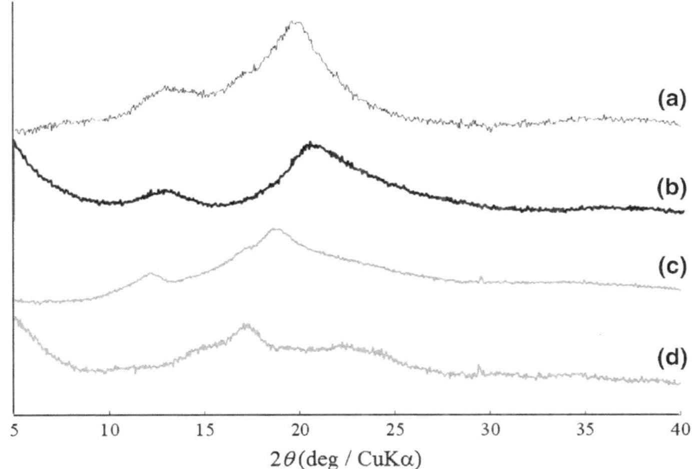

Fig. 4 XRD patterns of powdered samples obtained from hydrogels using PAA-Na-*g*-PVL (**a**) and CMC-*g*-PCL (**b**), supramolecular polymeric nanocomposite (**c**), and amylose (**d**)

inclusion nanocomplex, which was completely different from that of a sole amylose (Fig. 4d). This result indicated amylose induced complexation with PVL graft chains in the vine-twining polymerization system to form inclusion nanocomplexes in the intermolecular graft copolymers, which acted as the cross-linking points for the hierarchically formation of the nanocomposite hydrogel.

Because amylose is enzymatically hydrolyzed and produced by amylase and phosphorylase catalyses, respectively, the behavior of enzymatic disruption and reproduction of the nanocomposite hydrogel by the two enzymatic reactions was investigated. The pure nanocomposite hydrogel sample was first prepared by adding aqueous sodium acetate buffer to the aforementioned powdered sample. When the hydrogel was maintained in the presence of β-amylase under the enzymatic reaction conditions, it was transferred into a solution. This disruption behavior was owing to the disappearance of cross-linking points by hydrolysis of amylose components in the hydrogel by β-amylase catalysis.

When the phosphorylase-catalyzed enzymatic polymerization in the resulting solution was conducted after adding the enzyme and G-1-P, it turned into the hydrogel. Because maltooligosaccharides present in the solution, which were produced by the β-amylase-catalyzed hydrolysis of amylose components in the above experiment, acted as a primer for the enzymatic polymerization, the enzymatically produced amylose induced complexation with the PVL graft chains in the intermolecular graft copolymers to form the cross-linking points. It was further confirmed that the disruption-reproduction cycle was repeatable. Accordingly, it should be noted that the present supramolecular nanocomposite hydrogel shows enzymatically recyclable behavior by the two enzymatic reactions.

The author also investigated the hierarchically fabrication of a supramolecular nanocomposite film through hydrogelation by the vine-twining polymerization using a newly designed graft copolymer (Kadokawa et al. 2013). To provide the film form of the amylosic supramolecular nanocomposite, carboxymethyl cellulose (CMC) was employed as the new main-chain of the graft copolymer because of its water solubility and film formability. Accordingly, CMC-*graft*-poly(ε-caprolactone) (CMC-*g*-PCL) was chemically synthesized by the condensation of amine-terminated PCL with carboxylates in CMC. The vine-twining polymerization was then carried out by the phosphorylase-catalyzed enzymatic polymerization using CMC-*g*-PCL in sodium acetate buffer to obtain the supramolecular nanocomposite hydrogel (Fig. 5). The XRD data of a powdered sample, which was provided by lyophilization of the hydrogel, showed the typical diffraction pattern assignable to

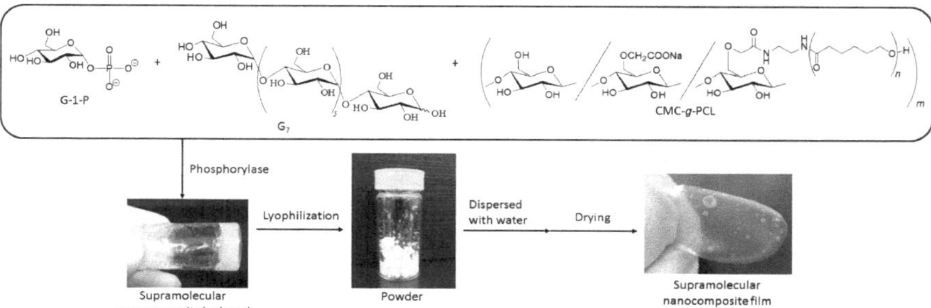

Fig. 5 Preparation of supramolecular nanocomposite film through hydrogelation by vine-twining polymerization using CMC-*g*-PCL

an amylosic inclusion nanocomplex as shown in Fig. 4b. This result supported that gelation took place owing to complexation by amylose toward graft PCL chains in the intermolecular CMC-*g*-PCL for the cross-linking. It was found that a supramolecular nanocomposite film was fabricated by adding water to the powdered material, followed by drying. The several nanocomposite films were fabricated by the same procedure from the hydrogels composed of the different amounts and molecular weights of the amylose molecules because these values could be controllably changed by the conditions in the phosphorylase-catalyzed enzymatic polymerization. The mechanical properties of the resulting films composed of amyloses with moderate amounts and molecular weights under tensile mode were superior to those of a CMC film. This result suggested that the supramolecular cross-linking structure in the nanocomposite films contributed to enhancing the mechanical properties.

3 Hierarchically Fabrication of Amylosic Supramolecular Polymeric Nanocomposite by Vine-Twining Polymerization Approach Using a Primer-Guest Conjugate

Taking the vine-twining propagation manner around a guest polymer into account, as aforementioned, the author investigated to hierarchically fabricate a supramolecular polymeric nanocomposite, which was composed of amylose-PLLA inclusion nanocomplexes by using a G_7-functionalized PLLA (G_7-PLLA) as a primer-guest conjugate substrate (Tanaka et al. 2013). This approach has been based on the following strategic manner, in which in this system, a propagating amylose chain started from G_7 in the conjugate substrate by phosphorylase catalysis potentially includes a PLLA segment in the other substrate and such inclusion complexation among the substrates successively takes place, leading to a supramolecular polymeric nanocomposite.

The key substrate, G_7-PLLA conjugate, was easily synthesized from maltoheptaosyl azide and acetylene-terminated PLLA by cupper(I)-catalyzed click chemistry. The resulting G_7-PLLA conjugate was suspended in sodium acetate buffer and the phosphorylase-catalyzed enzymatic polymerization of G-1-P was conducted in this system for the vine-twining polymerization (Fig. 6). The precipitated product was collected by filtration. Its solubility and morphology were completely different from those of the general amylose-PLLA inclusion nanocomplex previously obtained by the vine-twining polymerization using a sole PLLA as a guest (Kaneko et al. 2011b). The XRD pattern of the isolated product (Fig. 4c) was similar as that of the general amylose-PLLA inclusion nanocomplex. It should be further noted that the XRD peaks of the product were slightly sifted to lower angles (12° and 19°) compared with those detected in the XRD pattern of the aforementioned supramolecular nanocomposites composed of amylose and PVL or PCL (Fig. 4a, b). The difference

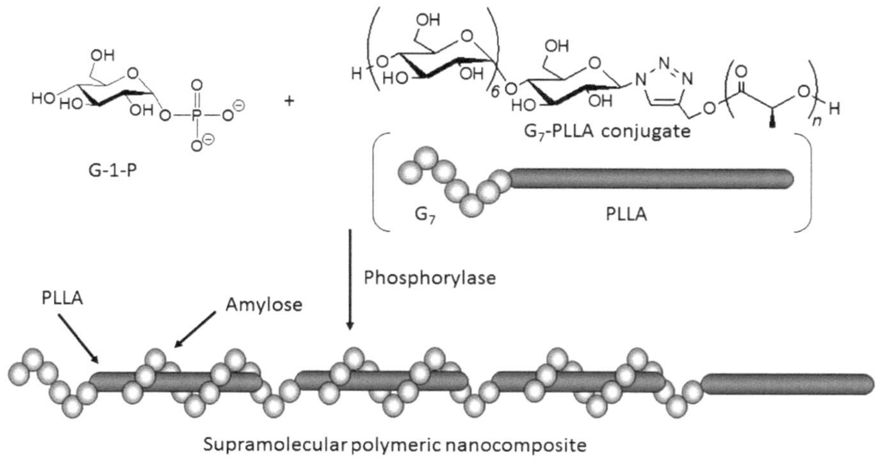

Fig. 6 Preparation of supramolecular polymeric nanocomposite by vine-twining polymerization using G_7-PLLA as primer-guest conjugate

is attributed to the longer diameter of the amylose-PLLA inclusion nanocomplex in the present supramolecular polymeric nanocomposite owing to a bulky structure of PLLA. The ^{1}H NMR spectrum of the soluble fraction in DMSO-d_6 showed signals due to both the amylose and PLLA. These analytical results indicated that amylose was produced by the enzymatic polymerization in this system, and which induced complexation with the PLLA segment.

Because the above analytical results did not sufficiently support the supramolecular polymeric nanocomposite structure, however, the estimation of molecular weights was investigated by the GPC measurement. The GPC peak of the product with DMSO eluent was detected at higher molecular weight region than that of its amylose part obtained by the alkaline hydrolysis, where the PLLA part in the product was completely hydrolyzed and separated off. This result strongly indicated that the product had the higher molecular weight because of the supramolecular polymeric nanocomposite structure.

4 Conclusions

In this chapter, an overview of hierarchically fabrication of eco-friendly supramolecular nanocomposites by means of inclusion complexation by amylose in the vine-twining polymerization was presented. The method was achieved by the phosphorylase-catalyzed enzymatic polymerization in the presence of the designed graft copolymers with guest polymeric chains. Inclusion complexation of amylose chains with guest chains in the intermolecular graft copolymers took place with the progress of the polymerization to construct the supramolecular nanocomposite hydrogel and film. Moreover, the vine-twining polymerization, which was performed in the presence of the primer-PLLA conjugate, gave the supramolecular polymeric nanocomposite.

Because the vine-twining polymerization based on the enzymatic polymerization provides amylosic supramolecular nanocomposites with well-defined structures, which are hardly produced by conventional chemical synthetic method, the present approach will be extended to the systems using newly designed guest polymer components with hierarchically ordered structure. Accordingly, the approach has a potential to fabricate additional eco-friendly supramolecular nanocomposites. Furthermore, the eco-friendly supramolecular nanocomposites based on amylose are practically applicable to the biomedical and environmentally benign materials in the future.

References

Frampton MJ, Claridge TDW, Latini G, Brovelli S, Cacialli F, Anderson L (2008) Amylose-wrapped luminescent conjugated polymers. Chem Commun 2797–2799

Fujii K, Takata H, Yanase M, Terada Y, Ohdan K, Takaha T, Okada S, Kuriki T (2003) Bioengineering and application of novel glucose polymers. Biocatal Biotransform 21:167–172

Ikeda M, Furusho Y, Okoshi K, Tanahara S, Maeda K, Nishino S, Mori T, Yashima E (2006) A luminescent poly(phenylenevinylene)-amylose composite with supramolecular liquid crystallinity. Angew Chem Int Ed 45:6491–6495

Kadokawa J (2011) Precision polysaccharide synthesis catalyzed by enzymes. Chem Rev 111:4308–4345

Kadokawa J (2012) Preparation and applications of amylose supramolecules by means of phosphorylase-catalyzed enzymatic polymerization. Polymers 4:116–133

Kadokawa J (2013) Architecture of amylose supramolecules in form of inclusion complexes by phosphorylase-catalyzed enzymatic polymerization. Biomolecules 3:369–385

Kadokawa J (2014) Chemoenzymatic synthesis of functional amylosic materials. Pure Appl Chem 86:701–709

Kadokawa J, Kaneko Y (2013) Engineering of polysaccharide materials by Phosphorylase-catalyzed enzymatic chain-elongation. Pan Stanford Publishing Pte Ltd, Singapore

Kadokawa J, Kobayashi S (2010) Polymer synthesis by enzymatic catalysis. Curr Opin Chem Biol 14:145–153

Kadokawa J, Kaneko Y, Nakaya A, Tagaya H (2001a) Formation of an amylose-polyester inclusion complex by means of phosphorylase-catalyzed enzymatic polymerization of α-D-glucose 1-phosphate monomer in the presence of poly(ε-caprolactone). Macromolecules 34:6536–6538

Kadokawa J, Kaneko Y, Tagaya H, Chiba K (2001b) Synthesis of an amylose-polymer inclusion complex by enzymatic polymerization of glucose 1-phosphate catalyzed by phosphorylase enzyme in the presence of polyTHF: a new method for synthesis of polymer-polymer inclusion complexes. Chem Commun 449–450

Kadokawa J, Kaneko Y, Nagase S, Takahashi T, Tagaya H (2002) Vine-twining polymerization: amylose twines around polyethers to form amylose-polyether inclusion complexes. Chem Eur J 8:3321–3326

Kadokawa J, Nakaya A, Kaneko Y, Tagaya H (2003) Preparation of inclusion complexes between amylose and ester-containing polymers by means of vine-twining polymerization. Macromol Chem Phys 204:1451–1457

Kadokawa J, Nomura S, Hatanaka D, Yamamoto K (2013) Preparation of polysaccharide supramolecular films by vine-twining polymerization approach. Carbohydr Polym 98:611–617

Kaneko Y, Beppu K, Kadokawa J (2007) Amylose selectively includes one from a mixture of two resemblant polyethers in vine-twining polymerization. Biomacromolecules 8:2983–2985

Kaneko Y, Beppu K, Kadokawa J (2008a) Preparation of amylose/polycarbonate inclusion complexes by means of vine-twining polymerization. Macromol Chem Phys 209:1037–1042

Kaneko Y, Beppu K, Kadokawa J (2009a) Amylose selectively includes a specific range of molecular weights in poly(tetrahydrofuran)s in vine-twining polymerization. Polym J 41:792–796

Kaneko Y, Beppu K, Kyutoku T, Kadokawa J (2009b) Selectivity and priority on inclusion of amylose toward guest polyethers and polyesters in vine-twining polymerization. Polym J 41:279–286

Kaneko Y, Fujisaki K, Kyutoku T, Furukawa H, Kadokawa J (2010) Preparation of enzymatically recyclable hydrogels through the formation of inclusion complexes of amylose in a vine-twining polymerization. Chem Asian J 5:1627–1633

Kaneko Y, Kadokawa J (2005) Vine-twining polymerization: a new preparation method for well-defined supramolecules composed of amylose and synthetic polymers. Chem Rec 5:36–46

Kaneko Y, Kyutoku T, Shimomura N, Kadokawa J (2011a) Formation of amylose-poly (tetrahydrofuran) inclusion complexes in ionic liquid media. Chem Lett 40:31–33

Kaneko Y, Saito Y, Nakaya A, Kadokawa J, Tagaya H (2008b) Preparation of inclusion complexes composed of amylose and strongly hydrophobic polyesters in parallel enzymatic polymerization system. Macromolecules 41:5665–5670

Kaneko Y, Ueno K, Yui T, Nakahara K, Kadokawa J (2011b) Amylose's recognition of chirality in polylactides on formation of inclusion complexes in vine-twining polymerization. Macromol Biosci 11:1407–1415

Kida T, Minabe T, Nakano S, Akashi M (2008) Fabrication of novel multilayered thin films based on inclusion complex formation between amylose derivatives and guest polymers. Langmuir 24:9227–9229

Kida T, Minabe T, Okabe S, Akashi M (2007) Partially-methylated amyloses as effective hosts for inclusion complex formation with polymeric guests. Chem Commun 1559–1561

Kitaoka M, Hayashi K (2002) Carbohydrate-processing phosphorolytic enzymes. Trends Glycosci Glycotechnol 14:35–50

Kobayashi S, Makino A (2009) Enzymatic polymer synthesis: an opportunity for green polymer chemistry. Chem Rev 109:5288–5353

Lenz RW (1993) Biodegradable polymers. Adv Polym Sci 107:1–40

Nakai H, Kitaoka M, Svensson B, Ohtsubo K (2013) Recent development of phosphorylases possessing large potential for oligosaccharide synthesis. Curr Opin Chem Biol 17:301–309

Nomura S, Kyotoku T, Shimomura N, Kaneko Y, Kadokawa J (2011) Preparation of inclusion complexes composed of amylose and biodegradable poly(glycolic acid-co-ε-caprolactone) by vine-twining polymerization and their lipase-catalyzed hydrolysis behavior. Polym J 43:971–977

Ohdan K, Fujii K, Yanase M, Takaha T, Kuriki T (2006) Enzymatic synthesis of amylose. Biocatal Biotransform 24:77–81

Rouilly A, Rigal L (2002) Agro-materials: a bibliographic review. J Macromol Sci Polym Rev C42:441–479

Singha AS, Thakur VK (2009a) Fabrication and characterization of H. sabdariffa fiber-reinforced green polymer composites. Polym-Plast Technol Eng 48:482–487

Singha AS, Thakur VK (2009b) Physical, chemical and mechanical properties of Hibiscus sabdariffa fiber/polymer composite. Int J Polym Mater 58:217–228

Singha AS, Thakur VK (2009c) Fabrication and characterization of S. cilliare fibre reinforced polymer composites. Bull Mater Sci 32:49–58

Singha AS, Thakur VK (2009d) Grewia optiva fiber reinforced novel, low cost polymer composites. J Chem 6:71–76

Singha AS, Thakur VK (2009e) Synthesis, characterization and analysis of Hibiscus Sabdariffa fibre reinforced polymer matrix based composites. Polym Polym Compos 17:189–194

Sarko A, Zugenmaier P (1980) Crystal structures of amylose and its derivatives. In: French AD, Gardner KCH (eds) Fiber diffraction methods, vol 141. American Chemical Society, Washington DC, pp 459–482

Schuerch C (1986) Polysaccharides. In: Mark HF, Bilkales N, Overberger CG (eds) Encyclopedia of polymer science and engineering, vol 13. 2nd edn. Wiley, New York, pp 87–162

Seibel J, Jördening H-J, Buchholz K (2006) Glycosylation with activated sugars using glycosyltransferases and transglycosidases. Biocatal Biotransform 24:311–342

Shogren RL (1993) Complexes of starch with telechelic poly(ε-caprolactone) phosphate. Carbohydr Polym 22:93–98

Shogren RL, Greene RV, Wu YV (1991) Complexes of starch polysaccharides and poly(ethylene-co-acrylic acid)—structure and stability in solution. J Appl Polym Sci 42:1701–1709

Takata H, Takaha T, Okada S, Takagi M, Imanaka T (1998) Purification and characterization of α-glucan phosphorylase from *Bacillus stearothermophilus*. J Ferment Bioeng 85:156–161

Tanaka T, Sasayama S, Nomura S, Yamamoto K, Kimura Y, Kadokawa J (2013) An amylose-poly (L-lactide) inclusion supramolecular polymer: enzymatic synthesis by means of vine-twining polymerization using a primer–guest conjugate. Macromol Chem Phys 214:2829–2834

Thakur VK, Thakur MK, Gupta RK (2014a) Review: raw natural fiber-based polymer composites. Int J Polym Anal Charact 19(3):256–271

Thakur VK, Thakur MK, Raghavan P, Kessler MR (2014b) Progress in green polymer composites from lignin for multifunctional applications: a review. ACS Sustain Chem Eng 2(5):1072–1092

Thakur VK, Vennerberg D, Kessler MR (2014c) Green aqueous surface modification of polypropylene for novel polymer nanocomposites. ACS Appl Mater Interf 6:9349–9356

Thakur VK, Vennerberg D, Madbouly SA, Kessler MR (2014d) Bio-inspired green surface functionalization of PMMA for multifunctional capacitors. RSC Adv 4:6677–6684

Thakur VK, Thunga M, Madbouly SA, Kessler MR (2014e) PMMA-g-SOY as a sustainable novel dielectric material. RSC Adv 4:18240–18249

Yanase M, Takaha T, Kuriki T (2006) α-Glucan phosphorylase and its use in carbohydrate engineering. J Sci Food Agric 86:1631–1635

Ziegast G, Pfannemüller B (1987) Phosphorolytic syntheses with di-, oligo- and multi-functional primers. Carbohydr Res 160:185–204

Mechanical Properties of Eco-friendly Polymer Nanocomposites

Asim Shahzad

Abstract Biopolymers are an alternative to petroleum-based synthetic polymers that are renewable, do not contribute to environmental pollution, and are biodegradable. However, some of their properties like tensile strength, impact strength, thermal stability, and permeability are not of sufficiently high standard and must be improved. One way to improve the properties of biopolymers and thus enhance their commercial potential is to incorporate nano-sized bio-based reinforcements in the polymers. The composites thus formed are called eco-friendly polymer nanocomposites. The research in these composites has increased substantially in the last few years with a corresponding increase in research papers. These composites are finding applications in various fields like medicine, packaging, electronics, the automotive sector, and the construction industry. Polysaccharide polymers that are abundant in nature are increasingly being used for this purpose. The biopolymers most commonly used in these composites are thermoplastic starch (TPS), polylactic acid (PLA), cellulose acetate, chitosan, polyvinyl alcohol (PVA), and epoxidized plant oils. Some examples of the bio-based reinforcements used in these composites are cellulose nanowhiskers, chitin whiskers, and starch nanoparticles (SNP). Extrusion and injection molding are the most widely used methods for manufacturing of these composites. Results show that incorporation of bio-based nanoreinforcements in biopolymers results in improvement in mechanical properties of these composites. These include tensile, flexural, and impact properties. Poor dispersion and agglomeration of nanoreinforcements in biopolymers and their poor interfacial bonding are issues which impose a limit on these composites' mechanical performance. Various physical and chemical methods for surface treatments of nanoreinforcements are used. These methods have been shown to result in improvements of mechanical properties of these composites. There are a number of other issues like sensitivity to moisture and temperature, expensive recycling processes, high variability in properties, nonlinear mechanical behavior, poor long-term performance, and low impact strength, which are hindering the

A. Shahzad (✉)
Materials Research Centre, College of Engineering, Swansea University, Swansea, UK
e-mail: mr_asim_shahzad@yahoo.com

V.K. Thakur and M.K. Thakur (eds.), *Eco-friendly Polymer Nanocomposites*,
Advanced Structured Materials 75, DOI 10.1007/978-81-322-2470-9_18

development of these materials. However, as the investment and research in these materials increase, they are expected to replace many conventional materials in optical, biological, and engineering applications.

Keywords Biopolymers · Nanocomposites · Cellulose · Chitosan · Starch · Mechanical properties

1 Introduction

This chapter aims to present an overview of the mechanical properties of eco-friendly polymer nanocomposites. The main constituents of these composites and their properties are discussed first, which is followed by a discussion of the mechanical properties of these composites.

1.1 Polymer Nanocomposites

Polymer nanocomposites are now a well-established class of materials which exhibit superior mechanical, thermal, and barrier properties at low reinforcement levels (typically 5 wt%), as well as better recyclability, transparency and low weight, compared with conventional polymer composites (Lin et al. 2011a, b, c; Thakur et al. 2012a). These nanocomposites are based on conventional thermoset and thermoplastic polymer matrices, most of which are not biodegradable (Thakur et al. 2014a, b, c, d, e, f). Biopolymers are a relatively new class of polymers which are biodegradable, environment friendly, and renewable (Singha and Thakur 2009a, b, c, d). Biopolymers, as stand-alone materials, exhibit poor mechanical properties, low thermal stability, and poor barrier properties (Thakur et al. 2012b, c, d, e). When such polymers are reinforced with natural fibers, their properties are considerably improved with simultaneous improvement in biodegradability. Reinforcement with nanoparticles and nanofibers, derived from naturally occurring polymers, results in further improvement in their properties. Such improvements include, but are not limited to, higher modulus and strength, decrease in gas permeability, increase in heat distortion temperature as well as improvement in the biodegradability of biopolymers.

The most widely used reinforcements in nanocomposites are carbon nanotubes, layered clay, and nanoparticles. The most widely used polymer matrices are polyamides, nonpolar polymers, polyesters, epoxies, and polyurethane. However, this chapter will focus on nanocomposites which are made from both biodegradable fiber/fillers and biodegradable matrices, since only these composites truly justify the definition of eco-friendly and green nanocomposites.

Natural polysaccharides, such as starch derived from corn, wheat, rice or potato, and cellulose and its derivatives, are the main biopolymers employed in the development of green nanocomposites. The biodegradable thermoplastic polyester PLA, starch, and epoxidized vegetable oils are other examples of polymers widely used in the development of reinforced biopolymers.

Cellulose nanofibers are as strong as steel and can be used as an alternative to glass fibers. With just 5–10 % of nanofiber loading, the tensile strength of the nanocomposites can increase by two to three times (Oksman et al. 2009). The composites thus formed are (theoretically at least) completely biodegradable, environment friendly, and renewable; hence they are called eco-friendly polymer nanocomposites. As these nanocomposites continue to find applications in various fields, the research in these composites has increased exponentially in the last decade or so. These applications include medicine, packaging, electronics, the automotive sector, construction, and other areas. Medical applications include artificial implants, wound dressing products, drug delivery, and medical devices. Automotive applications include car panels, door modules, and various load bearing parts (Oksman et al. 2009). A typical example of the research being conducted on these composites is the NanoCelluComp project, which is a consortium consisting of 11 partners from five European countries. This project aims to develop a novel process technology that utilizes the high mechanical performance of cellulose nanofibers combined with biopolymers for the manufacture of 100 % bio-derived high-performance composite materials (JEC Composites 2014). This process technology will be compatible with existing composite manufacturing processes to facilitate industry uptake. Simultaneously, the project is assessing the sustainability of the process and materials in terms of life cycle analysis (including environmental impacts) and costs compared to existing materials.

1.2 Fillers and Matrices

Fillers used as nanoreinforcement are, generally, of three types: isodimensional particles, elongated particles and layered particles, as shown in Fig. 1.

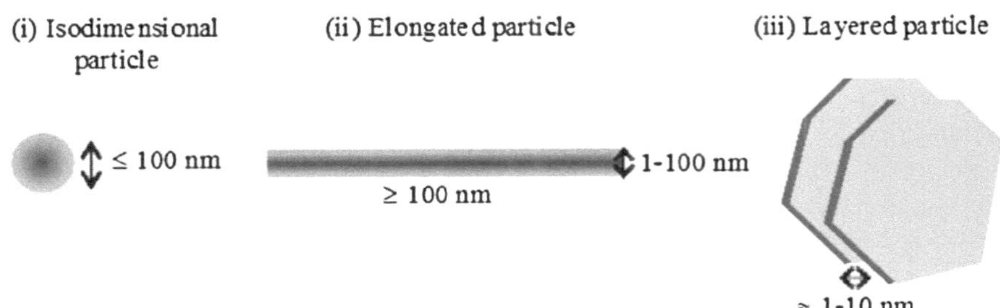

Fig. 1 Three types of fillers used as nanoreinforcement (Angellier and Dufresne 2013)

Table 1 Comparison between cellulose nanocrystals (CNC) and microfibrillated cellulose (MFC)

Cellulose nanocrystals	Microfibrillated cellulose
Consist of crystallites	Consist of crystalline and amorphous regions
Rod-like nanoparticles	Long and flexible nanoparticles
Diameter ~ 5–10 nm; length ~ 100–500 nm	Diameter ~ 10–100 nm; length ~ several microns
Young's modulus ~ 130–250 GPa	Young's modulus ~ 150 GPa

Isodimensional particles have the same size in all directions and their aspect ratio is usually close to one. Some examples of these particles are spherical silica, carbon black, and fullerenes. Elongated particles consist of fibrils with diameters ranging between 1 and 100 nm and lengths of several hundred of nanometers. Some examples of these particles are cellulose and chitin nanofibers and carbon nanotubes. Because of their higher aspect ratio, nanocomposites based on these fillers offer superior mechanical properties compared to other fillers. Layered particles have one dimension, often the thickness, ranging from several angstroms to several nanometers. Their examples are layered silicates and starch nanocrystals (SNs).

Cellulose fibers consist of smaller and stronger long and thin filaments called microfibrils. Nanoscale particles can be derived from microfibrils which are of two types: nanowhiskers, also called cellulose nanaocrystals (CNC) and microfibrillated cellulose (MFC). Mechanical shearing actions separate these fibrils into MFC, whereas longitudinal cutting of microfibrils through strong acid hydrolysis treatments results in nanowhiskers. Table 1 summarizes the differences between the two.

In order to produce composites with superior mechanical properties, the reinforcements used in them should have high strength and stiffness, preferably at low density. Nanoscale cellulose fibers and particles satisfy these requirements, with the added benefit of being biodegradable and abundantly available. With a Young's modulus of about 150 GPa (much greater than glass fibers and at par with Kevlar fibers) and a surface area of several hundred m^2/g (Kalia et al. 2011), nanofibers have the potential to significantly reinforce polymers at low filler loadings. For these reasons, Moon et al. (2011) consider them to be ideal materials on which to base a new biopolymer composites industry.

CNC is derived as nanowhiskers for use as reinforcement in biopolymers such as starch, PLA, and PHB. Apart from being biodegradable, these nanowhiskers have high specific strength and stiffness. Similarly, chitin nanowhiskers are synthesized from chitin, the second most abundant biopolymer after cellulose. These nanowhiskers have many desirable properties for use as reinforcements, like low density, nontoxicity, biodegradability, biocompatibility, easy surface modification, and functionalization. Starch nanoparticles (SNP) are synthesized from starch granules for use as reinforcement.

Owing to its unique properties, such as high mechanical strength, high crystallinity (above 60 %), and a highly pure nanofibrillar network structure, bacterial cellulose (BC), produced by *Acetobacter xylinum*, is becoming a popular biopolymer for use in nanocomposites. BC is available as a tridimensional network of

Table 2 Comparison of properties of cellulose nanaocrystals (CNC) with conventional reinforcements (Moon et al. 2011)

Material	Density (g/cm^3)	Tensile strength (GPa)	Elastic modulus (axial) (GPa)	Elastic modulus (transverse) (GPa)
Glass	2.5	3.4–4.3	72–87	–
Carbon	1.8	1.5–5.7	150–500	–
Kevlar-49	1.4	3.5	124–130	2.5
Steel wire	7.8	4.1	210	–
Clay nanoplatelets	–	–	170	–
Carbon nanotubes	–	11–63	270–950	0.8–30
Boron nanowhiskers	–	2–8	250–360	–
CNC	1.6	7.5–7.7	110–220	10–50

nano- and microfibrils with 10–100 nm width for use as reinforcement. Their structure is quite different from plant cellulose with high purity and high degree of polymerization. Their high degree of crystallinity imparts to them a tensile modulus of 114 GPa which is higher than that of synthetic glass (Lee et al. 2009). Their main disadvantage is their high hydrophilicity.

Table 2 compares the properties of CNC with various conventional reinforcements used in composites. It is readily apparent that their properties are comparable to, or even better than, other reinforcements. These superior properties will inevitably result in composites with superior properties.

Depending on their origins, biopolymers used as matrices in composite materials are divided into three main classes: agropolymer based (renewable sources), microbially derived, and chemically synthesized. Some authors also mention a fourth class, which consists of blends of the aforementioned three classes.

Agropolymer-based biopolymers are obtained from biomass by fractionation. They are further divided into two categories: polysaccharides, and proteins and lipids. Polysaccharides are derived from starches (wheat, potato, maize), lignocellulosic products (cellulose), pectins, chitosans, and gums (Thakur and Thakur 2014a, b, c). Proteins and lipids can be either animal derived (casein, whey, gelatin, collagen), or plant derived (soya, gluten) (Thakur and Kessler 2014a, b). These materials and their derivatives offer a wide range of properties and applications.

Microbially-derived biodegradable polyesters are obtained by microbial fermentation from genetically modified plants. For this reason, they are also referred to as bacterial biopolymers. Their best example is polyhydroxyalkanoate (PHA). PHAs occur naturally in a variety of organisms, but microorganisms can be employed to tailor their production in cells. PHAs can be produced by varieties of bacteria using several renewable waste feedstocks. The feedstocks include cellulose, vegetable oils, organic waste, municipal solid waste, and fatty acids depending on the specific PHA required. Poly(3-hydroxybutyrate) (PHB) is the main representative of PHAs and was, in fact, the first PHA discovered. PHB is similar in its material properties to thermoplastics, with a good resistance to moisture.

Chemically synthesized biodegradable polyesters are obtained by synthesis from monomers obtained from biomass. Their best example is polylactic acid (PLA) derived from corn. PLA shares some similarities with commodity polymers such as polyethylene terephthalate (PET). It has many good characteristics like good transparency, glossy appearance, high rigidity, and ability to tolerate various types of processing conditions.

2 Tensile Properties

Tensile properties are, by far, the most widely studied mechanical properties of eco-friendly polymer nanocomposites. Overall, the mechanical performance of CNC-reinforced composites depends on the aspect ratio, crystallinity, processing method, and CNC/matrix interfacial interaction. The mechanical properties are proportional to aspect ratio and crystallinity of nanoreinforcement and it has been shown that increase in aspect ratio and crystallinity results in increase in mechanical properties. Slow processing methods which encourage water evaporation result in composites with improved properties. This is because nanoparticles have sufficient time to interact and connect to form a continuous network, which is the basis of their reinforcing effect. Nanoreinforcement which is compatible with the biopolymer matrix also exhibits improved mechanical properties of the nanocomposites.

The mechanical properties of CNC nanocomposites have also been found to exceed the predicted values gained from classical models based on filler reinforced nanocomposites. This phenomenon, also called percolation effect, has been explained by the formation of a rigid network of whiskers which is responsible for the unusual reinforcing effect observed (Samir et al. 2005). Above the percolation threshold the cellulosic nanoparticles can connect and form a 3D continuous pathway through the nanocomposite film, cemented by hydrogen bonds. For example, for rod-like particles such as tunicin whiskers with an aspect ratio of 67, the percolation threshold is close to 1 vol%. The percolation approach has been found to fit satisfactorily the experimental data, especially at high filler loading. The similarity between predicted and experimental data demonstrates the major role of filler/filler interaction in the final mechanical behavior of the cellulose-based nanocomposites. Moreover, the percolation approach also accounts for the excellent thermal stabilization of nanocomposites' modulus at high temperatures. Thus the mechanical properties of CNC nanocomposites are mainly affected by factors that ensure or interfere with the formation of the percolated network such the aspect ratio of CNCs and their interfacial interactions (between them or with the host matrix).

There are two main issues with the use of cellulose nanoreinforcement in biopolymer matrices which ultimately affect the mechanical properties of the composites made from them. The first one is the noncompatibility of the polar fiber and the nonpolar matrix. This results in poor fiber/matrix interfacial adhesion which adversely affects the mechanical properties of composites. The second one is the poor dispersion of fibers in the matrices. The agglomeration of cellulose nanofibers is particularly an issue when used with thermoplastic matrix. Cellulose fibrils have high density of hydroxyl groups on the surface which try to bond with adjacent hydroxyl groups by weak hydrogen bonding. This results in agglomeration of nanofibers. This again results in poor mechanical properties of composites because the reinforcing effect of fibers is greatly diminished. Various surface treatments are used to overcome this issue. These include both physical treatments and chemical treatments. Physical treatments include surface fibrillation, electric discharge (corona and cold) plasma, irradiation, and electric currents. Chemical treatments include the use of silane coupling agents, maleated polypropylene/maleic anhydride grafted polypropylene (MAPP), adhesives, and surfactants/surface active agents. An example of good dispersion of nanocrystalline cellulose in thermoplastic starch (TPS) matrix is shown in Fig. 2. Cellulose crystals appear like uniformly dispersed white dots as shown in figure (b), their concentration being function of the cellulose

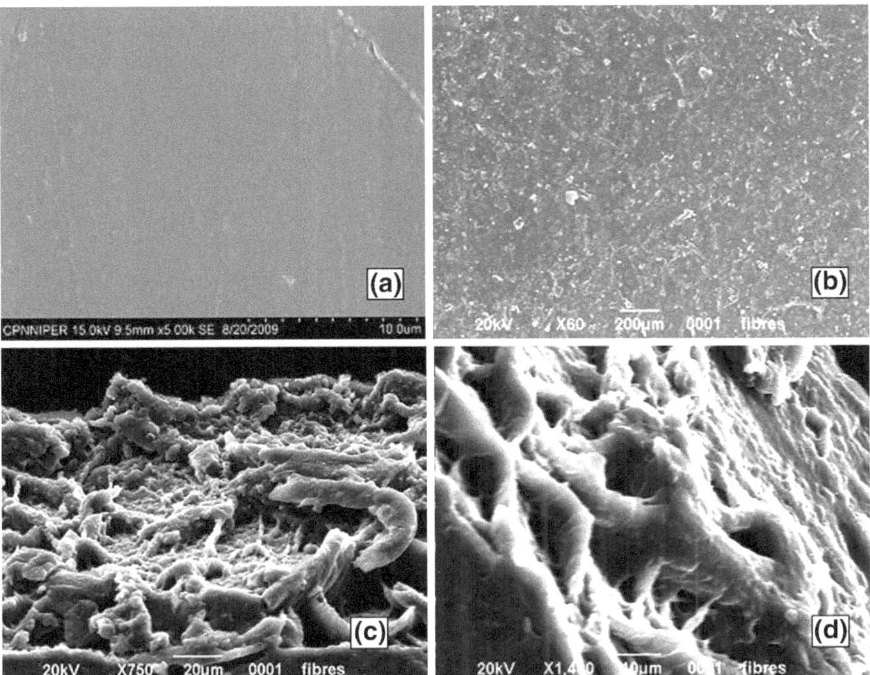

Fig. 2 SEM micrographs of **a** unreinforced TPS film, **b** nanocrystalline cellulose/TPS nanocomposites containing 8 % cellulose, and **c**, **d** fractured surface of nanocrystalline cellulose/TPS nanocomposites (Kaushik and Kumra 2014)

percentage. The figure shows a relatively uniform dispersion of nanocrystals in TPS and no aggregates on the fracture surfaces of the nanocomposites.

The problem of noncompatibility is not generally encountered with biopolymer matrices. However, even in these matrices, the strong interfibrillar hydrogen bonding still impedes proper dispersion of cellulose nanofibers.

The use of a particular manufacturing method of nanocomposites can also affect their mechanical properties. The most commonly used manufacturing methods for cellulose nanocomposites are solvent casting, impregnation of a fiber network or nanopaper with a polymer and melt compounding (Oksman et al. 2009). The solvent casting of nanocomposites entails dissolution of the polymer matrix and proper dispersion of nanocellulose in the same solvent. The impregnation of nanofiber paper/network of films requires the polymer matrix to be in the dissolved stage or have a low viscosity. In this method the final product shape is limited to films or flat products. Melt compounding needs to be done using a specific extruder where the nanocelluloses are mixed with a polymer melt. This is the most widely used method because of the possibility of scaling up the process, and because of the choice of injection molding or compression molding the products.

In one landmark study on the effects of manufacturing processes on mechanical properties of nanocomposites, Oksman et al. (2009) used solvent casting process to make aligned CNC/poly(vinyl alcohol) (PVA) nanocomposites. CNCs were aligned by using a strong magnetic field. The SEM analysis of the nanocomposites showed cellulose nanowhiskers aligned in the direction perpendicular to the magnetic field (Fig. 3). The dynamic mechanical properties of the nanocomposites were studied in the parallel and transverse direction. The results showed that the storage modulus below the T_g (at 25 °C) was remarkably higher (2 GPa) in the transverse direction (whiskers in parallel direction) than the parallel direction of the magnetic field (whiskers in perpendicular direction). This improvement is significant and can be considered as a direct impact of orientation of nanowhiskers in the PVA matrix. These results showed that aligned CNCs can have a similar reinforcing effect on the matrix as the synthetic fibers.

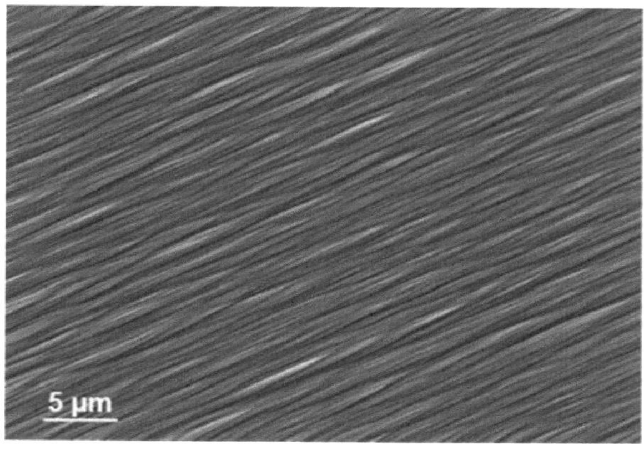

Fig. 3 SEM micrograph of CNC/PVA nanocomposite showing highly oriented structure (Oksman et al. 2009)

They also used modified impregnation method to make cellulose/cellulose nanocomposites. These nanocomposites also showed improved tensile and creep properties over neat CNCs. They then used melt compounding technique to manufacture CNC/poly(lactic acid) nanocomposites, where polyethylene glycol (PEG) was used as plasticizer and maleic anhydride (MA) was used as a coupling agent. Composites made with and without plasticizer both showed increase in tensile properties. However, the maximum improvement in tensile properties was observed for composites made without plasticizer. The tensile modulus of these composites for 5 wt% CNCs improved by about 35 %, the tensile strength by about 90 %, and elongation to break by about 35 %.

2.1 Chitosan

Chitin is a semicrystalline biopolymer which forms microfibrillar arrangement in living organisms like shrimp, crab, tortoise and insects, with diameters ranging from 2.5 to 25 nm. These fibrils can be separated as nanofibers and nanoparticles, also called chitosan, and used as reinforcements in nanocomposites. They have similar structure as nanocrystalline cellulose. Chitosan possesses many properties like biocompatibility, biodegradability, and nontoxicity which encourage their use as reinforcements and matrices in eco-friendly composites. They have been widely studied in inorganic fillers like hydroxyapatite, clay, carbon nanotubes, and grapheme oxide. However, their mechanical properties have been found to be short of expectations. Their use with organic fillers has been less extensively studied.

2.1.1 Chistosan as Filler

The studies have shown increase in tensile properties of nanocomposites for α-chitin whisker (ChW)/PVA (Junkasem et al. 2006; Uddin et al. 2012), α-ChW/chitosan films (Sriupayo et al. 2005), and α-ChW/soy protein isolate films (Lu et al. 2004). In these studies an optimum ChW loading was reported, above which there was negligible improvement in these properties. These values were 5.1 % for α-ChW/PVA, 2.96 % for α-ChW/chitosan films, and 20 % for α-ChW/soy protein isolate films.

Junkasem et al. (2006) reported that the tensile strength of ChW/PVA nanocomposites increased with initial addition of the chitin whiskers to reach a maximum value (5.7 ± 0.6 MPa) at the chitin whisker to PVA ratio of about 5.1 % and decreased with further increasing the whisker content. The presence of ChWs within the nanocomposites increased Young's modulus by about 4–8 times over that of the neat matrix.

Uddin et al. (2012) also reported an optimum loading of ChW in PVA matrix. At 5 wt% ChW loading, the optimum values of tensile strength (1880 MPa) and toughness (68 J/g) were obtained. However, tensile modulus continued to increase with increase in ChW loading, attaining a peak value of 50 GPa at 30 wt% loading.

In an interesting study, all chitosan nanocomposites were made of ChW reinforced chitosan films by Sriupayo et al. (2005). The tensile strength increased with initial increase in the whisker content to reach a maximum value at the whisker content of 2.96 wt% and decreased gradually with further increase in the whisker content, while the percentage of elongation at break decreased with initial increase in the whisker content, and leveled off when the whisker content was greater than or equal to 2.96 wt%.

Chang et al. (2010) studied the mechanical properties of chitosan nanoparticle CNP/TPS nanocomposites. When the CNP content increased from 0 to 6 wt%, the tensile strength increased from 2.8 to 10.8 MPa, while the elongation at break decreased from 59 to 23 %. This was attributed to the interfacial interaction between chitosan nanoparticles and the glycerol plasticized starch matrix because of the similar polysaccharide structures of CNP and starch. At CNP concentration of 8 wt%, the tensile strength showed deterioration, possibly due to the agglomeration of CNP. Thus CNP nanoparticles were also shown to have an optimum concentration in TPS, above which their mechanical properties decreased.

2.1.2 Chitosan as Matrix

Li et al. (2009) prepared bio-based nanocomposite films using cellulose whiskers as the reinforcing phase and chitosan as the matrix. The results showed that the whisker content had a profound effect on the mechanical properties of the composites as shown in Fig. 4. The tensile strength of the composite films in the dry state increased from 85 to 120 MPa with increasing filler content from 0 to 20 wt%, whereas the elongation at break decreased from 20 to 6 %. This indicated that the incorporation of cellulose whiskers into the chitosan matrix resulted in strong interactions between filler and matrix, which restricted the motion of the matrix. In the wet state, the tensile strength of the composite films increased from 9.9 to 17.3 MPa with an increase of whisker content from 0 to 15 wt%, and then decreased with further increasing filler content. The composites containing more than 20 wt% cellulose whiskers exhibited a decrease in both the tensile strength and elongation at break in both dry and wet state due to the microphase separation and the stiffness of cellulose whiskers. Thus optimum filler content was found to exist for cellulose whiskers in chitosan matrix.

Similar increases in mechanical properties have been reported for cellulose nanofiber reinforced chitosan films by Gallstedt and Hedenqvist (2006) and Nordqvist et al. (2007).

Khan et al. (2012) reported an optimum content of 5 wt% of CNC in chitosan matrix. The nanocomposites showed a 25 % increase in tensile strength and an 87 % increase in tensile modulus at this CNC content.

Fernandes et al. (2009) prepared nanocomposite films based on different chitosan matrices (two chitosans with different DPs and one water-soluble

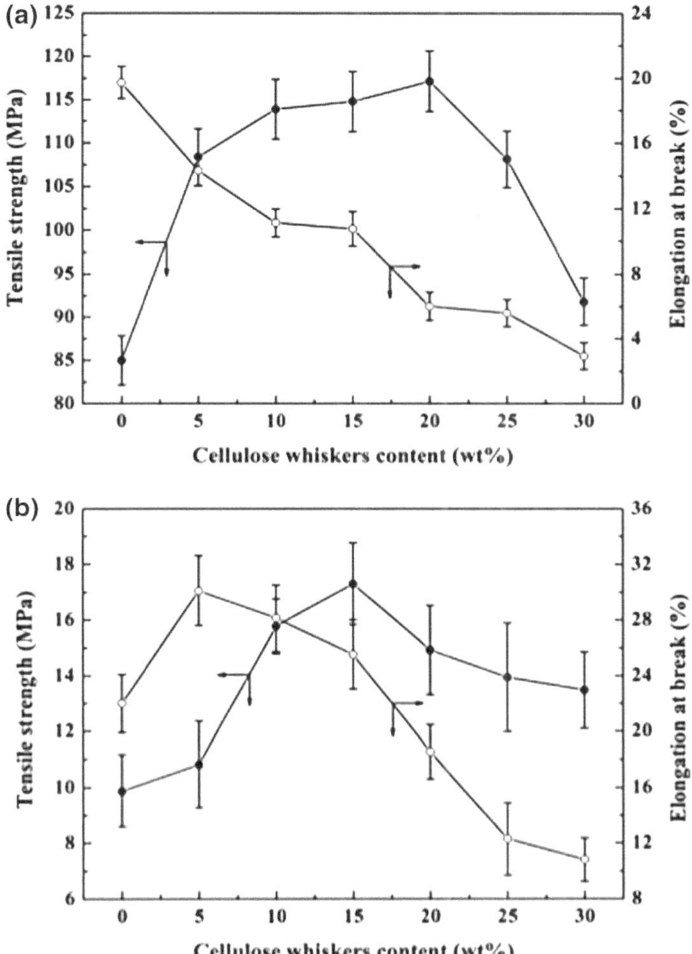

Fig. 4 Dependence of the cellulose whisker content on the tensile strength (●) and elongation at break (O) of the chitosan films in dry (**a**) and wet (**b**) states (Lee et al. 2009)

derivative) and BC nanofibrils. The incorporation of nanofibrils resulted in considerable increase in the strength and modulus of the nanocomposite films. Additionally, the presence of nanofibrils caused a significant decrease in the elongation at break which was more pronounced for higher cellulose contents. The superior mechanical properties of the films compared with those of the neat matrix confirmed the good interfacial adhesion and the strong interactions between the two components. These results were explained by the inherent morphology of BC with its nanofibrillar network and the similar structures of the two polysaccharides.

The use of BC nanofibers reinforced chitosan has also been explored for use in medical applications. Ciechanska (2004) reported improved mechanical properties and moisture holding capacity of these nanocomposites.

2.2 Starch

After cellulose, starch is the second most abundant biomass material. It is found in many plants' roots, stalks, crop seeds, and staple crops such as rice, corn, wheat, tapioca, and potato. It is a semicrystalline biopolymer. Because of its versatile properties and high availability at low price, it has great potential to be used in bionanocomposites. It can be used either as SNP or starch nanaocrystals (SNC) or as a matrix. The main issue is the immiscibility of starch with polymer matrices which results in processing issues and poor mechanical properties. This issue can be overcome by mixing with water or nonaqueous plasticizer (generally polyols, such as glycerol). The resulting material is called TPS and is more suitable for thermoplastic processing. However, TPS shows some drawbacks such as a strong hydrophilic character, lower T_g and melting point, poor mechanical properties (high strain to failure and low tensile strength) compared to conventional polymers and post-processing variation of the properties which reach equilibrium only after several weeks. To improve these material weaknesses, TPS is usually blended with other polymers like aliphatic polyesters. For a plasticizer content of up to 12 %, mechanical properties of TPS are similar to the anti-plasticization effect generally observed in synthetic polymers; that is, both the stress and strain at break decrease. For higher amounts of plasticizer, mechanical properties are controlled by the glass transition, which involves the molecular motions. This results in an increase in the strain at break and a decrease in both the strength and Young's modulus.

Nanocrystals extracted from starch are also used as fillers in polymer matrices. These nanoparticles not only increase the mechanical properties but also the physical properties such as permeability and fire retardancy of these composites. Unlike cellulose or chitin which exhibit needle-like nanocrystals, SNs occur as platelet-like nanoparticles. Thus starch has the versatility of being used both as a matrix and a filler.

2.2.1 Starch as Matrix

The use of CNC/MFC as filler in starch matrix has been shown to result in improvement in mechanical properties of the composites as shown in Table 3. This has been attributed to strong interfacial interactions between the cellulose crystallite surface and the starch matrix, and also to the formation of a rigid network of nanofibers connected by hydrogen bonds.

Cao et al. (2008a, b) studied the properties of plasticized starch-based nanocomposites reinforced with flax CNCs. The CNC content had a profound effect on

Table 3 Tensile properties of (untreated) cellulose CNC/MCF reinforced biopolymer composites

Matrix/fiber content (%)	Tensile strength (MPa)	Tensile modulus (MPa)	Strain (%)	References
TPS/15	6.8 (yield)	220	–	Kaushik et al. (2010)
TPS/40	6.9	480	13.6	Lu et al. (2006)
TPS/10	–	271	–	Alemdar and Sain (2008)
TPS/30	11.9	498	7.2	Cao et al. (2008a, b)
TPS/39	–	6000	–	Dufresne et al. (2000)
TPS/25	12	6000	1	Anglès and Dufresne (2001)
TPS/25	10.3	7200	1	Mathew et al. (2008)
TPS/30	3.1	10,000	0.45	Lu et al. (2005)
TPS/10.3	–	9000	3.3	Orts et al. (2005)
TPS/10	1.75	2450	–	Alemdar and Sain (2008)
TPS/30	2.95	2580	0.1	Cao et al. (2008a, b)
TPS/10	1.95	–	2.1	Chen et al. (2009)
TPS/20	1.4	1600	1.8	Teixeira et al. (2009)
PLA/?	30	–	2.5	Qu et al. (2010)
PLA/5	58	2600	2.8	Mathew et al. (2006)
PLA/5	71.1	2900	–	Wang and Sain (2007)
PLA/3-20	55–75	4700	1.6–3	Iwatake et al. (2008)
PLA/3-20	61–71	3800–5700	1.7–2.7	Suryanegara et al. (2009)
PLA/0-90	35–180	5000–13,000	1–3.3	Nakagaito et al. (2009)
PLA/5	71	3600	2.7	Jonoobi et al. (2010)
PCL/0-12	18–25	400–600	–	Siqueira et al. (2009)
Chitosan/20	120	–	6	Li et al. (2009)
PVA/5	82	400	–	Cho and Park (2011)
PVA/90	84	7.7	2	Leitner et al. (2007)
PVA/10	178	1010	–	Bhatnagar and Sain (2005)
PVA/10	108	6600	2.1	Wang and Sain (2007)
PVA/50	145.1	8490	–	Bruce et al. (2005)
PVA/15	62	5200	–	Lu et al. (2008)
PVA/10	160	8000	4.1	Kakroodi et al. (2014)
NR/10	4	17	–	Bras et al. (2010)

the mechanical properties. The tensile strength increased from 3.9 to 11.9 MPa with increasing filler content from 0 to 30 wt%, Young's modulus increased significantly from 31.9 to 498.2 MPa, and the elongation at break decreased from 68.2 to 7.2 %. This was attributed to the reinforcement effect of the homogeneously dispersed high-performance nanocrystals in the TPS matrix and the strong hydrogen bonding interaction between nanocrystals and TPS molecules.

Orts et al. (2005) studied the properties of cellulose microfibrils obtained from different sources of cellulose fibers added at low concentrations (2–10 % w/w) to starch gels and films as reinforcing agents. Significant changes in mechanical properties, especially maximum load and tensile strength, were observed for fibrils derived from cotton, softwood, and BC. For extruded starch plastics, the addition of cotton-derived microfibrils at 10.3 wt% increased Young's modulus by fivefold relative to a control sample with no cellulose reinforcement. Addition of microfibrils did not always change mechanical properties in a predictable direction. Whereas tensile strength and modulus were shown to increase following the addition of microfibrils to starch thermoplastic and a cast latex film, these parameters decreased when microfibrils were added to a starch–pectin blend, implying that complex interactions are involved in the application of these reinforcing agents.

Kaushik et al. (2010) showed that a 15 % cellulose nanofibers content in TPS matrix resulted in a considerable increase in properties over neat TPS. Yield strength increased from 4.7 to 6.8 MPa, whereas tensile modulus registered an increase from 78 to 220 MPa. This increase was again attributed to the formation of a rigid network of nanofibers connected by hydrogen bonds and also by mutual entanglements.

Mathew et al. (2008) studied properties of tunicin whiskers reinforced TPS nanocomposites at different relative humidity (RH) levels. The nanocomposites exhibited good mechanical strength due to the strong interaction between tunicin whiskers, TPS matrix, plasticizer (sorbitol) and water, and due to the ability of the cellulose filler to form a rigid three-dimensional network. An even distribution of whiskers (as determined by SEM) and plasticizer in the matrix contributed to the mechanical performance. For all RH levels, the modulus increased gradually with filler load, and above 5 % whiskers, a significant improvement was observed. The tensile strength and Young's modulus were high at lower RH levels, and elongation at break remained constant, irrespective of RH and filler content.

Dufresne et al. (2000) reported a 350 % increase in tensile modulus of MFC/starch nanocomposites at 50 wt% cellulose content compared to neat starch. However, at high humidity levels (75 % RH), the reinforcing effect was clearly diminished. This was attributed to the hydrophilic nature of both the starch and the cellulose, which resulted in the plasticization of starch and weakening of cellulose/starch interfacial adhesion. On the other hand, the addition of cellulose to starch resulted in a decrease of both water uptake at equilibrium and the water diffusion coefficient.

Alemdar and Sain (2008) investigated the properties of wheat straw nanofiber/TPS composites. At 10 % nanofiber loading, the tensile modulus increased from 111 to 277 MPa.

In a study on BC/TPS composites, Martins et al. (2009) exhibited superior mechanical properties of BC fillers compared to vegetable cellulose fillers. The Young modulus increased by 30 and 17 fold (with 5 % fibers), while the elongation at break was reduced from 144 % to 24 % and 48 % with increasing fiber content, respectively, for composites with bacterial and vegetable cellulose. SEM micrographs of fractured samples also corroborated these findings as shown in Fig. 5, which provided evidence of the strong interfacial adhesion between the cellulose fibers and the TPS matrix, as shown by their good dispersion within the matrix, without noticeable aggregates. It was observed that the characteristic nano- and

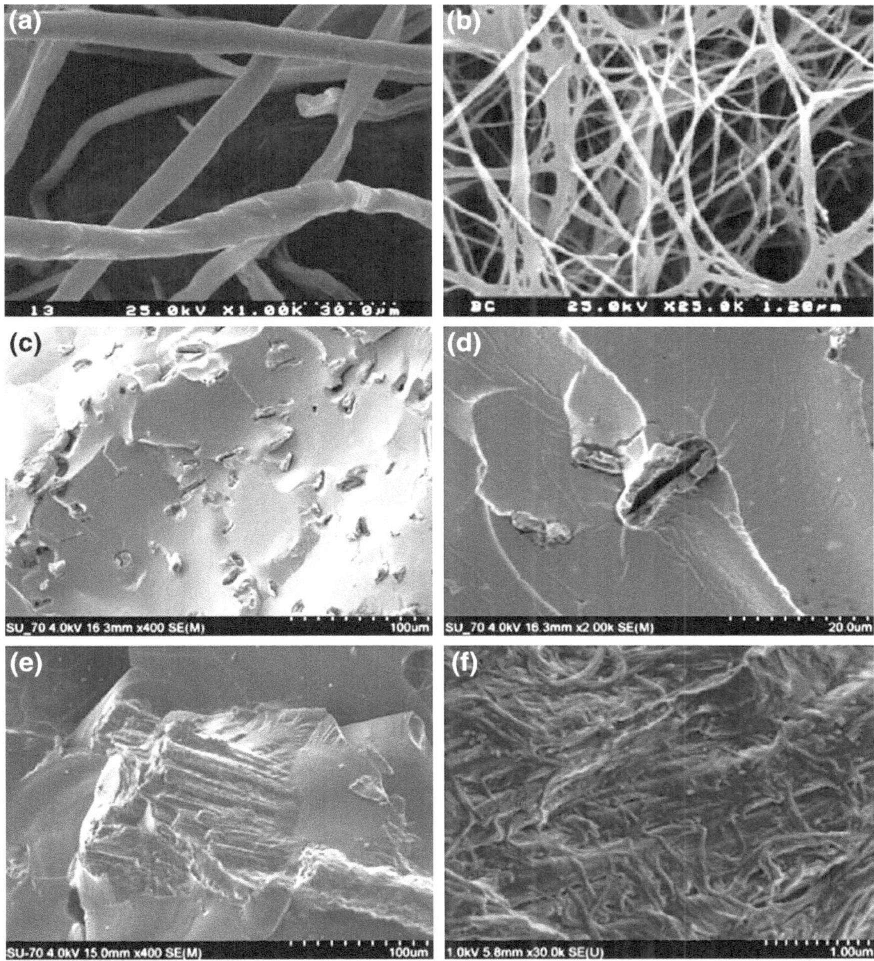

Fig. 5 SEM micrograph of vegetable cellulose (VC) and bacterial cellulose (BC), and the fractured surface of TPS filled with VC and BC: **a** neat VC, **b** neat BC, **c** and **d** TPS/VC composites (5 wt%); **e** and **f** TPS/BC composites (5 wt%) (Martins et al. 2009)

microfibril network of BC was maintained and totally impregnated with TPS. These results corroborated the superior mechanical properties of the BC-based composites compared to their vegetable cellulose counterparts.

Wan et al. (2009) studied BC reinforced TPS composites. At 22 wt% filler loading, the nanocomposites showed 137 % increase in tensile strength and 132 % increase in tensile modulus compared to neat BC. Thus BC was shown to be a good filler in TPS.

In a novel application of BC, Grande et al. (2009) added starch to the culture medium of cellulose-producing bacteria in order to introduce the granules into the forming network of cellulose which allowed the preservation of the natural ordered structure of cellulose nanofibers. Microscopic analysis revealed that starch acted as a matrix which filled the voids in the BC network. Using MCF as reinforcement, the nanocomposites showed considerable improvement in mechanical properties.

Lu et al. (2006) studied ramie CNC reinforced TPS composites. The results indicated that the synergistic interactions between fillers and between filler and TPS matrix play a key role in reinforcing the composites. The nanocomposites, conditioned at 50 % RH, showed increases in both tensile strength and Young's modulus from 2.8 MPa for TPS film to 6.9 MPa and from 56 MPa for TPS film to 480 MPa, respectively, with increasing nanocrystal content from 0 to 40 wt%.

The effects of various surfactants on the properties of starch-based nanocomposites have also been studied and most of the studies point to their positive effects on mechanical properties of nanocomposites. For example, glyceryl monostearate (GMS) was used by Mondragon et al. (2008) in MFC/TPS, which resulted in improvement in mechanical properties. Similar improvements in properties were reported by Takagi and Asano (2008a, b) for CNC/esterified starch nanocomposites.

2.2.2 Starch as Filler

SNs have been successfully used as filler in biopolymer matrices such as waterborne polyurethane (Chang et al. 2009; Chen et al. 2008a, b), starch (Angellier et al. 2006a, b; Viguié et al. 2007; Garci et al. 2009, 2011), pullulan (Kristo and Biliaderis 2007), PLA (Yu et al. 2008), PVA (Chen et al. 2008a, b), PCL (Habibi and Dufesne 2008), poly(butylene succinate) (PBS) (Lin et al. 2011a, b), and soy protein isolate (SPI) (Zheng et al. 2009), and the results have been very encouraging. Another advantage of using SNs is that they possess a highly reactive surface covered with hydroxyl groups, which provides the possibility of modification via chemical reaction. As discussed earlier, these treatments can be used to improve the properties of these composites.

Lin et al. (2011) compared the mechanical properties of PBS nanocomposites reinforced by CNCs and SNs. The results are shown in Fig. 6. With the addition of just 2 wt% CNC, the tensile strength of the nanocomposites improved from 26.2 MPa for the neat PBS sheet to 29.0 MPa, with a simultaneous increase in elongation to break from 6.33 to 7.43 %. When the CNC loading level exceeded 2 wt%, the tensile strength and elongation to break gradually decreased, whereas the Young' modulus of the nanocomposites sharply increased. In the case of PBS/SN nanocomposites, the addition of 5 wt% SN resulted in simultaneous increase in the strength, toughness, and rigidity of the nanocomposites. For SN content of greater than 5 wt%, the Young's modulus of these nanocomposites gradually increased at the expense of tensile strength and elongation to break, similar to the PBS/CNC nanocomposites. The greater improvements in PBS/SN composites compared to

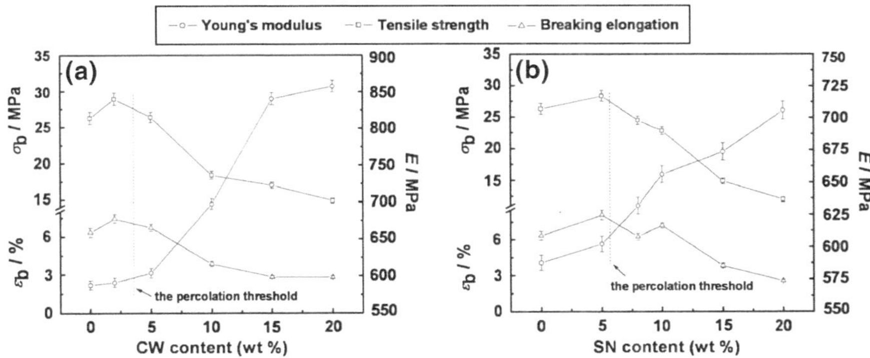

Fig. 6 Effects of CW and SN content on tensile strength, elongation at break and Young's modulus for the PBS/CW (**a**) and PBS/SN (**b**) nanocomposites (Lin et al. 2011)

PBS/CNC composites were ascribed to the percolation phenomenon as discussed earlier. Since the percolation threshold is inversely proportional to the aspect ratio of the nanofillers, it was calculated to be 3.6 wt% for rod-like CNC and 5.5 % for platelet-like SN, as shown in Fig. 6.

Habibi and Dufresne (2008) also compared the mechanical properties of CNC and SN reinforced nanocomposites. However, the matrix they used was PCL and both nanocrystals were grafted with PCL to improve compatibility with the matrix. The addition of unmodified CNC or SN in PCL resulted in improvement in their tensile modulus but, at the same time, reduced drastically all the other mechanical properties, especially the elongation at break. The elongation at break decreased from 637 % down to 8.0 % and 2.3 % at 20 wt% of cellulose and SNs, respectively. The addition of PCL-grafted CNC and SN nanocrystals resulted in continuous increase in the tensile modulus of composites, whereas the decrease in both the strength and the elongation at break was much more gradual. The elongation at break decreased from 637 % down to 420 % and 83 % at 20 wt% of PCL-grafted cellulose and SNs, respectively. This behavior was ascribed to the better dispersion of the fillers within the polymeric matrix, induced by the grafting of PCL chains on the surface of the nanoparticles. The possibility of chain entanglements between the grafted chains and the PCL chains from the matrix could also play an important role in this phenomenon.

SNs can be more efficient reinforcing agents than the native starch from which they are extracted. This was demonstrated for native pea starch (NPS)/PVA and pea starch nanocrystals (PSN)/PVA nanocomposites by Chen et al. (2008a, b). The PSN/PVA nanocomposites exhibited smoother fracture surfaces, higher light transmittance, higher tensile strength and elongation at break, and lower moisture uptake than the corresponding NPS/PVA nanocomposites. For example, the values of tensile strength and elongation at break for PSN/PVA composites containing 10 wt% of PSN were 40 MPa and 734 %, while those of the corresponding NPS/PVA composites were 35 MPa and 579 %, respectively. This was attributed to the smaller size of PSN compared to NPS which dispersed more homogeneously in PVA matrix, resulting in stronger interactions with PVA.

All-starch nanocomposites made of SN in a matrix of waxy maize starch plasticized with sorbitol were studied by Viguie′ et al. (2007). Adding SNs resulted in an increase of both the tensile modulus and strength and in a decrease of the strain at break regardless of the plasticizer content. However, highly plasticized films (35 wt% sorbitol) showed a higher relative reinforcing effect. For instance, the Young's modulus and the strength of the composites reinforced with 15 wt% SNs increased by a factor of 7 and 12, respectively, when plasticized with 35 wt% sorbitol, and only by a factor 2.7 and 4.2, respectively, when plasticized with 25 wt% sorbitol.

Similar to their use as a matrix, the reinforcing effect of SNs is generally ascribed to the formation of a hydrogen bonded percolating filler network above a given starch content corresponding to the percolation threshold. This threshold has been reported to be 10 % for natural rubber (NR) matrix (Angellier et al. 2005a). It was shown that with a content of up to 20 wt%, SN presented the advantage of effectively reinforcing the NR without significantly decreasing the elongation at break of the material. For example, the elongation at break decreased slightly from 303 to 277 %, whereas the strength increased sharply from 77.1 to 229.5 MPa at 20 wt% SN filler (Angellier et al. 2005b).

2.3 Cellulose

Cellulose nanofibers and nanoparticles have many characteristics which make them suitable for use as reinforcements in composites. These include high aspect ratio, low density, outstanding mechanical properties, and a reactive surface of hydroxyl side groups that facilitates grafting chemical species to achieve different surface properties (surface functionalization). Fillers with a high aspect ratio give the best reinforcing effect. Favier (1995) showed that tunicin whiskers with aspect ratio of 67 had greater effect on modulus increase than bacterial (aspect ratio 60) and Avidel (aspect ratio 10) whiskers.

Studies done by Kvien and Oksman (2007) and Gindl and Keckes (2007) have shown that CNCs have similar reinforcing effect on polymer matrices as synthetic fibers—they have optimum properties in the direction of alignment.

The mechanical properties of CNC reinforced composites have been shown to be dependent on the following factors (Moon et al. 2011):

(a) the CNC and matrix material properties,
(b) the degree of CNC in-plane orientation,
(c) the density of CNC–CNC contacting points within the CNC network,
(d) interfacial properties, either CNC–CNC and/or CNC–matrix,
(e) the CNC volume fraction,
(f) CNC morphology/size, and
(g) moisture content.

Factors b, d, and f are so crucial that they have been considered to be the main reason for the considerably lower tensile properties of CNC composites than those theoretically predicted. Because of low density of CNC, the specific strength and specific modulus of CNC-based composites becomes comparable to that of metals, ceramics, and conventional composites. This makes CNC composites a suitable material to be considered for application in automotive and aerospace industry.

However, there are some limitations to the use of nanoceluulose as reinforcement in polymer matrices, as pointed out by Lee et al. (2014). They studied data on tensile properties of BC and CNC reinforced polymer nanocomposites reported in literature and compared with the tensile properties of the benchmark matrix PLLA. They found that only 20 % of the nanocomposites reported in literature exceeded the tensile properties of PLLA. Moreover, a nanocellulose loading of greater than 30 vol% is required to produce cellulose reinforced polymer nanocomposites with tensile properties exceeding those of PLLA, indicating that low nanofibre content does not lead to dramatic property improvements in nanocellulose reinforced polymer composites. Whilst both CNC and BC serve as excellent nanoreinforcement for the production of high-performance nanocomposites, it was observed that BC outperforms CNC as reinforcement. Tensile modulus and strength values of 21 GPa and 320 MPa, respectively, were obtained for random-in-the-plane BC reinforced epoxy composites at 60 vol% BC, the highest values reported so far for cellulose reinforced nanocomposites.

All-cellulose nanocomposites (cellulose nanofillers in cellulose matrix) are an emerging and interesting class of nanocomposites. Shakeri et al. (2011) manufactured all-cellulose nanocomposite films consisting of dissolved cellulose reinforced with undissolved cellulose crystallites, using partial dissolution of MFC cellulose in ionic solution and subsequent film casting. The randomly oriented nanocrystallite reinforced films were transparent and of high strength and stiffness with regard to comparable cellulosic materials. This composite exhibited a good interface between the fibers and the surrounding matrix from the selectively dissolved/resolidified MFC-fibril surfaces, which resulted in excellent bonding, good mechanical and thermal performances, and optical transparency.

Surface functionalization of CNCs allows better dispersion in the polymer matrix and improves fiber/matrix interfacial bond strength, overcoming two main issues with the nanoreinforcements. This gives cellulose nanoreinforcements an edge over other nanoreinforcements. It is no wonder, then, that cellulose reinforced nanocomposite find wide range of applications in automotive, construction, packaging, display, water treatment, and medical fields. They are also the most widely studied of all the nanobiocomposites. Table 3 lists the tensile properties of some selected composites. The table is not comprehensive, but is intended to show representative values of cellulose nanocomposites with different biodegradable polymer matrices. Although these values point toward a great potential of these composites, they do not represent a competitive advantage over existing materials such as cellophane (Newman and Staiger 2008).

The various surfactants/treatments use include, but are not limited to, corona or plasma discharge, surface derivatization, graft copolymerization, acetylation, silylation, coupling agents, and grafting. The details of these treatments can be found elsewhere in this book. We shall concentrate on the applications of these processes and their effects on cellulose-based nanocomposites.

The most commonly used matrices with cellulose nanoreinforcement are PLA, PVA, PCL, and NR. We shall now discuss the mechanical properties of a selection of such composites.

2.3.1 Poly(lactic acid) PLA

Apart from its biodegradability, the most attractive feature of PLA is its similar properties to commodity plastics such as polypropylene. Main disadvantages of PLA are its brittleness, low thermal stability, relatively poor barrier properties, and high price. Blending and copolymerization with other polymer are used to overcome these issues.

Jonoobi et al. (2010) studied mechanical properties of CNF reinforced PLA composites. The tensile strength and modulus were improved with increased nanofiber contents. The modulus of the PLA was increased from 2.9 to 3.6 GPa with the addition of 5 wt% nanofibers, a 24 % increase. Similarly, a 21 % increase in tensile strength was observed for nanocomposites compared to neat PLA. On the other hand, strain to failure of nanocomposites was decreased with increase in nanofiber content. Classical models of Halpin–Tsai and Krenchel were used to compare the predicted theoretical data with the experimental data. It was found that experimental data were nearer the predicted value of Krenchel than Halpin–Tsai, which was a confirmation of the random distribution of nanofibers in the matrix, as hypothesized by Krenchel, rather than aligned in longitudinal direction, as hypothesized by Halpin–Tsai.

Bondesen and Oksman (2007) studied the feasibility of using PVA as compatibilizer to improve the dispersion of cellulose whiskers in PLA matrix. The hydroxyl groups on partial hydrolyzed PVA are expected to interact with the hydrophilic surfaces of the cellulose and the residual vinyl acetate groups with the hydrophobic PLA. Two feeding methods were used and evaluated: dry mixing with PLA prior to extrusion or pumping as suspension directly into the extruder. Due to immiscibility of the polymers, phase separation occurred with a continuous PLA phase and a discontinuous PVA phase. The whiskers were primarily located in the PVA phase and only a small amount was located in the PLA phase. This inadequate dispersion of whiskers in the PLA phase was probably the reason why no improvements in thermal properties were observed for the nanocomposites. The relatively small improvements in tensile properties of the nanocomposites also indicated that it was principally the PVA phase that was reinforced with whiskers rather than the PLA phase.

Oksman et al. (2006) studied a new processing technique for manufacturing CNC/PLA nanocomposites. MFC was treated with *N*,*N*-dimethylacetamide (DMAc) containing lithium chloride (LiCl) in order to swell the CNC and partly separate the cellulose whiskers. It was found that the extrusion process using liquid feeding of the CNC suspension is a promising way to produce cellulose nanocomposites. However, the mechanical results did not show improvements as expected when compared with pure PLA. A possible explanation given for such behavior was the combination of used additives and the high processing temperature. It was concluded that DMAc/LiCl is not suitable as swelling/separation agent for CNC if high temperature processing is used. Similarly, improvements in mechanical properties have been reported for CNC/PLA nanocomposites by Okobu et al. (2009) and Iwatake et al. (2008). As a result of uniform distribution, the Young's modulus and tensile strength of PLA increased by 40 and 25 %, respectively, without a reduction in yield strain at a fiber content of 10 wt%.

Petersson and Oksman (2006) compared the properties of PLA-based nanocomposites made of CNC and nonbiodegradable layered silicate bentonite. At 5 wt % loading, there was more evidence of agglomeration in CNC/PLA composites than betonite/PLA composites, as shown in Fig. 7. The PLA/bentonite nanocomposite showed a 53 % increase in tensile modulus and a 47 % increase in the yield strength compared to neat PLA. The PLA/CNC system on the other hand showed no increase in tensile modulus and only a 12 % increase in yield strength compared to neat PLA. These results were lower than expected. Two factors were ascribed for

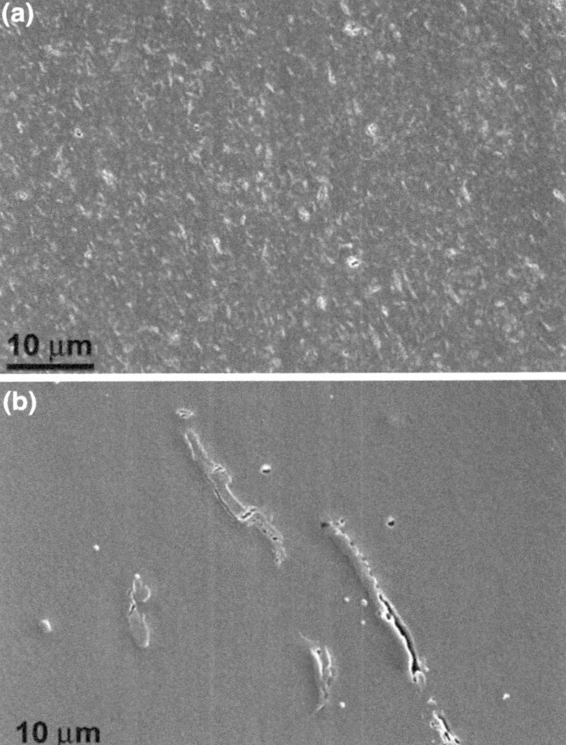

Fig. 7 SEM of the nanocomposites, **a** bentonite/PLA, **b** CNC/PLA (Petersson and Oksman 2006)

this discrepancy. First, the bentonite added to the PLA/bentonite material had theoretically twice the surface area than CNC. A larger surface area allowed the nanoreinforcement to interact with a larger amount of polymer chains and thereby having a larger effect on the mechanical properties. Secondly, the bentonite clay is organically modified to be compatible with polymers like PLA. Good interaction between betonite and PLA allowed for good stress transfer to take place in the composite which gave rise to large improvements in the mechanical properties of betonite/PLA. Authors concluded that better exfoliation of cellulose whiskers within the PLA matrix is expected to result in further improvements in properties of their composites.

The agglomeration of CNC in PLA was also reported by Haafiz et al. (2013). While there was some improvement in tensile modulus of nanocomposites at 5 wt% of CNC loading, both the tensile strength and the strain to failure showed gradual decrease with increase in CNC loading. This was attributed to agglomeration of CNC which resulted in poor interfacial adhesion between the CNC and the matrix. This was also confirmed by the SEM micrographs of the fractured surfaces.

2.3.2 Poly(vinyl alcohol) PVA

PVA is attractive as matrix because it is water soluble, has excellent chemical resistance, and is biocompatible and biodegradable. PVA has been studied in terms of its biomedical applications, like tissue reconstruction and replacement, cell entrapment and drug delivery, soft contact lens materials, and wound covering bandages for burn victims (Siro and Plackett 2010).

Leitner et al. (2007) reported a gradual increase in tensile properties of CNC reinforced PVA composites. At a cellulose content of 50 wt%, the tensile modulus of PVA increased by a factor of 20 and the tensile strength increased by a factor of 3.5. Both properties showed further improvements at cellulose contents of 70 and 90 wt% respectively. Similarly, Bruce et al. (2005) reported approximately five times higher tensile strength for PVA containing 50 wt% MFC when compared to the neat polymer.

In another study Bhatnagar and Sain (2005) made nanocomposites containing 10 % cellulose nanofibers obtained from various sources, such as flax bast fibers, hemp fibers, kraft pulp, and rutabaga. There was noticeable improvement in tensile properties with tensile modulus showing a four to five times improvement over neat PVA. Fortunati et al. (2012) reported that a 5 wt% cellulose content proved ideal to promote a direct mechanical interaction between the PVA and CNC. Zimmerman et al. (2004) reported up to three times higher tensile modulus and up to five times higher tensile strength containing 20 wt% MCF in PVA. Encouraging results have also been reported for BC/PVA nanocomposites for potential applications as medical devices by Wan et al. (2006) and Millon and Wan (2006).

Kakroodi et al. (2014) also reported significant improvements in tensile properties of PVA reinforced with cellulose nanofibers extracted from aloe vera rind. Their results showed that incorporation of even small amounts of nanofibers (as low as 2 wt%) had significant effects on both the modulus and strength of PVA. Tensile modulus and strength of PVA increased by 32 and 63 %, respectively, after adding 2 wt% of cellulose nanofibers. Samples with higher concentrations of nanofibers also showed improved mechanical properties due to a high level of interfacial adhesion and dispersion of fibers in the PVA matrix. Tensile modulus increased by 68 % at 7 wt% and by 88 % at 10 wt% of nanofibers. Similarly, tensile strength increased by 101 % at 7 wt% and by 125 % at 10 wt% of nanofibers. The most remarkable effect was on elongation at break of the composites, which decreased from 165 % for neat PVA to 4.1 % for composites at 10 wt% of nanofibers.

2.3.3 Poly(caprolactone) PCL

PCL is an oil-derived biodegradable, semicrystalline polymer. Among its good properties for use as matrix are good water resistance, low melting point, low viscosity, good processability, and high strain to failure. Its main disadvantages are low tensile strength and poor dispersion of cellulose nanoreinforcement due to its hydrophobic nature. Various surface treatments have been used to improve fiber/matrix compatibility of which grafting is the most popular method. Lonnberg et al. (2008) grafted MCF and Habibi et al. (2008) grafted CNC with different molecular weights of PCL, whereas Sequiera et al. (2009) grafted nanocellulose with N-octadecyl isocyanate for this purpose. All of these studies reported improvements in fiber/matrix adhesion following the treatments.

2.3.4 Natural Rubber

NR is a biodegradable material which has good mechanical properties and is an ideal matrix material for use in eco-friendly composites (Thakur et al. 2014a). The addition of the filler usually results in the improvement of its stiffness and hardness, and also of the resistance to abrasion, tear, cutting, and rupture. Traditionally, carbon black and silica have been used as the main fillers. However, cellulose nanocrystals have recently been used as filler in NR with encouraging results.

Sareena et al. (2012) used coconut shell powder (CSP) as filler in NR matrix. CSP was also subjected to alkalization treatment to improve particle/matrix adhesion. The optimum tensile strength was found at 10 parts per hundred (phr) of rubber CSP composites which was attributed to smaller particle size and uniform dispersion of filler in NR compounds. At more than 10 phr, the weak interaction and bonding between the filler particles and the NR matrix was responsible for the deterioration of tensile strength. Alkalized CSP composites exhibited improved tensile strength. Young's modulus and hardness of alkalized and non-alkalized CSP composites showed gradual improvement with increase in filler loading which was attributed to the stiffening effect of the filler particles and their proper dispersion in the matrix.

Bras et al. (2010) studied properties of bagasse CNC reinforced NR composites. Both the Young's modulus and the tensile strength significantly increased upon whisker addition to rubber while the strain at break decreased. The maximum increase of strength and modulus was 374 and 530 %, respectively, at 10 wt% of whiskers. This high reinforcing effect was ascribed to the mechanical percolation phenomenon of cellulose whiskers which forms a stiff continuous network of cellulosic nanoparticles linked through hydrogen bonding. As explained earlier, it strongly depends on the aspect ratio of the rod-like reinforcing particles and therefore on the origin of cellulose. Authors also compared the performance of bagasse whiskers as reinforcing agents in NR matrix to other nanoparticles isolated from different sources (starch, date palm tree, and Capim Dourado) reported by other authors. Bagasse nanocrystals had the lowest aspect ratio compared to the other whiskers isolated from the other sources which could be the reason for lower tensile properties of bagasse NR nanocomposites.

Similar improvements in mechanical properties of nanocellulose/NR and nanolignin/NR composites have been reported by Favier et al. (1995), Hajji et al. (1996), Abraham et al. (2007) and Thakur et al. (2014b). Due to the uniform dispersion of the nanocellulose in rubber latex, the composites showed improved stiffness and strength without any loss of its elastomeric nature, as shown in SEM micrograph of Fig. 8. The research on nanocellulose/NR composites is an emerging field and more work needs to be done in this area.

2.3.5 Epoxidized Vegetable Oils

Epoxidized vegetable oil-based bioresins are another option for making eco-friendly nanocomposites. The three most widely used oils for this purpose are soybean, canola, and linseed oils. However, the fact that these oils are a vital

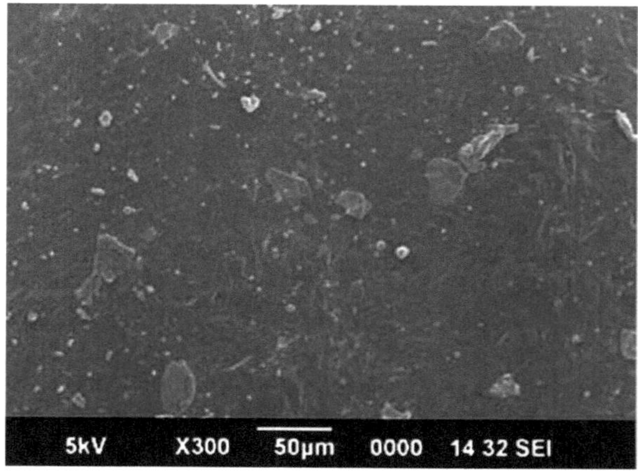

Fig. 8 SEM micrograph showing good dispersion of nanocellulose whiskers in natural rubber latex (Abraham et al. 2007)

ingredient of food items consumed throughout the world is a major drawback and a strong lobby exists against the use of these oils for nonfood applications.

2.4 The Effects of Surface Treatments/Surfactants/Additives

Fukuzumi et al. (2009) reported the positive effect of 2,2,6,6-tetramethylpiperidine-1-oxyl (TEMPO) oxidation treatment on softwood and hardwood nanofibrils in PVA matrix. Both the tensile strength and elastic modulus of PVA film were remarkably improved by 20 % nanofibrils addition. Atomic force microscopy confirmed the homogenous dispersion of nanofibrils in the matrix.

Xaio et al. (2012) grafted MFC with L-lactic acid oligomers (g-MC) in PLA matrix. The grafting resulted in improved compatibility between fibers and the matrix, and this was manifested in improved tensile properties of nanocomposites. The tensile strength increased from 30 MPa for non-grafted fibers to 70 MPa for grafted fibers at 30 wt% fiber content. Similarly, strain to failure increased from 0.8 % for non-grafted fibers to 3.2 % for grafted fibers.

The positive effects of grafting CNC with n-octadecyl isocyanate (ICN) on their interaction with PLA matrix were reported by Espino-Pérez et al. (2013). The incorporation of untreated CNC in PLA resulted in decrease in tensile properties. However, the incorporation of CNC-ICN in PLA showed improvements on tensile properties. The best results were achieved at 2.5 wt% CNC-ICN.

Lee et al. (2009) studied the effect of surface functionalization of BC nanofibrils and their use as reinforcement in polylactide (PLLA) matrix. BC was functionalized with various organic acids via an esterification reaction which rendered the otherwise hydrophilic BC hydrophobic and resulted in better compatibility between PLLA and BC. A direct wetting method, allowing the determination of the contact angle of polymer droplets on a single BC nanofiber, was developed to quantify the interfacial adhesion between PLLA and functionalized BC. It was found that the contact angle between PLLA droplets and functionalized BC decreased with increasing chain lengths of the organic acids used to hydrophobise BC. The mechanical properties of the surface functionalized BC reinforced PLLA nanocomposites showed significant improvements compared to neat PLLA and BC reinforced PLLA (tensile modulus by 50 % and tensile strength by 15 %). The thermal degradation and viscoelastic behavior of the nanocomposites were also improved over neat PLLA. Therefore, it was concluded that PLLA nanocomposites with overall improved properties can be fabricated through the surface functionalization of BC.

Similarly Juntaro et al. (2008) studied the effects of acetone-treated sisal fibers attached with BC nanofibrils and incorporated in PLLA matrix. Acetone treatment was used to improve the adhesion of BC/sisal fibers and sisal fiber/PLLA matrix. SEM images showed marked improvement in attachment of BC on sisal fibers following the treatment, as shown in Fig. 9. The interfacial adhesion between the (modified) fibers and the PLLA matrix was quantified using single-fiber pull-out

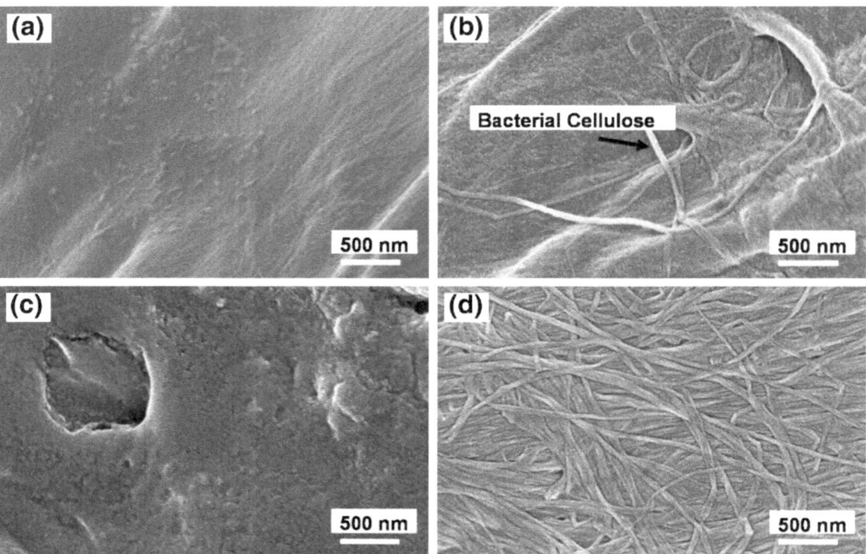

Fig. 9 SEM micrographs of sisal fiber surfaces: **a** natural sisal fiber; **b** sisal fiber after attachment with bacterial cellulose; **c** acetone-treated sisal; **d** acetone-treated sisal fiber after attachment with bacterial cellulose (Juntaro et al. 2008)

tests. The apparent interfacial shear strength of the unmodified sisal fibers was significantly improved from 12.1 to 14.6 MPa after the bacterial cellulose growth. The tensile strength and modulus of the composites also showed improvement of around 40–70 % following the growth of BC. However, the improvement was attributed more to the presence of BC rather than the acetone treatment.

3 Flexural and Impact Properties

The flexural and impact properties of eco-friendly nanocomposites have not been studied to the same extent as their tensile properties. Most of the studies done have been on cellulose nanofibers and nanoparticles-based composites. Takagi and Asano (2008a, b) studied the properties of CNFs in esterified starch at 70 wt% fiber loading. Two different processing conditions were used: vacuum-treated and extra-stirred samples. The latter method showing significantly higher flexural strength of nanocomposites. Nakahara (2008) grafted nanocellulose with either PVA or PLA, followed by dewatering and kneading with PLA. The nanocomposite exhibited high values of flexural modulus of 5.1–5.7 GPa, flexural strength of 100–105 MPa, and an impact strength of 26 J/m. Hashiba (2009) also reported high flexural strength (110 MPa) and modulus (4 GPa) of nanocellulose/PLA nanocomposites.

4 Conclusions

Eco-friendly polymer nanocomposites are exciting new materials which are finding niche in medicine, packaging, electronics, automotive, and construction industry. Their main advantages are biodegradability, renewability, and low carbon footprint. They are based on biopolymer matrices reinforced with nanofibers or nanoparticles, extracted from biopolymers. The exceptionally high tensile properties of biopolymer nanoreinforcement make them an ideal material for use as reinforcements in composites. However, there are some issues which hinder these composites from attaining their optimum properties. The two major issues are poor interfacial adhesion of reinforcements and the matrices because of noncompatibility, and poor dispersion/agglomeration of reinforcements in biopolymer matrices. At the moment a lot of research is going on in various parts of the world on overcoming these issues. Various physical and chemical treatments of reinforcements, compatibilizers, and surfactants have been used for this purpose.

It has been shown that if cellulose nanoreinforcement can be made to align in a biopolymer matrix, they exhibit optimum properties in the direction of alignment, just like in synthetic fiber composites. Chitosan is a semicrystalline biopolymer which is used both as a matrix and as nanoreinforcement in composites. The use of chitosan nanoreinforcements in biopolymer matrices results in improvement in properties of composites. However, an optimum loading limit has been observed, depending on the biopolymer matrix, above which the properties of nanocomposites start to decline. A similar optimum loading limit has been found to exist when CNCs are used as reinforcements in chitosan matrix.

Starch is another semicrystalline biopolymer which can be used both as nanoreinforcement and as a matrix in composites. When used as a matrix, starch is usually blended with plasticizer to improve its processing and mechanical properties. Studies have shown significant improvements in tensile properties of nanocomposites, whether starch is used as a reinforcement or a matrix. BC-based nanocomposites have been shown to exhibit superior properties than vegetable cellulose-based nanocomposites.

Cellulose is the most abundant biomass material and is also the most widely studied in terms of its use in nanocomposites. More importantly, CNCs exhibit optimum properties in the direction of alignment just like synthetic fibers. The most widely used biopolymer matrices with CNCs are PLA, PVA, PCL, and NR. All of these nanocomposites exhibit superior tensile properties. The use of surface treatments and surfactants has also been reported to have positive impact on the mechanical properties of these nanocomposites.

References

Abraham E, Pothen LA, Thomas S (2007) Preparation and characterization of green nano composites. In: Proceedings of the fibre reinforced composites conference, Port Elizabeth, South Africa, December 2007

Alemdar A, Sain M (2008) Biocomposites from wheat straw nanofibers: morphology, thermal and mechanical properties. Compos Sci Technol 68:557–565

Angellier H, Dufresne A (2013) Mechanical properties of starch-based nanocomposites. In: Dufresne A, Thomas S, Pothan LA (eds) Biopolymer nanocomposites: processing, properties, and applications. Wiley, Hoboken

Angellier H, Molina-Boisseau S, Dufresne A (2005a) Mechanical properties of waxy maize starch nanocrystals reinforced natural rubber. Macromolecules 38:9161–9170

Angellier H, Molina-Boisseau S, Belgacem MN, Dufresne A (2005b) Surface chemical modification of waxy maize starch nanocrystals. Langmuir 21:2425–2433

Angellier H, Molina-Boisseau S, Dufresne A (2006a) Waxy maize starch nanocrystals as filler in natural rubber. Macromol Symp 233:132–136

Angellier H, Molina-Boisseau S, Dole P, Dufresne A (2006b) Thermoplastic starch-waxy maize starch nanocrystals nanocomposites. Biomacromolecules 7:531–539

Anglès MN, Dufresne A (2001) Plasticized starch/tunicin whiskers nanocomposite materials. 2. Mechanical behavior. Macromolecules 34:2921–2931

Bhatnagar A, Sain M (2005) Processing of cellulose nanofiber reinforced composites. J Reinf Plast Compos 24:1259–1268

Bondeson D, Oksman K (2007) Polylactic acid/cellulose whisker nanocomposites modified by polyvinyl alcohol. Compos Part A-Appl Sci Manufact 38:2486–2492

Bras J, Hassan ML, Bruzesse C, Hassan EA, El-Wakil NA, Dufresne A (2010) Mechanical, barrier, and biodegradability properties of bagasse cellulose whiskers reinforced natural rubber nanocomposites. Ind Crops Products 32:627–633

Bruce DM, Hobson RN, Farrent JW, Hepworth DG (2005) High-performance composites from low-cost plant primary cell walls. Compos Part A-Appl Sci Manufact 36:1486–1493

Cao X, Chen Y, Chang PR, Muir AD, Falk G (2008a) Starch-based nanocomposites reinforced with flax cellulose nanocrystals. Polym Lett 2:502–510

Cao XD, Chen Y, Chang PR, Stumborg M, Huneault MA (2008b) Green composites reinforced with hemp nanocrystals in plasticized starch. J Appl Polym Sci 109:3804–3810

Chang PR, Ai F, Chen Y, Dufresne A, Huang J (2009) Effects of starch nanaocrystals-graft-polycaprolactone on mechanical properties of waterbone polyurethane-based nanocomposites. J Appl Polym Sci 111:619–627

Chang PR, Jian R, Yu J, Ma X (2010) Fabrication and characterization of chitosan nanoparticles/plasticized-starch composites. Food Chem 120:736–740

Chen G, Wei M, Chen J, Huang J, Dufresne A, Chang PR (2008a) Simultaneous reinforcing and toughening: new nanocomposites of waterbone polyurethane filled with low loading level of starch nanocrystals. Polym 49:1860–1870

Chen Y, Cao X, Chang PR, Huneault MA (2008b) Comparative study on the films of poly(vinyl alcohol)/pea starch nanocrystals and poly(vinyl alcohol)/ native pea starch. Carbohydr Polym 73:8–17

Chen Y, Liu C, Chang PR, Cao X, Anderson DP (2009) Bionanocomposites based on pea starch and cellulose nanowhiskers hydrolyzed from pea hull fiber: effect of hydrolysis time. Carbohydr Polym 76:607–615

Cho M, Park B (2011) Tensile and thermal properties of nanocellulose-reinforced poly(vinyl alcohol) nanocomposites. J Ind Eng Chem 17:36–40

Ciechanska D (2004) Multifunctional bacterial cellulose/chitosan composite materials for medical applications. Fibres Text East Eur 12:69–72

Dufresne A, Dupeyre D, Vignon MR (2000) Cellulose microfibrils from potato tuber cells: processing and characterization of starch-cellulose microfibril composites. J Appl Polym Sci 76:2080–2092

Espino-Pérez E, Bras J, Ducruet V, Guinault A, Dufresne A, Domenek S (2013) Influence of chemical surface modification of cellulose nanowhiskers on thermal, mechanical, and barrier properties of poly(lactide) based bionanocomposites. Eur Polym J 49:3144–3154

Favier V (1995) Etude de nouveaux mate´riaux composites obtenus a` partir de latex filmoge`nes et de whiskers de cellulose: effet de percolation me´canique. PhD thesis, Joseph Fourier University, Grenoble, France

Favier V, Canova GR, Cavaille JY (1995) Nanocomposite materials from latex and cellulose whiskers. Polym Adv Technol 6:351–355

Fernandes SCM, Oliveira L, Freire CSR, Silvestre AJD, Neto CP, Gandinia A, Desbrieres J (2009) Novel transparent nanocomposite films based on chitosan and bacterial cellulose. Green Chem 11:2023–2029

Fortunati E, Puglia D, Monti M, Santulli C, Maniruzzaman M, Kenny JM (2012) Cellulose nanocrystals extracted from okra fibers in PVA nanocomposites. J Appl Polym Sci. doi:10.1002/APP.38524

Fukuzumi H, Saito T, Wata T, Kumamoto Y, Isogai A (2009) Transparent and high gas barrier films of cellulose nanofibers prepared by TEMPO-mediated oxidation. Biomacromolecules 10:162–165

Gallstedt M, Hedenqvist MS (2006) Packaging-related mechanical and barrier properties of pulp-fiber-chitosan sheets. Carbohydr Polym 63:46–53

Garcia NL, Ribba L, Dufresne A, Aranguren M, Goyanes S (2011) Effect of glycerol on the morphology of nanocomposites made from thermoplastic starch and starch nanocrystals. Carbohydr Polym 84:203–210

Gindl W, Keckes J (2007) Drawing of self-reinforced cellulose films. J Appl Polym Sci 103:2703–2708

Grande CJ, Torres FG, Gomez CM, Troncoso OP, Canet-Ferrer J, Martínez-Pastor J (2009) Development of self-assembled bacterial cellulose–starch nanocomposites. Mater Sci Eng C 29:1098–1114

Haafiz MKM, Hassan A, Zakaria Z, Inuwa IM, Islam MS, Jawaid M (2013) Properties of polylactic acid composites reinforced with oil palm biomass microcrystalline cellulose. Carbohydr Polym 98:139–145

Habibi Y, Dufresne A (2008) Highly filled bionanocomposites from functionalized polysaccharide nanocrystals. Biomacromolecules 9:1974–1980

Habibi Y, Goffin AL, Schiltz N, Duquesne E, Dubois P, Dufresne A (2008) Bionanocomposites based on poly(epsilon-caprolactone)-grafted cellulose nanocrystals by ring-opening polymerization. J Mater Chem 18:5002–5010

Hajji P, Cavaille JY, Favier V (1996) Tensile behavior of nanocomposites from latex and cellulose whiskers. Polym Compos 17:612–619

Hashiba M (2009) Thermoplastic resin compositions containing cellulose nanofibers with good bending properties. PCT Int Appl 2008-JP58502; 2007-195163:24

Iwatake A, Nogi M, Yano H (2008) Cellulose nanofiber-reinforced polylactic acid. Compos Sci Technol 68:2103–2106

JEC Composites (2014) Nanocellulose-based composites. www.jeccomposites.com. Accessed 18 Aug 2014

Jonoobi M, Harun J, Mathew AP, Oksman K (2010) Mechanical properties of cellulose nanofiber (CNF) reinforced polylactic acid (PLA) prepared by twin screw extrusion. Compos Sci Technol 70:1742–1747

Junkasem J, Rujiravanit R, Supaphol P (2006) Fabrication of α-chitin whisker-reinforced poly (vinyl alcohol) nanocompositenanofibres by electrospinning. Nanotechnology 17:4519–4528

Juntaro J, Pommet M, Kalinka G, Mantalaris A, Shaffer MSP, Bismarck A (2008) Creating hierarchical structures in renewable composites by attaching bacterial cellulose onto sisal fibers. Adv Mater 20:3122–3126

Kakroodi AR, Cheng S, Sain M, Asiri A (2014) Mechanical, thermal, and morphological properties of nanocomposites based on polyvinyl alcohol and cellulose nanofiber from aloe vera rind. J Nanomater 2014:1–7

Kalia S, Dufresne A, Cherian BM, Kaith BS, Av′erous L, Njuguna J, Nassiopoulos E (2011) Cellulose-based bio- and nanocomposites: a review. Int J Polym Sci 2011:1–35

Kaushik A, Kumra J (2014) Morphology, thermal and barrier properties of green nanocomposites based on TPS and cellulose nanocrystals. J Elast Plast 46:284–299

Kaushik A, Singh M, Verma G (2010) Green nanocomposites based on thermoplastic starch and steam exploded cellulose nanofibrils from wheat straw. Carbohydr Polym 82:337–345

Khan A, Khan RA, Salmieri S, Le Tien C, Riedl B, Bouchard J, Chauvec G, Tand V, Kamal MR, Lacroix M (2012) Mechanical and barrier properties of nanocrystalline cellulose reinforced chitosan based nanocomposite films. Carbohydr Polym 90:1601–1608

Kristo E, Biliaderis CG (2007) Physical properties of starch nanocrystals reinforced pullulan films. Carbohydr Polym 68:146–158

Kvien I, Oksman K (2007) Orientation of cellulose nanowhiskers in polyvinyl alcohol (PVA). Appl Phys A Mater Sci Process 87:641–643

Lee K, Blaker JJ, Bismarck A (2009) Surface functionalization of bacterial cellulose as the route to produce green polylactide nanocomposites with improved properties. Compos Sci Technol 69:2724–2733

Lee KY, Aitomäki Y, Berglund LA, Oksman K, Bismarck A (2014) On the use of nanocellulose as reinforcement in polymer matrix composites. Compos Sci Technol. doi:10.1016/j.compscitech.2014.08.032

Leitner J, Hinterstoisser B, Wastyn M, Keckes J, Gindl W (2007) Sugar beet cellulose nanofibril-reinforced composites. Cellulose 14:419–425

Li Q, Zhou J, Zhang L (2009) Structure and properties of the nanocomposite films of chitosan reinforced with cellulose whiskers. J Polym Sci 47:1069–1077

Lin N, Yu J, Chang PR, Li J, Huang J (2011a) Poly(butylene succinate)-based biocomposites filled with polysaccharide nanocrystals: Structure and properties. Polym Compos 32:472–482

Lin M-F, Thakur VK, Tan EJ, Lee PS (2011b) Surface functionalization of $BaTiO_3$ nanoparticles and improved electrical properties of $BaTiO_3$/polyvinylidene fluoride composite. RSC Adv 1:576–578

Lin M-F, Thakur VK, Tan EJ, Lee PS (2011c) Dopant induced hollow $BaTiO_3$ nanostructures for application in high performance capacitors. J Mater Chem 21:16500–16504

Lonnberg H, Fogelstrom L, Malstrom E, Zhou Q, Berglund L, Hult A (2008) Microfibrillated cellulose films grafted with poly(e-caprolactone) for biocomposite applications. Nordic Polymer Days, 11–13 June, Stockholm

Lu Y, Weng L, Zhang L (2004) Morphology and properties of soy protein isolate thermoplastics reinforced with chitin whiskers. Biomacromolecules 5:1046–1051

Lu Y, Weng L, Cao X (2005) Biocomposites of plasticized starch reinforced with cellulose crystallites from cottonseed linter. Macromol Biosci 5:1101–1107

Lu Y, Weng L, Cao X (2006) Morphological, thermal and mechanical properties of ramie crystallites—reinforced plasticized starch biocomposites. Carbohydr Polym 63:198–2004

Lu J, Wang T, Drzal LT (2008) Preparation and properties of microfibrillated cellulose polyvinyl alcohol composite materials. Compos Part A 39:738–746

Martins IMG, Magina SP, Oliveira L, Freire CSR, Silvestre AJD, Neto CP, Gandini A (2009) New biocomposites based on thermoplastic starch and bacterial cellulose. Compos Sci Technol 69:2163–2168

Mathew AP, Chakraborty A, Oksman K, Sain M (2006) The structure and mechanical properties of cellulose nanocomposites prepared by twin screw extrusion. In: Oksman K, Sain M (eds) Cellulose nanocomposites: processing, characterization, and properties, vol 938. American Chemical Society, Washington DC, pp 114–131

Mathew AP, Thielemans W, Dufresne A (2008) Mechanical properties of nanocomposites from sorbitol plasticized starch and tunicin whiskers. J Appl Polym Sci 109:4065–4074

Millon LE, Wan WK (2006) The polyvinyl alcohol-bacterial cellulose system as a new nanocomposite for biomedical applications. J Biomed Mater Res Part B-Appl Biomater 79B:245–253

Mondragon M, Arroyo K, Romero-Garcia J (2008) Biocomposites of thermoplastic starch with surfactant. Carbohydr Polym 74:201–208

Moon RJ, Martini A, Nairn J, Simonsen J, Youngblood J (2011) Cellulose nonmaterial's review: structure, properties and nanocomposites. Chem Soc Rev 40:3941–3994

Nakagaito AN, Fujimura A, Sakai T, Hama Y, Yano H (2009) Production of microfibrillated cellulose (MFC)-reinforced polylactic acid (PLA) nanocomposites from sheets obtained by a papermaking-like process. Comp Sci Technol 69:1293–1297

Nakahara S (2008) Resin composite materials containing surface- treated microfibrillated cellulose (MFC) reinforcement, their manufacture, and their articles. Jpn Kokai Tokkyo Koho 2007-17153:10

Newman RH, Staiger MP (2008) Cellulose nanocomposites. In: Pickering KL (ed) Properties and performance of natural-fibre composites. Woodhead Publishing Limited, Cambridge

Nordqvist D, Idermark J, Hedenqvist MS (2007) Enhancement of the wet properties of transparent chitosan-acetic-acidsalt films using microfibrillated cellulose. Biomacromolecules 8:2398–2403

Oksman K, Mathew AP, Bondeson D, Kvien I (2006) Manufacturing process of cellulose whiskers/polylactic acid nanocomposites. Compos Sci Technol 66:2776–2784

Oksman K, Mathew AP, Sain M (2009) Novel bionanocomposites: processing, properties and potential applications. Plast Rubber Compos 38:47–61

Okubo K, Fujii T, Thostenson ET (2009) Multi-scale hybrid biocomposite: processing and mechanical characterization of bamboo fiber reinforced PLA with microfibrillated cellulose. Compos Part A-Appl Sci Manufact 40:469–475

Orts WJ, Imam JSSH, Glenn GM, Guttman ME, Revol J (2005) Application of cellulose microfibrils in polymer nanocomposites. J Polym Environ 13:301–306

Petersson L, Oksman K (2006) Biopolymer based nanocomposites: comparing layered silicates and microcrystalline cellulose as nanoreinforcement. Compos Sci Technol 66:2187–2196

Qu P, Gao Y, Wu G, Zhang L (2010) Nanocomposites of poly(lactic acid) reinforced with cellulose nanofibrils. BioResources 5:1811–1823

Samir MASA, Alloin F, Dufresne A (2005) Review of recent research into cellulosic whiskers, their properties and their application in nanocomposite field. Biomacromolecules 6:612–626

Sareena C, Ramesan MT, Purushothaman E (2012) Utilization of coconut shell powder as a novel filler in natural rubber. J Reinf Plast Compos 31:533–547

Shakeri A, Mathew AP, Oksman K (2011) Self-reinforced nanocomposite by partial dissolution of cellulose microfibrils in ionic liquid. J Compos Mater 46:1305–1311

Singha AS, Thakur VK (2009a) Grewia optiva fiber reinforced novel, low cost polymer composites. J Chem 6:71–76

Singha AS, Thakur VK (2009b) Synthesis, characterisation and analysis of Hibiscus sabdariffa fibre reinforced polymer matrix based composites. Polym Polym Compos 17:189–194

Singha AS, Thakur VK (2009c) Fabrication and characterization of H. sabdariffa fiber-reinforced green polymer composites. Polym-Plast Technol Eng 48:482–487

Singha AS, Thakur VK (2009d) Fabrication and characterization of S. cilliare fibre reinforced polymer composites. Bull Mater Sci 32:49–58

Siqueira G, Bras J, Dufresne A (2009) Cellulose whiskers versus microfibrils: influence of the nature of the nanoparticle and its surface functionalization on the thermal and mechanical properties of nanocomposites. Biomacromolecules 10:425–443

Siro I, Plackett D (2010) Microfibrillated cellulose and new nanocomposite materials: a review. Cellulose 17:459–494

Sriupayo J, Supaphol P, Blackwell J, Rujiravanit R (2005) Preparation and characterization of α-chitin whisker-reinforced chitosan nanocomposite films with or without heat treatment. Carbohydr Polym 62:130–136

Suryanegara L, Nakagaito AN, Yano H (2009) The effect of crystallization of PLA on the thermal and mechanical properties of microfibrillated cellulose-reinforced PLA composites. Compos Sci Technol 69:1187–1192

Thakur VK, Kessler MR (2014a) Free radical induced graft copolymerization of ethyl acrylate onto SOY for multifunctional materials. Mater Today Commun. doi:10.1016/j.mtcomm.2014.09.003

Thakur VK, Kessler MR (2014b) Synthesis and characterization of AN-g-SOY for sustainable polymer composites. ACS Sustain Chem Eng 2:2454–2460

Thakur VK, Thakur MK (2014a) Recent advances in graft copolymerization and applications of chitosan: a review. ACS Sustain Chem Eng 2:2637–2652

Thakur VK, Thakur MK (2014b) Recent trends in hydrogels based on psyllium polysaccharide: a review. J Clean Prod 82:1–15

Thakur VK, Thakur MK (2014c) Processing and characterization of natural cellulose fibers/thermoset polymer composites. Carbohydr Polym 109:102–117

Takagi H, Asano A (2008a) Effects of processing conditions on flexural properties of cellulose nanofiber reinforced "green" composites. Compos A 39:685–689

Takagi H, Asano A (2008b) Effects of processing conditions on flexural properties of cellulose nanofiber reinforced "green" composites. Compos Part A Appl Sci Manufact 39:685–689

Teixeira E, Pasquini D, Curvelo ASS, Corradini E, Belgacem MN, Dufresne A (2009) Cassava bagasse cellulose nanofibrils reinforced thermoplastic cassava starch. Carbohydr Polym 78:422–431

Thakur VK, Yan J, Lin M-F et al (2012a) Novel polymer nanocomposites from bioinspired green aqueous functionalization of BNNTs. Polym Chem 3:962–969

Thakur VK, Singha AS, Thakur MK (2012b) Biopolymers based green composites: mechanical, thermal and physico-chemical characterization. J Polym Environ 20:412–421

Thakur VK, Singha AS, Thakur MK (2012c) Graft copolymerization of methyl acrylate onto cellulosic biofibers: synthesis, characterization and applications. J Polym Environ 20:164–174

Thakur VK, Singha AS, Thakur MK (2012d) Modification of natural biomass by graft copolymerization. Int J Polym Anal Charact 17:547–555

Thakur VK, Singha AS, Thakur MK (2012e) Green composites from natural fibers: mechanical and chemical aging properties. Int J Polym Anal Charact 17:401–407

Thakur VK, Thakur MK, Gupta RK (2014a) Review: raw natural fiber-based polymer composites. Int J Polym Anal Character 19(3):256–271. doi:10.1080/1023666X.2014.880016

Thakur VK, Thakur MK, Raghavan P, Kessler MR (2014b) Progress in green polymer composites from lignin for multifunctional applications: a review. ACS Sustain Chem Eng 2:1072–1092

Thakur VK, Vennerberg D, Madbouly SA, Kessler MR (2014c) Bio-inspired green surface functionalization of PMMA for multifunctional capacitors. RSC Adv 4:6677–6684

Thakur VK, Thunga M, Madbouly SA, Kessler MR (2014d) PMMA-g-SOY as a sustainable novel dielectric material. RSC Adv 4:18240–18249

Thakur VK, Grewell D, Thunga M, Kessler MR (2014e) Novel composites from eco-friendly soy flour/SBS triblock copolymer. Macromol Mater Eng 299:953–958

Thakur VK, Vennerberg D, Kessler MR (2014f) Green aqueous surface modification of polypropylene for novel polymer nanocomposites. ACS Appl Mater Interfaces 6:9349–9356

Uddin AJ, Fujie M, Sembo S, Gotoh Y (2012) Outstanding reinforcing effect of highly oriented chitin whiskers in PVA nanocomposites. Carbohydr Polym 87:799–805

Viguié J, Molina-Boisseau S, Dufresne A (2007) Processing and characterization of waxy maize starch films plasticized by sorbitol and reinforced with starch nanocrystals. Macromol Biosci 7:1206–1216

Wan WK, Hutter JL, Millon LE, Guhados G (2006) Bacterial cellulose and its nanocomposites for biomedical applications. In: Oksman K, Sain M (eds) Cellulose nanocomposites. Processing, characterization, and properties. American Chemical Society, Washington DC

Wan YZ, Luo H, He F, Liang H, Huang Y, Li XL (2009) Mechanical, moisture absorption, and biodegradation behaviors of bacterial cellulose fiber-reinforced starch biocomposites. Compos Sci Technol 69:1212–1217

Wang B, Sain M (2007) The effect of chemically coated nanofiber reinforcement on biopolymer based nanocomposites. Bioresources 2:371–388

Xiao L, Mai Y, He F, Yu L, Zhang L, Tang H, Yang G (2012) Bio-based green composites with high performance from poly(lactic acid) and surface modified microcrystalline cellulose. J Mater Chem 22:15732–15739

Yu J, Ai F, Dufresne A, Gao S, Huang J, Chang PR (2008) Structure and mechanical properties of poly(lactic acid) filled with (starch nanaocrystals)-graftpoly (ε-caprolactone). Macromol Mater Eng 293:763–770

Zheng H, Ai F, Chang PR, Huang J, Dufresne A (2009) Structure and properties of starch nanocrystal-reinforced soy protein plastics. Polym Compos 30:474–480

Zimmermann T, Pohler E, Geiger T (2004) Cellulose fibrils for polymer reinforcement. Adv Eng Mater 6:754–761

Nanoclay/Polymer Composites: Recent Developments and Future Prospects

K. Priya Dasan

Abstract Clay can be counted among the most widely investigated and commercially high demand filler in the polymer industry. Recently there has been a growing interest for the development of polymer/clay nanocomposites due to their superior properties compared to conventional filled polymers even at a very low fraction of filler addition. The easy availability, processability, low cost, and non-toxicity of clay and the advancements in the processing of clay nanocomposites have invited a lot of commercial attention for these materials. The value-added properties enhanced without sacrificing of pure polymer properties make the clays more and more important in the modern polymer industry. Today it finds a wide range of applications ranging from household items to aerospace to medicine. This chapter looks at the chemical and physical aspects of this wonderful material, clay/polymer nanocomposite processing techniques, and the commercial importance of these nanocomposites.

Keywords Nanoclay · Nanocomposite · Intercalation · Exfoliation

1 Introduction

Clay has been in use in the polymer industry for more than five decades. Besides the cost-effective factor, this natural filler offer marked improvement in physical and other engineering properties. Clay can be produced at a high enough purity at large scale for critical manufacturing applications. Most of the commercially high demand polymers are hydrophobic in nature. Since hydrophilic clays and hydrophobic polymers are not compatible in their virgin state, many of the applications required surface modification of clay. With nanotechnology becoming one of the most popular areas of research in almost all disciplines, there was a spurt of

K.P. Dasan (✉)
Material Chemistry Division, SAS, VIT University, Vellore 632014, India
e-mail: priyajeenetd@gmail.com

V.K. Thakur and M.K. Thakur (eds.), *Eco-friendly Polymer Nanocomposites*,
Advanced Structured Materials 75, DOI 10.1007/978-81-322-2470-9_19

research in clay/polymer nanocomposites. The morphology of clay which facilitates easy surface modification and nanolevel dispersion in polymer matrix has bought a revolution in the application of nanoclays in various forms for reinforcements in polymer matrix.

One of the most popular and earlier applications of clay nanocomposites for high-end applications was by Toyota Motors where clay was originally added as filler to a nylon 6 matrix in order to increase the strength of the panels to be used in the interior of Toyota Camry cars (Usuki et al. 1993). The product showed dramatic improvements in mechanical and physical properties and heat distortion temperature at very low content of clay. This was followed by many investigative reports by researchers from various parts of the world on the outstanding performance of clay as nano-filler for various polymers. The reports indicated superior mechanical, chemical, thermal, electrical, barrier, and fire retardancy for clay nanocomposites (LeBaron et al. 1999; Thostenson et al. 2005; Mittal 2007; Ray and Okamoto 2003; Alexandre and Dubois 2000; Kiliaris and Papaspyrides 2010; Giannelis 1998; Pavlidou and Papaspyrides 2008; Yano et al. 1993). These clay polymer nanocomposites developed found potential use in a wide range of applications such as electronics, food packaging, barrier materials, etc. (Fig. 1) (www.bccresearch.com). Clay nanocomposites currently find applications in automobile as bumpers, step-assists, gas tanks, fuel pumps, interior and under-bonnet parts, body panels, electrical parts and appliances, power tool housings, packaging and building components, shock absorbers, coatings, etc. (Fig. 2). The major advantages of using nanocomposites in automobiles are greater safety, increased comfort, better driveability, and reduction in vehicle weight, which in turn leads to fuel saving. The

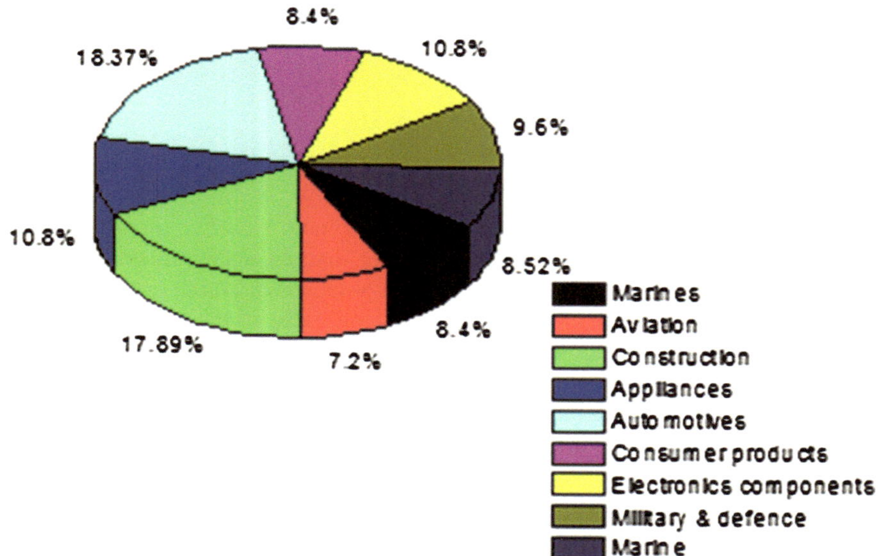

Fig. 1 Market share of clay nanocomposites in different applications

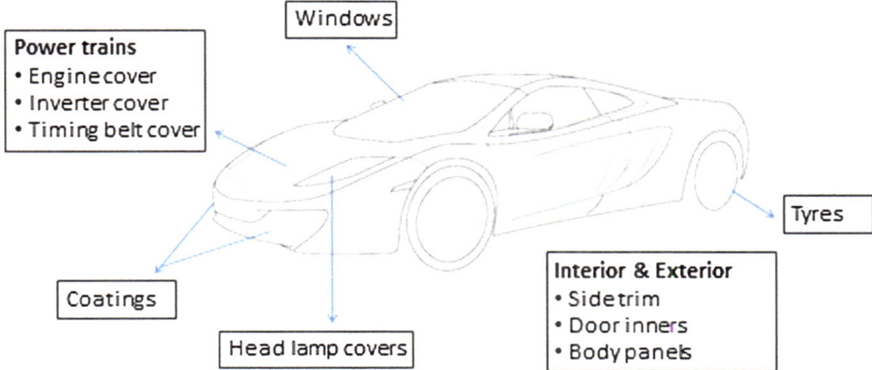

Fig. 2 Potential applications of clay nanocomposites in automobiles

use of polymer nanocomposites continues to grow in the automotive sector, thanks to their exclusive properties. The Toyota research team later came out with various other types of clay nanocomposites based on polymers such as polystyrene, acrylic, polyimides, epoxy resin, and elastomers for application in automobiles (Okada and Usuki 2006). Unitika Co. of Japan introduced nylon-6 nanocomposite for engine covers on Mitsubishi GDI engines 3, which showed around 20 % weight reduction and excellent surface finish. In 2002, General Motors came out with a step-assist automotive component made of polyolefin reinforced with 3 % nanoclays for GM's Safari and Chevrolet Astro vans, followed by the application of these nanocomposites in the doors of Chevrolet Impalas. Aerospace is expected to be a potential area for nanocomposites in the coming era. A recent study indicates that existing and new opportunities for nanocomposites can make up as much as a quarter of the total tonnage of interiors' components (www.researchandmarkets.com). The existing fields of applications are mainly restricted to interior seating and flooring which can be expanded further to include brackets, trays and clips, cockpit flooring, and seat rails. A market research report, "Nanocomposites-A Global Strategic Business Report," states that the world nanocomposites market is forecast to reach 1.3 billion lbs by the year 2015, and growth in the nanocomposites market will be driven by robust demand outlook in the emerging application possibilities in the automotive market (www.electronics.ca). Analysts forecast the Global Nanocomposites market to grow at a CAGR of 17.69 % over the period 2013–2018 (www.researchandmarkets.com). Another area of growing interest is food packaging which requires polymeric systems with high-barrier properties. Mitsubishi Gas Chemical and Honeywell Specialty Polymers use nanoclay/nylon composites as barrier layers in multilayer PET bottles and films for food packaging. MGC's MXD6 nylon nanocomposite, called Imperm N, is used commercially in Europe in multilayer PET bottles for beer and other alcoholic beverages. Recently, LG Chem Ltd. of South Korea developed high-barrier, monolayer blow molded containers of HDPE with 3–5 % nanoclay for handling toluene and light hydrocarbon fluids.

The U.S. military and NASA, in conjunction with Triton Systems, Inc., Chelmsford, Mass., were looking into nanoclay as a barrier enhancer for EVOH in long-shelf-life packaging. The increasing interest in clay nanocomposites has resulted in many companies coming up with surface modified or nanostructured clay in market. The market for nanoclay is dominated by companies such as southern clay products Inc (Cloisite series). Nanocor (Nanomer series), Elementis specialities (Bentone series), Sud Chemie Inc.

2 Polymer Nanocomposites

Global consumption of nanocomposites was an estimated 118,768 metric tons with a value of over $800 million in 2010. In 2011, the market reached around 138,389 metric tons and $920 million and by 2016, it is expected to be 333,043 metric tons and $2.4 billion (www.bccresearch.com). This is a 5-year compound annual growth rate (CAGR) of 19.2 % in unit terms and 20.9 % in value terms. The properties of composites that can be improved many times further even at very low volume fraction of fillers has been the most intriguing part of polymer nanocomposite technology. In polymer nanocomposites, the mixing of the filler phase is achieved at the nanometer level, so that at least one dimension of the filler phase is less than 100 nm. The nanoscale dispersion of the filler phase in the polymer matrix leads to tremendous interfacial contacts of the nanoparticles with the polymer matrix which in turn leads to synergistic improvements in the composite properties. Besides the interfacial surface area, the other factors which directly influence the composite properties are filler volume fraction, the aspect ratio, alignment in the composite, and other geometric considerations. Combination of filler nanoscale dimension and high aspect ratio with its nanoscale dispersion within polymer matrix leads to significant improvements in the polymer properties at very low filler volume fractions (Lin et al. 2011a, b; Thakur et al. 2012, Thakur et al. 2014a, b, c, d). Due to lower filler loading, the macroscopic homogeneity, low density, and opacity of primary polymer gets retained in the final nanocomposite system. Some of the very commonly used nanofillers are layered silicates, carbon nanotubes and nanofibers, inorganic nanoparticles, expandable graphite, new entrants such as cellulose nanowhiskers or cellulose nanofibers, etc. Among these fillers, clay is one of the most abundant and cost-effective fillers. The easy availability, abundance, easy processibility, and possibility of surface modification have made it much more popular.

3 Health and Environmental Issues Related to Nanoclay

One major concern in product development in today's time is the environmental issues related to the processing and end-use applications. No environmental issues have been raised or reported so far regarding the extraction or application of this filler in any field. Though this naturally occurring material is viewed as a nontoxic and eco-friendly material, the safety concern during the extraction, development, need to be studied. Information on occupational exposure to clay particles in mines, processing plants, and user safety data are limited. However, exposure to airborne fine (0.1–2.5 lm) and ultrafine (<0.1 lm) particles can result in lung damage. Natural-clay particles are smaller than 0.004 mm in diameter; however, manufactured nanoclay particles are in the ultrafine-size range and have shown toxic properties (Opportunities and Risks of Nanotechnologies, Allianz Report in Cooperation with the OECD International Future Program). Studies have shown that exposure to these particles can result in increase in morbidity and mortality. Most of the industries dealing with clay and clay-related products have developed their own standard protocol and safety measures at the workplace. Therefore, the introduction of proper studies and regulations on the effect on airborne clay nanoparticles need to be carried out with much urgency.

4 Structure of Clay

Based on the origin, clay is of mainly two types—residual clay and transported clay (or sedimentary clay). As the name suggests residual clays are found in its place of origin and are produced from the surface weathering of rock or shale and are generally found in the place of origin. Residual clays are generally produced naturally by the chemical decomposition of rocks. Chemically clays are polyhydrosilicates with sheet-like geometry (Fig. 3). The structure of clay particles is perceived in layers; each layer is composed of two types of structural sheets: octahedral and tetrahedral. The tetrahedral sheet is composed of silicon-oxygen tetrahedra linked to neighboring tetrahedra by sharing three corners, resulting in a hexagonal network. The remaining fourth corner of each tetrahedron forms a part to adjacent octahedral sheet. The octahedral sheet is usually composed of aluminum or magnesium in six-fold coordination with oxygen from the tetrahedral sheet and with hydroxyl. The two sheets together form a layer, and several layers may be joined in a clay crystallite by interlayer cations, van der Waals force, electrostatic force, or by hydrogen bonding.

The variety of clay minerals can be described by the arrangement of tetrahedral and octahedral sheets, i.e., 1:1 clay mineral would have one tetrahedral and one octahedral sheet per clay layer; 2:1 clay mineral would contain two tetrahedral sheets and one octahedral sheet sandwiched between the two tetrahedral sheets. Based on the variation in the layered structure, clay minerals can be classified mainly into two types which can be further classified as described below. The main classifications of natural clays are kaolinite, montmorillonite/smectite, and illite group. The kaolinite group has three subgroups which are polymorphs with a general chemical formula of $Al_2Si_2O_5(OH)_4$. These subgroups known as kaolinite, dickite, and nacritevary in their structure are used as fillers in ceramics, paint, rubber, paper, and plastics. They have silicate sheets (Si_2O_5) bonded to aluminum

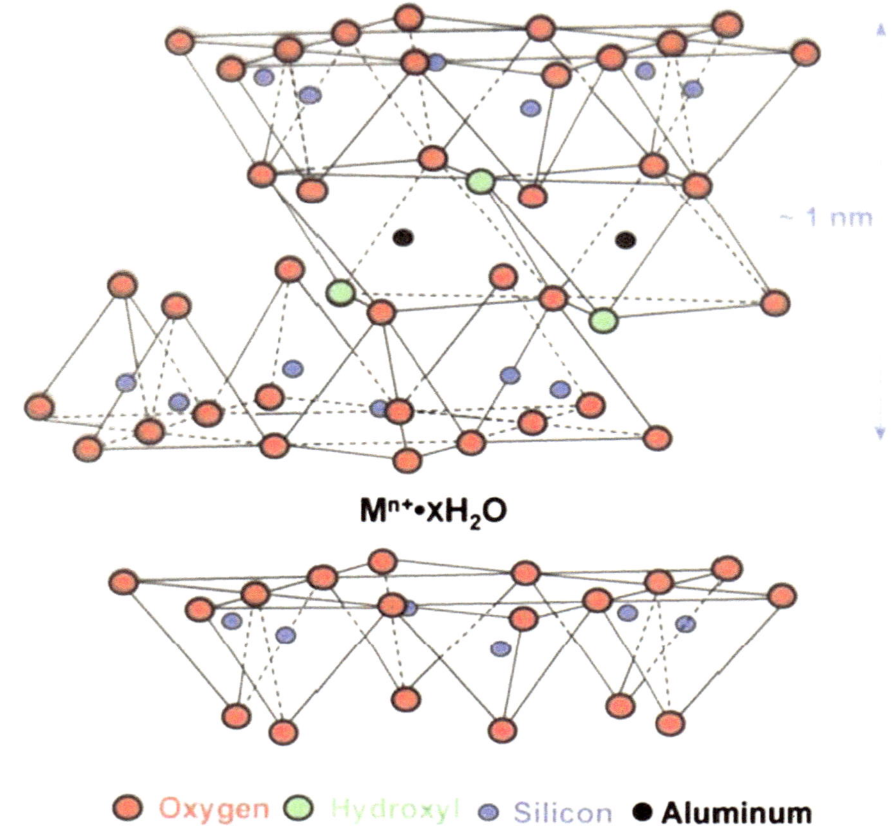

Fig. 3 General structure of clay

oxide/hydroxide layers ($Al_2(OH)_4$) with the two types of layers tightly bonded. Smectite clay group includes montmorillonite, talc, pyrophyllite, saponite, and nontronite which are differentiated in their chemical characteristics. The general formula of this group is $(Ca,Na,H)(Al,Mg,Fe,Zn)_2(Si,Al)_4O_{10}(OH)_2XH_2O$ where silicate layers sandwich an aluminum oxide/hydroxide layer ($Al_2(OH)_4$). The important difference among the members of this group is seen in the chemical characteristics. These types of clays are widely used in in paints, plastics, rubbers, as plasticizer in molding sands, in drilling muds, and as electrical, heat, and acid-resistant porcelain. One of the members, talc, has traditional presence in facial powder. The illite group consists of the mineral illite having a general formula $(K, H)Al_2(Si,Al)_4O_{10}(OH)_2XH_2O$ and is generally used in drilling mud. The structure of this group is similar to the montmorillonite group with silicate layers sandwiching an aluminum oxide/hydroxide layer in the same stacking sequence.

The presence of charge in tertrahedral and octahedral sheets influences the layered structure of clay which results in its usefulness in many applications. The versatility of this material is the possibility of ionic substitution in the sheet structure, resulting in useful modifications. Ions, i.e., Fe^{3+} and Al^{3+}, are small enough to enter the tetrahedral coordination with oxygen and substitute Si^{4+}. Similarly, cations such as Mg^{2+}, Fe^{2+}, Fe^{3+}, Li^{1+}, Ni^{2+}, and Cu^{2+} can be substituted for Al^{3+} in the octahedral sheet. Large sized cations like K^{+}, Na^{+}, and Cs^{+}, can be located between the layers and, therefore, are referred to as interlayer cations. Anionic substitution is also possible, for example, the hydroxyl ion (OH^{-}) can be substituted by F^{-}. The possibility of isomorphous substitution in clay minerals helps in modifying or developing charges on clay without changing its chemical structure. Replacement of Si^{+4} with Al^{3+} in tetrahedral coordination, replacement of Al^{3+} by Mg^{2+} are very commonly adopted. Therefore, the cation exchange capacity or CEC plays a major role while selecting the clay for end-use modification. It is the capacity of a material to hold cations (generally Al^{3+}, Ca^{2+}, Mg^{2+}, Mn^{2+}, Zn^{2+}, Cu^{2+}, Fe^{2+}, Na^{+}, K^{+}, and H^{+}) and is described as the quantity of positively charged ions held by the negatively charged surface of clay minerals. It can be expressed as centimol positive charge per Kg of material or milliequivalent (meq) of positive charge per 100 g of material. In early times, CEC of clays were of priority for applications such as soil fertility, plant nutrient retention capacity, preventing cation contamination in groundwater, etc. More recently, CEC became an important aspect in studying the development of nanoclays through the modification in clay sheet structure.

Among the various clays discussed above, MMT plays a major role commercially. It is widely used as reinforcement for the polymer–clay nanocomposite synthesis because it is environmentally friendly, readily available in large quantities at a relatively low cost, and its intercalation chemistry is well understood (Brindley and Brown 1980; Moore and Reynolds 1997). MMT is extremely fine-grained, does not form macroscopic crystals, and swells on addition of water or organic liquids. MMT clay is the most widely used material for preparing polymer nanocomposites due to its high aspect ratio and economic advantages. The mineral platelet thickness of MMT is only 1 nm, although its dimensions in length and width can be measured in hundreds of nanometres, with a majority of platelets in 200–400 nm range after purification. Due to very small size and thickness of the platelets, a single gram of clay contains over a million individual platelets. It has a highly modifiable structure and the cation substitution creates a charge imbalance that allows the chemical composition to vary. Due to this reason exact theoretical formula is never seen in nature for Montmorillonite clay and is generally represented as without the structural substitution as $(OH)_4Si_8Al_4O_{20}H_2O$. However, the occurrence in nature in any form consists of water molecules. Commercially available nanoclays with over 98 % montmorillonite have variable colors, which are the result of substitution of interlayer cations by iron, titanium, and manganese within the lattice structure and depend on the level of substitution and valence state of cations. Because the metals have central coordination positions in the layer structure, the process leading to their removal is not economically viable. Organic-cation substitution can be used in producing organophilic montmorillonite nanoclay.

5 Nanoclay Preparation

The first criteria required for making the clay more compatible with polymer is to make them less hydrophilic or more organophilic. This is generally achieved through a cation exchange process by the replacement of sodium and calcium cations present in the inter layer space or clay galleries by alkylammonium or alkylphosphonium (onium) cations (Ahmad et al. 2010) (Fig. 4). An exchange of inorganic cations with organic cations renders the clay organophilic and hydrophobic, and lowers the surface energy of the clay layers. It then becomes possible for the organic polymer to diffuse between the clay layers and to delaminate the clay platelets to individual layers. This technology has been widely developed as reported by many researchers (Theng 1974; Lagaly 1986; Pinnavaia and Farzanch 1983; Giannelis 1996). Long-chain alkyl ammonium salts have been widely used for exchanging the inorganic cations because they increase the basal spacing of the clay to a large extent, apart from lowering the surface energy, which can further be helpful in achieving exfoliation of the clay layers in the polymer matrix. The insertion of alkylammonium or alkylphosphonium cations into the galleries results in increasing the inter layer spacing, which promotes the following intercalation of polymer chains into the galleries during nanocomposite preparation. Also, the alkylammonium or alkylphosphoniumcations provides functional groups which are compatible with polymer chains or initiate the polymerization which results in increased interfacial interactions.

Clay particles or tactoids dispersion in polymer can be found in three typical morphologies (Fig. 5). The separated morphology is obtained when the polymer is unable to intercalate between the silicate sheets. In the intercalated morphology, one or more extended polymer chain is intercalated between the silicate layers resulting in a well-ordered multilayer morphology built up with alternating polymeric and inorganic layers. The fully exfoliated morphology is obtained when the silicate layers lose their parallelism and they are completely and uniformly dispersed in a

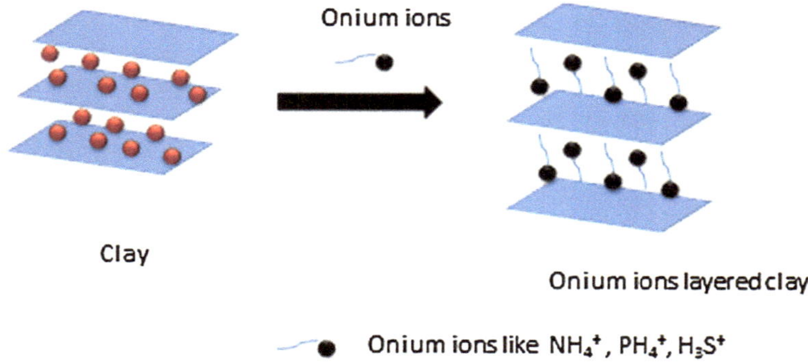

Fig. 4 Schematic representation of cation exchange process

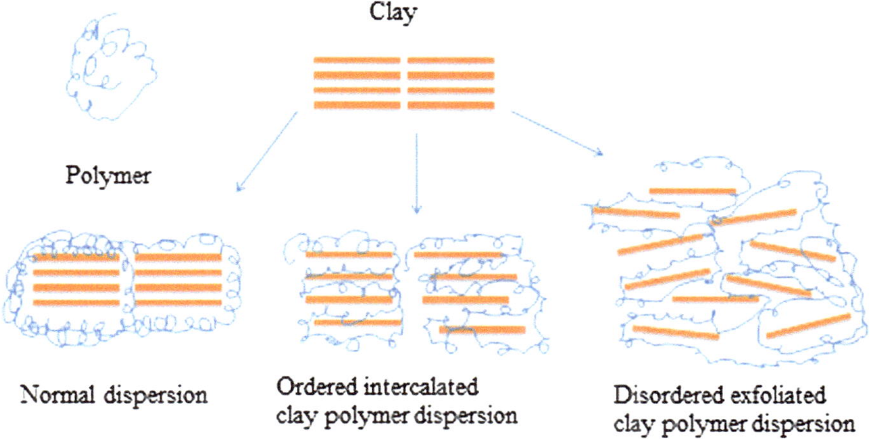

Fig. 5 Clay morphology in polymer matrix

continuous polymer matrix. The most desirable morphology is exfoliated followed by intercalated structure. However, the attainment of this structure requires separation of the tactoids from the primary particle, followed by the destruction of the order of the clay platelets within the tactoids. However, a balance between an exfoliated and intercalated structure often results in desired property enhancements (Alexandre and Dubois 2000). For attainment of these structures, the following processing methods are generally adopted (Fig. 6).

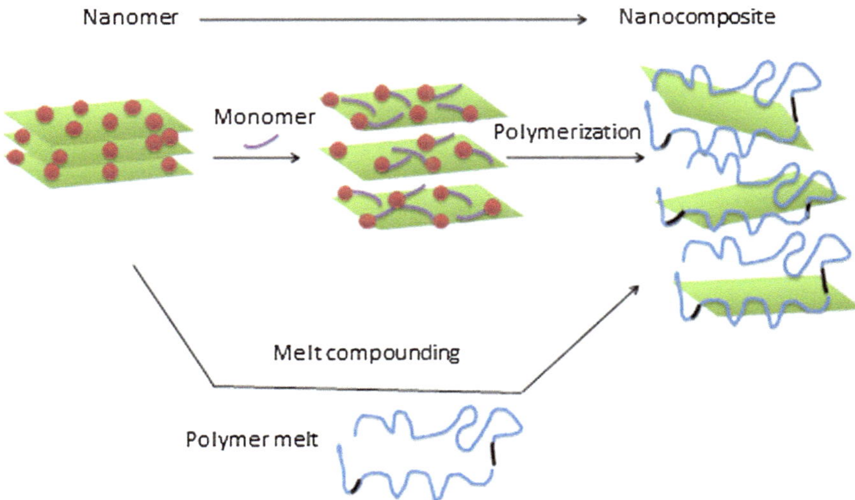

Fig. 6 Processing techniques in nanoclay/polymer composites **a** Template and intercalation method **b** in situ polymerization techniques

5.1 In Situ Template Synthesis

The advantage of this method is that the clay layers get well dispersed within the polymer matrix without the use of onium cations. In this method, the polymer and clay are dissolved in an aqueous solution followed by refluxing at high temperature. The end product is obtained on washing and drying. This results in nucleation and growth of clay layers on the polymer chains and the polymer chains are trapped in the clay interlayers. However, this process requires very high temperature and results in decomposition of polymers. Generally hectorite clay is used in this method as it requires lower temperature for processing. The possibility of agglomeration is a major drawback of this method.

5.2 Solution Intercalation

In this method the polymer or prepolymer is dissolved in a solvent and the clay is dispersed in the same solution. Generally used solvents are water, acetone, and chloroform. The clay is swollen in the solvent and the polymer chains intercalate between the layers. The intercalated nanocomposite is obtained by solvent removal through vaporization or precipitation. During solvent evaporation, the entropy gained by the exit of solvent molecules from the interlayer spacing allows the polymer chains to diffuse between the layers and sandwiching. The growing environmental concern and cost of solvents and the economy involved in the phase separation of products has prompted industries to look at water as solvent. There are also health and safety concerns associated with the application of this technology. The nanocomposite preparation by emulsion polymerization, with the clays dispersed in the aqueous solution, is also categorized as solvent intercalation method (Rehab and Salahuddin 2005).

5.3 In Situ Intercalative Polymerization

In this method the organoclay is swollen in monomer liquid or monomer solution. The monomers thus diffused into the inter layer spacing are polymerized by heat or radiation, by the diffusion of an initiator or by the organic initiator present on the organic modifier of clay (Hussain et al. 2006). The polymerization is carried out within the clay galleries as well as extra galleries. The growth of polymer chains results in the exfoliation and formation of disordered structure. This method is suitable for the preparation of thermoset/clay nanocomposites and has been widely used for the epoxies and styrene polymer nanocomposites (Lan et al. 1995). The initial working this area was carried out by the Toyota Research Group to produce clay/nylon-6 nanocomposites. This method is capable of producing well-exfoliated nanocomposites and has been applied to a wide range of polymer systems. The technology is suitable for raw polymer manufacturers to produce clay/polymer nanocomposites in polymer synthetic processes and is also particularly useful for thermosetting polymers.

5.4 Melt Intercalation

Clay is mixed within the polymer matrix at polymer molten temperature followed by extrusion and injection molding. Prior to mixing, clays are organically modified and polymer chains are surface modified with more polar functional groups. This in turn enhances their compatibility and therefore promotes exfoliation. The efficiency of intercalation using this method may not be as high as that of in situ polymerization and often the composites produced contain a partially exfoliated layered structure. The fact that the composite can be manufactured with conventional polymer processing techniques makes it commercially very important.

The above mentioned processes are in use with much success in the polymer industry for a long time. A few other methods are reported by researchers, which has been successful in certain systems. These include solid intercalation, co-vulcanization and the sol–gel method. Some of these methods are in the early stages of development and have not yet been widely applied.

6 Processing Parameters

Processing conditions have a significant influence on the delamination and dispersion of clay in the matrix. The beneficial effect obtained by using nanoclay reinforcement has been observed, even for low nanoclay content. However, many researchers have observed lowering of properties with increasing nanoparticle percentage after an initial improvement at very low loading. The main cause for this is the nonuniform dispersion of fillers in the matrix. Achieving a uniform dispersion without heterogeneity requires systematic investigation of the system. Pressure, curing temperature, stirring time, and speed affect the dispersion properties of nano-fillers inside the composites to form exfoliated or intercalated composites. Many times the non-optimal processing conditions result in the formation of heterogeneity due to the presence of air bubbles trapped during the sample preparation. Sometimes, agglomeration of nanoclay, formation of nanoclay clusters, and uncured resin occur if large amount of nanoclay are added into the resin. Resin viscosity, sonication time plays a major role in the fabrication of nanoclay composites. Extra energy imposed into the mixture under sonication causes early curing of the resin, which results in brittling of resultant composites. Okamoto et al. (2001) Investigation of flow-induced structure formation of nanocomposites shows that it ultimately influences the mechanical behavior of the composite. According to their claim, addition of nanoclay eliminates the surface melt fracture and elevates the critical shear rate for onset of melt fracture. Higher melt viscosity is known to impose higher shear stresses to platelets during melt mixing in extruder. The higher shear stress in the extruder breaks the organoclay particles into stacks of platelets ortactoids, which can be subsequently sheared apart into smaller platelets. The influence of various parameters on the structure of epoxy-clay nanocomposites was investigated in detail by researchers (Kormann et al. 2001; Lu et al. 2001; Chin et al. 2001). It was found that higher cure temperature and lower curing agent concentration were propitious to the formation of exfoliated epoxy-clay nanocomposites. SEM analysis of a polystyrene/clay nanocomposite was carried out to investigate the morphology of PS/montmorillonite nanocomposites by Fan et al. (2002). Monodispersed spherical particles of about 200 nm in diameter were observed when PS/montmorillonite powder was dispersed in water, whereas planar silicate sheets were found for cetyltrimethyl ammonium bromide exchanged montmorillonite. Jeon et al. (2003) investigated the effects of varying VA content (from 3–28 wt%) and four different types of clay on the structure of EVA/clay nanocomposites. They found the interlayer distance of montmorillonite (MMT) to increase with an increase in VA content. However, when the VA content was beyond 15 wt%, there was no further increase in interlayer distance.

7 Characterization of Composites

The most important criteria to be evaluated in case of polymer nanocomposites are the morphology of the clay tactoids or platelets in polymer matrix. This in turn provides information on the interfacial interaction and the ultimate properties of composites. The degree of intercalation/exfoliation is generally analyzed using XRD and SEM. The layered clay structures show a characteristic peak in XRD analysis due to their regular layered structures. The peak is indicative of the platelet separation or d-spacing in clay structure. Using the peak width at half maximum height and peak position (2θ) in the XRD spectra, the interlayer space can be calculated utilizing Bragg's law. Any change in the interlayer or d-spacing of a clay lattice by organic modification or polymer intercalation results in the shifting of peak position, its broadness, and intensity in the XRD spectra. According to Bragg's law, increasing of d-spacing results in broadening and shifting of related XRD peak toward lower diffraction angles (2θ). By monitoring the position (2θ), shape, and intensity of the characteristic peak for organoclay in nanocomposite structure, it is possible to determine the degree of intercalation/exfoliation. It becomes necessary to use wide-angle X-ray diffraction (WAXD) patterns in the range of $1° < 2\theta < 10°$ to identify the fully exfoliated structures which correspond to the d-spacing of at least 6 nm. WAXD helps in determining the d-spacing in the pure clay as well as in nanocomposite structure, within 1–4 nm, using the position, broadness, and intensity of characteristic peak in WAXD diffractogram. However, the disappearance of such a peak is not a conclusive evidence for a highly exfoliated structure in nanocomposite. TEM micrographs allow a qualitative understanding of the internal structure, exfoliation, or spatial distribution of layers within the polymer matrix and their structural defects. In the TEM micrographs, the darker lines in the brighter matrix show the clay layers because of the presence of heavier elements including Al, Si, and O in the composition of clay sheets or layers compared to lighter atoms such as C, H, N, and Na present in the polymer matrix or inter layer spacing of clay sheets. Therefore, the distance between darker liner sections presented in the TEM micrographs can qualitatively show the d-spacing and dispersion status. Indirect methods such as melt flow viscosity and DMA can also be used to evaluate the interaction between the polymer and the nanoclay structure. TGA can be used to assess the amount of organic matter exchanged on the clay surface during the surface modification process. High-resolution TGA can be used to find the presence of any excess of surface modification molecules present as a pseudo bilayer, but unbound to the surface.

8 Properties of Nanocomposites

A few works exist on polymer blends reinforced with nanoclay for enhanced properties (Azizi et al. 2011; Lotti et al. 2008). Good tear strength (Gatos and Karger-Kocsis 2007), higher dynamic mechanical properties (Venkatesh et al. 2012), increased flammability resistance (Tabuani et al. 2011; Zanetti and Costa 2004; Zhang et al. 2009) reduced coefficient of thermal expansion; and enhanced thermal stability (Hemati and Garmabi 2011; Durmus et al. 2007; Xue et al. 2011; Bertini et al. 2006) have been reported as some of the achievements of using nanoclay as reinforcement in polymers. Polymer/organoclay nanocomposite formation was studied by many researchers (Alexandre and Dubois 2000; Zanetti 2000; Lagaly 1986) and were found to exhibit unique property combinations. Timmerman et al. examined the nanoclay reinforcement effects on cryogenic micro cracking of epoxy/carbon fiber composites. The matrices of carbon fiber/epoxy composites were modified with layered inorganic clays and a traditional filler to determine the effects of particle reinforcement, both at micro and nanoscale, on the response of these materials to cryogenic cycling. The incorporation of nanoclay reinforcement in the proper concentration resulted in laminates with micro crack densities lower than those seen in the unmodified or macro-reinforced materials as a response to cryogenic cycling. The decomposition behavior of EVA nanocomposites was studied by Hull et al. (2003). The incorporation of nanoscale clay fillers into EVA appeared to reinforce the protective layer. Poly(vinyl alcohol)/sodium montmorillonite nanocomposites of various compositions were prepared and their characterization was done by Strawhecker and Manias (2000). Young's modulus tripled, whereas water permeability reduced by 60 % with nanofiller incorporation. Furthermore, due to nanoscale dispersion of the filler, the nanocomposites retained their optical clarity. Synthesis and characterization of novel segmented polyurethane/clay nanocomposites were done by Chen et al. A two-fold increase in tensile strength and a three-fold increase in elongation were found for 1 % benzidine-montmorillonite in polyurethane compared to pure polyurethane. The nanocomposite exhibited lower water absorption properties than pure polyurethane.

The measurements of free volume property by positron annihilation lifetime spectroscopy (PALS) for nanocomposite materials consisting of SBR and layered silicate clay of rectorite and conventional composite materials with SBR/N326 (carbon black) have been carried out. The PALS and differential scanning calorimetry (DSC) results showed that the layered rectorite had a stronger effect than carbon black on restraining polymer chain mobility, which resulted in the decrease of fraction of free volume and gas permeation. The use of octadecylamine modified montmorillonite as a substitute for carbon black in NR compounds was studied by Arroyo et al. (2003). Rubber samples with 10 phr of pristine clay and octadecylamine modified montmorillonite were separately compared with 10 and 40 phr carbon black loaded systems. The organoclay and carbon black accelerated the vulcanization reaction and, gave rise to a marked increase in torque, indicating a higher degree of crosslinking which was also confirmed by swelling measurements. The vulcanization rate and torque value of the organoclay containing composite was found to be sensibly higher than the carbon black composite even at high contents (40 phr) of carbon black.

The viscoelastic and thermal properties of fully and partially cured DGEBA epoxy resin composites were studied modified with montmorillonite nanoclay exposed to UV radiation. Samples were fabricated and cured to 80 % conversion (partially cured) based on isothermal cure kinetic studies. Influence of 1–3 wt% loading of montmorillonite nanoclay on the cure behavior and development of physical properties of these composites were evaluated. Results of the study revealed that for optimization of modified epoxy composite properties, a different curing cycle was necessary due to interaction of different amounts of nanoclay and epoxy molecules. Addition of nanoclay increased the viscoelastic properties, storage modulus and activation energy of decomposition of partially cured samples evolved over exposure time, while fully cured samples degraded over the same period.

The significant improvement in properties with nanoclay prompted researchers to evaluate the interfacial characteristic of clay/polymer systems. Fourier Transform Infrared spectroscopy (FTIR) and X-ray photoelectron spectroscopy (XPS) were conducted to analyze the chemical composition between epoxy matrix and nanocomposite (Chan et al. 2011). These experiments revealed that a chemical bonding at an interface between the matrix and nanoclay of the composites did exist. A four phase model of nanocomposites which includes the effective interface between the nanoplatelets and polymer, as well as interplatelet and outer phases, was used in the simulations by Dai and Mishnaevsky (2013). Different crack growth criteria were compared, including the 3D Benzeggagh and Kenane law (BK law) criterion, the 3D Wu and Reuter law (power law) criterion, and the Reeder law criterion. The effects of the platelelet aspect ratio, clustering, and orientation effects on the crack propagation were studied in numerical experiments by them. It was observed that the increasing aspect ratio leads to the increasing Young modulus, but decreasing strength. The clustering of disks had an adverse effect, meaning increased strength and lower stiffness. In the simulations, damage mechanisms such as crack deflection and delamination were observed. Chan et al. (2011) evaluated the mechanism of reinforcement in a nanoclay/polymer composite based on tailor-made experiment setup. It was found that the Young's modulus and tensile strength of a composite with 5 wt% of nanoclay increased up to 34 and 25 % respectively, as compared with a pristine sample. Images obtained from SEM and results extracted from TEM proved that interlocking and bridging effects did exist in the composites. Nanoclay clusters with diameter of 10 nm could enhance the mechanical interlocking inside the composites, thus breaking up the crack propagation. The authors stated that formation of boundaries between the nanoclay clusters and epoxy can refine the matrix grains and further improve the flexural strength of the composites.

Arash et al. (2014) studied the effect of nanoclay loading (0, 2, 4 and 6 %) on long-term physical properties and withdrawal strength of fasteners before and after saturation for polypropylene/medium density fiberboard MDF dust composites. Sanding dust of MDF was used as lignocellulose material and polypropylene as the thermoplastic material. The results showed that long-term water absorption (WA) and thickness swelling (TS) decrease as the nanoclay loading increased. Also, WA and TS both increase with increasing MDF dust content. Maximum withdrawal strengths of fasteners (screws and nails) were obtained in the samples reinforced with 2 wt% nanoclay. Lau et al. (2006) in his review has explained in detail the growth of nanotubes from nanoclay substrates to form exfoliated nanotube/nanoclay polymer composites. Mechanical properties and dimensional stability of organo-nanoclay modified biofiber polymer composites has been studied by Chen and Yan (2013). Hydrophobic Kraft fibers were obtained by organo-nanoclay modification. The treated fibers were used as reinforcements for improving mechanical performance and dimensional stability of HDPE composites. After the organo-nanoclay treatment, Kraft fibers had a more uniform dispersion in the HDPE matrix and the resulting composites had higher Young's modulus and thermal stability than composites containing untreated fibers. The water absorption properties of the organo-nanoclay treated Kraft fiber–HDPE composites decreased after adding the compatibilizer. An interesting review of recent times shows the application of clay materials along with natural fibers as very high performing packaging materials for the food industry (Majeed et al. 2013). The review describes the works done by a few researchers on the influence of clay as a ternary component in fiber reinforced polymer. Nanoclay has been observed to be a good reinforcing agent for reed flour/PP Najafi et al. (2011) as well as rice husk HDPE Kord and Kiakojouri (2011b) systems. For the reed flour/clay/PP system, the incorporation of coupling agent was found to enhance the mechanical properties of the composites system. The same trend has been observed for wood flour/HDPE composite system with compatibilizers (Zhong et al. 2007). As for the HDPE system, the addition of nanoclay to the rice husk flour reinforced HDPE composite increased crystallization temperature, crystallization enthalpy, and crystallinity level. The author suggested that a fully exfoliated morphology can be obtained by enhancing the compatibilizer loading. Kord (2011a) reported that the incorporation of nanoclay increase the stiffness and flexibility of the PP-based hybrid composites. This study also reported a significant decrease in water absorption and thickness swelling with the increase in nanoparticles loading. However, an improvement was reported in the dynamic mechanical behavior, dimensional stability, and fire retardancy of the hybrid composites. Hybridization of 5 wt% of nanoclay with microcrystalline cellulose increased Young's modulus from 1.040 to 1.240 GPa of the microcrystalline cellulose reinforced ethylene–propylene (EP) co-polymer (Pratheep Kumar and Pal Singh 2007). The addition of nanoclay to the cellulose containing composites resulted in 15 % decrease of water absorption of the composites. Furthermore, the addition of nanoclay increased their thermal degradation onset temperature indicative of an enhanced thermal stability (Kord and Kiakojouri 2011). Najafi et al. (2011) reported an improved mechanical, barrier properties for polypropylene-wood flour composites reinforced with nanoclay. The addition of 4 phc nanoclay with MAPP as compatibilizer to the reed flour/PP biocomposites increased the tensile strength and Young's modulus from 14.6 and 1390 MPa to 28.7 and 2630 MPa, respectively. In addition to improvement in tensile properties, hybridization of nanoclay with reed flour also lowered the water absorption and thickness swelling.

9 Conclusion

A large number of fillers are being introduced into the polymer industry every year. The use of these fillers has helped to achieve properties of polymer and thereby fine-tune the specifications as per requirement. Fillers have played a major role in the development of polymer technology and making the polymer industry a billion dollar industry. The advancements in nanotechnology have taken it further to very high performing materials. Also it introduced new areas of potential applications for polymers. Clay has been used from the start of the polymer industry and still holds its position as the most consumed filler in this industry. The possibility of surface modification and dispersion at nanolevel has helped in making this material more versatile.

References

Ahmad I, Hussain M, Seo K-S, Choa Y-H (2010) J Appl Polym Sci 116:314

Alexandre M, Dubois P (2000) Polymer–layered silicate nanocomposites: preparation, properties and uses of a new class of materials. Mater Sci Eng R Rep 28:1–63

Arash Chavooshia Z, Navib Mohammad, Madhoushia Mehrab, Mohammad Yousef A (2014) bareshic MDF dust/PP composites reinforced with nanoclay: morphology, long-term physical properties and withdrawal strength of fasteners in dry and saturated conditions. Constr Build Mater 52:324–330

Arroyo M, Manchado MAL, Herrero B (2003) Organo-montmorillonite as substitute of carbon black in natural rubber compounds. Polymer 44:2447–2453

Azizi S, Yunus WMZW, Ahmad M (2011) Effect of polyethylene-grafted maleic anhydride on properties of high-density polyethylene and polystyrene blend/layered silicate nanocomposites. J Reinf Plast Comp 30:1649–1654

Brindley GW, Brown G (1980) crystal structures of clay minerals and their X-ray identifications. Mineral Soc, London, pp 1–123

Bertini F, Canetti M, Audisio G, Costa G, Falqui L (2006) Characterization and thermal degradation of polypropylene–montmorillonite nanocomposites. Polym Degrad Stabil 91:600–605

Chana M-L, Lau K-T, Wonga TT, Cardonab F (2011) Interfacial bonding characteristic of nanoclay/polymer composites. Appl Sur Sci 258:860–864

Chan M-L, Lau K-T, Wong T-T, Ho M-P, Hui D (2011) Mechanism of reinforcement in a nanoclay/polymer composite. Composites: Part B 42:1708–1712

Chen X, Guo Q, Mi Y (1998) Bamboo fiber-reinforced polypropylene composites: a study of the mechanical properties J Appl Polym Sci 69:1891–1899

Chin IJ, Thomas TA, Kim HC, Thomas PR, Wang J (2001) Polymer 42:5947–5952

Christoph L Opportunities and risks of nanotechnologies, Allianz report in cooperation with the OECD international future program clay analysis of clay minerals, vols 4 and 10. Oxford University Press, Oxford, pp 104–120

Dai G, Mishnaevsky Jr L (2013) Damage evolution in nanoclay-reinforced polymers: a three-dimensional computational study. Compos Sci Technol 74:67–77

Durmus A, Woo M, Kas_goz A, Macosko CW, Tsapatsis M (2007) Intercalated linear low density polyethylene (LLDPE)/clay nanocomposites prepared with oxidized polyethylene as a new type compatibilizer: structural, mechanical and barrier properties. Eur Polym J 43:3737–3749

Fan J, Liu S, Chen G, Qi Z (2002) SEM study of polystyrene/clay nanocomposite. J Appl Polym Sci 83:66–69

Gasification studies of polymer layered silicate nanocomposites. Chem Mater 14:881–887

Giannelis EP (1996) Polymer layered silicate nanocomposites. Adv Mater 8:29–35

Giannelis EP (1998) Polymer-layered silicate nanocomposites: synthesis, properties and applications. Appl Organo met Chem 12:675–680

Gatos KG, Karger-Kocsis J (2007) Effect of the aspect ratio of silicate platelets on the mechanical and barrier properties of hydrogenated acrylonitrile butadiene rubber (HNBR)/layered silicate nanocomposites. Eur Polym J 43:1097–1104

Hull TR, Price D, Liu Y, Wills CL, Brady J (2003) An investigation into the decomposition and burning behaviour of Ethylene-vinyl acetate copolymer nanocomposite materials. Polym Degrad Stab 82:365–371

Hemati F, Garmabi H (2011) Compatibilised LDPE/LLDPE/nanoclaynanocomposites: I. Structural, mechanical, and thermal properties. Can J Chem Eng 89:187–196

Hussain F, Hojjati M, Okamoto M, Gorga R.E (2006) Review article: polymer-matrix nanocomposites, processing, manufacturing, and application: an overview. J Compos Mater, 40(17):1511–1571

Jeon CH, Ryu SH, Chang YW (2003) Optical properties of plasticized polycarbonate. Polym Int 53:153–155

Jieming CX, Ning Y (2013) Mechanical properties and dimensional stability of organo-nanoclay modified biofiber polymer composites. Composites: Part B 47:248–254

Jyh MH, George J, Jiang, Zong MG, Wei X, Wei PP (2002) The characterization of organic modified clay and clay-filled PMMA nanocomposite. J Appl Polym Sci, 83:1702–1718

Lau K, Gu C, Hui D (2006) A critical review on nanotube and nanotube/nanoclay related polymer composite materials. Composites: Part B 37:425–436

Kiliaris P, Papaspyrides CD (2010) Polymer/layered silicate (clay) nanocomposites: an overview of flame retardancy. Prog Polym Sci 35:902–958

Kojima Y, Usuki A, Kawasumi M, Okada A, Kurauchi T (1993) Kamigaito O (1993) Synthesis of nylon 6–clay hybrid by montmorillonite intercalated with $\in$-caprolactam. J Polym Sci Part A: Polym Chem 31:983–986

Kormann X, Lindberg H, Berglund LA (2001) Synthesis of epoxy–clay nanocomposites. Influence of the nature of the curing agent on structure. Polymer 42:4493–4499

Kord B, Kiakojouri SMH (2011) Effect of nanoclay dispersion on physical and mechanical properties of wood flour/polypropylene/glass fibre hybrid composites. Bioresources 6:1741–1751

Kord B (2011a) Effect of nanoparticles loading on properties of polymeric composite based on hemp fiber/polypropylene. J Thermoplast Compos 2:4

Kord B (2011b) Nanofiller reinforcement effects on the thermal, dynamic mechanical and morphological behavior of HDPE/rice husk flour composites. Bioresources 6:1351–1358

Lagaly G (1986) Development in ionic polymers. Appl Sci Publ, London

Lan T, Kaviratna PD, Pannavaia T (1995) Mechanism of clay tactoid exfoliation in epoxy-clay nanocomposites. J Chem Mater 7:2144–2150

LeBaron PC, Wang Z, Pinnavaia TJ (1999) Polymer-layered silicate nanocomposites: an overview. Appl Clay Sci 15:11–29

Lu JK, Ke YC, Qi ZN, Yi XS (2001) Studies of intercalation and exfoliation behavior of organoclays in epoxy resin. J Poym Sci Part B: Polym Phys 39:115–120

Lotti C, Isaac CS, Branciforti MC, Alves RMV, Liberman S, Bretas RES (2008) Rheological, mechanical and transport properties of blown films of high density polyethylene nanocomposites. Eur Polym J 44:1346–1357

Lin M-F, Thakur VK, Tan EJ, Lee PS (2011a) Dopant induced hollow BaTiO3 nanostructures for application in high performance capacitors. J Mater Chem 21:16500–16504

Lin M-F, Thakur VK, Tan EJ, Lee PS (2011b) Surface functionalization of BaTiO3 nanoparticles and improved electrical properties of BaTiO3/polyvinylidene fluoride composite. RSC Adv 1:576–578

Majeed K, Jawaid M, Hassan A, Abu Bakar A, Abdul Khalil HPS, Salemae AA, Inuwa I (2013) Potential materials for food packaging from nanoclay/natural fibres filled hybrid composites. Mater Des 46:391–410

Matti P, Nam PH, Okamoto M, Kotaka T, Hasewaga N, Usuki A (2002) Influence of crystallization on inercalation, morphology and mechanical properties of polypropylene/clay nanocomposites. Macromol, 35:2043–2058

Mittal V (2007) Polypropylene-layered silicate nanocomposites: filler matrix interactionsand mechanical properties. J Thermoplast Compos Mater 20:575–599

Moore DM, Reynolds RC (1997) X-ray diffraction and the identification and analysis of Najafi A, Kord B, Abdi A, Ranaee S (2011) The impact of the nature of nanoclay on physical and mechanical properties of polypropylene/reed flour nanocomposites. J Thermoplast Compos

Okada A, Usuki A (2006) Twenty years of polymer composites. Macromol Mater Eng 291: 1449–1476

Okamoto M, Nam PH, Maiti P, Kotaka T, Hasegawa N, Usukin A (2001) A house of cards structure in polypropylene/clay nanocomposites under elongational flow. Nano Lett 1:295–298

Pavlidou S, Papaspyrides CD (2008) A review on polymer-layered silicate nanocompopsites. Prog Polym Sci 32:1119–1198

Pinnavaia TJ, Farzanch F (1983) Metal complex catalyst interlayered in smecitite clay, hydroformylation of 1 hexene with rhodium complexes ion exchanged into hectorite. Inorg. Chem, 22:2216–2220

Pratheep Kumar A, Pal Singh R (2007) Novel hybrid of clay, cellulose, and thermoplastics. I. Preparation and characterization of composites of ethylene–propylene copolymer. J Appl Polym Sci 104:2672–2682

Ray SS, Okamoto M (2003) Polymer/layered silicate nanocomposites: a review from preparation to processing. Prog Polym Sci 28:1539–1641

Rehab A, Salahuddin N (2005) Nanocomposite materials based on polyurethane intercalated into montmorillonite clay. Mater Sci Eng A399:368–376

Strawhecker KE, Manias E (2000) Structure and properties of poly(Vinyl alcohol)/ Na-montmorillonite nanocomposites. Chem Mater 12:2943

Tcherbi-Narteh A, Hosur M, Triggs E, Owuor P, Jelaani S (2014) Viscoelastic and thermal properties of full and partially cured DGEBA epoxy resin composites modified with montmorillonite nano clay exposed to UV radiation. Polymer Degradation and Stability. Polym Degrad Stab 101:81–91

Theng BKG (1974) The chemistry of clay organic reactions. Wiley, Adam Hilger, London
Thostenson ET, Li C, Chou TW (2005) Nanocomposites in context. Compos Sci Technol 65: 491–516
Tabuani D, Ceccia S, Camino G (2011) Polypropylene nanocomposites, study of the influence of the nanofiller nature on morphology and material properties. Macromol Symp 301:114–27
Timmerman JF, Hayes BS, Seferis JC (2002) Nanoclay reinforcement effects on the cryogenic microcracking of carbon fibre/epoxy composites.Compos Sci Technol 62:1249
Tjong SC (2006) Structural and mechanical properties polymer nanocomposites. Mater Sci Eng R 53:73–197
Thakur VK, Yan J, Lin M-F et al (2012) Novel polymer nanocomposites from bioinspired green aqueous functionalization of BNNTs. Polym Chem 3:962–969
Thakur VK, Vennerberg D, Kessler MR (2014a) Green aqueous surface modification of polypropylene for novel polymer nanocomposites. ACS Appl Mater Interfaces 6:9349–9356
Thakur VK, Thakur MK, Raghavan P, Kessler MR (2014b) Progress in green polymer composites from lignin for multifunctional applications: a review. ACS Sustain Chem Eng 2:1072–1092. doi:10.1021/sc500087z
Thakur VK, Vennerberg D, Madbouly SA, Kessler MR (2014c) Bio-inspired green surface functionalization of PMMA for multifunctional capacitors. RSC Adv 4:6677–6684
Thakur VK, Thunga M, Madbouly SA, Kessler MR (2014d) PMMA-g-SOY as a sustainable novel dielectric material. RSC Adv 4:18240–18249
Usuki YK, Okada A, Karauchi T, Kamigaito O (1993) J Polym Sci Part A: Polym Chem 31:2493
Venkatesh GS, Deb A, Karmarkar A, Chauhan SS (2012) Effect of nanoclay content and compatibilizer on viscoelastic properties of montmorillonite/polypropylene nanocomposites. Mater Des 37:285–291
Wang Z, Pinnavaia TJ (1998) Hybrid organic–inorganic nanocomposites: exfoliation of Wang ZF, Wang B, Qi N, Zhang HF, Zhang LQ (2005) Influence of fillers on free volume and gas barrier properties in styrene-butadiene rubber studied by positrons. Polymer 46:719–724
Xue B, Zhang P, Jiang Y, Sun M, Liu D, Yu L (2011) Preparation and characterization of linear low-density polyethylene/dickitenanocomposites prepared by the direct melt blending of linear low-density polyethylene with exfoliated dickite. J Appl Polym Sci 120:1736–1743
Yano K, Usuki A, Okada A, Kurauchi T, Kamigaito O (1993) Synthesis and properties of polyimide/clay hybrid. J Polym Sci Polym Chem 31:2493–2498
Zanetti M, Costa L (2004) Preparation and combustion behaviour of polymer/layered silicate nanocomposites based upon PE and EVA. Polymer 45:4367–4373
Zanetti M, Lomakin S, Camino G (2000) Macromol Mater Eng 279:1
Zhang J, Hereid J, Hagen M, Bakirtzis D, Delichatsios MA, Fina A et al (2009) Effects of nanoclay and fire retardants on fire retardancy of a polymer blend of EVA and LDPE. Fire Saf J 44:504–513
Zhong Y, Poloso T, Hetzer M, De Kee D (2007) Enhancement of wood/polyethylene composites via compatibilization and incorporation of organoclay particles. Polym Eng Sci 47:797–803